Introduction to Coastal Processes and Geomorphology

Written for anyone interested in coastal geomorphology, this is the complete guide to the processes at work on our coastlines and the resulting features seen in coastal systems across the world. Accessible to students from a range of disciplines, the quantitative approach of this book helps to build a solid understanding of wave and current processes that shape coastlines. From sandy beaches to coral reefs, the major coastal features are related to contemporary processes and to sea level changes over the past 25 000 years. Key equations describing these processes and standard methods and instrumentation used to collect measurements are all presented in this wide-ranging overview. Designed to support a one- or two-semester course and grounded in current research, this Second Edition has been substantially updated and rewritten – featuring cutting-edge new topics, insights from new models and technologies, additional global examples and an enhanced package of online teaching materials.

Robin Davidson-Arnott has been a Professor in the Department of Geography at the University of Guelph since 1976. He was a member of the Task Force of the International Joint Commission (Canada/USA) Great Lakes Water Levels Reference Study Phase 1 (1987–9), and was seconded as a Scientist to the Ontario Ministry of Natural Resources Development of Ontario Shoreline Management Policy and Technical Guideline (1992–5), and to the International Joint Commission (Canada/USA) Upper Great Lakes Water Level Regulation Study (2007–11). He has worked as a consultant for a number of studies for Ontario Conservation Authorities and Parks, Canada, and been awarded the R.J. Russell Award from the Coastal and Marine Specialty Group of the Association of American Geographers in 2000. His research interests are in coastal geomorphology: on beach and nearshore processes on sandy coasts, nearshore erosion of cohesive coasts, coastal saltmarshes, aeolian sediment transport and coastal dunes. He has received continuous support in his research from the Natural Sciences and Engineering Research Council of Canada for over 30 years. He has authored and co-authored many books and journal articles on the subject, including a contribution to *Geomorphology and Global Environmental Change* (2009, Cambridge University Press).

Bernard Bauer is a process geomorphologist with research and teaching experience in coastal, aeolian, and fluvial environments. He is the recipient of the HydroLab Award from the International Association for Great Lakes Research, the R.J. Russell Award from the Association of American Geographers, and a Presidential Young Investigator Award from the US National Science Foundation (NSF). His research has been conducted in Canada, USA, Australia, New Zealand and Ireland, leading to peer-reviewed publications in major international journals. He has held administrative positions as Department Chair (University of Southern California), Faculty Dean and Associate Provost (University of British Columbia), and he also served as a Programme Director at the US National Science Foundation. He now devotes considerable time as a volunteer member of committees dealing with water sustainability issues in the Okanagan, British Columbia.

Chris Houser is a coastal geomorphologist with a focus on barrier island response and recovery to storms and sea level rise, physical and social dimensions of the rip current hazard, and scale interaction in coastal systems. His research has been conducted in Canada, the United States, Australia, Denmark and Costa Rica, leading to peer-reviewed publications on coastal geomorphology, geology, coastal management and beach safety. He has held academic and administrative positions at the University of West Florida, Texas A&M University and the University of Windsor, and is currently the Dean of Science at Windsor.

Given the impending challenges at the coast under the impacts of global climate change, it is heartening to encounter this well-presented text on the interaction of processes and sediment supply that provide the characteristics of the present coastal features. The writing style is that of being in the classroom and being exposed to the lectures on the topic, replete with background, a variety of perspectives, and areas of continued need for research. It is the essence of communication for the purpose of generating understanding and appreciating the vectors of change at many levels now and into the future. BRAVO!!

Norbert Psuty, Rutgers University

This substantially updated second edition is a well-balanced and authoritative introduction to a wide range of coastal systems, usefully supported by examples drawn from across the globe. Written by acknowledged coastal experts, the text is grounded in morphodynamics yet styled to allow easy access to a wide spectrum of readers, reaching out beyond coastal scientists to include those who manage the coast. The use of special interest boxes serves as an elegant device linking an understanding of morphodynamics to key coastal-management issues. This book is an authoritative key text for all those with an interest in coastal processes and geomorphology.

Jim Hansom, University of Glasgow

There has been a significant increase in the number and quality of studies devoted to coastal geomorphology over the last three decades. Much of this research has been driven by advances in technology and by the development of novel approaches, especially in instrumentation and modelling. At the same time our coasts are now affected by major changes driven by natural and human factors that call for an increasingly cross-disciplinary approach. The book by Robin Davidson-Arnott, Bernard Bauer and Chris Houser is more than a remarkable introduction to coastal geomorphology, building up with team synergy on the first edition published a few years ago by R. Davidson-Arnott. Written by three scientists who have contributed substantively, and still do, to the understanding of coasts and the geomorphic processes involved in the construction, shaping and reworking of coastal landforms, the second edition of this book finely crystallizes cutting-edge research in coastal geomorphology. The book is cast in a quantitative, yet easily comprehensible, format complemented by online teaching materials that will be appealing to students and scholars from a range of disciplines. The text is well-balanced with good, clear diagrams and figures, and each chapter backed by up-to-date references and supplementary information on coastal processes in a box format. This book deserves to appear on your shelves. I will certainly recommend it to students and to colleagues.

Edward Anthony, Centre Européen de Recherche et d'Enseignement des Géosciences de l'Environnement, Aix-Marseille University; Editor-in-Chief of *Marine Geology*

... combines an accessible yet scholarly treatment of the underlying processes with a broad range of interesting case studies. .... *Introduction to Coastal Processes and Geomorphology* would certainly be my current choice for a course text in this field.

Geological Magazine

This excellent book is both well-written and illustrated ... it will more than satisfy undergraduate coastal geomorphology students yet it is also clear and accessible enough to be of great use to students from a range of disciplines ... one of the best coastal geomorphology texts available.

Jim Hansom, University of Glasgow

... an excellent, modern synthesis of key concepts and literature that also provides a sound knowledge base for advanced studies and researchers. Supplemented with ... excellent online resources, including field data sets and presentation-quality figures and video clips. This text, and its related resources, is a must have for anyone interested in coastal geomorphology.

Ian J. Walker, University of Victoria

Introduction to Coastal Processes and Geomorphology

Second Edition

Robin Davidson-Arnott
University of Guelph

Bernard Bauer
University of British Columbia

Chris Houser
University of Windsor

CAMBRIDGE
UNIVERSITY PRESS

University Printing House, Cambridge CB2 8BS, United Kingdom

One Liberty Plaza, 20th Floor, New York, NY 10006, USA

477 Williamstown Road, Port Melbourne, VIC 3207, Australia

314–321, 3rd Floor, Plot 3, Splendor Forum, Jasola District Centre, New Delhi – 110025, India

79 Anson Road, #06–04/06, Singapore 079906

Cambridge University Press is part of the University of Cambridge.

It furthers the University's mission by disseminating knowledge in the pursuit of education, learning, and research at the highest international levels of excellence.

www.cambridge.org
Information on this title: www.cambridge.org/9781108424271
DOI: 10.1017/9781108546126

First edition published 2010
Second edition published 2019

Printed in the United Kingdom by TJ International Ltd. Padstow Cornwall

A catalogue record for this publication is available from the British Library.

ISBN 978-1-108-42427-1 Hardback
ISBN 978-1-108-43986-2 Paperback

Brief Contents

Preface *page* xi
Acknowledgements xv

Part I Introduction 1

1 Introduction 3

2 Coastal Geomorphology 9

Part II Coastal Processes 29

3 Sea Level 31

4 Wind-Generated Waves 75

5 Wave Dynamics 109

6 Surf Zone Circulation 157

7 Coastal Sediment Transport 183

Part III Coastal Systems 227

8 Beach and Nearshore Systems 229

9 Coastal Sand Dunes and Aeolian Processes 280

10 Barrier Systems 343

11 Saltmarshes and Mangroves 395

12 Coral Reefs and Atolls 444

13 Cliffed and Rocky Coasts 476

Index 517

Contents

Preface *page* xi
Acknowledgements xv

Part I Introduction 1

1 Introduction 3

1.1 Humans and the Coastal Zone 3
1.2 Approaches to the Study of Coasts 5
1.3 Information Sources 6

2 Coastal Geomorphology 9

2.1 Definition and Scope 9
2.2 The Coastal Zone 11
2.3 Controls on Coastal Form and Evolution 18
2.4 Shoreline Mapping and Technological Innovation 19
2.5 Understanding and Modelling Coastal Dynamics 21
2.6 Uncertainty in Predicting Coastal Evolution 25

Part II Coastal Processes 29

3 Sea Level 31

3.1 Synopsis 31
3.2 Defining Mean Sea Level 32
3.3 Changes in Mean Sea Level 36
3.4 Astronomical Tides 43
3.5 Short-Term Dynamic Changes in Sea Level 55
3.6 Climate Change and Sea Level Rise 63

4 Wind-Generated Waves 75

4.1 Synopsis 75
4.2 Wave Definition and Description 76
4.3 Wave Generation 80
4.4 Wave Measurement and Parameterisation 86
4.5 Wave Prediction 101
4.6 Wave Climate 105

5 Wave Dynamics 109

5.1 Synopsis 109
5.2 Wave Theory 110

5.3 High-Order Wave Theories 118
5.4 Wave Transformations in Intermediate Water Depths 122
5.5 Wave Breaking 128
5.6 Wave Groups and Low-Frequency Modes of Energy 148

6 Surf Zone Circulation 157

6.1 Synopsis 157
6.2 Undertow 158
6.3 Rip Cells 164
6.4 Longshore Currents 174
6.5 Wind and Tidal Currents 177

7 Coastal Sediment Transport 183

7.1 Synopsis 183
7.2 Sediment Transport Mechanisms, Boundary Layers and Bedforms 183
7.3 Cross-Shore Sediment Transport 194
7.4 Longshore Sand Transport 200
7.5 Littoral Sediment Budget and Littoral Drift Cells 211

Part III Coastal Systems 227

8 Beach and Nearshore Systems 229

8.1 Synopsis 229
8.2 Beach and Nearshore Sediments and Morphology 230
8.3 Nearshore Morphodynamics 249
8.4 Beach Morphodynamics 263

9 Coastal Sand Dunes and Aeolian Processes 280

9.1 Synopsis 280
9.2 Morphology and Structure of Coastal Dunes and Dune Fields 282
9.3 Foredune Morphodynamics and Maintenance 295
9.4 Aeolian Processes on Beaches and Dunes 302
9.5 Flow Modification and Wind Steering by Topography 317
9.6 Geometric Controls on Sand Delivery to Foredunes 324
9.7 Prediction of Long-Term Sediment Delivery to Foredunes 327
9.8 Long-Term Foredune Evolution and Beach–Dune Interaction 328

10 Barrier Systems 343

10.1 Synopsis 343
10.2 Barrier Types and Morphology 344

10.3 Barrier Dynamics: Overwash and Inlets 350
10.4 Barrier Spit Morphodynamics 370
10.5 Barrier Islands 375
10.6 Anthropogenic Impacts and Natural Hazards 381

11 Saltmarshes and Mangroves 395

11.1 Synopsis 395
11.2 Saltmarsh and Mangrove Ecosystems 395
11.3 Saltmarshes 398
11.4 Mangroves 430

12 Coral Reefs and Atolls 444

12.1 Synopsis 444
12.2 Corals and Reef Formation 445
12.3 Geomorphology and Sedimentology of Coral Reefs 450
12.4 Impacts of Disturbance on Coral Reefs 466

13 Cliffed and Rocky Coasts 476

13.1 Synopsis 476
13.2 Cliffed Coast Morphology 477
13.3 Cliffed Coast Erosion System 480
13.4 Erosion of Soft Rock Cliff Coasts 488
13.5 Hard Rock Coasts 504

Index 517

Preface

This textbook is designed for an upper-level undergraduate course in coastal geomorphology, but it would also be appropriate for an entry-level graduate course. The approach adopted by the authors is process oriented rather than morphology based, privileging morphodynamics over landform description. The primary objective is to provide students with a fundamental appreciation of how erosion, transportation and deposition of sediments in the coastal zone act in concert to produce the vast variety of features found along marine, estuarine and lacustrine shorelines around the world. Nevertheless, the discussion necessarily includes the broader context within which these myriad coastal features evolve, including the tectonic and geologic controls that influence the erodibility of coasts, the isostatic, eustatic and meteorologic factors responsible for regional-scale, long-term trends in sea-surface fluctuations, and the biologic and human influences that affect coastal systems. The intent is to provide a comprehensive and authoritative treatise that includes sufficient information for the reader to be able to understand the broad dynamics of coastal systems, that will then enable a more detailed foray into the primary sources found in refereed journal articles and advanced texts on topics such as wave dynamics, nearshore circulation, tides, rip currents and sediment transport mechanics in water and air.

It is assumed that students reading this textbook will have backgrounds in earth sciences, physical geography, or coastal/ocean engineering, although it is written at a level that accommodates those from allied disciplines (e.g., marine biology, environmental science, geography, geology). Ideally, students will have already completed an introductory course in geomorphology or oceanography, and will also have some comfort with mathematical relations and basic physics. However, the text is written in a style that focuses on the meaning of the terms in the equations so as to facilitate a basic understanding of the concepts rather than the mathematical derivation. This is particularly true for Chapter 5 on wave dynamics. The course instructor can further explain how the equations are used in applied ways through laboratory exercises and homework assignments, as appropriate. The hope is that the book will serve as a useful reference beyond the requirements of a single course, and that coastal managers and other scientists and social scientists interested in the coastal zone will also find it useful.

The origin of the book stems from a fourth-year course taught by Dr Robin Davidson-Arnott at the University of Guelph during a career spanning over 30 years. The first edition was released in 2010, and it was adopted widely for many university courses across the world. The current (second) edition of the text is substantially updated and re-written, reflecting the input from Dr Bernard Bauer and Dr Chris Houser, who are close colleagues of Professor Davidson-Arnott. Each of the co-authors has made extensive contributions to the scientific literature on coastal geomorphology and, even though their research interests overlap considerably, they bring special expertise and different perspectives to the second edition of this book. The valuable comments from six anonymous reviewers of the proposal to revise the book has also been useful in informing the overall tone and philosophy of the text, helping to identify weaknesses and gaps in the first edition, and pointing to useful resources that have enhanced the effectiveness of our treatment of the topic.

Nevertheless, the content of the book, by necessity, reflects our personal experiences and biases. For example, each of us is a field-oriented scientist believing that conceptual theories and numerical models are simplifications of reality that should always be tested against what can be measured, observed or experienced in the real world. Such external realities reveal the truth, unlike a computer code that mimics our limited understanding of the truth. In this context, it is important to note that our experiences, although extensive and geographically varied, are hardly globally representative. Many of our research projects have been carried out in Canada (Atlantic, Pacific and Great Lakes coasts), the United States of America (Atlantic, Pacific and Gulf coasts), western Europe (Ireland, England, Denmark), Australia (New South Wales), New Zealand (South Island), South America (Brazil) and the Caribbean. Our professional and leisurely travels are more extensive, but there are still coastal systems, such as high-latitude ice-dominated coasts, estuaries, deltas, muddy coasts and fjords, with which we have limited experience and expertise. The reader is encouraged to seek out coverage of these coastal topics in texts and scholarly manuscripts that focus on them.

The book comprises 13 individual chapters that can be read in any order without much loss of comprehension. However, it will be useful to understand that the chapters are loosely organised into three major sections. The first section (**Part I**) provides introductory material that positions coastal geomorphology among the sciences and introduces basic ideas and terminology. Chapter 1 makes the essential point that the history of human civilisation has been intricately linked to coasts, and that the majority of humans live close to the coast. Thus, there are compelling reasons to understand coastal processes, not the least of which is the future fate of the majority of global mega-cities under the prospect of sea level rise driven by human-induced climate change. Chapter 2 defines the scope of coastal geomorphology for the purposes of this textbook, and introduces essential terminology used by coastal geomorphologists. In this second edition of the book, the discussion of coastal classification has been expanded considerably and sub-sections added on coastal mapping and modelling.

The second section (**Part II**) has five chapters dealing with coastal processes that are most prevalent in oceans but also lakes and bays. Chapter 3 discusses changes in sea level over a range of time scales, but most importantly due to the impacts of climate change since the last glaciation. Tides and storm surges are also covered in this chapter with respect to their dynamic origins as well as the implications for coastal hazard management. Chapters 4 and 5 provide a comprehensive introduction to waves, and these chapters have been completely rewritten in this second edition of the text. Chapter 4 deals with wave description and wave measurement, and then discusses the factors that control their formation. Analytical techniques such as spectral analysis are introduced and related to the statistical properties of wave fields, and the concept of a wave climate is discussed. Chapter 5 is focused on the dynamic aspects of wave motion, beginning with a summary of linear wave theory and many derived quantities such as wave energy and wave power. Higher order wave theories are touched upon briefly so as to make students aware of the limitations of linear wave theory and to show how more complex theories are needed to describe how waves transform as they propagate into shallow water. The chapter ends with a treatment of the integral properties of waves (e.g., radiation stress) and a discussion of the importance of wave groups and infragravity energy in the nearshore circulation. Chapter 6, on surf zone circulation, builds on Chapter 5 by showing how rip currents, undertow, and longshore currents are the necessary manifestation of mass and momentum conservation in the nearshore (i.e., driven by radiation stress gradients due to shoaling and breaking waves). The discussion of rip currents has been updated with an extensive treatment of the hazards they pose. Finally, Chapter 7 draws on the water motion described in the previous three chapters to describe the processes of sediment transport, both cross-shore and alongshore. The basics mechanics of sediment transport are discussed and several models of on–offshore and alongshore transport are summarised, demonstrating how beaches evolve and are maintained in quasi-equilibrium depending on the wave climate. Updated discussions of offshore transport in rip channels and transport modelling are new to this second edition.

The third section of the book (**Part III**) describes a range of different coastal systems.

Sand beaches, barred profiles and nearshore morphodynamics are discussed at length in Chapter 8. New sections on oil spills, satellite monitoring of morphologic change, and morphology-process feedbacks are included in this chapter. Chapter 9 on aeolian processes and forms was completely re-written and re-organised. The basic dune forms are introduced, stressing the unique relationship between foredune evolution and vegetation succession along sandy shorelines. The mechanics of aeolian sediment transport are discussed as a complement to the concepts covered in Chapter 7 related to waves and currents. Beach–dune interaction is treated at the end of the chapter in an attempt to demonstrate the intricate linkages between nearshore processes driven by waves and currents and aeolian processes on the subaerial beach. This leads nicely to a discussion of barrier systems in Chapter 10, which covers the evolution and structure of spits, bay mouth bars, and barrier islands. New sections on barrier island response to sea level rise, on eco-geomorphic feedbacks on barrier systems, barrier island recovery following hurricanes, and anthropogenic influences are included in this second edition. Biologically dependent coasts are treated in Chapter 11 (saltmarshes and mangroves) and Chapter 12 (coral reefs). Both of these have been updated, and there is increased consideration of the response of these systems to sea level rise. Recent understanding of the structure, evolution and sensitivity of coral reefs to storms, ocean warming, acidification and over-fishing are discussed. The book concludes in Chapter 13 with a discussion of processes and forms of cliffed systems, beginning with an examination of the controls on erosion of soft rock coasts and then extending this to hard rock coasts.

As mentioned earlier, there are gaps in coverage in this book related to particular coastal systems and there is no chapter devoted to coastal management. A more comprehensive treatment would have expanded the book considerably and increased the production and purchase costs, as well as taking us outside the domain of our collective expertise. Nevertheless, we felt compelled to add several special interest boxes throughout the book on topics related to coastal management, and most of these are new. Intense media coverage of the impact of major hurricanes and the increasing frequency and intensity of coastal storms recently served to focus public attention on the vulnerability of coastal infrastructure to a range of hazards. There is growing acknowledgement among coastal scientists that the threat of sea level rise and climate change is immediate and substantial. The coastline will respond, but it is not evident that human society will act quickly enough to mitigate the adverse impacts or be able to adapt to the anticipated rapid rates of change.

A variety of supplementary material is available online to complement the material presented in the book. This includes colour versions of all photographs and diagrams and several videos. Virtual field trips providing examples of the coastal environments described in Part III are also available. Finally, data from field experiments that can be used in laboratory exercises for students are included in separate spreadsheets.

Acknowledgements

When I was asked by Cambridge if I would like to produce a second edition of this book I realised immediately that, while I was still active in research, I would need help from people who are still teaching the subject, in addition to doing research. I would therefore like to begin by thanking my colleagues and co-authors Bernie Bauer and Chris Houser for agreeing so readily to join me in this task. Their participation has not only helped to update the text, but it has brought new perspectives and increased the breadth and rigour of the material covered. I am also grateful to my colleagues in what is now the Department of Geography, Environment and Geomatics at the University of Guelph for continuing to provide me with support for ongoing research and for the production of the book. In particular I am grateful that the department has allowed me continued access to the services of our Cartographer Marie Puddister who has once again compiled all the figures for the book, including colour versions for the web, and created many new figures. I have continued to benefit from interacting with colleagues in the field and at conferences and I would like to acknowledge especially Patrick Hesp, Ian Walker, Jeff Ollerhead, and Irene Delgado-Fernandez who have contributed so much to our field experiments in Prince Edward Island. Finally, I would like to thank my family for all the good times we have, and my wife Sharon for sharing my life and for continuing to put up with my ongoing work in retirement.

RD-A

First and foremost I would like to thank Robin Davidson-Arnott for giving me the opportunity to participate in the writing of the second edition of his textbook – I was honoured by the invitation because the first edition was already well known around the world, but more so because I respect Robin immensely. He and I have collaborated fruitfully on a range of research projects spanning 30 years, and I continue to be impressed by his broad knowledge, keen insights, sharp intellect, Puritan work ethic, positive attitude, and overall good nature. He is gentlemanly in a way that is far too uncommon these days, and I have yet to meet anyone that had a disparaging word to say about Robin. We share many common and highly valued colleagues (including Patrick, Ian, Jeff and Irene, as noted above), to which I would also wish to acknowledge Norb Psuty, Karl Nordstrom, Paul Gares, Bill Nickling, Cheryl McKenna Neuman, Steve Namikas and Rob Brander (among others). My interest in coastal processes was stimulated by excellent mentoring and research opportunities provided by Brian Greenwood, which coincided with the initiation of a life-long collaboration with my close colleague and friend Doug Sherman (thanks, dude!). Along the way, we prematurely lost Bill Carter and Jim Allen, both of whom were enthusiastic and unique coastal characters with much more to offer. I gratefully acknowledge Mike Hilton for hosting me as a William Evans Visiting Fellow at the University of Otago during my last sabbatical, which allowed me time to focus on the book and to witness some of the world's most dynamic process geomorphology in action. It is no surprise that the New Zealand coastline figures prominently in the photos of the revised text. The University of British Columbia Okanagan continues to pay my salary, and I can't believe how lucky we are as university professors during tumultuous times of global political, economic, and religious unrest. My bedrock, however, has always been my wife, Bea, and my family.

BOB

I was first introduced to this book as an undergraduate student in the Department of Geography at the University of Guelph in 1996. Robin used an early draft of this text in his Coastal Geomorphology class and it sparked my interest in coastal systems. While I did foray into the desert for several years with Bill Nickling, I returned to the coast with Brian Greenwood, and I am grateful to both for providing me with a strong background in process geomorphology. This foundation was strengthened through collaborations with Doug Sherman, Rob Brander, Cheryl Hapke, Christian Brannstrom, Sarah Trimble, Brad Weymer and Phil Wernette. None of this would have been possible without my family and their support for my time in the field.

CH

We would all like to thank Matt Lloyd, Earth Sciences Editor, for initiating the project and guiding us through it as well as the many other helpful staff at CUP who have looked after us along the way. Special thanks to Marion Moffatt who did a wonderful job of editing the manuscript. We are also all grateful to Marie Puddister for her outstanding job on the graphics. Thanks also to all of our colleagues who have let us make use of their photographs and research.

Part I

Introduction

1

Introduction

1.1 Humans and the Coastal Zone

The coastal zone is a dynamic environment influenced by atmospheric, oceanographic and terrestrial processes. The combination of processes operating in these systems shapes the coastal zone and determines whether dunes, cliffs or marshes are the primary landform along a particular section of coast. These coastal features evolve in response to variations in natural factors such as sea level, wave climate and sediment transfers between the land and ocean, as well as those due to human activities such as harbour construction and sand mining. In some areas, the accumulation of sediments or sea level fall may cause the land to advance seaward, while areas experiencing a net loss of sediment or a rising sea level may be eroded. Our changing climate means that coastal systems will experience significant change over the next century and there is a need to predict the nature and extent of coastal alternation based on a sound understanding of the physical and biological processes involved.

This book describes and explains the physical processes, and some biological ones, that act to shape our coast, and the unique landforms that develop in response to those processes. As in any other branch of applied science, these process–form interactions can be studied for their own interest. However, there are often aspects of this study which are of particular importance to human life and activities. For example, a large proportion of the world population is concentrated in the coastal zone, including almost all of the major cities such as New York, Tokyo, Amsterdam and Shanghai. The coastal zone is used for fishing, transportation, recreation, waste disposal, cooling and drinking water, and is a source of energy from wave and tidal power. Many of these activities pose an environmental threat to coastal systems, both physical and biological, through pollution, siltation, dredging, infilling and a host of other activities that alter the way natural systems operate. In recent years there has been increasing pressure from leisure activities focused on water sports, and recreation at the seashore (Figure 1.1). In addition, natural processes often pose a hazard to human occupation and utilisation of the coastal zone through wave action, flooding, storm surge, tsunami inundation, as well as through coastal erosion and sedimentation. Because of the threats to human life and activities posed by both environmental impact and natural hazards, there is a strong economic incentive to improve our understanding of processes operating in the coastal zone so that the effects of these hazards can be minimised. Such knowledge is also invaluable in the development of comprehensive coastal zone management planning.

Each maritime country has a unique perspective of their coastline, shaped by history and culture, and by the physical and biological nature of the coast itself. There are commonalities among great differences; for example, the people

Figure 1.1 Examples of recreational pressures on the coast: A. Beach, promenade and sea front shops and apartments, Malo les Bains, Dunkirk, France. Development of the seafront in many coastal towns in Britain, France and western Europe began in the late nineteenth century with the advent of cheap rail travel. Small seafront guest houses are now being replaced by apartments that are used for weekends and holidays; B. Resort development, Frigate Bay, St Kitts, West Indies in March, 2014. The advent of cheap air fares from northern parts of the US, Canada and western Europe has fuelled resort development on a massive scale in Florida and much of the Caribbean and Mexico. Developments here include a large five-star hotel and golf course, other smaller hotels, time-share and condominium apartments and individual houses that are privately owned or rented.

of the Netherlands and of the Maldives both face a similar threat posed by a dense coastal population and rising sea level, even though one nation is situated on a large delta that has had a significant proportion of its area reclaimed by dyking, while the other sits on a small coral atoll. In popular tourist locations around the world and in particular in the Caribbean, the coast is quickly being developed and altered leading to new and unprecedented challenges for countries that are particularly vulnerable to sea level rise but lack the resources or the appropriate management agencies. In contrast, in the United States the US Army Corps of Engineers, a federal agency, has played a key role in coastal development and the management of coastal hazards and they have been in the forefront of applied research on coastal processes and engineering. In Canada there is no equivalent federal agency and the relatively small population and limited resources has left a much greater proportion of the coast relatively pristine.

Canada has one of the longest marine coastlines in the world and within it four distinct regions can be recognised (Figure 1.2). Almost all of the population of Canada lives within 50 km of one of these coasts, and more than half along the Great Lakes–St Lawrence system. The Pacific coast is dominated by swell waves and is generally ice free, while the Arctic coast is dominated by the presence of ice year-round and, in the eastern Arctic by ongoing post-glacial isostatic uplift. The east coast experiences strong mid-latitude storms as well as the effects of one or two hurricanes a year, and much of it is influenced by a seasonal ice cover. Along this coast, the tidal range is < 1 m in parts of the Gulf of St Lawrence and may be over 15 m in the Bay of Fundy. Finally, the Great Lakes are freshwater, but act as small seas, with tides being replaced by seasonal and long-term water level fluctuations. Like the Atlantic coast, seasonal ice foot development occurs in all the lakes and there is considerable surface ice cover on Lakes Erie and Huron.

The potential impact of oil exploration and exploitation off the Arctic and Atlantic coasts, destruction of coastal wetlands and interference with longshore sediment transport, as well as the effects of coastal erosion and storm-wave damage are examples of some of the conflicts that exist in the Canadian coastal zone and that provide a stimulus for developing an improved

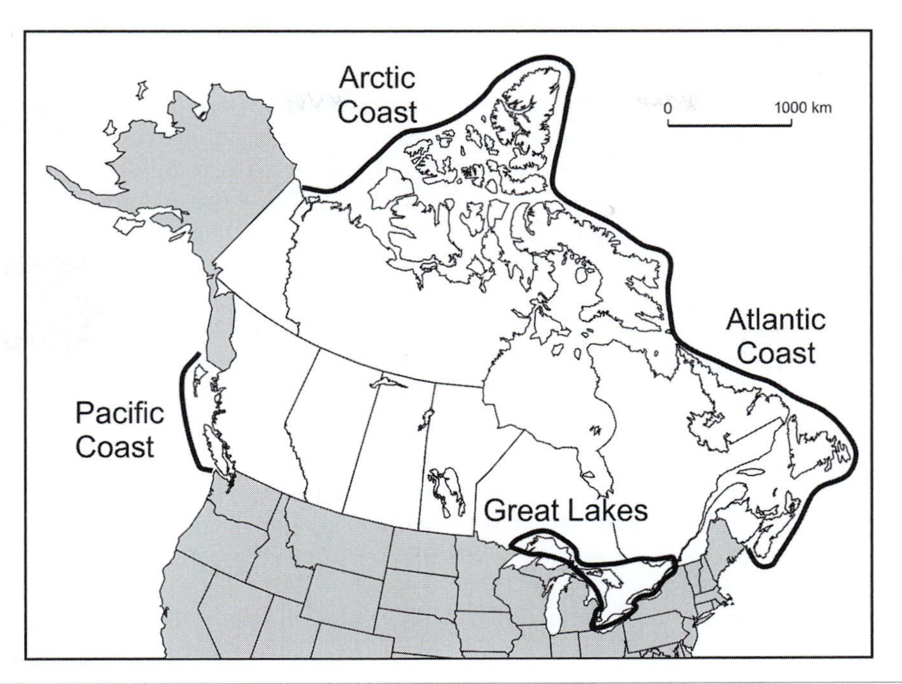

Figure 1.2 Primary divisions of the coasts of Canada (Owens, 1977).

knowledge and understanding of the features and processes.

1.2 Approaches to the Study of Coasts

Coastal geomorphology focuses on the morphology of the coastal zone and on processes such as waves, tides and currents that act to shape features as disparate as high rock cliffs, low coral atolls, muddy deltas and sandy beach and dune systems. Many of these landforms are formed through a combination of coastal, fluvial and aeolian processes, requiring close collaboration, and the sharing of paradigms, instrumentation, field methodology and modelling approaches across those areas of geomorphology.

The coastal zone and coastal processes are also the subject of study by a number of other disciplines, each of which brings a different focus or approach. In particular, there is considerable overlap of interest between coastal geomorphologists, sedimentologists, coastal oceanographers and coastal engineers in the study of waves and currents, and coastal erosion and deposition. While the ultimate objectives of the different disciplinary groups may be somewhat divergent, they share a common interest in expanding our understanding of these physical processes on Earth. Indeed, some coastal scientists and engineers are working with astronomers to help interpret the landscape history of Mars (e.g., Parker *et al.*, 1993; Citron *et al.*, 2018; Goudge *et al.*, 2018). In recent decades, many of the artificial barriers that often separated these groups have disappeared. This is evident in the range of disciplines represented at international coastal conferences, in the groups of collaborators carrying out large projects, and in the contributors to most of the journals that appear in the reference lists at the end of each chapter in this book.

There are also areas of overlap between coastal geomorphologists and biologists studying the aquatic ecology of beaches, estuaries and marshes. For example, the development of coastal dunes is dependent on the presence and diversity of vegetation, which are in turn stressed by the waves and currents that accompany elevated water levels during large storms. Similarly, seagrass and marsh vegetation promote sediment deposition and substrate stability through the attenuation of waves and currents but are also susceptible to erosion by storm waves and currents. Estuaries and marshes play a significant role as nurseries for juvenile fish, and fisheries biologists have an interest in the functioning and conservation of these systems. The coastal zone provides habitat for many species of fish and shellfish, and some open water species may breed in coastal waters. Waves and currents are important for the dispersal of organisms and influence the presence and survival of shellfish, and a variety of other organisms that live in surface sediments of the sandy beach and nearshore environment. Coral reefs create unique and ecologically important environments because they attenuate swell and storm waves, but these systems are vulnerable to erosion by waves, sea level rise and ocean acidification. Globally, the loss of coral reefs will result in dramatic changes to the coast in tropical environments and a loss of ecological diversity.

1.3 Information Sources

There is a long history of the study of coastal processes and landforms. In the past 100 years or so there have been a number of textbooks, published in English, aimed at various levels of undergraduate and graduate instruction, and as resources for researchers of all kinds. Two books by D.W. Johnson (Johnson, 1919, 1925) provide a wealth of information about the coast of the United States and approaches to the study of coasts in the early twentieth century. An understanding of coastal geomorphology was important to the Allied invasion of Normandy in Operation Overlord during World War II. The position of the nearshore bars, influenced by the balance of storm and fair-weather waves, determined the distance that soldiers were exposed to the enemy as they stormed the beach. After World War II there was a rapid growth in studies of coastal geomorphology, marked by the appearance of the first edition of *Beaches and Coasts* by Cuchlaine King (1959) and a popular book by Willard Bascom (Bascom, 1964). Both of these highlighted the research that began in World War II. An English translation of a text by Zenkovitch (1967) provided access to a considerable body of literature from what was then the Soviet Union over the same period. This was followed in the early 1970s by the publishing of the *Coastal Engineering Manual* by the US Army Corps of Engineers through the Beach Erosion Board and later the Coastal Engineering Research Centre. This manual provided a background on coastal processes (particularly waves, wave hindcasting and sediment transport) that has guided coastal engineering and management. While designed primarily to support practising coastal engineers, it proved a useful source for people interested in physical processes in the coastal zone.

The past 40 years have seen a number of textbooks that provide a variety of different perspectives and many of these still provide a good source for information and insights on both processes and coastal landforms. Included in these are books by Davis, (1984), Carter (1988), Carter and Woodroffe (1994), Trenhaile (1997), Komar (1998), Short (1999), Bird (2000) and Woodroffe (2002), which were all generally aimed at senior undergraduates, graduate students, and researchers. Books by Pethick (1984), Masselink and Hughes (2003) and Davis and FitzGerald (2004) were aimed at providing an introduction to the subject that was accessible to undergraduates, both in terms of content and affordability. There are also a number of texts that are devoted to a specific aspect of coastal engineering (Kamphuis, 2000) or coastal geomorphology (Trenhaile, 1987; Sunamura, 1992; Nordstrom, 2000).

Much of the material in this book is drawn from articles published in journals and conference proceedings as well as some of the specialist texts noted above. While each chapter in this book can be read independently, one of our aims is to provide sufficient basic information on vocabulary, methods and processes to make exploring this literature much easier. Almost all the journals are now available online, and provide access to issues that go back to the journal inception. Increasingly, they provide a number of routes to access related publications. For example, the *Journal of Coastal Research* provides broad coverage of all the material covered in this book and includes physical and biological processes, aspects of coastal management and case studies from around the world. There is also considerable coverage in *Marine Geology*, *Continental Shelf Research*, *Coastal Engineering* and the *Journal of Estuarine, Coastal and Shelf Science*. Both *Geomorphology* and *Earth Surface Processes and Landforms* encourage papers on coastal geomorphology. Useful updates can be found in *Progress in Physical Geography* and substantial reviews often appear in *Earth Science Reviews*, *Annual Review of Fluid Mechanics* and *Annual Review of Earth and Planetary Sciences*. Google Scholar provides an easy way to search journal papers and conference proceedings to explore almost every aspect of coastal geomorphology.

Conferences provide a major forum for the exchange of information and ideas, and published conference proceedings still provide a useful source of new information. The *Coastal Engineering Conferences* sponsored by the American Society of Civil Engineers (ASCE) began just after World War II and are held every two years, with additional and specialised conferences such as *Coastal Sediments* and *Coastal Dynamics* held every four years. The Coastal Education Research Foundation, which sponsors the *Journal of Coastal Research*, also sponsors an International Coastal Symposium (ICS), which is held every two or three years in countries around the world. In Canada the first Canadian Coastal Conference was held in Halifax in 1978 (McCann, 1980) under the auspices of the Geological Survey of Canada. Beginning in 1980, Canadian Coastal Conferences were held every two or three years, sponsored initially through a committee of the National Research Council and later by its successor the Canadian Coastal Science and Engineering Association (CCSEA). In the past two decades meetings of Coastal Zone Canada have highlighted most work on coastal management and processes in Canada. In addition to these specialised conferences, coastal geomorphology has become an important focus with dedicated sessions at the annual conferences of the Geological Society of America (GSA), American Association of Geographers (AAG), and the American Geophysical Union (AGU). Aeolian processes are discussed at the International Conference on Aeolian Research, which is now held every two years, and aeolian manuscripts are published in *Aeolian Research*, the *Journal of Geophysical Research*, and most other geomorphology journals.

There is, of course, a vast amount of material available on the Internet, from real-time access to data from wave buoys and cameras set up at various beaches, to data and information provided by a host of government departments and agencies, and from websites of individual organisations and researchers. A good search engine will open up a huge range of possibilities and the problem is to determine what is relevant and what is not. This is why texts like this are important to training the generation of coastal geomorphologists and scientists who will need to respond to a rapidly changing coastal system. The intense media coverage of natural disasters in the coastal zone such as the December 2004 tsunami in the Pacific and Indian oceans, and Hurricane Katrina (2005) in the United States focused our attention on vulnerability and adaptation to these and other coastal hazards. Recent events such Hurricane Ike (2008), Superstorm Sandy (2012) and an increasing frequency of nor'easters along the Atlantic Seaboard of the United States have maintained and even elevated this focus on coastal geomorphology, particularly given the threat of rapid sea level rise predicted over the next century. Given that several hundred million people live along, at or close to the coast, sea level rise and storm erosion pose a significant socio-economic threat in the future. As a consequence, there is a growing acknowledgement of the need for some comprehensive system of coastal zone management to facilitate adaptation to natural hazards and to reduce human impact on natural coastal systems. It is hoped that the material presented in this book can be used to provide coastal managers with background on the physical processes and features of the coastal zone which need to be considered in developing effective management strategies and plans to ensure resiliency of the coastal environment.

References

Bascom, W. 1964. *Waves and Beaches: the Dynamics of the Ocean Surface*. New York: Anchor Books, Garden City, 267 pp.

Bird, E.C.F. 2000. *Coastal Geomorphology: An Introduction*. Chichester: John Wiley & Sons, 332 pp.

Carter, R.W.G. 1988. *Coastal Environments*. London: Academic Press, 617 pp.

Carter, R.W.G. and Woodroffe, C.D. (eds.), 1994. *Coastal Evolution: Late Quaternary Shoreline Morphodynamics*. Cambridge: Cambridge University Press, 517 pp.

Citron, R.I., Manga, M. and Hemingway, D.J. 2018. Timing of oceans on Mars from shoreline deformation. *Nature*, **555**(7698), 643.

Davis, R.A., Jr (ed.), 1984. *Coastal Sedimentary Environments*, 2nd edition. New York: Springer Verlag, 716 pp.

Davis, R.A. Jr and FitzGerald, D.M. 2004. *Beaches and Coasts*. Oxford: Blackwell, 419 pp.

Goudge, T.A., Mohrig, D., Cardenas, B.T., Hughes, C.M. and Fassett, C.I. 2018. Stratigraphy and paleohydrology of delta channel deposits, Jezero Crater, Mars. *Icarus*, **301**, 58–75.

Johnson, D.W. 1919. *Shore Processes and Shoreline Development*. New York: Prentice Hall, 584 pp.

Johnson, D.W. 1925. *The New England Acadian Shoreline*. New York: Wiley, 608 pp.

Kamphuis, J.W. 2000. *Introduction to Coastal Engineering and Management*. Singapore: World Scientific, 437 pp.

King, C.A.M. 1959. *Beaches and Coasts*. London: Edward Arnold, 403 pp.

Komar, P.D. 1998. *Beach Processes and Sedimentation*, 2nd edition. Englewood Cliffs, NJ: Prentice Hall, 544 pp.

Masselink, G. and Hughes, M.G. 2003. *Introduction to Coastal Processes and Geomorphology*. London: Edward Arnold, 354 pp.

McCann, S.B. (ed.) 1980. *The Coastline of Canada, Littoral Processes and Shore Morphology*. Proceedings of a Conference held in Halifax, 1–3 May, 1978, Geological Survey of Canada, paper 80-10, 439 pp.

Nordstrom, K.F. 2000. *Beaches and Dunes of Developed Coasts*. Cambridge: Cambridge University Press, 338 pp.

Owens, E.H. 1977. Coastal Environments of Canada: The Impact and Cleanup of Oil Spills. Fisheries and Environment Canada. Economic and Technical Review, EPS-3-EC-77 Report, 413 pp.

Parker, T.J., Gorsline, D.S., Saunders, R.S., Pieri, D.C. and Schneeberger, D.M. 1993. Coastal geomorphology of the Martian northern plains. *Journal of Geophysical Research: Planets*, **98**(E6), 11061–78.

Pethick, J. 1984. *An Introduction to Coastal Geomorphology*. London: Edward Arnold, 260 pp.

Short, A.D. (ed.) 1999. *Handbook of Beach and Shoreface Morphodynamics*. Chichester: John Wiley & Sons, 379 pp.

Sunamura, T. 1992. *Geomorphology of Rocky Coasts*. Chichester: John Wiley & Sons, 302 pp.

Trenhaile, A.S. 1987. *The Geomorphology of Rock Coasts*. Oxford: Oxford University Press, 384 pp.

Trenhaile, A.S. 1997. *Coastal Dynamics and Landforms*. Oxford: Clarendon Press, 366 pp.

Woodroffe, C.D. 2002. *Coasts: Form, Process and Evolution*. Cambridge: Cambridge University Press, 623 pp.

Zenkovitch, V.P. 1967. *Processes of Coastal Development*. Edinburgh: Oliver & Boyd, 738 pp.

2

Coastal Geomorphology

2.1 Definition and Scope

Geomorphology has been defined as the area of study leading to an understanding and appreciation for landforms and landscapes (Bauer, 2004). Coastal geomorphology is, therefore, the branch of geomorphology that is focused on processes and forms in the coastal zone, whether adjacent to oceans, seas, estuaries or lakes. The methods and theories of coastal geomorphology are strongly informed by those in other branches of geomorphology (e.g., fluvial, aeolian, glacial, tectonic), although the interests of coastal geomorphologists also overlap substantially with those of coastal engineers, oceanographers and marine geologists as well as other scientists interested in the environmental, biological, meteorological and sedimentological aspects of coastal regions. Coastal geomorphology is therefore a hybrid science that has both pure and applied aspects.

Coastal systems vary greatly in their dynamics and according to the types of controls that are imposed on their evolution through time. At one end of the continuum, the response of a loose, sediment bed to wave or swash action varies on the scale of fractions of seconds as water particles are accelerated up and down the foreshore or across a bar face with each wave cycle (Figure 2.1A). Individual grains of sand are mobilised, transported and deposited as a consequence of the fluid motion, and this leads to changes in the morphology at scales of minutes to hours (Figure 2.1B). The wave field will continue to evolve during a passing storm, and therefore the potential for sediment transport (hence, morphological adjustment) is altered continuously. The locus of energy dissipation (or work) on the shoreline will also shift up and down the beach face according to tidal stage as well as storm surge. At the other end of the continuum, cliffs and platforms sculpted in resistant rock such as dolomite or granite may show no observable morphological change over decades or centuries despite being acted upon incessantly by waves and tides (Figure 2.2). Tectonic and eustatic changes in relative sea level are important considerations, as are the fracturing patterns and weathering processes that break down the rocks. Often there are biological or chemical components that complicate matters considerably.

Coastal geomorphology, much like the other branches of geomorphology, evolved from highly descriptive, non-theoretical interpretations of landscapes to much more rigorous, scientific investigations of form-process interactions (Goudie, 2004). Field observation during these early stages of the discipline focused mainly on macro features that were easily observable from land, especially the effects of sea level rise in producing partially submerged coastal features such as fjords (drowned glaciated valleys) and rias (drowned river valleys) or the influence of tectonic uplift in revealing the shape of marine terraces and rock platforms that were once submerged. The action

Figure 2.1 Small scale coastal features. A. Large swash ripples on a shallow sloping foreshore, Doughboy Bay, Stewart Island, New Zealand. Ripples are altered after each up-rush and back-wash cycle. Note heavy (black) mineral lines and patterns of seaward drainage between ripple bifurcations. B. Active construction of a sand berm by repeated swash action and overwash, yielding upward growth and landward progradation (near Port Lincoln, Australia).

Figure 2.2 Large-scale coastal features. Rocky shoreline with very resistant materials near Bodega Bay, California. Note old marine terrace being uplifted due to tectonic instability on this coast.

of waves and currents, however, were rarely measured – rather they were believed to operate in a manner that flattened submerged relief (i.e., marine planation theory) and straightened the coastline by trimming rocky headlands and building sandy barriers across bays and inlets (Woodroffe, 2002). However, the advent of new methodologies for coring and geophysical remote sensing, combined with new dating techniques, led to key developments in sedimentology that transformed the way coastal geomorphologists and geologists thought about coastal evolution.

In particular, detailed histories of the glacial and post-glacial evolution of coasts, involving isostatic compensation and the changing nature of sediment supply from drainage basins, shoreline erosion and littoral drift have revolutionised coastal science and engineering.

Much of the research in coastal geomorphology in the past few decades concentrated on the zone influenced by waves, currents and tides, as the primary driving forces for landform evolution along the coast. These forces, mediated by the strength of materials that they act upon, yield a broad range of erosional and depositional features that define the shape of the shoreline and provide the basis for coastal classification. A major stimulus for these scientific efforts was the need during World War II to predict where and when to land personnel-carrying water craft safely so as to facilitate the invasion of hostile territories in Europe and the Pacific theatre. Early efforts in this regard were not always successful, and the cost was major loss of life, as recorded in historical accounts and re-created in Hollywood movies. However, rapid advances in the technologies and instruments available for measuring fluid motion, sediment transport and morphological change (Davidson-Arnott, 2005a) have enabled coastal scientists to better understand the morphodynamics of coasts, which involves the simultaneous measurement of process dynamics (the forcing functions) and morphological change (the system response) in order to decipher cause, effect and feedback relationships. Studies on sandy beaches and barrier islands are especially conducive to this process-based approach because of the rapid adjustments that occur in these high-energy environments. Similar process studies on rock platforms, cliffed shorelines and coral reefs are increasingly commonplace. There have also been revolutionary advances in the capacity to monitor and analyse shoreline change using remote-sensing technologies, precise digital geo-referencing systems and geographic information systems (GIS). These are especially critical at this time in the twenty-first century because of the anticipated impacts of global climate change on the frequency and intensity of major storms superimposed on accelerated sea level rise, which are likely to have significant consequences for human occupation and utilisation of the coastal zone.

2.2 The Coastal Zone

As scientific disciplines mature, the degree of sophistication with which its practitioners approach the subject matter evolves significantly. Initially, there is a phase of observation and description in which a glossary or lexicon of terms is assembled that can be used to describe what is being observed, ideally with reasonable precision and some semblance of common comprehension. After a lengthy phase of description, the practitioners are able to make conceptual sense of the phenomena under investigation (i.e., they begin to theorise about logical order and structure). A first step in this process of theorisation involves proposing a classification system (i.e., taxonomy) that enumerates the broad spectrum of possible states in a rational scheme. There may be weaknesses in the logical underpinnings of the proposed classification system, or perhaps someone discovers new instances that don't fit nicely into the categories or classes, which usually leads to new and different classification systems being proposed with new (and perhaps different) terminology. It should be appreciated that there is no single 'correct' classification system because each may have a different purpose and orientation – i.e., they provide insight into how scientists were thinking about the coastal zone at the time of development of the classification scheme, and they provide a general perspective on how the coastal zone can be conceptualised and understood.

2.2.1 Zonation and Terminology

There are many terms used to describe coastal features and processes, and sometimes they are defined inaccurately or used incorrectly. Differences in meaning are also hampered by the tendency for different disciplines (or sub-disciplines) to work in isolation or towards different objectives for which an existing lexicon may be inadequate. Thus, coastal scientists in biology, engineering or geology sometimes adopt terms

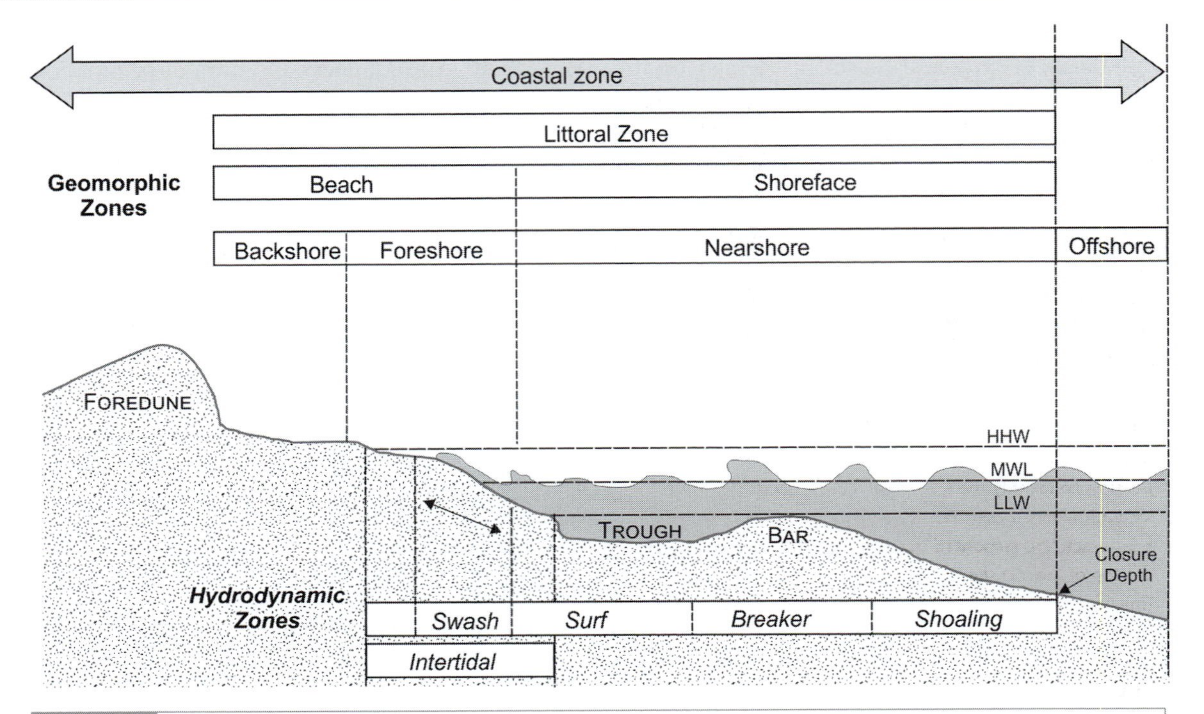

Figure 2.3 Definition sketch of the coastal zone with geomorphic (upper) and hydrodynamic (lower) sub-divisions in the cross-shore direction. HHW refers to high-high water line and LLW refers to low-low water line during spring tides, whereas MWL refers to mean water line in the presence of waves. The swash zone moves up and down the foreshore according to tidal stage while its width depends on wave type and foreshore slope.

that differ from those used by geomorphologists. What follows is an introduction to the nomenclature used commonly in coastal geomorphology and, more importantly, in this book (Figure 2.3). The list is not exhaustive, and additional definitions related to waves or to specific coastal environments (e.g., rocky cliffs, coral reefs, mangrove swamps) are introduced in later chapters, as needed.

2.2.1.1 *Geomorphic Zones*

Coastal zone – The narrow outer margin of a continent or island that is dynamically dependent on the interaction of marine/lacustrine and terrestrial influences, due to its location close to the land–sea boundary (shoreline). The landward and offshore extents of the coastal zone are imprecisely defined deliberately. The limit on land, for example, may be only a few hundred metres from a cliff top or foredune ridge or it may be several kilometres inland if there are extensive sand dunes or intertidal marsh flats. The seaward limit is usually thousands of metres away from the shoreline, but may extend to the edge of the continental shelf or be defined according to shipping routes and national economic interests or by biological criteria predicated on the penetration of sunlight to the sea bed.

Shoreline – A general term used to indicate the location where land meets sea. Given the dynamic nature of this boundary, which depends on wave conditions, tidal stage and storm surge, it is best to provide an operational definition being adopted for any particular study rather than assume that the meaning is implicitly understood by all. Short-term studies of swash dynamics, for example, might be interested to monitor the varying position of the advancing and retreating motion of the waves on the foreshore, which is arguably the actual shoreline position. However, long-term studies of beach change might define a semi-permanent shoreline

Figure 2.4 Varying definitions of a shoreline (photo from near Port Lincoln, Australia). Depending on purpose, it is possible to identify the shoreline with: (i) the time-varying location where water meets the land (far left), (ii) the high tide wrack line associated with the high tide (brown ring at the top of the foreshore), (iii) the eroded toe of the foredune (middle of photo by the break in slope between the beach and the dune), or (iv) the line of permanent vegetation (at the top of the foredune toe). To the right of the photo is a large scarp that demarcates a transition from a wave-dominated erosional period to an aeolian building phase in which the active foredune system in the foreground was emplaced. This was likely stimulated by regional tectonic uplift of the coast.

according to the cross-shore position of the foredune toe or the line of permanent vegetation on the foredune ramp (Figure 2.4). The latter definition of the shoreline lends itself to GIS analysis using remotely sensed imagery and freely available add-in extensions to ArcGIS such as the digital shoreline analysis system (DSAS).

Offshore Zone – The outer, deep-water section of the coastal profile where there is no appreciable sediment transport due to ordinary wave action. The boundary with the nearshore is sometimes defined by the 'closure depth' (Nicholls *et al.*, 1998) which is the depth of water at which only the largest waves during extreme storms are able to influence the bed. This boundary is defined in a relative sense as the transition to intermediate water depths is less than half the wavelength of large storm waves. Note that this depth depends on the wave climate (see Chapter 4), and therefore is linked to the typical size of waves encountered along a specific section of coast. The landward boundary may also be defined (especially by coastal engineers) as the outer limit of the breaker zone.

Littoral zone – The portion of the coastal profile where sediment is transported frequently by wave action and by wave-generated currents. The seaward limit is the closure depth, and the landward limit is the position on the beach where normal wave action ceases to be relevant. On most sandy beaches, this limit is defined by a storm berm or wave-cut scarp (either cut into the beach or into the foredune toe), although it can also be identified by a semi-permanent wrack line or by large woody debris. On cliffed coasts and rocky shores, there may be a wave notch that indicates the upper limit of wave action, and usually there is a biological transition from marine or intertidal species to subaerial species that can be used to demarcate the limit.

Beach – The portion of the coastal profile that is influenced both by wave action and by subaerial processes because it is periodically exposed to air or episodically inundated by large waves. The beach comprises the foreshore and backshore (see definitions below), and is also referred to, simply, as the shore (Figure 2.5). Sometimes the terms backbeach and midbeach are used by aeolian geomorphologists to differentiate among zones of the subaerial portion of the beach.

Shoreface – The seaward dipping region dominated by wave action, extending offshore from the low tide line (or from the lower limit of the beach face). The term is used primarily by geologists/sedimentologists and may be divided into an upper and lower shoreface marked by a distinct change in slope on sandy coasts at a depth around 6–10 m (Cowell, 1999). The seaward limit is ill-defined but coincides with the transition to the continental shelf, which has a gentler slope than the shoreface.

Nearshore zone – Similar to the shoreface, with a landward limit at the base of the foreshore but a seaward limit defined by the closure depth (as with the littoral zone). It is the zone of wave action and sediment redistribution, not including the beach zone. Note that if the offshore zone is defined as beginning at the seaward edge of the breaker line rather than the closure depth, then

Figure 2.5 Examples of beaches. A. Wide, flat beach backed by small foredune and cliffs (near Crescent City, California). Wheel tracks mark the break in slope between the flat backshore or berm (left) and the steeper foreshore (right). B. Shingle–gravel beach backed by eroding, soft-sediment scarp (north of Oamaru, New Zealand). The break in slope from the steeper foreshore (left) to the backshore (right) is marked by the transition from gravel to cobble-sized shingle. Storm surge during high tide rearranges the foreshore and berm, and occasionally attacks the cliff at the rear of the beach.

the extent of the nearshore zone should be adjusted accordingly.

Foreshore – A subsection of the beach that is subject to the action of swash motion (Figure 2.6). The seaward limit is defined by the lowest extent of swash activity, often indicated by a break in slope or the presence of a pronounced beach step. The upper limit is defined by the highest extent of swash activity, which is demarcated by a line between dry and saturated sand or by morphologic features such as a berm crest, scarp or recent wrack line. The width (and height) of the foreshore may vary from season to season depending on the wave climate, and it is related to tidal range. The steepness of the foreshore varies according to grain size and wave conditions. The foreshore is usually planar in profile geometry although sometimes with cusps alongshore, and it is sometimes referred to as the beach face.

Figure 2.6 Example of a foreshore with large cuspate features on a popular swimming beach south of Sydney, Australia. Swash motion is semi-rhythmic alongshore with greater landward excursions in the cusp embayments, as indicated by the zone of saturated (dark) sand. Most of the people are sitting on the flat backshore, which lies above the transition to the foreshore, as defined by the gradient from dry to moist sand (and also by zone of human footprints).

Backshore – The most landward section of the beach, extending from the top of the foreshore to the foredune toe or cliff face. This is the section of the coast that most people would refer to as the 'beach' because it is usually dry, relatively flat, and conducive to recreational activities. It comprises the berm or sometimes multiple berms. It is often subject to aeolian activity during wind events, but may be inundated by wave action during major storms, especially during high tides.

2.2.1.2 *Hydrodynamic Zones*

Shoaling zone – The region of intermediate water depth in which waves begin to experience transformations because of the constraints imposed by limited water depth on the wave dynamics (see Chapters 4 and 5). The seaward limit aligns roughly with the closure depth whereas the landward limit is the zone of wave breaking. Both limits are wavelength dependent, so this zone varies in width accordingly. Shoaling involves an increase in wave height and a decrease in wavelength as well as changes in wave shape to more asymmetric and skewed profiles, but not to the extent that the crests become unstable and break.

Breaker zone – The zone in which pronounced wave breaking takes place due to instability in the wave form. It is ill-defined and depends on beach geometry (slope, presence of nearshore bars) and deep-water wave characteristics (wavelength, height), therefore it can be either narrow and clearly visible (e.g., with plunging breakers) or wide and diffuse (e.g., with spilling breakers). It is best to think of the breaker zone as being the general location where wave shoaling reaches its ultimate unstable stage, and significant amount of white, turbulent water is produced. Sometimes there are multiple breaker lines, especially in nearshore systems with multiple bars.

Surf zone – The zone of surf waves (bores) extending from the breaker zone to the foreshore. After waves shoal and break, they often reform as smaller, solitary waves that propagate to shore. Depending on beach slope and incident wave conditions, the surf zone may contain a dozen or more bores (e.g., on shallow gradient beaches with incident waves breaking far offshore) or there may not be a surf zone at all (e.g., on steep beaches where wave breaking occurs directly on the foreshore).

Swash zone – A very dynamic region of the coast where incident waves expend their remaining energy, either via wave breaking directly on the foreshore or as wave run-up (swash uprush) and wave run-down (backwash). The width of the swash zone varies with foreshore slope and incident wave conditions. On lake shorelines with active storm waves, the swash zone is essentially the same size as the foreshore, extending from the top of the step to the top of the berm crest. However, when the storm dies away, the swash motion generated by the small residual waves will only occupy the lower section of the foreshore. On ocean shorelines, the swash zone migrates up and down the foreshore according to the tidal cycle, and the swash zone itself can be narrow or extensive depending on how large the incident waves are.

Intertidal zone – The section of the shoreline that is cyclically exposed and submerged by low and high tides, respectively (Figure 2.7). It is the horizontal expression of the tidal (vertical) range, usually referenced to the low-low water (LLW) and high-high water (HHW) levels. On shallow gradient beaches, the intertidal zone can be hundreds of metres wide consisting of a single bar-trough system or multiple ridge-and-runnel systems whereas on rocky coasts the low tide often exposes shore platforms that are inhabited by numerous intertidal organisms.

2.2.2 Classification of Coasts

The characteristics of the coastal zone vary considerably from place to place, and they may do so over relatively short distances. The most distinguishing feature(s) of a specific coast may be due to large-scale tectonic factors (e.g., a sequence of exposed marine terraces), local geology and stratigraphy (e.g., sea stacks and arches), dominant process dynamics (e.g., recurve spit), or the effects of recent changes in relative sea level (ria coast). It is possible to imagine classification systems having multiple bases upon which the categorisation scheme is founded.

Visual descriptions – A visual basis for a classification system simply asks 'what does the coast look like?' In answering this question, you might come up with material descriptors such as sandy, rocky, or volcanic, but you might equally identify the environment (e.g., ocean, lake, bay, estuary) and determine that the dominant processes are associated with waves, currents, tides or some terrestrial origin such as permafrost or slope instability. The advantage with visual descriptions is that they are easy to understand and intuitive. The disadvantage is that the terminology is often vaguely defined, and often there is no rational basis for selecting categories and membership criteria. Further, there is some likelihood that the system will not be comprehensive (i.e., inclusive of all possibilities).

Figure 2.7 Intertidal zone. Aerial photo of a rocky shoreline with sand patches exposed at low tide in a macrotidal environment (south of Calvert Island, British Columbia). The intertidal zone extends approximately from the current position of the waves almost to the line of large conifer trees. In the left-centre and middle part of the photo, adjacent to the forest, are two small pocket beaches backed by large woody debris (LWD), which was floated there during recent high tide storm events. In this part of British Columbia, most beaches have fringes of LWD that are frequently remobilised during storms, and are thought to play a role in the morphodynamics and ecology of the beaches.

Genetic (evolutionary) pathways – There has been renewed interest in understanding landform evolution over long time scales, especially in light of rapid sea level rise and its consequences for shorelines around the world. For example, old theories regarding coastal landforms passing through various stages of evolution, such as Darwin's ideas about coral atolls, are still quite relevant (see Chapter 12). Coasts for which relative sea level is rising (submergent coasts) have a much different look than emergent coasts for which the rate of uplift is more rapid than the rise in global sea level. A classification system based on these notions must therefore consider tectonic, isostatic, eustatic and steric factors, which are most appropriate for coastal evolution at large spatial and temporal scales.

Mass budget states – The end member states of a classification system based on sediment budgets would be depositional (prograding) versus erosional (retrograding) coasts. Each class can be further subdivided according to the nature of the mass budget (positive vs negative, relative magnitude) as well as the characteristic forms that might be prevalent (e.g., bay mouth bars and spits versus rocky platforms and sea stacks).

Process dominance – A classification system based on processes (or agency) would identify what sort of process suites are the most active on the coast, and these might include waves, tides, currents, ice, permafrost, slope instability, rock weathering and a range of biological processes involving plants and animals (e.g., coral reefs, mangroves).

Energy conditions – Some coasts are known as high energy (e.g., open, exposed coasts in the major oceans) versus low energy (e.g., estuaries, bays, small lakes). The challenge is to identify the proper metrics to differentiate the categories, which might include mean and maximum wave heights, mean and maximum tidal range, or some cumulative energy value derived from the wave climate. It seems reasonable to tailor the range in high–low energy conditions to the type of environment, thereby establishing one set of norms for oceans and large seas versus another for lakes and small bays.

Morphodynamic states – Since the 1970s, it has been appreciated that sandy coastlines can be classified according to their morphodynamic state, which uses terms such as reflective, intermediate, and dissipative. This will be discussed in greater detail in Chapter 8.

The multiplicity of these approaches, as well as the myriad factors controlling coastal evolution across the globe, has given rise to a number of different classification schemes. Finkl (2004) provides an excellent and comprehensive review of coastal classification systems, so only a brief summary of a few of the most well-known systems will be offered here, mainly to illustrate the nature of the challenge associated with classifying coasts.

2.2.2.1 *Shepard's Classification System*

Shepard (1937, 1963) was one of the first to propose a comprehensive coastal classification scheme that was updated several times into the 1970s. According to this scheme, coasts can be divided into one of two categories: **primary** coasts, which are controlled by terrestrial agents such as the tectonic setting or geological materials; and **secondary** coasts, which are shaped by marine agents such as waves and currents. Thus, the categorisation is predicated on whether the look of the coast is due to the dominant influences deriving from the land or alternatively from the ocean. Both the primary and secondary categories were subdivided further to distinguish between other major controls. For example, **primary** coasts encompass:

- land erosion coasts, which were shaped by subaerial or fluvial erosion and then drowned (e.g., rias, fjords)
- subaerial deposition coasts, which include deltas, drowned alluvial plains, and glacial deposition
- volcanic coasts
- faulted and folded coasts
- ice-dominated coasts.

Secondary coasts follow a parallel logic, and include:

- wave erosion coasts (e.g., wave-cut benches, wave-straightened cliffs)
- marine deposition coasts (e.g., barrier islands, cuspate forelands, saltmarshes, mudflats)
- organic coasts that are dominated by biotic activity (e.g., coral coasts, oyster reefs, mangroves, marsh grass embayments).

The use of the term 'primary' in association with land-based influences (and 'secondary' in reference to water-based influences) may reflect the reality that this scheme was proposed during a period when there was relatively little technological capacity to make observations from or below the sea. In the present-day context, it is inaccurate to interpret the term 'primary' as implying dominance, in the literal sense, over 'secondary' effects.

2.2.2.2 *Inman and Nordstrom's Classification System*

Despite long-standing appreciation that there were tectonic influences on coastlines, it wasn't until the mid 1960s that the theory of plate tectonics was proposed with sufficient empirical support to convince the scientific community of its explanatory power. The theory gained almost immediate and widespread acceptance, and it is perhaps not surprising that the ideas and terminology used in plate tectonics were usurped for the purposes of a coastal classification scheme. Inman and Nordstrom (1971) devised a two-tier classification system in which the first-order features are related to continental scales and are based on the plate tectonic context of the coast. The scales are of the order of 1000 km long, 100 km wide, with heights of a few kilometres. In this scheme, coasts are divided into three major groups:

Collision coasts: (1) continental margins characterised by cliffed shorelines, narrow coastal plains and deep ocean trenches, with high mountains just inland; (2) island arc collision coasts, which are similar to continental margins in general form but lack the associated major mountain ranges and large rivers.

Trailing edge coasts: (1) neo-trailing edges formed near the beginning of plate separation (e.g., the Red Sea) with little or no continental shelf development; (2) Afro-trailing edges characterised by hilly or plateau coasts; and (3) Amero-trailing edges with a wide continental shelf and large rivers draining the interior of the continent.

Marginal sea coasts: coasts fronting on marginal seas and protected from the open ocean (Mediterranean, China).

Inman and Nordstrom (1971) also defined a broad range of second-order features, which enhance the classification scheme using attributes based on morphology and processes. The reader is referred to the original paper for details.

2.2.2.3 *Valentin's Classification System*

Valentin (1952) proposed a classification system that focused on the direction of long-term movement of the shoreline (relative to fixed reference grid), and the two primary classes were: **advancing** coasts and **retreating** coasts. It is notable that this work came in advance of the plate tectonic theory, although Valentin recognised that there were both tectonic and eustatic factors leading to advance or retreat. Specifically, he suggested that advancing coasts could arise because the sea floor was emerging out of the ocean (i.e., via tectonic or isostatic uplift) or because the coastline was prograding due to sediment accretion on the coast or because of biological factors (e.g., coral growth, mangrove trees trapping mud). Retreating coasts, on the other hand, might be due to submergence (e.g., eustatic sea level rise, tectonic downwarping) or because of retrogradation, which implies erosion of the coast by wave action and other forces. Valentin's framework was used by Sherman and Bauer (1993) to develop a generic process–response classification scheme that explicitly includes micro, meso and macro scales of adjustment.

2.3 Controls on Coastal Form and Evolution

The three systems of coastal classification presented above are each internally consistent and easy to understand, yet they are very different in how they categorise (and choose to organise) the types of coast across the world. At this point, it is not unreasonable to ask 'what is the utility of a coastal classification system?' Arguably, the process of categorisation within a relatively young, but maturing scientific discipline such as coastal geomorphology is an essential step leading toward a more complete understanding of the phenomena being studied. Indeed, implicit to any classification system is an underlying sense of order, which is then subject to scrutiny using the scientific method. In particular, it is useful to query the basic logic of the classification system, and to ask whether we truly understand how different coasts evolved to their current state. This demands an intricate knowledge of the many controls on coastal evolution as well as the complex interactions that ultimately lead to what we are able to see and describe on the coast.

Fairbridge (2004) suggested that a minimalist description of coasts requires information on three categories of attributes, which can be interpreted as factors or controls on coastal evolution. These are:

Type of material – The two end-members are hard versus soft materials (Figure 2.8). Hard refers to rocks with considerable internal strength (e.g., granite) whereas soft refers to unconsolidated sediments with little internal cohesion or easily dissolved rocks such as limestone. The type of material (i.e., its relative hardness) dictates the strength or resistance to change, and therefore the speed at which geomorphic evolution can take place.

Agency or process – *Physical* processes involve the motion of water (waves, currents, tides) and air (wind) in ways that lead to erosion, transportation, and deposition of materials in the coastal zone. *Chemical* processes are ones that alter the surface or internal strength of

Figure 2.8 The role of material strength is critical to coastal morphology. A. A hard rock island off the coast of California near Bodega Bay. Note the craggy, steep sided features as well as the wave-cut notch at the base of the island, just above current water level. B. Wave cut cliff in soft, sandy aeolionite with a more resistant cap rock (near Port Lincoln, Australia). As the base of the cliff is attacked by waves during storms, the soft materials are eroded and transported alongshore to become part of the beach, whereas the cap rock collapses and breaks apart to form lag deposits at the base of the cliff (bottom foreground of photo). These slabs are easily abraded and disintegrated, so they do not accumulate as a talus slope that might offer protection to further cliff erosion.

the materials, thereby making them more susceptible to degradation, mobilisation, and reconsolidation. *Biological* processes are similar to chemical processes, but they operate on a biological basis which often involves biochemical processes. Note that agency can be either destructive (erosional) or constructive (depositional).

History or context – This attribute category includes broad-scale controls (i.e., context) such as the geo-tectonic framework (e.g., type of plate boundary, propensity for faults and fractures, volcanic activity, etc.), isostatic adjustments (e.g., glacial unloading), eustatic and steric impacts (due to changes in global water volume in the oceans), and human modifications to the coasts, whether planned or inadvertent.

A somewhat different, but parallel approach is that by Davies (1972), who also focused on broad controls on coastal evolution (Table 2.1). He recognised three categories of factors, but organised them according to: (1) physical factors of the land; (2) physical factors of the sea; and (3) biological factors. It is easy to see how Davies' approach is adapted from the combined earlier works of Shepard, Valentin and Inman and Nordstrom, because it includes many of the same concepts but is organised in a different way. This suggests that, perhaps, the era of classification reached its pinnacle in the 1970s, because there have been relatively few innovations since (cf. Finkl, 2004). Rather much of the recent effort has been devoted to mapping shorelines using advanced technologies, not for the purposes of classification, but for resource management and for understanding shoreline dynamics for scientific and engineering purposes.

2.4 Shoreline Mapping and Technological Innovation

Mapping of the coast was originally done for the purposes of navigation, and this effort continues today with ever-increasing sophistication, although for much different and varied objectives. Early maps prior to the nineteenth century were derived from sketches, coastal surveys and ship-based soundings. The manual effort involved was extraordinary, so only the locations of greatest commercial or strategic naval importance were mapped – the majority of the coast

Table 2.1 Factors controlling coastal evolution and form

Factors of the land	
Geological structure	Plate tectonics, mountain ranges, continental shelf width, relative relief
Local geology	Rock type, fractures and fissures
Geomorphic processes	Fluvial incision, deltas, sediment supply
Isostatic sea level change	Tectonic isostasy, glacial isostasy
Factors of the sea	
Eustatic sea level change	Glacial eustasy, geoidal changes
Wave climate	Water body size, orientation of the coast, wind and storm climatology
Tides	Tidal range, tidal type, tidal currents
Ice effects	Shorefast ice, winter ice cover
Local erosion and deposition	Shoreline transport and deposition processes
Biological effects	Mangroves, saltmarshes, coral reefs, seagrass beds, coastal dune vegetation

Source: Davies (1972).

remained unmapped and undocumented. By the early twentieth century, aerial imagery was becoming available, and it provided the basis for drawing very accurate maps of the shoreline, even along coasts that were remote and largely unexplored. Black and white photography was replaced by colour photography in the 1950s, which made identification and interpretation of shoreline features much easier. Eventually, infrared photography became common as a complement to colour imagery, but this has mostly been replaced by digital photography, which has revolutionised the collection of spectral information from the landscape and sea surface. Digital cameras with the capacity to measure across a broad range of spectral wavelengths are mounted on airplanes, satellites, and most recently drones or unmanned autonomous vehicles (UAVs), providing global coverage at extraordinary resolution. Global positioning systems (GPS) are used to georeference these images with respect to national or international reference grids, allowing for repeat surveys to be compared and contrasted with great accuracy. It is therefore possible to monitor shoreline change across long stretches of coastline and assess the impact of the driving forces such as hurricanes and mid-latitude storm systems. It is also possible to map sensitive shorelines and wetlands, especially on large lakes that are subject to increasing development and recreational pressures in the form of seawalls, docks and boating activity.

Since the 1990s, light detection and ranging (LiDAR) has been used intensively to survey the coast at a resolution that is unprecedented, thereby providing detailed renditions of the coast that were largely unimaginable only a few decades ago. As a consequence, LiDAR imagery is used for purposes of shoreline change detection, storm damage assessment, vegetation surveys and infrastructure inventories with repeat flights showing before and after conditions that reveal the impact of natural disasters as well as the imprint of human activities. A special issue of the *Journal of Coastal Research* (Brock and Purkis, 2009) was devoted to the emerging role of LiDAR in coastal research and management, and it provides a comprehensive introduction to the many applications of this important technology. The United States Geological Survey (USGS) used LiDAR to assess barrier island dynamics along the Gulf of Mexico, and this yielded important information on the impact of hurricanes and the consequences for the insurance industry and federal disaster management. Significant portions of the US and European shoreline have now been mapped using airborne LiDAR methods. LiDAR has also been used by researchers interested in cliff instability on rocky coasts to assess fracture patterns and miniscule movements along planes of weakness. Aeolian geomorphologists have used terrestrial LiDAR to provide information on changes to foredune morphology and on moisture conditions on beaches.

With the widespread use of digital technologies came the need to store, manipulate, exchange, analyse and visualise huge amounts of data. Enhanced computing capacities, especially high-speed processors became essential, but so too did the requirement for advanced software. The rapid development and adoption of geographic information system (GIS) software packages in the late 1980s, with a proliferation of freely available add-ins and specialised analysis modules, has been critically important in advancing the effort to map the coastal zone for a broad range of management, engineering and scientific purposes. The data bases that are utilised in GIS models are capable of portraying information on many more variables than can be incorporated in a simple coastal classification system. Moreover, recent data sets contain immense spatial and temporal detail that can be utilised by computer-based learning algorithms to accomplish classification in an objective manner according to rule sets established by the end user. Thus, it is now possible to produce tailor-designed classifications that might focus on the intersection of data layers that represent geology, sedimentology, slope, vegetation cover, tectonic processes, hydrodynamic variables, property value, infrastructure type, population density and any other attribute that can possibly be mapped and represented alongside a digital elevation model (DEM). Shoreline mapping and classification using GIS was carried out for the entire shoreline of the Great Lakes on both the US and Canada shores in the early 1990s as part of an International Joint Commission study of the impacts of high lake levels and storm events on coastal properties. Similar GIS data bases are available for coasts along many parts of the world and are proving to be invaluable for local authorities charged with coastal management. They can be adapted to address a whole range of issues from modelling the potential impact of oil spills to addressing some of the challenges posed by global climate change and sea-level rise (McFadden *et al.*, 2007), such as hazard zone mapping, infrastructure protection and community relocation.

2.5 Understanding and Modelling Coastal Dynamics

As is the case in other areas of geomorphology, a reductionist paradigm has guided much of the field and laboratory experimentation in coastal geomorphology (Bauer and Sherman, 1999; Walker *et al.*, 2017). Reductionism refers to the tendency for scientists to examine processes at smaller and smaller scales in the hope of being able to assert control over a range of extraneous factors that might have a bearing on the problem (Rhoads and Thorn, 1996; Harrison, 2002). For example, a coastal geomorphologist interested in the work of waves or blowing wind on a loose sediment bed might devise experiments in a wave tank or wind tunnel so that the grain size and fluid forcing can be selected and controlled with great accuracy. This eliminates much of the uncertainty inherent to field experimentation for which there is no control over the waves, the wind or the sediment. The laboratory experiments can be repeated to verify the outcomes, and new simulations can be performed using different combinations of grain size, wave height, wave period or wind speed to determine the influence of each factor in isolation. The ultimate goal of this approach is to build sufficient understanding at the micro-scale so as to explain what happens in the real world, and ideally to develop computer models of coastal processes based on the fundamental physics (or chemistry or biology) that will emulate natural processes. A numerical model is therefore an abstraction or simplification of reality that is useful for revealing what might happen in the near or distant future (or past) given a set of initial conditions and circumstances. Field experimentation will always be necessary in order to verify the model predictions, to assess the range of validity and applicability of the model, and to provide insight into the inherent complexity of the real world. In this respect it is critically important to appreciate the limitations of any model so that it is not used beyond the scope for which it was designed.

The example above, of predicting sediment transport by waves or wind, is chosen purposely because it reveals much about the distinction between what is understood about coastal processes relative to what can be effectively modelled. Many decades of careful measurements have shown that sediment motion on a bed of non-cohesive sediments is first initiated when the combined lift and drag forces exerted by the fluid on a single particle exceed the buoyant weight of the particle, tending to keep it on the bed. The equations are easily written and coded. However, two major issues hamper any attempt to operationalise these equations in the form of a sediment transport law. First, the exact position of the particle with respect to its neighbours cannot be described for every single particle on the bed. This is a critical concern because how a particle sits on the bed dictates whether it may lift, roll or slide out of position. Thus, it is not sufficient simply to quantify the lift and drag forces, but there is a need to apply the method of moments, which involves lever arms and centroids (e.g., Komar, 1998). Moreover, as soon as one particle begins to move, there is a cascading effect on all neighbouring particles because the packing geometry of the localised bed has been altered. So the problem of sediment transport evolves into one that requires keeping track of the continually changing geometry of the bed in addition to the trajectories of the particles in transport (and their interaction with the bed).

Second, the fluid motion is characteristically turbulent, which means that the fluid forces on the bed are subject to change instantaneously even if the overall mean flow conditions are constant. Both issues imply that further reductionism is needed to sort out the details, or alternatively, to adopt alternative approaches to operationalise the basic equations in a way that facilitates the prediction of sediment transport rate (e.g., using stochastic functions or constitutive relations). It is enlightening to realise that most predictive equations of sediment transport are based on simplified parameterisations that incorporate some of the basic physics (e.g., transport rate is proportional to the cube of the fluid speed) and also one or more empirical terms or functions intended to account for unknown complexities (see Chapter 9). Not surprisingly, the art of sediment transport prediction is quite uncertain despite a sound understanding of the basic physics involved.

The reductionist approach is ideal for yielding new and novel insights into basic processes, but it has largely failed on its promise to deliver enhanced predictive capacity. Physicists have long understood the challenges of upscaling, which involves using micro-scale knowledge (e.g., at the scale of sand grains and fluid particles) to predict phenomena occurring at larger scales or higher levels (e.g., the rhythmic pattern of beach cusps or the offshore increase in nearshore bar size and spacing). Many people have argued that reductionism is not a viable approach to the study of landscapes (e.g., Harrison, 2002) because landforms are emergent features – i.e., arising from complex interactions that are not predictable because of the stochastic, if not chaotic, nature of morphodynamics. There is now a general recognition that modelling of large-scale coastal evolution (temporal scale of decades and spatial scale of tens to hundreds of kilometres) may require a different approach than simply running the small-scale models for larger areas and over longer periods of time (Cowell *et al.*, 1995; Ashton *et al.*, 2001; Harvey, 2007; Baquerizo and Losada, 2008; Kroon *et al.*, 2008). Walker *et al.* (2017) try to reconcile the reductionist and emergent approaches based on a decade-long monitoring study of a beach–dune system in Prince Edward Island, and they provide an interesting discussion of the challenges. The key message from these sort of investigations is that upscaling from the smallest scales of inquiry to the landscape scale can be fraught with pitfalls, and it is critically important that investigators understand the broader context in which their experimental results should be interpreted.

Regardless of the scale of inquiry, most geomorphologists adhere to concepts embedded in general systems theory, which includes such ideas as dynamic equilibrium, thresholds and feedback loops (Chorley and Kennedy, 1971; Rhoads and Thorn, 1996). Key among these is that the landforms visible on the landscape are in a state of mutual adjustment with the suite of processes that act upon them. In other words, there should be a direct association between the processes active along the shoreline and the characteristic forms that evolve as a consequence of those processes. In this way, coastal geomorphologists have come to understand how a beach berm differs from a sea stack, and why each of these forms has taken very different evolutionary pathways to their current state. Many coastal engineers assert that these dynamic coastal systems can be modelled reliably as long as there is sufficient empirical data to calibrate the models and to provide robust boundary conditions.

One of the most popular numerical models used by coastal engineers is SBEACH, which is an empirically based, short-term processes model used to estimate beach–dune profile response during storms. The model domain extends from the landward limit of wave run-up to the offshore closure depth. It was developed by researchers at the US Army Corps of Engineers (USACE) for sandy beaches with uniform grain sizes in the range of 0.2 to 0.42 mm, and the developers recommend that users calibrate the model with survey data from the study site for multiple before-and-after storm periods. It is used extensively to predict the fate of beach nourishment schemes for different cross-sectional designs and storm scenarios. Although SBEACH used to be freely available through the USACE, it is now sold commercially (with educational discounts) as a module within a comprehensive collection of coastal engineering design and analysis software (CEDAS) that also includes ACES, which is a wave prediction module, and several other modules dealing with longshore sediment transport, breakwater design, wave overtopping, inlet hydraulics and vessel generated waves. A similar model, XBEACH, is an open-source numerical code developed by Deltares in collaboration with UNESCO-IHE and TU Delft. It also computes the nearshore hydrodynamic conditions and morphodynamic response of the beach–dune profile during storm events, but it must be purchased commercially. XBEACH has been used for coastal planning, engineering design and scientific research in many countries, but enjoys its greatest popularity in Europe, especially the Netherlands.

Although every computer model of coastal evolution (of which there are a great number) differs in the details, they all share some commonalities. Typically, they are modular in the sense that they contain several different subroutines that deal with different aspects of the problem semi-independently. Basic components include:

Bathymetric or topographic grid – This module allows the user to establish the physical boundaries of the model, usually in the form of a digital elevation model (DEM) that is measured directly (e.g., via surveying) or extracted from some prior product (e.g., contour map). Often, the grid is nested, with a main grid that has large spatial coverage extending from the beach to the continental shelf but with sparse details, and a high-density grid providing great detail in the area of concern, whether it be the surf zone, a tidal inlet or the area surrounding a breakwater. The extent of the grid depends on the problem of interest and depends very much on whether the outer boundary conditions to the high-density grid are to be simulated (e.g., using a wave shoaling model) or prescribed (i.e., based on a known wave conditions at the break point).

Hydrodynamic module(s) – In this component of the model, the hydrodynamic conditions are predicted for the modelling domain.

The modelling domain is a 3D mesh grid that is bounded by the bathymetry on the bottom, by the landward, seaward and alongshore limits imposed by the user or defined by topographical controls (e.g., headlands, submarine canyons) or by variable limits such as the closure depth or breaking zone, and the sea surface. Note that the grid defining the sea surface is a dynamic one that must accommodate wave motion, tidal stage and wave set-up and set-down. Complex wave models also simulate the energy input into the sea surface by wind, which is either measured or modelled (using global circulation models and regional downscaling algorithms). The shoreline boundary is similarly complex because it must accommodate wave run-up and run-down as well as groundwater dynamics (e.g., infiltration and exfiltration). The hydrodynamic models are usually founded on conservation principles of mass, momentum and energy as codified, for example, in the Navier–Stokes equations using a range of assumptions about how turbulent energy is created, transferred and dissipated within the water column. The hydrodynamic modules can be extraordinarily complex or quite simplified depending on the modelling objective (e.g., forces on floating oil platforms in the North Sea; effects of sea ice on wave attenuation and sediment transport along the coast; prediction of sediment transport in longshore currents).

Sediment transport module(s) – Geomorphologists are interested mainly in how coastal landforms evolve, so the details of the hydrodynamic modules may not be of great interest as long as they accurately predict parameters that are needed to simulate sediment transport. In particular, it is critical to know the time-varying shear stress at every location on the sea bed, which depends on the size of waves, the degree to which waves interact with currents (e.g., longshore, tidal, rip cells, undertow), and the changes in water depth due to tidal stage and storm surge. Sediment transport is often sub-divided into: (1) initiation of motion; (2) bed-load transport; and (3) suspended load transport. The potential for sediment transport depends critically on grain size, which varies across the nearshore profile, so a detailed mapping of grain size distributions is required and often incorporated in the bathymetric grid. Moreover, the sea bed is typically not flat but contains bedforms of varying scale (e.g., ripples, megaripples, bars), and these serve to impart form drag or friction to the hydrodynamic processes. This indicates how there can be critical feedback between the modules that must be accommodated during the simulations as conditions change throughout the progression of a storm.

Morphologic adjustment – The output from the sediment transport module is an instantaneous map of the transport flux at every node in the model domain. In order to calculate the impact on the morphology, the spatial difference in transport rate (i.e., the sediment flux divergence) is needed. The basic relation involves mass conservation in the form of the simplified Exner equation (Paola and Voller, 2005):

$$\frac{\partial z}{\partial t} = -\frac{1}{1-p} \nabla \cdot q_s \tag{2.1}$$

where z is the bed elevation, t is time, p is sediment porosity and q_s is the sediment volume flux. This relation shows that the spatial difference in transport flux between grid nodes is proportional to erosion or deposition of sediment from the sea bed across those nodes. Account needs to be taken of temporary sediment storage in the water column as suspended sediment as well as losses and gains across the exterior boundaries of the modelling domain. Note also that these are vector quantities that don't necessarily align with the orientation of the original bathymetric grid, so the model needs to keep track of the varying geometries involved.

A model simulation is initiated by setting the appropriate boundary conditions and time steps, and then solving the hydrodynamic equations for waves and currents, using the hydrodynamic parameters as input into the sediment transport module, calculating the sediment flux at every node and feeding the results into the Exner equation (2.1) to predict the morphologic response. Once the distribution of erosion and deposition has been quantified, an adjustment is made to the original bathymetric grid, and another loop through the modules is performed for every subsequent time step. After a pre-determined simulation time (e.g., 1 hour or 1 day), the final results are mapped, analysed and assessed.

2.6 Uncertainty in Predicting Coastal Evolution

There continues to be healthy debate within the discipline about the extent to which numerical models can be used reliably for the purpose of predicting shoreline evolution far into the future. There is a strong recognition that there can be great uncertainty in coastal evolution modelling, and hence there is a powerful rationale for incorporating stochastic approaches that incorporate statistical probabilities, random sampling and multiple simulations to determine model sensitivity and outcome likelihoods. There is even a sense that coastal geomorphologists are increasingly receptive to ideas borrowed from chaos theory, involving non-linear dynamics, fractal dimensions and self-organisation (Phillips, 1995; Werner, 2003; Baas, 2007).

The need for robust models of coastal evolution has been heightened in the past two decades because of the anticipated deleterious consequences of global climate change. Predictions of increased frequency and magnitude of major coastal storms superimposed on increasing rates of eustatic sea level rise (Stive, 2004; Cowell *et al.*, 2006; Slott *et al.*, 2006) are believed to have major consequences for coastal communities across the globe, especially in low-lying areas like the Netherlands, Bangladesh, the Maldives and the US Gulf Coast. As with the science of weather prediction, model forecasts of coastal evolution are only accurate over short time intervals (e.g., hours to days). It is not yet feasible to predict with any degree of reliability (or within current computational capacities) what might happen to a specific coast over periods of decades or centuries. As a consequence, relatively simply conceptual models of shoreline change due to sea level rise such as the Bruun Rule (Schwartz, 1967; Rosati *et al.*, 2013) continue to be used for engineering and shoreline management purposes (Atkinson *et al.*, 2018) despite the many critiques of its robustness (Le Cosannet *et al.*, 2016) and the availability of alternatives (Davidson-Arnott, 2005b).

The history of sea-level change due to the cyclic advance and retreat of glaciers over the past 100 000 years is especially problematic for modelling coastal evolution because the assumed tendency toward equilibrium between processes and characteristic landforms is transient. Different coastal landforms evolve at quite different rates depending on the controls discussed above, especially the relative hardness of the materials and the degree of consolidation. Sandy coasts, for example, are especially dynamic, and characteristic shoreline features such as spits, bay mouth bars and barrier islands will respond to individual storms but may grow or disappear entirely on time scales of centuries to millennia if sediment supply is altered or if sea level changes. On the other hand, erosion of hard bedrock coasts occurs very slowly. Erosional platforms associated with most modern coasts could not have formed during the 6000 years that sea level has been close to its present level. They are inheritance features, evolved from a complex suite of multi-scalar processes that have acted upon the coast over several glacial–interglacial cycles, if not over many millions of years. It would be foolhardy to believe that a model based on the mechanics of rock abrasion and hydraulic pressurisation due to wave action alone could reasonably simulate the evolution of rock platforms over such extensive time periods without also taking into account the effects of sea level fluctuations, tectonic instability, geologic structure and altered climatic conditions.

As is described in greater detail in the next chapter, sea level in the previous interglacial period was roughly the same as that today. Over the last glacial period, sea level trended downward with the growth of continental ice sheets, culminating in a low stand of about 130 m lower than present at around 30 000 to 25 000 years BP. During this time, subaerial processes that are common on continents today would have been active on much of the continental shelf. The base level of rivers is tied to mean sea level, and therefore they would have eroded channels into the continental shelf and delivered sediment to coastal areas that are now entirely submerged. The vestiges of this geomorphic work are clearly evident on bathymetric maps of the shoreface. Subsequent to the last glacial maximum, relatively rapid melting and retreat of the ice margin produced general submergence of most of the global shorelines, especially in the tropics and mid latitudes. The history is complicated in higher latitudes because most of the coast and parts of the continental shelf were covered in ice during the glacial period. Post-glacial isostatic adjustments have caused the shoreline to rise at the same time that sea level was rising, so whether a section of coast is experiencing submergence or emergence depends critically on the rate of isostatic rebound, in addition to any other tectonic motions that characterise the region. The consequence for places that were glaciated, like Canada, Russia, Fenno-Scandinavia, and parts of New Zealand, is that the coastal features viewed on the landscape today are relatively young in a geologic sense – they are still evolving. This draws into question whether any sort of equilibrium modelling effort based on waves, currents and tides alone can realistically predict coastal evolution over scales that are relevant to the need for society to understand the long-term ramifications of building and living close to shore.

References

Ashton, A., Murray, A.B. and Arnault, O. 2001. Formation of coastal features by large-scale instabilities induced by high angle waves. *Nature*, **414**, 296–300.

Atkinson, A.L., Baldock, T.E., Birrien, F. *et al.* 2018. Laboratory investigation of the Bruun rule and beach response to sea level rise. *Coastal Engineering*, **136**, 183–202.

Baas, A.C.W. 2007. Complex systems in aeolian geomorphology. *Geomorphology*, **91**, 311–31.

Baquerizo, A. and Losada, M.A. 2008. Human interaction with large scale coastal morphological evolution: an assessment of the uncertainty. *Coastal Engineering*, **55**, 569–80.

Bauer, B.O. 2004. Geomorphology. In Goudie, A. (ed.), *Encyclopedia of Geomorphology*, vol 1, A to I. London: Routledge, 428–35.

Bauer, B.O. and Sherman, D.J. 1999. Coastal dune dynamics: problems and prospects. In Goudie, A.S., Livingstone, I. and Stokes, S. (eds.), *Aeolian Environments, Sediments and Landforms*. Chichester: John Wiley & Sons, Ltd., pp. 71–104.

Brock, J.C. and Purkis, S.J. 2009. The emerging role of LiDAR remote sensing in coastal research and resource management. *Journal of Coastal Research*, **SI 53**, 1–5.

Chorley, R.J. and Kennedy, B.A. 1971. *Physical Geography: A Systems Approach*. London: Prentice Hall, 370 pp.

Cowell, P.J. 1999. The shoreface. In Short, A.D. (ed.), *Handbook of Beach and Shoreface Morphodynamics*. Chichester: John Wiley & Sons, 39–71.

Cowell, P.J., Roy, P.S. and Jones, R.A. 1995. Simulation of LSCB using a morphological behaviour model. *Marine Geology*, **126**, 45–61.

Cowell, P.J., Thom, B.G., Jones, R.A., Everts, C.H. and Simanovic, D. 2006. Management of uncertainty in predicting climate-change impacts on beaches. *Journal of Coastal Research*, **22**, 232–45.

Davidson-Arnott, R.G.D. 2005a. Beach and nearshore instrumentation. In Schwartz M.L. (ed.), *Encyclopedia of Coastal Science*. Dordrecht: Springer, 130–8.

Davidson-Arnott, R.G.D. 2005b. Conceptual model of the effects of sea-level rise on sandy coasts. *Journal of Coastal Research*, **21**, 1166–72.

Davies, J.L. 1972. *Geographical Variation in Coastal Development*. Edinburgh: Oliver & Boyd, 204 pp.

Fairbridge, R.W. 2004. Classification of coasts. *Journal of Coastal Research*, **20**, 155–65.

Finkl, C.W. 2004. Coastal classification: systematic approaches to consider in the development of a comprehensive system. *Journal of Coastal Research*, **20**, 166–213.

Goudie, A. (ed.), 2004. *Encyclopedia of Geomorphology*, vol 1, A to I. London: Routledge, Taylor & Francis Group, 1200 pp.

Harrison, S. 2002. On reductionism and emergence in geomorphology. *Transactions of the Institute of British Geographers*, **26**, 327–39.

Harvey, A.M. 2007. Geomorphic instability and change: introduction: implications of temporal and spatial scales. *Geomorphology*, **84**, 153–8.

Inman D.L. and Nordstrom, C.E. 1971. On the tectonic and morphologic classification of coasts. *Journal of Geology*, **79**, 1–21.

Komar, P.D. 1998. *Beach Processes and Sedimentation*, 2nd edition. Upper Saddle River, NJ: Prentice Hall, 544 pp.

Kroon, A., Larson, M., Möller, I. *et al.* 2008. Statistical analysis of coastal morphological data sets over seasonal to decadal time scales. *Coastal Engineering*, **55**, 581–600.

Le Cosannet, G., Oliveros, C., Castelle, B. *et al.* 2016. Uncertainties in sandy shorelines evolution under the Bruun rule assumption. *Frontiers in Marine Science*, **3**, 49, DOI: 10.3389/fmars.2016.00049.

McFadden, L., Nicholls, R.J., Vafeidis, A. and Tol, R.S.J. 2007. A methodology for modeling coastal space for global assessment. *Journal of Coastal Research*, **23**, 911–20.

Nicholls, R.J., Larson, M., Copobianco, M. and Birkemeier, W.A. 1998. Depth of closure: improving understanding and prediction. *Proceedings, Coastal Engineering 1998*, 2888–901.

Paolo, C. and Voller, V.R. 2005. A generalised Exner equation for sediment mass balance. *Journal of Geophysical Research*, **110**, F04014.

Phillips, J.D. 1995. Nonlinear dynamics and the evolution of relief. *Geomorphology*, **14**, 57–64.

Rhoads, B.L. and Thorn, C.E. 1996. *The Scientific Nature of Geomorphology*. Chichester: John Wiley & Sons Ltd., 481 pp.

Rosati, J., Dean, R. and Walton, T. 2013. The modified Bruun rule extended for landward transport. *Marine Geology*, **340**, 71–81.

Schwartz, M.L. 1967. The Bruun theory of sea-level rise as a cause of shore erosion. *Journal of Geology*, **75**, 76–92.

Shepard, F.P. 1937. Revised classification of marine shorelines. *Journal of Geology*, **45**, 602–24.

Shepard, F.P. 1963. *Submarine Geology*, 2nd edition. New York: Harper and Rowe, 412 pp.

Sherman, D.J. and Bauer, B.O. 1993. Dynamics of beach-dune systems. *Progress in Physical Geography*, **17**, 413–47.

Slott, J.M., Murray, A.B., Ashton, A.D. and Crowley, T.J. 2006. Coastline responses to changing storm patterns. *Geophysical Research Letters*, **33**, L18404, DOI: 10.1029/2006GL027445.

Stive, M.J.F. 2004. How important is global warming for coastal erosion? *Climatic Change*, **64**, 27–39.

Valentin, H. 1952. Die Küsten der Erde. *Petermanns Geographisches Mitteilungen Ergänzungsheft*, **246**, 118.

Walker, I.J., Davidson-Arnott, R.G.D., Bauer, B.O. *et al.* 2017. Scale-dependent perspectives on the geomorphology and evolution of beach–dune systems. *Earth-Science Reviews* **171**, 220–53.

Werner, B.T. 2003. Modeling landforms as self-organised, hierarchical dynamic systems. In Wilcock, P.R. and Iverson, R.M. (eds.), *Prediction in Geomorphology*. Washington, DC: American Geophysical Union, Geophysical Monograph Series, vol. 135, pp. 133–50.

Woodroffe, C.D. 2002. *Coasts: Form, Process and Evolution*. Cambridge: Cambridge University Press, 623 pp.

Part II

Coastal Processes

3

Sea Level

3.1 Synopsis

Earth is unique among the planets of our solar system because 70 per cent of the surface is covered in liquid water, mostly as an interconnected system of vast oceans. Coastal geomorphologists need to understand certain aspects of ocean dynamics, specifically the trends and fluctuations in sea-surface levels across a range of time scales. Dynamic changes in sea level on the order of hours to a few decades (i.e., periods longer than surface gravity waves) are a reflection of the response of the water surface to meteorological and oceanographic processes as well as to tides. At even longer periods (decades to millenia), changes in sea level forced by glacial periods and tectonic movements can lead to vertical shoreline excursions of up to 100 m with vertical translations extending tens of kilometres and more. In subsequent chapters, we will find that the dominant process variable controlling changes in coastal form and evolution is the incident wave climate, but it is critical to appreciate that long-term changes in the relative level of the water along the shoreline can greatly influence the way in which coastal evolution occurs. The relative position of the water on the shore will determine where wave action is applied, and fluctuations in sea level will have an effect on coastal erosion, transportation and deposition. In addition to tides produced by the gravitational force of the Moon and the Sun, short-term fluctuations in sea level occur as a result of storm surge, seasonal variations in pressure and wind patterns, and changes in weather patterns on a scale of years to decades (e.g., the El Niño–Southern Oscillation (ENSO) cycle in the Pacific). These fluctuations are also extremely significant ecologically, both directly through exposure and coverage of the intertidal zone and indirectly because of the forced movement of water and nutrients.

Changes in the relative position of the land and sea on a time scale of thousands to millions of years has caused inundation (transgression) or exposure (regression) of the land. Eustatic changes result from changes in the global volume of water in the ocean basins, with the most significant of these being the effects of glacier and ice sheet growth and decay during the Pleistocene. During the last glacial period, sea level reached its lowest point around 25 000 BP at an elevation of about 130 m below the present level. The succeeding Holocene transgression produced a rapid rise in sea level to the present level about 5000 years ago. Isostatic loading and unloading due to the growth and decay of the ice sheets complicates the response of the coast in mid and high latitudes. Over long time periods tectonic forces have produced relative changes in the elevation of the coast locally and changes in the geometry and volume of the ocean basins.

One consequence of human-induced global warming is an increase in ocean volume due to thermal expansion. In addition, the potential for increased melting of glaciers and snow fields, particularly in Greenland and Antarctica where the largest reservoirs of fresh water are located, is critically important to future coastal evolution.

3.2 Defining Mean Sea Level

It is convenient to think of sea level changes occurring roughly on three time scales (Fairbridge, 1983): (1) long term, (2) medium term, and (3) short term. Long-term changes are on the order of 10^6–10^9 years, where interest is focused on eustatic cycles of a few million years and on large-scale plate tectonics. Medium-term changes are on the order of 10^3–10^6 years, where interest is primarily on the effects of cycles of glaciation in the Quaternary through crustal loading and unloading, and through eustatic changes associated with exchanges of water between the oceans and continental glaciers and ice sheets. Local change in elevation of the land due to tectonic forces also occurs on this time scale. Short-term changes are on the order of hours to 10^3 years, where attention is focused on astronomical tides, residual isostatic effects of the last glaciation, and on changes due to meteorological, oceanographic and climatic factors. Modern coastal features and processes are most affected by short-term sea level changes and by medium-term changes related primarily to the past 25 000 years – from the time of the last glacial maximum through melting of the continental ice sheets and the Holocene transgression.

3.2.1 The Geoid as a Frame of Reference

There are a large number of causes of changes in sea level that vary in their magnitude and on the time scale over which they operate (Table 3.1). However, in order to evaluate these changes, a reference system is needed from which the deviations from mean sea level can be measured. Conceptually, it is convenient to think of mean sea level as the average level of the sea with respect to a point on the coastline. Consequently, fluctuations due to tides, meteorological conditions and short-term variations in ocean temperature and currents can be seen as departures above and below this level. Over the medium and long term, changes in mean sea level relative to an adjacent land mass can occur because of uplift or subsidence of the land or because of changes in the height of the ocean itself, for example, due to changes in the volume of water in the oceans. However, if we are to distinguish between these two causes a constant, universally acceptable reference frame is needed – one that accommodates the reality that there are no truly fixed reference points on the Earth given plate tectonics and the dynamic state of the lithosphere. This is now made possible by referencing mean sea level to a known global reference framework such as the North American Datum of 1983 (NAD83), the North American Vertical Datum of 1988 (NAVD88), or the European Terrestrial Reference System of 1989 (ETRS89).

The NAD83 and the ETRS89 are idealised Earth models based on a mathematically smooth ellipsoid that estimates the position of the mean surface across the entire globe. This approach is similar to imagining that all topography on Earth is eliminated by filling in the ocean basins with material moved from continents and submarine mountain chains until a featureless surface is achieved. The ellipsoid model provides a conceptually simple reference level for mean sea level, and it sees widespread use for the relative ease in which it can be implemented. With the development of satellite-based global positioning systems the Earth ellipsoid can be defined with great precision and accuracy, and the World Geodetic System (WGS84) provides the reference coordinate system for use with GPS.

The challenge associated with using the ellipsoid model for determining sea level is that it does not truly reflect the equipotential surface based on gravity (i.e., the height and orientation of a surface with respect to the Earth's core where the gravity vector has the same value everywhere across the globe). Differences in material density in the lithosphere and asthenosphere lead to situations in which gravity is not always pointed directly 'downward' with exactly the same magnitude. Scientists have measured and modelled such an equipotential surface – termed the geoid.

Table 3.1 Major causes of sea level change, characteristic magnitude and time history

Name	Cause	Magnitude	Time scale	Time history
Surf beat	Wave groups	0.05–0.5 m	30–300 seconds	periodic
Seiche	Oscillations in basin	0.02–0.75 m	seconds to hours	episodic periodic
Tides	Gravitational force of Moon and Sun	0.05–15 m	semi-diurnal or diurnal	periodic
Wave set-up (and set-down)	Wave momentum flux	0.05–0.75 m	hours–days	episodic
Wind set-up and storm surge	Wind stress barometric pressure	0.05–3.0 m	hours–days	episodic
Dynamic ocean temperature, currents	Expansion and contraction due to temperature changes	0.05–0.5 m	days to months seasonal	episodic
Astronomical cycle effects on tidal range	Sun, Moon and Earth alignment, orbit, axis	0.05–5.0 m	14 days to 10^3 years	periodic
Eustatic – changes in volume of ocean water – glacial eustasy	Growth and decay of glaciers, lake and groundwater storage	metres to 10^2	10^2–10^4 years	continuous episodic (periodic?)
Isostasy – glacio-isostasy, hydroisostasy	Loading and unloading of crust	up to 150 m	10^3–10^4 years	continuous
Geoidal	Earth revolution, gravity changes	metres	10^3–10^6 years	continuous
Eustatic – changes in ocean basin volume	Plate tectonics	up to 10^2 metres	10^4–10^6 years	continuous
Tectonic uplift and downwarping	Local tectonics, plate tectonics, volcanism	metres to 10^3 metres	10^1–10^6 years	continuous

Note: The magnitudes of change and time period are general and in some cases may fall outside the ranges indicated. Periodic and episodic changes result in a temporary deviation from mean sea level whereas continuous changes result in a change in the relative position of the sea and land.

It is also referred to colloquially as the lumpy potato model of Earth because it looks more like a potato (geoid) than a smooth egg (ellipsoid). Depending on location, the geoid surface can be either above or below the ellipsoid surface by up to 100 m or so, which leads to two different ways of calculating elevation depending on whether the reference frame is an ellipsoid (e.g., NAD83) or a geoid, such as the Earth Gravitational Model of 2008 (EGM08).

The geoid is particularly relevant to ocean dynamics and estimating sea level because water flows according to gravitational attraction. Consider a situation for which there are no ocean surface fluctuations due to tides, currents and meteorological factors. The ocean surface would become perfectly still, but it would not be perfectly flat. In fact, the water surface would have topography (i.e., mounds and depressions) because the pull of gravity at any point on the surface of the Earth is controlled by the distribution of mass beneath it (Pugh and Woodsworth, 2014). The density of the material in the asthenosphere and mantle can be quite variable, which will affect the strength of the regional gravity vector, thereby leading to sea-surface topography on the order of centimetres to tens of metres depending on location. These mounds and depressions are not noticeable to an observer on a ship because they are horizontally extensive, but sophisticated satellite-based measurements have demonstrated that sea-surface topography, on average, aligns well with the geoid model.

The geoid itself is always deforming or adjusting to gravitational and rotational changes on the Earth so that changes in mean sea level over time (due to eustatic changes, for example) cannot be thought of as a simple worldwide linear shift up or down (Figure 3.1). Instead, sea level change at a point on a coastline over a long period

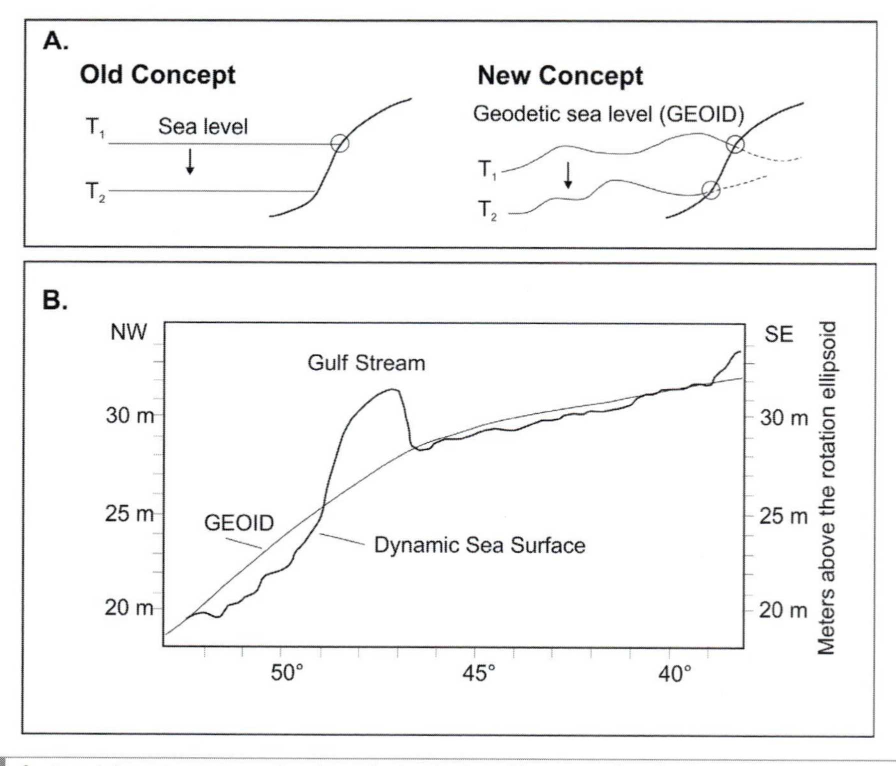

Figure 3.1 Static and dynamic concepts of sea level: A. sketch of an old concept of sea level as a simple uniform surface and sea level change expressed as a movement of this up or down versus the new concept of a dynamic geodetic surface; B. profile across the Gulf Stream showing dynamical deviations in the sea surface from the geodetic sea surface resulting from surface water temperature (after Morner, 1987).

of time reflects changes in the geoid as well as eustatic and isostatic changes on local, regional and global scales. When all these factors are taken into account, the net outcome is referred to as the relative sea level (RSL) change. This is the truest indication of sea level change, and can be readily compared to other locations across the globe.

3.2.2 Sea Level Measurement

Changes in sea level can be measured directly at locations along the coast using a pressure transducer or float positioned in a stilling well (Pugh and Woodworth, 2014). This approach has been used at major harbours around the world for over 100 years and longer in some cases. The stilling well is necessary to dampen high-frequency fluctuations associated with the passage of waves and often uses a narrow conical opening within the well pipe that restricts the rise and fall of water in the pipe so that only the longer, sustained changes in water level are sensed inside the well. Additional dampening may be provided by using a small diameter opening for the return flow to the sea at the base of the pipe or by having the opening far below the water surface so that the action of small waves is filtered out due to depth attenuation (see Chapter 5).

The first tide gauges used a stilling well and a float connected to a chart recorder and later to data loggers. Float systems require a lot of maintenance and are not easily interfaced with an electronic transmission system that can send the signal continuously to a central station. In recent years they have largely been replaced by acoustic transducers that measure the travel time of an acoustic signal between the sensor and the water surface in the well. The major limitation of these is that the travel time of the acoustic pulse is highly sensitive to the temperature gradient in the air between the sensor and the water surface. This can be overcome by replacing the acoustic sensors with radar or laser sensors (Miguez *et al.*, 2008). All of these gauges are deployed off marine structures such as harbour piers or oil platforms. They are subject to being discontinued or moved when redevelopment occurs, thus shutting down the historical record or possibly introducing errors in the datum if they are moved.

All gauges require a local datum to which their elevation can be tied, and there are a number of restrictions or limitations to their accuracy due to inherent limitations in the instrumentation and to the effects of local hydraulic conditions. One problem is to correct the gauges for any local vertical motion of the land at the site, which can be done using GPS benchmarking (Wöppelmann *et al.*, 2007). Tide gauges provide a record of sea level changes over the past century or so, but more recently satellite altimetry is being used to assess sea level fluctuations across all ocean surfaces (Church *et al.*, 2006; Berge-Nguyen, 2008). The Global Sea Level Observing System (GLOSS) provides a common set of criteria for installation of tide gauges and standards of accuracy, as well as a repository for tide gauge data from all over the world. Elevations can now be transmitted in real time and are available for some sites over the Web.

In the past four decades measurement of the elevation of the ocean surface from satellite altimetry has provided large spatial coverage of the elevation of the oceans and in particular the response to changing temperature, winds and pressure. The joint USA/France TOPEX/Poseidon satellite, launched in 1992, and the Jason series (Jason 1 2001–12; Jason 2 2008–; and Jason 3 launched in 2016) are dedicated solely to altimetry (Pugh and Woodworth, 2014). They orbit the Earth every 112 minutes at an altitude of 1336 km and provide continuous coverage every 9.9 days of the area equatorward of 66° latitude. They transmit a short radar pulse and measure the return pulse that, with corrections and averaging, can provide elevation with an accuracy of 2–4 cm. The radar reflections are strongest from portions of the waves that are perpendicular to the pulse direction, and given the characteristic shape of ocean waves (Chapter 4), the return signal is biased to the trough region of the wave. The bias increases with increasing wave height and therefore corrections have to be made in order to get an accurate estimate of the mean water level. The accuracy of the readings requires precise knowledge of the altitude and inclination of the satellite as well as a number of other factors that affect the transmission of the signal through the atmosphere (Pugh and Woodworth, 2014). Despite the complexities, errors in measurement are usually of the order of only a few centimetres.

Practically, it is still not easy to define mean sea level accurately because fluctuations in sea level occur on a range of time scales (Table 3.1). Thus, the question of an appropriate averaging interval is a complex one. If we have a record that extends over one year only, then we have little choice but to average over the period of record. However, if we have a record of measurements that extends several decades, would it be reasonable to average over the entire record or simply certain segments? The situation becomes even more complex given that the length of records varies widely from country to country across the globe, and recognising that fluctuations in sea level due to changes in atmospheric pressure, winds, ocean currents and sea-surface temperature can extend over years, decades or longer. Thus, mean sea level can truly only be defined for a given period of interest, and it should be understood and interpreted in this context only.

3.2.3 Estimating Past Sea Levels

The historic record of sea level derived from instrumented measurements extends back less than 200 years. Determination of past sea level change over millenia requires two major components: (1) the identification of some indicator or proxy that can be tied to sea level; and (2) the dating of that indicator directly or indirectly (Khan *et al.*, 2017). Common indicators are in situ shells of organisms such as clams and barnacles, saltmarsh diatoms, and coral micro-atolls found in lagoons, shore platforms, the base of beach ridges coastal sand dunes, mangroves and saltmarshes (e.g., Goodwin and Harvey, 2008). In each case the relationship of the indicator to mean sea level should be known and care must be taken to account for factors such as tidal range and the effect of exposure on the height of wave run-up (Figure 3.2). There are numerous methods available for dating sediments and organic materials (e.g., Lamb *et al.*, 2006). In the past, dune and beach sand have been dated primarily by radiocarbon (^{14}C) dating of carbonate shells or driftwood, with the assumption that they were emplaced at the time of deposition of the sediments. Recent developments in thermoluminescence, and especially optically stimulated luminescence dating, now permit the dating of the time of burial of quartz and feldspar grains within the dune and beach deposits, and these methods have proven to be extremely valuable in a number of fields in the geosciences (Roberts and Liam, 2015). Optical luminescence has a precision of a few years to decades and can be used to date sediment over a range of a few years to over 200 000 years (Bateman *et al.*, 2017).

Once the age and elevation of the indicator has been determined, these can be compared to modern sea level. Determination of the age and elevation of a sequence of such indicators, for example a series of beach ridges (e.g., Argyilan *et al.*, 2005) or uplifted coral reefs (e.g., Dechnic *et al.*, 2017), allows for a local sea level curve to be plotted. This approach can work reasonably well for Holocene and even Quaternary shorelines where the evidence is still relatively intact and where assumptions can be made about the stability of the land surface. It becomes increasingly problematic beyond the Quaternary, except for purposes of sorting out stratigraphic sequences (Moucha *et al.*, 2008; Murray-Wallace and Woodroffe, 2014).

3.3 Changes in Mean Sea Level

It is convenient to conceive of changes in the relative mean sea level at a point on the coastline as arising from two principal causes: (1) eustatic changes in the level of the oceans due to volume changes such as those caused by growth and decay of ice sheets; and (2) changes in the elevation of the continents such as those related to plate tectonics or isostatic adjustments. However, eustatic and tectonic processes operate simultaneously and it is not always possible to isolate a single, primary cause of measured changes in sea level at one location or even one region. For example, an increase in ocean volume due to glacial melt will exert increasing pressure on

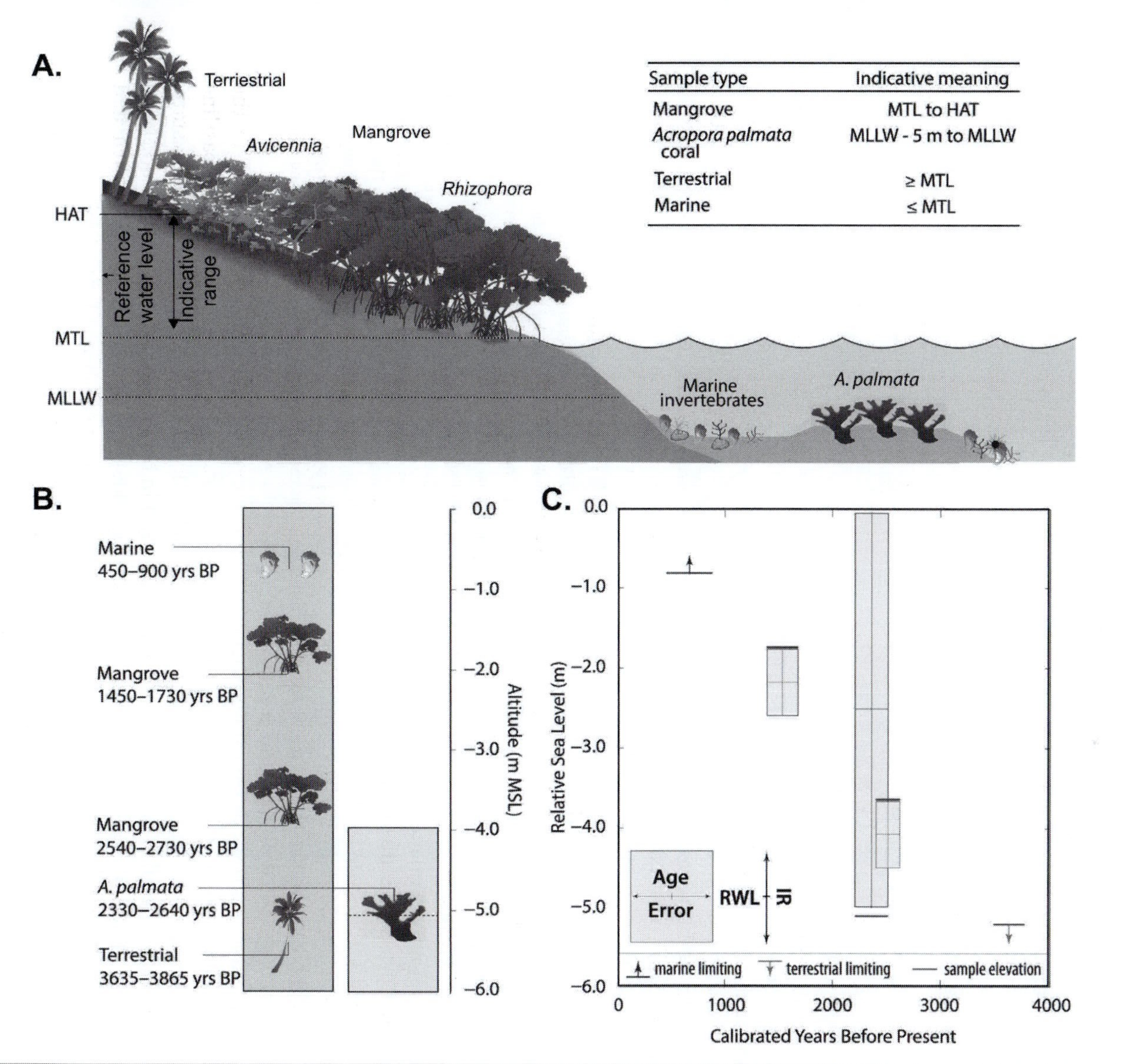

Figure 3.2 Schematic application of several sea level indicators associated with tropical coasts to reconstruct RSL change based on dating of indicators in situ or in cores: A. vertical distribution of mangrove and coral zones with respect to the tidal frame for the Caribbean region. Each indicator species occupies a defined range of elevation relative to a measure of sea level (e.g., highest astronomical tide or mean tide level) and distinguishes between terrestrial, intertidal and subtidal organisms; B. altitude of dated samples in hypothetical cores taken in mangrove peats and sediments (left) and from coral reefs (right); C. production of sea level index points and limiting dates. RSL is reconstructed by subtracting a sample's RWL from its altitude in a core. RWL = reference water level; IR = indicative range; RSL = relative sea level; MLLW = mean lower low water; MTL = mean tide level; HAT = highest astronomical tide (From Khan *et al.*, 2017).

continental shelves thereby forcing isostatic compensation, which will lead to the continental margins sinking slightly. So the relative rise in sea level measured along the coast will have both eustatic and isostatic components. In this section, the main eustatic and tectonic controls on sea level change are examined along with the history of sea level change since the last interglacial about 100 000 years BP. The potential impacts of human activities and global climate change on sea level are examined in Section 3.6.

3.3.1 Eustatic Changes in the Level of the Sea

The term eustasy is applied to all changes of sea level resulting from changes in the volume and distribution of water in the ocean basins. These changes can arise from: (1) changes in the volume and distribution of water in the global ocean, most notably those resulting from the growth and decay of ice sheets (glacial eustasy); (2) changes in the volume of the ocean basins (tectono-eustasy) arising from plate tectonics and continental drift as well as sediment infill and hydro-isostasy; and (3) changes in ocean mass/level distribution arising from changes in the Earth's rotation, tilt and gravitational distribution (geoidal eustasy).

The greatest impact on modern shorelines has come from fluctuations during the Quaternary associated with the growth and decay of ice sheets and glaciers (Murray-Wallace and Woodroffe, 2014). Throughout the Quaternary sea level fell during periods when continental ice sheets grew in size and it rose during interglacial periods when the ice sheets melted, driven by cycles in the orbit and tilt of the Earth (Milankovitch cycles – for a detailed description of these controls see e.g., Ruddiman, 2014). Over the past 0.9 million years the cycles have been about 100 000 years in length, reflecting dominance of modulation of the precession cycle by the eccentricity of the Earth's orbit (Figure 3.3A). During the warmest part of the last interglacial around 128–116 ka BP (corresponding to Marine Isotope Stage 5), sea level was about 5–7 m above the present level (Dutton and Lambeck, 2012). In contrast, during the last glacial advance (about

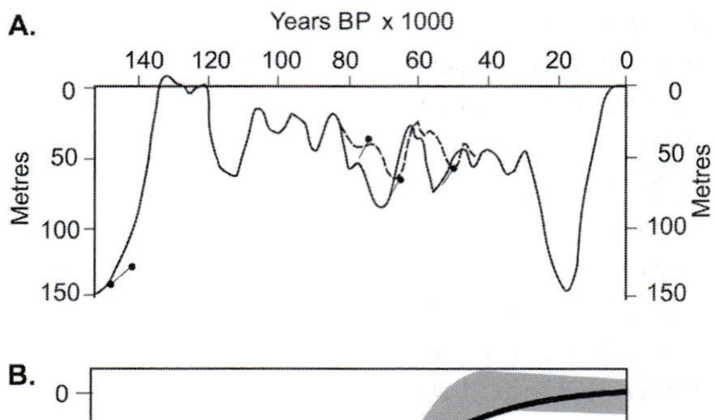

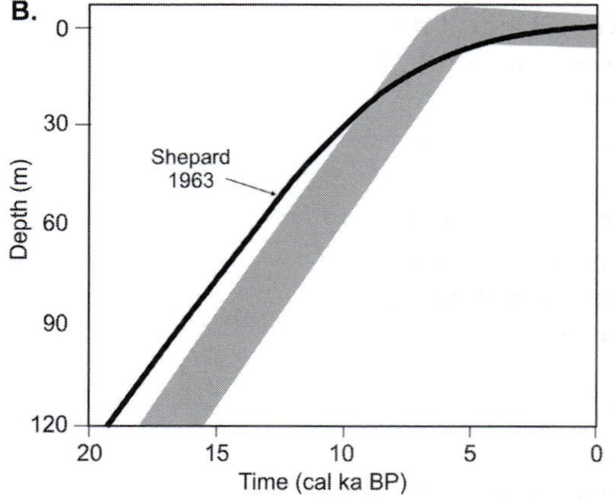

Figure 3.3 Changes in sea level: A. sea level changes over the past 150 000 years in the Australian region based on New Guinea shoreline data (continuous line) and deep sea cores (dots and dashed line) (Chappell, 1987); B. a generalised form of post-glacial eustatic sea level (stippled band) showing a steady rise due to glacial melting until about 5.5 ka BP and then small changes around the present level reflecting redistribution of water over the earth surface as a result of glacial-isostatic adjustments and ocean siphoning). The general rise is based on data compiled for far-field sites such as Barbados and the Huon Peninsula (see, e.g., Woodroffe, 2003). This can be compared to the curve developed from much earlier work by Shepard (1963).

30–20 ka BP), global sea level was on the order of 110–130 m below present levels (Peltier and Fairbanks, 2006). The impact of this was particularly important in areas beyond the limits of glaciation where large areas of the continental shelves were exposed to terrestrial processes. Melting of the continental ice sheets, which began about 25 000 years BP, led to rapid increases in the volume of water in the oceans and produced the Holocene transgression, which ended with sea level at, or very close to, the present level between 7000 and 4000 years BP (Figure 3.3B).

It is likely that fluctuations in the rate of sea level rise occurred on several time scales as a result of climatic cycles so that the sea level curve should show oscillations, as suggested by Fairbridge, rather than the smooth form of the Shepard curve (Figure 3.3B). Large pulses of water leading to rapid sea level rise may have played a significant role in the drowning of coral reefs and barrier islands on the outer edge of continental shelves (Chappell *et al.*, 1996).

Over the past 2000 years, the actual eustatic contribution to sea level rise through further melting of ice sheets and glaciers was small, probably < 2 m (Woodroffe *et al.*, 2015). As a result, RSL changes in the past 4000–5000 years are likely to have been influenced more by regional variations in tectonic processes, including changes in the geoid, than by eustatic processes.

Evidence for these general eustatic changes during the Pleistocene is derived primarily from far-field areas – those far away from the influence of the relatively rapid isostatic loading and unloading by the growth and decay of ice sheets. Often the best locations are in places where older shorelines are preserved as a result of local tectonic uplift, which can be measured and dated with a high degree of accuracy and precision (e.g., Isla and Angulo, 2016). Such locations include the Huon Peninsula in New Guinea (Chappell, 1974; Chappel *et al.*, 1996) and the island of Barbados in the Caribbean (Schellmann *et al.*, 2004; Fairbanks *et al.*, 2005) The actual sea level history along any specific coast will vary because of other factors such as isostatic uplift and local tectonic factors (examined in the next section), as well as changes in the geoid. However, most modern shorelines in the middle and low latitudes reflect the impact of a large rise in sea level, beginning about 20 000 years ago, that brought with it transgression of the shoreline accompanied by a reworking of sediments deposited on the exposed shelf. These sediments are now stored in large barrier and dune systems such as those found along the coasts of North America and Australia.

Evidence from a number of places in the equatorial regions of the Pacific and possibly elsewhere suggests that sea level toward the end of the Holocene transgression 5–6 000 years ago may have been 1–2 m above the present mean sea level. A subsequent drop in MSL may reflect the effects of local adjustments in the geoid and equatorial siphoning produced by the collapse of a bulge in the surface of the Earth that formed on the periphery of the continental ice sheets as material in the mantle was displaced under their weight (Mitrovica and Peltier, 1991). The lowering of the ocean bed in these regions would draw off water from far-field areas close to the equator and produce a regional lowering of RSL of 1–2 metres. However, recent work on the coast of Mauretania, a tectonically stable area, found little evidence of a sea level high stand (Certain *et al.*, 2018). They suggest that if siphoning occurred it may have been offset by a continuing eustatic contribution from ongoing deglaciation.

The landward displacement of the shoreline associated with the Holocene transgression means that most coastlines worldwide are relatively young. Significantly, adjustments to the transgression are still taking place, which implies that geomorphic processes along the coast are actively creating new landscapes although the rates of change are slowing. There is evidence of a decrease in the amount of sediment in the littoral system as it becomes locked up in dunes and barriers. River deltas have changed locations and estuaries are gradually infilling, while in tropical areas coral reef growth has largely stabilised.

The rapid rise in sea level to about 5000 years BP must have had an impact on Palaeolithic humans living along the coast over much of the tropical and temperate world. Some of the impacts of rapid transgression and the separation of Britain from the continental mass of Europe are now quite well documented (e.g., Turney and Brown, 2007). Recent unravelling of the sea level history of the Sea of Marmara (Eriş *et al.*, 2007) and the Black Sea (Ryan *et al.*, 1997; Ballard *et al.*, 2000) and the discovery of drowned settlements along paleo-shorelines have provided exciting opportunities for linking coastal processes and human history, and may provide a plausible origin for legends of the Great Flood.

3.3.2 Tectonic Changes in the Level of the Sea

Tectonic changes in sea level occur because of crustal movements related to local, regional and global tectonic activity. Over long periods of geologic time, uplift or subsidence can occur, particularly along plate margins. One area where this is currently pronounced is in the eastern Mediterranean along the Hellenic Arc (Pirazzoli, 1987). Along the western part of the island of Crete periods of rapid subsidence between 4000 and 1700 years BP led to the formation of a series of stepped shorelines. Around 1530 years BP sudden uplift of about 10 m occurred in a single event leading to emergence of all of the older Holocene shorelines (Pirazzoli, 1987). On the Pacific coast of central and southwestern Japan, which is located on the subduction zone at the contact between the Philippine Sea plate and the Asian plate, the last interglacial shoreline dated around 125 000 years BP has been uplifted as much as 200 m and the post-glacial shoreline dated at 6000 years BP is around 30 m above present sea level (Ota and Machida, 1987). In active tectonic zones rapid uplift or subsidence can occur along relatively short segments of fault lines associated with individual Earthquake events. Thus, Stephenson *et al.* (2017) document uplift of about 1 m for shore platforms located on the Kaikoura Peninsula on the east coast of the South Island of New Zealand following a magnitude 7.8 earthquake on 14 November 2016 (Figure 3.4). In some areas of Kaikoura the uplift reached 2 m, and at Waipapa Bay north of Christchurch, up to 6 m of uplift occurred during the same earthquake. The greatest rates of uplift have been recorded for the Huon Peninsula in New Guinea where a series of raised reefs have been dated to yield uplift rates of 0.07–0.33 m per century (Bloom *et al.*, 1974; Chappell, 1974).

Of particular interest in glaciated areas and areas close to the glacier margins is the effect of isostatic adjustment of the crust due to loading and unloading by glacial- and hydro-isostasy. The basic concept of isostatic adjustment is well known, particularly with respect to the growth and melting of Pleistocene ice sheets. In the simplest sense, the weight of ice in a large glacier or ice sheet depresses the crust locally causing material in the upper mantle to flow outward away from the area of depression. When the ice melts, the overburden weight is removed and material in the mantle flows back beneath the original centre of depression. This forces the crust upward due to the differences in material density between the crust and mantle. The amount of depression and subsequent isostatic compensation is related directly to the thickness

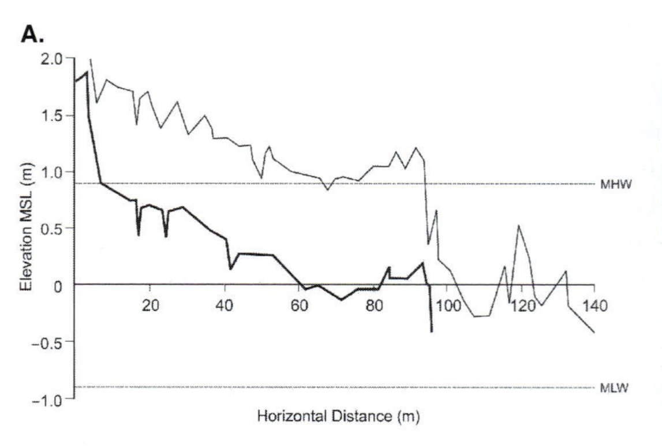

Figure 3.4 Uplift of the Kaikoura shoreline, South Island, New Zealand following the magnitude 7.8 earthquake of 14 November 2016: A. cross-shore profile KM2 measured in April 1994 and after the November earthquake (Stephenson *et al.*, 2017 figure 5); and B. photo looking alongshore to the west at high tide from the top of profile KM2 after the earthquake. The pre-earthquake high-tide level was along the gravel beach on the left of the photo. The boulders visible in the middle left are part of a revetment put in place to protect a road. Photo courtesy Wayne Stephenson.

of the overlying ice, and the rate of rebound can be predicted from models based on the rheological properties of the upper mantle (e.g., Tushingham and Peltier, 1991; Dyke, 1998; Hagendoorn *et al.*, 2007).

Associated with crustal loading and depression under an ice sheet is a lateral movement of mantle material that is displaced sideways beneath the crust beyond the immediate area of glaciation. A pressure bulge is created beyond the margins of the ice sheets, producing uplift adjacent to the glaciated area. Such bulges may have extended 800–1000 km south of the main glacial limit along the east coast of North America and the west coast of Europe. When the ice sheets melted and retreated, the bulge collapsed leading to subsidence along these marginal regions, and this continues today in areas such as the continental shelf off Nova Scotia (Scott *et al.*, 1987). The situation may be complicated further in areas with a wide continental shelf due to hydro-isostasy resulting from depression of the shelf and adjacent coast under the weight of rising sea level. This can occur in areas close to the location of ice sheets as well as in areas such as Australia that are distant from major ice sheet development.

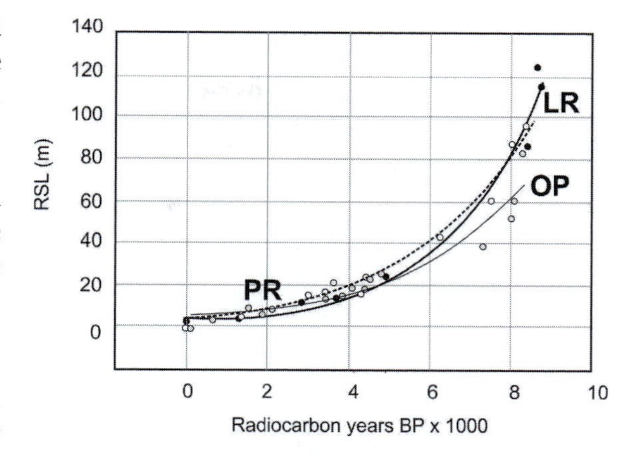

Figure 3.5 Relative sea level data for three sites on Devon Island, eastern Arctic Canada showing exponential decrease in the rate of isostatic uplift: Lyell River (LR), Port Refuge (PR) and Owen Point (OP). After Dyke (1998).

The rate of rebound associated with unloading of the crust as the ice melts is typically very rapid in the first few thousand years, with very little rebound occurring beyond 10 000 years. Plots of uplift versus time for shorelines on Devon Island in the Canadian eastern Arctic (Dyke, 1998) generally show a negative exponential form to the present (Figure 3.5), though in areas where there were major fluctuations in the rate of ice sheet melting, or even periods of significant advance, the curves are more complex (Lohne *et al.*, 2007).

In areas near the southern limits of the continental ice sheets in North America and western Europe, isostatic uplift occurred simultaneously with the most rapid phase of sea level rise during the Holocene transgression. As a result, there was a rough balance between isostatic uplift and eustatic drowning of the coast, creating a relatively stable sea level position along the shoreline. However, in northern areas such as the eastern Canadian Arctic, Scotland and northern Scandinavia, where major ice cover persisted past 10 000 years BP, much of the isostatic uplift occurred near the end of the Holocene transgression (Ullman *et al.*, 2016). This resulted in the preservation of beaches at altitudes >100 m above the present shoreline. In these areas isostatic uplift is still occurring at rates >0.25 m per century (e.g., Dyke, 1998; Smith *et al.*, 2012). Isobase maps for eastern North America (Andrews, 1970) show the extent of uplift that has occurred in the area around Hudson Bay and the Canadian Arctic archipelago over the past 6000 years (Figure 3.6). In eastern Canada areas such as the Scotian Shelf and Halifax have experienced continuous submergence, reflecting both the Holocene rise in sea level and the collapse of the forebulge under the present coast and continental shelf. The northern Gulf of St Lawrence, where there was considerable ice loading, has experienced continuous emergence. However, in the central Gulf of St Lawrence around the Magdalen Islands the situation is more complicated and the history of RSL reflects the effects of all three major controls (Rémillard *et al.*, 2017). Instead of simple curves reflecting continuous emergence or submergence, RSL here produces a J-shaped curve (Figure 3.7). There is initial emergence following ice retreat, but this slows after several thousand

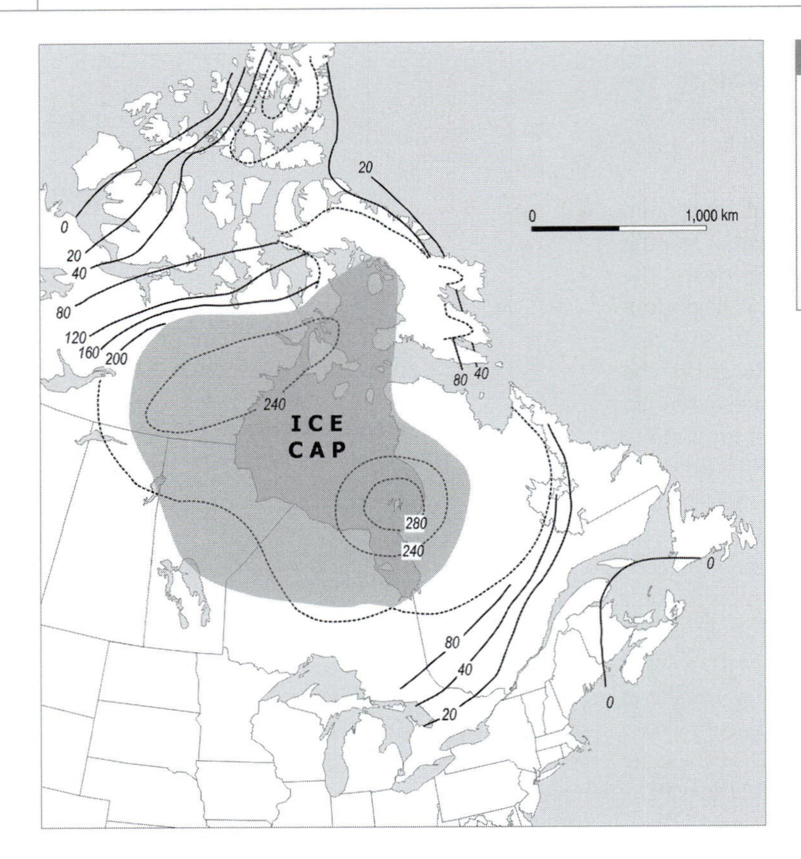

Figure 3.6 Isobase map for the region around Hudson Bay showing emergence (m) since 6000 years BP (solid lines). The shaded area marks the estimated extent of the ice cap at 6000 years BP and the dashed lines indicate the estimated uplift in that region following melting of the ice cap. After Andrews (1970).

years and uplift switches to subsidence as a result of the collapse of the forebulge and likely some effect from ongoing hydro-isosostacy (Figure 3.7).

The North American Great Lakes region provides many examples of the effects of isostatic adjustment on post-glacial shoreline evolution. The presence of numerous dated lake shorelines and a network of lake level gauging stations have provided a database for assessment of rheological response of the Earth to deglaciation. Present rates of uplift in the Great Lakes area (Tushingham, 1992; Figure 3.8) show an increase to the north and east, reflecting closer proximity to the centre of ice accumulation in Hudson Bay as well as the decreased time for unloading following retreat of the Wisconsinan ice sheet. A key factor for ongoing shoreline evolution in each of the lakes is the relationship between the present rate of isostatic uplift of a point on the shoreline relative to the rate of uplift at the exit to the lake. On Lake Superior, for example, the contours of isostatic uplift are aligned northwest–southeast so that areas to the west and south of Thunder Bay, including most of the US shoreline, are rising more slowly than the exit at Port Iroquois. Thus, the southwestern shorelines are gradually drowning whereas the northeastern shores, including all of the Canadian shorelines, are remaining stable or are gradually emerging (relative to the outlet). The same situation is occurring on Lake Michigan with the outlet at the north experiencing uplift whereas the southern shores are being depressed and are slowly drowning.

The effects of isostatic adjustment, as well as the timing of exposure of drainage outlets from the lakes, during retreat of the last ice sheet has produced a complex series of lake levels in the basins over the past 14 000 years. The lake level curve for Lake Huron/Michigan (Figure 3.9) shows initial inundation associated with glacial Lake Algonquin followed by a rapid fall to a level about 100 m below the present level as the lake

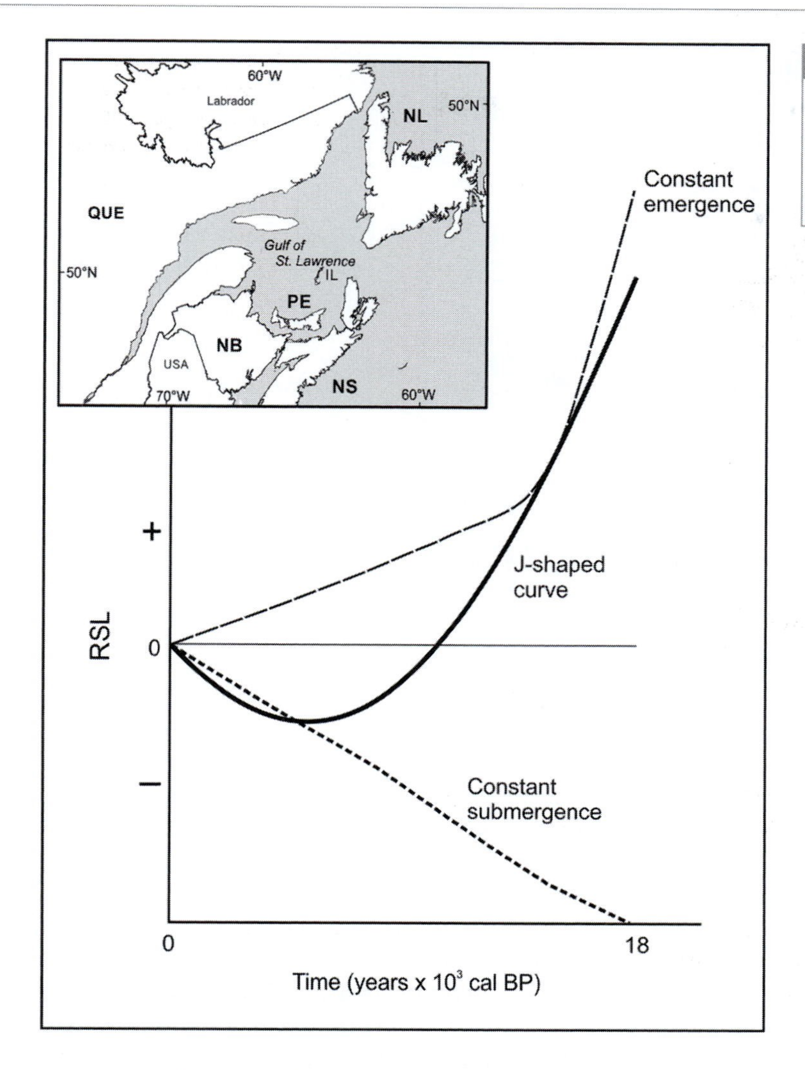

Figure 3.7 Schematic representation of the three major theoretical post-glacial relative sea level curves in formerly glaciated areas such as the Gulf of St Lawrence (inset). After Rémillard *et al.* (2017).

drained to the east, which was still isostatically depressed (Lewis, 1969). Between 10 500 BP and 5000 years BP, the lake level rose about 100 m to a high stand that was a few metres above the present level. This event, known as the Nipissing transgression, thus resembles, in scale, the eustatic sea level rise in the oceans. Notably, the high stand is marked by transgressive barriers and high dune systems along many parts of Lake Huron and Lake Michigan (Davidson-Arnott and Pyskir, 1989; Hansen *et al.*, 2010).

Local depression of the crust along the coast can also occur as a result of the accumulation of large amounts of sediment such as occurs in deltaic areas. Evidence from the Mississippi Delta suggests that subsidence due to isostatic loading of the sediments deposited onto the continental shelf accounts for 1–2 mm a^{-1} (Woolstencroft *et al.*, 2014). However, as in all deltaic environments the effects of compaction of the sediments must be taken into account as well, and this likely contributes most of the rest of the total rate of about 10 mm a^{-1} (Törnqvist *et al.*, 2008).

3.4 Astronomical Tides

3.4.1 Tides and Coastal Processes

Along ocean coasts there is a regular daily rise and fall of sea level that may range from a few decimetres to as much as 15 m in a few places. These

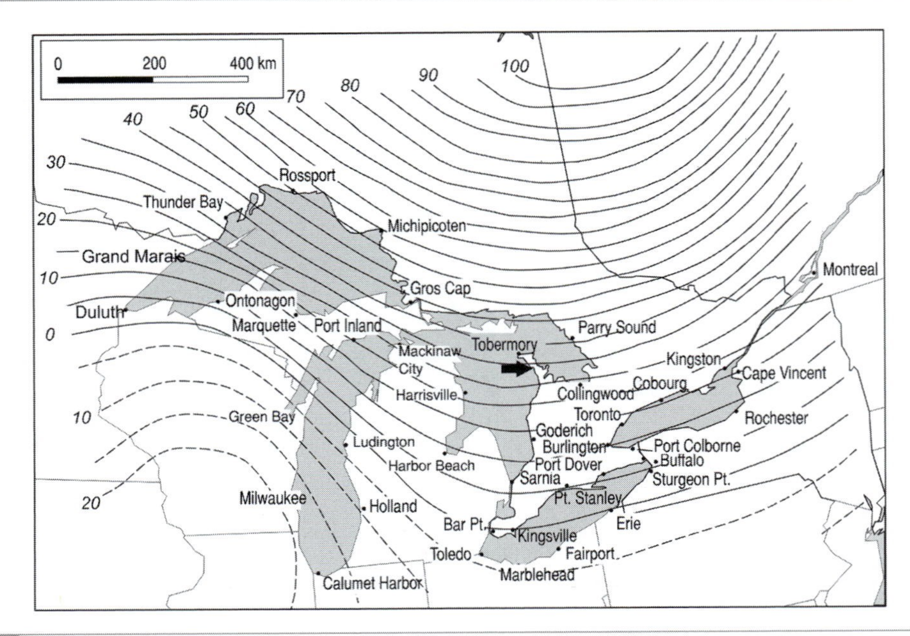

Figure 3.8 Present-day post-glacial uplift (cm/century) for the Great Lakes region as predicted by the model ICE-3G (Tushingham, 1992). Dashed lines indicate negative uplift (subsidence).The arrow marks the location for the point graphed in Figure 3.9.

fluctuations are termed astronomical tides because they are produced by the gravitational influence of the Moon and Sun. Tides are distinguished by their regularity and predictability from other short-term changes in sea level produced, for example, by strong winds or changes in barometric pressure. The latter are sometimes termed meteorological tides. Tidal fluctuations and the currents they produce are very important in all aspects of the coastal zone, affecting: (1) physical processes such as the shoreward extent of wave action and the flushing of waters in estuaries, lagoons and bays; (2) biological activities such as the zonation of plants and the feeding activities of birds, fish and other marine organisms; and (3) chemical processes such as those associated with the wetting and drying of intertidal rock surfaces. The intertidal zone – the zone located between the high and low tide lines – may be only a few metres wide on steep coasts with a small tidal range, and hundreds of metres wide on gently sloping coasts with a large tidal range. The tidal range greatly affects the form and width of sandy beaches, and thus the source area for coastal sand dunes. The flow of water into and out of inlets connecting lagoons and bays to the open ocean maintains the openings and permits the exchange of water and nutrients. The rise and fall of the tides across the intertidal zone creates stresses for some organisms but at the same time the variability creates a variety of rich and diverse habitats such as intertidal pools and saltmarshes.

The tides ultimately owe their origin to the gravitational forces exerted by the Moon and the Sun. These affect all objects on the Earth, but the response is most apparent in the oceans because the deformation of the surface is large enough to be seen by eye. Tides can be measured in large lakes but the response is only a matter of a few millimetres up to a couple of centimetres for large lakes such as Lake Superior. The association of the regular rise and fall of the tides with the Moon, in particular, has long been recognised – Pytheas, a Greek navigator and astronomer, wrote about the relationship between the position of the Moon and the height of the tide around 300 BCE. In western Europe tidal predictions for the time of high tide at particular ports

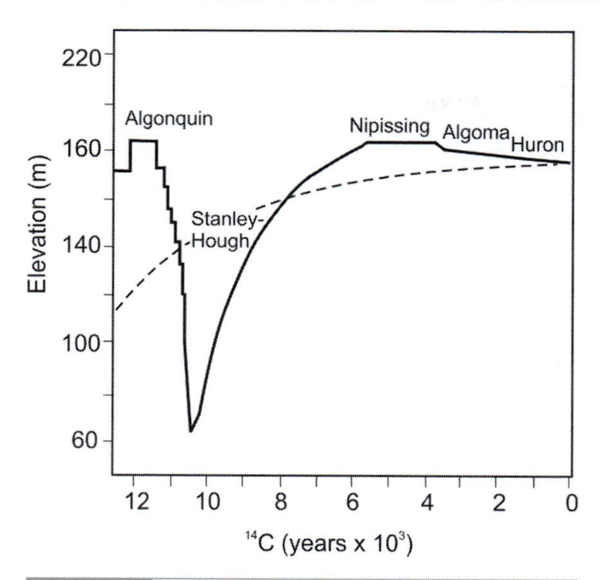

Figure 3.9 Post-glacial lakes in the Huron Basin (after Lewis, 1969). The dashed line shows the continuous isostatic uplift of a point on the present shoreline of Lake Huron (see Figure 3.8 for location). The uplift rate has slowed over the past 12 000 years and is now very small. The solid line indicates the elevation of lake level in the Huron Basin over the same period. This reflects the effects of the exposure, as ice retreated to the NE, of isostatically depressed outlets to the lake some 10 000–11 000 years ago that led to a fall in lake level by about 100 m. Subsequently, continuous uplift of the outlets resulted in a rise in lake level to about 5500 years ago when the present outlet to Lake Erie was established. The names refer to recognised glacial (Algonquin) and post-glacial lake levels.

were available as early as the thirteenth century (Macmillan, 1966). In the seventeenth century Isaac Newton, building on developments in astronomy and his own development of gravitational theory, developed the equilibrium theory of tides that related the tide generating forces to the gravitational pull of the Moon and the Sun.

The equilibrium theory of tides provides a rough explanation of tides and the ability to predict their timing and height. Some of the observed discrepancies between observations and predictions can be attributed to the effects of land masses and the shape and slope of the coastline. However, some of the discrepancies arise from the need to account for the forces involved in the actual movement of the fluid mass of water in the tidal waves – that is, the dynamical movement of the ocean – and accounting for this as well as the initiating gravitational forces gives rise to the modern dynamic theory of tides.

The general characteristics of measured tides are examined in Section 3.4.2, followed by a description of the equilibrium theory of tides. Section 3.4.3 describes the dynamic theory of tides and the behaviour of tidal waves in shallow water. A much more detailed but very readable description of tides and tide generating forces can be found in Pugh and Woodworth (2014).

3.4.2 Characteristics of Tides

Tides are regular daily oscillations in water level with the highest level being termed high tide and the lowest level termed low tide. The range of oscillation varies spatially across the globe and temporally at any point on the coast. This is evident from a comparison of the predicted tides for four locations over a one-month period (Figure. 3.10). At each station the key characteristics are: (1) the tidal type or the number of cycles in a day (e.g., diurnal, semi-diurnal); (2) the spring–neap cycle or the variation in tidal elevation over a two-week period; and (3) the tidal range or the difference in elevation between high and low tide.

The Moon orbits around the Earth in the same direction as the Earth's sense of rotation and therefore advances slightly in every 24 hour period (an Earth day). As a result the Moon rises 50 minutes later every day, and the lunar day is thus 24 hours and 50 minutes long. Because the gravitational effect of the Moon is significantly greater than the Sun, it is the lunar day that is critical to the timing of the tides and is the major determinant of tidal type.

In many locations, there are two tidal cycles every day, giving rise to a semi-diurnal tidal type such as that at Immingham (Figure 3.10). In some locations one of the tides is greatly suppressed, thus producing a diurnal tidal type with a single cycle over the lunar day (Figure 3.10 – Doson). In other areas elements of both the diurnal and semi-diurnal type are present, producing a composite tidal form. These are termed mixed

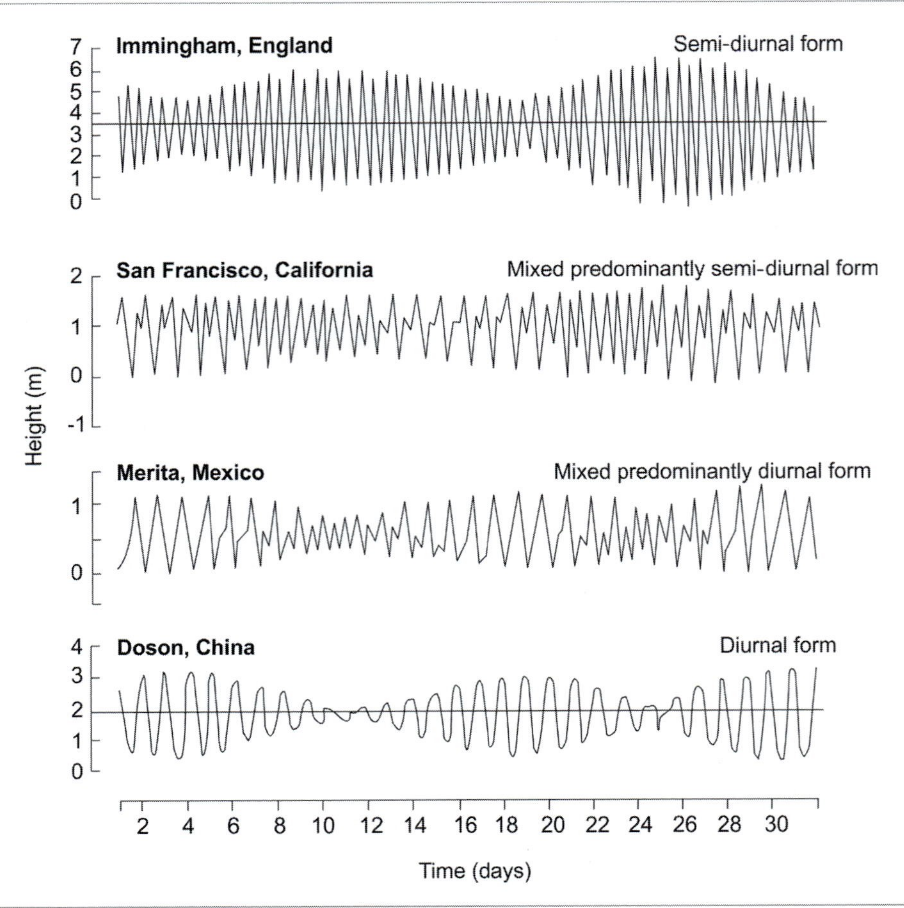

Figure 3.10 Examples of tidal curves for one month showing each of the major tidal patterns. After Defant (1962).

semi-diurnal tides where the semi-diurnal component is dominant or, alternatively, mixed diurnal tides where the diurnal component dominates (Figure 3.10, San Francisco and Merita, respectively). Diurnal tides occur along the coast of Antarctica and the Indian Ocean and in parts of the eastern Arctic archipelago. Semi-diurnal tides are common along much of the Atlantic and Arctic coasts and mixed tides are common in the northern Pacific.

The difference between the elevation of high and low tide is the tidal range and it controls the vertical and horizontal excursion of the water line on the coast. The greatest range occurs during spring tides when the Sun and Moon are in alignment thereby exerting greater gravitational pull on the oceans. The average tidal range at spring tides can be used as an important shoreline descriptor and it ranges from <0.5 m in enclosed seas to over 15 m in a few elongate estuaries and embayments. On the basis of the spring tidal range, Davies (1964) classified shorelines into:

Microtidal <2 m **Mesotidal** 2–4 m **Macrotidal** >4 m

Figure 3.11 shows the worldwide distribution of tidal range. Microtidal and mesotidal ranges generally occur on the open coasts of the world oceans as well as enclosed seas such as the Mediterranean, Baltic and Gulf of St Lawrence. Macrotidal areas occur primarily in embayments and areas with shallow continental shelves such as around the British Isles, parts of the Canadian

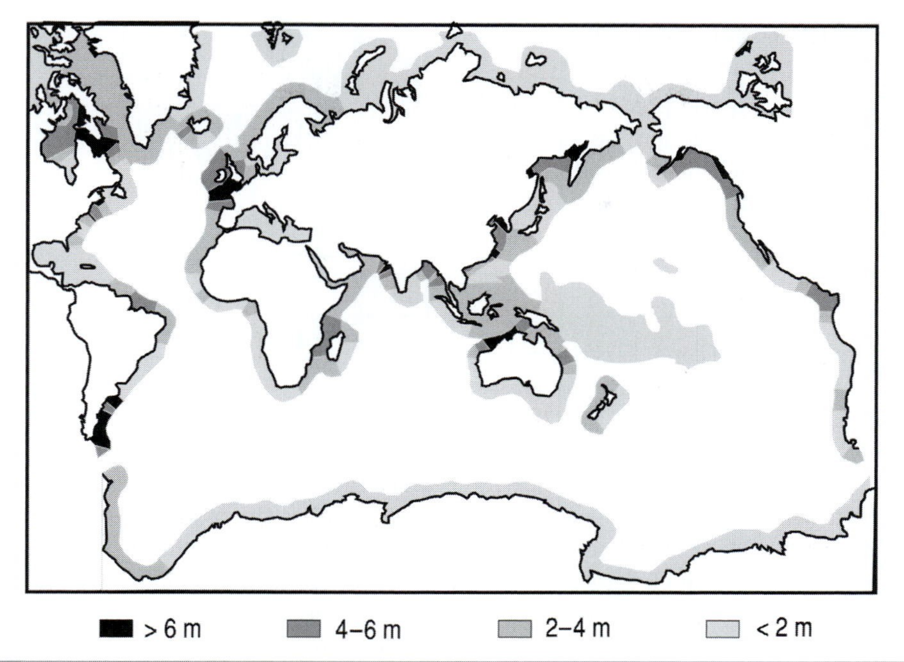

Figure 3.11 World distribution of tidal range. After Davies (1964).

eastern Arctic archipelago and northwest Australia. In a few locations, such as the Bristol Channel and Severn Estuary in England, the Gulf of St Malo in France and the Bay of Fundy in eastern Canada, spring tidal range can exceed 12 m. Because the macrotidal designation encompasses such a wide range it is useful to distinguish between macrotidal areas with a range of 4–6 m and hypertidal areas with a range of >6 m (Archer, 2013). Archer (2013) goes on to subdivide the hypertidal range from 6 to 16 m into five 2 m classes labelled hypertidal A–E. While restricted to a few areas of the world, the large tidal range and associated strong tidal currents in hypertidal estuaries produce unique geomorphic and sedimentary features and are attractive for their potential for power generation.

The frequency distribution of sea level elevation related to astronomical tides can be derived from records at an individual tide gauge, and a number of descriptors can be identified (Figure 3.12). These descriptors include: the mean elevation of spring high and low tides (MHWS and MLWS); the mean elevation of neap high and low tides (MHWN and MLWN); and the highest and lowest predicted astronomical tide levels (HAT and LAT). Meteorological forces can result in absolute water levels that are higher than the HAT or lower than the LAT.

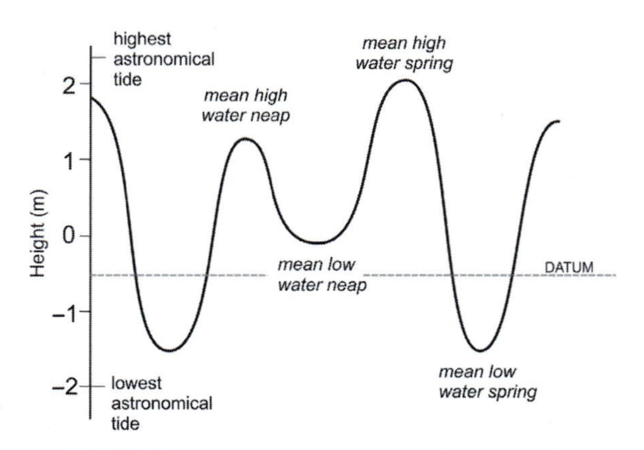

Figure 3.12 Descriptors of the various water levels associated with tides measured at one location. MHWS and MLWS are mean high water and mean low water for spring tides; MHWN and MLWN are the equivalents for neap tides. HAT and LAT are the highest and lowest forecast tidal elevations at the station. Tide levels at the gauge station are referenced to a surveyed datum.

3.4.3 Equilibrium Theory of Tidal Generation

Newton's theory of gravity predicts that the gravitational force of attraction between two bodies is a function of their masses and inversely proportional to the distance between them:

$$F = \frac{Gm_1m_2}{R^2} \quad (3.1)$$

where F is the gravitational force, G is the universal gravitational constant, m_1, m_2 are the masses of the two bodies, respectively, and R is the distance between the two bodies.

Equation 3.1 demonstrates that the gravitational attraction between the Sun and Earth is 0.46 times that of the Moon and Earth because the Sun, despite having a much larger mass, is much farther from Earth. The gravitational effects of the Sun and Moon on tide generation will vary with astronomical factors that control the phase relationships between the Sun and the Moon, variations in their distance from the Earth, and variations in their declination or position over the Earth. Thus, the equilibrium theory of tides is based on a complete understanding of these astronomical factors that control the gravitational forces.

If we consider a fictional Earth covered entirely by water, the water particles everywhere on the surface will be acted on by the combination of the Earth's gravitational force, the gravitational force of the Moon, and the centripetal or inertial force generated by the rotation of the Earth–Moon system. The Earth–Moon system rotates about a common centre of mass that, because of the much greater mass of the Earth, is actually located within the Earth (Figure 3.13). This point is known as the barycentre, and it is located on average about 4671 km from the Earth's centre (radius of 6378 km). The small inequality of the vertical gravitational pull of the Moon, when directly overhead, relative to the gravitational pull of the Earth acting in the opposite direction, would have only a negligible effect on the water surface. However, everywhere else on Earth, the tangential orientation of Moon's gravitational pull is not opposed by any

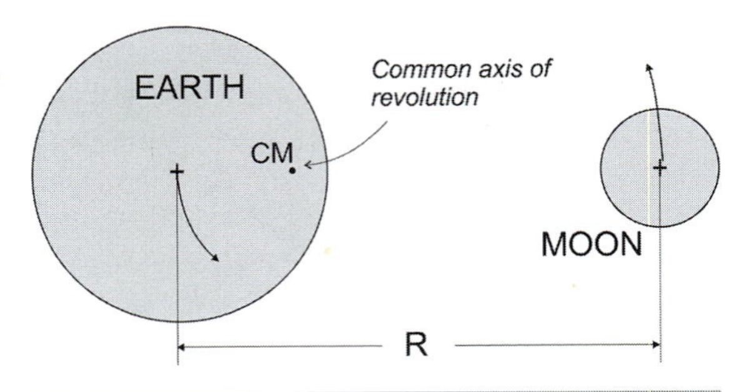

Figure 3.13 The Earth–Moon system and centre of rotation. CM refers to the centre of mass.

counterbalancing force vector, and thus tends to accelerate the water surface towards a point directly opposite the Moon. This generates a sloping water surface that would tend to grow until the pressure gradient associated with the sloping surface of the tidal wave balances the tangential acceleration (Figure 3.14). The effect of the tangential force is depicted as a bulge facing the Moon though the actual wave height in the open ocean is only about 0.5 m. On the side of the Earth opposite to that of the Moon there is a similar tangential stress produced because the gravitational pull of the Moon is slightly smaller as a result of the increase in distance, equal to the diameter of the Earth. This gives rise to an excess tangential centripetal force that produces the second bulge in the water surface on the side opposite the Moon.

Rotation of the Earth on its axis leads to the movement of the tidal bulges around the Earth in the form of ('true') tidal waves (Figure 3.14). Similar reasoning can be applied to the gravitational effect of the Sun, but the tidal wave formed by the Sun is much smaller than that of the Moon because of its smaller gravitational force. On an Earth uniformly covered with water, both the Moon and the Sun will generate tidal waves that travel around the world from east to west simply because the sense of rotation of Earth is west to east beneath the celestial bodies that cause the tides. These waves each have an associated amplitude, period and phase and so, based on the initial work of Lord Kelvin, it is possible to extract from a tidal record at any point on the Earth simple cosine waves related to each of the

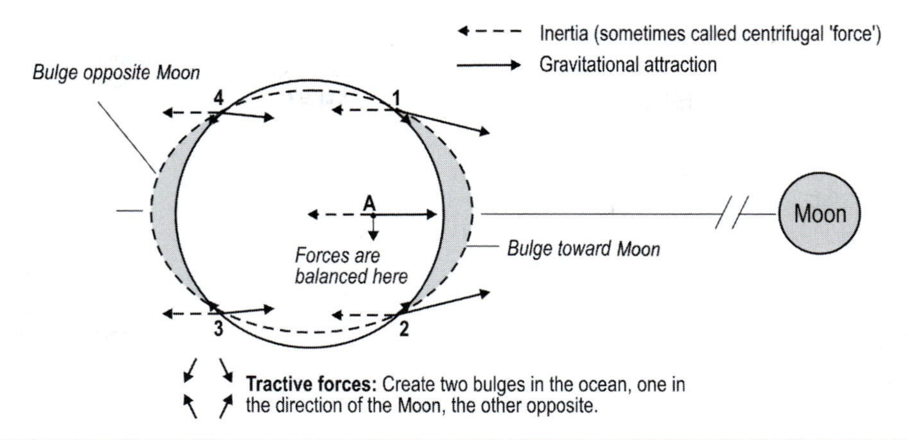

Figure 3.14 Inertial and gravitational forces acting on the Earth surface to produce tides. On the side facing the Moon the tangential gravitational forces exceed the inertial forces resulting in a convergence of water towards a point in line with the Moon. On the opposite side of the Earth the gravitational forces are slightly weaker than the inertial forces resulting in a convergence to a point opposite the Moon. Rotation of the Earth results in the rotation of these two 'bulges' producing two high and two low tides every sidereal day of 24 hours and 50.47 minutes. A similar reasoning can be applied to the gravitational force of the Sun.

main tide generating forces and to combine these to forecast the tides at that point.

Because the Moon is the most important of the tide-generating controls, it is the lunar day that is the primary control on the tidal cycle and the principal lunar semi-diurnal tidal constituent M_2 has a period of 12.42 hours giving rise to two waves in one lunar day (Figure 3.15A). Similarly we can identify the main solar semi-diurnal constituent S_2 that has a much smaller amplitude and a period of 12 hours. When the two waves are exactly in phase they reinforce each other so that the high tide is higher and the low tide is lower than for the lunar constituent only. However, because the solar wave has a slightly shorter period, the peak of the S_2 wave gradually occurs earlier and earlier than that of the M_2 wave so that after about 180 hours it arrives half a lunar day earlier than the lunar peak. The two waves are now exactly out of phase and their combination produces a tide with lower highs and higher lows compared to when they were in phase (Figure 3.15A). Spring tides occur when the Sun and Moon are aligned (full and new Moons) and neap tides are associated with the first and third quarter. There are thus two spring–neap cycles in a lunar month of 29.5306 days. The relationship to phases of the Moon is also shown in a plot of the tides for Halifax (Nova Scotia) and Rustico (Prince Edward Island) for the month of January 2018 (Figure 3.16A, B). Spring tides occur a few days after full and new moons while neap tides occur just after the first and last quarters.

The declination of the Moon relative to the equator changes over a nodical month of 27.2122 days (Figure 3.15B). When it is close to the equator the tidal forces will be symmetric at any point on the Earth, producing semi-diurnal tides with equal amplitude, termed equatorial tides (Figure 3.16). The orientation of the forces changes as the declination increases, producing an inequality in the range of the semi-diurnal tides which depends on the latitude of the point of observation. These tides are termed tropical tides (Figure 3.16). Variations in the distance between the Earth and the Moon or Sun also produce variations in the tides. The distance between the Earth and the Moon varies by about 25 000 km between their closest point (perigee) and their farthest point (apogee), producing a 30 per cent variation in the lunar component of the semi-diurnal tide (Figure 3.15C). Because one complete cycle from perigee to succeeding perigee takes 27.6 days (an anomalistic month) the heights of the two sets of spring tides within one

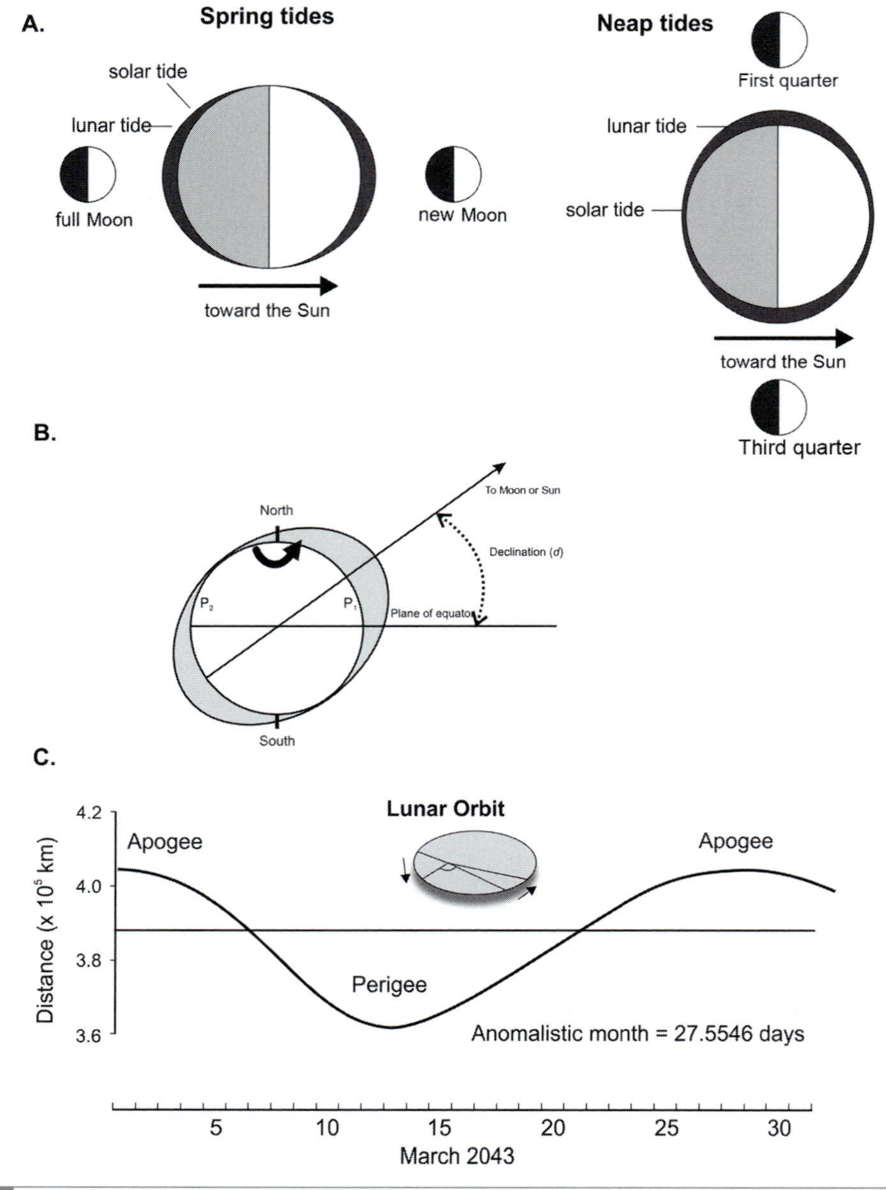

Figure 3.15 The three principal lunar cycles controlling tidal patterns: A. relative positions of the Sun, Moon and Earth over one lunar (synodic) month giving rise to the spring–neap cycle; B. variation in lunar declination with respect to the equator over one nodical month; C. variation in the distance of the Moon over one anomalistic month. Figures B and C modified from Pugh and Woodworth (2014, figure 1.2).

month will usually be different and this is clear in the plots for Halifax and Rustico (Figure 3.16).

Any other factors that affect the relative position of the Sun and Moon with respect to the Earth's surface or the distance from the two bodies will result in variations in the gravitational force and thus in the tidal range. Thus, the spring tidal range tends to be greatest around the equinoxes (21 March and 21 September). Over an 18.6 year nodal period the maximum lunar monthly declination varies from 18.3° to 28.6°. Maximum and minimum values occurred in

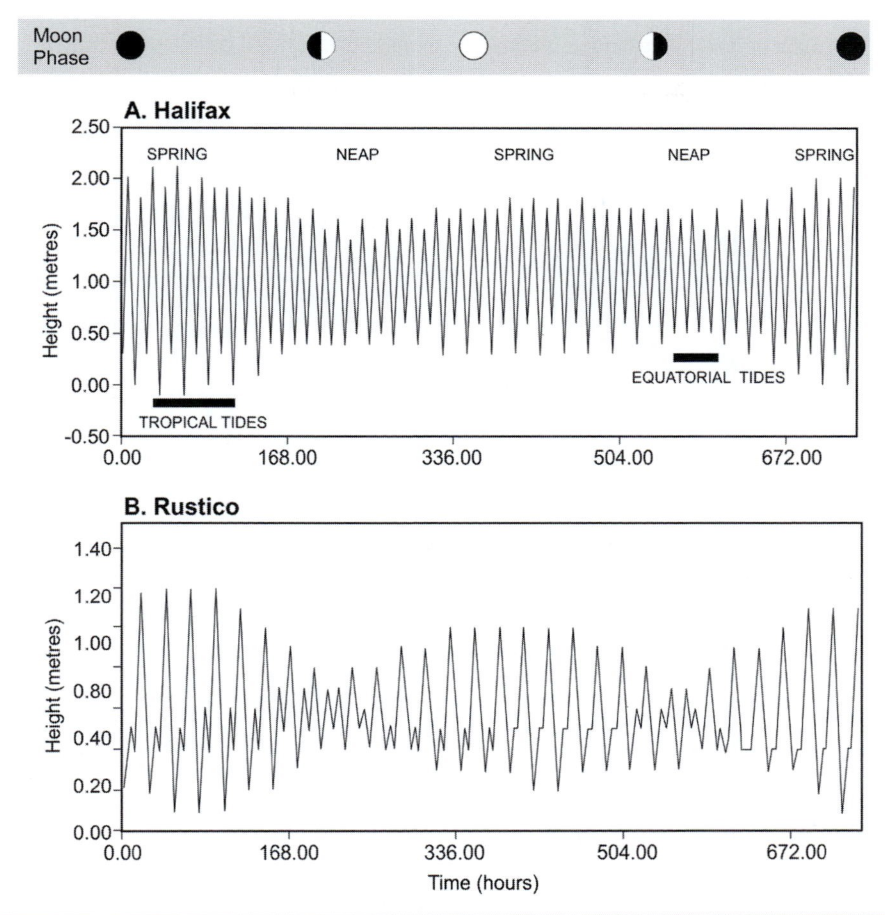

Figure 3.16 Graph of the predicted tides for January 2018 for: A. Halifax, Nova Scotia, on the Atlantic coast of Canada, which is characterised by semi-diurnal tides with a range of about 2 m; and B. Rustico on the north shore of Prince Edward Island, Gulf of St Lawrence, which is characterised by mixed diurnal tides with a range of about 1 m. Here the second daily tide almost disappears during spring tides. At both locations there is a distinct spring–neap cycle with the second set in the month (equatorial tides) being lower than the first set (tropical tides).

June 2006 and October 2014 and the next maximum is forecast for January 2022. The effect of the cycle on tidal amplitude is about 3.7 per cent with increased tides associated with minimum declination values. It may even be that multiple harmonics of the nodal signal can be identified in the sea level record (Hansen *et al.*, 2015). The effect of this is proportional to tidal range and therefore the effect on extreme water levels is generally only significant for hypertidal areas. Over a period of time there are occasions when extreme tidal forces occur as a result of the coincidence of several factors that produce large tides – when the Sun and Moon are in line with the Earth and at their closest respective distances. For maximum semi-diurnal tides both the Moon and the Sun should have zero declination.

Each of the astronomical factors described above produces a periodic variation in tidal forces and thus in tidal amplitude. The period or frequency is determined by return frequency of the particular condition and the amplitude will vary with the strength of the force and with location on the Earth's surface. The phase and amplitude of each will vary over the Earth's surface but can be extracted from harmonic analysis of tide gauge

Table 3.2 The main periodic contributions to tidal generation

Species	Notation	Period (hours)	Relative amplitude (%)	Description
Semi-diurnal	M2	12.42	100	Main lunar
	S2	12.00	46.6	Main solar
	N2	12.66	19.1	Moon distance
	K2	11.97	12.7	Moon and Sun relative distance
Diurnal	K1	23.93	58.4	Soli-lunar
	O1	25.82	41.5	Lunar diurnal
	P1	24.07	19.3	Solar diurnal
Fortnightly	Mf	330	17.2	Lunar phase
Monthly	Mm	661	9.1	Lunar monthly
Solar semi-annual	Ssa	4385	8.0	Seasonal solar
Solar annual	Sa	8759	1.3	Annual solar
Nodal		163 024	0.1	Moon orbital

Source: Carter (1988).

records at a station and then used for future predictions. The most important harmonic tide generating components are given in Table 3.2.

3.4.4 Dynamic Theory of Tidal Generation

Tide gauge measurements often reveal differences between the time and amplitude of the measured tides at a location and those predicted solely from consideration of astronomical factors (i.e., the equilibrium theory of tides). Leaving aside sea level changes due to meteorological factors, these differences arise from a number of factors that influence the propagation of the tidal waves around the oceans, including the irregular distribution and shape of land masses and ocean basins, the interaction of the tidal waves with the ocean bed and land margins, the effects of inertia of the water mass, and the effect of a rotating Earth in an inertial reference frame (i.e., the Coriolis 'force').

In the open ocean tidal waves have a wavelength on the order of 20 000 km and a period of a bit over 12 hours – these waves behave dynamically as shallow water waves (see Section 5.3), because the depth of the ocean is, on average, less than 0.05 of the wavelength. Frictional drag on the bed leads to loss of energy, whereas shoaling and refraction affects the speed and direction of propagation as the tidal wave travels across the ocean floor. The situation changes dramatically when waves reach the continental shelf with a depth of about 200 m compared to 4000 m for the deep ocean. The sudden change in depth results in significant reflection of wave energy, that interacts with the next incoming wave to produce a standing wave. If we consider a tidal wave within a closed rectangular basin (Figure 3.17A), the wave will travel across the basin, reflect off the far wall and then interact with other advancing tidal waves. The result is to produce a standing wave with the highest vertical motion, or amplitudes of oscillation, at the boundaries (antinode) and a nodal point near the centre where there is no vertical movement. Also associated with the standing wave is a horizontal movement of water that has its maximum excursion at the nodal point and minimum at the boundaries (Figure 3.17A).

The interaction of the water motion associated with the standing wave and the Coriolis

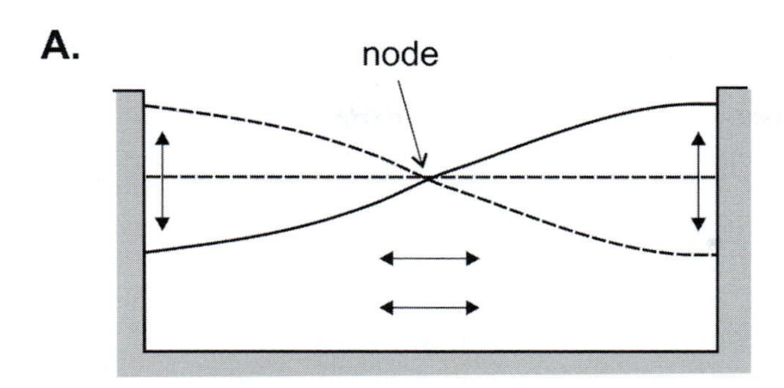

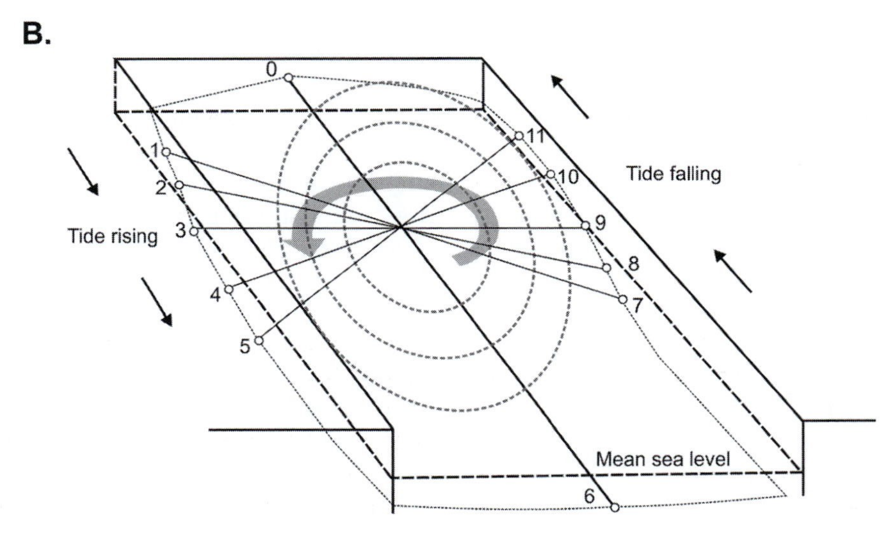

Figure 3.17 Tidal amphidromes: A. vertical and horizontal motions associated with a standing wave in a closed basin; B. sketch of the progression of a tidal wave anticlockwise around an amphidrome in the northern hemisphere. The radial lines are co-phase lines which mark equal arrival times while the circles are co-range lines which mark areas with equal amplitude. The crest of the tidal wave (high tide) is 0. At points 1–5 the tide is rising and the trough of the wave is at point 6. At points 7–11 the tide is falling.

effect results in the generation of a circular rotation similar to that of atmospheric cyclones and this follows an anticlockwise rotation about the nodal point or amphidrome in the northern hemisphere (Figure 3.17B). The movement and amplitude of the wave are depicted by co-phase lines that radiate outward from the centre of the amphidrome and by co-range lines that form concentric circles around it. The co-phase lines mark points of equal arrival time of the tidal stage and they can be depicted in terms of time or degrees around a circle. The co-range lines join places having the same tidal amplitude and they increase in magnitude away from the centre.

There are a number of amphidromic points in the oceans that control large-scale tidal motion. In the Atlantic Ocean there are two large amphidromes associated with the M_2 tide, one in the North Atlantic with its centre roughly a third of the way between Nova Scotia and Britain and the other in the South Atlantic centred midway between Brazil and southern Africa. On the continental margins smaller amphidromic systems come into existence as a result of the complex interaction of tidal waves associated with these large systems and coastal topography that produces patterns of reflection and refraction. Where the shape of the water body is complex a series of amphidromes may form, as is the case for the European continental shelf around Britain, the North Sea and English Channel. An example of the smaller amphidromic systems can be seen in the Gulf of St Lawrence on the east coast of Canada (Farquharson, 1970). The tidal wave from the Atlantic Ocean enters the Gulf primarily through the 110 km wide Cabot Strait between Cape Breton (Nova Scotia) and Newfoundland (see Figure 3.7 inset), and moves anticlockwise around the coast leading to the generation of an amphidromic point near the Magdalen Islands (Figure 3.18). The tidal range increases from < 0.25 m around the islands to about 2 m at the eastern end of the Northumberland Strait. A further example of the complexity of the tides in many coastal situations can be seen in the Northumberland Strait (Figure 3.18). Here the crest of the tidal wave (high tide) propagating through the eastern entrance to the strait meets the succeeding trough (low tide) propagating through the western entrance and produces a quasi-amphidromic point just off West Point, Prince Edward Island. Water motion here is not completely dampened, but the tidal range in the vicinity is < 1 m.

In elongate embayments the effects of constriction of the tidal wave, shoaling and harmonic amplification can produce extremely large tidal ranges. The Bay of Fundy is one such location and it has the highest measured spring tidal range in the world (over 16 m). The tide at the entrance to the bay is already quite high (>3 m) because of amplification over the wide continental shelf. The bay itself is about 270 km long with an average depth of about 60 m, giving rise to a natural period of oscillation around 12.5 hours. This is very close to the semi-diurnal

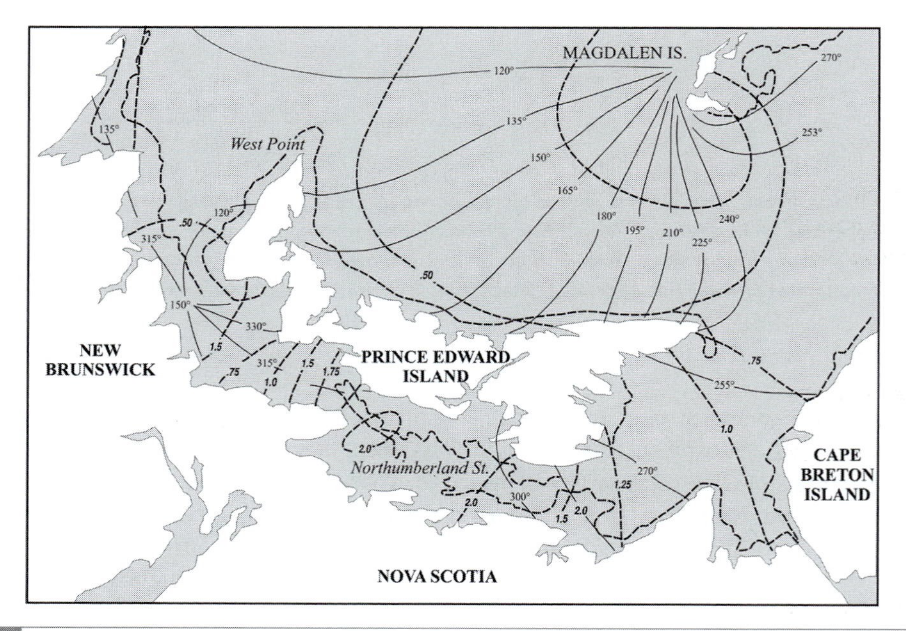

Figure 3.18 Amphidromic points and co-tidal lines for the southern Gulf of St Lawrence. The main amphidromic point off the Magdalen Islands is produced by the tidal wave which enters the Gulf through the Cabot Strait between Cape Breton and Newfoundland. A secondary amphidromic point occurs at the eastern end of the Northumberland Strait where the tide propagating around Prince Edward Island and eastward through the strait is out of phase with the next tide propagating westward through the straight. After Farquharson (1970).

tide and thus resonance results in an increase in the tidal range up the basin to extreme spring tidal ranges over 15 m in the Minas Basin and Cumberland Basin. However, some portion of the increase in tidal range is due to the narrowing of the bay which results in a convergence of wave energy and thus in the height of the tidal wave.

In estuaries, and river mouths entering bays with a large tidal range, a tidal bore may develop on the rising tide. The bore appears as a breaking wave that travels up river at speeds of 20–25 km hr^{-1} and is accompanied by a rapid increase in water level following its passage. Bores occur on the Peticodiac River and several other rivers draining into the upper Bay of Fundy as well as on the Severn River in England and the Qiantang River, China (Pugh and Woodworth, 2014).

3.5 Short-Term Dynamic Changes in Sea Level

Changes in sea level on a time scale of hours to decades can result from changes in sea surface temperature, reflecting seasonal patterns of heating and cooling, as well as decadal scale changes in global pressure and wind systems. Large scale events such as those associated with ENSO events can change ocean surface temperatures and the strength and position of major ocean currents, thus leading to areas of elevated or depressed sea level (Figure 3.19). Short-term meteorological processes such as winds and pressure changes can also produce elevated or depressed water levels along coasts due to the effects of wave shoaling and breaking – these changes are dealt with in Chapter 5.

3.5.1 Effects of Temperature, Pressure and Ocean Currents

There are a number of factors that contribute to dynamic fluctuations in sea level over a period of months to a decade or so (Komar and Enfield, 1987). These fluctuations (Figure 3.19) are

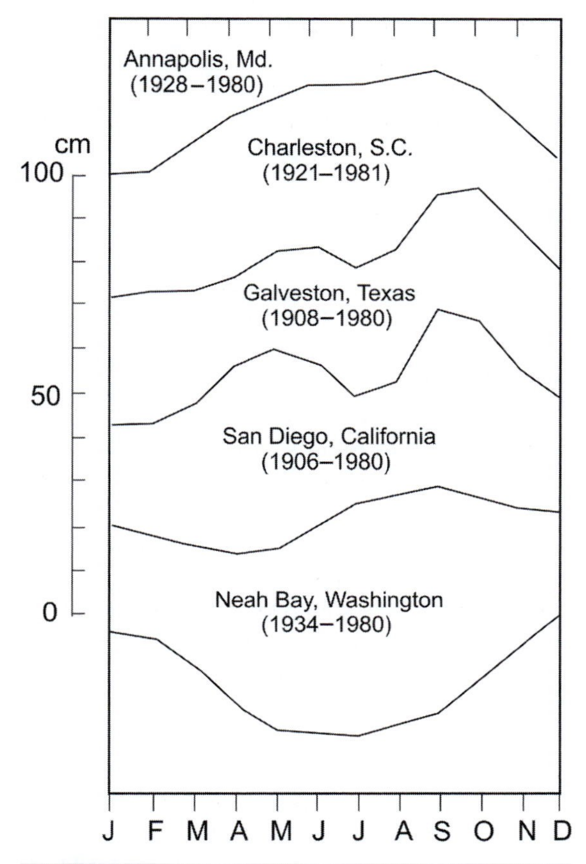

Figure 3.19 Examples of annual cycles of sea level changes determined from tide gauges. The dates in brackets give the period over which the data were averaged (Komar and Enfield, 1987).

generally about 0.1–0.3 m but in some cases may range up to 1 m. Seasonal cycles occur as a result of fluctuations in average barometric pressure, sea surface temperatures and changes in location of ocean currents. There is a change in sea level of about 1 cm for every 1 millibar (roughly 0.1 kPa) in atmospheric pressure. As a result, sea level responds not only to pressure changes associated with weather systems but also to seasonal changes in pressure such as those associated with the movement of the Azores and Hawaiian high pressure zones and the shifting of the Intertropical Convergence Zone north and south.

Sea level changes in the Pacific occur as a result of changes in pressure, winds and ocean

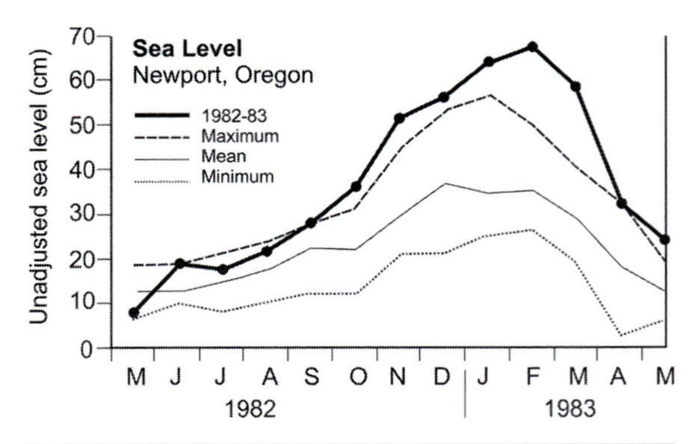

Figure 3.20 Monthly average sea level at Newport, Oregon, during the 1982–3 El Niño event compared to the maxima, average and minima previously measured (Komar and Enfield, 1987).

currents collectively associated with El Niño events. The 1982–3 El Niño event produced an increase of sea level of 0.4–0.5 m that travelled from west to east across the Pacific and affected many coastal areas of North, Central and South America. Typically El Niño events result in a rise of 0.1–0.2 m in sea level along the California coast and northward into Oregon (Komar and Enfield, 1987; see Figure 3.20). They also increase coastal vulnerability to storms and increased sea levels over much of the coast of the Pacific Ocean (Barnard *et al.*, 2015).

Much of the early work on these effects relied on data from tide gauges which were located primarily in harbours. Over the past few decades, satellite imagery of sea surface temperatures and sea levels have given us a much more complete picture of the dynamics of the ocean surface. These have shown that regional variations in sea level can persist for up to a decade or more. Changes in mean sea level over the period 1993–2016 from Topex and Jason-1, 2 and 3 satellite altimetry show decadal scale dynamic variations superimposed on a long-term trend of sea level rise in regional oceans and seas (Figure 3.21). Variations in sea level in the Atlantic Ocean are on the order of 10–20 mm over periods of 2–5 years (Figure 3.21A). Averaging over the whole of the Atlantic Ocean tends to suppress some of the extreme fluctuations. In the much smaller Caribbean Sea, the fluctuations are on the order of 50–100 mm and over a period between 2002 and 2004 sea level varied over 80 mm (Figure 3.21B). Here variations in sea surface temperature (Glenn *et al.*, 2015) and in regional atmospheric pressure likely account for the majority of this change. There are also clear connections between climatic factors such as the amount and distribution of rainfall in the region and ENSO events in the Pacific (Glenn *et al.*, 2015).

3.5.2 Storm Surge

Changes in sea level during storms occur as a result of wind stress on the water surface and changes of atmospheric pressure leading to positive displacement of the water level (storm surge or set-up) and negative displacement (negative surge or set-down). Wind stress on the water surface results in wind drift, a slow flow of surface water in the direction of the wind. Close to the shoreline, the nearshore slope and beach provide a barrier to the surface drift, resulting in a set-up of water at the shoreline and a compensating return flow offshore below the surface (Figure 3.22A). The return flow will be retarded by frictional forces, and the greater this retardation the greater will be the equilibrium set-up. The highest storm surges thus tend to occur in shallow, gently sloping coastal areas and in semi-enclosed bays and estuaries where local wind set-up is superimposed on the general coastal one. Offshore winds can also force a displacement of water offshore thereby depressing the mean water level at the shore, but this tends to be much smaller than a corresponding surge because the effects are dissipated into the open ocean. However, in shallow, enclosed water bodies such as lagoons and lakes, set-down on the upwind coast can be of the same order as the surge on the downwind coast (Figure 3.22B). Storm surge is sometimes referred to as a meteorological tide, with the increased sea level being superimposed on the oscillations associated with astronomical tides. The predicted water level changes associated with astronomical tides are usually subtracted from tide gauge records in order to separate out the storm surge effect. Because of the potential impact on shoreline features and on human activities, more

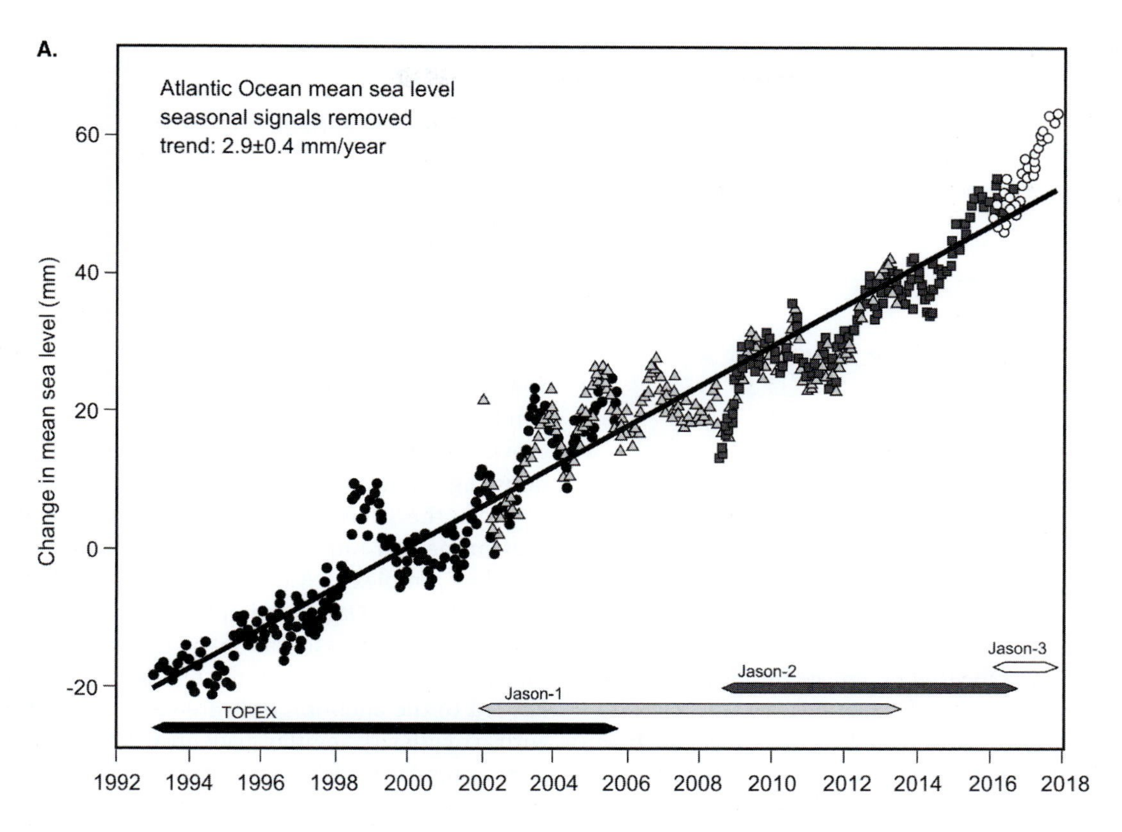

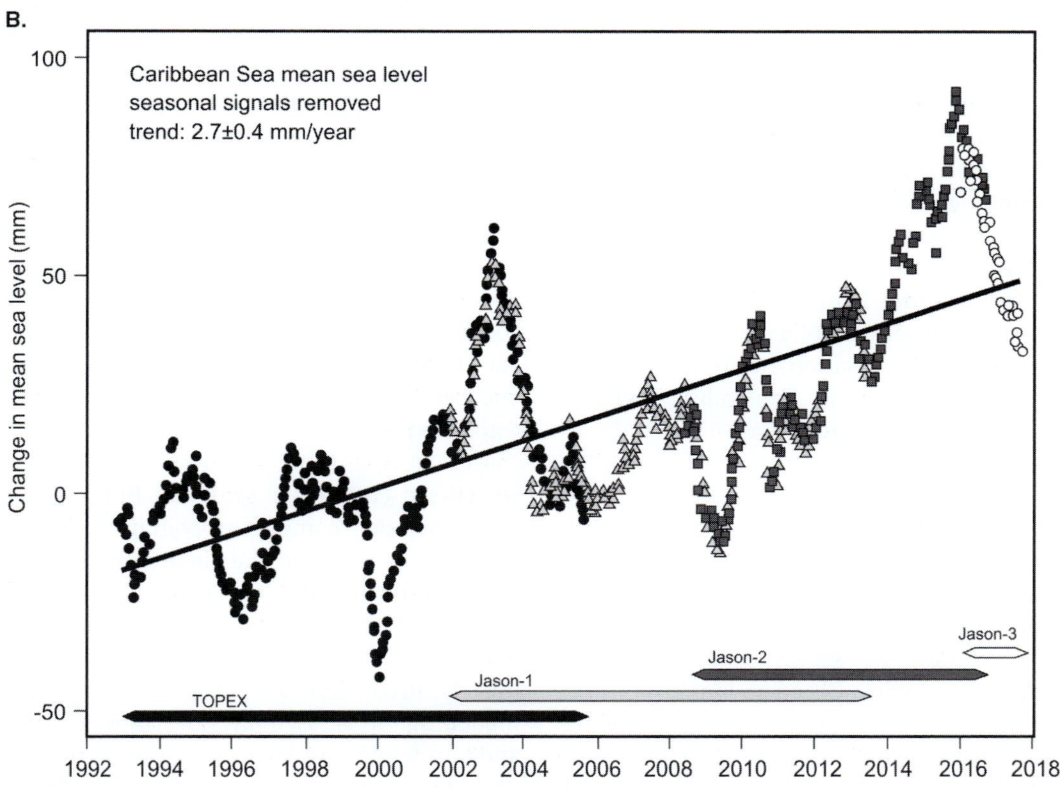

Figure 3.21 Variations in mean sea level from satellite altimetry for the period 1993–2017 for: A. the Atlantic Ocean; and B. the Caribbean Sea. Note the much large variations in height for the smaller water body of the Caribbean Sea. (NOAA – Laboratory for Satellite Altimetry).

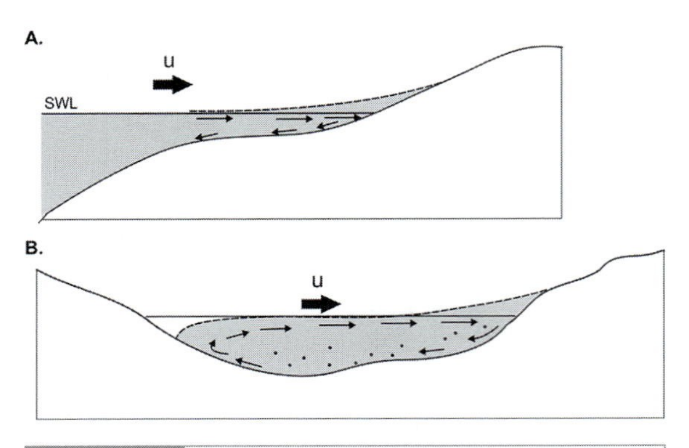

Figure 3.22 Storm surge generated by onshore winds: A. surge effects on an open coast; B. positive and negative surge in a lagoon or lake (SWL = still water level).

attention is focused on storm surges rather than set-down.

A storm surge can be considered as being generated by two components – a static component related to the atmospheric pressure and a dynamic component associated with the tangential wind stress on the water surface generated by the wind field. The static component, or inverse barometer effect, arises because of the low atmospheric pressure associated with both tropical and extratropical cyclones that are responsible for generating significant storm surges. A 1 hectopascal (hPa) reduction in atmospheric pressure near the centre of a storm produces an increase on the water level of about 1 cm and generally accounts for about 10–15 per cent of a storm surge (World Meteorological Organisation, 2011).

The key parameters controlling dynamic storm surge can be expressed as:

$$S = K \cdot \frac{w^2}{h} \tag{3.2}$$

where S is the storm surge amplitude, w is wind speed, h is water depth and K incorporates several other factors such as bottom stress, stratification, surface roughness and wind angle (World Meteorological Organisation, 2011). Prior to 1970, storm surge estimation was carried out using relatively simple empirical equations or statistical models. These still provide a reasonable estimate of the maximum surge, particularly for simple coastal situations such as large lakes and semi-enclosed basins. Quite a lot of the early work on modelling storm surge was in fact done in the North American Great Lakes because of the virtual absence of tides, the simple lake geometry and water depth, the presence of water level gauges at many points along the shoreline and the existence of a relatively large number of meteorological stations at the upwind and downwind ends of each of the lakes. Large lakes still provide useful testing sites for empirical equations (e.g., Chittibabu and Rao, 2012).

Since the 1970s most storm-surge prediction uses numerical models of increasing sophistication and complexity (Gonnert *et al.*, 2001; World Meteorological Organisation, 2011). The need for improved representation of the bed topography and the complexity of the shoreline form has led to the adoption of finite element models with irregular triangular grids instead of square or rectangular grids, and this is also the case for modelling of wave transformation and tidal currents. The models are also increasingly demanding of the meteorological and tide forecast models that provide the input into real time storm surge predictions. Many countries have their own dedicated models, in part because of the ease of access to real time meteorological data and the need to link this with a hazard warning system – for example CS3X in Britain (Pugh and Woodworth, 2014) and the Hirlam and DCSM in the Netherlands (WMO 2011). In the USA the forecast model SLOSH has been in use for more than 40 years and has been used extensively to predict surges associated with hurricanes on the Gulf and East coasts. It is a finite difference model with a structured curvilinear grid that limits the ability to model complex topography near the coast, for example in the vicinity of tidal inlets. The ADCIRC model developed by the Army Corps of Engineers is a finite element model (Luettich *et al.*, 1992) that uses a highly flexible, unstructured grid allowing for very detailed modelling. It is used by government agencies and academic researchers among others (e.g., Chittibabu and Rao, 2012; Farhadzadeh and Gangai, 2017).

It is useful to distinguish between storm-surge events associated with tropical cyclones (hurricanes, typhoons) and those generated by mid-latitude cyclones (Dolan and Davis, 1994; Pugh and Woodsworth, 2014). Tropical cyclones are relatively small in extent and originate at sea. They are generated and driven by the energy released by condensation of moisture evaporated from a warm ocean surface and thus they decay as they move into higher latitudes over cooler ocean waters and die out rapidly over land (Emanuel, 2003). Hurricanes have a pressure gradient of about 20 mb 100 km^{-1} and this produces intense winds that exceed 32 m s^{-1} for the least intense tropical cyclones. At landfall they can produce very high surges but these are usually confined to a few tens of kilometres of shoreline and occur over a time period of one or two tidal cycles or less. Tropical cyclones are usually absent from regions between 5° north and south of the equator because the Coriolis effect is weak and insufficient to generate significant rotation. In the northern hemisphere tropical cyclones form in the Caribbean and off the Pacific coast of Mexico and in a broad zone across the Indian and Pacific oceans, affecting the mainland coast of Asia from India to China as well as the islands of Indonesia and Japan. In the southern hemisphere they are generally absent from the Atlantic, but in the Pacific they are formed in a broad zone from west of Australia and New Guinea to the southwest coast of Africa. The effects of wind dominate over those of pressure, though the extreme low pressure at the centre of the storm can account for a rise in sea level of up to 1 m.

The surge produced by hurricanes generally increases with increasing intensity (Xia *et al.*, 2008). Hurricane Hugo, which came ashore near Charleston, North Carolina on 21–22 September 1989 was one of the most intense Caribbean

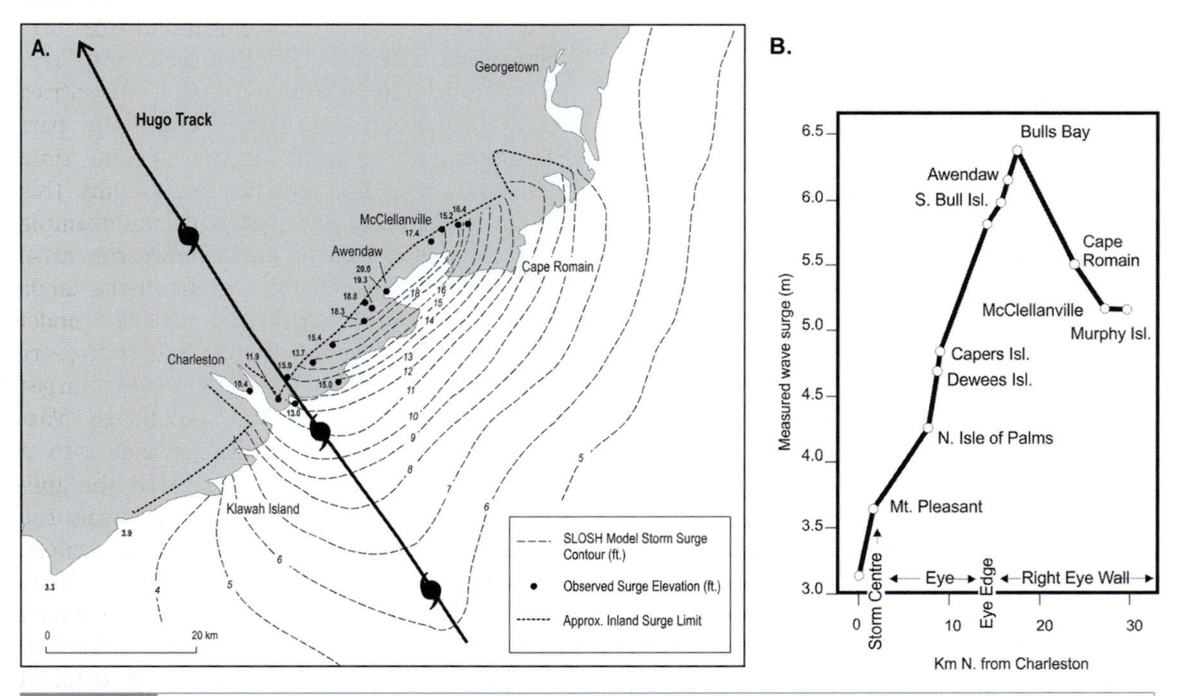

Figure 3.23 Storm surge associated with Hurricane Hugo 22 September 1989; A. hurricane path, and storm surge values predicted by SLOSH (Sea, Land and Overland Surge from Hurricanes); B. plot of observed surge values on the South Carolina coast. Source: Adapted from Coch and Wolff (1991). Reproduced with permission from the Coastal Education and Research Foundation, Inc.

hurricanes of the twentieth century, producing a maximum surge of about 6.1 m just to the right of the eye (Coch and Wolfe, 1991; Figure 3.23A). However, at Kiwah Island just 24 km to the south (left) of the eye the surge was only 1.3 m and the major impact of the surge only extended along about 100 km of the shoreline (Figure 3.23B). This illustrates a key feature of storm surge, and to a lesser extent wave heights, associated with tropical cyclones when they approach land. Because of the small diameter of the centre around which the winds move, the winds are onshore to the right of the eye (in the northern hemisphere) and they generate large waves moving onshore. Winds to the left of the path blow alongshore or offshore and thus the surge is greatly reduced and the effect of waves generated by the storm out to sea tends to be reduced by the winds blowing offshore just before it makes landfall. Thus, while the height of the surge tends to increase with increasing intensity, the track of the storm, angle of approach to the shoreline trend and configuration of the coastline greatly influence the actual surge level at a point on the coast (Xia *et al.*, 2008; Lindner and Neuhauser, 2018).

Cyclones in the Bay of Bengal affect low-lying coasts in India, Bangladesh and Burma. They can produce very high surges because of the configuration of the bay, low coastal gradients and the intensity of the storms (Murty and Flather, 1994). The impacts of surges are felt particularly along the low-lying areas of the Ganges River delta. A severe cyclone in November, 1970 produced a storm surge of more than 9 m on the coast north of Chittagong and one in 1990 generated a storm surge ranging from 5 to 10 m, leading to the loss of over 100 000 lives. Two or three hurricanes and post-tropical storms bring large waves to the Atlantic coast of Canada every year, and on average one crosses the coast every two or three years. In the Pacific typhoons affect the Philippines other islands as well as much of the coast of China, Japan and northern Australia.

Extratropical storms are formed largely along the boundary between warm and cold air masses and their track is controlled by large-scale pressure systems and by waves in the upper atmosphere jet stream. They extend over many hundreds of kilometres and the pressure gradient is about 5 mb 100 km^{-1}. They affect large areas of the coastline and their effects may take place over a number of tidal cycles. The east coasts of the US and Canada are affected particularly by depressions that track northeastward off the coast and produce northeasters which commonly occur from late fall through the winter (Dolan and Davis, 1994; Forbes *et al.*, 2004).

These systems often track eastward across the north Atlantic, together with others generated off Labrador, and affect the coast of western Europe (Betts *et al.*, 2004; see Figure 3.24). The impacts are felt particularly on the North Sea coasts where the coastal geometry and shallow water often enhance storm surge elevations. A storm surge in 1953, which coincided with spring tides and was accompanied by waves of 7–8 m, breached dykes along much of the coast of Holland and did extensive damage to low lying areas in other coastal communities (Figure 3.25). There were more than 1800 lives lost in the Netherlands and 300 in England and it stimulated research into better protection and better forecasting in both countries. In the past few years western Europe has been hit by a number of intense storms that have had a major impact on the coasts of Britain, and northwest Europe (e.g., Dissanayake *et. al.*, 2014; Castelle *et al.*, 2015; Spencer *et al.*, 2015; Brooks *et al.*, 2017). Spencer *et al.* (2015) show the calculated storm surge for a number of towns along the North Sea (east) coast of England where surge levels increased from about 1 m at Aberdeen in the north to 2 m at Lowestoft in the south (Figure 3.26). These surges were similar to those of the 1953 storm in magnitude but wave height was considerably lower.

Extratropical storms are also responsible for most periods of significant wave action and storm surge generation in the Great Lakes. An intense storm on 2 December 1985 resulted in extensive overwash and damage to property along much of the coastline of eastern Lake Erie and over 80 cottages were destroyed at Long

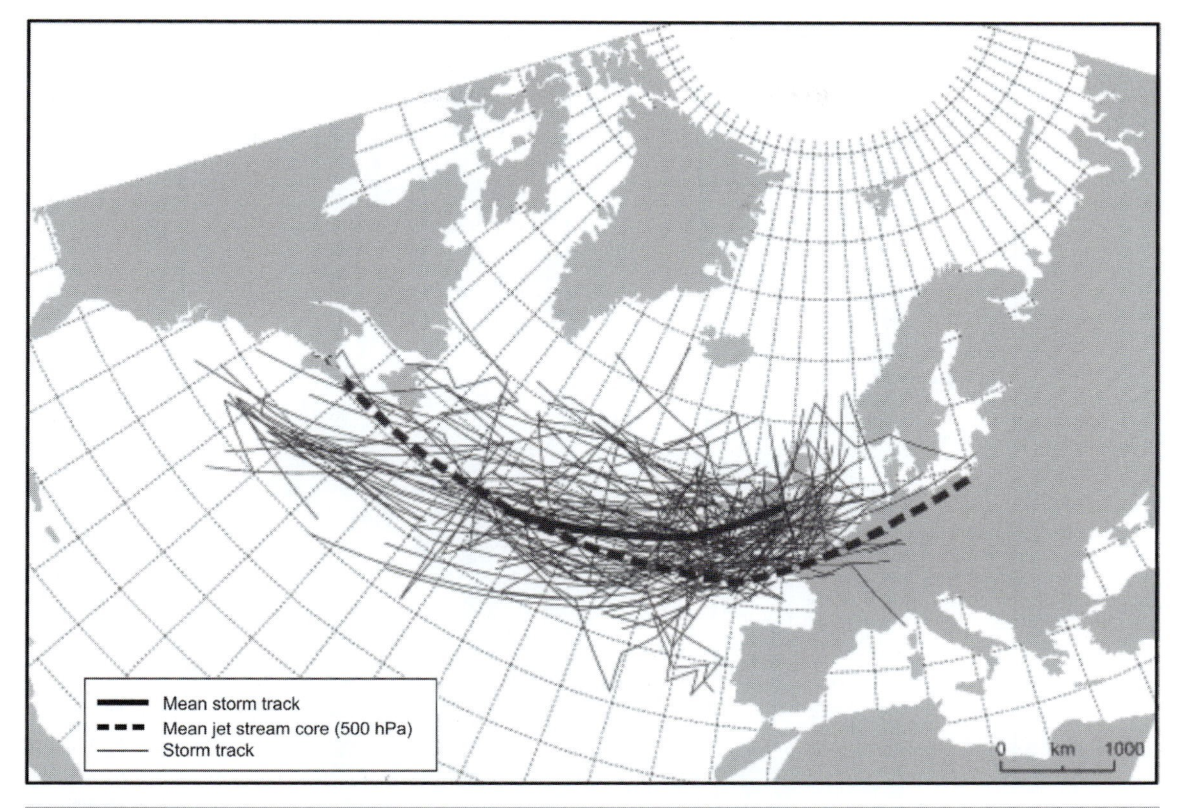

Figure 3.24 Position of individual storm tracks promoting extreme surge events (top 1%) at Brest for the period 1950–92. The mean jet stream core (500 hPa level) and mean storm track position are also shown. After Betts *et al.* (2004).

Point alone (see Figure 3.27). The maximum surge recorded at Port Colborne near the east end of the lake was just over 2 m and the set-down at Kingsville towards the western end was over 1.5 m (Figure 3.27). At Port Stanley, near the centre of the lake, water level fluctuations were < 0.5 m.

3.5.3 Seiches and Other Long Waves

In enclosed bodies of water such as lakes, lagoons, bays and harbours, long period standing waves termed seiches (pronounced sayshh) can be generated by processes such as changes in atmospheric pressure and winds, and by the release of a storm surge following a change in wind direction (Miles, 1974). Seiches in Lake Erie, as an example, are generated by storm surge set-up on one shoreline and the matching surge set-down at the opposite shoreline (see Figure 3.27). As the storm system that generated the surge passes on and the wind direction shifts, the water level differential is no longer maintained and the water is released from the high side towards the low side. As a result, a series of water level oscillations with decreasing amplitude are generated as the water sloshes back and forth. The longest natural period of oscillation, T_o, depends on the length of the water body and its depth, and is given by:

$$T_o = \frac{2L}{\sqrt{gh}} \quad (3.3)$$

where: T_o is the period of oscillation, L is the length of the basin, h is the average depth of basin.

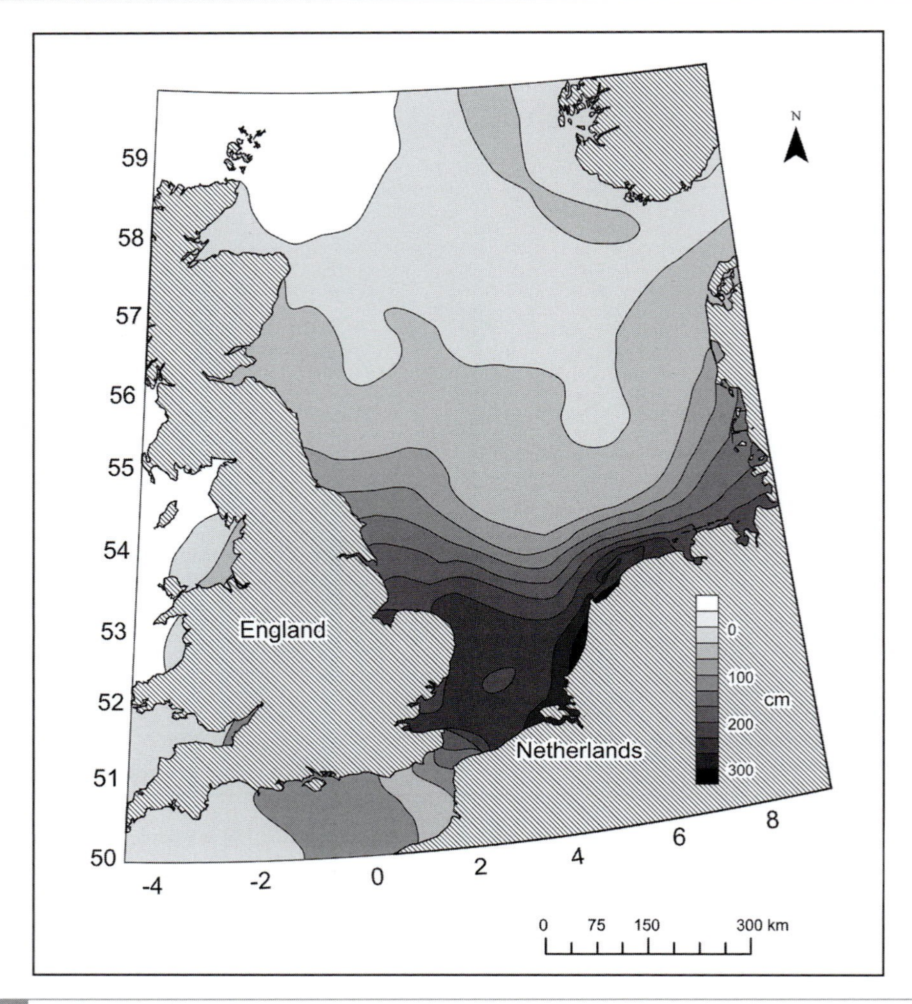

Figure 3.25 Maximum surge (cm) during the 1953 North Sea storm computed using the POL CS3 model. Note how the surge elevation increases as the gap between the coast of England (left) and the European shoreline of Germany, Holland and Belgium (right) becomes narrower and narrower. After Wolfe and Flather (2005).

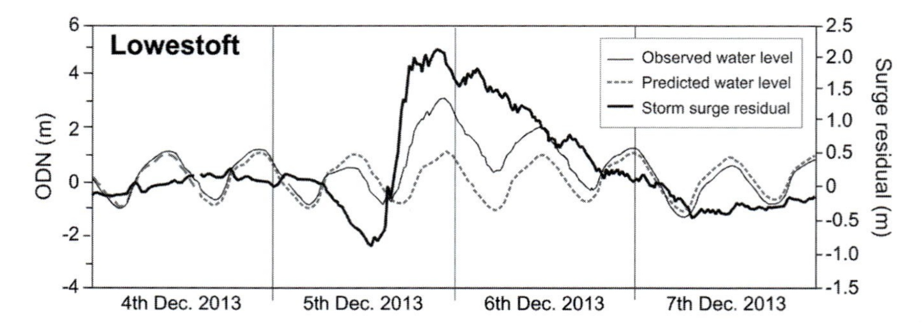

Figure 3.26 Storm surge for Lowestoft, SE England, 4–7 December 2013 from the UK National Tide Gauge Network. The surge is calculated as the residual arrived at by subtracting the forecast tidal elevation from the actual measured elevation. Note the negative surge or set-down (observed level less than the predicted tide) in the middle of the day on 5 December and then the rapid set-up towards the end of the day through the end of the day on 6 December. After Spencer *et al.* (2015).

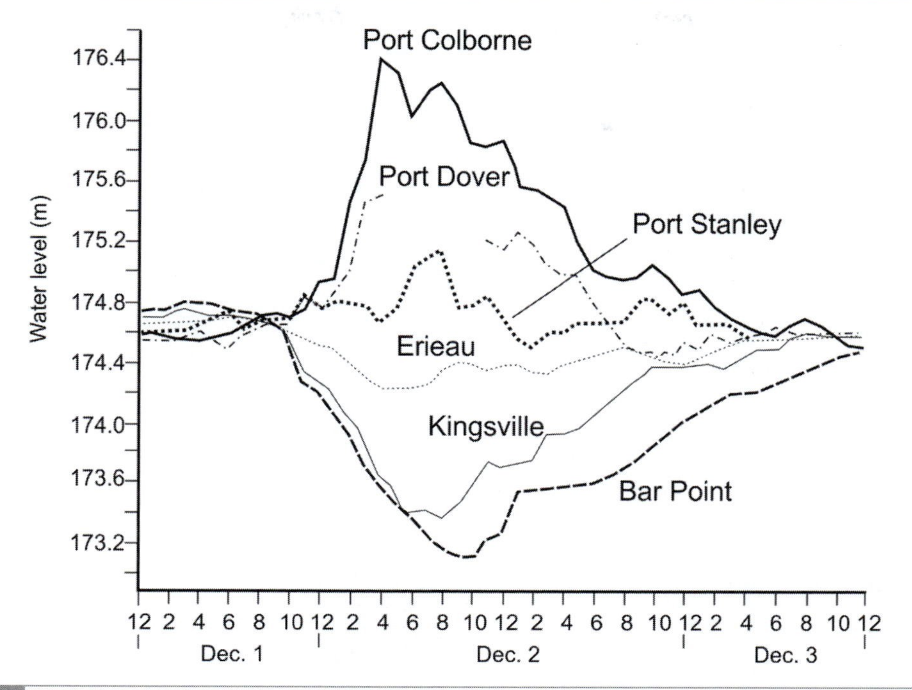

Figure 3.27 Water levels recorded at several gauges on Lake Erie during the storm of 2 December 1985. Note the positive surge at the eastern end of the lake (Port Colborne) and corresponding negative surge at the upwind, western end (Kingsville, Bar Point). Source: Environment Canada.

Note that the term $(gh)^{0.5}$ is the formula for the speed of a shallow water wave and $2L$ is the distance from one shore to the opposite shore and back again. The period of a seiche along the length of Lake Erie (about 350 km) is just over 9 hours.

3.6 Climate Change and Sea Level Rise

The results of a large number of studies worldwide suggest that over the past 1000 years sea level averaged globally rose at a rate of < 2 mm a^{-1} (Fleming *et al.*, 1998). There are a number of components that control changes in sea level and various models have been used to estimate the likely contribution of each of these components over the past century (see Table 3.3). The two major contributors to present sea level rise are: (1) steric height variations produced by the expansion or contraction of the water in the oceans as a result of changing temperature and density; and (2) variations in the mass (volume) of water in the oceans due to melting of glaciers and snow packs worldwide, and particularly melting of the Greenland and Antarctic ice sheets. We can use measured trends in global sea level to extrapolate future sea levels (e.g., Wake *et al.*, 2006) but, because melting of ice sheets and warming of the oceans tends to lag behind increases in atmospheric temperature, present trends may not be a good predictor of future levels. In particular, there is still considerable uncertainty about the present contribution from the two major ice sheets of Greenland and Antarctica (Church et al., 2013).

3.6.1 Measured Rates of Mean Sea Level Rise

Most data on rates of sea level rise for periods prior to the twentieth century come from a variety of proxy measures, with considerable

Table 3.3 Global mean sea level budget (mm a^{-1}) over different time intervals from observations and from model-based contributions

Source	1901–90	1971–2010	1993–2010
Observed contributions to global mean sea level rise			
Thermal expansion	–	0.8 (0.5 to 1.1)	1.1 (0.8 to 1.4)
Glaciers except Greenland and Antarctica[a]	0.54 (0.47 to 0.61)	0.62 (0.25 to 0.99)	0.76 (0.39 to 1.13)
Glaciers in Greenland[a]	0.15 (0.10 to 0.19)	0.06 (0.03 to 0.09)	0.10 (0.07 to 0.13)[b]
Greenland ice sheet	–	–	0.33 (0.25 to 0.41)
Antarctic ice sheet	–	–	0.27 (0.16 to 0.38)
Land water storage	–0.11 (–0.16 to –0.06)	0.12 (0.03 to 0.22)	0.38 (0.26 to 0.49)
Total of contributions	**–**	**–**	**2.8 (2.3 to 3.4)**
Observed GMSL rise	**1.5 (1.3 to 1.7)**	**2.0** (**1.7** to **2.3**)	**3.2** (**2.8** to **3.6**)
Modelled contributions to GMSL rise			
Thermal expansion	0.37 (0.06 to 0.67)	0.96 (0.51 to 1.41)	1.49 (0.97 to 2.02)
Glaciers except Greenland and Antarctica	0.63 (0.37 to 0.89)	0.62 (0.41 to 0.84)	0.78 (0.43 to 1.13)
Glaciers in Greenland	0.07 (–0.02 to 0.16)	0.10 (0.05 to 0.15)	0.14 (0.06 to 0.23)
Total including land water storage	**1.0 (0.5 to 1.4)**	**1.8 (1.3 to 2.3)**	**2.8 (2.1 to 3.5)**

Notes: Uncertainties are 5 to 95%. The Atmosphere–Ocean General Circulation Model (AOGCM) historical integrations end in 2005; projections for RCP4.5 are used for 2006–10. The modelled thermal expansion and glacier contributions are computed from the CMIP5 results, using the model of Marzeion *et al.* (2012) for glaciers. The land water contribution is due to anthropogenic intervention only, not including climate-related fluctuations.

[a] Data for all glaciers extend to 2009, not 2010. [b] This contribution is not included in the total because glaciers in Greenland are included in the observational assessment of the Greenland ice sheet. [c] Observed GMSL rise – modelled thermal expansion – modelled glaciers – observed land water storage.

Source: Church *et al.* (2013) table 13.1.

uncertainty regarding their precision and accuracy. In the past century, tide gauge data were collected at an increasing number of stations worldwide and these provide measurements with a precision on the order of 2 cm. Gauges with a record of more than 50 years offer an opportunity to filter out annual- and decadal-scale dynamic variations as well as cyclic variations associated with the 18.6 year tidal oscillation, and thus to determine recent trends in sea level. An example of this is the record for Newlyn on the southwest coast of Britain (Arújo and Pugh, 2008) where a linear trend through the annual mean sea level (MSL) values shows sea level increasing at a rate of 1.77 ± 0.12 mm a^{-1} (Figure 3.28). This is consistent with a computed value of 1.8 ± 0.3 mm a^{-1} for the period 1950–2000 using a global combined tide gauge and satellite altimetry data base (Church *et al.*, 2004) and with a value of 1.74 ± 0.16 mm a^{-1} for nine tide gauges worldwide for the 100 year period from 1904–2003 (Holgate, 2007). Use of the tide-gauge data requires considerable effort to identify and remove errors from the data recording system, as well as the tidal components and the non-tidal residuals (Arújo and Pugh, 2008). It also requires that corrections be made for any vertical changes in the elevation of the land at the location of the tide gauge. Caution also has to be exercised in using tide-gauge measurements to estimate sea level rise globally because of the uneven distribution of gauges worldwide (Pugh and Woodsworth, 2014).

As was shown in Section 3.2.2, beginning in 1993 satellite altimetry has provided a high precision record of sea level over the world oceans between 66° latitude north and south. Continued analysis and updating of the altimetry data has shown both the need for calibration against established gauge sites, and to make corrections for atmospheric conditions and small variations in satellite orbits (Nerem *et al.*, 2007). The data collected to date, after calibration and corrections for glacial isostatic adjustment, are consistent with tide gauge curves and show sea level rising at an average of 3.0 ± 0.4 mm a^{-1} (Figure 3.29A). The largest contribution to sea level rise comes from added water from glacier melting followed by an increase in volume due to warming of the ocean surface layer (Figure 3.29B). The wealth of new data from the oceans paradoxically have shown the dynamic variability of the ocean surface due to decadal scale atmospheric and ocean current changes, particularly in the Caribbean (see Figure 3.21B) and the tropical Pacific and Indian oceans (Church *et al.*, 2006). There is thus a need to acquire longer records in order to smooth these dynamic changes and provide a robust estimate of long-term trends.

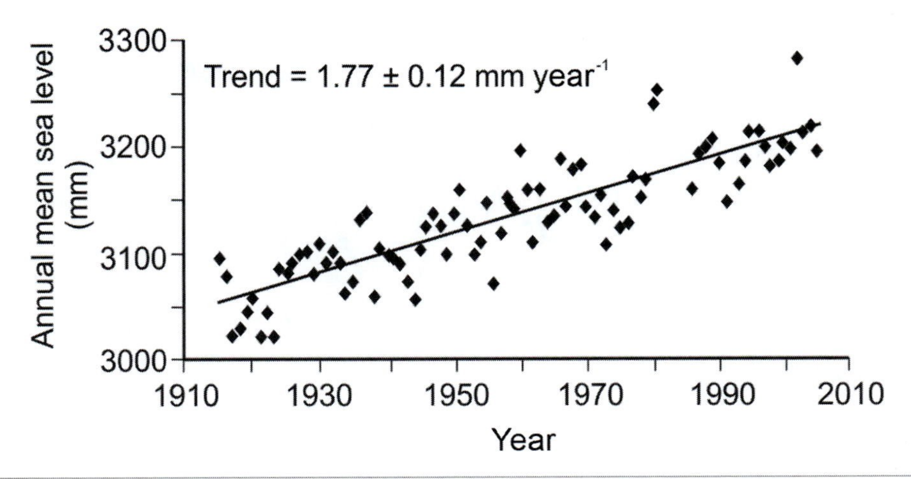

Figure 3.28 Annual mean sea level values for Newlyn, England from 1915 to 2005. Source: Arujo and Pugh (2008). Reproduced with permission from the Coastal Education and Research Foundation, Inc.

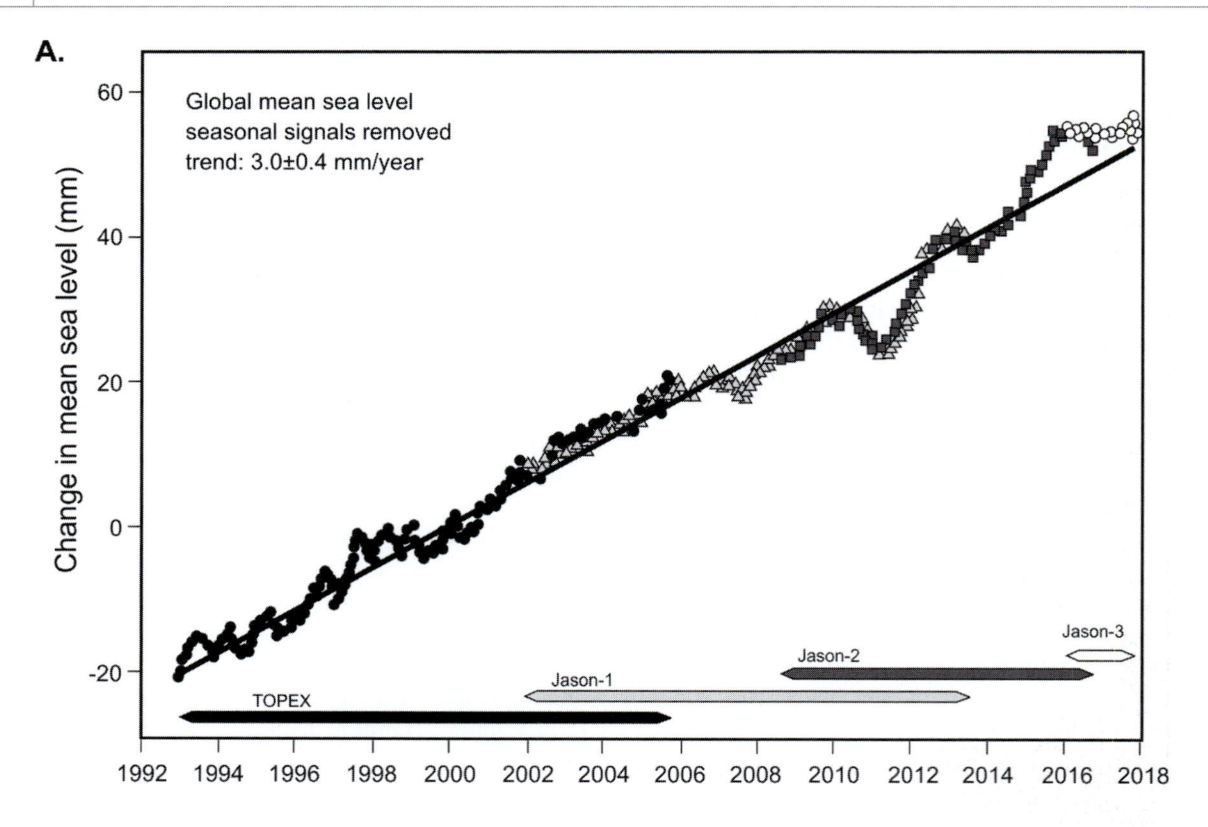

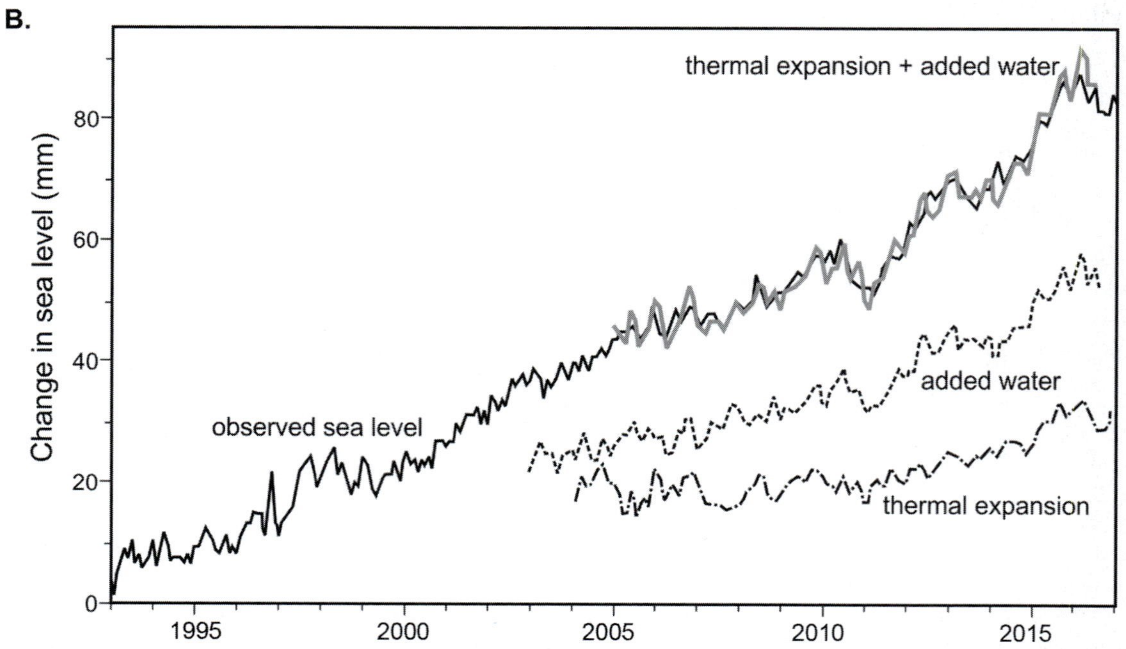

Figure 3.29 Recent sea level rise: A. Global mean sea level variations from TOPEX and Jason altimeter measurements 1993–2017, showing sea level rising at a rate of 3.0 ± 0.4 mm a^{-1}. (NOAA – Laboratory for Satellite Altimetry). The rate increases by about 0.3 mm a^{-1} after correction for the effects of glacial isostatic adjustment (Nerem *et al.*, 2007); B. contributions of ocean warming and glacier melt to recent sea level rise. After Church *et al.* (2013).

Box 3.1 Coastal Management Application

Prediction of Sandy Shoreline Recession Due to Sea Level Rise

The forecast increase in the rate of eustatic RSLR as a result of global warming has stimulated interest in the response of coasts generally, and sandy beach systems in particular, to this. In the latter half of the twentieth century the dominant paradigm was based on the 2D model of Bruun (1962) and the use of the 'Bruun rule' as outlined by Schwartz (1967). The Bruun model predicts that a rise in sea level leads to erosion and offshore transport of sediment which is deposited on the nearshore profile to a thickness equal to the change in sea level, and the Bruun rule provides a simple means of calculating the horizontal recession of the shoreline associated with this. While no significant efforts were made to substantiate the model itself, the Bruun rule was used to predict recession of sandy beach systems for decades, despite a number of publications which were critical of it (e.g., SCOR Working Group, 1991; Pilkey and Cooper, 2004; Davidson-Arnott, 2005; Ranasinghe *et al.*, 2012). There were also some modifications to the Bruun rule as shortcomings became more evident, culminating in the work of Rosati et al. (2013).

The best evaluation of the Bruun model and rule is probably the recent series of wave tank tests carried out by Atkinson *et al.* (2018), because the tests come closest to conforming to the assumptions of the model and to the step increment in water level. Atkinson *et al.* used a total of seven different experimental runs, four of them with a barred profile and three with a planar profile and berm. In each case the profile was subject to wave action until an equilibrium profile was established, the water level was raised by a step increment, and the profile again subject to wave action until equilibrium was established. Recession of the profile was then measured (Figure 3.30; **Mean**) and compared to predictions based on the original Bruun rule (**Bruun**), the Rosati *et al.* modified rule (**Rosati**) and a profile translation model developed by Atkinson *et al.* (**Trans**). It can be seen in Figure 3.30 that, while there are small differences in the performance of each of the predictive measures for individual experiments, there is no significant difference in the predictive ability of any of them. Moreover, when the data for the initial profile slope (**Slope**) provided in Atkinson *et al.* (2018) are used to project the recession distance based on the step increment

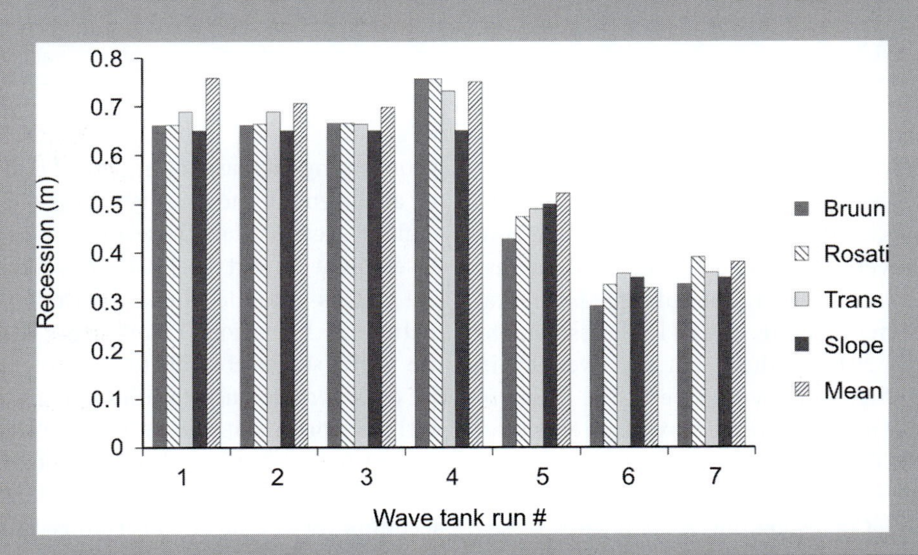

Figure 3:30 Measured shoreline recession (Mean) for seven wave tank runs compared to predicted recession using the Bruun rule (Bruun), the Rosati *et al.*, 2013 modification of the Bruun rule (Rosati), and a profile translation model developed by Atkinson *et al.* (2018) (Trans). Data from Atkinson *et al.* (2018, table 2). Also graphed is the predicted recession assuming upward translation of the profile over a height equal to the change in water level assuming maintenance of the initial profile slope for each of the wave tank runs.

Box 3.1 (cont.)

in water level there is no appreciable difference in the predicted recession compared to that of the Bruun rule or the Rosati *et al.* modification of it. This is in agreement with the findings of the SCOR Working Group (1991), who concluded that for most natural sandy coasts the Bruun rule was indistinguishable from predictions based on the profile slope.

Davidson-Arnott (2005) produced a conceptual model of the response of sandy coasts to RSLR that conformed closely to the assumptions of the Bruun model, with the exception that it was assumed that sediments eroded from the beach and dune were ultimately moved landward rather than offshore (Figure 3.31) and that the nearshore profile was translated landward and upward following the original profile form and slope. On relatively gentle slopes characteristic of most sandy coasts there is now increasing support for the landward movement of the nearshore profile as envisaged by the RD-A model (e.g., Aagaard and Sorensen, 2012; Aagaard, 2014) and evidence from a number of sites for erosion (ravinement) and landward translation of sediments under rising sea level (Goff *et al.*, 2014; Houston and Dean, 2014; Dean and Houston, 2016).

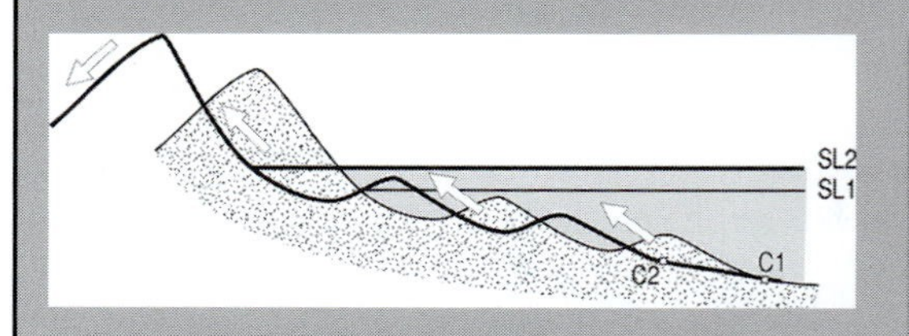

Figure 3:31 Schematic illustration of the Robin Davidson-Arnott (RD-A) model of shoreline recession on a sandy coast showing erosion and landward migration of the nearshore profile and transgression of the beach and foredune under an incremental increase in RSL. Source: Davidson-Arnott (2005). Reproduced with permission from the Coastal Education and Research Foundation, Inc.

There are two key points for coastal management that arise from this: (1) beach sediments are generally not lost offshore and beaches, dunes and barrier islands are preserved and translated landward under rising sea level; and (2) prediction of coastal evolution under RSLR requires evaluation of sediment budgets (littoral and dune) and modelling of the coastal processes (covered in Part II of this book) as well as an appreciation of the controls on the coastal system within which the impact of RSLR is being evaluated (covered in Part III of this book).

3.6.2 Future Rates of Sea Level Rise

There is still considerable uncertainty as to forecasts of global temperature change, and consequently, even greater uncertainty in how fast the oceans will respond. The Intergovernmental Panel on Climate Change (IPCC) Fifth Assessment Report from 2013 predicts that sea level rise in the twenty-first century will be of the order of 0.26 to 0.55 m for representative concentration pathway (RCP) 2.6 and 0.45–0.82 m for RCP 8.5 (the most and least conservative scenarios, respectively). Corresponding annual rates of sea level rise increase to about 4.5 mm a^{-1} until the middle of the century for RPC 2.6, and both RPC 4.6 and RPC 6.6 scenarios also reach a maximum before the end of the century. The RPC 8.5 shows a continuing increase in the rate of rise to 11 mm a^{-1} at the end of the century. These estimates are slightly greater than those made by the Fourth Assessment Report in 2007. However, many of the models of sea level change are based on prediction of the various components that contribute to the sea level budget (see Table 3.3). There are considerable uncertainties in both the quality of the historical data input for testing the models and on forecasting the contributions from each of them in the future (Church et al., 2013). This accounts for the large uncertainty in these predictions (Figure 3.32).

Clearly, rising sea level and the potential for an acceleration of this rate have implications for natural coastal systems as well as human

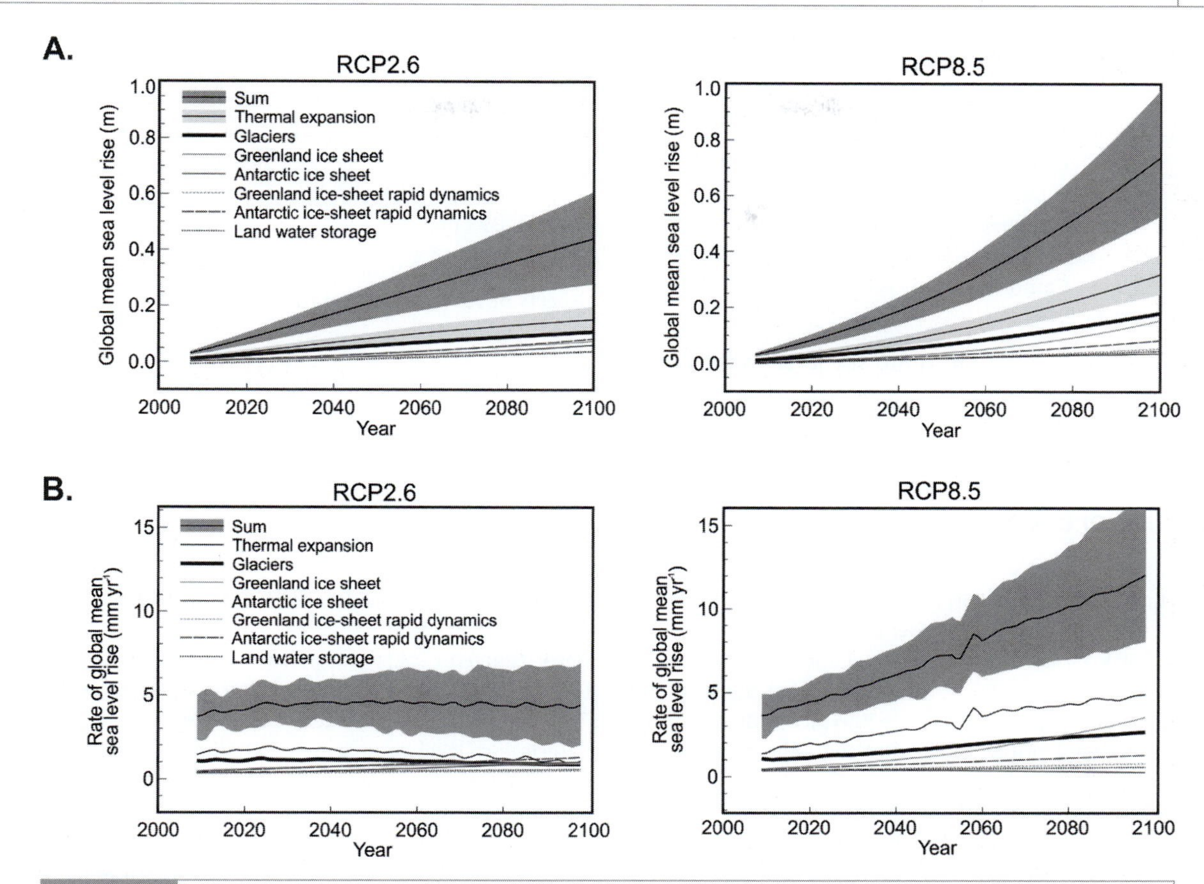

Figure 3.32 Predicted sea level rise A. and rate of sea level rise B. to 2100 based on IPCC Scenario 2.6 (least greenhouse gas inputs) and Scenario 8.5 (most greenhouse gas inputs). After Church *et al.* (2013).

economic activities (e.g., Le Cozannet *et al.*, 2014). Early scenarios focused on the extent of inundation of low-lying areas such as Florida, Louisiana and Bangladesh as well as the threat to Pacific island atoll nations, where most of the land base is only a few metres above present sea level. Some of these scenarios simply took the highest projected sea level and spread this over the present land mass without any allowance for dynamic coastal processes, with the result that most low-lying regions disappeared. This approach is still to be found in some assessments of vulnerability (McGranahan *et al.*, 2007). However, with the continued downward revision in the 'best guess' estimates of the rate of sea level rise, more attention is now being focused on the response of different shoreline geomorphic types and ecological communities to the change in sea level, and to assessing the impact of the rise on coastal zone management strategies (Vafeidis *et al.*, 2008; Lemmen *et al.*, 2016). The hysteria of the early projections has waned and been replaced by a more sober treatment of sea level rise as simply one factor among many to be considered within an integrated coastal management plan.

Further Reading

Church, J.A., Clark P.U., Cazenave, A. *et al.* 2013. Sea level change. In Stocker, T.F., Qin, D., Plattner, G.-K. *et al.* (eds.), *Climate Change 2013: The Physical Science Basis. Contribution of Working Group I to the Fifth Assessment Report of the Intergovernmental Panel on Climate Change*. Cambridge and New York: Cambridge University Press.

The focus here is on future sea level scenarios, but there is much on the science of measuring and modelling sea level change.

Pugh, D. and Woodworth, P. 2014. *Sea Level Science: Understanding Tides, Surges, Tsunamis and Mean Sea-Level Changes*. Cambridge: Cambridge University Press, 393 pp.

This is an excellent and comprehensive review of all aspects of sea level, including tides.

References

Aagaard, T. and Sorensen, P. 2012. Coastal profile response to sea level rise: a process-based approach. *Earth Surface Processes and Landforms*, **37**, 354–62.

Andrews, J.T. 1970. *A Geomorphological Study of Post-glacial Uplift with Particular Reference to Arctic Canada*. London: Institute of British Geographers Special Publication no. 2, 156 pp.

Archer, A.W. 2013. World's highest tides: Hypertidal coastal systems in North America, South America and Europe. *Sedimentary Geology*, **284–5**, 1–25.

Argyilan, E.P., Forman, S.L., Johnston, J.W. and Wilcox, D.A. 2005, Optically stimulated luminescence dating of late Holocene raised strandplain sequences adjacent to Lakes Michigan and Superior, Upper Peninsula, Michigan, USA. *Quaternary Research*, **63**, 122–35, DOI:10.1016/j.yqres.2004.12.001.

Arújo, I.B. and Pugh, D.T. 2008. Sea levels at Newlyn 1915–2005: analysis of trends for future floooding risk. *Journal of Coastal Research*, **24**(4c), 203–312.

Atkinson, A.L., Baldock, T.E., Birrien, F. *et al.* 2018. Laboratory investigation of the Bruun rule and beach response to sea level rise. *Coastal Engineering*, **136**, 183–202.

Ballard, R.D., Coleman, D.F. and Rosenberg, G.D. 2000. Further evidence of abrupt Holocene drowning of the Black Sea Shelf. *Marine Geology*, **170**, 253–61.

Barnard, P.L., Short, A.D., Harley, M.D. *et al.* 2015. Coastal vulnerability across the Pacific dominated by El Niño/Southern Oscillation. *Nature Geoscience*, **8**, 801–7, DOI: 10.1038/NGEO2539.

Bateman,M.D., Rushby,G., Stein, S. *et al.* 2017. Can sand dunes be used to study historic storm events? *Earth Surface Processes and Landforms*, **43**(4), 779–90, DOI: 10.1002/esp.4255.

Berge-Nguyen, M., Cazenave, A., Lombard, A., Llovel, W., Viarre, J. and Cretaux, J.F. 2008. Reconstruction of past decades sea level using thermosteric sea level, tide gauge, satellite altimetry and ocean reanalysis data, *Global and Planetary Change*, **62**, 1–13.

Betts, N.L, Orford, J.D., White, D. and Graham, C.J. 2004. Storminess and surges in the south-western approaches of the eastern North Atlantic: the synoptic climatology of recent extreme coastal storms. *Marine Geology*, **210**, 227–46.

Bloom, A.L., Broeker, W.S., Chappell, J., Mathews, R.K. and Mesolella, K.J. 1974. Quaternary sea-level fluctuations on a tectonic coast: new ^{230}Th ^{234}U dates from New Guinea. *Quaternary Research*, **4**, 185–205.

Brooks, S.M.,Spencer, T. and Christie, E.K. 2017. Storm impacts and shoreline recovery: mechanisms and controls in the southern North Sea. *Geomorphology*, **283**, 48–60.

Bruun, P. 1962. Sea-level rise as a cause of shore erosion. *Journal of Waterways and Harbours Division*, **88**(1), 117–30.

Castelle, B., Marieu, V., Bujan, S., Splinter, K.D., Robinet, A., Senechal, N., Ferreira, S. 2015. Impact of the winter 2013–2014 series of severe Western Europe storms on a doublebarred sandy coast: beach and dune erosion and megacusp embayments. *Geomorphology*, **238**, 135–48.

Certain, R., Dia, A., Aleman, N., Robin, N., Vernet, R., Barusseau, J-P. and Olivier Raynal, O. 2018. New evidence of relative sea-level stability during the post-6000 Holocene on the Banc d'Arguin (Mauritania). *Marine Geology*, **395**, 331–45.

Chappell, J. 1974. Geology of coral terraces, Huon Peninsula, New Guinea: a study of Quaternary tectonic movements and sea-level changes. *Geological Society of America Bulletin*, **85**, 553–70.

Chappell, J. 1987. Late Quaternary sea-level changes in the Australian region. In Tooley, M.J. and Shennan, I. (eds.), *Sea Level Changes*. Institute of British Geographers Special Publication **20**, Oxford: Blackwell, pp. 296–331.

Chappell, J., Omura, A., Esat, T. *et al.* 1996. Reconciliation of late Quaternary sea levels derived from coastal terraces at Huon Peninsula with deep sea oxygen isotope records. *Earth and Planetary Science Letters*, **141**, 227–36.

Chittibabu, P. and Rao, Y.R. 2012. Numerical simulation of storm surges in Lake Winnipeg. *Natural Hazards*, **60**, 181–97, DOI: 10.1007/s11069-011-0002-7.

Church, J.A., White, N.J., Coleman, R., Lambeck, K. and Mitrivica, J.X. 2004. Estimates of the regional distribution of sea level rise over the 1950–2000 period. *Journal of Climate*, **17**, 2609–25.

Church, J.A., White, N.J. and Hunter, J.R. 2006. Sea-level rise at tropical Pacific and Indian Ocean islands. *Global and Planetary Change*, **53**, 155–68.

Church, J.A., Clark P.U., Cazenave, A. *et al.* 2013. Sea level change. In Stocker, T.F., Qin, D., Plattner, G.-K. *et al.* (eds.), *Climate Change 2013: The Physical Science Basis. Contribution of Working Group I to the Fifth Assessment Report of the Intergovernmental Panel on Climate Change*. Cambridge and New York: Cambridge University Press.

Coch, N.K. and Wolff, M.P. 1991. Effects of Hurricane Hugo storm surge in coastal South Carolina. *Journal of Coastal Research*, **SI 8**, 201–26.

Davidson-Arnott, R.G.D. 2005. A conceptual model of the effects of sea-level rise on sandy coasts. *Journal of Coastal Research*, **21**, 1166–72.

Davidson-Arnott, R.G.D. and Pyskir, N.M. 1989. Morphology and formation of an Holocene coastal dune field, Bruce Peninsula, Ontario. *Geographie Physique et Quaternaire*, **42**, 163–70.

Davies, J.L. 1964. A morphogenetic approach to world shorelines. *Zeitschrift fur Geomorphologie*, **8**, 127–42.

Dean, R.G. and Houston, J.R. 2016. Determining shoreline response to sea level rise. *Coastal Engineering*, **114**, 1–8.

Dechnik, B.,Webster, J.M., Webb, G.E., Nothdurft, L. and Zhao, J-X. 2017. Successive phases of Holocene reef flat development: evidence from the mid- to outer Great Barrier Reef. *Palaeogeography, Palaeoclimatology, Palaeoecology*, **466**, 221–30.

Defant, A. 1962. *Physical Oceanography*, vol. II. New York: Macmillan, 598 pp.

Dissanayake, P., Brown, J. and Karunarathna, H. 2014. Modelling storm-induced beach/dune evolution: Sefton coast, Liverpool Bay, UK. *Marine Geology*, **357**, 225–42.

Dolan, R. and Davis, R.E. 1994. Coastal storm hazards. *Journal of Coastal Research*, **SI 12**, 103–114.

Dutton, A. and Lambeck, K. 2012. Ice volume and sea level during the last interglacial. *Science*, **337**(6091), 216–19, DOI: 10.1126/science.1205749.

Dyke, A.S. 1998. Holocene delevelling of Devon Island, Arctic Canada: implications for ice sheet geometry and crustal response. *Canadian Journal of Earth Sciences*, **35**, 885–904.

Emanuel, K. 2003. Tropical cyclones. *Annual Review of Earth and Planetary Science*, **31**, 75–104.

Eriş, K.K., Ryan, W.B.F., Çaðatay, M.N. *et al.* 2007. The timing and evolution of the post-glacial transgression across the Sea of Marmara Shelf south of Istanbul. *Marine Geology*, **243**, 57–76.

Fairbanks, R.G., Mortlock, R.A., Chiu, T.-C. *et al.* 2005. Radiocarbon calibration curve spanning 0 to 50,000 years BP based on paired 230Th/234U/238U and 14C dates on pristine corals. *Quaternary. Science. Reviews*, **24**, 1781–96.

Fairbridge, R.W. 1983. Isostasy and eustasy. In Smith, D.E. and Dawson, A.G. (eds.), *Shorelines and Isostasy*. Institute of British Geographers Special Publication **16**, London: Academic Press, pp. 3–25.

Farhadzadeh, F. and Gangai, J. 2017. Numerical modelling of coastal storms for ice-free and ice-covered Lake Erie. *Journal of Coastal Research*, **33**, 1383–96.

Farquharson, W.I. 1970. *Tides and Tidal Currents in the Gulf of St Lawrence*. Ottawa: Canada Department of Energy, Mines and Resources, Surveys and Mapping Branch.

Fleming, K., Johnston, P., Zwartz, D., Yokoyama, Y., Lambeck, K. and Chappell, J. 1998. Refining the eustatic sea-level curve since the last glacial maximum using far- and intermediate-field sites. *Earth and Planetary Science Letters*, **163**, 327–42.

Forbes, D.L., Parkes, G.S., Manson, G.K. and Ketch, L.A. 2004. Storms and shoreline retreat in the southern Gulf of St. Lawrence. *Marine Geology*, **210**, 169–204.

Glenn, E., Comarazamy, D., González, J.E. and Smith, T. 2015. Detection of recent regional sea surface temperature warming in the Caribbean and surrounding region. *Geophysical Research Letters*, **42**, 6785–92, DOI: 10.1002/2015GL065002.

Goff, J.A. 2014. Seismic and core investigation of Panama City, Florida, reveals sand ridge influence on formation of the shoreface ravinement. *Continental Shelf Research*, **88**, 34–46.

Gönnert, G., Dube, S.K., Murty, T. and Siefert, W. 2001. *Global Storm Surges: Theory, Observations and Applications*. Holstein, Die Kueste, 623 pp.

Goodwin, I.D. and Harvey, N. 2008. Subtropical sea-level history from coral microatolls in the southern Cook Islands, since 300 AD. *Marine Geology*, **253**, 14–25.

Hagedoorn, J.M., Wolf, D. and Martinec, Z. 2007. An estimate of global mean sea-level rise inferred from tide-gauge measurements using glacial-isostatic models consistent with the relative sea-level record. *Pure and Applied Geophysics*, **164**, 791–818.

Hansen, E.C., Fisher, T.G., Arbogast, A.F., Bateman, M.D. 2010. Geomorphic history of low-perched, transgressive dune complexes along the southeastern shore of Lake Michigan. *Aeolian Research* **1**, 111–27.

Hansen, J.M., Aagaard, T. and Kuijpers, A. 2015. Sea-level forcing by synchronization of 56- and 74-year oscillations with the Moon's Nodal Tide on the Northwest European Shelf (Eastern North Sea to Central Baltic Sea). *Journal of Coastal Research*, **31**, 1041–56.

Holgate, S.J. 2007. On the decadal rates of sea level change during the twentieth century. *Geophysical Research Letters*, **34**, L01602, DOI:1029/2006GL028492, 4 pp.

Houston, J.R. and Dean, R.G. 2014. Shoreline change on the east coast of Florida. *Journal of Coastal Research*, **30**, 647–60.

Isla, F.I. and Angulo, R.J. 2016. Tectonic processes along the South America coastline derived from Quaternary marine terraces. *Journal of Coastal Research*, **32**, 840–52.

Khan, N.S., Ashe, E., Horton, B.P. *et al.* 2017. Drivers of Holocene sea-level change in the Caribbean. *Quaternary Science Reviews*, **155**, 13–36.

Komar, P.D. and Enfield, D.B. 1987. Short-term sea-level changes and coastal erosion. In Nummedal, D., Pilkey, O.H. and Howard, J.D. (eds.), *Sea-Level Fluctuations and Coastal Evolution*, SEPM Special Publication no. 41. Tulsa, OK: SEPM, pp. 17–27.

Lamb, A.L., Wilson, G.P. and Leng, M.J. 2006. A review of coastal palaeoclimate and relative sea-level reconstructions using $\delta^{13}C$ and C/N ratios in organic material, *Earth-Science Reviews*, **75**, 29–57.

Le Cozannet G., Garcina, M.,Yates, M., Idier, D. and Meyssignac, B. 2014. Approaches to evaluate the recent impacts of sea-level rise on shoreline changes. *Earth-Science Reviews*, **138**, 47–60.

Lemmen, D.S., Warren, F.J., James, T.S. and Mercer Clarke, C.S.L. (eds.) 2016. *Canada's Marine Coasts in a Changing Climate*. Ottawa: Government of Canada, 274 pp.

Lewis, C.F.M. 1969. Late Quaternary history of lake levels in the Huron and Erie basins. *Proceedings Twelfth Conference on Great Lakes Research*, Ann Arbor, MI: International Association for Great Lakes Research, pp. 250–70.

Lindner, B.L. and Neuhauser, A. 2018. Climatology and variability of tropical cyclones affecting Charleston, South Carolina. *Journal of Coastal Research*, **34**, 1052–64.

Lohne, Ø, Bondevik, S., Mangerud, J. and Svendsen, J. 2007. Sea-level fluctuations imply that the Younger Dryas ice-sheet expansion in western Norway commenced during the Allerød. *Quaternary Science Reviews*, **26**, 2128–51.

Luettich, R.A. Jr, Westerink J.J. and Scheffner, N.W. 1992. ADCIRC: an advanced three-dimensional circulation model for shelves, coasts and estuaries. Report 1: theory and methodology of ADCIRC- 2DDI and ADCIRC-3DL. Technical Rep. DRP-92–6, US Army Corps of Engineers, Waterways Experiment Station, Vicksburg, Mississippi, USA.

MacGranhan, G., Balk, D. and Anderson, B. 2007. The rising tide: assessing the risks of climate change and human settlements in low elevation coastal zones. *Environment and Urbanisation*, **19**, 17–37.

Macmillan, D.H. 1966. *Tides*. New York: Elsevier, 240 pp.

Miguez, B.M., Le Roy, R. and Wöpelmann, G. 2008. The use of radar tide gauges to measure variations in sea level along the French coast. *Journal of Coastal Research*, **24**, 61–8.

Miles, J.W. 1974. Harbor seiching. *Annual Review of Fluid Mechanics*, **6**, 17–34.

Mitrovica, J.X. and Peltier, W.R. 1991. Free air gravity anomalies associated with glacial isostatic disequilibrium: load history effects on the inference of deep mantle viscosity. *Geophysical Research Letters*, **18**, 235–8.

Morner, N-A. 1987. Models of sea-level changes. In Tooley, M.J. and Shennan, I. (eds.) *Sea Level Changes*. Institute of British Geographers Special Publication no. 20, Oxford: Blackwell, pp. 332–55.

Moucha, R., Forte, A.M., Mitrovica, J.X. *et al.* 2008. Dynamic topography and long-term sea-level variations: there is no such thing as a stable continental platform. *Earth and Planetary Science Letters*, **271**, 101–8.

Murray-Wallace, C.V. and Woodroffe, C.D. 2014. *Quaternary Sea-level Changes: A Global Perspective*. Cambridge: Cambridge University Press, 504 pp.

Murty, T.S. and Flather, R.A. 1994. Impact of storm surges in the Bay of Bengal. *Journal of Coastal Research*, **SI 12**, 149–61.

Nerem R.S., Cazenave, A., Chambers, D.P., Fu e, L.L., Leuliette, E.W. and Mitchum, G.T. 2007. Comment on 'Estimating future sea level changes from past records' by Nils-Axel Morner. *Global and Planetary Change*, **55**, 358–60.

Ota, Y. and Machida, H. 1987. Quaternary sea level changes in Japan. In Tooley, M.J. and Shennan, I. (eds.), *Sea Level Changes*. Institute of British Geographers Special Publication no. 20, Oxford: Blackwell, pp. 182–224.

Peltier, W.R. and Fairbanks, R.G. 2006. Global glacial ice volume and Last Glacial Maximum duration from an extended Barbados sea level record. *Quaternary Science Reviews*, **25**, 3322–37.

Pilkey, O.H., Cooper, J.A.G. 2004. Society and sea level rise. *Science*, **303**(5665), 1781.

Pirazzoli, P. A. 1987. Sea level changes in the Mediterranean. In Tooley, M.J. and Shennan, I. (eds.), *Sea Level Changes*. Institute of British Geographers Special Publication no. 20, Oxford: Blackwell, pp. 154–81.

Pugh, D. and Woodworth, P. 2014. *Sea Level Science: Understanding Tides, Surges, Tsunamis and Mean Sea-Level Changes*. Cambridge: Cambridge University Press, 393 pp.

Ranasinghe, R, Callaghan, D. and Stive, M.J.F. 2012. Estimating coastal recession due to sea level rise: beyond the Bruun rule. *Climatic Change*, **110**, 561–74, DOI: 10.1007/s10584-011-0107-8.

Rémillard, A.M., St-Onge, G., Bernatchez, P. *et al.* 2017. Relative sea-level changes and glacio-isostatic adjustment on the Magdalen Islands archipelago (Atlantic Canada) from MIS 5 to the late Holocene. *Quaternary Science Reviews*, **171**, 216–33.

Roberts, R.G. and Lian, O.B. 2015. Dating techniques: illuminating the past. *Nature*, **520**(7548), 438–9.

Rosati, J.D., Dean, R.G. and Walton, T.L. 2013. The modified Bruun rule extended for landward transport. *Marine Geology*, **340**, 71–81.

Ruddiman, W.F. 2014. *Earth's Climate: Past and Future*. New York: Macmillan, 464 pp.

Ryan W.B.F., Pittman, W.C.L., Major, C.O. *et al.* 1997. An abrupt drowning of the Black Sea Shelf. *Marine Geology*, **138**, 119–26.

Schellmann, G., Radtke, U., Potter, E.K., Esat, T.M. and McCulloch, M.T. 2004. Comparison of ESR and TIMS U/Th dating of marine isotope stage (MIS) 5e, 5c, and 5a coral from Barbados: Implications for palaeo sea-level changes in the Caribbean. *Quaternary International*, **120**, 41–50.

Schwartz, M.L. 1967. The Bruun theory of sea-level rise as a cause of shore erosion. *Journal of Geology*, **75**, 76–92.

SCOR Working Group 89. 1991. The response of beaches to sea level changes: a review of predictive models. *Journal of Coastal Research*, **7**, 895–921.

Scott, D.B., Boyd, R. and Medioli, F.S. 1987. Relative sea-level changes in Atlantic Canada: observed level and sedimentological changes vs. theoretical models. In Nummedal, D., Pilkey, O.H. and Howard, J.D. (eds.), *Sea-Level Fluctuations and Coastal Evolution*. SEPM Special Publication no. 41, Tulsa, OK: SEPM, pp. 87–96.

Shepard, F.P. 1963. *Submarine Geology*, 2nd edition. New York: Harper and Rowe, 412 pp.

Smith D.E., Hunt, N., Firth, C.R. *et al.* 2012. Patterns of Holocene relative sea level change in the north of Britain and Ireland. *Quaternary Science Reviews*, **54**, 58–76.

Spencer, T., Brooks, S.M., Evans, B.R., Tempest, J.A. and Möller, I. 2015. Southern North Sea storm surge event of 5 December 2013: water levels, waves and coastal impacts. *Earth-Science Reviews*, **146**, 120–45.

Stephenson,W.J., Dickson, W.E. and Denys, P.H. 2017. New insights on the relative contributions of coastal processes and tectonics to shore platform development following the Kaikōura Earthquake. *Earth Surface Processes and Landforms*, **42**, 2214–20.

Törnqvist, T.E., Wallace, D.J., Storms, J.E.A. *et al.* 2008, Mississippi Delta subsidence primarily caused by compaction of Holocene strata. *Nature: Geoscience*, **1**, 173–6, DOI:10.1038/ngeo129.

Turney, C.M. and Brown, H. 2007. Catastrophic early Holocene sea level rise, human migration and the Neolithic transition in Europe. *Quaternary Science Reviews*, **26**, 2036–41.

Tushingham, A.M. 1992. Postglacial uplift predictions and historical water levels of the Great Lakes. *Journal of Great Lakes Research*, **18**, 440–55.

Tushingham, A.M. and Peltier, W.R. 1991. ICE-3G: a new global model of late Pleistocene deglaciation based upon geophysical predictions of post-glacial relative sea level change. *Journal of Geophysical Research*, **96**, 4497–523.

Ullman, D.J., Carlson, A.E., Hostetler, S.W. *et al.* 2016. Final Laurentide ice-sheet deglaciation and Holocene climate-sea level change. *Quaternary Science Reviews*, **152**, 49–59.

Vafeidis, A.T., Nicholls, R.J., McFadden, L. *et al.* 2008. A new global coastal database for impact and vulnerability. *Journal of Coastal Research*, **24**, 917–24.

Wake, L., Milne, G. and Leuliette, E. 2006. Twentieth century sea-level change along the eastern US: unravelling the contributions from steric changes, Greenland ice sheet mass balance and late Pleistocene glacial loading. *Earth and Planetary Science Letters*, **250**, 572–80.

Wolstencroft, M., Shen, Z., Törnqvist, T.E., Milne, G.A. and Kulp M. 2014. Understanding subsidence in the Mississippi Delta region due to sediment, ice, and ocean loading: Insights from geophysical modelling, *Journal of Geophysical Research: Solid Earth*, **119**, 3838–56, DOI:10.1002/2013JB010928.

Woodroffe, C.D. 2003. *Coasts: Form, Process and Evolution*. Cambridge: Cambridge University Press, 623 pp.

Woodroffe, S.A., Long, A.J., Milne, G.A., Bryant, C.L. and Thomas, A.L. 2015. New constraints on late Holocene eustatic sea-level changes from Mahé, Seychelles. *Quaternary Science Reviews*, **115**, 1–16.

Wolf, J. and Flather, R.A. 2005. Modelling Waves and Surges during the 1953 Storm. *The Royal Society, Philosophical Transactions: Mathematical, Physical and Engineering Sciences: The Big Flood: North Sea Storm Surge*, **363** (1831), 1359–75.

Wöppelmann, G., Martin M.B., Bouin, M-N. and Altamimi, Z. 2007. Geocentric sea-level trend estimates from GPS analyses at relevant tide gauges worldwide, *Global and Planetary Change*, **57**, 396–406.

World Meteorological Organisation. 2011. Guide to storm surge forecasting. WMO Report No. 1076. Geneva: World Meteorological Organisation, 120 pp.

Xia, M, Xie, L., Pietrafesa, L.J. and Peng, M. 2008. A numerical study of storm surge in the Cape Fear River estuary and adjacent coast. *Journal of Coastal Research*, **24**(4C), 159–67.

4

Wind-Generated Waves

4.1 Synopsis

Wind-generated waves are the primary force leading to long-term modification of the coast. Waves carve out erosional features and create depositional landforms. Waves are also the dominant source of energy that drives nearshore circulation in the form of longshore currents, rip currents and undertow. Thus, wind-generated waves are responsible for the reconfiguration of the shoreline through time. Their effectiveness is largely because the ceaseless forcing of wind and pressure systems that encircle the globe easily disturbs the sea surface. The energy imparted to the water surface by wind leads to wave creation and growth, and the resultant waves travel long distances toward the shore, where their energy is ultimately dissipated.

An ocean wave is a vertical displacement of the water surface that is periodic in space and time. As with any other waveform, water waves are characterised by their height (vertical distance between the crest and trough), wavelength (horizontal distance from crest to crest or trough to trough), and period of oscillation. They are referred to as progressive waves because a series of waves in succession will travel in the same direction, carrying their energy in the direction of propagation. Wave energy is moved from the area of wave generation to distant locations due to the oscillatory motion of the waves, which includes both potential energy (due to the upward and downward displacement from the still water level) and kinetic energy (associated with the orbital motion of water molecules in the water column beneath the wave). The movement of water beneath surface waves has the capacity to disturb the sea bed, but only in relatively shallow water.

While the exact mechanisms of energy transfer to waves by wind are not completely understood, there is enough knowledge and experience to permit forecasting of wave growth based on wind conditions as well as the subsequent transformation and interaction of waves as they travel across a lake or ocean. Often, these wave models involve statistical descriptions of wave fields that are derived from extensive measurements using a range of technologies such as wave rider buoys and satellite imagery. The models can also be used to hindcast wave conditions at specific locations based on historical weather data, which is quite useful when actual wave measurements are not available. In this way, it is possible to build up the wave climate for different coastal regions of interest or strategic importance, which includes measures such as monthly average wave height and wave period or the directional energy spectra for different seasons. This type of summary data is critical for effective coastal zone management as well as the design and construction of breakwaters, jetties, sea walls and harbours.

4.2 Wave Definition and Description

The main features of surface waves are illustrated in Figure 4.1, whereas Table 4.1 summarises the nomenclature and symbols that are commonly used to describe waves. A wave is a periodic undulation of the water surface above and below the mean water level (MWL), which refers to the average level of the water surface in the presence of the waves. The MWL is slightly different than the still water level (SWL) – the level that the water surface would assume in the absence of waves – because the dynamic action of waves exerts a force on the water surface such that MWL is typically a bit lower than the SWL, except close to shore. In the open ocean or in the middle of a large lake, the mean water level and the still water level are effectively the same. It would require extremely sensitive instrumentation to measure the difference, and the influence of atmospheric pressure variations would affect the mean sea surface elevation. However, as waves approach the shoreline and begin to experience shoaling transformations, the difference between the MWL and the SWL can be pronounced, due to the action of the waves. This is most evident in lakes where the water surface is often flat and glassy, and where one can actually measure the SWL, unlike on an ocean coast where the presence of waves is continuous. Additional details are provided in Chapter 5 where the topic of wave set-down and wave set-up is discussed. For the purpose of Figure 4.1, it is reasonable to assume that SWL = MWL, and that the water surface is defined as $z = 0$, where z is the vertical coordinate (with negative values below the water line).

The shape of a simple wave in deep water is very close to that of a sinusoid with symmetry above and below the MWL (Figure 4.1A). Indeed, the mathematical descriptions of surface gravity waves arising out of linear wave theory uses sine and cosine functions along with other trigonometric functions that are harmonic. The height, H, of the wave is equal to the distance from the trough to the crest, whereas the amplitude, a, is usually taken as one-half the wave height. This is valid for simple waves, but more complex waves such as those in the surf zone are skewed, meaning that the displacement of the water surface above the MWL is not the same as that for the displacement below the MWL. Thus, there can be one amplitude for the upper portion of the wave above MWL, a_1, and another amplitude, a_2, for the lower portion of the wave. The sum of these two amplitudes yields the wave height ($H = a_1 + a_2$), whereas the difference is proportional to the wave skewness. The asymmetry of a wave (not to be confused with wave skewness) refers to the difference in geometry or slope of the front face of a wave relative to its rear face, which can be pronounced during wave breaking. A pure sinusoid, as represented in Figure 4.1A is neither skewed nor asymmetric.

The wavelength, L, is the crest-to-crest distance or, alternatively, the distance between any two similar points along successive waves (e.g., trough to trough). Wave period, T, is the time taken for the passage of two such points past a fixed location, usually expressed in seconds. Wave frequency is simply the reciprocal of the period ($f = 1/T$) and it is expressed as cycles per second or Hz (after the physicist Heinrich Hertz). These fundamental attributes of surface gravity waves are used to derive other wave properties such the wave celerity,

$$C = \frac{L}{T}. \tag{4.1}$$

The celerity or speed at which the waveform advances across the water surface, C, is also referred to as the phase speed or speed of propagation. A characteristic of surface gravity waves is that the period does not change appreciably while they travel toward the shoreline, therefore the celerity only changes when the wavelength changes. This happens when waves enter shallow water and they experience significant transformations that lead to increases in wave height and ultimately wave breaking.

Figure 4.1B indicates that the orbits of water particles beneath a wave are circular, but only in

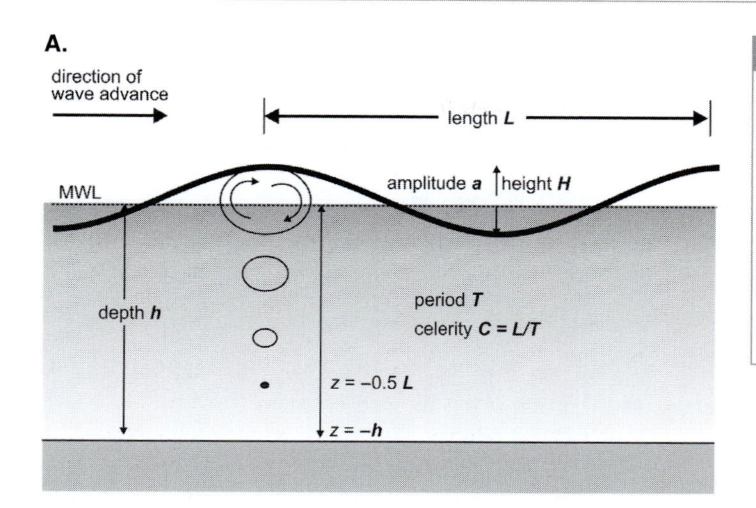

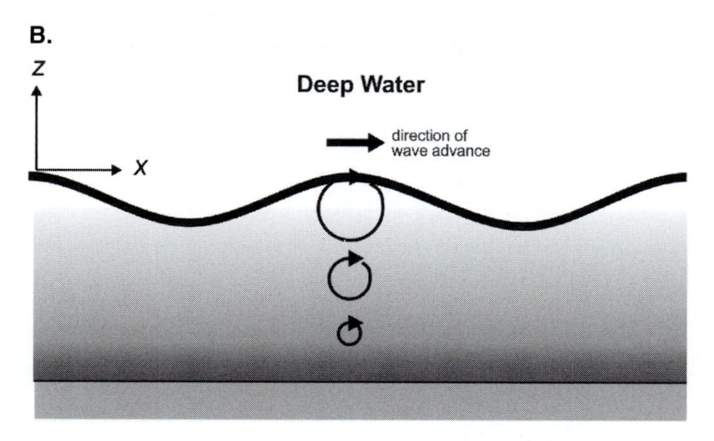

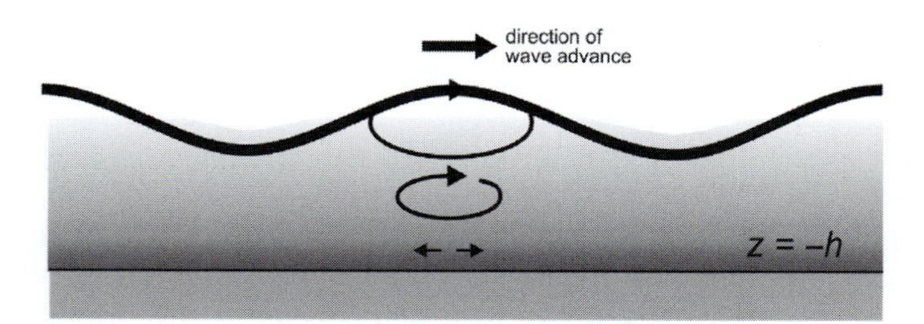

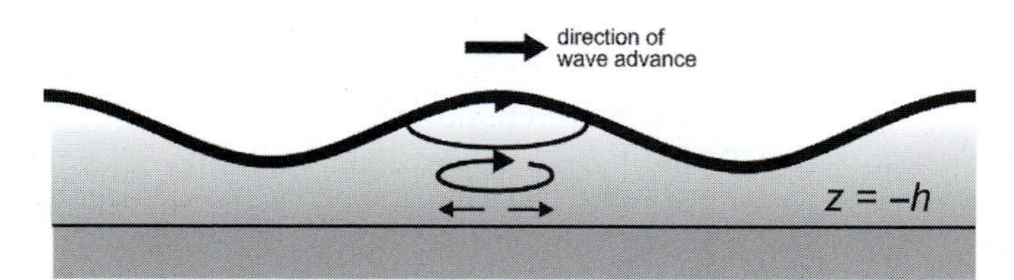

Figure 4.1 Simple wave descriptors: A. definition sketch of a progressive, deep water wave showing the exponential decay of the circular orbital motion with depth. Note that water depth ($z = -h$) is below wave base ($z = -0.5L$); B. characteristic shape of wave orbits in deep, intermediate and shallow water. For intermediate and shallow conditions, the wave base is truncated by the sea bed, which alters the dynamics of the wave.

Table 4.1 Common notation for wave parameters

Symbol	Name	Description
x	Horizontal distance	Coordinate variable used to indicate horizontal position in the direction of wave propagation
z	Vertical distance	Coordinate variable used to indicate vertical position relative to SWL or MWL ($z = 0$)
$\eta(t, x)$	Water surface elevation	Time-dependent position of the water surface at any horizontal location, x
SWL	Still water level	Average water surface in the absence of waves
MWL	Mean water level	Average water surface elevation in the presence of waves
H	Wave height	Vertical distance between crest and trough
a	Wave amplitude	Vertical distance from crest to MWL or trough to MWL
L	Wavelength	Horizontal distance between successive crests or troughs
T	Wave period	Time between successive crests or troughs
f	Frequency	$1/T$
σ	Radian frequency	$2\pi/T$
C	Wave celerity	speed of movement of individual wave form
C_g	Wave group velocity	speed of movement of a group of waves
k	Angular wavenumber	$2\pi/L$
h	Total water depth	vertical distance from water surface to bed; $z = -h$ at the bed
o	(subscript) Deep water	$h > 0.5\ L_o$
s	(subscript) Shallow water	$h < 0.05\ L_o$
b	(subscript) Break point	breaking conditions

deep water. On the front face of the wave, the water particles move upward and then forward as the crest approaches, whereas on the backside of the wave, the particles move downward and then backward as the trough approaches. At four points in this orbit, the motion is either purely horizontal (onshore at the crest or offshore at the trough) or purely vertical (upward or downward at the zero crossings). At all other points in the orbit there is a combination of these motions, yielding a symmetric rotating vector field.

In intermediate and shallow water, however, the dynamics of the wave begin to be constrained by the presence of the bottom, and therefore the orbital diameters become elliptical rather than circular. Indeed, in very shallow water, there is virtually no vertical component to the orbital motion, yielding a simple to-and-fro oscillation. In addition, the diameter of the orbital motions decreases exponentially with depth and becomes negligible at $z = -0.5L$ (Figure 4.1A). This provides a convenient definition of a deep water wave (i.e., when water depth is greater than one-half the wavelength; $h > 0.5\,L_o$, where L_o is the length of a deep water wave). In Chapter 5, the theoretical basis for this definition will be presented as well as for the complementary definition for a shallow water wave ($h < 0.05\ L_o$).

Wind-generated waves generally have relatively short periods, typically less than about 15 seconds (Figure 4.2). The wind-generated wave regime includes capillary waves (usually less than 0.25 seconds) and surface gravity waves that can be divided further into ordinary sea waves (about 1 to 10 seconds) and long-period swell waves (10 to 15 seconds). The cut-offs are not firm but rather are defined by the scales of the system under investigation (e.g., a small lake versus the Pacific Ocean), as well as the dynamics of the waves. Capillary waves can also be superimposed on much larger waves, giving the water surface a temporarily rippled look as a strong wind gust passes through and disturbs the wavy surface.

In the area where waves are actively generated by wind, the sea surface appears somewhat chaotic with waves that are irregular and short crested (Figure 4.3). Typically, there are also whitecaps (white horses) associated with localised wave breaking and there may be spindrift, which refers to the tops of wave crests being blown off by strong winds. Waves of this type are called sea waves because they are found in the open ocean or some distance offshore in relatively deep water. The disorganised state of the sea surface indicates that there is a superpositioning of many different waves across a range of periods, lengths and heights.

Figure 4.3 Partly arisen sea with active energy input by wind, showing chaotic nature of the sea surface. Widespread whitecapping is due to the interaction of different waves of different lengths traveling in slightly different directions, yielding short-term instabilities in the wave crests. This type of rough sea surface modifies the wind boundary layer in a way that facilitates further energy input into the waves.

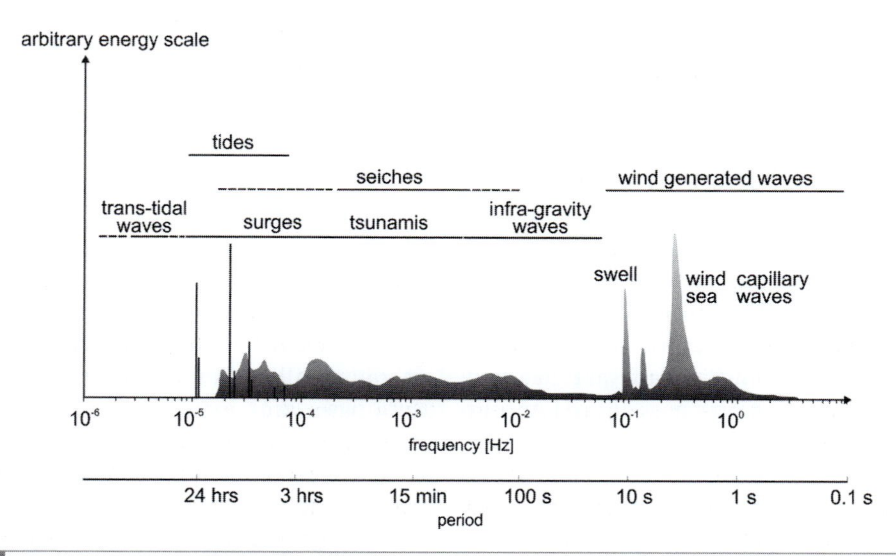

Figure 4.2 Energy content of the ocean surface according to the period (or frequency) of oscillations, each of which has a different dynamical origin. Wind generated waves tend to contain the most energy, on average, although tides have peak energy levels that are similar. Tides occur in a constrained frequency band, whereas most other oscillations occur across wide bands. (From Holthuijsen, 2007; after Munk, 1950.)

As will be shown in Chapter 5, there is a direct relationship between the wavelength of a deep water wave and its period. Specifically, $L_o = gT^2/2\pi$, which indicates that the wavelength increases as the square of the wave period. Substitution in Equation 4.1 for wave celerity shows that the speed at which a deep water wave moves across the surface is therefore proportional to the wave period. In other words, longer period waves travel faster. The implication is that locations distant from where the waves are being generated will be characterised by more orderly wave fields because the waves have sorted themselves out. The longest swell waves (those with large wave period and long crests) will arrive first on the shore, followed by decreasingly shorter waves in succession.

The energy spectrum shown in Figure 4.2 suggests that there can be many other types of oscillation in lakes and oceans outside the frequency band occupied by wind generated waves.

Infragravity waves, for example, are long-period waves (i.e., 30 seconds to 4 minutes or more) that are associated with the interaction of wind generated waves leading to low-frequency oscillations that are either tied dynamically to wave groups or are resonantly amplified due to wave–wave interaction. Surface oscillations on large bodies of water such as oceans, seas and large lakes can also be generated by wind stress (e.g., seiches), earthquakes (e.g., tsunamis), or the gravitational force of the Sun and Moon (e.g., tides). Trans-tidal waves have periods longer than the tidal cycle (approximately 24 hr and 50 minutes) and are generated by low-frequency fluctuations in Earth's crust or by large-scale pressure fluctuations in the atmosphere. There are also types of oscillations known as internal waves that exist below the water surface and are generally associated with density interfaces due to temperature differences (e.g., thermoclines) or salinity differences (e.g., haloclines). Figure 4.2 shows that there are specific frequency ranges where these types of oscillation appear to be most energetic, and in many instances, there are energy gaps between them. The primary concern of this chapter, however, is with surface gravity waves created by wind blowing across a water surface.

4.3 Wave Generation

When wind in the atmosphere blows across a surface, whether land or sea, there is a transfer of momentum to the surface. Over land, the frictional resistance of the surface causes a slowing down of the near-surface wind, which yields a boundary layer or zone where the wind speed is much slower than higher up (see discussion in Chapter 9). A similar process occurs over water, except that water does not offer nearly as much frictional resistance as land (especially when there are boulders, trees and buildings) because water can be easily deformed and is, in effect, a dynamically slippery surface. As a consequence, very little of the wind energy goes into creating surface currents (referred to as wind drift). However, the transfer of wind energy into surface waves is very effective once there are waves on the surface, and the question that has perplexed ocean scientists for a long time is 'how does this happen?' The physics are quite complex, so the following treatment is only a brief, qualitative summary of the primary controls on wave growth derived from theoretical modelling and empirical measurements. The interested reader is referred to some of the advanced literature on this topic, beginning with the text by Phillips (1966) as well a thorough review by Komen *et al.* (1994).

4.3.1 Stages and Processes of Wave Growth

Observations on small ponds and lakes show that the water surface remains smooth and glassy until a wind gust passes through and disturbs the surface. If the wind has sufficient speed and energy, small ripples known as capillary waves or cat's paws are created on the surface that are about one centimetre in height with wavelengths of only a few centimetres and wave periods less than about 0.25 seconds (Figure 4.4). Capillary waves are often clustered in surface patches that are forced along the water surface with the wind gust, although close inspection will reveal that individual waves move out in an arc from the centre of the patch. If the wind dies away after

Figure 4.4 Wind blowing across a pond (toward the camera) creating capillary waves (cat's paws). Near the upwind boundary (upper part of photo by the shoreline) there is a fringe of flat water because the wind coming off the land is not strong enough to generate waves. However, the waves grow in size away from the shoreline (and toward the camera) as more energy is put into surface oscillations.

a few tens of seconds, the ripples will quickly dissipate and the water surface will become calm again. The rapid dissipation of wave energy in capillary waves is due to the effect of surface tension, which quickly restores the water surface to its lowest energy state. A flat water surface has stronger molecule-to-molecule bonds (cohesion) at the air–water interface than a wavy surface, which is being stretched. Capillary waves are short-lived because the forcing during a wind gust is not prolonged enough to impart a significant amount of energy to the water surface to counteract the influence of surface tension. This dynamical interaction between wind stress and surface tension is only effective when the waves are of very short length and period. When the wave period exceeds about 0.25 s, the wave form becomes too large to be affected by surface tension to any substantial degree. Of course, surface tension is always present on the water surface regardless of its state, but the magnitude of the force becomes irrelevant to the dynamics of larger waves. Instead, the force of gravity becomes the restoring force, and hence the class of waves that includes ordinary progressive waves and ocean swell is referred to as surface gravity waves.

If the wind continues to strengthen above the threshold governed by surface tension, then the capillary waves do not die away but continue to grow in height and length. This is most apparent at the upwind margins of ponds and lakes as the wind moves from land to water (Figure 4.4) where there is a transition from no waves, to small capillary waves, and eventually to larger surface gravity waves. On larger water bodies, such as lakes and seas and with sustained wind, the waves will grow in height and length from the upwind shoreline to the downwind shoreline, where the waves will ultimately break. Numerous observations of this process of wave growth have led to the identification of three major controls on the size of waves at any particular location on a water body, namely: (1) wind speed; (2) the fetch length (distance) over which the wind blows; and (3) the duration of the wind event. Each of these controls imparts a limitation on the maximum size of wave that is possible on a given lake or sea for a given wind event. For example, it is readily apparent that a wave with H = 1 m is quite common on the ocean, but a wave of this size would be impossible on a mud puddle or small pond, even if the pond was infinitely deep. This is referred to as a fetch-limited situation because the water body is not long enough to accommodate a large wave. Similarly, a short-lived wind gust (duration limited) or a gentle breeze (speed limited) is not able to create a large wave even if the fetch is unlimited.

The rate of energy transfer to the water surface varies depending on the difference between the air and water temperatures and with the stability of the atmosphere. In general, the rate of energy transfer increases as the air temperature becomes cooler relative to the water surface, ostensibly because cold air is denser than warm air. In mid-latitude regions with strong seasonality, smaller waves are expected in late spring because the air temperature tends to be much warmer than the cold water. Similarly, larger waves are expected in the late fall when the water temperature is relatively warm in comparison to the air temperatures. The effect of air temperature is diminished for very strong winds, but may still be 5–10 per cent greater for cold winds in the fall compared to similar winds in the spring.

So, how exactly does the wind create waves? There is general agreement that the growth of waves is due primarily to pressure fluctuations rather than the simple frictional drag of the wind across the water surface. Near-surface wind is highly turbulent, which implies that there is a range of eddy sizes with different frequencies and spatial scales. These eddies have vector components that are oriented up and down as well as forward and backward. The turbulent action of the passing eddies stresses the surface in ways that lead to minor depressions of the water surface. As a consequence, the relative roughness of the water surface increases and this, in turn, produces feedback between the changing surface topography and the character of the boundary layer of the wind. Most models of wave growth use this basic idea, which was formalised in a theory introduced by Miles (1957). Wave growth occurs because of resonant interactions between the pressure fluctuations associated with blowing wind and the disturbances generated on the water surface, which eventually become waves. The very existence of even a small wave distorts the wind field leading to an asymmetry in the dynamic action on the windward and leeward faces of the wave. Pressure is at a maximum on the windward side of the crest and a minimum on the leeward slope (Figure 4.5). The pressure imbalance feeds energy into the wave by reinforcing the upward movement beneath the crest and the downward movement in the trough. As is the case with flow over other types of wave forms, such as aeolian sand dunes in deserts or subaqueous sand dunes in large rivers, the effect is enhanced when the wave form is large because the pressure asymmetry is greater. When the wave is sufficiently large, flow separation occurs at the crest and an eddy recirculation zone occupies the lee side of the wave, which is actually the front face of the wave given that the wind blows from behind the wave.

As the wave field evolves, more and more energy is transferred to the sea surface. The energy content of a wave field is often represented by a graph known as a wave energy spectrum, which is similar to Figure 4.2 but with a much shorter frequency range (e.g., 1 s to 300 s). Such a graph reveals that the wave field can comprise many waves of different frequency, each of which contains a fraction of the total energy content of the wave field. The waves (or band of waves) that contain the most energy are referred to as the peak waves because they are

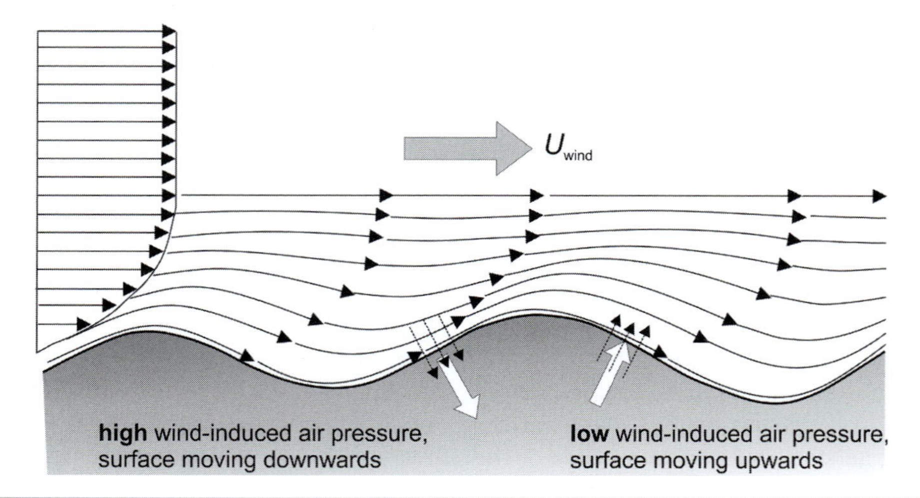

Figure 4.5 Air flow patterns over waves, showing the locations where the wind field exerts dynamic pressure into the growing waves. The pressure asymmetry across the front and rear faces of the wave crest facilitates energy input into the wave (from Holthuijsen, 2007).

defined by the largest peak in the energy spectrum. Peak waves are usually identified by their period, T_p, referred to as the peak period in the spectrum.

Observations of evolving wave fields have demonstrated that as more and more energy is transferred to the sea surface, the peak waves become more energetic. There is also a shift or cascade of energy to lower frequencies (i.e., longer periods). This may occur as a consequence of an increase in the wind speed through time (Figure 4.6A) or when the wind is sustained for a long period of time over longer distances (Figure 4.6B). Usually both effects occur simultaneously as a storm strengthens in intensity. Why the peak energy shifts to lower frequencies is an area of active investigation. However, the mechanism most certainly involves wave–wave interactions with wave triads (three waves) or quadruplets (four waves). The basic idea is that two waves or wave pairs will bleed some of their energy into low-frequency waves that have periods defined by the difference in frequencies of the original short-period waves. This yields a resonant transfer of energy in which there is a positive coupling between the low-frequency wave growth and the periodic input of energy from the dynamic interaction of the short-period waves. The process of resonant growth is similar to pushing a child on a swing in ever-increasing back-and-forth arcs – the push needs to happen at exactly the right time so as to enhance the swinging motion. In this example, the swinging child gains energy that is proportional to the energy expended by the person doing the pushing. In the same way, the low-frequency waves (the child on the swing) gain energy at the expense of the short-period waves (the person pushing the swing). Thus, the transfer of energy to lower frequencies is conservative in the sense that it does not increase or decrease the total energy content of the wave field – it simply shifts the existing energy at higher frequencies to lower frequencies. So, what exactly are the mechanisms that limit the growth of energy in a wave field?

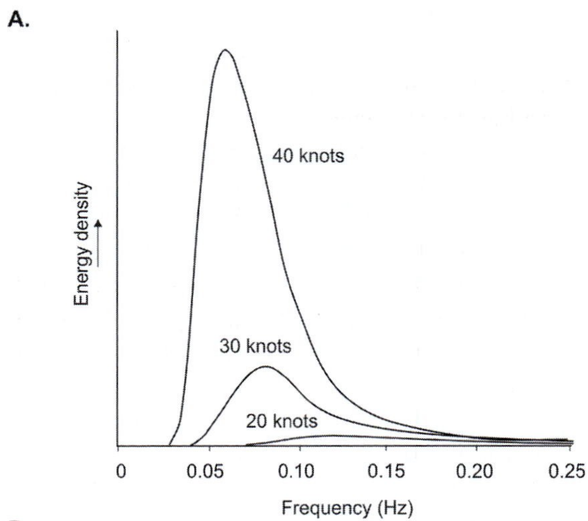

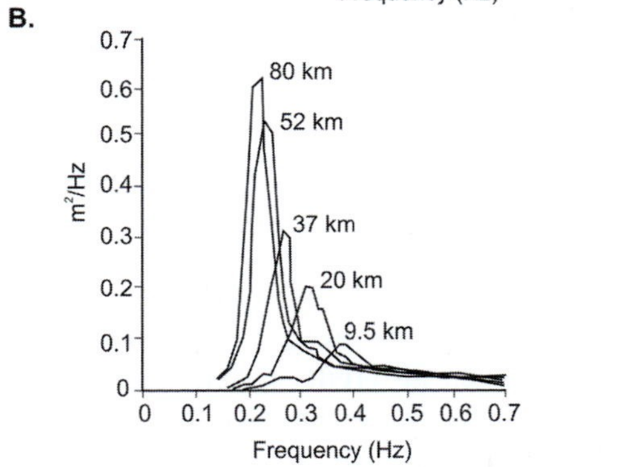

Figure 4.6 Changes in spectral shape due to the gradual evolution of a wave field in time and space. A. Increases in wind speed (knots) yield larger spectral peaks that are shifted to lower frequencies. B. A similar shift in spectral peaks occurs with increased fetch distance from the shoreline (km) while wind speed is held constant. From Joint North Sea Wave Project (JONSWAP) experiments in the North Sea (Hasselman *et al.*, 1973).

4.3.2 Limits to Wave Growth

On the open ocean where the wind is actively generating new waves, the energy spectrum is usually quite broad, implying that there is a wide range of wave periods all of which have similar energy content. New waves are constantly being generated while older waves grow in height and period. These somewhat chaotic conditions in the area of active wave generation are termed sea waves, and a young sea is referred to as a partly arisen sea (PAS). Observations indicate that there are limits to wave growth, and when the high-frequency band is saturated (can gain no more energy) it is referred to as an old sea or fully arisen sea (FAS). At this point in the evolution of the wave field there is a dynamic equilibrium in which the rate of energy transfer from the wind to the water surface is balanced by the rate of energy dissipation by wave breaking. In addition, there are dynamic transfers of energy among the waves, which maintain the spectral peak as well as the shape of the wave spectrum.

At the smallest scales, fluid motion is damped by the action of viscosity, and this effect is always there. Surface tension is also effective in damping the action of capillary waves, as discussed above. However, viscosity and surface tension are only effective in the context of very high-frequency (short-wavelength) waves. The primary mechanism of energy dissipation in open ocean waves is believed to be turbulence in the form of whitecapping (Figure 4.7). Wave theory (Chapter 5) shows that there is a critical wave steepness ($H/L = 0.142$) beyond which an oscillatory wave becomes unstable and begins to break. Water begins to spill down the face of the wave crest, thereby entraining air and creating random eddies that disrupt the orderly orbital motions beneath the wave form. The turbulent action near the zone of whitecapping yields a cascade of energy to smaller and smaller eddies, and ultimately to energy dissipation by viscous action and heat loss. There is also an associated pressure pulse that counteracts the upward motion of water beneath the wave crest, which further extracts energy from the wave.

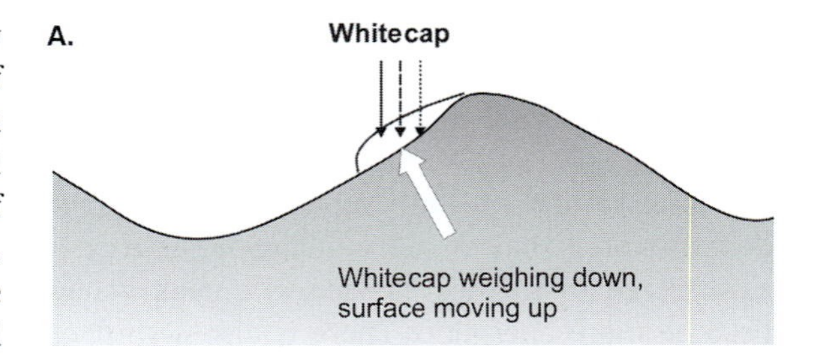

Figure 4.7 Whitecapping on sea waves. A. Unstable white water sliding down the front of the wave from the crest yields a pressure pulse that counteracts the upward motion of water in advance of the crest. B. Whitecapping due to a larger wave moving through a field of smaller waves, leading to localised instabilities of the smaller wave crests.

Whitecapping is most evident during very strong wind events and when there is a fully arisen sea. The reason for this is that the over-steepening of waves usually happens when two waves interact so that their wave crests reinforce each other, leading to localised regions of very high water surface elevation. What circumstances might lead to such a situation? One can imagine two waves of similar period and height moving in opposite directions, and every time the wave crest of one wave interacts with the wave crest of the opposite wave, the combination will create a wave crest that is double the height of either of the single waves. If the doubling of the height leads to the steepness criterion being exceeded, the wave will break. However, this situation is not very common in the open ocean because waves outside the eye of the storm usually travel in the same direction. Instead, there will be waves of different period and height propagating in the same approximate direction (Figure 4.7B). It will be shown in Chapter 5 that surface gravity waves are dispersive, which means that long-period waves travel faster than short-period waves. Thus, larger waves will overtake smaller waves and pass over (or through) them. This provides the main mechanism by which wave crests will interact to create the conditions leading to wave over-steepening and whitecapping. Thus, whitecapping preferentially affects the high-frequency waves in the spectrum, causing them to reach their limits to growth first, while energy is simultaneously being transferred to lower frequency waves (Figure 4.8).

The complexity of wave interactions in the area of wave generation is enhanced by the fact that wave generation takes place not just in the direction that the wind blows, but in an arc up to 45° on either side of the mean wind direction. The potential for whitecapping increases because waves of all sizes are more likely to cross at oblique angles, producing local areas of over-steepening that leave white, turbulent patches of water on the surface (Figure 4.3).

The limits to wave growth imposed by white-capping leads to a qualitative description of a fully arisen sea (FAS) as one in which the high-frequency portion of the energy spectrum is fully saturated (i.e., contains as much energy as it can hold). Detailed experiments have demonstrated that such saturation conditions are characterised by an energy spectrum with a downward slope from the peak wave frequency to the higher frequencies waves proportional to f^{-4} (if wave energy is represented on a logarithmic axis). Under this situation, the energy input to the waves by wind is balanced, on average, by energy dissipation due to whitecapping as well as transfer to other frequencies bands, as shown in Figure 4.8. Thus, the equilibrium state is very much a dynamic one involving active inputs and outputs of energy within every frequency band. Recall that there is also energy being extracted from the higher frequency waves and transfer to the lower frequency portion of the

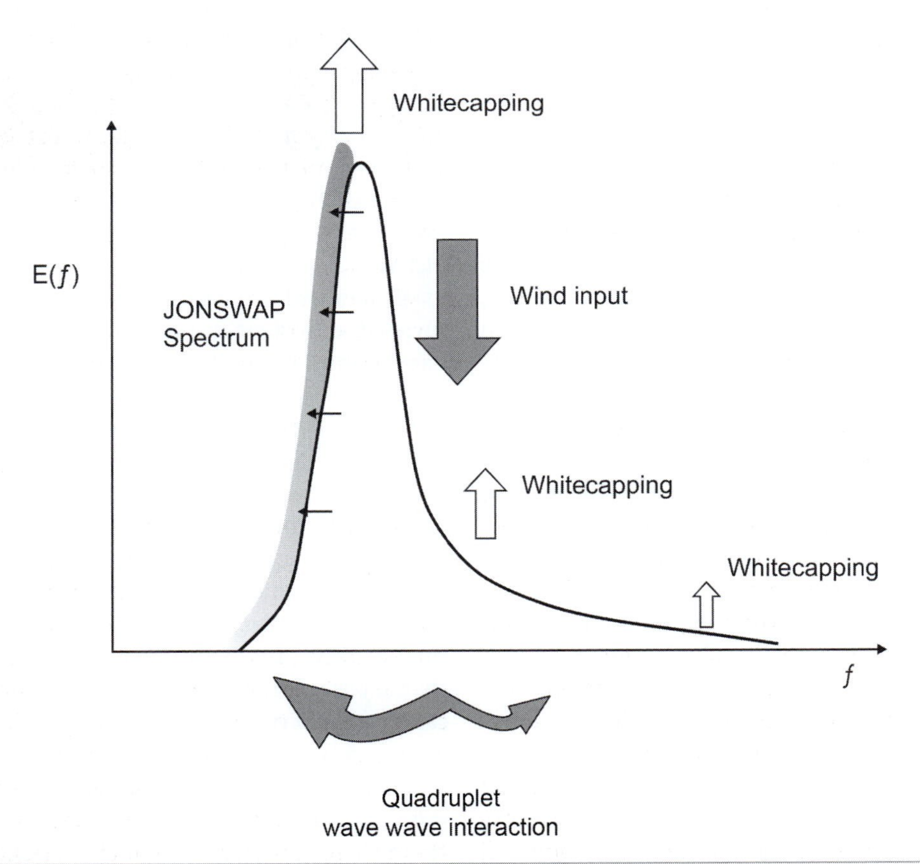

Figure 4.8 Graphical representation of the dynamic nature of a wave energy spectrum showing (i) input of energy by wind; (ii) dissipation of energy by whitecapping in the frequencies bands that have reached saturation; and (iii) the transfer of energy to lower and higher frequency bands due to wave–wave interaction. Based on the JONSWAP spectrum (after Holthuijsen, 2007).

spectrum by resonant interactions. The slope of the energy spectrum from the peak wave frequency to lower frequencies is usually quite steep, reflecting a rapid drop-off in energy at periods longer than can be generated by the particular combination of wind speed and duration that characterises the storm. The time it takes to reach the condition of a FAS may be only a few hours for moderate wind speeds, but can be as long as several days for very intense storms such as nor'easters and tropical hurricanes. This reinforces the notion that FAS is a relative term, with very different energy spectra and peak frequencies depending on the character of the storm that created the wave field.

To this point, it has been assumed that wave generation takes place in the open ocean with no restrictions on the space and time scales leading to wave growth or the limitations imposed by shallow water depth. The form of the spectrum, however, will be affected by factors such as the initial sea surface condition (i.e., pre-existing waves), the fetch length over which the wind blows, the duration of the storm, the water depth beneath the waves, and the presence of near-surface water currents upon which the wave field rides. Fetch length is controlled either by the size of the water body (e.g., ocean, lake, bay, pond) or by the lateral dimensions of the storm (e.g., mid-latitude depression, tropical hurricane, frontal disturbance, gust). Because wind generated waves are progressive, fetch length acts to restrict the distance (or equivalently, the length of time) over which energy can be transferred to the wave group. Thus, fetch can be thought of as equivalent to a duration limit although the actual spectral shape at the downwind margin may be somewhat different (Figure 4.6B).

The effect of shallow water is to increase the rate of energy loss through bottom friction and wave breaking, and hence, to limit the maximum energy content of waves. Often, the wave-breaking process leads to new wave motions at high frequencies inside the surf zone. Shallow water also leads to complications as regards wind energy input because wave celerity is reduced, and this affects the resonance between the pressure fluctuations induced by the wind and the wave form. Both the growth of total energy and the transfer of energy to low frequencies are less efficient than in deep water conditions (Young and Verhagen, 1996). The prediction of the shallow water wave energy spectrum is further complicated by the effects of coastline irregularity leading to complex shoaling and refraction patterns. Nevertheless, there is a great deal of interest in modelling the transition from a deep water energy spectrum to the shallow water condition because of the relevance to sediment transport, nearshore dynamics, coastal engineering and boating safety.

4.4 Wave Measurement and Parameterisation

4.4.1 Wave Measurement

Measurements of waves are essential for understanding coastal erosion, sediment transport, harbour design and marine ecosystem dynamics. Moreover, models of wave dynamics need to be calibrated and verified against actual wave measurements if they are to be used reliably for predictive purposes. Measurements of tides and sea level changes have been available for several centuries due to their importance for shipping and navigation, but routine measurements of waves have only become available in the past 50 years or so. The network of measuring stations is still much smaller than, for example, meteorological stations or river gauging stations. Recently, it has become possible to measure ocean waves with a high degree of accuracy using satellite-based sensors, and this has yielded an immense flow of data for oceanographers and coastal scientists. Importantly, satellite imagery provides extensive spatial coverage, which complements and significantly enhances site-specific measurements that are sporadically distributed in space and time.

4.4.1.1 *Time Scales of Wave Measurements*

As with other branches of the earth sciences, coastal geomorphology examines processes and landscape features across a wide range of spatial and temporal scales (Walker *et al.*, 2017).

In measuring and describing waves in the coastal zone, it is appropriate to consider the scales that are most relevant to the particular problem or application. For example, if the interest lies in explaining the evolution and maintenance of beach cusps, then measurements at frequencies of perhaps 1 Hz over the course of several hours during the cusp formation process would be sufficient. However, if the interest lies in coastal resource management or infrastructure planning, then a wave climate over many years to decades is required, with the hour-to-hour variations being less relevant. Every application has different requirements and therefore measurement strategies.

Holthuijsen (2007) considered four spatial/temporal scales – from smallest/shortest to largest/longest – each of which can be associated with a different methodological approach: phase resolving, phase averaging, event averaging and wave climate. At the smallest scale there is the need to describe wave parameters over a few wave periods or wave lengths (i.e., on the order of several seconds and tens of metres). Data collected at this scale make it is possible to develop and test numerical models that describe, in great detail, the motion of the water surface, the velocity of water particles and wave-induced pressures. This is termed the **phase resolving approach** since it has a resolution that is fine scale and appropriate for the study of wave dynamics within a single wave period or across wave groups.

Medium-scale studies deal with forms and processes on the order of a hundred to several hundred wave periods or wavelengths – i.e., a few minutes to about 30 minutes and covering a distance on the order of a few kilometres. This period is short enough for conditions to be considered homogeneous or stationary but long enough to provide robust statistics. Data recorded from a wave staff or buoy is typically broken up into packets of this length for analysis. At this scale, the objective is not to resolve every aspect of water motion (as at the small scale), but rather to characterise statistical properties of the wave field during the measurement time interval. Such an approach facilitates a comparison of similar recordings from other locations along the coast or from different times through a storm event at the same location. This is termed the **phase averaging approach**.

At larger scale – the **event averaging approach** – there is often the need to deal with time intervals equal to the duration of a prolonged storm and thus covering periods of days and distance scales on the order of hundreds to a thousand kilometres. This permits examination of the spatial and temporal variability of the wave-energy spectrum and therefore an accurate characterisation of the passage of waves from their area of generation across large distances until they finally reach the coast.

Finally, for the **wave climate approach**, if the concern is with coastal evolution and nearshore morphodynamics (e.g., the evolution of a barrier spit or the potential impact of constructing a harbour breakwall), then there is a need to consider average wave statistics over periods of tens of years or several decades. Such an approach leads to the development of the wave climate for a specific section of coast.

4.4.1.2 *Instrumentation*

Two general categories of wave-measuring device (Figure 4.9) can be identified: (1) instruments designed to measure properties of waves at a single point, typically requiring in-water deployment; and (2) remote sensing instruments such as radar or LiDAR that yield an image of wave conditions over a relatively large area of the sea surface from a distance and require placement on some sort of platform (e.g., plane, satellite). The latter group is able to provide information only of the surface expression of the waves (e.g., height, wavelength, steepness) although sequences of images with appropriate wave-tracking analysis can also yield insights into wave dynamics (e.g., periods and direction of travel). The former group, in contrast, reveals information on specific properties of the waves, such as the time series of surface elevation fluctuations, pressure and orbital velocity, from which can be derived other wave parameters (e.g., energy density). However, these properties are measured at only one location, and multiple

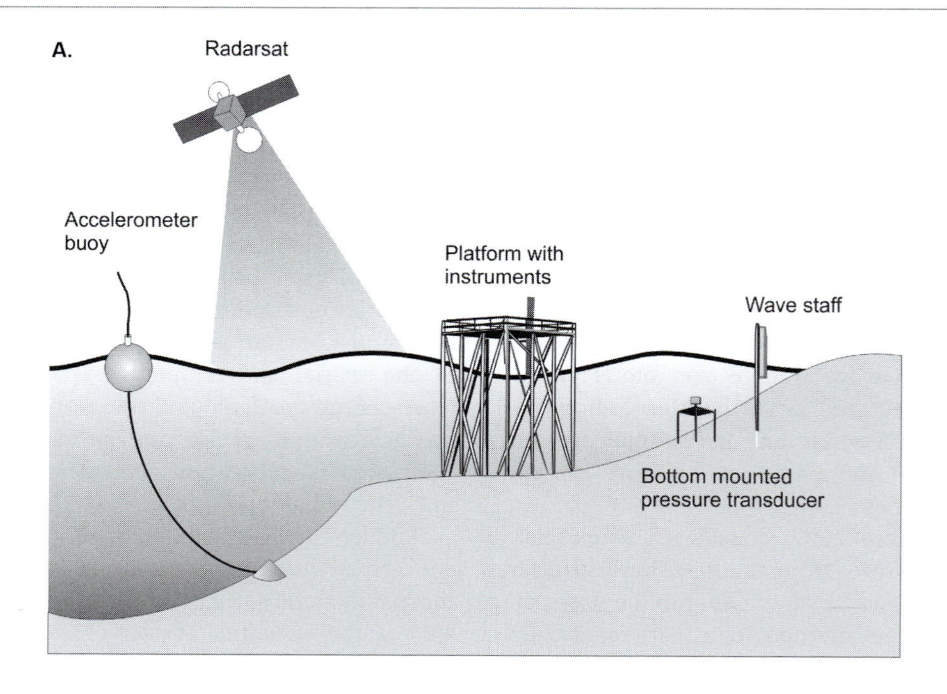

Figure 4.9 Wave measuring approaches. A. Schematic of several technologies used to measure wave parameters, including fixed instruments in the surf zone and mobile, satellite-based instruments covering the global oceans. B. Fixed instrument platform used in a study of wave generation in Lake Ontario (Donelan *et al.*, 1985; photo courtesy of Mark Donelan).

instruments need to be deployed to ascertain spatial trends. Thus, the two categories of instruments yield different information about the wave field that is highly complementary.

Instruments that measure surface elevation fluctuations can be divided into three groups: (1) graduated staffs; (2) electrical/electronic staffs; and (3) accelerometer buoys. Perhaps the simplest form of wave gauge is the graduated staff, which consists of a pole or long ruler that has centimetre markings and is fixed to the bed or other supporting structure such as a wharf pier. The design was originally intended for measuring tidal stage in harbours and shipping inlets, and it has been in existence for hundreds of years. The rise and fall of the water surface associated with waves is much more rapid than the tidal cycle, so rather than estimating and recording the water level manually, it is possible to use a video camera to secure the time series of surface elevation fluctuations. Often a grid on a board is used to scale the image of the wave profile (e.g., in wave tanks), especially when it is critical to know how wave breaking takes place or whether waves are asymmetric in profile. However, this type of recording is tedious to process and requires a great deal of operator oversight to extract reliable data. Its use tends to be restricted to highly specialised research studies rather than for routine data collection.

Wave staffs employing some form of electrical circuit have been used in numerous studies since the 1940s. These systems make use of the fact that water, especially sea water, is a highly conductive medium for electrical currents. The circuitry of the wave staff measures either the change in electrical resistance or capacitance of the system as water rises and falls over a length of non-insulated (bare) cable, which is part of the electrical circuit. The change in resistance or capacitance can be conditioned electronically to produce a variation in the type of output, which may be a direct current or a frequency signal. The output is saved on a device such as a strip-chart recorder, data logger or computer. Calibration of the instrument allows for conversion of the recorded signal (e.g., ohms) to variation in water surface elevation (e.g., centimetres) with time. The sensor itself may be fixed to a support that can be anchored into the bed (Figure 4.10A and B) or attached to some physical structure such as a jetty or platform (see Figure 4.9A). Wave staffs have been deployed at an inclined angle very close to the surface of the foreshore in order to measure wave run-up on moderately steep shorelines. The run-up spectrum determined by this method can be supplemented with information from video records (Holman and Guza, 1984) and this has proven particularly useful in areas such as Oregon and California where surf conditions make deployment of instrumentation in the breaker zone very difficult.

The advantage of surface-piercing wave staffs is that they provide a direct measure of water surface fluctuations and therefore wave height (or run-up excursion) and wave skewness or asymmetry. The disadvantage is that they are cumbersome to deploy and calibrate. Wave staffs are also subject to damage during storms because of the extreme wave forces on the staff (Figure 4.10C). Wave staffs were used in many studies of nearshore sediment transport, as well as for measurements of wave field growth during the 1970s and 1980s (e.g., in the North Sea or in the Great Lakes) although they are less popular now.

Wave parameters are now routinely measured with pressure sensors. Although there are a range of technologies (e.g., crystal wafers, diaphragms), there are two fundamental types depending on whether the rear of the sensing element is sealed or vented. In a sealed pressure sensor, the enclosure to the rear of the diaphragm is air-tight and occupied by air (or gas) of known pressure, typically standard atmospheric pressure at sea level (101.3 kPa or 760 mm of Hg), although in some varieties there is a total vacuum. When the pressure on the open side of the diaphragm increases or decreases relative to the pressure in the enclosure, a positive or negative signal is generated, respectively. It is critical to understand that these pressure variations are indexed to the enclosure pressure, and therefore the measurement is known as gauge pressure. If such a pressure sensor were

Figure 4.10 Resistance wave staffs for measuring surface elevation fluctuations: A. cross-shore array used in an experiment in Georgian Bay, Canada to determine wave shoaling transformations; B. close-up of wave staffs with mounting hardware and finned anchors that are jetted into the sandy bed; C. damage to wave staff mounts after a major storm during the 1985 Canadian Coastal Sediment Study, Prince Edward Island, Canada.

sitting in air on an ocean beach, it should measure zero gauge pressure assuming that the ambient atmospheric pressure was 101.3 kPa. Note, however, that as atmospheric pressure increases or decreases, the recorded pressure will also change. Thus, when measuring waves, it is necessary to compensate for these fluctuations in atmospheric pressure. It is best to use two identical pressure sensors (one in air and the other in the water) to facilitate this correction, although it is also feasible to use meteorological records if there is a weather station close by. Over short deployments of only a few hours, the need for such compensation is unnecessary because atmospheric pressure is not expected to change drastically.

To avoid this issue of atmospheric compensation, some manufacturers of pressure sensors have opened up the rear enclosure behind the diaphragm by connecting a venting tube that is mated with the electrical cable that goes back to shore. Since the vent is exposed to air, there is an automatic compensation for changes in atmospheric pressure (applied to the rear as well as the front of the diaphragm), and therefore the diaphragm responds only to changes in water pressure on the front of the diaphragm. A practical concern with vented pressure sensors is that the venting tube may allow moist air to enter into the rear enclosure, which may degrade the electronic components. As a precaution, it is advisable to fit a tube of desiccant to the vented end, thereby extracting any moisture in the humid air. Self-contained pressure sensors with internal batteries and data acquisition hardware (i.e., without cabling), such as those intended for deep water or long-term deployments, are typically of the gauge pressure variety.

The time series of pressure beneath an individual wave can be used to assess the period and wavelength of the wave, as well as the surface elevation fluctuations, using simple or complex wave theories. The procedure is discussed in Chapter 5, however, the key considerations involve accounting for the influence of changes in atmospheric pressure and mean water level due to tides, storm surge, and wave set-up or set-down, which will all influence the mean hydrostatic pressure measured by the pressure sensor. As will be shown in Chapter 5, there is an exponential decay in pressure (and wave orbital motion) beneath waves that is referred to as depth attenuation. The effect is wavelength dependent, such that the pressure signal from short period (small L) waves decays very rapidly with depth (see Figure 4.1), essentially filtering out any pressure signal from the measured time series if the pressure sensor is located in relatively deep water. Only the signal from the longer period waves will be captured in the measurements. This is a key consideration when planning instrument deployment schemes, especially if there is interest in the high-frequency bands of the wave spectrum. Because of the depth–wavelength dependency of the pressure signal, the analysis of a pressure time series is not a trivial task. However, these disadvantages are offset by the ease with which pressure sensors can be deployed and by their longevity and relatively cheap price. The cost of pressure sensors increases quickly according to increased sensitivity (accuracy and precision) and increased depth range.

Individual wave staffs or pressure transducers provide a one-dimensional (1D) picture of the variations in water surface elevation through time – i.e., they give information on wave height, wave period (or length), and wave asymmetry but not on the direction of travel. If information on the 2D attributes of waves are required (e.g., the directional energy spectrum), the deployment of a multiple-instrument array is necessary in order to compare travel times between various sensors. Alternatively, it is possible to take simultaneous measurements of surface elevation (with either a wave staff or pressure sensor) and the orbital motion beneath the wave using a current meter to determine travel direction based on anticipated phase relationships that wave theory reveals.

The electromagnetic current meter was widely used in the 1970s through to the 1990s to measure the velocity field beneath waves. The probe puts out a symmetric magnetic field within the water column, which becomes distorted in the presence of water motion. The distortion induces a voltage that can be sensed by the sensing elements embedded in the probe, and the voltage fluctuations are amplified and recorded. Electromagnetic current meters have been largely supplanted by the advent of acoustic Doppler velocimeters (ADV) that became increasingly more affordable and sophisticated in the past 20 years. Originally, these instruments were developed for laboratory use because of their delicate design and the need for computer processing power. However, now there are robust versions made for field deployment, and further advances in technology have yielded a profiling version (ADP) that samples the water column at a range of depth bins and records the velocity measurements in each of those bins separately. The main advantage of these sophisticated instruments is the capacity to take very high-frequency measurements (up to at least 25 Hz) of all three velocity components (u, v, w) almost simultaneously in the water column, allowing the characterisation of the velocity field beneath waves at every stage in the orbital cycle. The insights that have been revealed with respect to wave shoaling and wave breaking have been extraordinary. However, the disadvantage is that these instruments are relatively expensive and they require a high degree of technical proficiency to operate and maintain.

Wave staffs and pressure transducers are usually only deployed in relatively shallow water because of cable limitations, power requirements (i.e., the need for service access), and the constraints imposed by instrument sensitivity under large mean water pressures in deep water. Self-contained pressure sensors and current meters are often deployed on the sea bed to measure bottom pressures and currents due to tides and other long-period oscillations, but depth attenuation of the surface wave motion is a limiting constraint if the intent is to characterise the surface wave field. As a consequence, surface wave motion in deep water locations such as the continental shelf is typically measured using a wave rider buoy (Figure 4.11). An accelerometer is housed in the buoy, and it measures the surface displacement over time. The output is the one-dimensional energy spectrum. Heave–pitch–roll

Figure 4.11 Example of a wave rider buoy deployed along the California coast by the Scripps Institution of Oceanography. The aluminum frame attached to the buoy is a non-standard addition installed for the purpose of holding other instruments for specific experiments, but has since been used as a perch by marine wildlife. The presence of the frame (and the sea lion) will influence the performance of the buoy, which must be taken into account during data processing. (Photo courtesy of the Coastal Data Information Program, University of California San Diego.)

buoys, which incorporate several accelerometers mounted on gimbals and a compass that resolves the changing gimbal angles, permit estimation of the directional energy spectrum. Changes in the water surface slope due to waves are used to define the height and period of the wave, as well as the direction of travel for every frequency band. The buoy usually incorporates electronics for recording, processing and storing the data, as well as some form of radio communication link to shore or through a satellite channel. Solar panels are incorporated into the design to provide a source of energy. Free-floating buoys provide the best description of the water surface, but for practical reasons most buoys must be tethered to the bottom by an anchor system. The data from anchored buoys must be corrected for the effects of the mooring – not a trivial task. Buoys of this type are now being used to provide routine wave data in many parts of the world. There are extensive networks of buoys maintained by the National Oceanic and Atmospheric Administration (NOAA) on all the mainland coasts of the USA and there is also a good network around the coast of west Europe. Canada maintains wave buoys on all three coasts as well as the Great Lakes. Descriptions of the networks and, in many cases, access to the data being recorded by individual buoys can be readily accessed on the World Wide Web and a particularly useful site for acquiring buoy data is maintained by the US NOAA (www.ndbc.noaa.gov/).

The state of the sea surface (waves, ice cover, etc.) is now routinely measured using non-intrusive or remote sensing techniques such as radar altimetry, synthetic aperture radar (SAR), and light detection and ranging (LiDAR) technologies. The actual sensing equipment is mounted on satellites, airplanes, ships or observation towers on land (see Figure 4.9). The main advantage of these advanced technologies is that they provide a detailed description of the sea state over a large area of the ocean and therefore the output is much different than that obtained from individual instruments fixed in one location. The remotely sensed data are particularly useful in providing real-time information on the evolution of wave fields, interaction of waves and currents, modulation of wave action by ice cover, or transformation of waves in shallow water. In addition, this real-time information can be streamed directly into hydrodynamic models used to predict the fate of oil spills and the propagation of wave groups across the ocean, which is a boon to surfers who wish to track large, long-period waves to their favourite surfing location.

4.4.2 Analysis of Wave Records

Real wave records show complex patterns with a range of wave heights and periods, unlike the regular undulations used to conceptualise and theorise them. A 30-minute record from a wave staff or pressure transducer will typically have on the order of 400–1000 waves in the record, and no two waves will be exactly the same. In order to characterise the waves in the recording, two approaches are usually adopted: (1) calculation of the statistical properties of the raw data trace (time domain analysis); or (2) development of a spectral representation of the wave record based on the energy contained in various frequency bands (frequency domain analysis). These two approaches, although seemingly different, can be reconciled under most conditions.

4.4.2.1 *Time Domain Analysis*

Determination of the wave period (T) or wave height (H) from a time series of surface elevation fluctuations ($\eta(t)$) is relatively straightforward. Before the advent of digital wave records, this was done manually from a strip-chart trace of the waves, similar to the one shown in Figure 4.12. The trace shows that individual waves will vary considerably in height, and there appear to be several small waves that are superimposed on larger waves. This begs the question – when is a wave not a wave? Alternatively, how big does an undulation on the water surface have to be before it is considered to be a wave for the purpose of enumeration? Depending on which waves one decides to count as true waves, there will be implications for the total number of waves counted in the time series as well as the frequency or wave period. Clearly, inclusion of even the smallest undulations in the analysis will yield smaller and smaller average wave heights and shorter periods.

Based on the recommendation of a task force looking into this issue (IAHR, 1989), wave height and period are defined with respect to zero-crossings, which refer to positions where the trace of surface elevation fluctuations, $\eta(t)$, crosses the mean water line (defined by $z = 0$), either upward or downward (Figure 4.13). The beginning of a wave is defined as the instant when the trace crosses $z = 0$ in either the downward (down crossing method) or upward (up crossing method). The end of that wave occurs when the trace again crosses the MWL in the same orientation, either downward or upward, respectively. Every wave will therefore include exactly one large wave crest (highest point above MWL) and one large wave trough (lowest point below the MWL), while all other smaller oscillations are ignored. The wave height, H, is defined as the elevation difference between the highest and lowest points for each individual wave (i.e., $H = \eta_{max} - \eta_{min}$). The period is simply the time difference between the zero-crossings at the

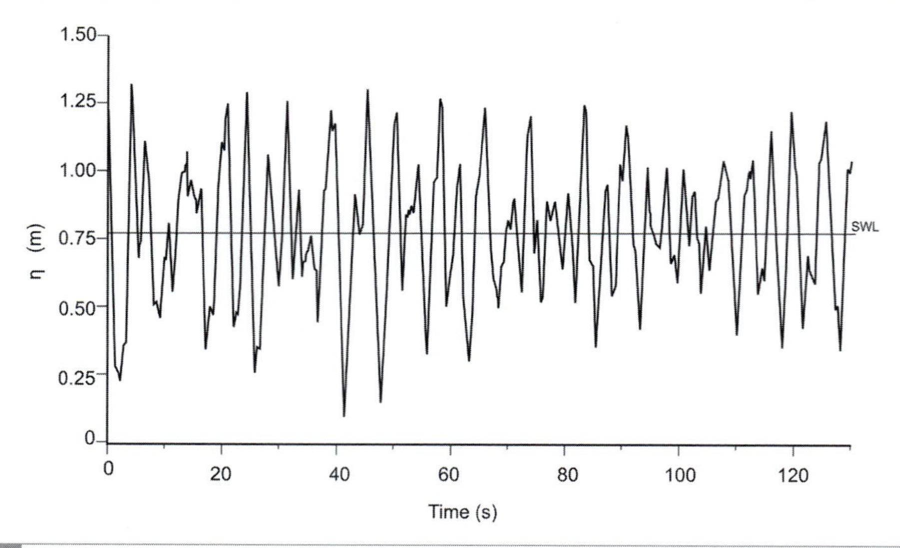

Figure 4.12 Two-minute time series of surface elevation fluctuations measured with a wave staff. Note the irregular nature of the waves, especially the minor oscillations superimposed on the longer period waves. This is typical of measurements inside the surf zone. The SWL is indicated by the solid line, which would have been marked on the wave staff during a calm period.

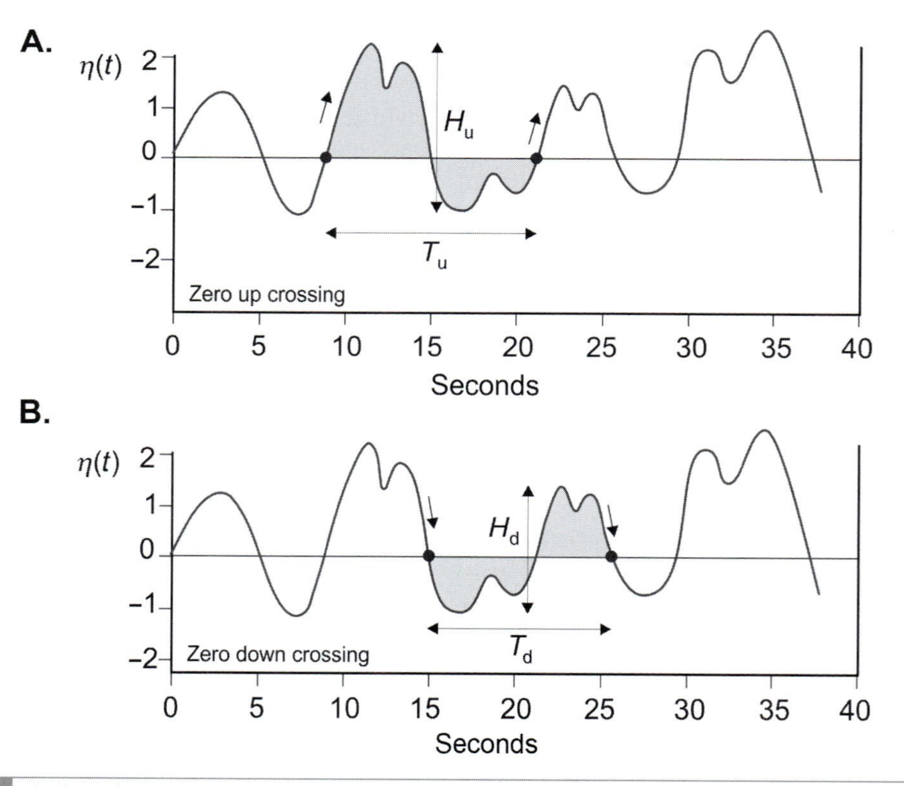

Figure 4.13 Analysis of wave parameters using the manual zero crossing method: A. zero up crossing approach; B. zero down crossing approach. Note that individual waves (H, T) will differ between the two approaches, but the statistics from the overall time series will be similar.

beginning and end of the wave. Note that the up crossing and down crossing methods use different start and end times for every individual wave, and therefore a different crest will be paired with a different trough, so the properties of each wave will differ. Nevertheless, these differences will average out across the entire wave record, as long as the wave record is sufficiently long, thereby providing robust statistical values.

The analysis of a wave record using the zero-crossing method will produce a large number of discrete values for H and T, and these can be represented in the form of a histogram or frequency distribution (Figure 4.14). A range of statistical quantities can be derived from the histogram data that describe the entire wave record. The most obvious is the mean (or average) wave height or period,

$$\bar{H} = \frac{1}{N}\sum_{i=1}^{N} H_i \tag{4.2}$$

where N refers to the number of waves in the time series, each of which has height, H_i, denoted by the subscript i, which ranges from the first wave (i = 1) to the last wave (i = N). If the wave heights and periods are normally distributed (i.e., Gaussian), as is generally the case in deep water, then the mean height is a robust statistical estimate of the wave field. The median wave height (with 50 per cent values being larger and smaller) will be identical to the mean wave height in this

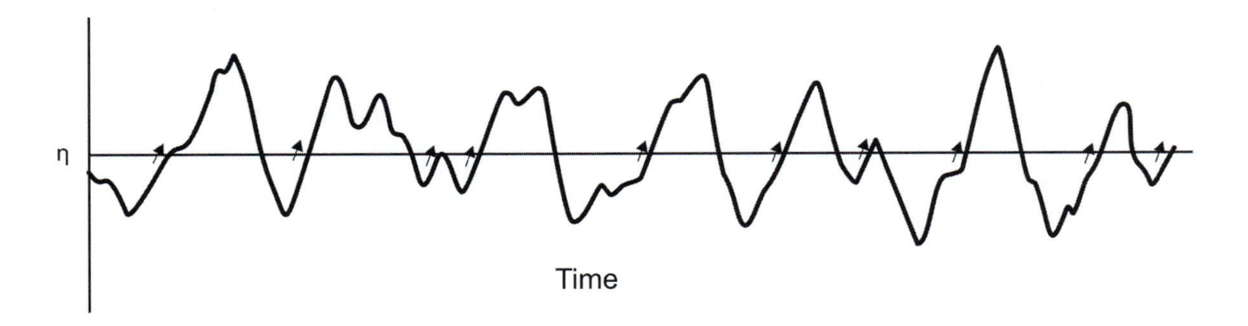

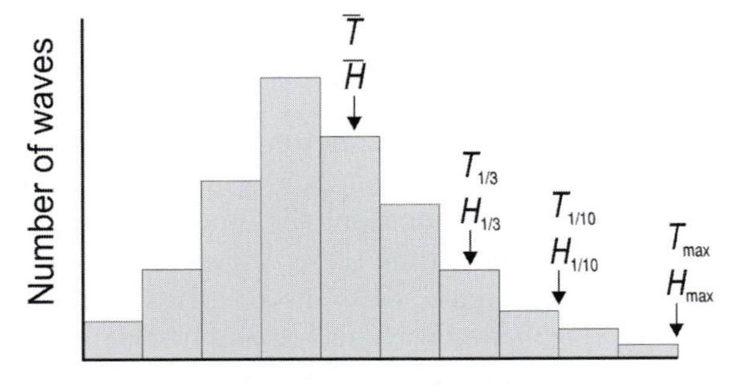

Figure 4.14 Analysis of a wave record. A. Portion of the wave record showing the up crossing method. B. Histogram of wave heights (or periods) derived from the wave record in A, showing mean, maximum and percentiles.

instance of a Gaussian distribution, but not for highly skewed distributions.

A measure of the variability or spread of the waves about the mean height is the standard deviation of H, also known as the root-mean-square wave height,

$$H_{\mathrm{rms}} = \sqrt{\frac{1}{N}\sum_{i=1}^{N} H_i^2}. \tag{4.3}$$

Often wave height distributions are not Gaussian, but rather there can be large numbers of relatively small waves and a smaller proportion of much larger waves. These larger waves are of special interest because the amount of energy contained in a wave is proportional to the square of wave height (see Chapter 5 for details). The largest waves have implications for the design of engineering structures that must survive the extreme forces generated by a wave field (rather than the average energy levels) or to prevent overtopping of a sea wall that protects valuable real estate or civil infrastructure. In this context, it is important to know how large the maximum wave height is, H_{max}. However, the largest wave is an unreliable measure of the distribution because it is a singular event, which could be exceeded at some other time. Therefore, it is more common to estimate the height of a nominal wave that represents a certain fraction of the largest waves. For example, $H_{1/10}$ refers to the wave height that represents the average of the largest 1/10 or 10 per cent of the wave heights in the distribution. Similarly, $H_{1/3}$ refers to the wave height that is the average of the largest 33 per cent of all waves in the distribution, and it is of special relevance because a rapid visual assessment of ocean waves by the human eye naturally focuses on the waves that are larger than the mean. Experiments have shown that this aligns roughly with the highest 1/3 of the waves, and therefore $H_{1/3}$ is formally known as the significant wave height. Note that this convention of using percentiles to identify a fraction of the largest waves and then calculating the average within that sub-sample is different than doing a frequency analysis (e.g., of sediment size or flood recurrence). For the latter, the convention is to identify a value (the Xth percentile) for which X per cent of the total sample is greater than that value – no averaging is done.

4.4.2.2 *Frequency Domain Analysis*

The zero-crossing technique of wave analysis is useful for providing insight into some of the challenges of defining a wave and extracting basic information from a wave record. However, it is a tedious process that yields only summary statistics on the wave field. Analysis of the full time series of $\eta(t)$ using computer programs can reveal a great deal more about the waves. In particular, it is now routine to conduct a spectral analysis of the time series, which involves a transformation of the data, originally recorded in the time domain, into the frequency domain. Many of the early developments for time series analysis came from audio electronics and signal processing (Cooley and Tukey, 1965; Jenkins and Watts, 1968), but the methods were easily adapted to the investigation of wave and current records. The basic approach is to decompose the complex record of surface elevation fluctuations (or pressure or velocity) into a series of sinusoidal waves each of which has a different frequency (period), energy content (height or amplitude), and phase relationship to the other waves.

It has been known for a long time that a complex trace or even a square wave can be reproduced using a linear combination of pure harmonic oscillations (Figure 4.15). Thus, the surface elevation fluctuations can be written as

$$\eta(t) = \sum_{i=1}^{\infty} a_i \cos\left(2\pi f_i t + \alpha_i\right), \tag{4.4}$$

where a_i is the amplitude ascribed to the sinusoid of frequency f_i, α_i is the phase, and t is time (or position along the time series). Consider the complex time series shown in the bottom panel of Figure 4.15. The objective is to take this measured wave record and determine that it is made of exactly three waves the same as those shown in the upper three panels, each with a different frequency and amplitude. The challenge is to pick exactly the right combination of such waves to emulate the measured wave trace and ignore

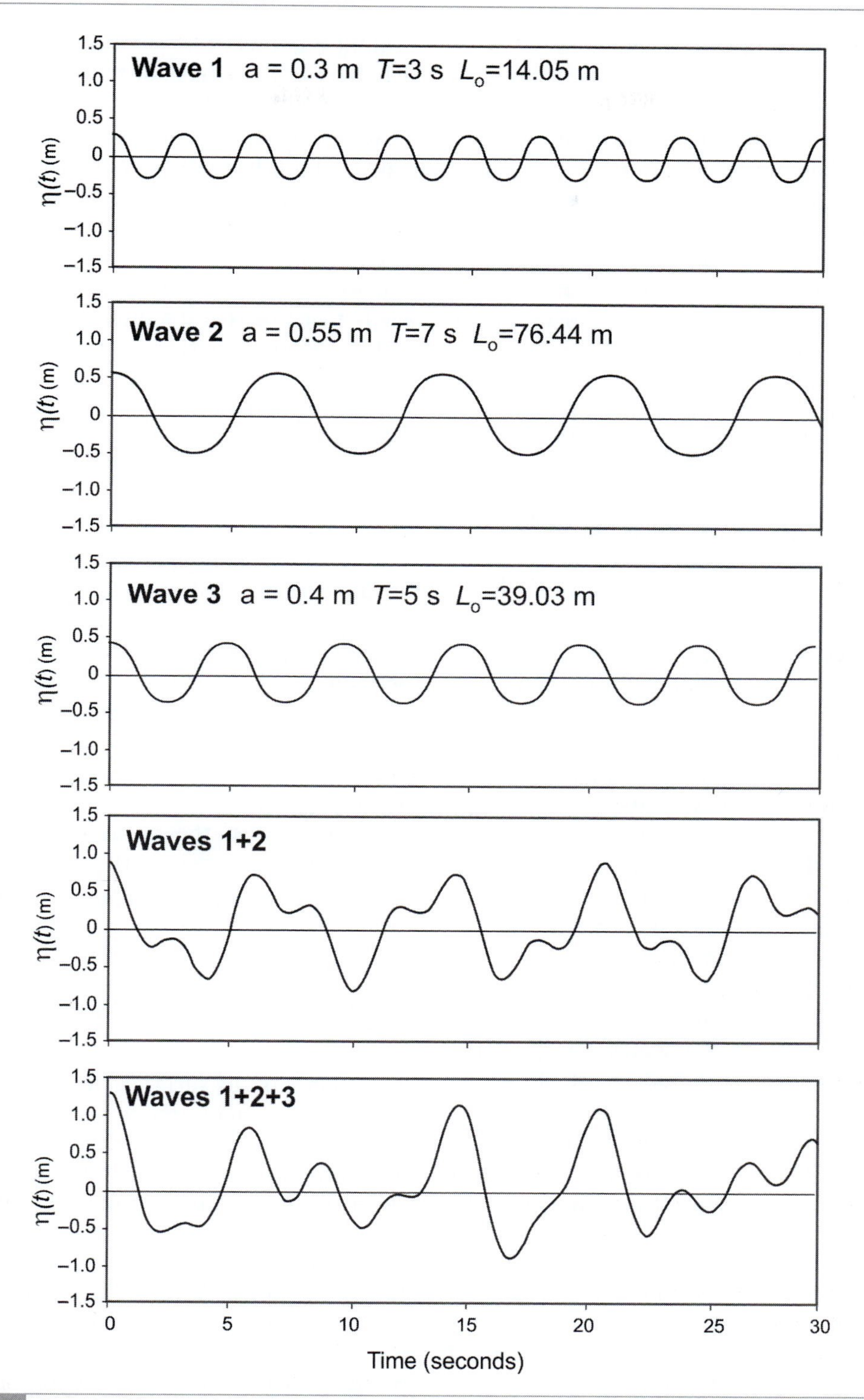

Figure 4.15 Example of how simple sinusoidal wave forms with differing periods and amplitudes (top three panels) can be combined to create complex surface elevation fluctuations (bottom two panels) by simple linear addition.

the infinite number of other possibilities. Fortunately, there are mathematical techniques that facilitate solving this inverse problem, and key among them is the fast Fourier transform (FFT).

There are numerous software packages (e.g., R, SigmaPlot, Matlab) with sub-routines that will perform Fourier spectral analysis to varying degrees of complexity and sophistication. The first step is typically to produce a periodogram, which is a plot with numerous spikes often referred to as a Fourier line spectrum or an amplitude spectrum (Figure 4.16A). Each spike represents a discrete sinusoid at a specific frequency represented along the horizontal axis and with amplitude proportional to the height of the spike on the vertical axis. The amplitude in this instance is not necessarily the actual wave amplitude but rather the size of the coefficients in Equation 4.4 attached to the sinusoid of a specific frequency in the Fourier analysis. For example, if the time series were of orbital velocity, $u(t)$, rather than surface elevation fluctuations, $\eta(t)$, the amplitude represented in the periodogram would have units of metres per second rather than metres. The number of spikes, and hence the frequency intervals between spikes, is defined in part by the user but also depends on the length of the time series as well as the sampling rate, which indirectly limit the shortest and longest periods (frequencies), respectively. The amplitude spectrum reveals how much of the overall variance in the time series is due to individual waves of different frequencies. If there were only two waves of unequal height in the wave field, then the amplitude spectrum would have only two spikes with different heights.

Wave energy is proportional to the square of wave height (H_i^2), so it is often more convenient to produce a variance spectrum (Figure 4.16B), which provides an overview of the relative energy content contained at different frequencies. The vertical axis in this graph is proportional to the amplitude (or velocity) squared, and therefore indicates the energy associated with waves of different frequencies, or alternatively the proportion of the variance in the time series that can be ascribed to a specific frequency. Since numerous frequencies are present in the ocean at one time, the variance associated with each frequency is then distributed over a frequency interval or band centred at f_i with width Δf_i to create the variance density spectrum (Figure 4.16C). The main advantage of this representation is that it is 'variance preserving' meaning that a summation of all the vertical bars in the graph will return a value equal to the total variance of the original time series. The disadvantage is that this is still a discrete representation of the smoothly continuous time series in the measured wave record. In order to overcome this shortcoming, numerous smoothing algorithms are available to yield a continuous variance density spectrum (Figure 4.10D). An integration beneath the curve will yield the variance of the original wave record.

The wave energy spectrum provides one of the most useful and complete descriptions of the wave field. Most modern wave prediction models will simulate the evolution of the wave spectrum through time. Nevertheless, it is possible to reconcile the time domain and frequency domain approaches to wave analysis. Consider a water surface elevation time series, $\eta(t)$, similar to the one shown in the lower panel of Figure 4.15, with the mean surface elevation having been adjusted to the MWL such that $\overline{\eta(t)} = 0$. The standard deviation, σ_η, is given by

$$\sigma_\eta = \eta_{\text{rms}} = \sqrt{\frac{1}{N} \sum_{i=1}^{N} \eta_i^2}. \tag{4.5}$$

The variance of the surface elevation time series, $\sigma_\eta{}^2$, is proportional to the total area beneath the curve of the continuous variance (energy) density spectrum. This suggests that the variance contained in any frequency band (of width Δf_i) is but a fraction of the total variance in the wave field, which is precisely what the energy spectrum is intended to portray – the relative importance of different frequency bands to the overall motion in the sea surface. An energy density spectrum with a distinct and narrow peak indicates that most of the energy content of the wave field is contained in a narrow range of frequencies, as might occur with clean swell waves.

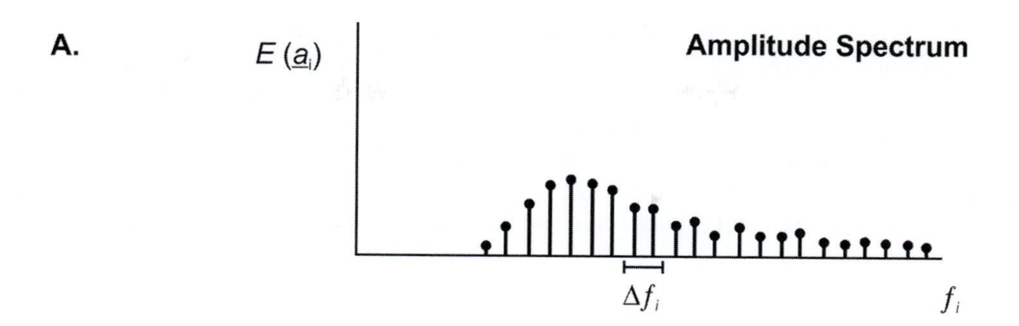

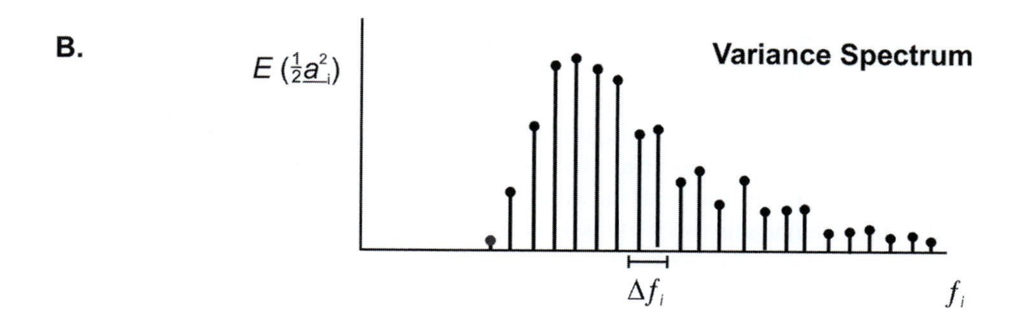

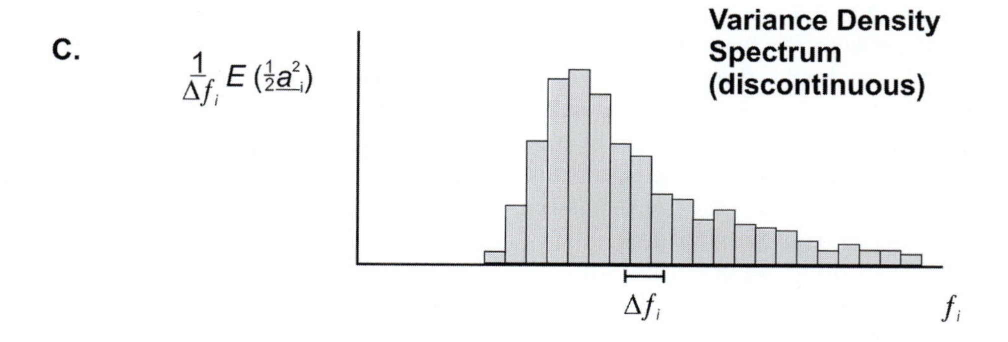

Figure 4.16 Transformation of the discrete amplitude spectrum into the continuous variance (energy) density spectrum, which provides a complete statistical description of surface elevation fluctuations (Holthuijsen, 2007).

In contrast, a broad, flat spectrum would represent wave motions smeared across a wide range of periods, perhaps because an old wave field was being modified by fresh winds. A spectrum measured in the surf zone with two distinct peaks might indicate that there are both incident waves (at high frequencies) and infragravity waves (at low frequencies).

Interpretation of energy spectra is assisted by the following relationships:

$$\bar{H} = \sqrt{2\pi\sigma_\eta} \quad (4.6)$$

$$H_{1/3} \approx \sqrt{2}H_{\mathrm{rms}} \approx 4\sigma_\eta \quad (4.7)$$

which provide a direct linkage between the time domain and frequency domain methods, under the assumption that the surface elevation fluctuation time series is Gaussian (or that the wave height distribution is Rayleigh distributed). It is essential to appreciate the difference between the statistical quantities that apply to the raw surface elevation time series, $\eta(t)$, such as σ_η, versus those that apply to sampled quantities derived from $\eta(t)$, such as H_i and T_i and the statistics that apply to the sample distribution (e.g., $\bar{H}$ or $H_{1/3}$). The frequency corresponding to the peak of the spectrum in the wind wave region is designated the peak frequency or peak period, T_p, which is approximately equivalent to $T_{1/3}$, the significant wave period derived from the zero-crossing method (within 5 per cent for wind sea).

It is important to remember that spectral analysis is a statistical technique, and in particular that the energy associated with each frequency is an estimate of the 'true' energy in that band. The periodogram provides estimates of this energy at discrete frequencies, but there is considerable statistical uncertainty, hence the confidence limits associated with each of the amplitude estimates are usually very large. Smoothing of the discrete spectra by a weighted local averaging scheme (referred to as windowing) to produce the continuous variance density spectrum (Figure 4.16D) eliminates the spiky nature of the periodogram and improves the statistical reliability of the estimates within specific frequency bands. But, too much smoothing ends up smearing the energy content in neighbouring frequency bands thereby hiding potentially important information about the existence of frequency specific motion. There is a trade-off between increasing the statistical confidence in the estimate of spectral energy (as more bands are grouped) and a reduction in the information contained therein because even large energy content in narrow frequency bands will be averaged away.

There are several things to be aware of when applying spectral analysis to wave records, and detailed treatments of the technique can be found in Jenkins and Watts (1968), Hegge and Masselink (1996) or Chatfield (2004). For example, the fast Fourier transform (FFT) is widely used because it is computationally efficient, but it requires that the number of data points in the time series is equal to $N = 2^x$, where x is an integer (e.g., N = 512, 1024, 2048, 4096, etc.). The length of the record and the sampling interval places restrictions on the frequencies that can be resolved. Thus, if the water surface elevation is sampled twice per second (0.5 Hz), the greatest frequency that can be resolved is 1 Hz. This high-frequency cut-off is known as the Nyquist frequency. Likewise, the lowest frequency that can be resolved is equal to $1/N$ although this will have considerable statistical uncertainty. A wave recording may consist of measurements made over a finite sampling period or it may be a sub-sample of a much longer record. In either case the sampling frequency (or digitising frequency for a continuous record), and the length of the record are important considerations. In general, a minimum of 6–8 data points should be used to define an individual wave so that a sampling frequency of 6–8 Hz is advised if there is a need to resolve waves with a period of 1 second. Of course, this presumes that one wishes to resolve most aspects of every wave. Slower sampling rates are acceptable if one is only interested in a random sampling of the sea surface fluctuations over a long time interval (i.e., to estimate the distribution). When there is interest in low-frequency water motion with periods on the order of 30 seconds to several minutes, a sampling time of 20–30 minutes is required. Beyond this time frame, the wave field tends to change in character, and then there are issues associated with non-stationarity. Fourier analysis is based on the assumption that the time series is stationary (i.e., with constant mean and variance), although it may be possible to de-trend a time series, to remove the tidal signal for example. Recent developments in wavelet analysis allow these assumptions to be relaxed and thus permit full time-frequency representation of the time series (Donelan *et al.*, 1996; Massel, 2001). Wavelet analysis is now used commonly, to the extent that it is supplanting the FFT, but the basic principles are similar.

Spectral analysis, whether by FFT or wavelets, provides an indication of how much energy is in the peak waves, as well as in other frequency bands. Low-frequency energy is often important in the design of harbours because of wave resonances at the scale of the harbour basin, which can be amplified and lead to unwanted oscillations that are dangerous to moored ships and dock pilings. On the beach there may be interest in determining the presence of standing, long waves and distinguishing them from edge waves. This can be accomplished using co-spectral analysis, which is a procedure by which two time series (say, pressure and velocity) are compared in a way to reveal the phase relationships between the motions at specific frequencies and to determine how coherent the motions are (Huntley, 1980). Co-spectral analysis has been used extensively to examine the impact of infragravity waves in the evolution of nearshore features such as crescentic bars and swash cusps (see Chapter 8). A full treatment of these advanced methods is beyond the scope of this introductory textbook, and the interested reader is referred to the published literature for detailed discussions.

4.5 Wave Prediction

A variety of techniques can be used to forecast (predict forward in time) or to hindcast (retrodict backward in time) the expected wave conditions at a particular location over a given time interval. Wave forecasting is similar to weather prediction, and typically relies on meteorological data such as wind speed and direction that are derived from predictive models of atmospheric circulation. The skill attributed to these models is very good for periods of hours to days, but decays rapidly beyond about one week. The widespread availability of satellite data has not only enhanced the spatial coverage but has provided critical information on observed wave conditions in real time, which is essential for providing the wave models with boundary (initial) conditions and verifying their output. Wave forecasting is used for a variety of purposes, mostly related to shipping, fishing and boating interests.

Wave hindcasting utilises the same predictive models as wave forecasting with one major difference. The models are driven by past meteorological measurements rather than weather predictions into the future, so there is typically greater accuracy. This is quite advantageous in areas with restricted fetch lengths where waves are locally generated rather than influenced by broad-scale regional wind and pressure patterns. Wave hindcasting is used to inform the design of engineering installations such as seawalls (i.e., determining the largest historical waves in the area) and to undertake forensic analysis of marine disasters in which ships have been damaged or sunk (i.e., due to rogue waves, for example). Similarly, wave hindcasting can be used to produce wave climates that provide insight into how coastal erosion has taken place through time and to drive models of long-term shoreline change. For most geomorphological and engineering work, hindcasting is used much more commonly than forecasting, but the same basic principles apply. Most wave models deal with wave generation in deep water because friction and shoaling will not affect waves in deep water, and therefore the shape of the resultant spectrum is easy to determine. In order to predict conditions close to shore, the deep water spectrum must be transformed to accommodate shoaling and breaking effects, and this is a complex undertaking. Fortunately, there are a large number of models that have this capacity.

Waves in the open ocean are generated by the transfer of energy from wind to the water surface, as described earlier in the chapter. Basic wave characteristics such as wave height and wave period are controlled by wind speed, duration and fetch length. The simplest models of wave prediction are based on this empirical knowledge having been put into practice by way of charts referred to as nomograms. The first operational version of a wave nomogram was assembled by Sverdrup and Munk (1947) who did their work in support of the Allied effort to invade Europe during World War II. Their model consisted of a series of empirical equations that were used to predict the significant wave height, $H_{1/3}$, and significant wave period, $T_{1/3}$, for given combinations of wind speed, duration and fetch length. The results across a range of values were plotted in one convenient graph, which was the nomogram. The equations were revised on several occasions by Bretschneider (1952, 1958) and therefore the general method is referred to as the Sverdrup–Munk–Bretschneider (SMB) method. The SMB equations were adopted by the US Army Corps of Engineers and widely propagated in their *Shore Protection Manual* (SPM; 1984), which was first distributed in the 1970s. Nomograms were particularly useful before the widespread use of computers because coastal engineers, scientists, and even ship captains could predict wave conditions with relatively simple data and no requirement to undertake complex calculations. The SMB method was widely used until the early 1980s but has been dropped from recent versions of the *Shore Protection Manual*, although it still includes a simple method for wave prediction under rather restricted conditions. Based on developments in wave theory and wave measurements on Lake Ontario, Donelan (1980) produced simple nomograms for wave period and wave height that can be used to explore wave generation in areas of restricted fetch (Figure 4.17).

The major weakness of wave nomograms is that they predict single values of wave height and wave period, and assume that the direction of wave propagation is parallel to the wind. More realistically, wave generation involves a full spectrum of wave motions, so the focus since the 1980s has been to develop models that predict the energy spectrum, from which can be derived simple statistics such as wave height and period. The first spectral model was that of Pierson, Neumann and James (1955; PNJ model), which dates back to 1955 and was based almost entirely on empirical data. Now, there are a large number of numerical models predicting the spectra of wind generated waves, and they may be divided into two types: (1) true spectral models; and (2) parametric models.

True spectral models are based on theoretical concepts supplemented by empirical observations. They predict the evolution of the spectrum from a consideration of energy inputs, losses and transfers among the many frequency components in the wave field (e.g., Resio, 1981). The theories of wave generation resulting from the work of Miles (1957), Phillips (1957) and Hasselmann *et al.* (1973) are at the base of these models, and the solutions to the differential equations are constrained by experimental data. In contrast, parametric models (e.g., Hasselmann *et al.*, 1976) solve the same energy or momentum transport equations but make assumptions about the shape of the spectrum so that the problem is reduced to prediction of a few non-dimensional parameters. This approach is computationally more efficient, and the models are easier to apply. The empirical results produced by the Joint North Sea Wave Project (JONSWAP) and its successors have provided a wealth of insight into wave generation and spectral evolution in actively growing seas, and these experimental results figure prominently in the literature. Additional research has been conducted in the Great Lakes (e.g., Schwab *et al.*, 1984; Donelan *et al.*, 1985), an environment that is ideal for studying wave generation because of the restricted fetches, the tendency for storms to be discrete events (with quiet conditions during inter-storm periods), and the relative abundance of meteorological stations to provide data to calibrate and test the model output. A useful summary of early work on wave prediction models appears in Bishop and Donelan (1989).

Newer versions of wave prediction models are known as second- and third-generation models, and they tend to be increasingly complex because

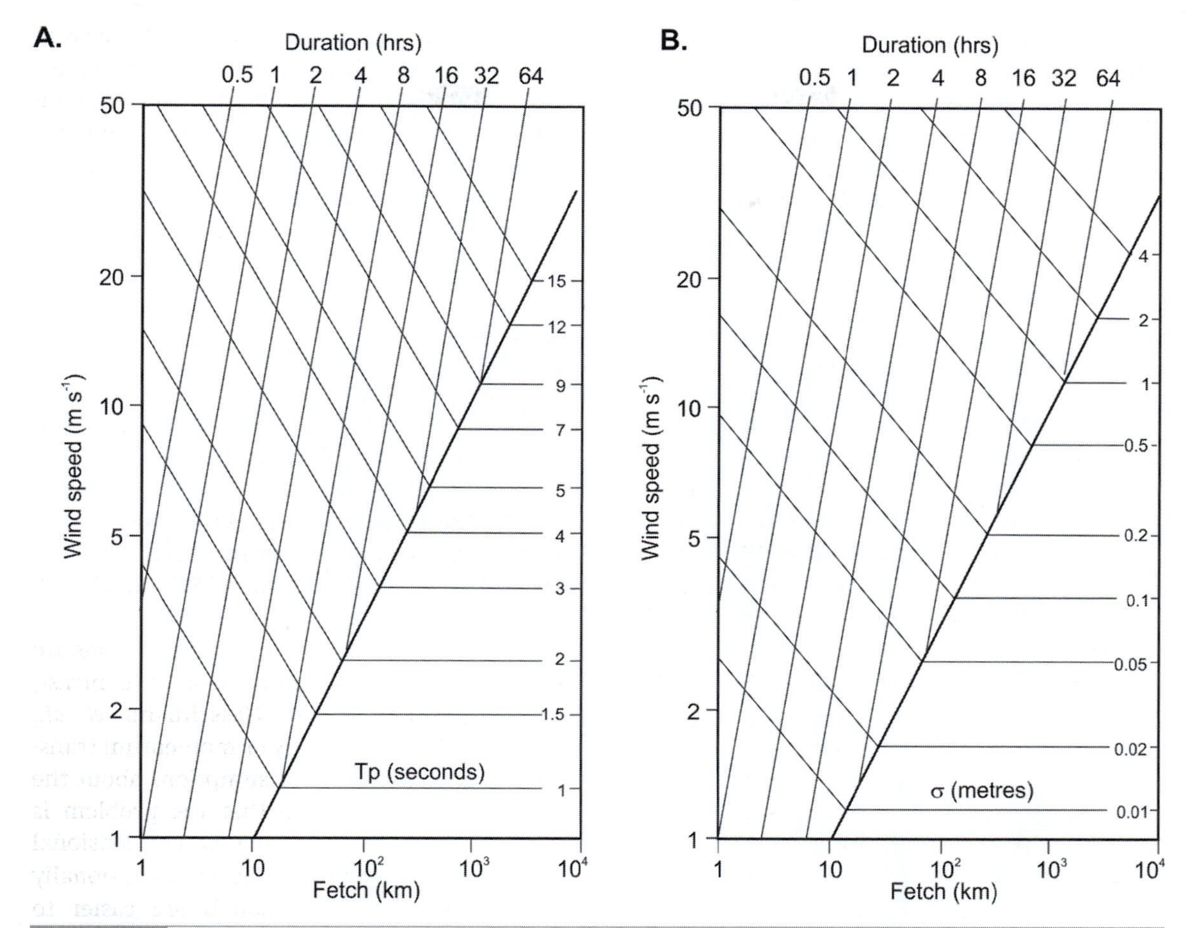

Figure 4.17 Wave nomograms for forecasting A. peak wave period, T_p, and B. wave height, σ, in areas of restricted fetch (Donelan, 1980). Users usually begin with known wind speed (vertical axis), and therefore enter the graphs from the left and move horizontally until the first limiting condition is reached, either a fetch restriction (bottom horizontal axis; no vertical lines) or a duration restriction (top horizontal axis; dotted diagonal lines). The point of intersection of wind speed and the limiting condition defines the wave period A. or wave height B., which can be estimated using the solid grey lines labelled inside the right of each diagram. The zone to the right-bottom corner of each diagram, delineated by the bold black line, indicates full wave development where no further growth of the wave is possible for the specific combination of wind speed and limiting conditions. The wave height provided in B. is the root mean square wave height H_{rms}. Note that $H_s \approx 1.4H_{rms}$.

they include more and more of the detailed dynamics of ocean waves. For example, models such as the WAM wave model (Hasselmann *et al.*, 1988; Komen *et al.*, 1994), and the SWAN – simulating waves nearshore – model (Booij *et al.*, 1999; Holthuijsen, 2007) compute quadruplet wave–wave interactions and do not impose any a priori shape on the spectrum. Models such as SWAN, as well as MIKE 21 SW (spectral wave module) and MIKE 21 BW (Boussinesq wave module) from the Danish Hydraulics Institute (DHI Group), incorporate refraction, diffraction and wave–current interactions and can model wave transformations into shallow water, thus providing extremely powerful tools for coastal design (Lin *et al.*, 2002). The third generation of predictive models are capable of providing highly accurate forecasts of the generation of waves within a storm area, the dispersion of those waves outside the area of generation, and the ultimate shoaling transformations that occur close to shore. Versions of such complex models are used for operational purposes by agencies in most countries across the globe, such as the US National Oceanic and Atmospheric Administration in their WAVEWATCH III model, which does a global ocean forecast as well as regional forecasts, the European Centre for Medium-Range Weather Forecasts in their HRES-WAM and HRES-SAW models, and the Bureau of Meteorology (Australia) in the AUSWAVE model. These model simulations are important for hurricane and storm predictions, and they are used routinely in general atmospheric and climate models that depend on accurate assessments of air–sea interactions.

Box 4.1 Direction of Wave Approach

A key element in any wave climate is the direction from which waves approach the coast. In mid-latitude locations, each season of the year will tend to have dominant waves coming from different directions as governed by prevailing winds and storm tracks. There are three controls to be considered: (1) the magnitude and frequency of winds in different directional bands; (2) the orientation of the coastline relative to the wind direction; and (3) wave refraction and diffraction patterns governed by offshore bathymetry and islands. Donelan *et al.* (1985) provide a useful discussion of many aspects of wave direction spectra, and Donelan (1980) points out the relevance of basin shape on peak wave direction in large lakes and enclosed seas (Figure 4.18).

Wave generation occurs not only in the direction that the wind blows, but in an arc that spans several degrees on either side of the primary direction. Wave generation in any off-wind direction can be related to the primary wind using the following simple predictive relationship:

$$U_\theta = U_p \cos\theta \qquad (4.8)$$

where U_p is the speed of the primary wind and U_θ is the speed of the effective wind in the direction designated by θ, which is the angle between the primary wind direction and the effective wind direction. Thus, the effective wind decreases in speed as the angle from the primary direction increases, and therefore it is less effective in creating waves.

Since wave height, wavelength and wave period increase with increasing wind speed, the largest waves are normally associated with the primary wind direction. However, on restricted water bodies such as lakes, there are often fetch limitations depending on which direction the wind blows from. Figure 4.18 shows an example of wind blowing from the east-southeast across an oval-shaped lake. The directional spread of wind from points A, B and C will yield a directional wave spectrum at the observation point, O. In this case, the effect of increasing fetch length along the more southerly track (A–O; aligned with θ_2) on wave size will more than compensate for the decrease in the effective wind speed ($U_{\theta 2}$), thereby creating large waves at O (as indicated in the lowermost cross section). However, this is only effective up to some critical angle from the primary wind direction beyond which the influence of increasing fetch length is less than that produced by the decrease in effective wind speed. Consequently, waves originating from the southernmost shoreline of the lake, in this example, would be very small. Waves from the more easterly directions (θ_1) centred around point C will be smaller than those associated with the primary wind direction (point B) because of the combined effect of decreasing fetch and decreasing effective wind speed.

In closed basins such as the Great Lakes, the Caspian Sea or Lake Victoria, the effect of this fetch and wind arc phenomenon will vary with the shape of the water body and the location of the point of interest on that water body. For elongated lakes, the difference between the wind direction and peak wave direction can be as much as 50°. A simple method for calculating the direction of peak wave approach for fetch limited areas is provided in Donelan (1980), and most wave models have algorithms that accommodate this effect.

Box 4.1 (cont.)

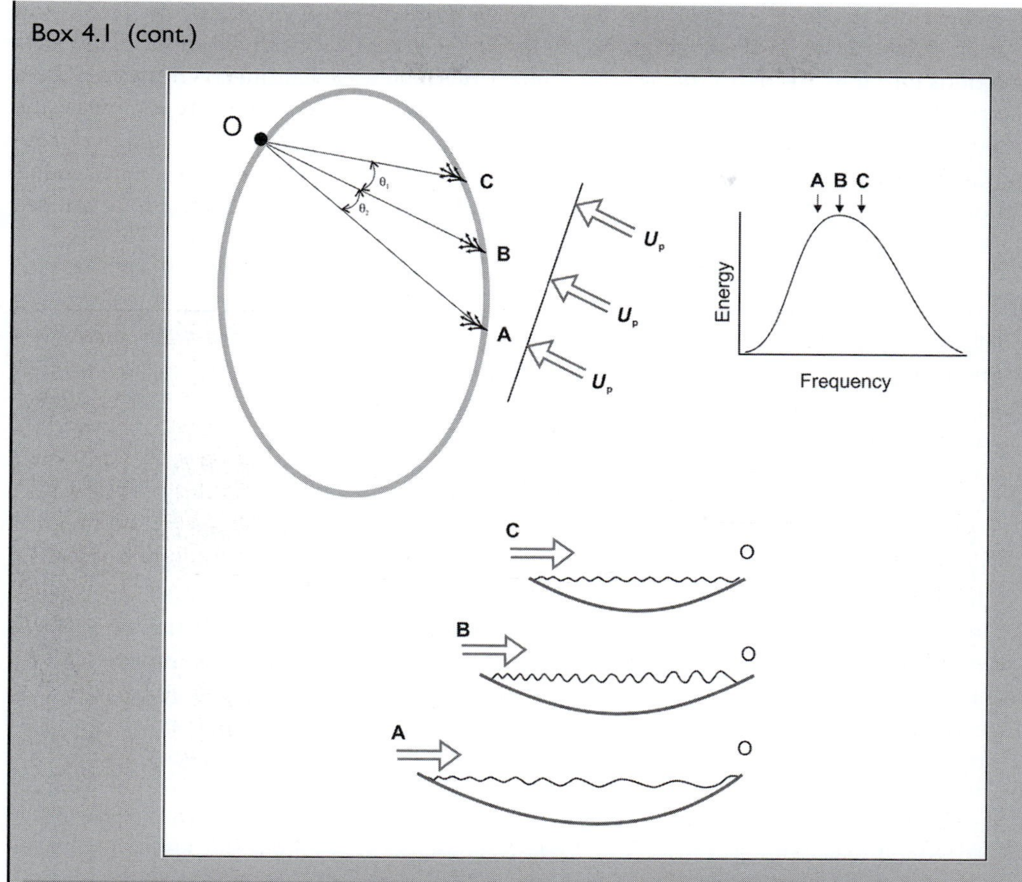

Figure 4.18 Schematic of wave energy spectrum dependency on the combined effect of fetch and wind speed in an enclosed, fetch-restricted basin or lake. Wind approaching from the east–southeast has a primary wind speed of U_p but a directional spread in effective wind speed of $U_\theta = U_p\cos\theta$, as shown by the barbs at locations A, B and C. The measurement point, O, receives wave energy generated by effective winds arriving from multiple locations everywhere along the opposite shore, but represented by specific points A, B and C. The wind blowing from B to O is parallel to the primary direction and therefore the energy input into the waves is at its maximum, thereby creating relatively large waves at O. The wind blowing from C to O is less effective with reduced wind speed, $U_{\theta 1}$, and the fetch is also reduced, therefore creating smaller waves with shorter periods at O. In contrast, the wind blowing from A to O has a longer fetch than B–O but is at an angle from U_p and therefore is less effective with reduced wind speed, $U_{\theta 2}$. The longer fetch compensates for the less effective wind, and as a consequence relatively large waves with longer periods are propagated to O (after Donelan, 1980).

4.6 Wave Climate

When someone describes the climate of a region, they are usually conveying information about long-term atmospheric conditions with respect to air temperature, humidity, air pressure and precipitation. Climate is often referred to as average weather, where the term weather is used exclusively in reference to the hour-to-hour and day-to-day conditions experienced or anticipated in the near future (i.e., a weather prediction). Climate provides information such as mean monthly temperature, average annual precipitation or heating degree days per year. Many locations also report wind roses, which show the mean (and maximum) sustained wind speeds (i.e., longer than 6 or 12 hours) from different compass directions (often in 10° or 15° bands). It is possible to think of coastal regions having climates with respect to wave conditions in exactly the same way as atmospheric conditions, thereby providing information on the long-term wave conditions along a stretch of shoreline. Typical measures include monthly or annual mean wave height and wave period, often in the form of a wave rose that includes directional information. Wave roses are usually reported as annual averages for significant wave height (or period) but it is not uncommon to find them at seasonal or monthly intervals.

In order to develop a wave climate for a specific location, there is a need for extensive data that adequately characterise the area and provide statistical confidence in the estimates. There are some coastal locations (usually near major ports) where such data have been collected for decades, but a surprisingly large portion of the global coastline has no measurements whatsoever. There is a gradually increasing compilation of ship-based observations because most commercial craft are required to maintain accurate logs of sea conditions, but most coastal wave climates are determined by wave hindcasting. Hindcast wave climates are usually based on a minimum of five years of recorded wind data, and more commonly 10 to 20 years. Clearly, the longer the record, the greater the statistical reliability associated with the estimates of wave parameters. However, there is also the issue of non-stationarity, and with very long records it is possible to determine whether there are gradual shifts in marine conditions due to climate change, sea level rise, and the changing nature of the shoreline (natural or engineered). In areas of restricted fetch, wave hindcasting is relatively simple (Davidson-Arnott and Pollard, 1980), but on exposed ocean coasts the wave climate will be influenced by storm events taking place a long distance away (Adams *et al.*, 2008). This is another reason to use long records to determine wave climates so that a representative sample of all events is included.

Wave climate data are used for a variety of purposes, most commonly as input into modelling coastal erosion, sediment transport calculations, and the design of coastal structures (Resio *et al.*, 2002). They also provide information for general beach conditions that may be critical to the management of recreational activities and for vacation planning (e.g., surfing, boating). Often it is necessary to develop the deep water wave climate first, and then transform the waves across the nearshore to produce the equivalent wave climate close to the beach. Detailed information on wave shoaling transformations is provided in Chapter 5.

Further Reading

Demirbilek, Z. and Vincent, C.L. 2002. *Water Wave Mechanics*. Chapter II-1, EM 1110-2-1100 (Part II), US Army Corps of Engineers, II-1-1-121.

This is the most recent edition of the *Shore Protection Manual* produced by the US Army Corps of Engineers. It is available on the Web as a PDF file and comes with a number of worked examples: www.publications.usace.army.mil/USACE-Publications/Engineer-Manuals/.

Holthuijsen, Leo H. 2007. *Waves in Oceanic and Coastal Waters*. Cambridge: Cambridge University Press, 387 pp.

This is a very readable description of wave generation, wave properties and wave modelling by one of the researchers involved in the development of the SWAN model.

References

Adams, P.N., Inman, D.L. and Graham, N.E. 2008. Southern California deep-water wave climate: characterisation and application to coastal processes. *Journal of Coastal Research*, **24**, 1022–35.

Bishop, C.T. and Donelan, M.A. 1989. Wave prediction models. In Lakhan, V.C. and Trenhaile, A.S. (eds.), *Applications in Coastal Modeling*. Amsterdam: Elsevier, pp. 75–105.

Booij, N., Ris, R.C. and Holthuijsen, I.H. 1999. A third-generation wave model for coastal regions, Part 1, model description and validation. *Journal of Geophysical Research*, **104**, 7649–66.

Bretschneider, C.L. 1952. Revised wave forecasting relationships. *Proceedings of the Second Coastal Engineering Conference*, New York: American Society of Civil Engineers, pp. 1–5.

Bretschneider, C.L. 1958. Revisions in wave forecasting: deep and shallow water. *Proceedings of the Sixth Coastal Engineering Conference*, New York: American Society of Civil Engineers, pp. 30–67.

Chatfield, C. 2004. *The Analysis of Time Series: An Introduction*. London: Chapman and Hall, 6th edition, 333 pp.

Cooley, R.J.W. and Tukey, J.W. 1965. An algorithm for the machine calculation of complex Fourier Series. *Mathematics and Computing*, **19**(90), 297–301.

Davidson-Arnott, R.G.D. and Pollard, W.H. 1980. Wave climate and potential longshore sediment transport patterns, Nottawasaga Bay, Ontario. *Journal of Great Lakes Research*, **6**, 54–67.

Donelan, M.A. 1980. Similarity theory applied to the forecasting of wave heights periods and directions. *Proceedings Canadian Coastal Conference*, Ottawa: National Research Council of Canada, pp. 47–60.

Donelan, M.A., Hamilton, J. and Hui, W.H. 1985. Directional spectra of wind-generated waves. *Philosophical Transactions Royal Society of London*, A, **315**, 509–52.

Donelan, M.A., Drennan, W.M. and Magnusson, A.K. 1996. Nonstationary analysis of the directional properties of propagating waves. *Journal of Physical Oceanography*, **26**, 1901–14.

Hasselman, K., Barnett, T.P., Bouws, E. *et al.* 1973. Measurement of wind-wave growth and swell decay during the Joint North Sea Wave Project (JONSWAP). *German Hydrographic Series* A8 (Suppl.), **12**, 1–95.

Hasselmann, K., Ross, D.B., Muller, P. and Sell, W. 1976. A parametrical wave prediction model. *Journal of Physical Oceanography*, **6**, 201–28.

Hasselmann, S., Hasselmann, K., Janssen, P.A.E.M. *et al.* 1988. The WAM model: a third generation ocean wave prediction model. *Journal of Physical Oceanography*, **18**, 1775–810.

Hegge, B. J. and Masselink, G. 1996. Spectral analysis of geomorphic time series: auto-spectrum. *Earth Surface Processes and Landforms*, **21**, 1021–40.

Holman, R.A. and Guza, R.T. 1984. Measuring run-up on a natural beach. *Coastal Engineering*, **8**, 129–40.

Holthuijsen, Leo H. 2007. *Waves in Oceanic and Coastal Waters*. Cambridge: Cambridge University Press, 387 pp.

Huntley, D.A. 1980. Edge waves in a crescentic bar system. In McCann, S.B. (ed.), *The Coastline of Canada, Littoral Processes and Shore Morphology*. Proceedings of a Conference held in Halifax, 1–3 May, 1978, Geological Survey of Canada, paper 80-10, pp. 111–21.

IAHR Working Group on Wave Generation and Analysis, 1989. List of sea state parameters. *Journal of Waterway, Port, Coastal and Ocean Engineering*, ASCE, **115**, 793–808.

Jenkins, G.M. and Watts, D.G. 1968. *Spectral Analysis and its Applications*. San Francisco: Holden-Day, 525 pp.

Komen, G.J., Cavaleri, L., Donelan, M., Hasselmann, K. Hasselmann, S. and Janssen, P.A.E. 1994. *Dynamics and Modelling of Ocean Waves*. Cambridge: Cambridge University Press, 532 pp.

Lin, W., Sanford, L.P. and Suttles, S.E. 2002. Wave measurement and modelling in Chesapeake Bay. *Continental Shelf Research*, **22**, 2673–86.

Massel, S.R. 2001. Wavelet analysis for processing of ocean surface wave records. *Ocean Engineering*, **28**, 957–87.

Miles, J.W. 1957. On the generation of surface waves by shear flows. *Journal of Fluid Mechanics*, **3**, 185–204.

Munk, W.H. 1950. Origin and generation of waves. *Proceedings of the First Coastal Engineering Conference*. New York: American Society of Civil Engineers, pp. 1–4.

Phillips, O.M. 1957. On the generation of waves by turbulent wind. *Journal of Fluid Mechanics*, **2**, 417–45.

Phillips, O.M. 1966. *The Dynamics of the Upper Ocean*. Cambridge: Cambridge University Press, 336 pp.

Pierson, W.J., Neumann, G. and James, R.W. 1955. *Practical Methods for Observing and Forecasting Ocean Waves by Means of Wave Spectra and Statistics*. US Navy Hydrographic Office Publishing, 603, 284 pp.

Resio, D.T. 1981. The estimation wind-wave generation in a discrete spectral model. *Journal of Physical Oceanography*, **11**, 510–25.

Resio, D.T., Bratos, S.M. and Thompson, E.F. 2002. Meteorology and wave climate. Chapter 2 in EM 1110-2-1100 Part 2, *Coastal Engineering Manual*, US Army Corps of Engineers, 69 pp.

Schwab, D.J., Bennett, J.R., Liu, P.C. and Donelan, M.A. 1984. Application of a simple numerical wave prediction model to Lake Erie. *Journal of Geophysical Research*, **89**, 3586–92.

Sverdrup, H.U. and Munk, W.H. 1947. *Wind, Sea and Swell: Theory of Relations for Forecasting*. US Navy Hydrographic Office, publication **601**, 44 pp.

US Army Corps of Engineers, 1984. *Shore Pr/otection Manual*, 2 volumes. Vicksburg, MS: Dept. of the Army, Waterways Experiment Station, Corps of Engineers, Coastal Engineering Research Centre.

Walker, I.J., Davidson-Arnott, R.G.D., Bauer, B.O. *et al.* 2017. Scale-dependent perspectives on the geomorphology and evolution of beach-dune systems. *Earth-Science Reviews*, **171**, 220–53, DOI: doi.org/10.1016/j.earscirev.2017.04.011

Young, I.R. and Verhagen, L.A., 1996. The growth of fetch limited waves in water of finite depth. Part 1: Total energy and peak frequency. *Coastal Engineering*, **29**, 47–78.

5

Wave Dynamics

5.1 Synopsis

Waves generated by storms over the open ocean move across vast distances until their energy is finally expended on the coast. Although the processes are complex, the general principles are fairly well understood, and this chapter provides a basic introduction to the fundamentals. Wave theories are mathematical descriptions of the form and dynamics of surface waves, extending from deep water to the breaker line and across the surf zone. It is important to appreciate that no single theory is universally applicable or completely reliable.

When waves move from deep water toward the coast, water depth decreases. This has significant ramifications for the dynamics of waves propagating across the ocean surface. Every individual wave in sequence slows down as it enters intermediate water depth, and therefore the wavelengths become shorter, the wave groups begin to bunch up, and the wave height increases. These collective processes are referred to as wave shoaling. If waves approach the shoreline at an oblique angle, different parts of the wave experience different depths of water, which leads to wave refraction or a bending of the wave crests. In this way, wave crests tend to align with the shape of the depth (bathymetric) contours. Wave refraction leads to a concentration of wave energy on headlands, and a divergence of energy in embayments.

As waves travel into shallow water, they often break, which involves the front face of the wave becoming very steep. The details of wave breaking are poorly understood, and there is, as yet, no comprehensive theory that describes all aspects of the breaking process. However, empirical observations over many decades have demonstrated that the type (or form) of breaking wave depends, in part, on slope steepness, leading to collapsing, surging, plunging and spilling breaker types.

The overall change in wave dynamics due to shoaling and breaking involves time-averaged spatial gradients in cross-shore momentum and energy that ultimately lead to an increase in the water level against the beach, known as wave set-up. In turn, this set-up yields a pressure gradient that leads to water circulation in the nearshore zone by way of undertow currents or rip currents. These currents are an essential outcome of the need to conserve momentum and mass (i.e., water moved toward the coast by wave action must ultimately move back out to sea in the form of currents). For the coastal geomorphologist, it is precisely this interaction of waves and currents that explains how sediments can be redistributed within the nearshore zone to create such interesting features as sand bars, longshore troughs, and cuspate forms on the beach.

5.2 Wave Theory

A wave theory is a mathematical simplification of reality. Its purpose is to describe various aspects of water waves such as their shape, their movement across the sea surface, the transport of wave energy and the orbital motion of water particles (including velocity and acceleration) beneath the wave. Coastal scientists may also be interested to know the instantaneous pressure changes induced by the wave at some critical water depth, perhaps because of potential impacts on a submerged structure but more pragmatically because pressure sensors are often used to measure wave dynamics (as discussed in Chapter 4). Therefore, there is a critical need to know how the pressure field relates to the orbital motion as well as to the surface expression of the wave. Wave theories describe all this.

The wave theories presented in this chapter apply to fairly simple, regular waves, and thus, are not strictly applicable to the complex motion found in irregular (partly arisen) seas where short waves are superimposed on longer waves. The more complex the wave form, the more complex the equations necessary to describe it. However, sometimes the higher degree of complexity is not necessary or justified by the problem, so the focus here is on the simplest wave theories available. Figure 5.1 presents graphical images of the most widely used wave theories in coastal geomorphology as well as their domain of application. Detailed descriptions of each theory can be found in Mei (1983), Komar (1998) or Demerbilek and Vincent (2002).

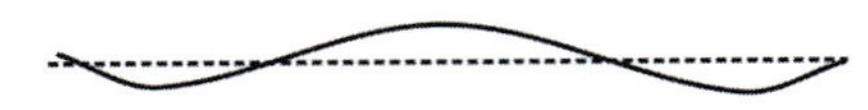

Figure 5.1 Four classes of wave theory with typical surface wave form and application domain.

5.2.1 Airy Wave Theory

Airy (1845) developed a simple theory for waves travelling over an ideal ocean or lake surface. The class of waves described in this theory is referred to as a surface gravity wave because the wave form appears on the water surface rather than at the submerged interface of two water layers with different densities, which is referred to as an internal wave. The term 'gravity' wave indicates that the dominant restoring force is Earth's gravitational field, which acts to suppress the indefinite growth of waves on the ocean surface due to wind energy inputs. The theory is referred to as 'linear' because only linear, first-order terms (i.e., H^x where $x = 1$) are retained in the equations that describe the wave dynamics, in contrast to more complex treatments that are non-linear or higher-order theories (i.e., in addition to H, H^x terms are used, where $x > 1$).

Airy wave theory makes many simplifying assumptions in setting up the problem, as do most other wave theories. It is assumed that the bottom is horizontal, uniform and non-porous so that it can be treated by a simple boundary condition such as no net movement of water into or out of the bed. It is further assumed that the fluid is 'ideal' in the sense that it is homogeneous, incompressible, inviscid (without viscosity) and irrotational (lacking vorticity). This simplifies the mathematics immensely. Airy theory also assumes that the surface form of the wave is sinusoidal (two dimensional) and that the wave heights are small relative to the wave length. This last assumption is necessary to constrain the problem to situations in which the water surface slope is small, and therefore Airy wave theory is also known as small amplitude wave theory. One might wonder whether all these simplifying assumptions restrict the usefulness of the solutions to situations that are too simple and of little interest to coastal geomorphologists. Fortunately, the utility of Airy wave theory has been demonstrated to be broadly applicable, although not universally so. The range of applicability will be discussed after the theory is introduced in more detail.

5.2.2 Wave Potential Function

The general approach taken to derive a particular wave theory is to list the range of assumptions that are deemed applicable to the problem, and then to set out a series of governing equations, usually in the form of coupled, partial differential equations with explicit boundary conditions that adequately formalise the problem. The solution to the set of governing equations can sometimes be derived explicitly but often require numerical codes to solve. In the case of Airy wave theory, there is an explicit solution to the main governing equation (the Laplace equation) in the form of the wave potential function,

$$\Phi = \frac{gH}{\sigma 2} \frac{\cosh(k(z+h))}{\cosh(kh)} \cos(kx - \sigma t), \qquad (5.1)$$

where Φ is the wave potential function, g is gravitational acceleration, x is horizontal distance, t is time, H is wave height, h is total water depth, z is vertical distance below the mean water line, k is the wave number ($2\pi/L$), σ is wave radian (or angular) frequency ($2\pi/T$), L is wavelength, T is the wave period and cosh() is the hyperbolic cosine function, which applies to all terms in closed parentheses. Total water depth, h, is considered a simple scalar quantity (as is wavelength), whereas z is a negative distance below the mean water line (taken as $z = 0$). Thus, the bottom is defined as $z = -h$.

The wave potential function is a complete description of the surface gravity wave arising from linear theory, including its surface expression and the horizontal and vertical velocities of water particles beneath the surface wave. Note that there are four components on the right-hand side (RHS) of Equation 5.1. The first component, g/σ, defines the scale of the wave (i.e., its relative size) through the radian frequency or wave period relative to the gravitational field of the Earth. An alternative version of the wave potential function uses the wave number (i.e., the wavelength) rather than the radian frequency, but there is no difference in terms of wave dynamics because wavelength and wave period are intricately related. The second component, $H/2$, defines the amplitude of the wave, which is not directly related to wavelength or wave period, and therefore is an independent parameter. The third component (involving the cosh functions) describes the depth attenuation of the wave dynamics or the rate at which the influence of the surface wave decreases with increasing depth in the water column. And the fourth component, $\cos(kx - \sigma t)$, is a general descriptor of the sinusoidal nature of the wave at the mean water surface ($z = 0$), which has both spatial and temporal dimensions.

5.2.3 Wave Profile at the Surface

The wave profile of an Airy wave is sinusoidal. It is instructive to consider the dynamics of the water surface as a wave cycle passes a single point. The elevation of the water surface, $\eta(x, t)$, will move upward as the wave crest approaches and then downward as the following wave trough approaches. A sensor such as a wave staff will register only this up-and-down motion (i.e., $\eta(t)$). Another sensor located close by will register a similar up-and-down motion, but not in perfect phase with the first sensor. The difference in phase will depend on the spacing between the sensors as well as the wavelength of the wave. A general expression describing the fluctuations in surface water elevation can be derived from the wave potential function by differentiating Equation 5.1 with respect to time, while constraining our interest to only the surface (i.e., $z = 0$), to yield:

$$\eta(x,t) = \frac{\partial \Phi}{\partial t} = \frac{H}{2} \cos\left(\frac{2\pi x}{L} - \frac{2\pi t}{T}\right). \qquad (5.2)$$

The continuous variables x and t define the spatial or temporal coordinates relative to the full wave form (specified by L and T). For example, if t is held constant, and x is varied incrementally from

0 to 2π, an entire cycle of the sinusoid is retrieved. This is similar to freezing the sea surface at one instant in time and tracing a line from one wave crest to the next. Some form of this relation occurs in all the equations used to describe instantaneous water particle motion and serves to locate the point of interest on the wave profile. Note that $\eta(x, t)$ is a variable quantity that depends on time and space, unlike H (wave height), which is a fixed quantity for any specific wave that prescribes the maximum range that η can assume. Wave height, H, is always a positive quantity, defining the maximum vertical distance between the crest and trough of a specific wave, whereas η varies between $-H/2$ and $+H/2$ in the case of a perfect sinusoid.

5.2.4 Wave Dispersion

One of the interesting characteristics of surface gravity waves is that they are dispersive, which is an attribute that arises from the dynamic boundary condition of the free water surface. The direct relationship between wave period and wave length is expressed in the wave dispersion equation,

$$\sigma^2 = gk\tanh(kh), \tag{5.3}$$

where tanh() is the hyperbolic tangent function (applied to terms in closed parentheses). Substituting for $k = 2\pi/L$ and $\sigma = 2\pi/T$, yields an alternative form of the wave dispersion relationship that makes explicit the dependency of wave length, L, on wave period, T:

$$L = \frac{gT^2}{2\pi}\tanh\left(\frac{2\pi h}{L}\right). \tag{5.4}$$

Equation 5.4 cannot be solved explicitly because L appears on both sides of the equals sign, however, the expression can be easily solved numerically or by iteration or with the use of wave tables (Wiegel, 1954).

Recalling (Equation 4.1) that wave celerity, C, is defined as the ratio of wavelength to wave period (L/T), yields the following:

$$C = \frac{L}{T} = \frac{gT}{2\pi}\tanh\left(\frac{2\pi h}{L}\right) \tag{5.5}$$

or alternatively,

$$C = \sqrt{\frac{gL}{2\pi}\tanh\left(\frac{2\pi h}{L}\right)}. \tag{5.6}$$

Equations 5.5 and 5.6 show that the speed at which a surface gravity wave propagates across the water surface is proportional to the wave period or the wavelength, respectively. Specifically, long waves (which also have long periods) travel faster than shorter waves. As a consequence, a fully arisen sea in the open ocean with a complete spectrum of wave frequencies will tend to sort itself out as the waves travel away from the generating area toward the coast. The longest period waves, which travel the fastest, will reach the coast first. Along the coast of California, for example, these long-period swell waves originating from storms in the Pacific Ocean typically provide the best opportunities for surfing, well in advance of the arrival of the migrating storm and the complex wave field associated with it.

5.2.5 Approximations for Deep and Shallow Water

The wave dispersion relation, in the form of Equation 5.3 or 5.4 (or the alternative forms 5.5 and 5.6), is a very powerful tool in wave analysis, but it is not easy to solve explicitly. Fortunately, there are simplified expressions that provide approximate solutions when considering 'deep' water or 'shallow' water. The quotation marks indicate that the reference is to relative depth, which means that the ratio of water depth to wavelength is large or small, respectively.

Consider Equation 5.4, and note that the ratio of water depth to wavelength ($kh = 2\pi h/L$) appears in the tanh() term. Recall that the function tanh (X) is equal to zero when $X = 0$; it increases linearly for values up to about $X = 1$ such that $\tanh(X) \sim X$; and for large values of X greater than 1.5, $\tanh(X) \rightarrow 1$ asymptotically. So when h/L is large (i.e., deep water), kh is large, and therefore $\tanh(kh)$ approaches 1. Equation 5.4 reduces to,

$$L_o = \frac{gT^2}{2\pi}, \tag{5.7}$$

where the subscript 'o' on wavelength (L_o) indicates that this relation applies specifically and exclusively to the deep water approximation. When using metric values (g = 9.81 m s^{-1}), Equation 5.7 reduces to $L_o = 1.56\ T^2$, which shows how the deep-water wavelength of a surface gravity wave depends only on the square of the wave period. This is the simplest and most useful outcome from the wave dispersion relation because it provides a very easy way to assess the maximum possible wavelength of a wave of known period. Wave period is simple to measure using only a watch while standing on the shoreline, and wave period does not change right up until wave breaking occurs.

Rearranging Equation 5.7 slightly also provides an expression for wave celerity in deep water (C_o):

$$C_o = \frac{gT^2}{2\pi}. \tag{5.8}$$

What exactly does 'deep' water mean in real terms? Guidance comes from the condition that Equation 5.7 will provide reasonably accurate solutions to the wave dispersion relation (Equation 5.4) when tanh(kh) approaches 1. By convention, coastal scientists have adopted a definition of deep water based on $h/L_o \geq$ ½ (i.e., $h \geq L_o/2$ or $kh \geq \pi$), which means tanh(kh) ≥ 0.996, implying that Equations 5.7 and 5.8 reproduce the results of the general wave dispersion relation (Equations 5.3 to 5.6) to within about 0.5 per cent for the deep-water case.

Shallow water conditions are defined, by convention, as $h/L \leq 0.04$ (i.e., $h \leq L_o/25$ or $kh \leq \pi/12.5$). Some coastal scientists prefer to use $h/L \leq 0.05$ or $kh \leq \pi/10$, which is a less restrictive condition that leads to slightly greater error. Using similar reasoning as with the deep-water case, it can be shown that tanh($2\pi h/L$) ~ $2\pi h/L$ when the value of $2\pi h/L \leq 0.3$. Substituting $2\pi h/L$ for tanh($2\pi h/L$) in Equation 5.4 produces:

$$L_s = T\sqrt{gh}. \tag{5.9}$$

Where the subscript 's' on wavelength (L_s) indicates that this relation applies specifically and exclusively to the shallow water approximation. Dividing both sides by the wave period, T, provides an expression for wave celerity in shallow water:

$$C_s = \sqrt{gh}. \tag{5.10}$$

Equations 5.9 and 5.10 demonstrate that the dynamics of shallow water waves are strongly influenced by water depth, which is very different from the deep-water situation where the sea bed plays no role in governing the wave dynamics. A summary of the general expressions is provided in Table 5.1.

By taking the general form of the expression for the wave dispersion relation (Equation 5.4 or 5.6) and dividing through by the deep water wave expression (Equation 5.7 or 5.8), it can also be shown that:

$$\frac{C}{C_o} = \frac{L}{L_o} = \tanh\left(\frac{2\pi h}{L}\right). \tag{5.11}$$

This expression is useful for examining wave shoaling effects as the wave transitions from deep to intermediate water depths. For example, from Equation 5.11, it is apparent that $L = L_o$ tanh (kh), which shows that the wavelength of a shoaling wave will always be equal to or shorter than L_o depending on water depth.

5.2.6 Orbital Motion Beneath Waves

To this point, the focus has been mainly on the water surface expression of Airy waves. However, the wave form is intricately related to the motion of water particles beneath the surface, which circumscribe closed circular or elliptical orbits that involve to-and-fro as well as up-and-down motion (see Figure 4.1). For sinusoidal waves in deep water, the orbits are exactly circular and the diameter of the circle at the surface is equal to the wave height. The wave potential function (Equation 5.1) provides exact solutions for the orbital diameters and velocity fields at various depths. Differentiating Equation 5.1 with respect to x (horizontal distance) and z (vertical distance) yields:

$$u = -\frac{\partial\Phi}{\partial x} = \sigma\frac{H}{2}\frac{\cosh\left(k(z+h)\right)}{\sinh\left(kh\right)}\cos\left(kx-\sigma t\right) = \frac{gk}{\sigma}\frac{H}{2}\frac{\cosh\left(k(z+h)\right)}{\cosh\left(kh\right)}\cos\left(kx-\sigma t\right) \tag{5.12}$$

Table 5.1 Summary table of linear wave equations and the net drift velocity U_D from Stokes' second order wave theory

Parameter	General Expression	Simplification
Surface elevation, η (x, t)	$\eta = a\cos(kx - \sigma t)$	
Phase velocity, C	$C = \frac{gT}{2\pi}\tanh\frac{2\pi h}{L}$	$C_o = \frac{gT}{2\pi}$ $C_s = \sqrt{gh}$
Wavelength, L	$L = \frac{gT^2}{2\pi}\tanh\frac{2\pi h}{L}$	$L_o = \frac{gT^2}{2\pi}$ $L_s = T\sqrt{gh}$
Horizontal orbital velocity, u	$u = \frac{\pi H}{T}\frac{\cosh(k(z+h))}{\sinh(kh)}\cos(kx - \sigma t)$	at the bed ($z = -h$) $u_{max} = \frac{\pi H}{T\sinh(kh)}$
Horizontal orbital diameter, A	$A = H\frac{\cosh(k(z+h))}{\sinh(kh)}$	at the bed ($z = -h$) $A_s = \frac{HT}{2\pi\sqrt{\frac{g}{h}}}$
Net drift velocity (Stokes) U_D	$U_{D(z)} = \left(\frac{\pi H}{L}\right)^2\frac{C}{2}\frac{\cosh\left(\frac{4\pi(z+h)}{L}\right)}{\sinh^2\left(\frac{2\pi h}{L}\right)}$	at the bed ($z = -h$) $U_D = \left(\frac{\pi H}{L}\right)^2\frac{C}{2\sinh^2\left(\frac{2\pi h}{L}\right)}$
Wave height, H	$H = H_o\left(\frac{\cosh^2 kh}{0.5(\sinh(2kh) + kh)}\right)^{0.5}$	

Note: $k = 2\pi/L$ and $\sigma = 2\pi/T$, $a = H/2$, h = water depth.

$$w = -\frac{\partial\Phi}{\partial z} = \sigma\frac{H}{2}\frac{\sinh(k(z+h))}{\sinh(kh)}\sin(kx - \sigma t)$$
$$= \frac{gk}{\sigma}\frac{H}{2}\frac{\sinh(k(z+h))}{\cosh(kh)}\sin(kx - \sigma t). \quad (5.13)$$

Two versions of the velocity equations for each of u and w are presented because the literature uses both, and there is often ambiguity about which is the correct form. Both versions are correct, and the reader can derive one from the other by a simple substitution from the wave dispersion relation for σ or k, and by realising that tanh(X) = sinh(X)/cosh(X). Appropriate care is required when working with these equations so that the correct hyperbolic functions are used. Although Equations 5.12 and 5.13 appear imposing at first, the interpretation of their meaning follows exactly the same logic as with the wave potential function described earlier. Specifically, the first part provides scaling information on the wave (i.e., wave period or wavelength); the second part is the wave amplitude; the third part with the hyperbolic functions provides information on how the wave dynamics change with depth below the surface, and the fourth part is simply the sinusoidal motion of the overall wave form. Equations for particle accelerations can be derived by differentiating Equations 5.12 and 5.13 with respect to time.

Often Equations 5.12 and 5.13 are given in a form that takes advantage of the fact that the horizontal and vertical orbital diameters, A and B respectively, at any depth can be parameterised by:

$$A = H\frac{\cosh(k(z+h))}{\sinh(kh)} \tag{5.14}$$

and

$$B = H\frac{\sinh(k(z+h))}{\sinh(kh)}. \tag{5.15}$$

In deep water, the orbits are circular everywhere beneath the wave. Thus, $A_o = B_o = H\exp(kz)$. The orbital paths are at their maximum at the water surface ($z = 0$; $A_o = B_o = H$), and the orbits decrease in diameter with depth according to the hyperbolic functions. The decrease is exponential. At a depth equal to $L_o/2$, the values A_o and B_o have been reduced by $\exp(-\pi)$, and therefore the orbital excursions are about 4 per cent of their surface values. This fact is well known to experienced scuba divers who rarely float on the surface during wavy conditions, preferring the calm conditions at depth. It is also of critical importance to sediment transport in the nearshore zone because it defines the depth to which a given wave is likely to disturb the bottom. In shallow water, the presence of the bed strongly influences the wave dynamics. The orbital paths are constrained vertically because there is less water depth for the wave to move within, so the orbits become elliptical with $A > B$. In very shallow water, there is a simple to-and-from motion with virtually no vertical movement ($A \gg B$).

In geomorphological studies, particularly in relation to bedforms, the horizontal orbital diameter at the bed is often an important parameter. Recognising that on the bed, $z = -h$, and substituting in Equation 5.14, yields:

$$A_{(bed)} = \frac{H}{\sinh(kh)}. \tag{5.16}$$

In shallow water the orbital diameter can be approximated by:

$$A_{s(bed)} = \frac{HT}{2\pi}\sqrt{\frac{g}{h}}. \tag{5.17}$$

Another useful wave parameter related to bedforms and sediment transport is the maximum horizontal velocity at the bed, which occurs under the wave crest (in the positive direction) and the wave trough (in the negative direction). Equation 5.12 reduces to:

$$u_{\max(bed)} = \sigma\frac{H}{2}\frac{1}{\sinh(kh)}. \tag{5.18}$$

By knowing wave height and wave period, this expression can be used to provide an estimate of the maximum orbital velocity on the bed at any water depth, thereby giving a rough indication of whether sediment particles on the bed might be entrained and transported.

5.2.7 Dynamic Pressure Beneath Waves

The pressure field beneath waves is generally of passing interest to coastal geomorphologists because pressure does not enter into many equations dealing with sediment transport. However, it is very common to measure wave parameters using pressure sensors, and in order to interpret the measurements it is critical to understand how the pressure signal varies with depth.

If the sea surface is flat and undisturbed by waves, a pressure sensor deployed at some arbitrary depth, z, beneath the water surface (not necessarily at the sea bed) will measure the hydrostatic pressure due to the weight of the column of water above the sensor:

$$p = -\rho g z, \tag{5.19}$$

where p is the hydrostatic pressure, ρ is the fluid density and g is gravitational acceleration. The negative sign is necessary in this expression because z is considered a negative value since the water surface is $z = 0$ and the bed is located at $z = -h$. Thus, the deeper the position of the sensor in the water column, the greater will be the hydrostatic pressure (a positive value). Note that Equation 5.19 refers only to the pressure generated by the column of water above the sensor and assumes that variations in atmospheric pressure have been accounted for. This can be done by either using a pressure transducer that is vented to the surface (i.e., total pressure – atmospheric pressure) or by deploying a second pressure transducer in the air above the water surface, and then subtracting the atmospheric pressure fluctuations from the total pressure measured by the first pressure transducer in the water column, as described in Chapter 4. Typically the atmospheric corrections are relatively small over the short term, but for long-term deployments (e.g., weeks to months) the pressure changes associated with storms and hurricanes can be quite important.

The water surface is generally not flat but is usually disturbed by the presence of surface gravity waves. In this situation, account must be taken of the effect of the waves as the water surface elevation, η, fluctuates above and below the mean water level (MWL), which is defined approximately by $z = 0$. When a wave crest passes over the pressure sensor, η is positive and there is an additional pressure contribution above the hydrostatic pressure due to the water column at MWL. Conversely, when a wave trough passes over the pressure sensor, η is negative and there is a reduction in total pressure relative to the hydrostatic pressure at the MWL. A component of these pressure differences is due to the incremental addition or subtraction of 'static' water on top of the hydrostatic water column, but there is also a dynamic pressure component due to the fact that the water particles are being accelerated through their orbital paths. To first order, the pressure field beneath waves at arbitrary depth, z, is given by,

$$p_z = -\rho g z + \rho g \frac{H}{2} \frac{\cosh\left(k(z+h)\right)}{\cosh\left(kh\right)} \cos\left(kx - \sigma t\right), \tag{5.20}$$

where p_z is the total wave pressure measured by a pressure sensor deployed at z beneath the mean water line (presuming that atmospheric pressure effects have been compensated for). There are two components to the total wave pressure appearing on the right-hand side of Equation 5.20. The first component is the hydrostatic pressure relative to the MWL as expressed by Equation 5.19. The second component is the wave-derived pressure fluctuation, to be interpreted in the same way as the wave potential function and the orbital velocity components.

Equation 5.20 is often represented using a pressure response factor (K_p):

$$K_p = \frac{\cosh\left(k(z+h)\right)}{\cosh\left(kh\right)} \tag{5.21}$$

and by inserting the expression for η from Equation 5.2, to yield,

$$p_z = -\rho g z + \rho g \eta K_p = \rho g \left(\eta K_p - z\right). \tag{5.22}$$

Rearranging Equation 5.22 to isolate η provides a useful expression for interpreting pressure transducer signals to derive the time series of surface elevation fluctuations,

$$\eta = \frac{p_z + \rho g z}{\rho g K_p}. \tag{5.23}$$

Note that p_z is the total wave pressure measured by the pressure transducer (absent atmospheric pressure effects) and that z is a negative quantity, which indicates that the hydrostatic pressure relative to the MWL is subtracted to yield only the dynamic pressure due to the wave. Before this expression can be solved, it is necessary to calculate wavelength, L, from the dispersion relation (on the basis of wave period and total water depth) because L appears in the pressure response factor, K_p. For a simple wave field with only one wave period, the process is straightforward, but for a spectrum of waves, the calculations need to be done for every frequency component because every different wave will be depth attenuated differently. The shortest waves may not even register on the pressure sensor. In addition, if there is a background current, the frequency spectrum will be shifted and this needs to be taken into account by using $(C-U_c)^2 = g/k(\tanh(kh))$ where U_c is the mean current speed in the direction of wave propagation (Dean and Dalrymple, 1984, pp. 68–9).

5.2.8 Wave Energy

The energy in water waves has two sources: (1) kinetic energy, E_k, associated with the orbital motion of the water particles; and (2) potential energy, E_p, resulting from displacement of the water surface away from the mean water level. According to Airy wave theory, if potential energy is determined relative to the MWL and all waves are propagating in the same direction, the two types of energy are equal.

The wave energy density, $\bar{E}$, or average energy per unit area of sea surface (Joules per square metre) is given by:

$$\bar{E} = E_k + E_p = \frac{\rho g H^2}{16} + \frac{\rho g H^2}{16} = \frac{\rho g H^2}{8}. \quad (5.24)$$

The total energy per unit crest width, E_T, in a wave of length, L, is given by:

$$E_T = \bar{E}L = \frac{\rho g H^2 L}{8}. \quad (5.25)$$

The total energy in a wave field is transported at a rate proportional to the group velocity, C_g, which is different from the wave celerity, C, as defined in Equation 5.5 or 5.6. Wave celerity or phase velocity defines the speed of propagation of the wave form across the water surface (i.e., the travel speed of the disturbance), whereas C_g refers to the mean rate at which energy is transferred or carried by the waves in the direction of wave propagation. If a group of say five or six waves is generated in a long wave tank, as the waves travel down the tank, the first wave decreases in height until it disappears because it expends all its energy in deforming the undisturbed water surface in front of it. The second wave advances and becomes the leading wave, and so on (Figure 5.2). At the same time, a new wave develops behind the last wave, thus maintaining the same number of waves in the group. The process keeps repeating, and the effect is that the group of waves travels down the tank at a speed equal to one-half the speed of the individual waves (assuming they are in deep water). The individual waves move through the group, while the group itself maintains the energy, which is conserved to first order.

Understanding how wave energy is created and transferred is important for forecasting wave propagation and predicting the time it takes for storm waves to reach the coastline where the energy is finally expended. Energy is defined as the capacity to perform work (i.e., work rate), and in the nearshore zone wave energy is essential for driving sediment transport, rock abrasion and the generation of longshore currents. More specifically, it is the rate at which energy is transported and transformed into work that is critical, and this is known as wave power, P_w, given by,

$$P_w = \bar{E}C_g = \bar{E}nC, \quad (5.26)$$

where n is a wave function that will be described below. Note that this conventional definition of wave power has SI units equal to Watts per metre of wave crest, but is more correctly interpreted as the rate (metres per second) at which the average wave energy density (Joules per square metre) is transported across the water surface by the wave group. At any one instant in the wave cycle, the wave power can be greater or less than P_w, but when averaged over an entire wave cycle, P_w is

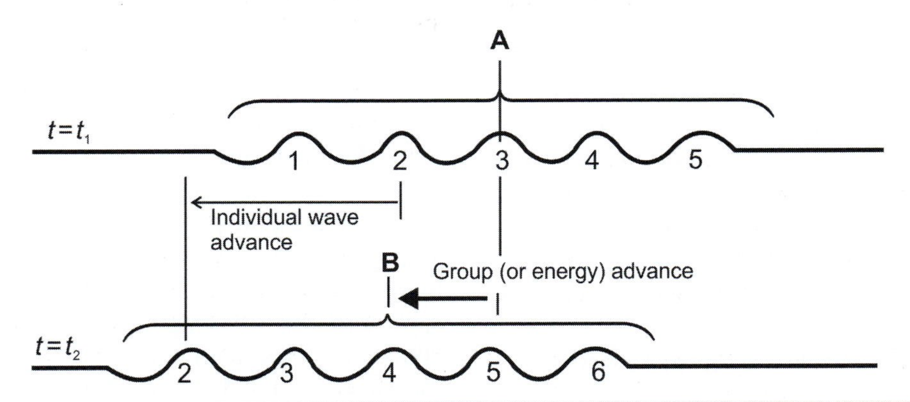

Figure 5.2 Idealised schematic showing the relationship between wave celerity (phase speed of the wave form) and the group velocity (speed at which wave energy is transported). By the time the wave group (i.e., entire packet of waves) has advanced from position A to position B (bold arrow), individual waves have advanced twice as far (fine arrow), as shown by comparing position of wave 2 at t_1 and t_2 (after Sunamura, 1992).

proportional to the mean rate of work expended or transferred per unit crest width.

As waves enter shallow water, the speed at which they propagate across the surface slows down. The rate of energy transfer must also decrease accordingly. From inspection of Equation 5.26 it is evident that $C_g = nC$, which suggests that the group velocity is proportional to the wave celerity in a way that relies on the wave function, n:

$$n = \frac{1}{2}\left(1 + \frac{2kh}{\sinh(2kh)}\right). \quad (5.27)$$

Substituting back into Equation 5.26 gives the full expression for wave power as,

$$P_w = \frac{1}{8}\rho g H^2 \frac{C}{2}\left(1 + \frac{2kh}{\sinh(2kh)}\right). \quad (5.28)$$

In deep water, $2kh$ becomes large, and in the limit, $2kh/\sinh(2kh) \rightarrow 0$, so that $n = 0.5 \times (1 + 0) = ½$. Therefore, wave energy is propagated at one-half of the speed of individual waves in deep water (i.e., $C_g = ½C$), as suggested in Figure 5.2. In shallow water $kh < 1$ and $\sinh(2kh) \rightarrow 2kh$, so that $n = 0.5 \times (1 + 1) = 1$. Therefore, the rate of energy propagation in shallow water is equal to the wave celerity. In other words, the energy is actually carried by the individual waves (bores) in shallow water.

Using these approximations yields simplified expressions for wave power in deep water, P_{wo},

$$P_{wo} = \bar{E}_o\ C_g = \bar{E}_o\ \frac{1}{2}\ C = \frac{1}{16}\rho g H^2 \frac{L_o}{T} \quad (5.29)$$

and in shallow water, P_{ws},

$$P_{ws} = \bar{E}_s\ C_g = \bar{E}_s\ C = \frac{1}{8}\rho g H^2 \sqrt{gh}, \quad (5.30)$$

where subscripts o and s refer to deep and shallow water conditions, respectively.

5.3 High-Order Wave Theories

Solution of the hydrodynamic equations for surface gravity waves can be improved over linear theory by retaining additional terms in the Taylor series expansions that provide high-order solutions to the governing equations and boundary conditions. These complex equations often produce better agreement between theoretical and observed wave behaviour than linear theory. Figure 5.3 shows a set of waves moving into shallow water that are no longer sinusoidal even though Airy wave theory might have been appropriate farther offshore. The wave crests are peaked and narrow whereas the troughs are broad and flat. Linear wave theory is unable to reproduce this skewed form of shoaling waves, so high-order theories are preferred. A very brief introduction to these high-order wave theories is presented in this section, and the interested reader is referred to any of several excellent texts dealing extensively with the topic (e.g., Mei, 1983; Dean and Dalrymple, 1984; Sleath, 1984; Demirbilek and Vincent, 2002).

Figure 5.3 Shoaling waves with narrow crests and broad, flat troughs (Moeraki, New Zealand) indicating the need for non-linear wave theories capable of describing the differences between the wave crests and troughs (i.e., wave skewness).

5.3.1 Stokes' Second-Order Wave Theory

Stokes (1847) used the same assumptions and approach as Airy (1845), including the small-amplitude approximation, but retained additional non-linear terms in his perturbation approach. Access to high-speed computers now allows expansion to fifth or higher orders (Demirbilek and Vincent, 2002), but Stokes' second-order theory is the simplest solution, retaining only the first sequence of non-linear terms in the Taylor series expansions (i.e., those in H^2).

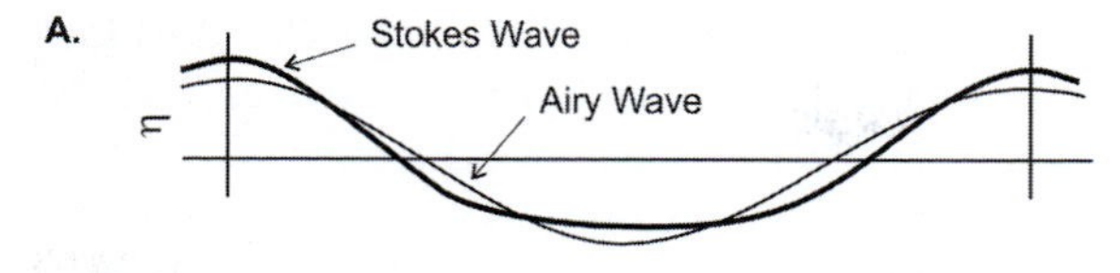

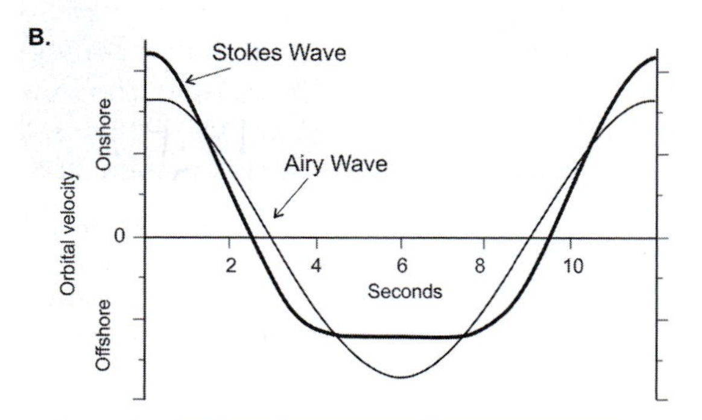

Figure 5.4 Contrasting surface elevation profiles A. and orbital velocities B. for Airy waves and Stokes waves (after Komar, 1998).

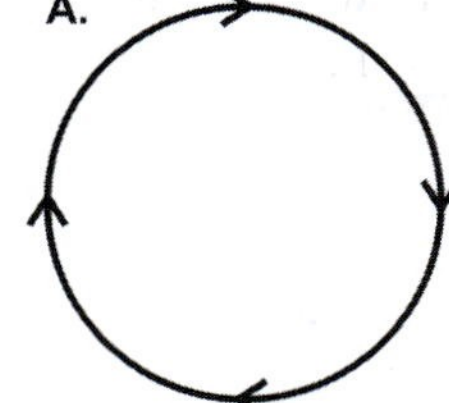

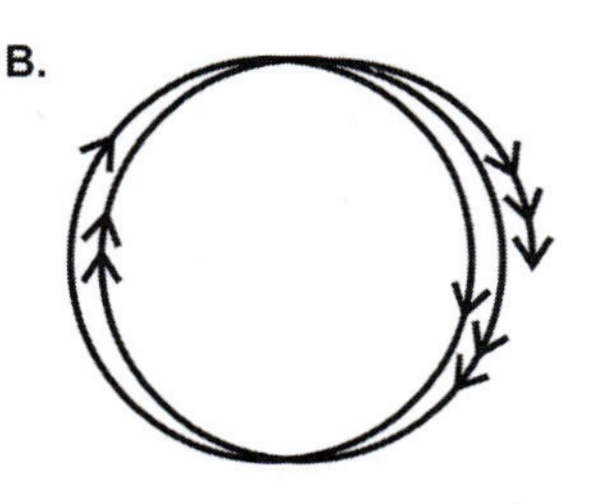

Figure 5.5 Linear wave theory predicts that wave orbits are closed, as in A., whereas non-linear theories are able to predict the incremental forward motion associated with subsequent wave cycles B., which is referred to as Stokes drift. Since the orbits are no longer closed, each subsequent wave moves the water particles forward by a slight distance, which when divided by the wave period provides the magnitude of the net drift velocity.

The water surface elevation predicted by Stokes' second-order theory is:

$$\eta = \frac{H}{2}\cos(kx - \sigma t) + \frac{\pi H^2}{8L}\frac{\cosh\left(\frac{2\pi h}{L}\right)}{\sinh^3\left(\frac{2\pi h}{L}\right)} \times \left(2 + \cosh\frac{4\pi h}{L}\right)\cos(2kx - 2\sigma t). \qquad (5.31)$$

The first part of the equation is identical to Equation 5.2 and represents the linear contribution to the surface form, while the remaining part is the second-order contribution. Third-order theory would involve additional sets of expressions including H^3 and so on. Figure 5.4A shows the difference in the surface profile of a Stokes second-order wave relative to a sinusoidal Airy wave, whereas Figure 5.4B contrasts the orbital velocity fields. It is evident that the Stokes profile more closely emulates the broad, flat trough and peaked crest typical of shoaling waves shown in Figure 5.3.

An important difference between linear and non-linear wave theories is that the orbital motion of water particles beneath the waves is no longer closed (Figure 5.5). As each wave passes a certain vertical slice of the ocean, each water particle advances a short distance in the direction of wave propagation. The existence of this net motion, called Stokes drift, is well known and it can be predicted from Stokes' second-order theory by:

$$U_{D(z)} = \left(\frac{\pi H}{L}\right)^2 \frac{C}{2}\frac{\cosh\left(\frac{4\pi(z+h)}{L}\right)}{\sinh^2\left(\frac{2\pi h}{L}\right)} \qquad (5.32)$$

where $U_{D(z)}$ is the net forward drift of water particles located at mean depth, z. In considering sediment transport and beach equilibrium, $U_{D(z)}$ is extremely important because it provides one mechanism by which sediment particles can be moved along the bed under the oscillatory action of waves. This is a challenge with linear theory because the orbital excursions are closed and the to-and-fro motion yields no net sediment transport unless there are currents superimposed on the wave field. Even though Stokes drift yields net motion that is of the order of less than a few millimetres per second, the much greater instantaneous velocities associated with the purely oscillatory portion of wave motion can easily entrain particles from the bottom. So Stokes drift has the effect of gradually shifting sediment along the bed in the direction of wave propagation when sediment transport is active.

At the bed $z = -h$ and $\cosh(4\pi (z + h)/L) = 1$, so Equation 5.32 reduces to:

$$U_{D(\text{bed})} = \left(\frac{\pi H}{L}\right)^2 \frac{C}{2\sinh^2\left(\frac{2\pi h}{L}\right)}. \quad (5.33)$$

The net Stokes drift velocity at the bed in deep water is very small in comparison to the maximum orbital velocities of the wave, but it increases in relative magnitude as the wave shoals into shallower water. Near the wave breakpoint, $U_{D(\text{bed})}$ can be substantial in relation to the wave orbital velocities at the bed.

5.3.2 Other High-Order Wave Theories

Close examination of the ocean surface indicates that waves are not perfectly sinusoidal as assumed in Airy wave theory. This is particularly true when waves move into intermediate and shallow water depth (Figure 5.3). Wave celerity decreases and conservation of mass requires wave height to increase. Thus, the small-amplitude assumption becomes less valid, and water particle accelerations increase. Wave theorists have devised a broad range of approaches to deal with these issues, all of which are computationally demanding. For example, one approach uses the trochoid rather than the sinusoid as the basis upon which to model the surface wave form (Figure 5.1). These types of 'rotational' approaches are referred to as trochoidal or 'Gerstner' theories and they create waves that have sharp, peaked crests and broad, wide troughs very much like high-order Stokes waves.

Another class of theories are referred to as 'shallow water' wave theories because they use the same basic assumptions as Airy did, with the exception that: (1) the wavelength is considered to be very long compared with water depth; and (2) the wave height is small relative to water depth. The most commonly known version is cnoidal wave theory, which is based on Jacobian elliptical functions and was developed by Kortweg and de Vries (1895). It is applicable in relative water depths of $h/L < 0.125$ and thus is best used close to the wave breakpoint. The properties of long waves close to the break point and particularly bores in the surf zone can be approximated by solitary wave theory, which is another shallow water theory. Solitary waves consist essentially of a wave crest riding above the mean water line (Figure 5.1), and theoretically they have infinite wave length and period. Solitary waves are a limiting case of cnoidal waves, and are often described as waves of translation since the water particles are displaced quite a distance in the direction of the wave advance. This translatory phenomenon can often be observed on sand bars at low tide when surf bores propagate across the bar surface and into the trough landward of it (see Chapter 8). Much of the recent modelling of waves in the surf and swash zones is now based on some form of Boussinesq model, which is the basis for the solitary wave approach (e.g., Madsen *et al.*, 1997; Bayram and Larson, 2000; Karambas and Koutitas, 2002). There are also a range of numerical approaches that include the exact solutions of Cokelet (1977) as well as the stream function and vocoidal theories proposed by Dean (1965) and Swart and Loubster (1978).

5.3.3 Range of Applicability of Wave Theories

Given the large number of wave theories available to the coastal geomorphologist, the challenge is to decide which one to implement for a given problem of interest. There are three questions to take into consideration: (1) how well does the theory apply to the situation under investigation; (2) how accurate do the predictions have to be; and (3) what is the cost (usually in terms of computational capacity) of using a more sophisticated wave theory?

Figure 5.6 shows the range of applicability for several theories. The horizontal axis is non-dimensional water depth (h/gT^2), increasing from left to right. The vertical axis is non-dimensional wave height (H/gT^2), increasing from bottom to top. Note that the variable grouping gT^2 is proportional to wavelength, L, by virtue of the dispersion relation. So the graph axes can be interpreted as relative water depth ($\sim h/L$) and relative wave height or wave steepness ($\sim H/L$) without loss of meaning. The range of applicability for specific theories is defined by the regions demarcated by solid lines. For example, Airy wave theory (linear theory) is generally applicable in the lower right-hand side of the diagram, which includes deep water and intermediate depth waves. However, given the small-amplitude assumption, linear theory does not extend very far toward the upper section of the diagram because this is the regime of relatively steep waves (i.e., greater H/L). When dealing with steep waves, many of the high-order theories are more applicable, including the Stokes solutions for deep-water waves. Intermediate water depths and shallow water conditions are best treated using cnoidal theory, stream functions or numerical solutions. Le Méhauté (1976) and Demirbilek and Vincent (2002) describe these limits in greater detail.

There are two very important points that arise from examination of Figure 5.6. The first is that there are upper limits to the theories that are related to wave-breaking conditions, which are not predicted explicitly by the theories. Physically, waves cannot exist in the upper regions of this diagram because the height of the wave relative to its length yields a steepness that is unstable. The wave will break. In deep water this is defined by a critical wave steepness of H_O/L_O = 0.14, whereas during wave shoaling in intermediate water there is a critical wave height to water depth ratio of about 0.78, based largely on empirical evidence. These upper limits are therefore imposed by the mechanics of wave breaking

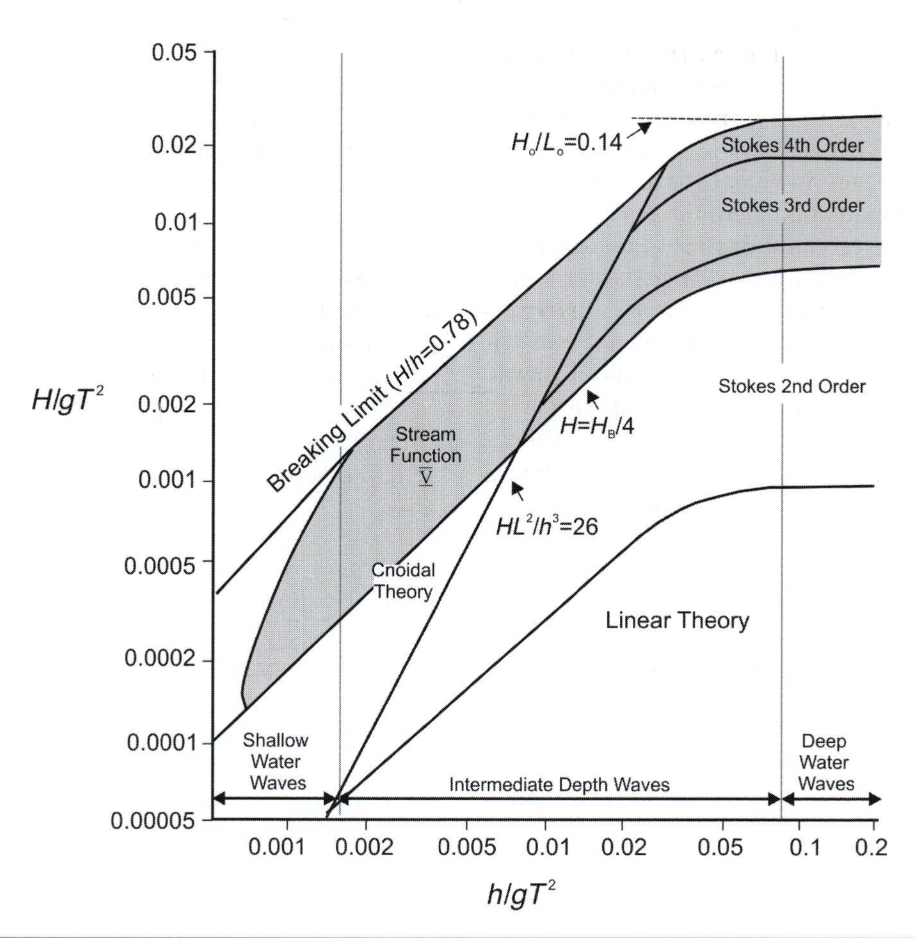

Figure 5.6 Range of applicability of different wave theories. Horizontal axis is non-dimensional water depth, and vertical axis is non-dimensional wave height. Upper limits on wave height are defined by the depth-determined breaking limit (H/h = 0.78) in shallow and intermediate water depths, and by critical wave steepness in deep water.

rather than any sort of inherent limit to the theory. The second, and equally important, point is that each of the theories will yield inappropriate answers if applied outside of their area of applicability. There is nothing in the theory that constrains the mathematical calculations to be restricted to a certain domain within the graph. So the coastal scientist needs to be vigilant about which theory is being used and whether it is actually applicable to the problem of interest. Linear theory, for example, does not perform well in providing wave predictions in the surf zone despite the ease with which it can be implemented in a computer to yield solutions, even for very shallow water conditions.

5.4 Wave Transformations in Intermediate Water Depths

Deep-water waves moving towards the shoreline will eventually encounter water depths at which the wave begins to interact with the bed. A whole suite of complex transformations begin to take place as a consequence of the need to conserve mass, momentum and energy, and these will engender changes in the length, speed, height and shape of the wave. The peak wave period, however, tends to remain constant, although for a spectrum of incident waves, there will be shifts in where energy is allocated within frequency space. The dominant processes that lead to wave transformations in intermediate water are wave shoaling and wave refraction, which are described in this section. In so doing, there is an inherent assumption that wave energy is conserved to first order although there may be energy inputs (e.g., from wind) and energy losses (e.g., due to bottom friction). At some point in the shoaling process, the water becomes so shallow as to induce wave breaking, which will be discussed in the succeeding section.

5.4.1 Wave Shoaling

Consider a two-dimensional wave of infinite width, and of deep-water wavelength, L_o, and height, H_o, moving toward the coast. Select any two points along the width of the crest, separated by distance, s, and follow these points as they propagate with the speed of the wave, C_o. These two points will define two wave rays as they trace parallel lines toward the shore, which are orthogonal to the wave crest at all times (hence, wave rays are also referred to as wave orthogonals). The wave rays are equivalent to streamlines, which have the property of not allowing any transfer of fluid or energy across them, to first order. Energy conservation between the wave rays requires that the wave power (rate of energy delivery) remains constant as the wave moves across the ocean surface. Applying Equation 5.26 for the segment of wave crest between the two arbitrary points separated by distance, s, the situation requires that,

$$P_w s = \bar{E} C_g s = \bar{E} n C s = \text{constant}$$

for all positions along the wave trajectory. If the wave in deep water (defined by L_o, H_o, C_o and s_o) travels to a new location in intermediate water where the spacing of the wave rays differs (defined by L_i, H_i, C_i and s_i), then

$$\bar{E}_o C_{go} s_o = \bar{E}_i n C_i s_i. \tag{5.34}$$

Substituting Equation 5.24 for average wave energy yields:

$$\frac{H_i}{H_o} = \left(\frac{1}{2n}\frac{C_o}{C_i}\right)^{0.5}\left(\frac{s_o}{s_i}\right)^{0.5}. \tag{5.35}$$

This expression predicts the change in wave height as a consequence of both shoaling and refraction effects. Note that the first term in parentheses on the RHS of this expression is simply the ratio of the group velocity of the waves in deep water ($C_{go} = ½C_o$) to the group velocity of the waves in intermediate water ($C_{gi} = nC_i$). A general form of this expression is referred to as the shoaling coefficient, $K_s = (C_{g1}/C_{g2})^{0.5}$, where the subscripts indicate two arbitrary locations outside the breaker zone (i.e., '1' replaces all 'o' subscripts and '2' replaces all 'i' subscripts on H, C and s in Equation 5.35).

As waves move into water of decreasing depth, the shape of the wave becomes distorted so that it is no longer symmetric above and below the MWL, as assumed by Airy wave theory. Stokes' second-order theory predicts that the wave crest becomes narrower and more peaked while the wave trough becomes flatter and wider (Figure 5.4A). The term 'wave skewness' has been adopted to define this difference, especially in reference to the unequal onshore and offshore orbital velocities (Malarkey and Davies, 2012). Wave skewness is calculated as the third moment about the mean from the velocity distribution. A similar concept, wave asymmetry, refers to the difference in accelerations between the leading part of the wave (in advance of the crest) versus the trailing part of the wave (behind the wave crest), and it is due to the different shape in front of and behind a breaking wave. The transition to wave skewness has important ramifications for the dynamics of a shoaling wave. For example, the horizontal orbital velocities under the wave crest increase but are of shorter duration in comparison to the wave trough, which has a longer phase of reduced velocity in the offshore direction (Figure 5.4B).

As waves shoal into shallow water, the wavelength decreases according to the wave dispersion relation (Equation 5.4), and this leads to an increase in wave steepness, H/L, in proportion to the relative water depth h/L. Changes in wave height due only to shoaling effects can be predicted using Equation 5.35 by assuming that the wave rays stay parallel to each other ($s_o = s_i$). The expression reduces to the following basic form:

$$H_i = H_o \left(\frac{\cosh^2(kh)}{0.5(\sinh(2kh) + kh)} \right)^{0.5}. \tag{5.36}$$

The term shoaling is used to describe the transformations in the length, height and celerity of the waves as they propagate from deep water into shallower water (Figure 5.7), but it does not include the actual wave breaking process. These transformations are predicted satisfactorily by Airy and Stokes wave theories for intermediate water depths. Changes in the ratios of C, L, H and C_g relative to the deep water values are shown in Figure 5.8. The values of n are also shown because this function (Equation 5.27) controls many of the shoaling transformations. In deep water, n = ½ whereas in shallow water n = 1. There is a zone of transition indicated by the dashed box, which corresponds to the region for which shoaling transformations are most

Figure 5.7 Long-crested waves shoaling across a shallow nearshore (near Invercargill, New Zealand). Note decrease in wavelength and increasing incidence of turbulence toward shore due to wave instability.

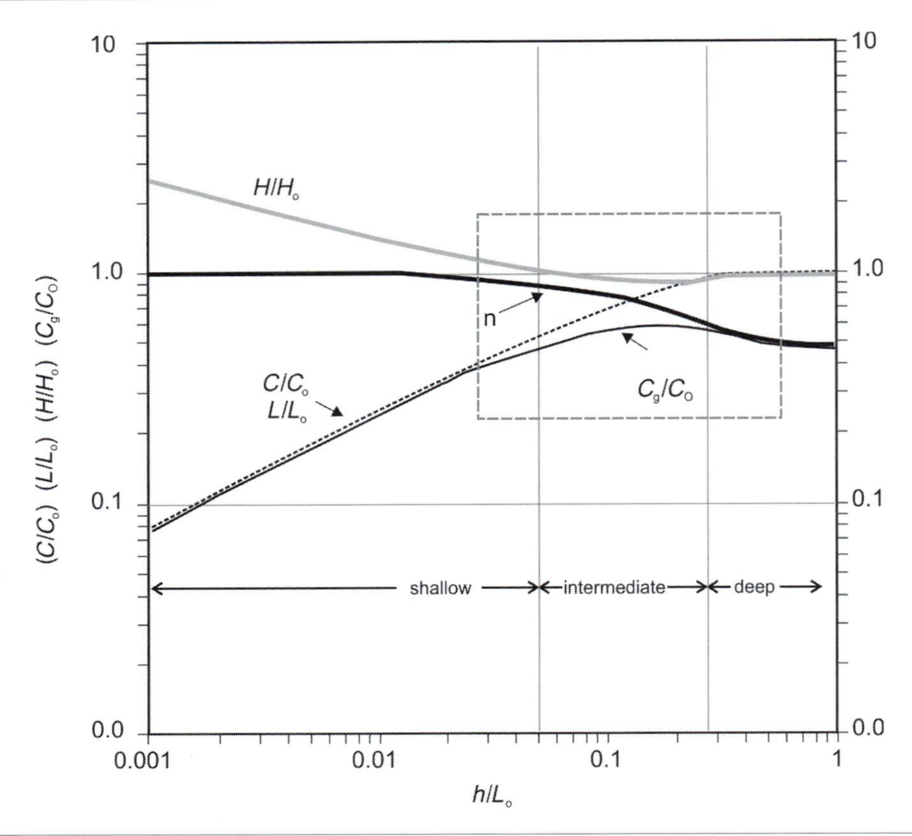

Figure 5.8 Theoretical changes in wave parameters during wave shoaling. All parameters are normalised by the deep-water wave condition such that values diverging away from a value of 1 indicate increases (as with wave height) or decreases (as with wavelength and celerity) relative to the incident waves outside the region of shoaling (indicated by dashed grey box). In very shallow water, the values are approximations, at best, because simple wave theory was assumed.

pronounced. As water depth decreases (from right to left on the diagram), the wave speed, C, and length, L, progressively decrease. In addition, the ratio of group velocity to deep-water celerity (C_g/C_O) also decreases but not before a small increase within the shoaling region. This suggests that wave energy is propagated at a rate slightly greater than in deep water during the initial stages of shoaling, and this is reflected in a slight initial decrease in wave height. At a relative depth of about $h/L_O = 0.04$, which defines the transition to shallow water, wave height increases rapidly as the wave continues to shoal into shallower water. These shoaling transformations are most readily apparent in long-period swell waves that interact with the bed a long distance offshore. Steep, short-period storm waves, characteristic of fetch-limited areas, do not feel the effects of the bed until they are in much shallower water and, because they are initially steeper, the transformations are more immediate and not as apparent visually.

5.4.2 Wave Refraction and Wave Diffraction

Waves entering shallow water often do so at an angle relative to the depth (bathymetric) contours. This may occur on a planar beach when the waves approach the shoreline obliquely (Figure 5.9A) or more commonly, when the offshore bathymetry is complex regardless of wave approach angle relative to the shoreline (Figure 5.9B and C). As a consequence, one

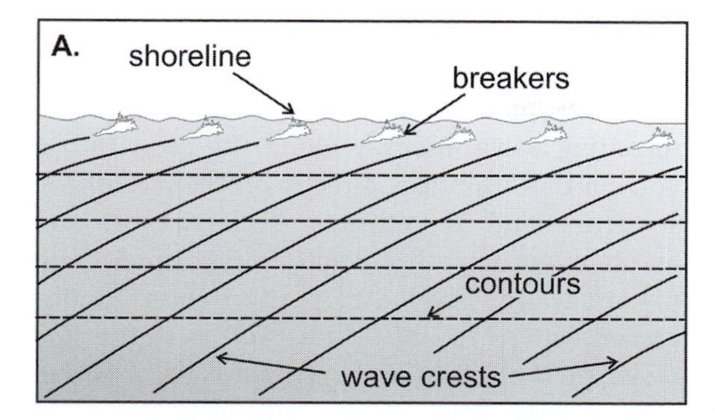

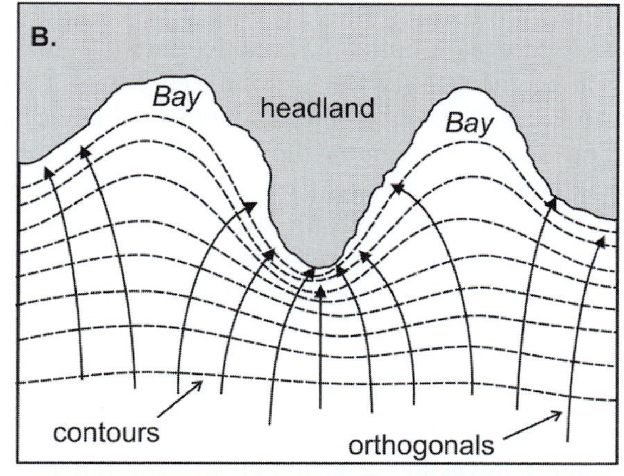

Figure 5.9 Wave refraction. A. Waves approaching a straight shoreline at a highly oblique angle across a planar nearshore, showing that refraction is most pronounced close to shore in shallow water. B. Waves approaching perpendicular to a complex shoreline with headlands and embayments will encounter different depths of water and therefore the wave rays (orthogonals – solid arrows) will converge or diverge according to the bathymetric contours (dashed lines). C. Small waves refracting around a shallow sandy deposit showing bending of wave crests. Criss-cross patterns are due to reflected waves moving offshore and alongshore at varying angles, intersecting with the incident waves.

section of the same wave will begin to interact with the bottom earlier than another section that is farther alongshore. Based on what was discussed about wave shoaling in the previous section, it is immediately evident that the section of wave that encounters shallow water first will experience transformations before other sections that are still in relatively deeper water. The section of wave in shallow water slows down and therefore the waves bunch up and increase their height, whereas the section in deeper water continues to propagate at its maximum velocity without any changes to wavelength or wave height. The consequence of this along-crest difference in shoaling leads to a gradual bending of the wave crests so as to align roughly with the bottom contours. This process is referred to as wave refraction.

Modelling wave refraction begins with Equation 5.35, which is the general expression for wave shoaling that accommodates both changes in depth and changes in spacing of wave orthogonals. However, unlike the case of a uniform decrease in water depth, where it was reasonable to assume that the wave orthogonals run parallel to each other ($s_o = s_i$), now there has to be allowance for divergence or convergence of the orthogonals (i.e., $s_o \neq s_i$). Note that this allowance simply increases or decreases the area over which wave energy is distributed, while the requirement that there can be no transfer of fluid or energy across the wave orthogonals still holds. In addition, it is necessary to assume that the changes in bathymetry are gradual and that the waves are regular, long-crested and of small amplitude.

Let the ratio $K_r = (s_o/s_i)^{0.5}$ be the refraction coefficient. It can be shown (Dean and Dalrymple, 1984, pp. 108–9) that

$$K_r = \left(\frac{\cos\theta_o}{\cos\theta_i}\right)^{\frac{1}{2}} \tag{5.37}$$

where θ is the angle of wave approach defined by the deviation of the wave orthogonal from a line that is directly on–offshore. The refraction process is analogous to the refraction of light waves, which is predicted by Snell's law for the constancy of wave number. A simple refraction theory for water waves based on this approach is provided in Box 5.1.

Wave refraction can be depicted by tracking the behaviour of wave crests, but in examining the effects of refraction on wave height and energy distribution along the shoreline it is often more useful to examine the behaviour of the wave orthogonals (Figure 5.9B). The convergence of orthogonals reflects a compression of the wave laterally on its path toward the shoreline, which results in a localised increase in energy density that manifests as an increase in wave height. Conversely, divergence of orthogonals represents a lateral stretching of the wave crest and produces a decrease in wave energy density and height. Refraction thus has the effect of increasing or decreasing the height of the wave as it reaches the shoreline and therefore also the longshore distribution of wave energy – a process that may be extremely important in determining the magnitude of wave erosion, the potential impact on harbour and shore protection structures, and the direction and magnitude of longshore sediment transport.

Along complex shorelines with numerous small islands and deep embayments bounded by irregular promontories, the wave energy that propagates toward the coast is often transferred laterally along the wave crests by a process known as wave diffraction. A long, offshore structure such as a breakwater will split an incoming wave laterally along the crest into a portion that impinges directly on the face of the structure and is reflected back to sea, and another portion that continues to propagate toward shore. This latter portion enters undisturbed water in front of it as well as to the side of it, in the lee of the structure. As a consequence, there is a discontinuity in wave height along the wave crest. The energy in the wave therefore leaks horizontally along the wave, into the undisturbed water behind the lee of the structure. Figure 5.10 shows a situation in which only short segments of the incoming waves (on the outside of the breakwater) survive through the narrow inlet beneath the bridge. As these wave segments move through the throat of the inlet, they come out the other side into an area of relatively undisturbed water in the shallow, tidal lagoon. These waves are then free to propagate across the calm water while also spreading laterally on either side of the inlet. Just as the leading waves in the group disturb the water in front of the wave group, so too does the energy leaking along the wave crest disturb the lateral margins of the wave group, which leads to

Figure 5.10 Wave diffraction in a small, shallow tidal lagoon (bottom of photo) after the incident waves (top of photo) have passed through a narrow inlet beneath the railway bridge (middle of photo) (near Dunedin, New Zealand).

curved waves emanating out in all directions. The wave diffraction process continues to expand the influence of the waves across a greater surface area as they propagate forward. A similar process occurs when waves pass by and around a small island, with wave diffraction providing one mechanism (in association with wave refraction) by which wave energy is delivered to the rear side of the island. Until recently, the relevance of these processes was not fully appreciated, but they offer a compelling explanation why the leeward sides of islands can experience considerable damage during major tsunami events.

Although simplified refraction models do not usually consider the effects of diffraction, it is now apparent that refraction effects alone would yield unreasonable solutions in situations where there is extreme focusing of energy on headlands or divergence of energy in embayments that have narrow throat regions (Figure 5.11). Thus, wave refraction and diffraction processes need to be included in those cases with complex bathymetry in order to assess more accurately the true energy levels along the shoreline.

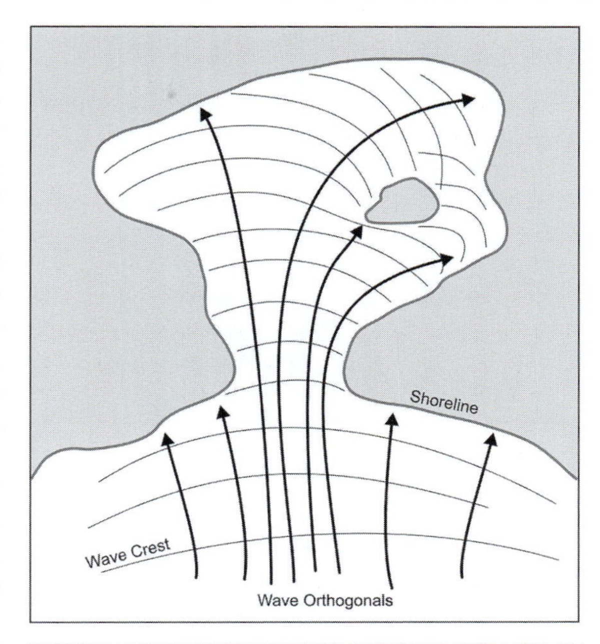

Figure 5.11 Patterns of wave crest propagation on a complex coast. Bold arrows are wave orthogonals (wave rays). For complex bathymetries in deep embayments with narrow inlets, islands and multiple headlands, it is necessary to include both refraction and diffraction to predict accurately the wave height at any specific location.

Box 5.1 Simple Wave Refraction Modelling

Changes in the approach angle of an oblique wave passing over relatively simple topography can be approximated using Snell's law as follows:

$$\sin \alpha_2 = \frac{C_2}{C_1} \sin \alpha_1 \tag{5.38}$$

where α is the angle between the wave crest and the depth contours, C is the wave celerity, and subscripts refer to two different positions along the shoaling trajectory with '1' referring to the offshore position (which might be deep water) and '2' referring to an onshore location prior to breaking. The geometry is illustrated in Figure 5.12. The wave approaches the coast at an angle $\alpha_1 = \alpha_o$ in deep water. At point B, the wave crest is at the contour line marking the transition from deep to intermediate water, while at point A the crest is still in deep water. Over one wave period the wave crest at A will travel a distance equal to the deep-water wavelength, L_o, reaching the bathymetric contour at which $h/L_o = 0.5$. However, the wave at point B will travel a slightly lesser distance, BD = L, that reflects a reduction in the wave celerity as the wave begins to shoal. For the triangle ABC, $\sin \alpha_o = AC/BC = L_o/BC$ and for triangle DBC, $\sin \alpha = L/BC$ (where $\alpha = \alpha_2$). Then,

$$\frac{\sin \alpha}{\sin \alpha_o} = \frac{L}{L_o} = \frac{C}{C_o} = \tanh\left(\frac{2\pi h}{L}\right). \tag{5.39}$$

Box 5.1 (cont.)

where the latter equalities in this expression are from the wave dispersion relation (Equation 5.11). This indicates how wave refraction is functionally dependent on water depth relative to the length of the wave.

The first, simple wave refraction diagrams were constructed manually using graphical techniques (Munk and Traylor, 1947), but eventually computer codes were developed in the 1960s (Wilson, 1966; Dobson, 1967). The program WAVENRG by May (1974) is a second generation program that was used in a number of studies in the US and Canada. These early programs did not incorporate wave diffraction, with the result that in areas of rapid convergence wave height grew very large and produced wave breaking in quite deep water. Advanced computer programs, such as MIKE 21 and STWAVE, are now available for simulating complex wave motion over irregular topography (see Chapter 4), and these have largely superseded the simpler approaches.

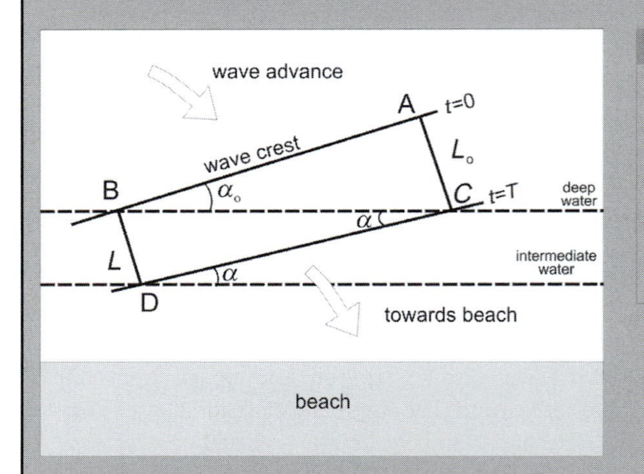

Figure 5.12 Simplified definition sketch for implementation of Snell's law for wave refraction. The wave crest in deep water (segment AB) is unaffected by the sea bed, but in intermediate-depth water only the inner portions of the wave crest (segment CD) are influenced by the bottom. As a consequence, the wavelength, L, defined by BD is shorter than L_o, defined by AC, and consequently the angle of the wave crest relative to the shoreline is altered progressively.

5.5 Wave Breaking

5.5.1 General Concepts

Wave breaking occurs when the wave form becomes unstable and the regular oscillatory motion beneath the wave is disrupted entirely. Very rapid changes in water surface elevation, orbital velocities and particle accelerations occur during the breaking process. The physics are highly non-linear and complex, but the net outcome is a transfer of energy from the oscillatory motion of the wave to chaotic turbulence in the water column following breaking. Wave breaking is accompanied by energy dissipation in terms of heat and work expenditure (e.g., bottom friction, sediment transport).

There are two primary situations that lead to wave breaking, one for deep water and the other for shallow water. As noted above, in deep water the maximum wave steepness is H_o/L_o = 0.142. This occurs when the interior angle of the wave crest is 120°. When wave steepness approaches this critical angle, the orbital velocities of water particles immediately beneath the crest become faster than the speed at which the wave form propagates across the water surface. As a consequence, the water particles 'move out' in front of the wave crest and begin to cascade down the front face of the wave (Figure 5.13). In deep water, this situation occurs frequently when waves of different wavelength (period) interact with each other, as they would in an area of active wave generation in the open ocean.

Figure 5.13 Wave breaking in deep water as a consequence of the wave form having reached critical steepness. Note foamy water spilling down the front of the wave (to left) and 'spindrift' due to wind blowing off the wave top (to right), leaving a heavy mist in the lee of the wave. Whitecapping is evident in other waves, as is the trail of turbulent water left behind by the previous broken wave (centre-left of photo).

Wave dispersion requires that longer waves travel faster than shorter waves, so a long wave often travels 'through' a short wave. The wave heights of the two waves are superimposed, leading to a local oversteepening of the combined wave. The result is wave breaking at the crest, and often this occurs over a distance of several metres or more, leading to a phenomenon known as whitecapping. A trail of highly turbulent surface water with a foamy appearance is left in the wake of the wave–wave interaction, with the net result that some of the wave energy is dissipated.

Waves moving into shallow water will also experience breaking, but as a final stage in the wave shoaling process. Recall that wave shoaling leads to a shortening of the length of the wave, and due to mass conservation, an increase in wave height. Combined, these shoaling transformations yield a progressive increase in wave steepness. The shallow water breaking condition, however, is different from its deep-water counterpart because of the control that water depth and beach slope play in the overall dynamics of shoaling and breaking waves. A number of expressions have been derived to predict the depth of water in which breaking takes place, h_b, and the wave height at breaking, H_b. McCowan (1894) was the first to suggest that wave height is limited to about three-quarters the water depth, whereas Munk (1949) used solitary wave theory to reaffirm that a shoaling wave will break when:

$$\gamma = \frac{H_b}{h_b} = 0.78, \tag{5.40}$$

where γ is a wave breaking index. Waves should break upon entering water depths that are only slightly greater than the wave height, and it has been shown from measurements in wave tanks and in the nearshore zone that γ can vary anywhere between 0.7 to about 1.1.

Breaker height was also predicted by Munk (1949) using a different approach:

$$\frac{H_b}{H_o} = \frac{1}{3.3\left(\frac{H_o}{L_o}\right)^{0.33}}. \tag{5.41}$$

This relationship predicts the breaking condition according to the deep-water wave height and wavelength. The approach assumes no losses of energy due to friction and no changes due to refraction. The key insight from this expression is that the wave height at breaking depends, in part, on the wave steepness in deep water (i.e., the character of the incident wave) and not just the water depth. H_b/H_o ranges from close to 1 for short period waves with large relative height, to 1.4–1.6 for long-period swell waves. Equation 5.41 is of value in understanding how wave height can potentially increase during the late stages of the shoaling process toward breaking, which clearly depends on how steep the wave was in deep water. In this regard, it serves a similar purpose as Equation 5.36, but it is of little value in predicting where in the nearshore zone wave breaking will take place.

5.5.2 Location of Wave Breaking

The location and consequences of wave breaking are controlled, in part, by water depth as well as the incident wave height and wavelength in deep water. However, the character of the nearshore and foreshore zones is equally important, especially beach slope. Three broad conditions can be recognised and are shown in Figure 5.14.

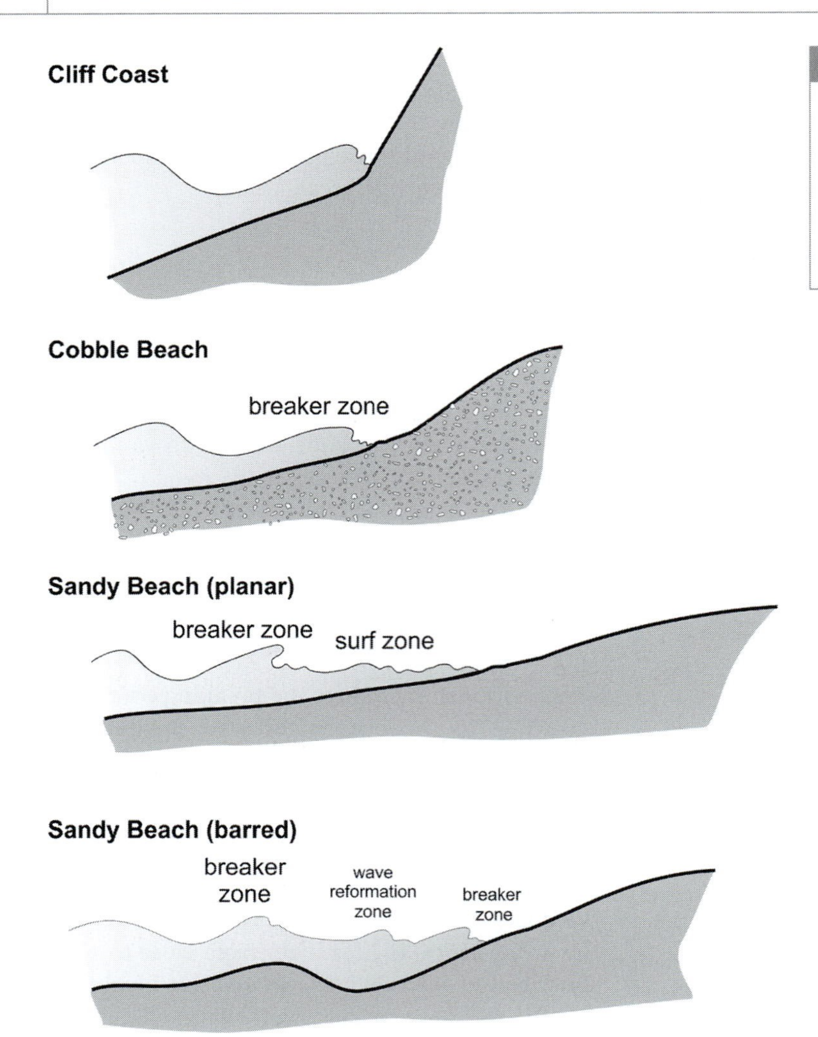

Figure 5.14 Wave breaking as controlled by nearshore topography. Steep beaches (cliffed and cobble beaches) usually have wave breaking at the shoreline whereas shallow sand and silt beaches have breakers some distance offshore, often induced by the presence of nearshore bars.

Very steeply sloping beaches These shorelines, often found on cliffed, rocky shores, have a steep nearshore and foreshore profile and are generally backed by a vertical face (e.g., cliffs, seawalls and other vertical structures; see Chapter 13). Waves propagating toward the coast do not encounter shallow water until they are very close to shore, if at all, and therefore they will not truly shoal across the nearshore zone. Rather, they will experience wave breaking essentially on the shore, and there is no surf zone. Often the wave motion is partly or completely reflected, especially if there is a vertical cliff or wall that extends well below the water surface. When waves break directly on the shore, a great deal of energy is expended over a localised area, and typically air compression occurs against the rocky cliff resulting in splash and spray reaching high into the air as well as physical fracturing of rocks when the air is trapped in cracks and crevices.

Moderately sloping beaches These shorelines are characterised by relatively steep accumulations of gravel or cobble. Wave breaking occurs either at the base of the foreshore or a short distance offshore. A small degree of wave reflection is common, but most of the wave energy is expended during the breaking process, which occurs over a short distance. Thus, the surf zone, if there is one, is usually quite narrow. The form of wave breaking can vary depending on tidal stage as well as on the incident wave steepness.

Waves breaking close to shore usually lead to significant swash motion on the foreshore.

Gently sloping beaches These shorelines are characterised by a shallow slope profile across the nearshore zone (primarily fine to medium sand beaches) with wave breaking occurring some distance offshore and giving rise to a breaker zone (or zones) and a wide surf zone between the line of first breaking and the beach. Where the profile is nearly planar, the surf zone is characterised by bores crossing the entire surf zone from the initial break point (Figure 5.15A). However, where the profile is characterised by the presence of nearshore bars with deeper water in the troughs landward of the bars, wave breaking may cease in the troughs and each bar crest may be associated with a separate breaker line (Figure 5.15B).

For planar beaches, it is possible to predict, in rough terms, where wave breaking should take place. Equation 5.35 gives an indication of the increase in wave height during shoaling (including refraction effects), and given a value for the wave breaking index, γ, it can be shown that the offshore position of wave breaking is (Dean & Dalrymple, 1984, p. 115):

$$x_b = \frac{h_b}{\tan\beta} = \frac{1}{\tan\beta \; g^{1/5} \; \gamma^{4/5}} \left(\frac{H_o{}^2 \; C_o \; \cos\theta_o}{2} \right)^{2/5}. \quad (5.42)$$

This relationship uses the deep-water wave characteristics and the beach slope to predict where wave breaking will occur, although there can be considerable uncertainty given that $0.7 < \gamma < 1.1$ due to the spectral nature of waves.

5.5.3 Breaker Types

Decades of field and laboratory observations have shown that waves adopt different forms during the breaking process. Three types of breaking wave are commonly recognised: spilling, plunging and surging, whereas Galvin (1968) identified a fourth – collapsing – and provided the definitive descriptions of each type (Figure 5.16). Galvin's work was based on high-speed filming of regular waves breaking on a plane laboratory slope, conditions that are simplified in comparison to the irregular waves and complex topography commonly found on beaches. Nevertheless, he provided valuable insight into the process of wave breaking and the controls on breaker type. Subsequent efforts by other researchers have been devoted primarily to devising classification criteria that allow separation of the different breaker types based on beach and wave properties.

Spilling breakers are found on wide, shallow beaches, and are characterised by a localised instability of the wave crest that is sustained as

Figure 5.15 Wave breaking on shallow gradient beaches. A. Wide, highly turbulent surf and swash zones on a gently sloping, planar beach with sequential bores rolling in from offshore (Mason Bay, Stewart Island, New Zealand). B. Plunging waves breaking on an outer nearshore bar dissipate a large proportion of their energy, but then reform as secondary waves that propagate across the alongshore trough only to break again on the shore and dissipate all remaining energy (near Port Lincoln, South Australia).

the wave shoals towards shore. Water from the crest moves slightly ahead of the wave form, and therefore slides down the front face of the wave. In so doing, a great deal of air is entrained, creating a line of foamy white water along the front face of the wave crest (Figure 5.17). Sometimes this line is thin and restricted to the wave top, whereas on slightly steeper beaches a thick plume of water extends down the entire front face of the wave. A turbulent wake is left in the lee of the wave form as it advances over the continually breaking front face. The wave form remains largely intact as the wave continues landward, and the wave height decreases slowly, which suggests that total energy expenditure is relatively small for spilling breakers (in contrast to plunging breakers) despite the intense levels of turbulence. On very wide, shallow beaches, there can be up to a dozen or so spilling breakers leading to a persistently turbulent and frothy sea surface out to several hundred metres offshore.

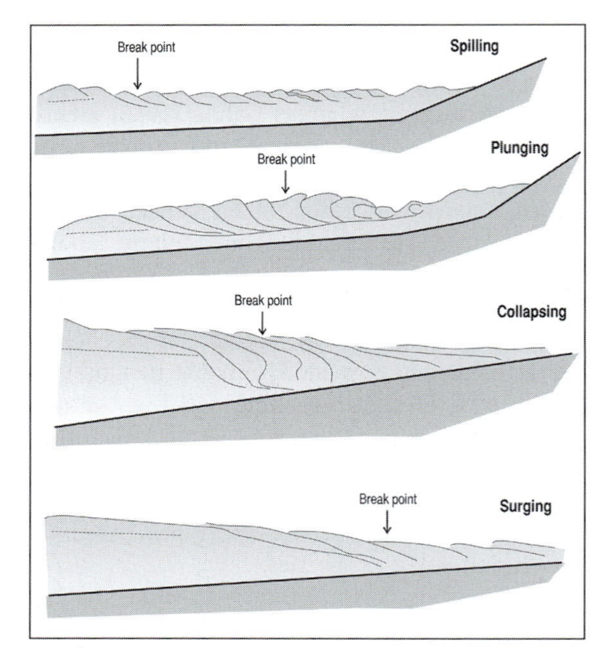

Figure 5.16 Types of wave breaking on a beach (after Galvin, 1968). See text for description.

As the nearshore slope steepens, there is a gradual shift from spilling to plunging types of breaker. The primary difference between these two breaker types is that the water leaving the wave crest of a plunging breaker no longer spills down the front face of the wave but actually 'leaps' out in front of the wave form (Figure 5.18A). This is often described as a 'jet' of water emanating from the crest, which then curls forward and downward. In essence, once the water particles are in the jet, they are no longer influenced by the orbital dynamics associated with the wave but come under the influence of gravity. The lip of the curl eventually reattaches to the water surface in advance of the wave, and thereby encloses an air pocket that surfers refer to as a 'tube'. An ideal surfing wave will 'peel'

Figure 5.17 Spilling breakers. A. Short-crested spilling wave moving from left to right has an unstable crest with white water spilling down the front of the wave. B. Long-crested spilling wave showing asymmetric profile from front to rear of wave crest. Section of wave in foreground is at the transition to plunging type breaking.

alongshore so that the surfer can ride the wave face just in front of the curling zone without being trapped inside the tube. In a two-dimensional situation where the wave comes directly onshore and does not peel, the curl closes on itself across the entire width of the breaking wave (Figure 5.18B–D). The wave collapses on the tube often making a thundering boom as the trapped air is compressed and then blasted out through the water above or to the rear face of the wave. A plunging wave expends most of its energy at the break point, although secondary disturbances in the form of new waves are generated when the wave crashes and disturbs the water ahead of it. The plunging process results in very rapid transformation of the wave form, and is quite distinct

Figure 5.18 Plunging breakers. A. Plunging wave breaking on the foreshore of a relatively steep beach showing the tube that develops as the water in the crest leaps in front of the wave. Note capacity of these waves to stir up sediment from the bed. B.–D. Time-lapse sequence of plunging wave from peak height to collapse. Note the suspended sediment in water and spindrift blowing off top of the crest by offshore wind.

from the spilling breaker even though there is a gradual transition between the two breaker types.

Surging breakers occur when small amplitude waves of relatively long wavelength reach a steep beach (Figure 5.19A). The resulting motion involves the rapid movement of the wave up the foreshore slope, thereby creating uprush or wave run-up. The uprushing water comes under the influence of gravity, decelerates and eventually comes to a halt at the maximum extent of run-up. The water then begins to accelerate back down the foreshore as a sheet of water on the backwash phase of the swash cycle. On beaches with coarse materials such as shingle or gravel, a significant portion of the water infiltrates into the beach, which leads to groundwater pumping in the foreshore sediments in time with the wave motion. The timing of the swash cycle (i.e., uprush and backwash) relative to the incident wave cycle is critical to understanding a range of phenomena such as the evolution of beach steps and cuspate forms. More importantly, whether the backwash reinforces or interferes with the incoming wave crests will dictate how high on the beach the next uprush will reach. This also influences the character of the surging breaker. At the extreme end of the continuum of interactions, there is no breaking at the base of the foreshore but only a simple back-and-forth movement of the water up and down the foreshore slope with minimal generation of turbulence and foam on the surface. At the other end, there is true breaking at the base of the foreshore, the generation of roller vortices and hydraulic jumps, and highly turbulent water moving up and down the foreshore usually carrying a great deal of sediment.

Collapsing breakers are often classified with surging breakers, and it is difficult to distinguish between the two. Collapsing breakers are so called because the wave form peaks up, as if to break, but rather than break, the wave form simply collapses in on itself (Figure 5.19B). As the collapsing breaker reaches the foreshore, the crest rises up, but then the front face of the wave advances ahead and stretches the wave, which creates space for the crest to collapse back down. There is a subsequent surging motion up the foreshore, but no true wave breaking has occurred. The surging motion of both spilling and collapsing types of breaker often generates small return waves that migrate offshore as the backwash vacates the foreshore.

It is important to recognise that the four breaker types represent idealised states within a continuum of transitional forms. Galvin (1968) observed that spilling and plunging breakers are formed when steep waves move across gentle beach slopes, whereas surging and collapsing waves occur when waves of relatively small steepness encounter moderate to steep slopes. He also noted that the sequence from spilling to

Figure 5.19 Surging and collapsing breakers. A. Typical surging wave with extensive wave uprush (near Port Elliot, South Australia). B. Collapsing breaker at the base of a steep foreshore (near Klamath, California, USA).

plunging and surging is observed if, for a given slope, period is held constant and wave height is decreased, or if height is held constant and period is increased. If height and period are held constant, an increase in slope should produce the same sequence. These observations suggested that it might be possible to use characteristic beach and wave parameters in non-dimensional form to predict the type of breaking wave that might usually be found on beaches of different type, as will be discussed next.

5.5.4 Breaker Indices (Surf Zone Similarity)

Galvin (1968) incorporated wave steepness and beach slope into two parameters used to predict breaker type:

$$B_o = \frac{H_o}{L_o \ \tan^2\beta} \tag{5.43}$$

$$B_b = \frac{H_b}{g \ \tan\beta \ T^2}, \tag{5.44}$$

where $\tan\beta$ is beach slope, B_o is an offshore index and B_b is a breaker index. Note that sometimes β is used for beach slope in these parameters, in which case it is implicit that β is not in degrees but rather a decimal that is the equivalent of $\tan\beta$. For example, a very steep beach might have a slope of about 0.1 (angle of 5.7°) whereas a relatively flat beach would have a slope of 0.001 to 0.0001. Based on his laboratory data, Galvin determined the values for the two equations that define the transitions between breaker types as follows:

Parameter	Surging–Plunging	Plunging–Spilling
B_o	0.09	4.8
B_b	0.003	0.068

A similar set of experiments carried out by Okazaki and Sunamura (1991) confirmed Galvin's results. On natural beaches, where slope is variable and waves are irregular, prediction of breaker type by a simple index is not always accurate (e.g., Weishar and Byrne, 1978). Nevertheless, a number of field and laboratory studies have shown that the breaker index is useful for the prediction of beach states and can provide some explanation for phenomena observed on the beach and in the surf zone. It is now common to classify beaches according to non-dimensional groupings of the key variables, referred to as surf similarity parameters. Bauer and Greenwood (1988) provide a concise summary of surf similarity parameterisation and their use to provide insight into whether wave energy will be primarily dissipated or reflected upon entering the nearshore.

The first use of a surf similarity parameter was that of Iribarren and Nogales (1949) who quantified the transition from non-breaking to breaking waves on a plane beach using:

$$\xi = \frac{\tan\beta}{\left(\frac{H}{L_o}\right)^{0.5}}. \tag{5.45}$$

Battjes (1974) derived the value of the critical condition, $\xi = 2.3$, which corresponds to a regime about halfway between complete reflection and complete breaking. He also proposed that the expression for ξ be termed the Iribarren number. Note that Galvin's offshore parameter (Equation 5.43) is the squared reciprocal of ξ.

Guza and Inman (1975) adopted a similar parameter based on earlier work by Carrier and Greenspan (1958) on non-linear, inviscid, shallow water waves. The surf scaling parameter, ε, is defined as:

$$\varepsilon = \frac{a \ \sigma^2}{g \ \tan^2\beta}, \tag{5.46}$$

where a is the wave amplitude, σ is the wave radian frequency, and $\tan\beta$ is the beach slope.

Wright and Short (1984) proposed that the surf scaling parameter, ε, could be used to distinguish the two extremes of their model of beach states. First, fully dissipative beaches with flat shallow profile and relatively large subaqueous sand storage; and second, highly reflective steep beaches with relatively small subaqueous storage (see Chapter 8 for a full description of this model). According to Wright and Short, values of $\varepsilon \leq 2.5$ will produce strong reflection, surging breakers, and resonance at subharmonic frequencies. Beaches with values for $\varepsilon \geq 2.5$ will be characterised by plunging breakers. When $\varepsilon \geq 20$, spilling breakers will occur, with increasing dissipation associated with larger values of ε.

All of these surf similarity parameters incorporate some measure of wave steepness and beach slope. The slope is easily defined on a planar laboratory beach but on any complex beach profile it is necessary to define the landward and seaward limits over which the slope is measured. As noted by Bauer and Greenwood (1988), the particular value for ε that is reported depends very much on where and when the measurements were taken, so there continues to be significant ambiguity about the precise values that define the transitions from wave reflection to wave dissipation.

Wave breaking on natural beaches is also influenced by wind speed and direction. Where the wind is onshore, waves tend to break earlier than would otherwise be predicted and the form of breaking shifts towards the spilling end of the continuum (Galloway *et al.*, 1989; Douglas, 1990). On the other hand, offshore winds tend to delay wave breaking and result in a shift towards the plunging end of the continuum. This provides some explanation for the spilling breakers commonly observed for waves that are generated by local winds and for plunging breakers on beaches with light or offshore winds. Onshore winds also enhance the onshore flow of water in the wave crest and surf bores and reduce this for offshore winds.

5.5.5 Dynamics of Breaking Waves in the Surf Zone

On gently-sloping sandy beaches, large waves tend to break some distance offshore, and a wide surf zone is generated between the point of breaking and the shoreline. Svendsen *et al.* (1978) divided the surf zone into three regions, each of which has quite different wave dynamics (Figure 5.20). The outer region is a zone of rapid wave transformations that includes the point of wave breaking, but not the gradual shoaling processes that occur seaward. The inner region includes the zone landward of initial breaking where the broken waves have either generated secondary waves at higher frequencies and smaller amplitude or reformed as smaller waves of translation (i.e., solitary waves or surf bores) that retain some of the residual momentum and energy of the original deep-water waves. These inner surf-zone waves carry energy towards the shore, where the final stages of energy dissipation take place as swash motion on the foreshore. The processes involved in wave breaking are extremely complex (Figure 5.21), and much work remains to be done to model the dynamics of breaking waves. Insightful descriptions of wave

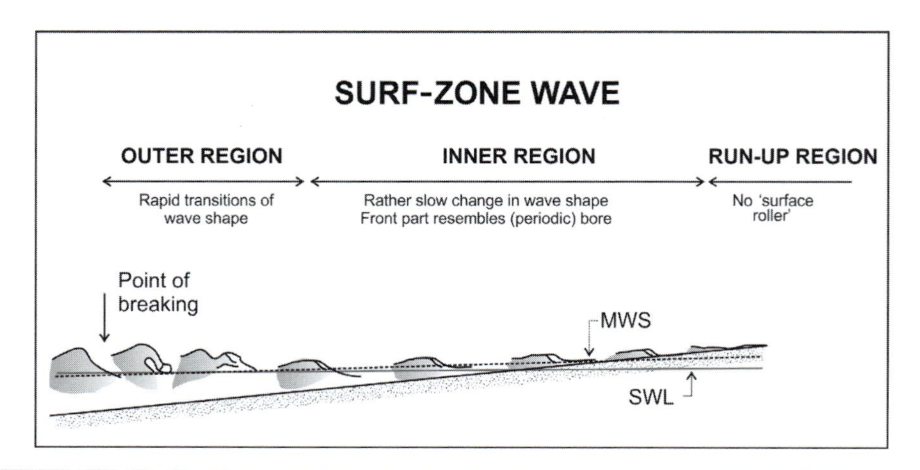

Figure 5.20 Regions of the surf zone as defined by Svendsen *et al.* (1978).

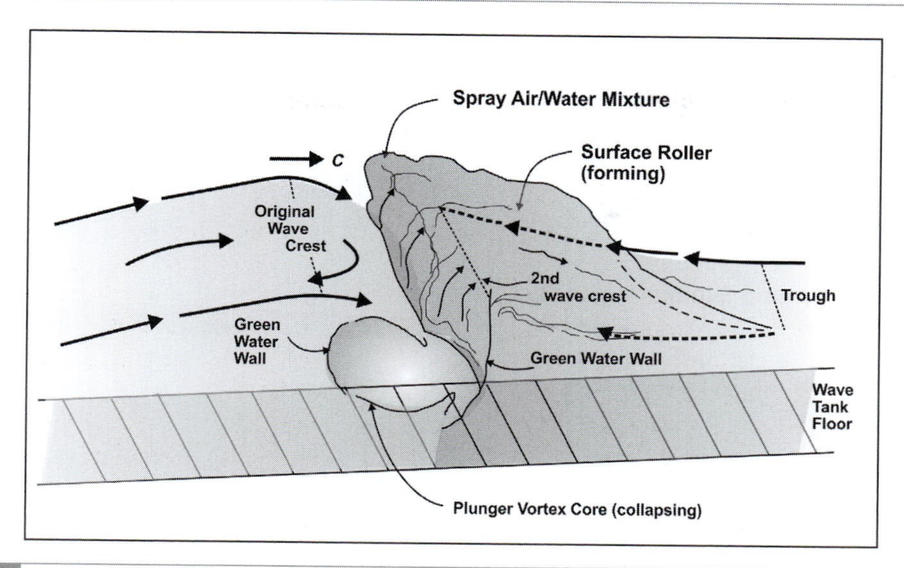

Figure 5.21 Schematic of the main features and relative water motion within a plunging wave that is breaking from left to right. Upon collapse of the vortex core, a secondary wave is created in front of the broken wave, which will propagate toward the shore. '*C*' refers to wave celerity (after Basco, 1985).

breaking are provided by Galvin (1968; 1972), Miller (1976), Peregrine (1983), Basco (1985) and Duncan (2001). Recent research uses computational flow dynamics (large eddy simulations) to investigate the vorticity dynamics in breaking waves (Kirby and Derakhti, 2019), which suggests that coherent flow structures created during the breaking process are central to understanding the creation of turbulence and the dissipation of wave energy.

Nevertheless, it is evident that the production and destruction of vortical motion across a range of scales is critically important to a deeper understanding of breaking waves. Basco (1985), for example, suggested that the main difference between a strongly plunging and a weakly spilling breaker is the strength of the vortex system within the primary breaking wave (Figure 5.21). In a plunging wave, the vortex that creates the tube is especially pronounced as the wave 'rolls over' upon itself in the classical sense of a surfing wave. The plunging jet of water that crashes into the wave trough in front of it creates a succession of secondary vortices in advance of the primary wave, and the full collapse of the primary wave contributes to the generation of secondary waves that propagate shoreward. However, most of the energy in the wave will have been dissipated as turbulent motion.

Spilling breakers are similar in kind but the vortex associated with the primary wave is not as pronounced and does not extend as deep into the water column (Figure 5.22). As the turbulent water from the crest cascades down the front face of the wave, a surface roller is generated in advance of the spilling breaker. However, a secondary wave is not produced by spilling breakers because there is no actual collapse of the wave and the turbulence remains high in the water column. Well-developed plunging breakers, in contrast, have extreme turbulence that is produced by the plunger vortex and by the surface roller, and the vortices penetrate deep beneath the wave and typically reaching the sediment bottom (Basco, 1985; Ting and Kirby, 1995). The dissipation of energy and reduction in wave height for the plunging breaker is rapid and pronounced because the plunging action takes place over a short distance. In contrast, the action of spilling breakers is less intense and continuous, and therefore the breaking takes place over a long distance, yielding a very wide surf zone

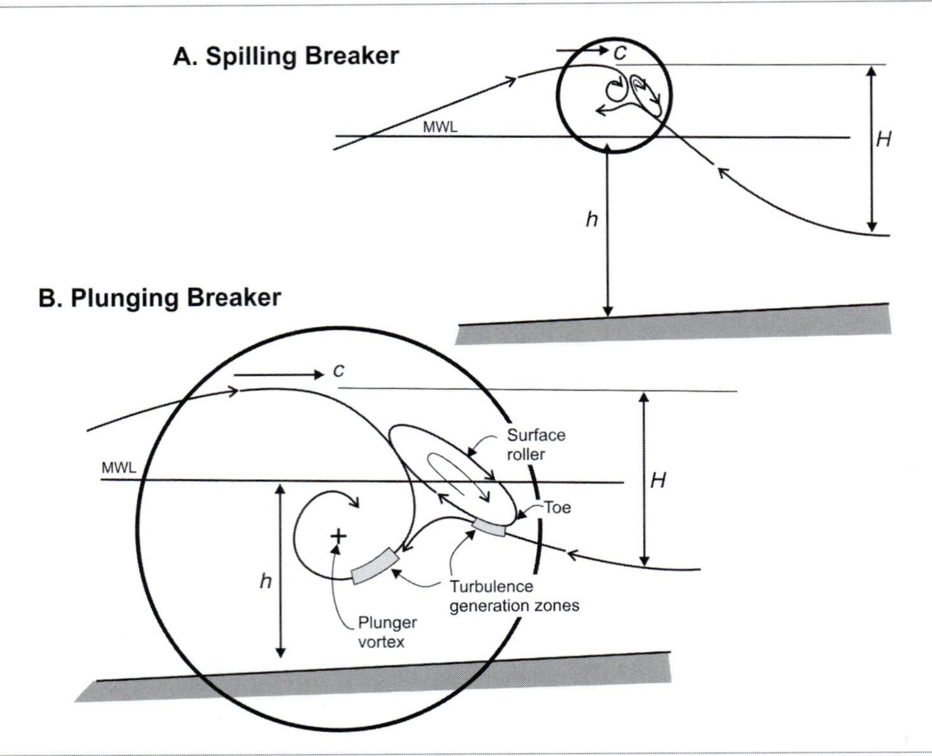

Figure 5.22 Schematic of the difference between the vortical motions in A. a spilling breaker for which the vortex and surface roller are relatively small in comparison the wave height, *H*, and B. a plunging breaker with a very pronounced plunger vortex and a large surface roller that scale approximately with *H*. *C* refers to wave celerity and *h* is average water depth relative to the mean water line (MWL).

rather than a distinct breaker line with reformed waves in the inner region of the surf zone.

The dynamics of wave breaking is of importance to nearshore processes and the equilibrium tendencies of various beach forms. For example, the greater depth of penetration of vortices produced by plunging breakers leads to the suspension of larger quantities of sediment near the break point (see Figure 5.18), and may influence the rate of sediment transport in the surf zone as well as the type of bedforms that occur on nearshore bars. Differences in the rate of energy dissipation between spilling and plunging breakers will also influence the pattern of wave set-up in the surf zone, and this will affect the manner in which offshore return flow develops as well as the strength of longshore currents.

After initial wave breaking has taken place within the outer region of the surf zone, wave reformation or secondary wave generation leads to a different type of wave that is more appropriately modelled using non-linear (high order) wave theories. Svendsen (1984) suggested that a surf bore can be considered to have two parts: (1) a surface roller, which consists of the turbulent, upper front face of the wave; and (2) the main body of the wave, which lies primarily below and behind the advancing front of the wave. Svendsen's (1984) simple two-box model of a surf bore considers flow in the upper roller portion to be directed strongly landward, while immediately beneath the surface roller the flow is directed uniformly offshore. In an actual surf bore, the situation is more complex (Figure 5.23) with upper onshore flows transitioning to lower offshore (undertow) flows. Toward the rear of the wave (after the crest has passed), the flows are dominantly onshore, whereas in the trough immediately preceding the bore, the flows are almost uniformly offshore.

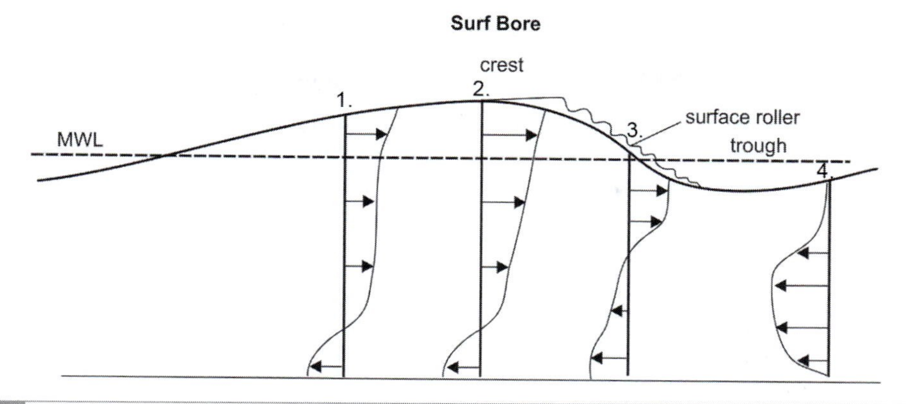

Figure 5.23 Characteristics of surf bores. Cross-sectional view showing the surface roller at the front face of the wave and the main body of the bore to the rear of the crest. Typical velocity profiles are shown for four positions along the bore: 1. seaward of the crest; 2. at the wave crest; 3. within and beneath the surface roller; and 4. in the wave trough immediately ahead of the bore (based on Svendsen, 1984).

Although there are several theoretical models that can be used to describe the dynamics of surf bores, in general, the wave height of surf bores in the inner surf zone on a planar beach is proportional to water depth:

$$H_{rms} = \lambda\ h, \qquad (5.47)$$

where H_{rms} refers to the root-mean-square wave height, which is a statistical measure used when there is a spectrum of wave heights rather than monochromatic waves (see Chapter 4). Therefore the depth-limiting condition for individual breaking waves (γ = 0.78: Equation 5.40) does not apply. Based on empirical data, the value of λ for bores ranges from 0.3 to 0.6 (e.g., Thornton and Guza, 1982) with larger values being associated with steeper beach slopes. As the bore moves across the surf zone into ever-decreasing water depth, progressive energy dissipation should result in a decrease in wave height (Figure 5.24), maintaining a constant relationship between wave height and water depth (Dally *et al.*, 1985). Energy dissipation comes primarily from losses due to turbulence in the roller vortex, but there are also some losses due to turbulence in the main body of the wave and to friction at the bed. Most modelling of energy dissipation and wave height decay in the surf zone begins with consideration of the surface roller, based on initial work by Duncan (1981) and Svendsen (1984), and makes use of the Boussinesq equations (Veeramony and Svendsen, 2000).

Figure 5.24 Complex surf zone with waves of different sizes breaking on an outer nearshore bar and reforming as short-crested bores moving at different angles as they approach shore (near Arcata, California, USA).

Regular waves in a wave tank will break repeatedly at the same location, and all incident waves will reach the same equilibrium height at breaking. After wave breaking takes place, the reformed waves propagate across the surf zone, and their height is proportional to water depth. This situation is known as a saturated surf zone because everywhere within the inner surf zone there is as much wave energy as is possible. However, on a natural beach with undulating or complex bathymetry, especially with a broad spectrum of incident waves, the location of wave breaking will vary. Larger, steeper waves will break farther offshore, while smaller waves will continue to move shoreward. Thus, at any one location there will be a combination of small, unbroken waves and secondary, reformed waves that are the residuals of larger waves that have already broken (Figure 5.25). Models of wave transformation in the surf zone must be capable of dealing with such semi-random wave action in order to predict the actual wave energy at any one location in the nearshore, based on knowledge of the incident wave spectrum (e.g., Lippman *et al.*, 1996; Elgar *et al.*, 1997). One consequence of this is that a surf bore produced by a large wave that has already broken offshore will often travel faster and catch up with smaller, unbroken waves that were far ahead of the large wave. The interaction of these waves of different wavelengths and heights produces localised increases and decreases in wave height, turbulence levels and sediment transport capacity depending on whether the phase interactions are constructive or destructive, as occurs with reflected and standing waves, for example.

On a barred nearshore, wave breaking is most pronounced on the bar crests, and this leads to distinct lines of wave breaking that coincide with the bathymetry. However, once wave breaking has taken place on the bar crest, the wave residuals or reformed waves move into locally deeper water that exists in the trough between two nearshore bars (see Figure 5.15B). Thus, the situation is more complicated than for a planar slope because partial breaking can take place on the bar crest, which then ceases when the wave enters the deeper water in the alongshore trough. These partially broken waves then shoal again onto another bar or finally on the beach. Such complications have been accounted for in more recent models of the transformation of random waves (e.g. Ruessink *et al.*, 2003), but the modelling is quite advanced and demanding. A further complication is the generation of secondary waves and the growth of harmonic energy (energy at higher frequencies than that of the incident waves), which is particularly evident on beaches with multiple nearshore bars. These harmonic waves appear to be bound to the incident waves and they ride along at roughly the same speed. In the field, the appearance of many more (often smaller) wave crests in the surf zone landward of the nearshore bar than seaward of it is a manifestation of this transfer of energy to secondary wave motion (Byrne, 1969; Davidson-Arnott and Randall, 1984; Masselink, 1998). It is common for wave spectra to transform from ones that have broad singular peaks offshore of the breaking zone, to ones that have two or three different frequency peaks inside the surf zone (Figure 5.25). Although the extent of wave breaking will depend on the spectrum of incident wave heights and on the depth of water over complex nearshore bathymetry, the tide will also play an important role. Where there is a significant tidal range, wave breaking will occur farther offshore and be more intense at low tide, whereas at high tide, wave breaking (and reflection) may only occur directly on the foreshore.

5.5.6 Wave Breaking and Wave Reflection at the Shoreline

Waves that have not experienced significant shoaling and wave breaking offshore, may reach the shoreline before expending their energy. This is particularly true when the beach slope is steep and the incident waves are of small height. Both conditions are relative, as can be deduced from the Iribarren number (Equation 5.45). Smaller values of ξ are associated with wave breaking offshore, whereas larger values of ξ indicate that wave breaking (and likely some degree of wave reflection) will occur at the shoreline. Wave breaking on the shoreline usually occurs as surging or collapsing breakers, but may also take place as plunging waves. Instead of a surface roller vortex being sustained as the wave propagates across the surf zone (as occurs with a spilling breaker) the breaking process on the shoreline leads to wave run-up on the foreshore. Sometimes this takes the form of a swash bore. On shallow gradient foreshores, these swash

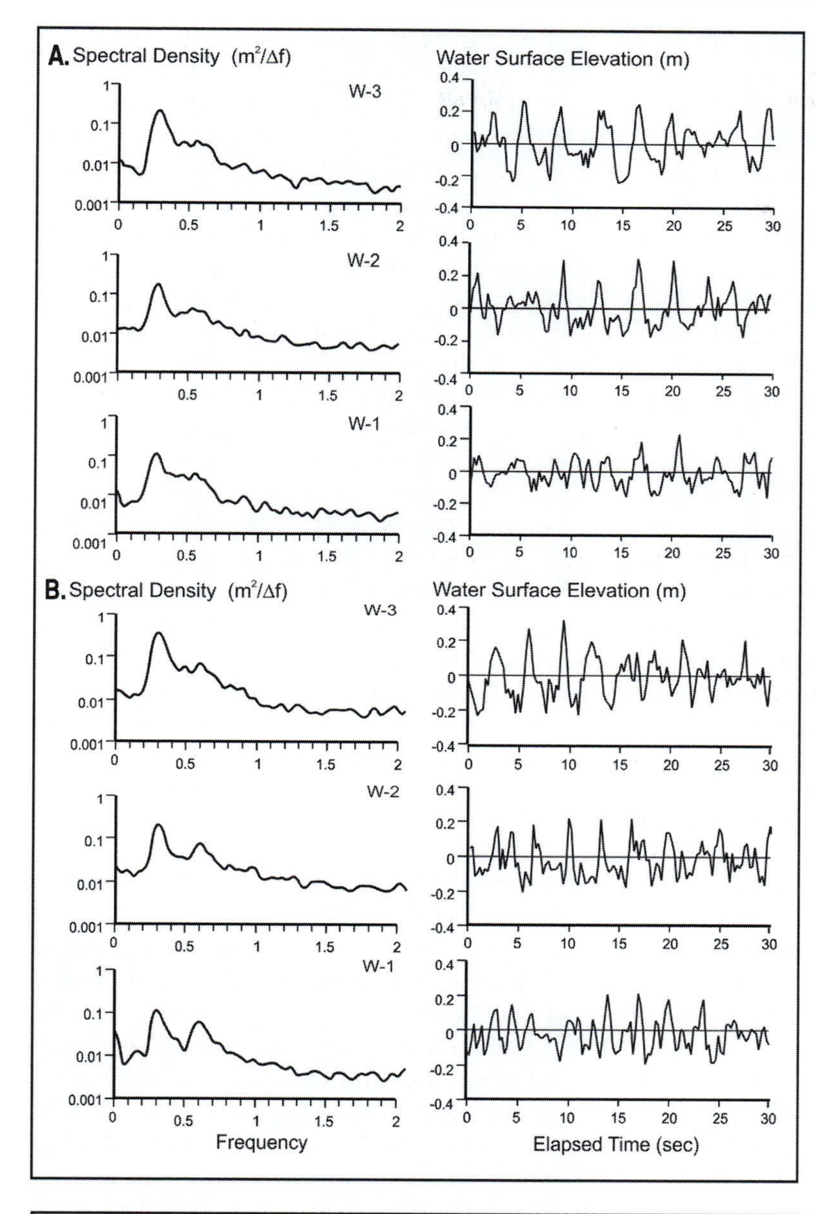

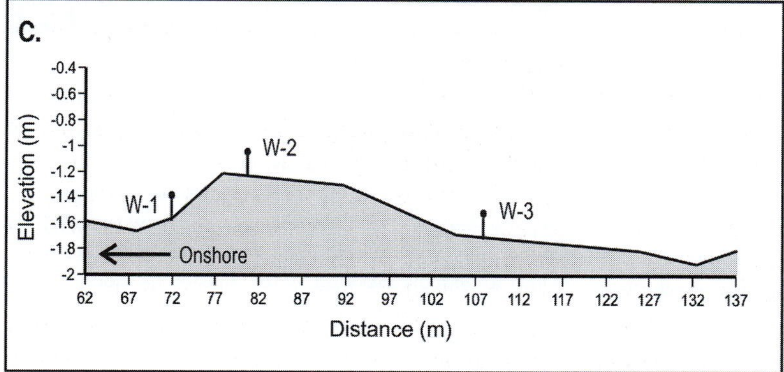

Figure 5.25 Changes in wave characteristics from offshore (W-3) to the nearshore bar crest (W-2) to the alongshore trough (W-1) during experiments at Linden Beach, Nova Scotia, during A. the rising tide and B. 95 minutes later at the high tide. Locations of wave staffs are shown in C. relative to the bar-trough bathymetry. The energy spectra (left panels) show the spectral energy density at different frequencies whereas the wave time series (right panels) show surface elevation fluctuations for the same wave groups moving through the instrument array. Note the growth in harmonic energy for positions closer to shore, especially evident in B. in the alongshore trough. The spectrum has evolved from one with a single energy peak at W-3 to one that has dual peaks at W-1. Source: Adapted from Dawson *et al.* (2002). Reproduced with permission from the Coastal Education and Research Foundation, Inc.

Figure 5.26 Swash motion on a planar foreshore at approximately mid tide, showing uprush (lower foreground) and backwash (upper background) interacting with incoming waves. The upper limit of the foreshore is approximately where the human footprints begin (upper left of photo) whereas the lower limit is demarcated by a prominent submerged beach step that undulates alongshore (Bondi Beach, Sydney, NSW, Australia).

bores look very much like surf bores or solitary waves. More often, they appear like advancing lobes of highly turbulent water, especially on steep beaches with coarse sediments. On gravel and cobble foreshores, an appreciable volume of water may infiltrate into the ground, but on fine-grained foreshores (and especially on the ebb tide when the sediments are saturated), most of the run-up returns to the sea as backwash. The interaction of backwash with the next incoming wave is critical to understanding the manner in which subsequent run-up phases evolve as well as to the evolution and maintenance of certain morphologies such as beach steps (Figure 5.26; e.g., Bauer and Allen, 1995).

On steep beaches and with small wave steepness, there can be a considerable amount of wave reflection rather than wave breaking. Wave reflection refers to the incident wave energy returning to the sea in the form of a wave that has similar period and wavelength as the incident waves. Under some circumstances, such as

Figure 5.27 Spilling breakers transitioning to surf bores (background of photo) and then to multiple swash bores arriving on a very shallow foreshore (foreground). A prior large bore has inundated the shallow foreshore, and the backwash returning to the sea is progressively thinning, thereby becoming unstable, leading to stationary surface waves and hydraulic jumps. The small bores moving onshore interact with the backwash, which impedes the shoreward progress of the swash bores (Doughboy Bay, Stewart Island, New Zealand).

a vertical cliff or sea wall with deep water at the toe (see Chapter 13), there can be almost perfect wave reflection, thereby creating a seaward propagating wave with a height equal to the incident wave. This situation is of particular interest to coastal engineers charged with designing sea walls and breakwaters. The interaction of an incident wave (height equal to H_1) with the reflected wave (height equal to H_2) will create situations in which the resultant height of the standing wave, H_{sw}, is doubled (i.e., $H_{sw} = H_1 + H_2$) if the wave periods are the same and the incident and reflected waves are phase locked. The term standing wave is in reference to the fact that the surface expression of the wave form does not appear to propagate in any particular direction, but rather stands in place. The consequence of this interaction is a sea surface that has regions in which the wave crests and wave troughs reinforce themselves, leading to extreme oscillations in water surface elevation (referred to as antinodes, where there is only up-and-down motion), and other regions where the crests and troughs cancel each other out, thereby yielding no surface elevation fluctuations whatsoever (referred to as nodes, where there is only lateral to-and-from motion). Beneath the nodes, the orbital velocities are horizontal and at their maximum with no vertical motion, whereas beneath the antinodes, the orbital velocities are vertical and at their maximum with no horizontal motion. If the antinode of a standing wave is positioned at the face of the sea wall, this can lead to bed scour and potentially undermine the structure if not built properly. In later chapters, it will become apparent how standing waves are also important to understanding nearshore morphology.

5.5.7 Wave Run-Up and Run-Down

The shoreward progression of wave energy across the nearshore zone terminates on the foreshore as swash motion, which involves wave run-up (or uprush) and wave run-down (or backwash). The upper and lower limits of uprush and backwash define the swash zone, which depends on the size of the incident waves, the beach slope, the grain size of the foreshore sediments and the tidal stage. Thus, the swash zone may occupy only a small portion of the foreshore (e.g., when the incident waves are small or in a macrotidal environment) or the entire foreshore (e.g., in non-tidal or microtidal environments with large waves). However, even in a macrotidal environment, the swash zone eventually transitions up and down the full extent of the foreshore because the wave action rides upon the mean water surface that is controlled by the tides.

Wave run-up is driven primarily by the forward momentum of the wave and is moderated by the beach slope and by infiltration into the foreshore sediments. The return flow or backwash is controlled by gravitational and frictional forces and thus depends on the beach slope, the surface roughness of the foreshore, and on the volume of water remaining to be transferred back to sea. The dynamics of the swash zone are complex, but critical to the exchange of sediment between the nearshore zone and the beach. During phases of onshore sediment transport, there is accretion on the foreshore and possibly berm building. During erosional phases, the foreshore is stripped of sediment, often leading to scarps, and sediment is moved offshore into the nearshore system.

Swash motion varies considerably along the continuum between steep and gently sloping beaches. On beaches with a gentle slope (Figure 5.27), waves arrive at the foreshore as surf bores, and the transition from bore to swash is indistinct – it can be envisaged as the depth at which turbulence in the surface roller extends to the bed. The swash zone is comparatively wide (tens of metres), the uprush velocities are small and the phases of flow acceleration gradual and long. As a result, the duration of a complete uprush–backwash cycle may be several multiples of the incident wave period and several bores may be present in the swash zone at one time. Often, there are phases of multiple bores moving progressively onshore atop a layer of deeper water, which is likely associated with the group structure of surf beat (described in the next section), followed by a lengthy period where much of the water has vacated the swash zone and only the largest surf bores are able to enter the lower foreshore regions. The water table is usually close to the surface so infiltration losses are negligible. The wave run-up distance will vary over a time scale of tens to hundreds of seconds, driven by the low-frequency oscillations of surf beat.

On steep beaches, the swash zone will be comparatively narrow, typically in the range of less than a metre to ten metres or so. The uprush and backwash velocities are greater than for gently sloping beaches, and the flow accelerations and decelerations are much more pronounced. Infiltration can be significant, especially with gravel and cobble beaches, and therefore the backwash is less voluminous than the uprush. The frequencies of run-up can be the same as that of the incident waves or somewhat longer depending on the excursion lengths and the time it takes for backwash to vacate the foreshore. However, only one or at most two waves are present in the swash zone simultaneously. Variation in run-up distance is driven by differences in the height of the breaking waves at the shoreline and by the interaction of incoming waves with the backwash. When the backwash reaches the bottom of the swash slope at the same time as another wave is just breaking on the shoreline, the interference results in a small swash excursion because the incident and returning actions tend to cancel each other. However, if the backwash phase ends before the next wave breaks, then maximum run-up will occur (see Box 5.2).

Box 5.2 Coastal Management Application

Tsunamis

A tsunami is an ocean wave of extremely long wave length and period that is generated by an impulsive disturbance that disrupts the water surface, such as an earthquake that leads to a thrust fault (vertical displacement) on the ocean floor or a submarine landslide initiated by a volcanic eruption (Fine *et al.*, 2004; Maramai *et al.*, 2005). Tsunamis are modelled as a long wave, which has different governing equations and boundary conditions than wind-generated surface gravity waves, and tsunami are therefore in a class of waves that includes seiches, tides and harbour waves. Indeed, the term 'tsunami' derives from the Japanese word for harbour wave, which is indicative of the importance of long waves to harbour safety and harbour design in seafaring nations. 'Tsunami' was adopted in the English scientific literature to avoid the common misusage of the term 'tidal wave' in reference to a tsunami. Tidal waves have a completely different origin.

Tsunamis have wavelengths measured in kilometres (rather than metres) and periods measured in tens of minutes (rather than a few seconds). Consequentially, they have travel speeds (wave celerity) of 500–800 km hr^{-1} in the open ocean (approximately half the speed of sound in air or the speed of a very fast commercial jetliner). Because of the excessively long wavelengths, they are essentially shallow-water waves even in the deep ocean, and their speed can be estimated using the shallow-water expression for wave celerity ($C = \sqrt{gh}$). Wave height in the open ocean may be less than 1 metre or so, but given the very long wavelength, they would not be noticeable on a ship. However, the extreme wavelength and their shallow-water quality imply that they are subject to shoaling and refraction processes even in the deep ocean. The tsunami wave will therefore experience significant transformations as it approaches the shoreline, such that different locations along the same coast may be subject to radically different impacts depending on the offshore bathymetry. Nevertheless, the tendency for waves to bunch up and increase their wave height is especially pronounced in a tsunami, leading to very destructive conditions along the coast with wave heights in excess of several tens of metres.

While tsunami waves can break as spilling or plunging breakers, the fact that they begin with very small wave steepness in the open ocean means that they tend to arrive at the coast as surging breakers with relatively little offshore dissipation of energy other than frictional losses at the sea bed. Catastrophic conditions on the shore are often initiated with a leading trough that vacates water from the nearshore and exposes typically submerged terrain. Unaware and curious beach-goers are lured to these unexplored areas to examine stranded sea creatures, but then are trapped by the ensuing wall of water associated with the first wave crest of the tsunami. A long period

Box 5.2 (cont.)

of highly turbulent onshore flow inundates the coast, ripping up trees, destroying buildings and mobilising rocky boulders the size of large cars (Imamura *et al.*, 2008; Goto *et al.*, 2010). This is evident in many of the photographs and videos of tsunami damage found on the web (Choowong *et al.*, 2008). Interestingly, small coral atolls sitting on seamounts with steep offshore slopes surrounded by very deep water may escape much of the potential impact because of the limited shoaling and refraction effects (Jackson *et al.*, 2005).

On 26 December 2004, an Indian Ocean tsunami was initiated by a rupture along a fault line west of northern Sumatra, Indonesia, which generated an earthquake measuring 9.1 M_W (on the moment magnitude scale) and produced a vertical displacement of the ocean floor of up to 15 m (Lay *et al.*, 2005). This kind of movement generates waves that travel in opposite directions away from the fault area (Figure 5.28), and in this instance, waves travelled westward across the Indian Ocean affecting the shoreline of India and Sri Lanka and places such as the Seychelles archipelago off the coast of Africa (Jackson *et al.*, 2005). They also travelled eastward away from the Sumatran earthquake site into the Andaman Sea affecting the coastline of Burma, Thailand and Malaysia. Approximately 280 000 people died during this tsunami, ranking it seventh in the list of the ten deadliest natural disasters on record, according to Wikipedia. While the tsunami associated with the 2004 Sumatran earthquake produced the largest loss of life on record, there are a number of other major tsunami events in recorded history, including the 1755 Lisbon earthquake (Chester, 2001), the 1964 Alaska earthquake which affected large parts of the Pacific Ocean and Hawaii, and the 1998 Papua New Guinea earthquake (Morton *et al.*, 2007). The 1929 earthquake off the Grand Banks of Canada triggered a large submarine slope failure that destroyed 12 transatlantic telegraph cables. The tsunami wave killed 28 people in Newfoundland (Fine *et al.* 2004). There has been growing recognition of the widespread occurrence of tsunamis and their potential effect on coastlines (Coleman, 1968; Young and Bryant, 1992; Clague *et al.*, 2000), especially through the investigation of sedimentary deposits associated with known tsunami events (e.g. Dawson and Stewart, 2007; Paris *et al.*, 2010; Umitsu *et al.*, 2007; Morton *et al.*, 2008). There remains some controversy over distinguishing between tsunami deposits and those due to storm events (Bryant, 2001; Felton and Crook, 2003; Kortekas and Dawson, 2007).

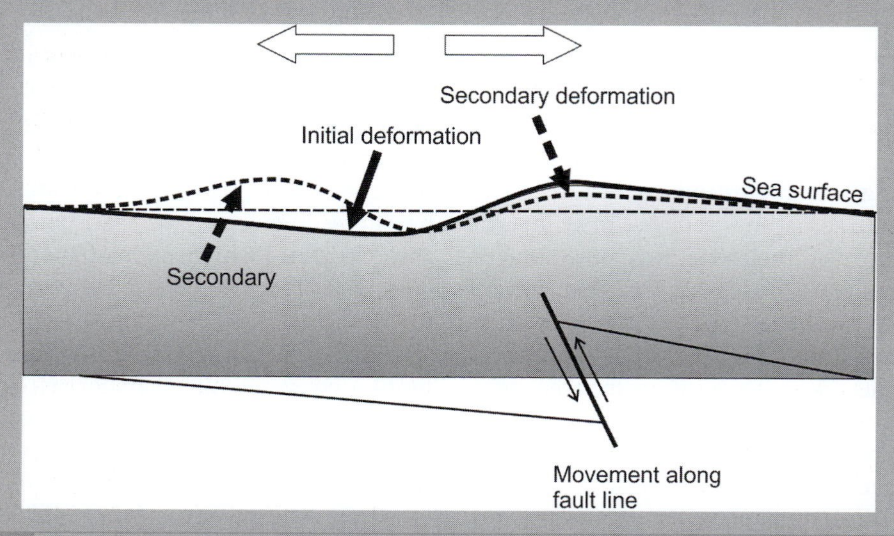

Figure 5.28 Tsunami generation by an earthquake rupture of the sea floor. Vertical thrusting motion on the sea bed yields a displacement of the ground that translates into a large deformation of the sea surface. The disturbance creates large waves that spread laterally away from the source in both directions.

Box 5.2 (cont.)

Figure 5.29 Examples of warning signs in tsunami-prone areas such as A. Timaru, south of Canterbury, New Zealand and B. north coast of California, USA.

The large number of tsunamis generated in the Pacific Ocean around the Ring of Fire prompted the US Government, through NOAA, to deploy a tsunami warning system in the region known as DART. This uses a sea-floor mounted pressure transducer system that can detect the motion associated with the long-period waves of the tsunami and is connected to a network of tsunami warning centres through a surface buoy linked to a satellite. Because of the celerity of tsunami waves, the system can only give useful warnings to coasts more than a few hundred kilometres from the point of generation. No such warning system existed in the Indian Ocean area in 2004. In many coastal regions prone to tsunami, there are warning signs posted along public access ways (Figure 5.29).

5.5.8 Radiation Stress, Mean Water Level and Wave Set-Up

Retaining non-linear terms (i.e., those with H^x, where $x > 1$) in the solutions of wave theories yields a range of interesting characteristics of waves that are sometimes referred to as integral properties because their effects require integration of the average velocity, momentum or energy, from the wave trough to the wave crest. This approach predicts a net forward movement of water (net drift velocity) associated with shoaling waves due to slight skewness and asymmetry in the velocity and acceleration fields beneath waves. When waves break, there is an additional translation of water shoreward, primarily in the crestal jet of plunging breakers or the roller vortex of spilling breakers and surf bores. These high-order effects are predictable from non-linear wave theories.

The landward translation of water with the waves is halted by the presence of the shoreline. Water cannot be 'piled up' on the shore to any substantial degree, and the principle of mass conservation requires that the water brought in by succeeding waves be transported back offshore in some fashion. Such offshore return flow does in fact occur, and it is, in part, responsible for the generation of unidirectional alongshore and offshore currents in the surf zone (see Chapter 6). However, while nearshore circulation (i.e., offshore currents) can be logically justified by the requirement that mass is conserved, a more detailed explanation of nearshore circulation is based on a consideration of the integral properties of waves, especially the flux of momentum carried by the waves.

An often-overlooked aspect of wave motion is that the presence of the waves leads to a slight reduction in the average surface water line. When the surface of the ocean or a lake is perfectly quiet, the average water surface elevation is referred to as the still water line (SWL). However, once the surface is disturbed by waves, the average surface is a bit lower than the SWL, and this new dynamic surface is referred to as the mean water line (MWL). From linear theory and with the use of the Bernoulli equation to describe the free surface, it can be shown that (Dean and Dalrymple, 1984, p. 287):

$$\bar{\eta} = -\left(\frac{H}{2}\right)^2 \left(\frac{k}{2\sinh(2kh)}\right), \quad (5.48)$$

where $\bar{\eta}$ refers to the mean water line beneath a wave field of height, H. This expression assumes that the reference sea-surface elevation is $\bar{\eta} = 0$ in deep water, despite the presence of waves, and that as these waves shoal into intermediate water depths, the MWL will be slightly depressed in proportion to the increase in wave height. This concept will be invoked below in relation to wave set-down and wave set-up.

Longuet-Higgins and Stewart (1964) showed that surface gravity waves exert a residual stress on the water column when averaged over one wave period, and this was likened to the force exerted by acoustic waves. The residual stress was termed the radiation stress, defined as the momentum flux per unit area due to the action of the waves. Radiation stress is a tensor quantity, and it has components in directions parallel to the wave advance (S_{xx}: the onshore flux of onshore momentum), parallel to the wave crest (S_{yy}: the alongshore flux of alongshore momentum), and tangential to the direction of wave advance (S_{xy}: the alongshore flux of onshore momentum). For waves approaching perpendicular to a gently-sloping planar beach there will only be one component, which is the onshore flux of onshore momentum that is approximated by:

$$S_{xx} = \bar{E}\left(\frac{1}{2} + \frac{2kh}{\sinh(2kh)}\right) = \bar{E}\left(2n - \frac{1}{2}\right). \quad (5.49)$$

In shallow water this becomes:

$$S_{xx} = \frac{3}{2}\bar{E} = \frac{3}{16}\rho g H^2. \quad (5.50)$$

For oblique wave approach angles, the expressions for the alongshore and tangential components are:

$$S_{yy} = \bar{E}\left(n - \frac{1}{2}\right) \quad (5.51)$$

and

$$S_{xy} = \frac{\bar{E}}{2}\ (n \ \sin 2\theta). \quad (5.52)$$

where θ is the angle of wave approach. As a wave enters intermediate water, and the wave height increases due to shoaling transformations, according to Equations 5.48–5.52 the momentum flux will increase progressively with wave height until the break point. Longuet-Higgins and Stewart (1964) showed theoretically and Bowen *et al.* (1968) verified experimentally that these spatial gradients in momentum flux and wave height lead to a progressive depression of the mean water level outside the breaker zone, which is referred to as wave set-down. In essence, the water level is tilted downward from the open ocean to the break point. However, at the break point, the momentum flux is released as the waves break and lose their energy (and height). The reformed waves in the surf zone will decrease in wave height from the break point to the shoreline, and therefore there is a consequent decrease in radiation stress. At the breakpoint the MWL is depressed whereas at the shoreline it is elevated, leading to an upward tilting of the MWL referred to as wave set-up inside the surf zone (Figure 5.30).

Radiation stress theory indicates that the slope of the MWL for a mildly sloping bottom is given by (Longuet-Higgins and Stewart, 1964):

$$\frac{d\bar{\eta}}{dx} = -\frac{1}{\rho g(h + \bar{\eta})}\frac{dS_{xx}}{dx}, \quad (5.53)$$

which shows that there is a change in the mean water surface wherever there are cross-shore gradients in the radiation stress. The expression

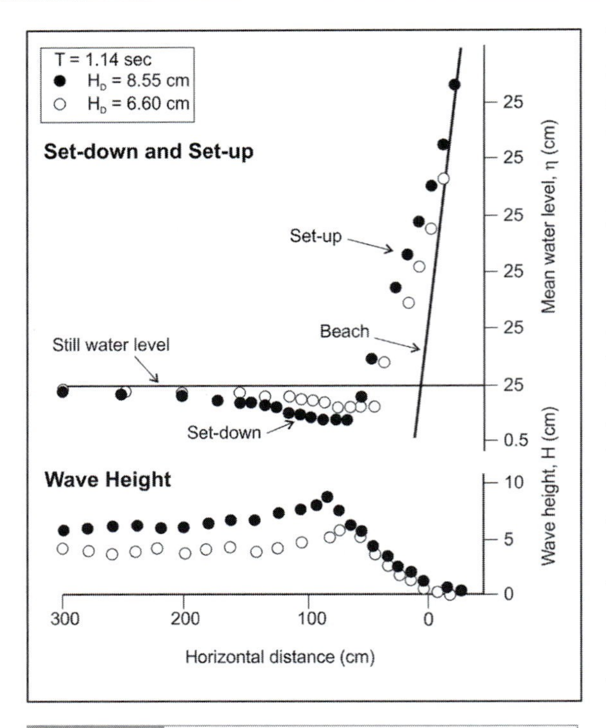

Figure 5.30 Wave set-down and set-up measured in a laboratory wave tank for two different wave heights (Bowen, 1969).

is not simple to solve because $\bar{\eta}$ appears on both sides of the equation, however, there are approximations that show that the wave set-down outside the breaker zone is typically less than about 5 per cent of the water depth at breaking (Dean and Dalrymple, 1984). Inside the breaker zone, the effect is more pronounced, and therefore wave set-up can be quite significant. With sustained action from the waves, the wave set-up is maintained as a gradual upward tilt of the water surface toward shore. However, this is inherently unstable because it yields a hydrostatic pressure gradient directed offshore. It is this pressure gradient (forced by wave set-up due to radiation stress) that ultimately provides the driving mechanism for nearshore circulation in the form of undertow and rip currents, which serve to satisfy mass conservation.

On planar beaches, the slope of the set-up under breaking waves is approximately the same regardless of incident wave height. The explanation is that larger waves will break farther offshore and thus the wave set-up occurs over a longer distance, which counteracts the fact that the set-down outside the breaker surface is larger. Set-down and set-up can be measured fairly easily in wave tanks using planar beaches (Bowen, 1969; Hsu *et al.*, 2006). Set-down and set-up have also been measured in the field (e.g. Holman and Sallenger, 1985), but conditions are complicated by irregular waves and varying profile geometries (including nearshore bars). It is also difficult to measure wave set-up when there is a wide surf zone or a relatively flat beach with a wide swash zone because the horizontal distances are very large relative to the vertical depression of the MWL. Also, the angle of wave approach under field conditions may vary considerably from shore perpendicular, especially for steep profiles where refraction effects are small. The amount of wave set-up will then depend also on the angle of wave approach at the shoreline (Hsu *et al.*, 2006). Nevertheless, wave set-up and set-down are key considerations in understanding nearshore circulation, and the concept of radiation stress is used widely to explain a range of coastal phenomena (Chapter 6).

5.6 Wave Groups and Low-Frequency Modes of Energy

Most surfers are familiar with the concept that 'every seventh wave is a big one'. Although not perfectly reliable, there is some truth to the adage. Observations and field measurements show that when individual waves arrive at the breaker zone, some appear larger while others are smaller. The larger waves break farther offshore, and often lead to more extreme wave run-up on the foreshore. These processes are not entirely random, but rather, there tend to be distinct groups of 5–10 larger waves followed by 5–10 smaller waves that are less noticeable. The phenomenon of waves arriving in groups is sometimes referred to as surf beat. Although its existence has been recognised for a long time, the details surrounding its origin and importance have only been revealed recently. In part, this is because there are several possible sources of low-frequency motion in the surf zone, all of which share the characteristic of having periods that are far in excess of the peak periods that define the incident wind-wave spectrum. In other words, rather than periods of 3–12 seconds, which define most wind-generated surface gravity waves, surf beat occurs at much longer periods, typically 20–200 seconds. These low-frequency oscillations are also referred to as infragravity motions because they exist at frequencies that are beyond (outside) the surface gravity wave band.

The actual frequency cutoff between gravity and infragravity waves is typically taken at about 25 seconds (0.04 Hz) for marine environments, but may be as short as 12–15 seconds (0.08 Hz) in fetch-limited areas such as the Great Lakes (Bauer, 1990). There are at least three types of low-frequency motion in the nearshore zone: (1) bound long waves; (2) free, leaky waves; and (3) edge waves (Figure 5.31). A concise summary is given in Hathaway *et al.* (1998). In addition, there are more exotic forms of low-frequency motions in the far infragravity (FIG) band, which are related to lateral shear instabilities of longshore currents leading to eddy vortex shedding (Bowen and Holman, 1989). These are beyond the scope of this introductory treatment of coastal geomorphology, although such motions appear to be most common on barred, nearshore profiles (Oltman-Shay *et al.*, 1989) and therefore may prove to be important to nearshore morphodynamics. Infragravity wave energy usually originates directly or indirectly from outside of the surf zone, as described by Munk (1949) and Tucker (1950) who referred to it as surf beat. Subsequent work identified it as a bound wave associated with the group structure of incident waves (Longuet-Higgins and Stewart, 1962; Huntley *et al.*, 1981; Symonds *et al.*, 1982; Haller *et al.*, 1999). Wave groups have their origin in the interaction of surface gravity waves of slightly different period. For example, a 6-second wave travelling across the water surface will move slightly faster than a 5-second wave, so the 6-second wave will overtake and pass through the 5-second wave. The combined action of the two different waves in sequence will yield locations (and times) on the sea surface where the wave crests and wave troughs reinforce each

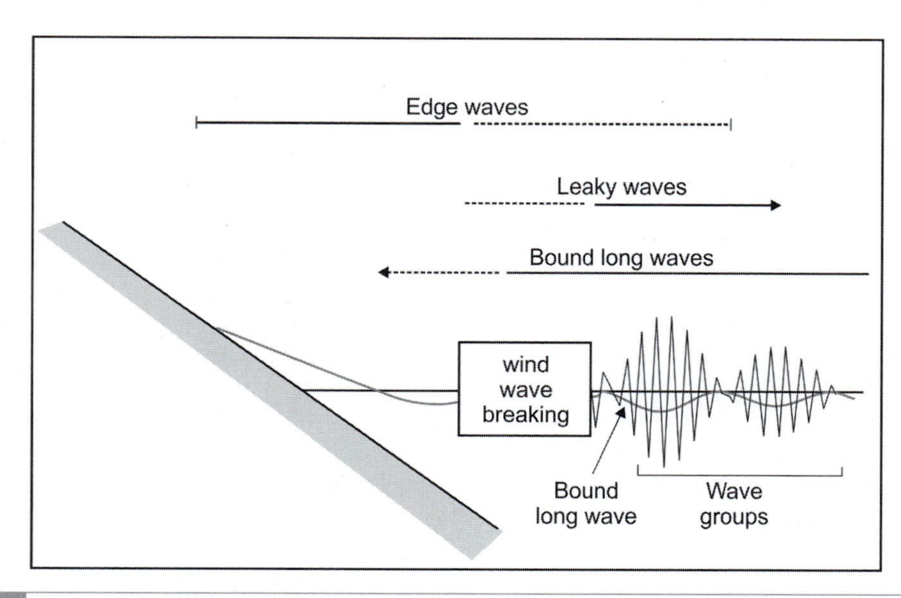

Figure 5.31 Types of infragravity wave motion in the nearshore zone. Solid lines indicate locations of greater prevalence and relative importance whereas dashed lines indicate regions of diminishing energy levels. The breaker zone for incident wind waves is indicated by the square box (after Huntley and Kim, 1984).

other and other locations where they counteract each other. It is possible to define a wave envelope that has an upper boundary that connects only the crests of a series of waves and a lower boundary that connects only the troughs. The wave envelope has a period (or wavelength) that is proportional to the difference in the frequencies of the two basic waves ($\Delta f = 1/T_1 - 1/T_2$), which in this example is 30 seconds (i.e., 0.2 – 0.167 = 0.033 Hz). In other words, the interaction of a wave field comprised of 6-second waves with a wave field of 5-second waves produces a group structure that has a period of 30 seconds. Recalling that the set-down under larger waves is greater than that under smaller waves suggests that the wave envelope must be associated with a variation in the mean water level with wavelength equal to the wave group, but 180° out of phase with it (Figure 5.32). This low-frequency oscillation of the sea surface is referred to as a bound long wave because it is dynamically forced by the wave group and travels with it at the speed of the wave group. The individual surface gravity waves that are responsible for the bound wave actually propagate through the wave group.

The bound long wave associated with the group structure of incident waves is released when the wave group is destroyed, typically at breaking. The bound wave is therefore transformed into a free mode of oscillation, much as any other progressive wave moving across the water surface, because the structure of the wave groups has been disrupted via the breaking process. The free long wave is able to propagate onshore and offshore, and it behaves much like a low-frequency surface gravity wave and is now subject to the same dynamical controls. Free long waves travelling to shore are unlikely to experience shoaling transformations because the wave heights are very small and the wavelengths are long. Such long waves usually experience reflection at the shoreline, and thus are referred to as leaky waves because energy is allowed to escape (leak) from the nearshore to the offshore (i.e., back out to sea). The interaction of a reflected long wave moving offshore and an incoming free long wave of similar period leads to the formation of a standing wave in the surf zone that is oriented parallel to the shoreline on a planar beach. These waves are of very small amplitude and are not visible to the eye, but they can be picked up by wave-measuring devices such as wave staffs, pressure sensors and current meters. If the motion is particularly

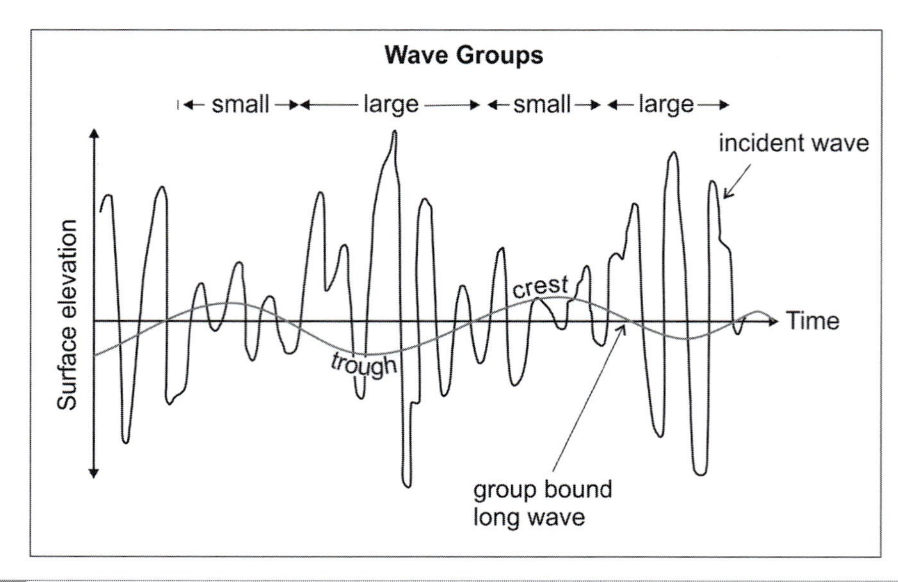

Figure 5.32 Schematic of the groupiness of incident waves and the associated bound long wave that travels with the groups (after Aagaard and Masselink, 1999).

pronounced, an energy peak will appear in the low-frequency band of the wave spectrum. Symonds *et al.* (1982) proposed a model that relates the time-varying breakpoint of the incident waves in a group to the generation of infragravity motion in the surf zone. Subsequent research has elaborated on these key ideas because of their relevance to nearshore circulation and the evolution of nearshore bars (Baldock *et al.*, 1997; Baldock *et al.*, 1999; Haller *et al.*, 1999; Svendsen and Veeramony, 2001; Karunarathna and Chadwick, 2007).

A third type of infragravity energy is referred to as an edge wave, which is very different from any of the waves discussed to this point. The central feature of edge waves is that they propagate alongshore rather than on–offshore. Although their theoretical existence was identified in the 1800s, they were thought to be nothing more than a mathematical peculiarity. It was not until the 1960s that the potential relevance of edge waves to rip currents and a range of rhythmic morphologies in the nearshore, such as beach cusps and cuspate bars, was contemplated. Edge waves are free modes of oscillation moving alongshore that are trapped to the shoreline – that is, their motion is most prevalent close to the beach rather than in the outer surf zone. They gain their energy from resonant interactions with other waves in the surf zone, usually involving triad interactions through which two waves of slightly different period will feed energy into low-frequency modes of motion at the difference frequency. In this way, it is possible for incident waves to transfer small amounts of energy into edge waves and thereby cause them to amplify over time. However, the resonant interactions are very frequency specific. If these are not maintained for lengthy time intervals, the energy will be quickly dissipated or pumped into neighbouring frequency bands. As a consequence, there may be a great deal of smearing of the energy across the infragravity band rather than the creation of a dominant wave mode that is clearly visible in an energy spectrum.

Edge waves propagate along the shoreline in sinusoidal fashion, similar to a progressive surface gravity wave if it were to travel perpendicular to the shore. The alongshore wavelength of an edge wave may be predicted for planar beaches from the following dispersion relation:

$$L_e = \frac{h}{2\pi} T_e^2 (2m - 1) \tan\beta, \qquad (5.54)$$

where L_e is the wavelength of the edge wave, T_e is the period, and m is the mode of the wave. The mode is an integer value (usually 0, 1, 2 or 3) that defines the on–offshore structure of the edge wave. The simplest mode edge wave ($m = 0$) has maximum amplitude along the shoreline and decreases exponentially in the offshore direction (Figure 5.33). Larger values of m indicate that there are nodal lines (zero crossings through the MWL) in the offshore direction, which define locations where there is no fluctuation in the sea surface due to the edge wave. The antinodal positions on either side of a nodal line will experience water level fluctuations, but the effect diminishes quickly in the offshore direction. The offshore decay of edge wave energy differentiates edge waves from low-frequency standing waves that may have the same long period but typically retain their energy in the offshore direction (Figure 5.33). Low-frequency standing waves arise because leaky waves are reflected from the shoreline and interact with incoming bound long waves to create the standing or partially standing structure. This fact is important when trying to differentiate high-mode edge waves from standing low-frequency waves in the field because they have very similar spatial structures in the cross-shore direction. However, because edge waves lose their energy quite rapidly in the offshore direction, it is unlikely that modes greater than 2 or 3 will play a significant role in nearshore morphodynamics farther offshore. Equation 5.54 indicates that the length of an edge wave is also dependent on water depth and beach slope.

The basic form of an edge wave is a progressive oscillation, meaning that it propagates alongshore in one direction or the other in sinusoidal fashion. On embayed beaches with rocky headlands or groynes at both ends, the progressive wave energy is easily reflected back in the opposite direction. In this way, a standing edge wave may be created,

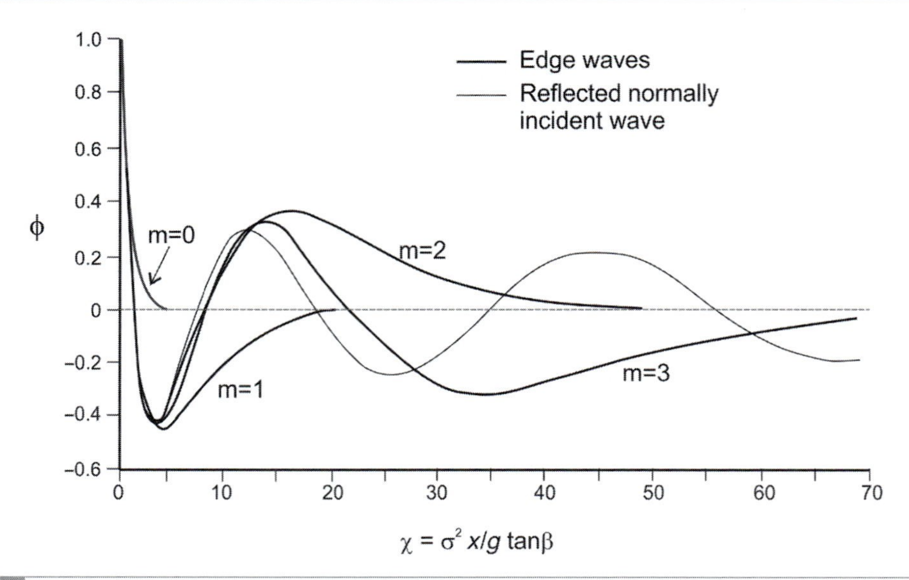

Figure 5.33 Cross-shore structure of edge waves. Both axes use non-dimensional quantities in order to generalise the diagram, with the horizontal axis showing normalised offshore distance from the shoreline and the vertical axis showing wave amplitude relative to the maximum amplitude at the shoreline. The first four modes (m = 0, 1, 2, 3) are shown, indicating that the mode number indicates how many zero crossings (nodal lines) there are in the offshore direction. High mode edge waves begin to show an offshore structure that is similar to a standing low-frequency wave (after Holman, 1983).

which is said to be 'standing' in the alongshore direction with nodal lines that extend offshore (not to be confused with the cross-shore nodal structure of higher mode edge waves). Standing edge waves are believed to provide the circulation template that allows for the creation of a range of rhythmic features on the foreshore and within the nearshore, and therefore they have been studied with great interest. A truly inspired theoretical treatment of this topic was given by Holman and Bowen (1982), but there has been relatively little field evidence for the importance of edge waves to sediment redistribution in the nearshore (cf. Bauer and Greenwood, 1990).

In practice, it requires a lot of instrumentation to determine the offshore and alongshore structure of low-frequency wave motion and to separate standing leaky waves from higher mode edge waves. Many field experiments have shown that infragravity energy in the surf zone is usually broad banded, without the selection or amplification of a particular cutoff frequency that would generate topography with a particular spatial pattern or form, as required in the model of Holman and Bowen (1982). When narrow-banded infragravity energy is detected in combination with certain types of rhythmic topography, it is not always clear that the edge waves created the topography rather than the topography having forced the low-frequency resonances that lead to the edge waves being generated. This remains one of the most intriguing chicken-versus-egg problems in nearshore morphodynamics.

Further Reading

Basco, D.R. 1985. A qualitative description of wave breaking. *Journal of Waterways, Port, Coastal and Ocean Engineering*, **111**, 171–88.

This paper provides an excellent description of the wave breaking process along with useful sketches and photographs.

Hathaway, K.K., Oltman-Shay, J., Howd, P. and Holman, R.A. 1998. Infragravity Waves in the Nearshore Zone. US Army Corps of Engineers, Technical Report **CHL-98-30**, 91 pp.

This technical report describes the range of low-frequency modes of energy often found in the near-shore zone and their potential relevance to water and sediment circulation.

References

Aagaard, T. and Masselink, G. 1999, The surf zone. In Short, A.D. (ed.), *Handbook of Beach and Shoreface Morphodynamics*. Chichester, UK: John Wiley & Sons, pp. 72–118.

Airy, G.B. 1845. Tides and waves. *Encyclopaedia Metropolitan*, **5**, 241–396.

Baldock, T.E., Holmes, P. and Horn, D.P. 1997. Low frequency swash motion induced by wave grouping. *Coastal Engineering*, **32**, 197–222.

Baldock, T.E., Huntley, D.A, Bird, P.A.D., O'Hare, T.O. and Bullock, G.N. 1999. Breakpoint generated surf beat induced by bichromatic wave groups. *Coastal Engineering*, **39**, 213–42.

Basco, D. R. 1985. A qualitative description of wave breaking. *Journal of Waterways, Port, Coastal and Ocean Engineering*, **111**, 171–88.

Battjes, J.A. 1974. Surf similarity. *Proceedings of the Fourteenth Coastal Engineering Conference*. New York: American Society of Civil Engineers, pp. 466–80.

Bauer, B.O. 1990. Assessing the relative energetics of infragravity motions in lakes and bays. *Journal of Coastal Research*, **6**, 853–65.

Bauer, B.O. and Allen, J.R. 1995. Beach steps: an evolutionary perspective. *Marine Geology*, **123**, 143–66.

Bauer, B.O. and Greenwood, B. 1988. Surf zone similarity. *Geographical Review*, **78**, 137–47.

Bauer, B.O. and Greenwood, B. 1990. Modification of a linear bar-trough system by a standing edge wave. *Marine Geology*, **92**, 177–204.

Bayram, A. and Larson, M. 2000. Wave transformation in the nearshore zone: comparison between a Boussinesq model and field data. *Coastal Engineering*, **39**, 149–71.

Bowen, A.J. 1969. Rip currents. 1. Theoretical investigations. *Journal of Geophysical Research*, **74**, 5467–78.

Bowen, A.J. and Holman, R.A. 1989. Shear instabilities of the mean longshore current: 1. Theory. *Journal of Geophysical Research*, **94**, 18023–30.

Bowen, A.J., Inman, D.L. and Simmons, V.P. 1968. Wave set-down and set up. *Journal of Geophysical Research*, **73**, 2569-77.

Bryant, E.A. 2001. *Tsunami: The Underrated Hazard.* Melbourne: Cambridge University Press, 350 pp.

Byrne, R.J. 1969. Field occurrences of induced multiple gravity waves. *Journal of Geophysical Research*, **74**, 2590–6.

Carrier, G.F. and Greenspan, H.P. 1958. Water waves of finite amplitude on a sloping beach. *Journal of Fluid Mechanics*, **4**, 97–109.

Chester, D.K. 2001. The 1755 Lisbon earthquake. *Progress in Physical Geography*, **25**, 363–83.

Choowong, M., Murakoshi, N., Hisada, K. et al. 2008. 2004 Indian Ocean tsunami inflow and outflow at Phuket, Thailand. *Marine Geology*, **248**, 179–92.

Clague, J.J., Bobrowsky, P.T. and Hutchinson, I. 2000. A review of geological records of large tsunamis at Vancouver Island, British Columbia, and implications for hazard. *Quaternary Science Reviews*, **19**, 849–63.

Cokelet, E.D. 1977. Steep gravity waves in water of arbitrary uniform depth. *Philosophical Transactions of the Royal Society A*, **286**, 183–203.

Coleman, P.J. 1968. Tsunamis as geological agents. *Journal of the Geological Society of Australia*, **15**, 267–73.

Dally, W.R., Dean, R.G. and Dalrymple, R.A. 1985. Wave height variation across beaches of arbitrary profile. *Journal of Geophysical Research*, **90**, 11917–27.

Davidson-Arnott, R.G.D. and Randall, D.C. 1984. Spatial and temporal variations in spectra of storm waves across a barred nearshore. *Marine Geology*, **60**, 15–30.

Dawson, A.G. and Stewart, I. 2007. Tsunami deposits in the geological record. *Sedimentary Geology*, **200**, 166–83.

Dawson, J.C., Davidson-Arnott, R.G.D. and Ollerhead, J. 2002. Low-energy morphodynamics of a ridge and runnel system. *Journal of Coastal Research*, **SI 36**, 198–215.

Dean, R.G. 1965. Stream function representation of non-linear ocean waves. *Journal of Geophysical Research*, **70**, 4561–72.

Dean, R.G. and Dalrymple, R.A. 1984. *Water Wave Mechanics for Engineers and Scientists*. Englewood Cliffs, NJ: Prentice-Hall, 353 pp.

Demirbilek, Z. and Vincent, C.L. 2002. Water Wave Mechanics. Chapter 1 in EM 1110-2-1100 Part 2, Coastal Engineering Manual, US Army Corps of Engineers, 121pp.

Dobson, R.S. 1967. *Some Applications of a Digital Computer to Hydraulic Engineering Problems*. Stanford, CA: Stanford University Department of Civil Engineering. Technical Report **80**, 7–35.

Douglas, S.L. 1990. Influence of wind on breaking waves. *Journal of Waterways, Port, Coastal and Ocean Engineering*, **116**, 651–63.

Duncan, J.H. 1981. An experimental investigation of breaking waves produced by a towed hydrofoil. *Proceedings of the Royal Society A*, **377**, 331–48.

Duncan, J.H. 2001. Spilling breakers. *Annual Review of Fluid Mechanics*, **33**, 519–47.

Elgar, S., Raubenheimer, B., Herbers, T.H.C. and Gallagher, E.L. 1997. Spectral evolution of shoaling and breaking waves. *Journal of Geophysical Research*, **102**, 15797–805.

Felton, E.A. and Crook, K.A.W. 2003. Evaluating the impacts of huge waves on rocky shorelines: an essay review of the book 'Tsunami: The Underrated Hazard'. *Marine Geology*, **197**, 1–12.

Fine, I.V., Rabinovich, A.B., Bornhold, B.D., Thomson, R.E. and Kulikov, E.A. 2004. The Grand Banks landslide-generated tsunami of November 18, 1929: preliminary analysis and numerical modeling. *Marine Geology*, **215**, 45–57.

Galloway, J.S., Collins, M.B. and Moran, A.D. 1989. Onshore/offshore wind influence on breaking waves: an empirical study. *Coastal Engineering*, **13**, 305–25.

Galvin, C.J. Jr. 1968. Breaker type classification on three laboratory beaches. *Journal of Geophysical Research*, **73**, 3651–9.

Galvin, C.J. Jr. 1972. Wave breaking in shallow water. In Meyer, R.E. (ed.), *Waves on Beaches and Resulting Sediment Transport*. New York; London: Academic Press, pp. 413–55.

Goto, K., Okada, K. and Imamura, F. 2010. Numerical analysis of boulder transport by the 2004 Indian Ocean tsunami at Pakarang Cape, Thailand. *Marine Geology*, **268**, 97–105, DOI: 10.1016/j.margeo.2009.10.023.

Guza, R.T. and Inman, D.L. 1975. Edge waves and beach cusps. *Journal of Geophysical Research*, **80**, 2997–3012.

Haller, M.E., Putrevu, U., Oltman-Shay, J. and Dalrymple, R.A. 1999. Wave group forcing of low frequency surf zone motion. *Coastal Engineering Journal*, **41**, 121–36.

Hathaway, K.K., Oltman-Shay, J., Howd, P. and Holman, R.A. 1998. Infragravity Waves in the Nearshore Zone. US Army Corps of Engineers, Technical Report CHL-98-30, 91 pp.

Holman, R.A. 1983. Edge waves and the configuration of the shoreline. In Komar, P.D. (ed.), *Handbook of Coastal Processes and Erosion*, Boca Raton, FL: CRC Press, pp. 21–33.

Holman, R.A. and Bowen, A.J. 1982. Bars, bumps, and holes: models for the generation of complex beach topography. *Journal of Geophysical Research*, **87**, 457–68.

Holman, R.A. and Sallenger, A.H. 1985. Set-up and swash on a natural beach. *Journal of Geophysical Research*, **90**, 945–53.

Hsu, T.W., Hsu, J.R-C., Weng, W-K., Wang, S-K. and Ou, S-H. 2006. Wave setup and setdown generated by obliquely incident waves. *Coastal Engineering*, **53**, 865–77.

Huntley, D.A. and Kim, C.S. 1984. Is surf beat forced or free? In *Proceedings of the Nineteenth Conference on Coastal Engineering*. New York: American Society of Civil Engineers, pp. 871–85.

Huntley, D.A., Guza, R.T. and Thornton, E.B. 1981. Field observations of surf beat. 1. Progressive edge waves. *Journal of Geophysical Research*, **86**, 6451–66.

Imamura, F., Goto, K. and Ohkubo, S. 2008. A numerical model for the transport of a boulder by tsunami, *Journal of Geophysical Research*, **7** (C01008), 1–12.

Iribarren, C.R. and Nogales, C. 1949. Protection des Ports. *Proceedings Seventeenth International Navigation Congress, Section II, Communication*, **4**, Lisbon, pp. 31–80.

Jackson, L.E. Jr, Barrie, J.V., Forbes, D.L., Shaw, J., Manson, G.K. and Schmidt, M. 2005. *Effects of the 26 December 2004 Indian Ocean Tsunami in the Republic of Seychelles*. Ottawa: Geological Survey of Canada, Open File 4539, 73 pp.

Karambas, T.V. and Koutitas, C. 2002. Surf and swash zone morphology evolution induced by non-linear waves. *Journal of Waterways, Port, Coastal and Ocean Engineering*, **128**, 102–13.

Karunarathna, H. and Chadwick, A.J. 2007. On low frequency waves in the surf and swash. *Ocean Engineering*, **34**, 2115–23.

Kirby, J.T. and Derakhti, M. 2019. Short-crested wave breaking. *European Journal of Mechanics/B Fluids*, **73**, 100–11, DOI: https://doi.org/10.1016/j.euromechflu.2017.11.001.

Komar, P.D. 1998. *Beach Processes and Sedimentation*, 2nd Edition, Upper Saddle River, NJ: Prentice Hall, 544 pp.

Kortweg, D.J. and de Vries, G. 1885. On the change of form of long waves advancing in a rectangular canal, and on a new type of stationary wave. *Philosophical Magazine*, **39**, 422–43.

Kotekaas, S. and Dawson, A.G. 2007. Distinguishing tsunami and storm deposits: An example from Martinhal, SW Portugal. *Sedimentary Geology*, **200**, 208–21.

Lawrence, P.L. and Davidson-Arnott, R.G.D. 1997. Alongshore wave energy and sediment transport on south-eastern Lake Huron, Ontario, Canada. *Journal of Coastal Research*, **13**, 1004–15.

Lay, T., Kanamori, H., Ammon, C.J. *et al.* 2005. The great Sumatra Andaman earthquake of 26 December 2004. *Science*, **208**, 1127–33.

Le Méhauté, B. 1976. *Introduction to Hydrodynamics and Water Waves*. New York: Springer-Verlag.

Lippmann, T.C., Brookins, A.H. and Thornton, E.B. 1996. Wave energy transformation on natural profiles. *Coastal Engineering*, **27**, 1–20.

Longuet-Higgins, M.S. and Stewart, R.W. 1962. Radiation stresses and mass transport in gravity waves with applications to surf beat. *Journal of Fluid Mechanics*, **13**, 481–504.

Longuet-Higgins, M.S. and Stewart, R.W. 1964. Radiation stress in water waves; a physical discussion with applications. *Deep Sea Research*, **11**, 529–63.

Madsen, P.A., Sorensen, O.R. and Schaffer, H.A. 1997. Surf zone dynamics simulated by Boussinesq type model. Part 1. Model description and cross-shore motion of regular waves. *Coastal Engineering*, **32**, 255–87.

Malarkey, J. and Davies, A.G. 2012. Free-stream velocity descriptions under waves with skewness and asymmetry. *Coastal Engineering*, **68**, 78–95.

Maramai, A., Graziani, L. and Tinti, S. 2005. Tsunamis in the Aeolian Islands (southern Italy): a review. *Marine Geology*, **215**, 11–21.

Masselink, G. 1998. Field investigation of wave propagation over a bar and the consequent generation of secondary waves. *Coastal Engineering*, **33**, 1–9.

May, J.P. 1974. WAVENRG: a computer program to determine the dissipation in shoaling water waves with examples from coastal Florida. In Tanner, W.F. (ed.), *Sediment Transport in the Nearshore Zone*. Tallahassee, FL: Florida State University Department of Geology: Coastal Research Notes, pp. 22–80.

McCowan, J. 1894. On the highest wave of permanent type. *Philosophical Magazine*, **38**, 351–8.

Mei, C.C. 1983. *The Applied Dynamics of Ocean Surface Waves*. New York: John Wiley & Sons, 740 pp.

Miller, R.L. 1976. Role of vortices in surf zone prediction: sedimentation and wave forces. In Davis, R.A. Jr and Ethington, R.L. (eds.), *Beach and Nearshore Sedimentation*. S.E.P.M. special publication **24**, pp. 92–114.

Morton, R.A., Gelfenbaum, G. and Jaffe, B.E. 2007. Physical criteria for distinguishing sandy tsunami and storm deposits using modern examples. *Sedimentary Geology*, **200**, 184–207.

Morton, R.A., Goff, J.R. and Nichol, S.L. 2008. Hydrodynamic implications of textural trends in sand deposits of the 2004 tsunami in Sri Lanka. *Sedimentary Geology*, **207**, 56–64.

Munk, W.H., 1949. The solitary wave theory and its application to surf problems. *New York Academy of Science*, **51**, 376–424.

Munk, W.H. and Traylor, M.A. 1947. Refraction of ocean waves; a process linking underwater topography to beach erosion. *Journal of Geology*, **55**, 1–26.

Okazaki, S-I and Sunamura, T. 1991. Re-examination of breaker type classification on uniformly inclined laboratory beaches. *Journal of Coastal Research*, **7**, 559–64.

Oltman-Shay, J., Howd, P. and Birkemeier, W.A. 1989. Shear instabilities of the mean longshore current: 2. Field Observations. *Journal of Geophysical Research*, **94**, 18031–42.

Paris, R., Fournier, J., Poizot, E. *et al.* 2010. Boulder and fine sediment transport and deposition by the 2004 tsunami in Lhok Nga (western Banda Aceh, Sumatra, Indonesia): a coupled offshore-onshore model. *Marine Geology* **268**, 43–54.

Peregrine, D.H. 1983. Breaking waves on beaches. *Annual Review of Fluid Mechanics*, **15**, 149–78.

Ruessink, B.G., Walstra, D.J.R. and Southgate, H.N. 2003. Calibration and verification of a parametric wave model on barred beaches. *Coastal Engineering*, **48**, 139–49.

Sleath, J.F.A. 1984. *Sea Bed Mechanics*. New York: John Wiley & Sons, 335 pp.

Stokes, G.G. 1847. On the theory of oscillatory waves. *Transactions Cambridge Philosophical Society*, **8**, 441–5.

Sunamura, T. 1992. *Geomorphology of Rocky Coasts*. Chichester, UK: John Wiley & Sons, 302 pp.

Svendsen, I.A. 1984. Wave heights and set-up in a surf zone. *Coastal Engineering*, **8**, 303–29.

Svendsen, I.A. and Veeramony, J. 2001. Wave breaking in wave groups. *Journal of Waterways, Port, Coastal and Ocean Engineering*, **127**, 200–12.

Svendsen, I.A., Madsen, P.A. and Buhr Hansen, J. 1978. Wave characteristics in the surf zone. *Proceedings of the Sixteenth Coastal Engineering Conference*, New York: American Society of Civil Engineers, pp. 520–39.

Swart, E.H. and Loubster, C.D. 1978. Vocoidal theory for all non-breaking waves. *Proceedings of the Sixteenth Conference on Coastal Engineering*. New York: American Society of Civil Engineers, pp. 467–86.

Symonds, G., Huntley, D.A. and Bowen, A.J. 1982. Long wave generation by a time-varying breakpoint. *Journal of Geophysical Research*, **87**, 492–8.

Thornton, E.B. and Guza, R.T. 1982. Energy saturation and phase speeds measured on a natural beach. *Journal of Geophysical Research*, **87**, 9499–508.

Ting, F.C.K. and Kirby, J.T. 1995. Dynamics of surf-zone turbulence in a strong plunging breaker. *Coastal Engineering*, **24**, 177–204.

Tucker, M.J. 1950. Surfbeats: sea waves of 1 to 5 minutes period. *Proceedings Royal Society London A*, **202**, 565–73.

Umitsu, M., Tanavud, C. and Patanakanog, B. 2007. Effects of landforms on tsunami flow in the plains of Banda Aceh, Indonesia, and Nam Khem, Thailand. *Marine Geology*, **242**, 141–53.

Veeramony, J. and Svendsen, I.A. 2000. The flow in surf-zone waves. *Coastal Engineering*, **39**, 93–122.

Weishar, L.L. and Byrne, R.J. 1978. Field study of breaking wave characteristics. *Proceedings of the Sixteenth Coastal Engineering Conference*, New York: American Society of Civil Engineers, pp. 487–506.

Wiegel, R.L. 1954. *Gravity Waves: Tables of Functions*. Berkeley, CA: University of California, Council of Wave Research, The Engineering Foundation, 300 pp.

Wilson, W.S. 1966. A method for calculating and plotting surface wave rays. US Army Corps of Engineers. CERC Technical Memorandum, **17**.

Wright, L.D. and Short, A.D. 1984. Morphodynamic variability of surf zones and beaches: a synthesis. *Marine Geology*, **56**, 93–118.

Young, R.W. and Bryant, E.A. 1992. Catastrophic wave erosion on the southeastern coast of Australia: impact of the Lanai tsunami ca. 105 ka? *Geology*, **20**, 199–202.

6

Surf Zone Circulation

6.1 Synopsis

As waves shoal, there is an onshore-directed transfer of momentum and mass as the wave height increases and the wavelength decreases. Through the zone of wave shoaling, this transfer results in the depression of the water surface called set-down. The transfer of mass resulting from Stokes drift reaches a maximum at the breakpoint, where wave breaking provides an additional transfer of mass. The landward transfer of mass by Stokes drift and breaking, leads to an increase in wave level through the surf zone and at the shoreline called set-up. Depending on the angle that the waves approach the shoreline, the onshore-directed transfer of mass results in some combination of alongshore and/or offshore-directed flow. If the waves are approaching at or near perpendicular to the beach the return flow takes the form of either a two-dimensional undertow or a three-dimensional rip cell circulation. When waves approach the shoreline at an angle, an alongshore-directed flow develops, with the sum of the alongshore and offshore-directed flows equal to the onshore-directed flow. The resulting combination of undertow, rip and alongshore currents partly determines the transport of sediment through the surf zone and the development of nearshore and beach morphology. These currents are also important for marine organisms living in the bed and in the water column.

If wave breaking is uniform alongshore, the onshore-directed transfer of mass is returned seaward as undertow, between the bed and the wave trough. Undertow, also known as bed return flow tends to be uniform alongshore and, because the cross-sectional area of discharge is large, the return flow is diffuse and current speeds are relatively small. The speed of the undertow current is < 0.05 m s^{-1} under low waves to > 0.20 m s^{-1} in intense storms. Because this flow takes place in the lower part of the water column where sediment concentrations are largest, it is an important process in the seaward transfer of sediment and for the development and migration of nearshore bars. If the nearshore bar is landward of wave breaking, the bar will migrate seaward with the undertow until it reaches the breakpoint where the offshore-directed flow of the undertow and the onshore-directed flow as Stokes drift are balanced.

If wave breaking is not uniform alongshore, the offshore-directed current becomes concentrated in narrow zones (rips) that periodically breach the breaker line alongshore. The offshore-directed flows in the rips are fed from two directions by alongshore currents in the surf zone called feeder currents. Because the cross-sectional area of discharge is relatively small, the current speeds are much faster than in undertow, on the order of 0.5 m s^{-1}, with flows exceeding 2 m s^{-1} in intense rips. The development of rip currents comes at the expense of the undertow current, leading to onshore-directed transport (and bar migration) between rips and offshore-directed transport within the rips. As a result, the bars tend to migrate landward with rips and offshore with undertow.

When and where waves approach the shoreline at an angle, a portion of the momentum flux and mass transport is directed alongshore, generating a current that flows alongshore current in the direction of wave approach. The speed of the longshore current strengthens with increasing wave height and increasing angle between the wave crest and the shoreline. Longshore currents coexist in the surf zone with undertow and rip cell circulation, and any increase in the longshore current comes at the expense of the speed of the cross-shore directed undertow and rip currents. The longshore current transports sediment alongshore, and increases or decreases in the strength of the current alongshore is responsible for erosion or accretion of the shoreline.

6.2 Undertow

Waves begin to shoal when they first 'feel' the bottom at half of their wavelength (L). As described in Section 5.4.1, wave shoaling leads to an increase in wave height and a decrease in wavelength and celerity. In turn, the deformation of the wave form produces a net drift velocity, Stokes drift, in the direction of wave advance which increases up to the breaker zone. The onshore-directed current increases in speed up to the breakpoint, and may still be evident through part of the surf zone if there are spilling breakers. An additional and substantial landward transfer of water occurs at the breakpoint, in which the water in the crest is thrown landward creating turbulent flow structures in the surf zone.

As we saw in Chapter 5, the initial breaking process can be quite mild, producing spilling breakers that result in a slow dissipation of energy and a gradual reduction in wave height, or violent with plunging breakers in which the overturning jet plunges into the trough ahead of the wave and there is a rapid dissipation of energy and reduction in wave height. On a simple planar profile (see Figure 6.1A), the breaking wave is transformed into a bore with a surface roller that decays relatively uniformly across the inner surf zone towards the swash zone. In the vertical, within the breaker and surf zones, a wave can be divided into three parts (Figure 6.1B): (1) the upper zone located primarily above the trough of the waves where the rapid transformation takes place (breaking) and where mass transport is primarily landward; (2) the middle zone located from the bottom of the wave trough to just above the bed where the turbulent flow structures generated in the upper layer break down and where the mean flow is directed offshore (as undertow); and (3) the boundary layer close to the bed where there is interaction between the waves, currents, and the bed resulting in a mean flow generally being directed onshore (Christensen *et al.*, 2002).

It is difficult to measure the mass flux in the area above the wave trough because instruments do not work well in the presence of bubbles generated by the turbulence in the roller or if they are periodically exposed. However, it is possible using first-order linear wave theory to estimate the mass transport above the level of the trough (Phillips, 1977; Dally and Dean, 1986; Masselink and Black, 1995) as:

$$Q_s = \frac{1}{8}\sqrt{\frac{g}{h}}H^2. \tag{6.1}$$

Each breaking wave adds to the water in the surf zone landward of the breaker line, and this excess water piles up against the beach and must eventually return offshore to balance the onshore flow. The mass transport offshore by undertow (Q_{undertow}) is equal to the sum of the mass transport by Stokes drift (Q_{drift}) and breaking waves (Q_b):

$$Q_{\text{undertow}} = Q_{\text{drift}} + Q_b. \tag{6.2}$$

A quick estimate of the speed of the undertow current (V_{undertow}) can be calculated assuming an average depth (h) through the surf zone:

$$V_{\text{undertow}} = \frac{Q_{\text{drift}} + Q_b}{h}. \tag{6.3}$$

While this mass balance approach may appear simple, it requires input of a larger number of parameters, and it becomes even more complex

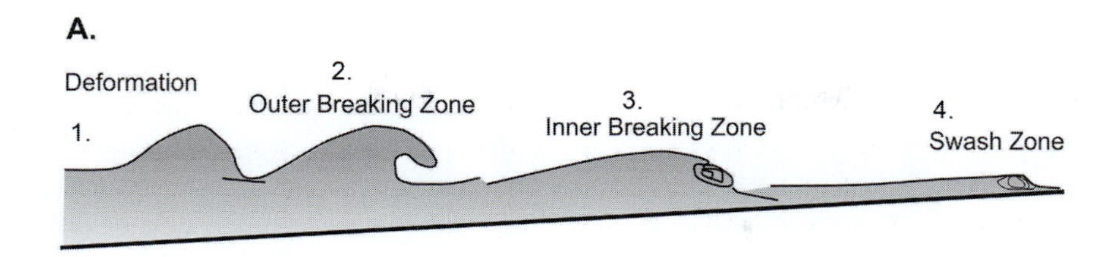

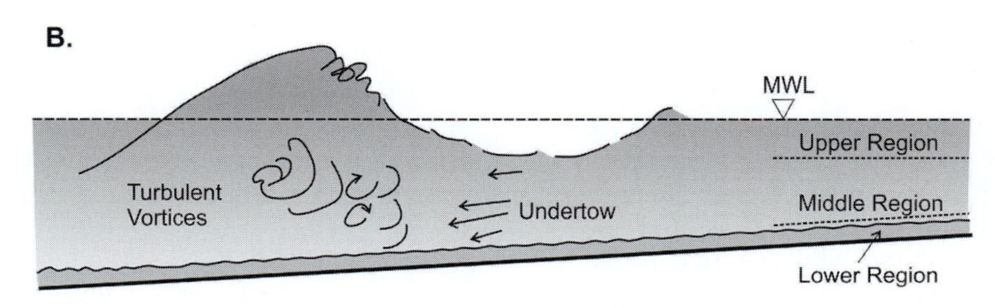

Figure 6.1 Zonation of the surf zone on a simple planar profile: A. four regions can be identified across the surf zone beginning with 1. the zone of strong wave deformation just before breaking; 2. the area of breaking and rapid transformation in the outer surf zone; 3. the inner surf zone characterised by slowly varying surf bores; and 4. the swash zone where wave run-up occurs. B. In the outer and inner surf zones the profile can be divided vertically into a region above the wave trough where breaking occurs, a middle zone dominated by turbulence generated by wave breaking and undertow, and a thin boundary layer between the middle zone and the bed.

if allowance is made for winds and angle of wave incidence (Haines and Sallenger, 1994). Recent models have built on the approach of Dally and Dean (1984) and Svendsen (1984a, b), and have incorporated this into 3D models of flow in the surf zone. Improvements to the modelling have also come with the application of Boussinesq equations to provide a better representation of the non-linear properties of wave motion at breaking (e.g., Musumeci *et al.*, 2005).

The forces necessary to drive undertow offshore comes from the set-up (or elevated water) at the shoreline, which produces an offshore-directed pressure gradient. The turbulent shear stresses created by the undertow current balances the difference between the momentum flux gradient and the pressure gradient (Ting and Kirby, 1994). While it is possible to model the set-up in terms of the mass transfers of water, the more general case is modelled based on the flux of momentum as described in Chapter 5 (Longuet-Higgins and Stewart, 1964). Note, however, that this approach to calculating the set-up only accounts for the momentum flux due to waves, and does not account for any additional set-up due to storm surge.

Box 6.1 Measurement of Surf Zone Currents

Measurement of currents on the upper shoreface, and especially in the surf zone, is made difficult by the rapid reversal of direction associated with wave oscillatory motion (Davidson-Arnott, 2005). Oceanographers routinely deployed mechanical current meters in deep and intermediate water depth to measure tidal and wind-driven

Box 6.1 (cont.)

circulation even prior to World War II. Similar in design to instruments used in rivers and air, these mechanical meters used some form of Savonius rotor to measure speed, and often a vane to measure the direction. When wave motion was present it was generally possible to assume that it was nearly uniform and to extract a mean value. However, in shallow water just outside the breaker zone, and in the surf zone, the oscillatory motion was too strong and too rapid to permit this type of instrument to be deployed. Most surf zone measurements in the 1950s and 1960s were average values obtained by tracking drogues or the movement of a cloud of dye.

While small, fast response ducted impellor meters were used successfully in Australia in the late 1970s (Wright *et al.*, 1982). It was the introduction of the electromagnetic current meter in the late 1960s that permitted measurement of currents at 4 Hz or faster and facilitated the isolation of both high- and low-frequency motions (Huntley and Bowen, 1973). The current meters make use of Faraday's law of electromagnetic induction in which the movement of a conductor (water) through a magnetic field generates an electromotive force that is proportional to the relative motion of the fluid and magnetic field. The current meters typically have two orthogonal pairs of sensors to measure the u and v components of flow in a plane. They are commonly paired in surf zone studies with an optical backscatterance probe (OBS) which provided a rapid measure of suspended sediment concentration in a small sampling volume (Downing *et al.*, 1981). Together these made it possible to measure sediment flux at various heights in the water column and spatially alongshore and cross-shore.

While electromagnetic current meters and OBS probes are still in use, they are now being replaced by acoustic Doppler velicometers (ADVs) and shallow-water acoustic Doppler current profilers (ADCPs). These instruments measure velocity over a depth range using the Doppler effect of sound waves reflected from particles and bubbles moving in the current at a single point (ADV) or at multiple points within the water column (ADCP), and can measure both velocity and sediment concentration to estimate the sediment transport. Further background on the theory that underlies the application of acoustic transducers and their ability to measure suspended sediment is summarised in Hay and Sheng (1992) and Thorne and Hurther (2014). Adoption of these instruments has significantly improved our understanding of flow structures and transport in rip channels (e.g., MacMahan *et al.*, 2005; Bruneau *et al.*, 2009; Orzech *et al.*, 2010; Houser *et al.*, 2013), transport over rippled beds (e.g., Masselink *et al.*, 2007), the exchange of sediment between offshore sand banks and the nearshore (e.g., Hequette *et al.*, 2009), transport within the swash zone (e.g., Houser, 2013; Puleo *et al.*, 2013), nearshore bar migration (e.g., List *et al.*, 2011; Cartier and Hequette, 2013) and longshore sediment transport (e.g., Cartier and Hequette, 2011). It is possible to measure sediment transport in the nearshore because it is reasonable to assume that sediment properties remain essentially constant in both time and space, but it is difficult to make this assumption in estuarine and deltaic environments where sediment and its scattering properties can vary significantly (e.g., Sassi *et al.*, 2012; Bradley *et al.*, 2013).

6.2.1 Laboratory Measurements

The offshore-directed return flow produced by wave advance and breaking can be readily demonstrated in a wave tank (flume), and observations of the flow have been described for a long time using dye tracers (Bagnold, 1940; Longuet-Higgins, 1983). In the past three decades, rapid advances in instrumentation beginning with small electromagnetic flow meters and more recently with acoustic Doppler velocimetry (ADV), laser Doppler velocimetry (LDV) and particle image velocimetry (PIV) have provided details of wave kinematics and permitted rapid advances in modelling of flows under waves and within the wave or wave–current boundary layer (e.g., Afzal *et al.*, 2015; Yuan and Madsen, 2015; Yuan, 2016). The general form of the flows is shown in Figure 6.2A and a comparison of early wave tank experiments with theoretical predictions from Stive and Wind (1986) is shown in Figure 6.2B.

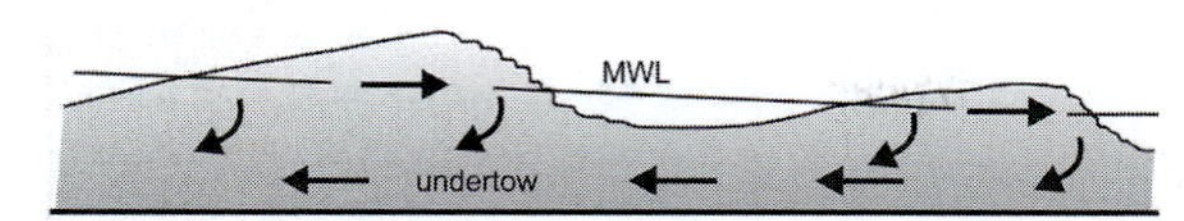

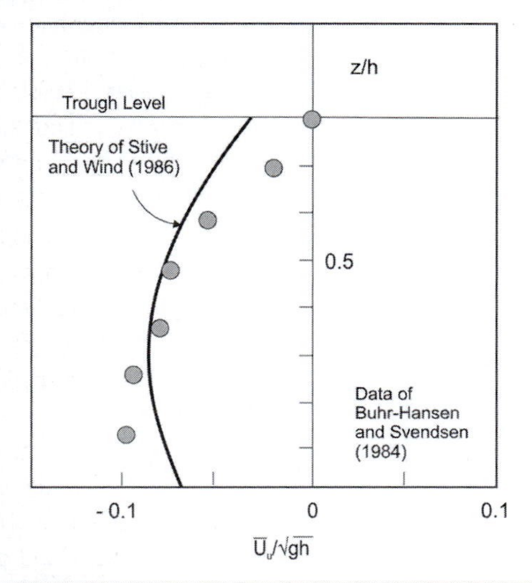

Figure 6.2 Flow velocity associated with undertow in a wave tank: A. schematic of the pattern of net flow associated with undertow (after Svendsen, 1984a); B. measurements of undertow in a wave tank (Komar, 1998; after Stive and Wind, 1986).

6.2.2 Field Measurements

In the 1950s through the 1970s, much of the focus of attention in field studies was on rip cell circulation based particularly on work carried out on the US west coast and on the east coast of Australia. There was a tendency to suggest that in the field all circulation was three dimensional and that 'undertow' was used by ignorant lay people who mistakenly labelled the offshore flow in a wave trough or in a rip current as undertow. However, the existence of persistent offshore flows at mid depth was reported in several field studies in the 1980s (Wright *et al.*, 1982; Greenwood and Sherman, 1984; Davidson-Arnott and McDonald, 1989; Haines and Sallenger, 1994 Plac). Since then there have been considerable advances in describing the characteristics of undertow observed in the field both on barred (e.g., Greenwood and Osborne, 1990; Masselink and Black, 1995; Aagaard *et al.*, 1997; Masselink, 2004; Goodfellow and Stephenson, 2005; Houser *et al.*, 2006), and on non-barred profiles (e.g., Miles and Russell, 2004; Reniers *et al.*, 2004). Because undertow occurs (relatively) uniformly along the beach, velocities are much lower than those associated with rips and are characteristically on the order of 0.05–0.5 m s^{-1}. The current usually does not extend very far beyond the breaker line.

Measurements using wave staffs and bi-directional electromagnetic current meters at Wasaga Beach in Georgian Bay (Lake Huron) carried out over the period 1983–5, showed that circulation over the straight outer bars of the multiple bar system was dominated by undertow and no evidence was found for the existence of rip currents (Davidson-Arnott and McDonald, 1989). Mean offshore flows at mid depth were characteristically < 0.1 m s^{-1} under moderate storm conditions and seldom exceeded 0.2 m s^{-1} (Figures 6.3, 6.4, 6.5). The set-up from wave breaking is complicated by the variable topography of the bars and by the effects of wind stress but some estimate of the magnitude can be obtained by comparing set-up at wave staff 5, just outside the main zone of breaking with that at wave staff 6 (Figure 6.3E). Mean flows are generally directed offshore and the magnitude of the flow reflects changes in significant wave height and the measure set-up through the storm. Measurements here and at Wymbolwood Beach 20 km to the east (Greenwood and Osborne, 1990; Osborne and Greenwood, 1992) show that the strength of the return flow is dependent on the height of the set-up, which in turn is controlled by the rate of energy dissipation (or breaking) over the bars, and particularly on the seaward slope of individual bars. Thus, mean offshore flows during a moderate storm tend to be greater over the second (innermost) bar where breaking is more intense compared to the outer bar where most of the waves shoal but do not necessarily break (Figure 6.4).

Measurements on the bar crest during one storm using a vertical array of current meters

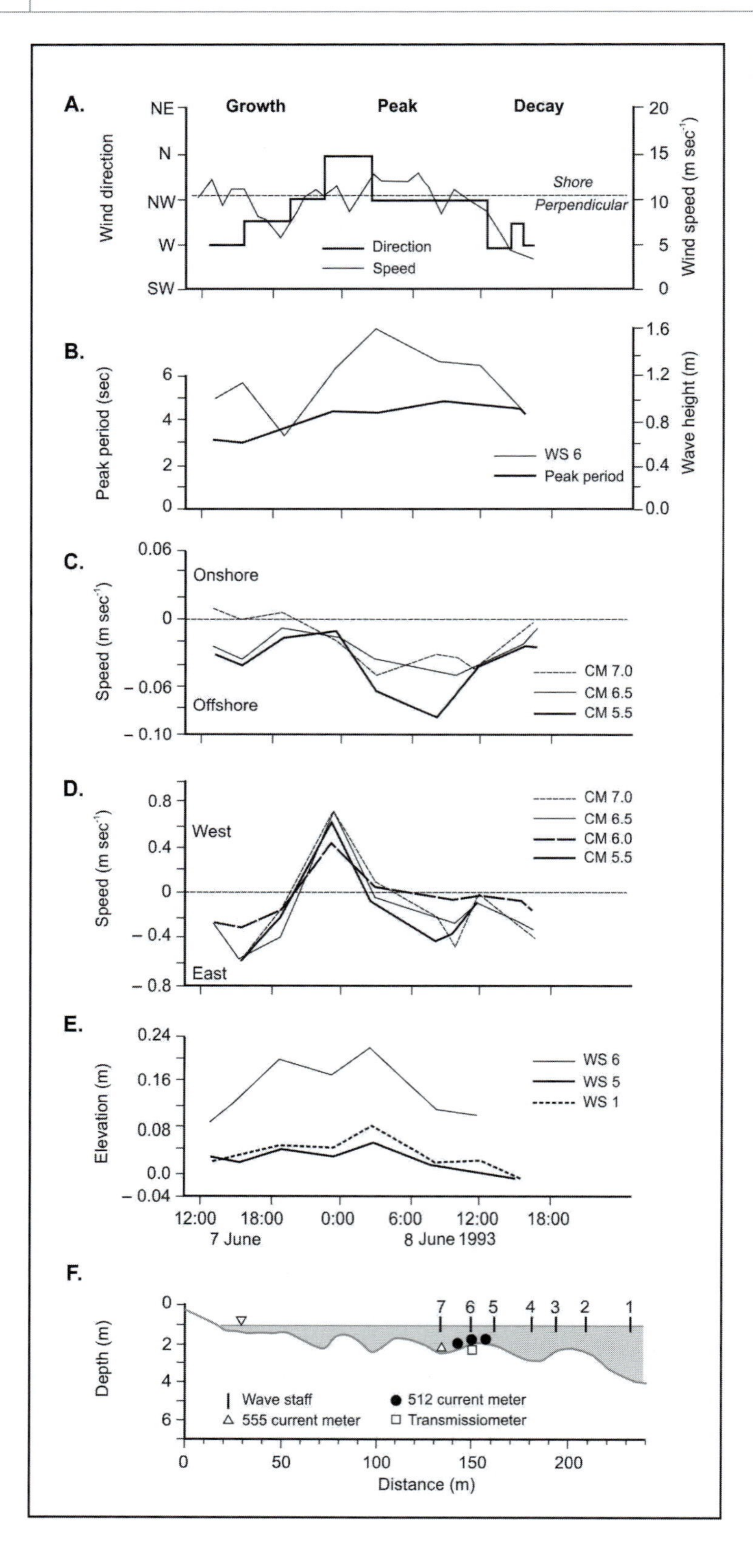

Figure 6.3 Mean flows over a straight nearshore bar at Wasaga Beach during a moderate storm: A. wind speed and direction measured at the site; B. variations in significant wave height and peak period at two wave staffs; C. mean on–offshore and D. mean alongshore speeds measured at three locations; E. mean water level measured at three wave staffs; F. profile of nearshore bar system and instrumentation location. The mean grain size is about 0.10–0.12 mm and tan β = 0.005 (Davidson-Arnott and MacDonald, 1989).

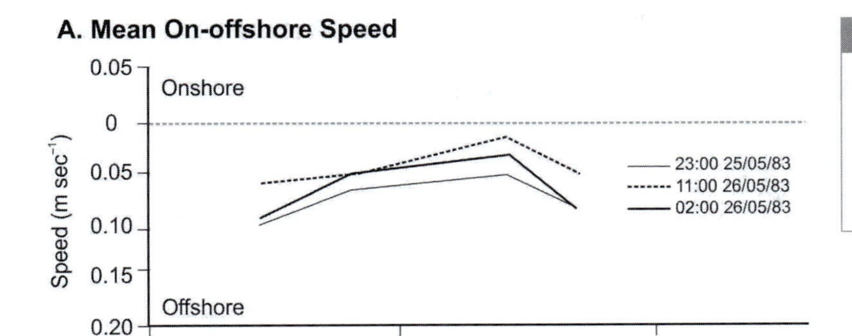

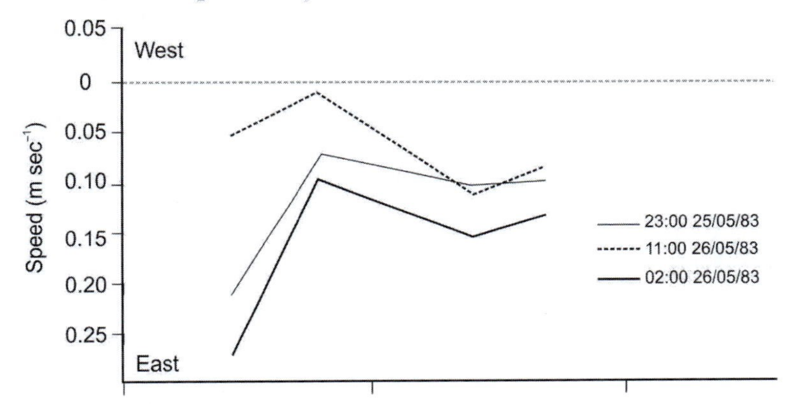

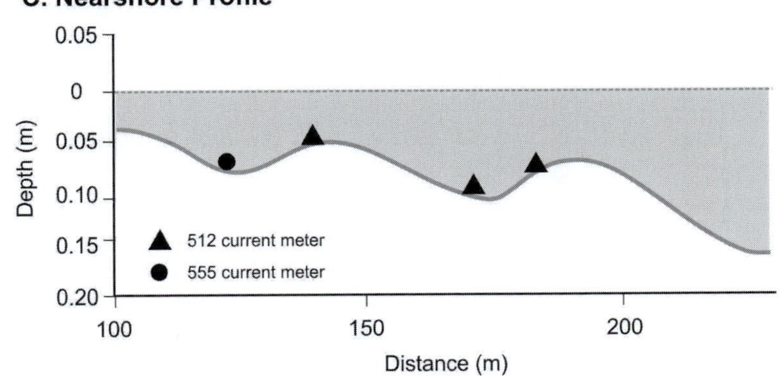

Figure 6.4 Variations in on-offshore and alongshore mean flows across two outer bars at Wasaga Beach during a moderate spring storm, 25–26 May 1983 (Davidson-Arnott and MacDonald, 1989).

showed that the maximum offshore flow was located at mid depth with flows decreasing towards the bed and towards the surface (Figure 6.5). This is in general agreement with wave tank measurements and predictions from theory, but there is insufficient vertical control to determine the form of the velocity profile (the vertical flow structure) explicitly. In part this is a function taking measurements at only three heights, but it also reflects the difficulties of field instrumentation where the upper instrument needs to be positioned below the wave troughs of irregular waves and where the height of the lower instrument above the bed may change because of scour or accretion during the storm event. Recent studies such as those at the Field

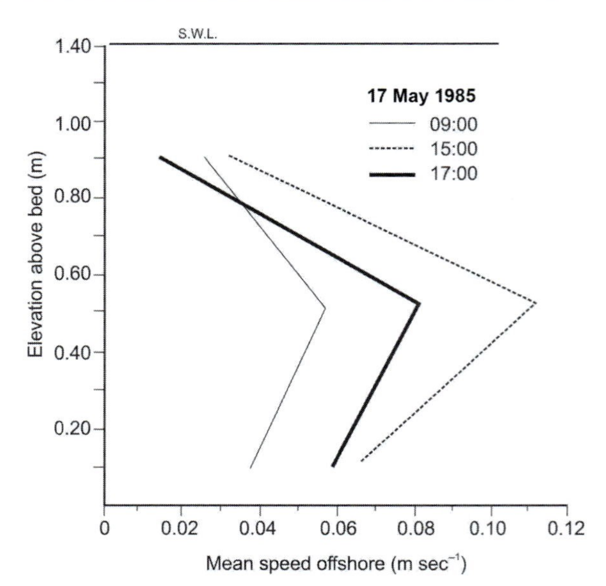

Figure 6.5 Vertical variations in mean flow over the crest of the second outer bar at Wasaga Beach during the growth, peak and decay of a moderate storm (Davidson-Arnott and McDonald, 1989).

Research Facility at Duck, North Carolina (Garcez Faria *et al.*, 2000; Reniers *et al.*, 2004) using a vertical array of eight current meters on a towed sled have provided a much better description of vertical profile. An alternative approach is to deploy the sensors at a site with a large tidal range and thus to make measurements when the instruments become fully submersed (e.g., Miles and Russell, 2004; Houser *et al.*, 2006).

6.3 Rip Cells

On some sandy beaches, the offshore return flow occurs as part of a complex three-dimensional system, termed a rip-circulation cell, that involves a strong narrow current called a rip current. These systems were first noted on the California coast by surfers, and the first observations and field measurements were provided by Shepard *et al.* (1941) and Shepard and Inman (1950). Rip currents (often called 'rips' or 'rip tides') are considered a public health hazard globally (Castelle *et al.*, 2016) and have been the focus of numerous studies in recent years through the cumulative efforts of theoretical, field, modelling and remote video monitoring studies. As a result, our understanding of the formation and behaviour of rip currents is relatively well developed (see Dalrymple *et al.*, 2011; Castelle *et al.*, 2016).

Unlike the vertical flow pattern associated with the undertow, the rip circulation system is described in the horizontal (Figure 6.6A). If waves approach nearly perpendicular to the beach, the idealised model of rip circulation involves the onshore mass transport of water due to breaking waves results in the formation of two converging longshore feeder currents close to the shoreline, which meet and turn seawards as a narrow, fast-flowing rip neck. The rip neck flows through the surf zone before decelerating and dissipating as an expanding rip head seaward of the surf zone. The nearshore circulation cell is completed when this water is transported shoreward again by the breaking waves. On long sandy beaches subject to swell waves, rip currents develop at quasi-regular intervals alongshore and are often visible as the 'quieter' water surface between breaking waves, by the foam, flotsam and suspended sediments associated with the rip head (Figure 6.6B), and a cuspate beach morphology (Figure 6.6A). Because of wave breaking, it is possible to outline the general nearshore topography by averaging the light intensity from a video sequence or series of high-frequency photographs (Holman *et al.*, 2006; Turner *et al.*, 2007). The bars and foreshore show up as light areas because of averaging of the white foam produced by wave breaking and the troughs and rip channels are dark because of the general absence of breaking there. An example of this application for a complex inner bar is shown in Figure 6.7, showing several quasi-permanent rip channels along the popular Casino Beach on Santa Rosa Island near Pensacola, Florida.

Rip cell circulation is typically associated with barred nearshore zones, and particularly with the three-dimensional or rhythmic topography that is characteristic of inner bar systems (Sonu, 1972; Greenwood and Davidson-Arnott, 1975; 1979; Short, 1979; Wright and Short, 1984;

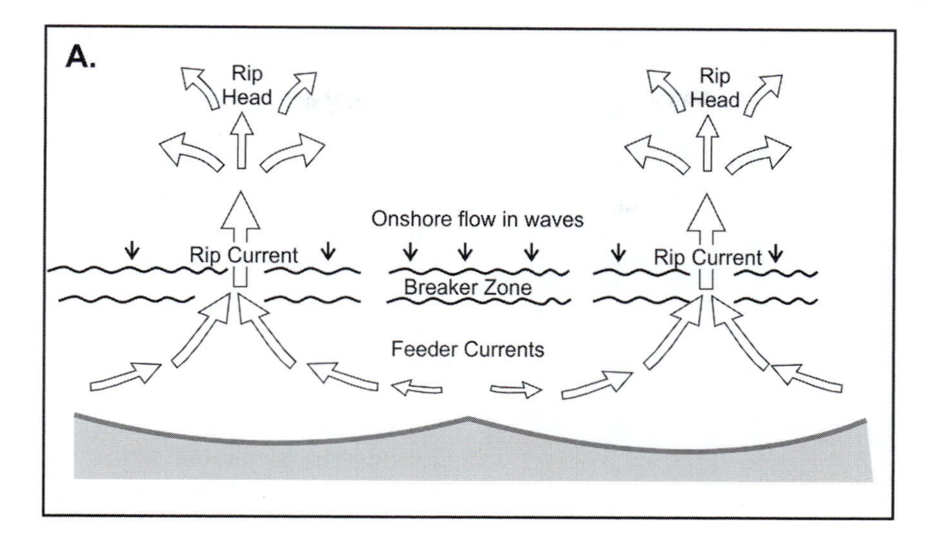

Figure 6.6 Rip cell circulation: A. plan view of rip cell circulation on an open coast; B. photograph of a rip current on the Caribbean coast of Costa Rica.

Bowman *et al.*, 1988; Gruszczynski *et al.*, 1993; Aagaard *et al.*, 1997; Brander, 1999a). The beach morphology takes on the form of giant cusps with the cusp horn located opposite the rip divide and the embayment located opposite the rip current (Figure 6.8A). The feeder currents occur in the trough landward of the innermost bar, and the rip current flows in a channel cut through the bar or between two transverse bars or shoals (Figure 6.8B). Under wave approach perpendicular to the shoreline, the rip systems may be nearly symmetric, while under oblique wave approach, the current becomes oblique to meandering (Sonu, 1972), and the rip

Figure 6.7 Quasi-permanent rip channels associated with a complex inner bar at Casino Beach on Santa Rosa Island near Pensacola, Florida. The bar morphology has been highlighted from a video record by averaging the light associated with foam produced by waves breaking on the bars.

channel becomes skewed alongshore. This has the effect of reducing the feeder current in the direction opposite to that of the wave approach and enhancing the flow in the direction of wave approach.

More recently, an alternative view of rip current circulation based on new field measurement techniques utilising GPS-equipped Lagrangian drifters (MacMahan *et al.*, 2009) has been proposed where rip current flow is confined primarily within the surf zone in semi-enclosed vortices with only episodic exits of flow offshore (MacMahan *et al.* 2010; Figure 6.9). This may explain the fact that the pattern in Figure 6.8B where the drogues were unable to cross the breaker line in the rip channels. As discussed by Castelle *et al.* (2016), there is a wide range of rip circulation systems associated with the different dominant physical forcing mechanism causing alongshore variability in wave breaking intensity. Each has

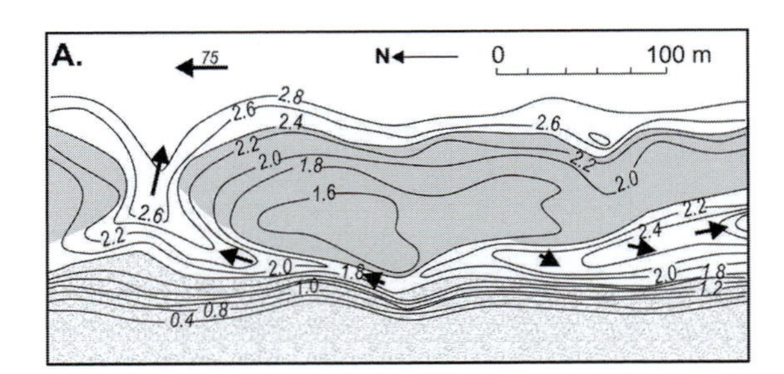

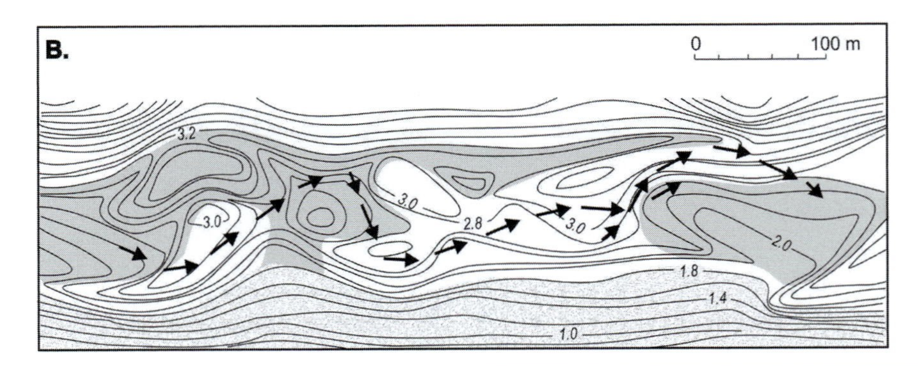

Figure 6.8 Rip cell circulation on a barred coast: A. current speeds and direction measured with drogues for perpendicular wave approach in an inner bar system, Kouchibouguac Bay, New Brunswick. Note the transverse bar/cusp horn on the left and the drainage divide situated in the middle of the straight section of the bar. The current flow to the right feeds another rip channel just beyond the measured area; B. current pattern traced by a drogue for complex topography under oblique wave approach producing a meandering pattern.

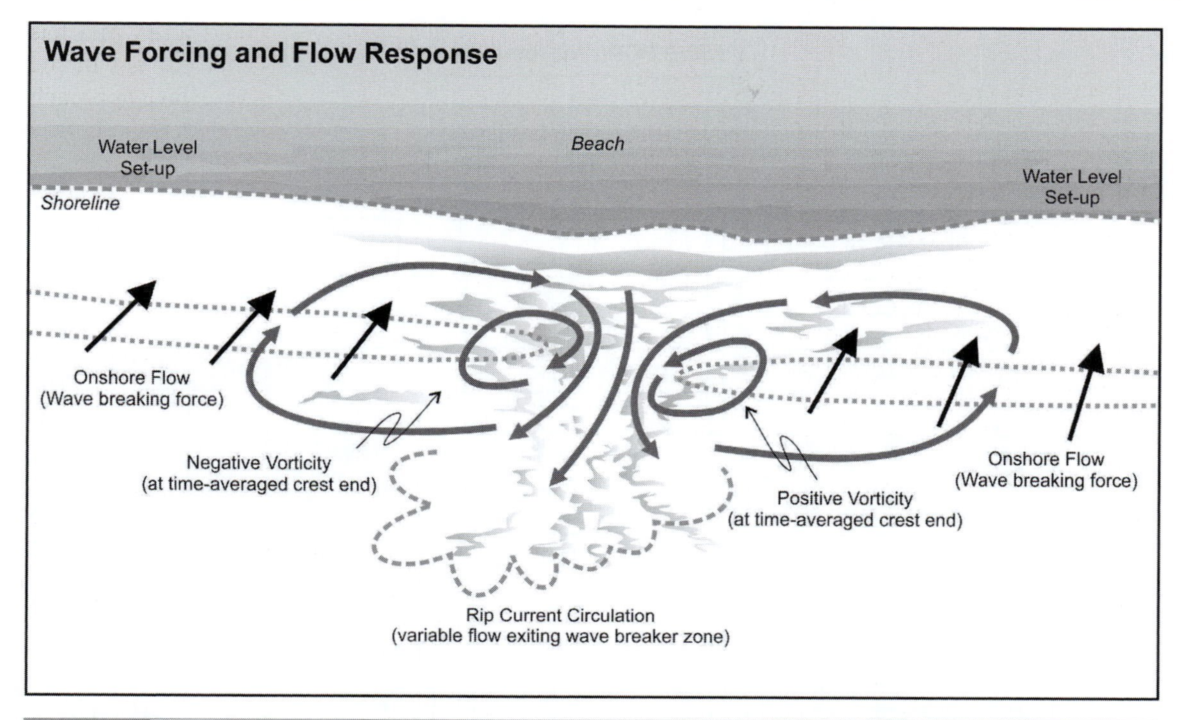

Figure 6.9 Model of time-averaged vortices and rip circulation in the nearshore resulting from the alongshore variation in breaking wave height (Castelle *et al.*, 2010, figure 5).

its own unique morphology, hydrodynamics and visual characteristics that can be used for identification (Figure 6.10).

Hydrodynamically-controlled rip currents are restricted to morphologically featureless, alongshore uniform beaches, on the seaward side of nearshore, or along planar sections of beaches, such as low tide terraces (Wright and Short, 1984; MacMahan *et al.*, 2005). In these environments, the rip current may erode a channel below the general level of the sand in the breaker and outer surf zones (MacMahan *et al.*, 2005). These rip systems are relatively unpredictable in space and time involving the episodic bursts of water flowing offshore associated with transient surf-zone eddies driven by wave breaking vorticity. These flash rips are visible as plumes of sediment and turbulent water seaward of the surf zone, are visible on the order of minutes, and are most common with plunging breakers and shore-normal wave incidence (Castelle *et al.*, 2016). Shear instability rip systems develop on alongshore uniform beaches in response to the longshore current shearing in the cross-shore direction. The resulting vortices have strong offshore flows.

Bathymetrically-controlled rip systems tend to be forced by the morphology of the nearshore bars or offshore features, and are relatively persistent in both space and time. The most common and best documented form develops in response to shore-normal waves breaking over complex three-dimensional nearshore morphology associated with intermediate beach states (Wright and Short, 1984). The resulting circulation system involves landward transport of sediment across the nearshore bar (via Stokes drift) that leads to the onshore migration of the bar, which in turn alter the strength of rip through changes in the intensity of wave breaking and changes in the width and depth of the rip channel. As the rip channels are present through a wide range of nearshore states, channel rips can be relatively stationary in position over temporal

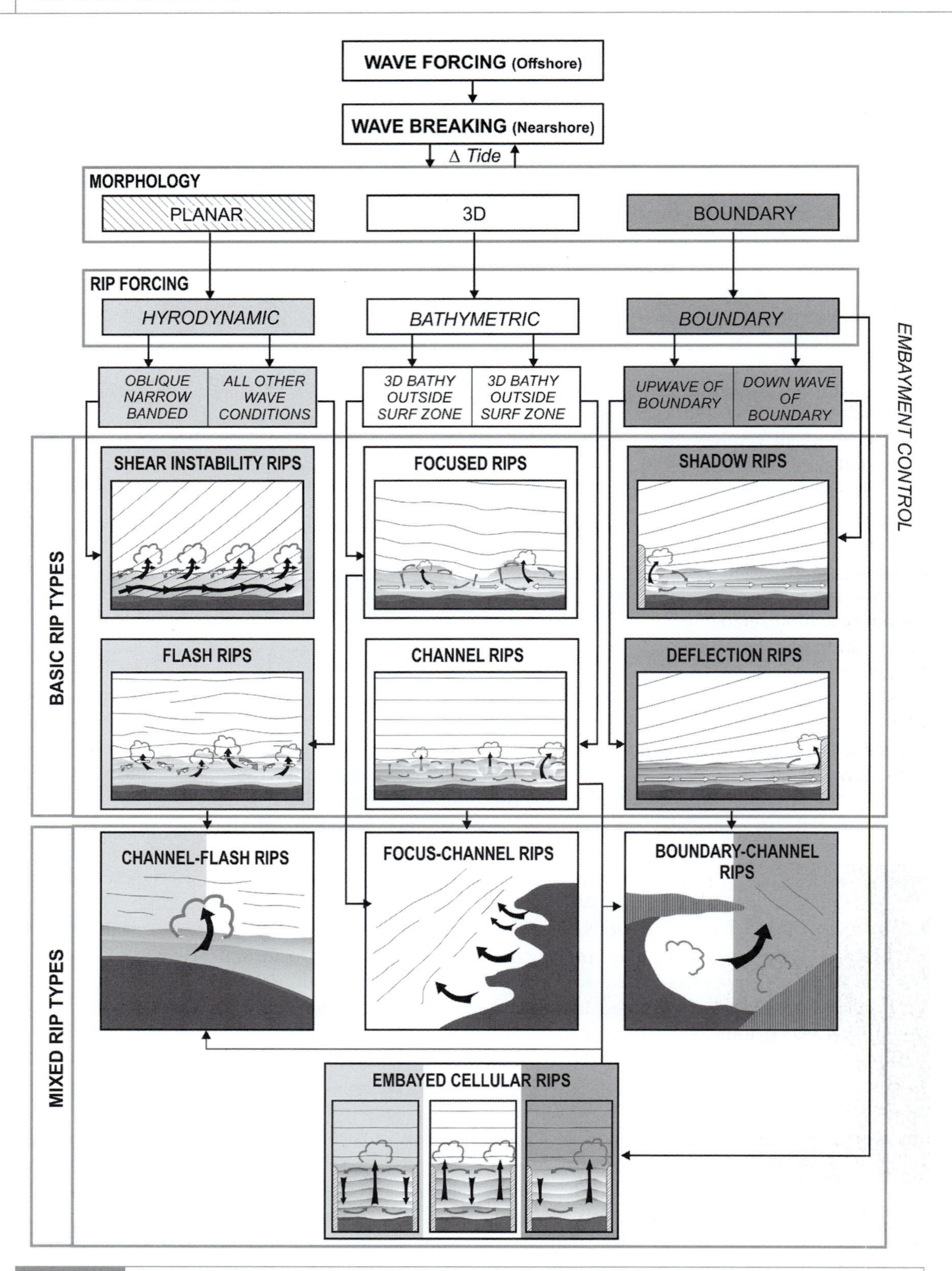

Figure 6.10 Summary diagram showing the classification framework of Castelle *et al.* (2016, figure 13) for rips associated with planar, 3D and boundary nearshore environments. See text for details of each rip type.

periods of days, weeks and months (MacMahan *et al.*, 2005). Channelised rip currents may also occur in quasi-fixed locations due to the presence of offshore bathymetric features such as seaward bars on multi-barred beaches, submarine canyons or transverse ridges. Specifically, wave refraction over these features results in opposing alongshore currents that deflect offshore as a rip current (Castelle *et al.*, 2016).

Boundary-controlled rip currents occur adjacent to natural and anthropogenic structures including headlands, rocky outcrops, groynes, jetties and piers (Figure 6.11). These persistent, often permanent features develop on the up-wave side of the boundary when strong alongshore currents driven by oblique waves are physically deflected offshore by the boundary. They also develop on the down-wave side due to the wave shadowing effect of the boundary that creates an alongshore variations in wave height and energy dissipation (Castelle *et al.*, 2016). The rip that develops on the up-wave side of the boundary exhibit relatively high exit rates, while the rip on the down-wave side tend to be characterised by a simple recirculation cell with few exits. A greater number of surf zone exits occur when and where the length of the boundary structure is greater than the width of the surf zone (Scott *et al.*, 2016), because the deflected current continues offshore beyond where the landward transfer by wave shoaling and breaking are at a maximum.

Figure 6.11 Example of a boundary rip associated with a rock groyne at Galveston Island, Texas. A deflection rip is visible on the left side of the groyne.

6.3.1 Rip Speeds

Because most of the offshore return flow is concentrated in a narrow zone, rip speeds are much higher than those associated with undertow, and typically range from < 0.3 m s^{-1} to > 1 m s^{-1} (Brander and Short, 2000), but can reach speeds greater than 2 m s^{-1}. Mean speeds measured over tens of minutes to hours in rip channels at several different sites tend to be quite low, ranging from 0.2 m s^{-1} to 0.65 m s^{-1} (MacMahan *et al.*, 2006). There have been suggestions that rip current speeds tend to pulsate, for example, at infragravity frequencies driven by wave groupiness, and this has been observed in several field experiments (Brander and Short, 2001; MacMahan *et al.*, 2005). This may not have a significant effect on the overall rate of sediment transport in the channel but, because transport rates are often u^3 or u^4 there is the potential for rates to be higher than under steady flows. It has generally been thought that mean flows increase with increasing wave energy (height), though the effect of this is limited by the fact that larger waves break in deeper water further offshore and so the maximum wave height on an inner bar is limited.

As is the case for the 2D undertow, the mass of water injected into the surf zone or landward trough must be returned seaward through the rip channel and thus it should be possible to compare measured discharge through the rip channel with predicted onshore flows:

$$V_{\text{rip}} = \frac{(Q_{\text{drift}} + Q_{\text{roller}})\lambda}{R_{\text{a}}}, \tag{6.4}$$

where V_{rip} is the mean velocity in the rip channel; Q_{drift} and Q_{roller} are the mean onshore discharge of water due to Stokes drift and the breaker roller; λ is the alongshore length between rip channels; and R_{a} is the cross-sectional area of the rip channel (after Aagaard *et al.*, 1997).

Equation 6.4 shows that there are three main controls on rip current velocity: (1) the input volume of water ($Q_{\text{drift}} + Q_{\text{roller}}$); (2) the spacing of rip channels (λ); and (3) the cross-sectional area of the channels (R_a). As the tide falls, the input volume of water increases (at least until the water depth over the bar crest limits the amount of water transferred landward), and the cross-sectional area decreases because of decreasing water depth. Consequently, we can expect quite large fluctuations in mean speed over time spans as short as individual tidal cycles. Over longer time scales or during major storms the channel dimensions and spacing can change, leading to either increases or decreases in the strength of the rip speed. An increase in the flow velocity due to increased wave input may lead to the scouring of a wider and/or deeper channel that ultimately weakens the rip speed (a negative feedback). Similarly, the interaction between the seaward flowing rip and incident waves may produce negative or positive feedbacks for flow velocity and discharge. Deeper channels lead to larger unbroken waves entering the rip channel and this in turn produces a radiation stress gradient that tends to counter the pressure gradients that drive the flow, leading to an even weaker current. When flow is fast, the wave-current interaction encourages wave breaking leading to a positive feedback on the rip flows (Haller *et al.*, 2002).

6.3.2 Controls on Rip Spacing

The spacing λ between rip currents or rip current channels varies considerably from beach to beach and often over time at any beach (McKenzie, 1958; Short, 1985). Since λ in effect characterises the collection area feeding the rip, it might be expected that this would decrease as wave height increases. However, observations by Short (1985) suggest the opposite – storm circulation may be dominated by a few mega rips with speeds up to 2 m s^{-1} during intense storms. In general, there is poor correlation between rip spacing and parameters such as incident wave height and period and surf zone width (Huntley and Short 1992; Turner *et al.*, 2007). This may be due in part to the persistence of rip channels and rhythmic topography over a range of wave conditions and to the difficulties of observing the spacing during formative high-energy events.

Much effort was expended over two or three decades (from 1970) trying to link rip current spacing, and the characteristic rhythmic topography associated with rip cells and crescentic nearshore bars, with some hydrodynamic mechanism that would produce periodic variations in incident waves or net drift velocities. Mechanisms of this type are termed template models (Blondeaux, 2001), in that the hydrodynamic mechanism imposes a fixed form on the rip cell and/or bar pattern. This can be contrasted with self-organisational models in which there is positive feedback between the mechanism and the development of the modelled circulation or morphology. The mechanism that attracted most attention, in terms of research devoted to testing it, was proposed by Bowen (1969a), linking rip current spacing to edge waves. Bowen (1969a) argues that longshore variations in wave height will give rise to longshore variations in momentum flux and set-up and that this should lead to a flow of water from areas of high waves towards areas of low waves where the rip current would flow seaward. Bowen and Inman (1969) then went on to show that the interaction of an edge wave with incoming waves would lead to longshore variations in wave height, with the height being greatest where the edge wave and incident wave are in phase and lowest where they are 180 degrees out of phase. The rip currents would thus be located where the edge wave and incident waves are 180 degrees out of phase and the breakers are smallest (Figure 6.12). As a result, the alongshore spacing for the rip currents is equal to the alongshore wavelength of the edge wave. Similar reasoning was also applied to the formation of rhythmic or crescentic bars (Bowen and Inman, 1971).

There is evidence that rips form where wave heights are lowest because of wave refraction over complex topography, but it is yet to be demonstrated in the field that the model put forward by Bowen and Inman (1971) works in terms of wave height variation, let alone the link to edge waves. Despite much searching, no clear link has

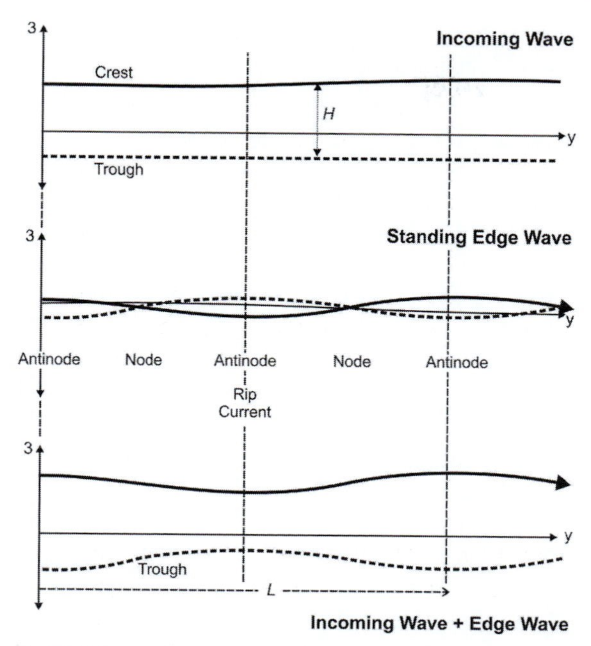

Figure 6.12 The addition of an incoming wave and a standing edge wave, producing an alongshore variation in breaking wave height (Bowen and Inman, 1969). Rip currents are located at the antinodes.

ever been established and it now seems that edge waves associated with rhythmic bars may be a product of the interaction between waves and existing topography rather than vice versa. Long-term studies, such as those using the Argus video system (Holman *et al.*, 2006; Turner, 2007), show that rip cell spacing is quite variable. It now seems much more likely that the location of a rip channel is random and that the generation of the circulation may be better modelled as a form of self-organisation (e.g., Murray *et al.*, 2003) with spacing moderated by wave conditions and pre-existing topography (Aagaard and Vinther, 2008).

Rip currents erode and transport sediment, particularly in the rip channel itself (Cook, 1970; Davidson-Arnott and Greenwood, 1974; 1976; Gruszczynski *et al.*, 1993; Sherman *et al.*, 1993; Aagaard *et al.*, 1997; Brander, 1999b) and there is evidently a complex feedback between topographical evolution and the rip current itself. Rip channels tend to persist for periods of weeks or months and over several storm events. The channel spacing is of the same order as the rhythmic topography and ranges from about 150 m to more than 600 m. Rip cell circulation is expressly included in the beach stage model developed by Short and Wright from their work on beaches on the east coast of Australia (Short, 1979; Wright and Short, 1984) – see Chapter 8 for more on this.

6.3.3 Rips and Undertow

Why is the circulation on some beaches always dominated by undertow and others by rip cell circulation? Circulation over the straight multiple parallel bars at Wasaga Beach (Davidson-Arnott and MacDonald, 1989) is always two-dimensional whereas beaches in California (MacMahan *et al.*, 2005) and New South Wales, Australia (e.g., Short, 1985) are generally characterised by well-developed rip cell systems. The fine sand and very gentle slope at Wasaga Beach combined with short-period wind waves results in wave breaking being dominantly spilling while swell waves and steeper nearshore slope in California and Australia tend to produce breakers in the plunging range. It is possible that the large vortices generated by plunging breakers may inhibit weak offshore flows associated with undertow, but undertow does occur under plunging breakers, particularly where there is no distinct trough and no preferential location for a breaker line. The existence of well-developed rip channels within barred systems is usually associated with the generation of rip cell circulation, but recent work by Aagaard and Vinther (2008) suggests that in some cases the system may be dominated by undertow even when distinct rip channels are present. They go on to suggest that morphodynamic adjustment may lag, changing wave and water level conditions and that the type of circulation may be controlled both by the nature of wave breaking that supplies water to the trough, and the morphology of the bar–trough system that determines the cross-sectional area of the trough.

Box 6.2 Rip Currents and Water Safety

Rip currents are a factor in hundreds of drownings each year and are now a recognised serious global public health issue (see Short, 1999; Klein *et al.*, 2003; Hartmann, 2006; Scott *et al.*, 2011; Brighton *et al.*, 2013; Woodward *et al.*, 2013; Kumar and Prasad, 2014; Houser *et al.*, 2017; Llopis *et al.*, 2018). In the United States, Australia and Costa Rica alone, an estimated 59, 21 and 49 rip-related drowning fatalities occur each year, exceeding the number of fatalities caused by most other natural hazards in those countries (Brander *et al.*, 2013; Arozarena *et al.*, 2015). Surf Life Saving Australia (SLSA) estimate that they make more than 22 000 rescues each year, of which 80 per cent are rip related (SLSA, 2010). Similarly, the United States Lifesaving Association (USLA) completes more than 88 000 beach rescues each year, of which the organisation estimates that a significant number are rip related (USLA, 2015). Rips are particularly dangerous for tourists and students studying abroad (see Houser *et al.*, 2016; Clifford *et al.*, 2018), who are not aware of the rip hazard on the beaches they visit.

Physically, rips do not pull a person under the water, but rather pull them away from shore, often into deeper waters. Drowning begins when the swimmer is no longer able to touch the bottom and keep their head consistently above water. The problem for non-swimmers and poor swimmers is twofold – it is very difficult to maintain a footing against waves and strong currents once the water depth is above the waist, and it is very difficult and tiring to swim against even weak currents. It is nearly impossible for strong swimmers to swim against typical rip current speeds of 0.3–1 m s^{-1}. Trained lifeguards combined with a good system of public education, both at the beach and away from it, can improve safety considerably (Short and Hogan, 1994). As described by LeDoux (1996), trying to fight against the current causes a person to become stressed and start experiencing a combination of: (1) increased adrenaline, (2) increased heart rate leading to raised blood pressure, (3) dilation of the bronchi causing rapid, shallow breathing, and (4) decreased blood flow and reduced function in the parts of the brain that produce logical, rational thinking and evaluation.

A person's risk of drowning or potential to need rescue depends to varying degrees on: the local nearshore hydrodynamic and bathymetric conditions, personal and group behaviours, and the beach safety and rip current knowledge of the individual beach user. Those with little to no experience tend to underestimate the danger compared to those with direct experience, which suggests that there is a potentially large at-risk population at the beach who do not understand the danger posed by risks despite warnings and direct experience (see Houser *et al.*, 2017). Klein *et al.* (2003) reported an 80 per cent reduction in drowning fatalities along a 100 km stretch of sandy beaches in Brazil following the introduction of a safety and training programme. Many beaches in the United States and around the world post the NOAA Break the Grip of the Rip® warning sign that informs beach users how to escape a rip current, and a simple illustration of a rip current from aerial perspective. Information is also distributed through the US National Weather Service rip current safety webpage (www.ripcurrents.noaa.gov/), brochures, videos, newspaper articles and public service announcements on television. Results from several survey-based studies suggest that the Break the Grip of the Rip® campaign has been successful in helping inform the public about rip current safety. Most respondents were familiar with the safety message of the campaign and could provide an accurate description of how to escape a rip current by swimming parallel and/or floating until the current weakened. Several studies suggest that only a small percentage of beach users (< 50%) recall observing rip current warning signs on or at the entrance to beaches (Caldwell *et al.*, 2013; Brannstrom *et al.*, 2014, 2015). Even if they do observe the signs some studies suggest that most beach users are unable to interpret the warning (see Brannstron *et al.*, 2014).

Even if there are warning signs or flags to warn of the rip hazard on a beach, some beach users will not take the appropriate actions to prepare for or avoid the hazard based on the behaviour of other beach users and their own past observations (whether accurate or not) of the hazard (Menard *et al.*, 2018). A study of university students studying abroad suggests that most did not select a beach based on safety or take appropriate precautions, and most selected a beach based on convenience or because of 'group think' (Houser *et al.*, 2016). Rips can be present

Box 6.2 (cont.)

and pose a threat to beach users under relatively small wave conditions. At Pensacola Beach, Florida there are semi-permanent and regular rip channels with flows reaching 0.5 m s^{-1} when waves are relatively small and do not appear to represent a potential danger.

Similarly, the beaches at Point Abino on the north shore of Lake Erie are characterised by the development of a rhythmic nearshore bar system with small rip channels at semi-regular intervals. Water depth in these channels is generally a metre or less and they are generally active only during high wave conditions. However, a fixed rip channel is formed at the junction between the beach and the limestone bedrock of the adjacent point, with particularly strong rips when there is a strong longshore current. Drownings at this location in 1982 and 1991 both occurred with waves < 1 m in height and winds from the southwest at < 10 m s^{-1}. In each case, children playing or swimming in the shallow water of the inner bar trough updrift of the point were carried alongshore and out into the rip channel. The adult who went to their rescue drowned because the depth in the channel was too great for them to stand and they quickly became exhausted trying to swim directly ashore against the current while trying to save lives. In both cases, had the adult known what was happening and been aware of the topography adjacent to Point Abino (Figure 6.13A), it would have been simple to swim sideways a few metres onto the rock platform (Figure. 6.13B) where they would have been able to stand.

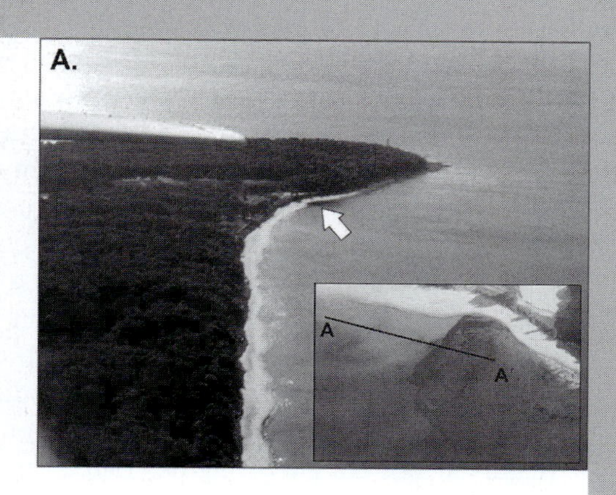

Figure 6.13 Rip channel at Point Abino on the north shore of Lake Erie: A. oblique aerial photograph looking east along the shoreline with the sandy beach and nearshore rhythmic bars in the foreground and the limestone headland of Point Abino in the background. The arrow points to the dark line marking the location of the rip channel. The inset shows a view onshore of the rip channel with the end of the inner sand bar on the left; B. profile parallel to the beach about 60 m offshore showing the drop-off from the nearshore sand bar into the rip channel trough and up onto the bedrock ledge of the point.

6.4 Longshore Currents

Where waves approach the shoreline at an angle, a portion of the momentum flux is directed alongshore and results in the generation of a longshore current that typically flows at a speed of 0.2–1.0 m s^{-1}. Longshore currents occur independently of rip cell circulation, but where rips are present the cellular circulation becomes skewed in the direction of the longshore current. Considerable effort has gone into attempts to measure the net longshore current (not always easy in the presence of strong oscillatory currents) and to develop theoretical and empirical models for predicting the mean flow. Longshore currents are a major control on longshore sediment transport, so successful modelling of net longshore transport in numerical models often requires modelling of the longshore current first. Initial modelling of longshore currents was based on waves alone but recent models also account for the effects of winds. Longshore current velocity can be modelled theoretically but there are several empirical equations that produce acceptable results for simple situations. Longshore current speeds tend to be nearly uniform through the upper half of the water column and this may extend close to the bed in the zone of strong wave breaking.

6.4.1 Longshore Currents on a Planar Beach

The work of Bowen (1969b) and Longuet-Higgins (1970a, b) showed that longshore currents generated by waves could be related to the longshore-directed component of radiation stress. As we saw in Chapter 4, radiation stress measures the excess momentum associated with waves and the gradient of this in the surf zone leads to wave set-up. When waves break at an angle to the shoreline the longshore-directed radiation stress will generate a flow alongshore. The longshore component of radiation stress is given by:

$$S_{xy} = E \sin\alpha \cos\alpha = \frac{1}{8}\rho g H^2 \sin\alpha \cos\alpha. \qquad (6.5)$$

Thus, the alongshore-directed momentum flux is a function of the wave energy density and the angle of the breaking wave to the shoreline (the $\sin\alpha$ term). The $\cos\alpha$ term is used to correct for the fact that as wave angle to the shoreline increases a unit length along the wave crest is distributed over a proportionately greater length alongshore (we will see this again in predictions of longshore sediment transport rates in Chapter 7). It is then possible to predict the steady longshore current velocity by balancing the momentum force with frictional forces at the bed and with eddy viscosity. The solutions produced by Longuet-Higgins (1970a, b) assume monochromatic (regular) waves, linear wave theory, a planar slope and a saturated surf zone:

$$V = \frac{5\pi}{8}\frac{\tan\beta}{C_{\mathrm{f}}} u_{\mathrm{m}} \sin\alpha_{\mathrm{b}} \cos\alpha_{\mathrm{b}}. \qquad (6.6)$$

where: V is the longshore current velocity; C_{f} is a friction coefficient on the order of 0.01; $\tan\beta$ is the surf zone slope; u_{m} is the maximum orbital velocity; and α_{b} is the wave angle to the shoreline at the breakpoint. Equation 6.6 produces a triangular decay in longshore current velocity from a maximum at the breaker line to zero at the beach (Figure 6.14), because the assumption of regular waves requires them to break at a single point. The longshore current distribution in the field does not have an abrupt cut-off at the breaker line but instead begins some distance offshore and produces a much smoother curve (Figure 6.14). Longuet-Higgins (1970a) modelled this by making allowance for lateral mixing which flattened the curve and extends it seaward of the breaker line with the shape of the distribution varying with the value assigned to the lateral mixing parameter P (Longuet-Higgins, 1970a). If irregular waves are considered, wave energy dissipation begins with breaking of the largest waves some distance offshore producing a smooth profile that approaches the form for regular waves in the inner surf zone, where almost all waves are broken and wave energy dissipation is controlled by water depth (Figure 6.14). Thornton and Guza (1986) introduced a formula for random waves on a planar beach that provides good comparison with measured data and numerical models now routinely include this (e.g., Larson and Kraus, 1991;

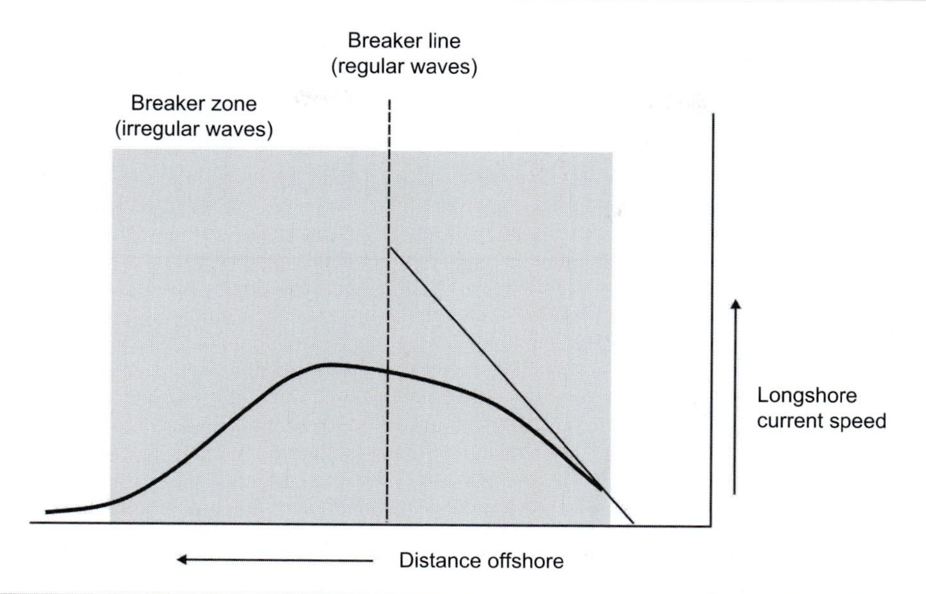

Figure 6.14 The general pattern of distribution of longshore current speeds across the surf zone on a planar profile for regular waves without lateral mixing (straight line and single breaker location) and for irregular waves (curved line and zone of wave breaking).

Grasmeijer and Ruessink, 2003), as well as the effects of wind.

It may not always be necessary or practical to use the more complex theoretical expressions or to set up and run a numerical model such as MIKE 21 (e.g., Jones *et al.*, 2007). An empirical estimate of the mean longshore current based on significant wave height (H_s) is given by Komar (1979):

$$V_l = 1.0\sqrt{gH_s}\sin\alpha\cos\alpha. \tag{6.7}$$

In simple conditions, such as a planar nearshore, this can be used to provide an order of magnitude estimate of the mean longshore current.

6.4.2 Longshore Currents on a Barred Beach

The presence of bars in the nearshore zone complicates the task of predicting longshore current velocity because of the variations in topography, wave breaking and set-up (Greenwood and Sherman, 1986; Larson and Kraus, 1991; Ruessink *et al.*, 2001). Maximum current speeds tend to occur on the landward edge of the bar crest or in the trough just landward of the bar. Measurements of longshore current speeds by Greenwood and Sherman (1984) across a profile with three bars at Wendake Beach, Georgian Bay, Canada showed that velocities were lower than would be predicted for a planar beach in the outer surf zone and higher than would be predicted in the inner surf zone. Measured current speeds at the height of the storm ranged from about 0.1 m s^{-1} to a maximum of about 0.4 m s^{-1} in the inner surf zone. The pattern of dimensionless mean current velocity is shown for the beginning and peak of a storm in Figure 6.15. These show that current speeds were lower just on the landward side of the bar crest and highest in the trough. There is no tidal modulation in the Great Lakes, though water depth over the bars does vary with wind and wave set-up. Recent work on modelling longshore currents on barred beaches by Ruessink *et al.* (2001) found reasonably good agreement between predictions of a 1D model and measured data from Egmond aan Zee on the Dutch coast and the FRF of the US Army Corps of Engineers at Duck, North Carolina, USA. The model predictions were improved by including a term for the wave roller during breaking which had the effect of shifting the location of the maximum current shoreward into the

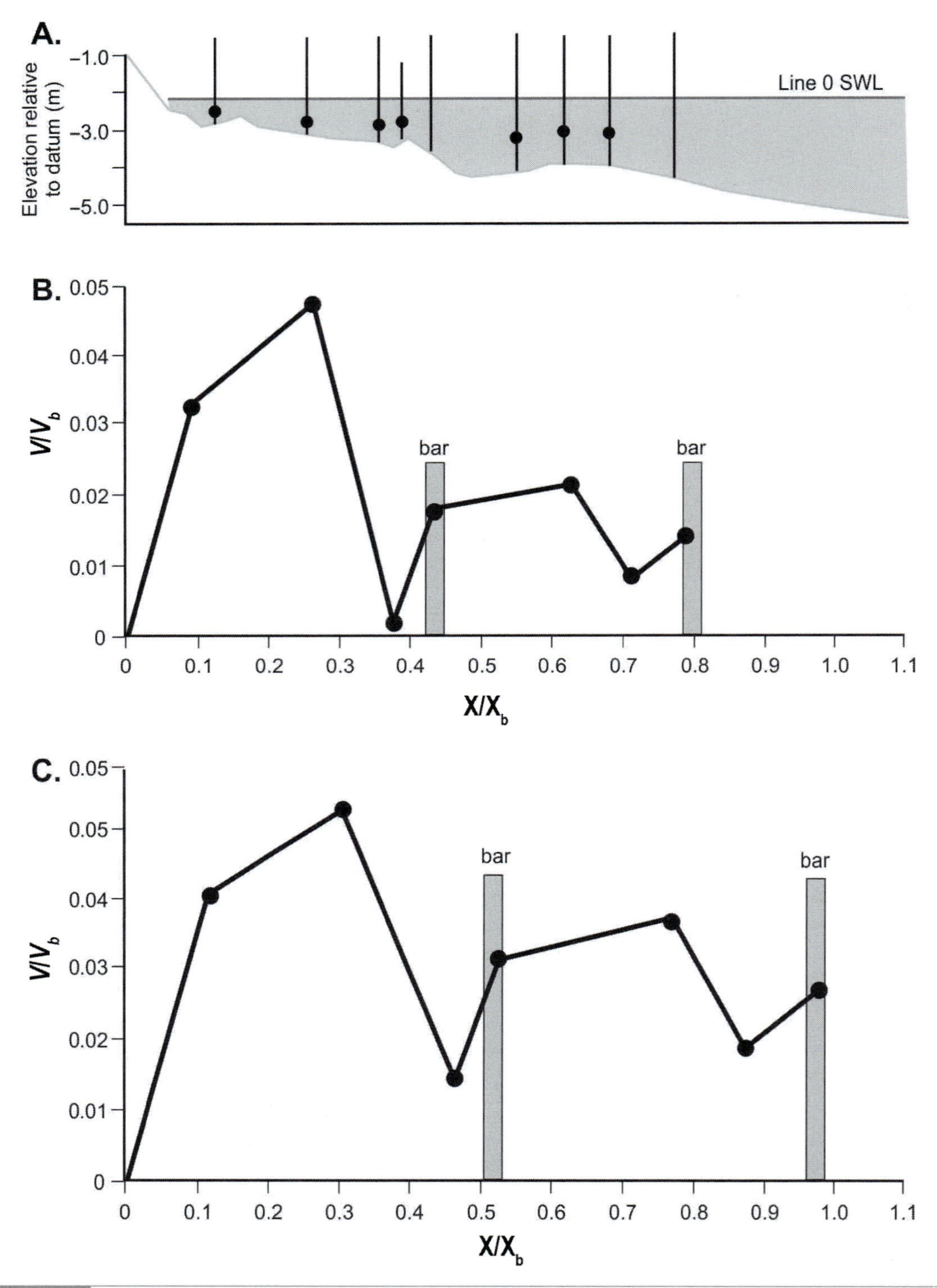

Figure 6.15 Measurements of longshore currents across a barred profile, Wendake Beach, Georgian Bay, Canada: A. profile across the nearshore showing topography and location of electromagnetic current meters and resistance wave staffs used in the experiment; B. dimensionless velocity profile near the beginning of the storm; and C. dimensionless velocity profile at the peak of the storm. Here, X is distance offshore; X_b is width of the surf zone; V is the measured mean longshore current at a point and V_b is the mean depth-integrated longshore current at the breaker line without lateral mixing (Longuet-Higgins, 1972; after Greenwood and Sherman, 1986).

trough. As we might expect from our earlier discussion of rip currents over barred topography (Aagaard *et al.*, 1997), the longshore current flow is tidally modulated and Ruessink *et al.* (2001) found that the improvement from including the wave roller was greatest at mid and high tide. Subsequent analysis of several data sets by Goda (2006) has supported the improvements produced in prediction of longshore current distribution and speed by including a roller term.

The presence of far infragravity oscillations (on the order of 200–400 seconds) in the spectra of longshore currents was first documented by Oltman-Shay *et al.* (1989), and these oscillations are now termed shear waves (Bowen and Holman, 1989). They result from instabilities in the horizontal shear of the longshore current and are likely to be better developed on barred shorelines where the longshore current flows in a well-defined trough. The effect is to produce a wave or meander in the longshore current that travels alongshore in the direction of the longshore current flow at a speed of about one-third that of the mean longshore current. The alongshore propagation of the shear wave will produce pulses in flow and these will also show up in the seaward flow in rip currents. Shear waves likely have a significant impact on the suspension of sediment and net sediment transport in the surf zone (Aagaard and Greenwood, 1995; Miles *et al.*, 2002).

In addition to the effects of tidal stage, there is now increased recognition of the role of winds and tidal currents in determining the strength and distribution of longshore currents (Nummedal and Finley, 1978; Whitford and Thornton, 1993; Masselink and Pattiaratchi, 1998). Wind forces become significant on all beaches as the wind angle approaches shore parallel and as wind speed increases (Whitford and Thornton, 1993). Castelle *et al.* (2006) in a study on a multiple bar system on the west coast of France found a wide scatter between measured and computed longshore currents for all conditions, but a very much stronger agreement when only conditions with weak winds were considered. On steep beaches with only a small wave angle the wind term may be negligible, but on gently sloping beaches the magnitude of the contribution from the wind may range from 0.1 to 1.0 of the contribution from waves (Whitford and Thornton, 1993). Some of this effect may result from the effect of wave refraction on gently sloping beaches which will reduce the wave angle at breaking. An example of this can be seen in the variations in measured longshore current speed and direction at Wasaga Beach over one storm (Figure 6.3D).

6.5 Wind and Tidal Currents

Swell waves reaching the coast of California or eastern Australia may have been generated by storms thousands of kilometres away and so are independent of local winds – in fact it is not uncommon on these coasts to have offshore winds with large waves. In this case, the alongshore and offshore flows may be generated only by wave set-up. Where swell waves reach the coast and winds are offshore the effect of the winds is to reduce the onshore flow near the surface and thus reduce the strength of the undertow or rip circulation. Local sea breeze effects can produce rapid changes in wind and wave conditions and greatly influence current strength, patterns and resulting morphological change (Sonu *et al.*, 1973; Masselink and Pattiaratchi, 1998; Torres-Freyermuth *et al.*, 2017). In fetch-limited areas and leeward coasts, large waves are almost always accompanied by strong onshore winds. The most significant effect of winds is to enhance set-up at the shoreline and thus to strengthen the nearshore circulation – the offshore directed flows in undertow or rip currents. Storm surge thus occurs in addition to wave set-up, and together they may exceed 2 m in mid-latitude gales and over 3 m in hurricanes.

Tidal currents are unidirectional currents that are primarily directed alongshore and are most important seaward of the breaker line. In straits and areas with large tidal ranges, tidal currents themselves may be strong enough to transport sediment and they can operate in depths where wave orbital motion is quite weak. On micro tidal coasts the currents are generally significant only near tidal inlets and estuaries.

Further Reading

Castelle, B., Scott, T., Brander, R.W. and McCarroll, R.J. 2016. Rip current types, circulation and hazard. *Earth-Science Reviews*, **163**, 1–21.

Svendsen, I.A. and Putrevu, U. 1996. *Surf zone hydrodynamics*. In Liu, P.L-F. (ed.), *Advances in Coastal and Ocean Engineering*, vol. 2, Singapore: World Scientific Press, pp. 1–78.

References

Aagaard, T. and Greenwood, B. 1995. Longshore and cross-shore suspended sediment transport at far infragravity frequencies in a barred environment. *Continental Shelf Research*, **15**, 1235–49.

Aagaard, T. and Vinther, N. 2008 Cross-shore currents in the surf zone: rips or undertow? *Journal of Coastal Research*, **24**, 561–70.

Aagaard, T., Greenwood, B. and Nielsen J. 1997. Mean currents and sediment transport in a rip channel. *Marine Geology*, **140**, 25–45.

Afzal, M.S., Holmedal, L.E. and Myrhaug, D. 2015. Three-dimensional streaming in the seabed boundary layer beneath propagating waves with an angle of attack on the current. *Journal of Geophysical Research: Oceans*, **120**, 4370–91.

Arozarena, I., Houser, C., Echeverria, A.G. and Brannstrom, C. 2015. The rip current hazard in Costa Rica. *Natural Hazards*, **77**, 753–68.

Bagnold, R.A. 1940. Beach formation by waves; some model experiments in a wave tank. *Journal Institute of Civil Engineers*, **15**, 27–52.

Blondeaux, P. 2001. Mechanics of coastal forms. *Annual Review of Fluid Mechanics*, **33**, 339–70.

Bowen, A.J. 1969a. Rip currents I: theoretical investigation. *Journal of Geophysical Research*, **74**, 5469–78.

Bowen, A.J. 1969b. The generation of longshore currents on a plane beach. *Journal of Marine Research*, **27**, 206–15.

Bowen, A.J. and Holman, R.A. 1989. Shear instabilities of the mean longshore current. 1 Theory. *Journal of Geophysical Research*, **94**, 18023–30.

Bowen, A.J. and Inman, D.L. 1969. Rip Currents II: laboratory and field investigations. *Journal of Geophysical Research*, **74**, 5479–90.

Bowen, A.J. and Inman, D.L., 1971. Edge waves and crescentic bars. *Journal of Geophysical Research*, **76**, 8662–71.

Bowman, D., Arad, D., Rosen, D.S. *et al.* 1988. Flow characteristics along the rip current system under low energy conditions. *Marine Geology*, **82**, 149–67.

Bradley, R.W., Venditti, J.G., Kostaschuk, R.A. *et al.* 2013. Flow and sediment suspension events over low-angle dunes: Fraser Estuary, Canada. *Journal of Geophysical Research: Earth Surface*, **118**, 1693–709.

Brander, R.W. 1999a. Field observations on the morphodynamic evolution of a low-energy rip current system. *Marine Geology*, **157**, 199–217.

Brander, R.W. 1999b. Sediment transport in low-energy rip current systems. *Journal of Coastal Research*, **15**, 839–49.

Brander, R.W. and Short, A.D. 2000. Morphodynamics of a large-scale rip current system at Muriwai Beach, New Zealand. *Marine Geology*, **165**, 27–39.

Brander, R.W. and Short, A.D. 2001. Flow kinematics of low-energy rip current systems. *Journal of Coastal Research*, **17**, 468–81.

Brander, R., Dominey-Howes, D., Champion, C., Del Vecchio, O. and Brighton, B. 2013. Brief communication: a new perspective on the Australian rip current hazard. *Natural Hazards and Earth System Sciences*, **13**, 1687.

Brannstrom, C., Trimble, S., Santos, A., Brown, H.L. and Houser, C. 2014. Perception of the rip current hazard on Galveston Island and North Padre Island, Texas, USA. *Natural Hazards*, **72**, 1123–38.

Brannstrom, C., Brown, H.L., Houser, C., Trimble, S. and Santos, A. 2015. 'You can't see them from sitting here': evaluating beach user understanding of a rip current warning sign. *Applied Geography*, **56**, 61–70.

Brighton, B., Sherker, S., Brander, R., Thompson, M., and Bradstreet, A. 2013. Rip current related drowning deaths and rescues in Australia 2004–2011. *Natural Hazards and Earth System Sciences*, **13**, 1069.

Bruneau, N., Castelle, B., Bonneton, P. *et al.* 2009. Field observations of an evolving rip current on a meso-macrotidal well-developed inner bar and rip morphology. *Continental Shelf Research*, **29**(14), 1650–62.

Caldwell, N., Houser, C. and Meyer-Arendt, K. 2013. Ability of beach users to identify rip currents at Pensacola Beach, Florida. *Natural Hazards*, **68**, 1041–56.

Cartier, A. and Héquette, A. 2011. Variation in longshore sediment transport under low to moderate conditions on barred macrotidal beaches. *Journal of Coastal Research*, **64**, 45–9.

Cartier, A. and Héquette, A. 2013. The influence of intertidal bar-trough morphology on sediment transport on macrotidal beaches, northern France. *Zeitschrift für Geomorphologie*, **57**, 325–47.

Castelle, B., Bonneton, P., Senechal, N. *et al.* 2006. Dynamics of wave-induced currents over an alongshore non-uniform multiple-barred sandy beach on the Aquitanian Coast, France. *Continental Shelf Research*, **26**, 113–31.

Castelle, B., Michallet, H., Marieu, V. *et al.* 2010. Laboratory experiment on rip current circulations over a moveable bed: drifter measurements. *Journal of Geophysical Research: Oceans*, **115**(C12). DOI: https://doi.org/10.1029/2010JC006343.

Castelle, B., Scott, T., Brander, R.W. and McCarroll, R.J. 2016. Rip current types, circulation and hazard. *Earth-Science Reviews*, **163**, 1–21.

Christensen, E.D., Walstra, D. and Emerat, N. 2002. Vertical variation of the flow across the surf zone. *Coastal Engineering*, **45**, 169–98.

Clifford, K., Brander, R., Houser, C. and Trimble, S. 2018. Beach safety knowledge of visiting international study abroad students to Australia. *Tourism Management*, **69**, 487–97.

Cook, D.O. 1970. The occurrence and geological work of rip currents off Southern California. *Marine Geology*, **9**, 1973–86.

Dally, W.R. and Dean, R.G. 1984. Suspended sediment transport and beach profile evolution. *Journal of Waterways, Port, Coastal and Ocean Engineering*, **110**, 15–33.

Dally, W.R. and Dean, R.G. 1986. Mass flux and undertow in a surf zone - discussion. *Coastal Engineering*, **10**, 289–307.

Dalrymple, R.A., MacMahan, J.H., Reniers, A.J.H.M. and Nelko, V. 2011. Rip Currents. *Annual Review of Fluid Mechanics*, **43**, 551–81.

Davidson-Arnott, R.G.D. 2005. Beach and nearshore instrumentation. In M.L. Schwartz (ed.), *Encyclopedia of Coastal Science*. Dordrecht, The Netherlands: Springer. pp. 130–8.

Davidson-Arnott, R.G.D. and Greenwood, B. 1974. Bedforms and structures associated with bar topography in the shallow-water environment, Kouchibouguac Bay, New Brunswick, Canada. *Journal of Sedimentary Petrology*, **44**, 698–704.

Davidson-Arnott, R.G.D. and Greenwood, B. 1976. Facies relationships on a barred coast, Kouchibouguac Bay, New Brunswick, Canada. In Davis, R.A., Jr and Ethington, R.L. (eds.), *Beach and Nearshore Sedimentation*. Tulsa, OK: Society of Economic Paleontologists and Mineralogists Special Publication no. 24, pp. 149–68.

Davidson-Arnott, R.G.D. and MacDonald, R.A. 1989. Nearshore water motion and mean flows in a multiple parallel bar system. *Marine Geology*, **86**, 321–38.

Downing, J.P., Sternberg, R.W. and Lister, C.R.B. 1981. New instrumentation for the investigation of sediment suspension processes in shallow marine environments. *Marine Geology*, **42**, 14–34.

Garcez-Faria, A.F., Tornton, E.B., Lippmann, T.C. and Stanton, T.P. 2000. Undertow over a barred beach. *Journal of Geophysical Research*, **105**(C7), 16999–7010.

Goda, Y. 2006. Examination of the influence of several factors on longshore current computation with random waves. *Coastal Engineering*, **53**, 157–70.

Goodfellow, B.W. and Stephenson, W.J. 2005. Beach morphodynamics in a strong-wind bay: a low energy environment? *Marine Geology*, **214**, 101–16.

Grasmeijer, B.T. and Ruessink, B.G. 2003. Modelling of waves and currents in the nearshore: parametric vs. probabilistic approach. *Coastal Engineering*, **49**, 185–207.

Greenwood, B. and Davidson-Arnott, R.G.D. 1975. Marine bars and nearshore sedimentary processes, Kouchibouguac Bay, New Brunswick, Canada. In Hails, J. and Carr, A. (eds.), *Nearshore Sediment Dynamics and Sedimentation: An Interdisciplinary Review*. Chichester, UK: John Wiley and Sons, pp. 123–50.

Greenwood, B. and Davidson-Arnott, R.G.D. 1979. Sedimentation and equilibrium in wave-formed bars: a review and case study. *Canadian Journal of Earth Sciences*, **16**, 312–32.

Greenwood, B. and Osborne, P.D. 1990. Vertical and horizontal structure in cross-shore flows: an example of undertow and wave set-up on a barred beach. *Coastal Engineering*, **14**, 543–80.

Greenwood, B. and Sherman, D.J. 1984. Waves, currents, sediment flux and morphologic response in a barred nearshore system. *Marine Geology*, **60**, 31–61.

Greenwood, B. and Sherman, D.J. 1986. Longshore current profiles and lateral mixing across the surf zone of a barred nearshore system. *Coastal Engineering*, **10**, 149–68.

Gruszczynski, M., Rudowski, S., Semil, J., Slominski, J. and Zrobek, J. 1993. Rip currents as a geological tool. *Sedimentology*, **40**, 217–36.

Haines, J.W. and Sallenger, A.H. 1994. Vertical structure of mean cross-shore currents across a barred surf zone. *Journal of Geophysical Research*, **99**, 14223–42.

Haller, M.C., Dalrymple, R.A. and Svendsen, I.A. 2002. Experimental study of nearshore dynamics on a barred beach with rip channels. *Journal of Geophysical Research*, **107**(C6), 3061.

Hansen, J.B. and Svendsen, I.A. 1984. A theoretical and experimental study of undertow. *Proceedings of the Nineteenth Coastal Engineering Conference*. New York: American Society of Civil Engineers, pp. 2246–62.

Hartmann, D. 2006. Drowning and beach-safety management (BSM) along the Mediterranean beaches of Israel: a long-term perspective. *Journal of Coastal Research*, **22**(6), 1505–14.

Hay, A.E. and Sheng, J., 1992. Vertical profiles of suspended sand concentration and size from multifrequency acoustic backscatter. *Journal of Geophysical Research: Oceans*, **97**(C10), 15661–77.

Héquette, A., Ruz, M.H., Maspataud, A. and Sipka, V. 2009. Effects of nearshore sand bank and associated channel on beach hydrodynamics: implications for beach and shoreline evolution. *Journal of Coastal Research*, **25**, 59–63.

Holman, R.A., Symonds, G., Thornton, E.B. and Ransinghe, R. 2006. Rip spacing on an embayed beach. *Journal of Geophysical Research*, **94**(C1), 995–1011.

Houser, C. (2013) Flow separation over a prograding beach step at Pensacola Beach, Florida. *Journal of Coastal Research*, **29**, 1247–56.

Houser, C., Greenwood, B. and Aagaard, T. 2006. Divergent response of an intertidal swash bar. *Earth Surface Processes and Landforms*, **31**, 1775–91.

Houser, C., Arnott, R., Ulzhöfer, S. and Barrett, G. 2013. Nearshore circulation over transverse bar and rip morphology with oblique wave forcing. *Earth Surface Processes and Landforms*, **38**, 1269–79.

Houser, C., Brander, R., Brannstrom, C., Trimble, S. and Flaherty, J. 2016. Case study of rip current knowledge amongst students participating in a study abroad program. *Frontiers: The Interdisciplinary Journal of Study Abroad Study Abroad*, **28**, 42–60.

Houser, C., Trimble, S., Brander, R. *et al.* 2017. Public perceptions of a rip current hazard education program: 'Break the Grip of the Rip!'. *Natural Hazards and Earth System Sciences*, **17**, 1003.

Huntley, D.A. and Bowen, A.J. 1973. Field observations of edge waves. *Nature*, **243**, 160–1.

Huntley, D.A. and Short, A.D. 1992. On the spacing between observed rip currents. *Coastal Engineering*, **17**, 211–25.

Jones, O.P., Petersen, O.S. and Kofoed-Hansen, H. 2007. Modelling of complex coastal environments: Some considerations for best practise. *Coastal Engineering*, **54**, 717–33.

Klein, A.H. da F., Santana, G.G., Diehl, F.L. and de Menezes, J.T. 2003. Analysis of hazards associated with sea bathing: results of five years work in oceanic beaches of Santa Catarina State, Southern Brazil. *Journal of Coastal Research*, **SI 35**, 107–16.

Komar, P.D. 1979. Beach-slope dependence of longshore currents. *Journal of Waterways, Port, Coastal and Ocean Division*, **105**, 460–4.

Komar, P.D. 1998. *Beach Processes and Sedimentation*, 2nd Edition. New Jersey: Prentice Hall, 544 pp.

Kumar, S.A. and Prasad K.V.S.R. 2014. Rip current-related fatalities in India: a new predictive risk scale for forecasting rip currents. *Natural Hazards*, **70**, 313–35.

Larson, M. and Kraus, N.C. 1991. Numerical model of longshore current for bar and trough beaches. *Journal of Waterways, Port, Coastal and Ocean Engineering*, **117**, 326–47.

LeDoux, J. 1996. *The Emotional Brain: the Mysterious Underpinnings of Emotional Life*. London: Simon and Schuster, 384 pp.

List, J.H., Warner, J.C., Thieler, E.R. *et al.* 2011. A nearshore processes field experiment at Cape Hatteras, North Carolina, USA. In *The Proceedings of the Coastal Sediments 2011*. Singapore: World Scientific, pp. 2144–57.

Llopis, I.A., Echeverria, A. G., Trimble, S., Brannstrom, C. and Houser, C. 2018. Determining beach user knowledge of rip currents in Costa Rica. *Journal of Coastal Research*, **34**(5), 1105–15.

Longuet-Higgins, M.S. 1970a. Longshore currents generated by obliquely incident sea waves 1. *Journal of Geophysical Research*, **75**, 6778–89.

Longuet-Higgins, M.S. 1970b. Longshore currents generated by obliquely incident sea waves 2. *Journal of Geophysical Research*, **75**, 6790–801.

Longuet-Higgins, M.S. 1972. Recent progress in the study of longshore currents. In Meyer, R.R. (ed.), *Waves on Beaches*. New York: Academic Press, pp. 203–48.

Longuet-Higgins, M.S. 1983. Wave set-up, percolation and undertow in the surf zone. *Proceedings of the Royal Society of London A*, **390**, 283–91.

Longuet-Higgins, M.S. and Stewart, R.W. 1964. Radiation stress in water waves: a physical discussion with applications. *Deep Sea Research*, **11**, 529–63.

MacMahan, J.H., Thornton, E.B., Stanton, T.P. and Reniers, A.J.H.M. 2005. RIPEX: Observations of a rip current system. *Marine Geology*, **218**, 113–34.

MacMahan, J.H., Thornton, E.B., and Reniers, A.J.H.M. 2006. Rip current review. *Coastal Engineering*, **53**, 191–208.

MacMahan, J., Brown, J. and Thornton, E. 2009. Low-cost handheld global positioning system for measuring surf-zone currents. *Journal of Coastal Research*, **25**, 744–54.

MacMahan, J., Brown, J., Brown, J. *et al.* 2010. Mean Lagrangian flow behaviour on an open coast rip-channelled beach: a new perspective. *Marine Geology*, **268**(1–4), 1–15.

Masselink, G. 2004. Formation and evolution of multiple intertidal bars on macrotidal beaches: application of a morphodynamic model. *Coastal Engineering*, **51**, 713–30.

Masselink, G. and Black, K.P. 1995. Magnitude and cross-shore distribution of bed return flow measured on natural beaches. *Coastal Engineering*, **25**, 165–90.

Masselink, G. and Pattiaratchi, C.B. 1998. The effect of sea breeze on beach morphology, surf zone hydrodynamics and sediment resuspension. *Marine Geology*, **146**, 115–35.

McKenzie, P. 1958. Rip current systems. *Journal of Geology*, **66**, 103–13.

Menard, A.D., Houser, C., Brander, R.W., Trimble, S. and Scaman, A. 2018. The psychology of beach users: importance of confirmation bias, action and intention to improving rip current safety. *Natural Hazards*, **94**(2), 953–73.

Miles, J.R. and Russell, P.E. 2004. Dynamics of a reflective beach with a low tide terrace. *Continental Shelf Research*, **24**, 1219–47.

Miles, J.R., Russell, P.E., Ruessink, B.G. and Huntley, D.A. 2002. Field observations of the effect of shear waves on sediment suspension and transport. *Continental Shelf Research*, **22**, 657–81.

Murray, A.B., LeBars, N. and Guillon, C. 2003. Tests of a new hypothesis for non-bathymetrically driven rip currents. *Journal of Coastal Research*, **19**, 269–77.

Musumeci, R.E., Svendsen, I.A. and Veeramony, J. 2005. The flow in the surf zone: a fully non-linear Boussinesq-type of approach. *Coastal Engineering*, **52**, 565–98.

Nummedal, D. and Finley, R.J. 1978. Wind-generated longshore currents. *Proceedings of the Sixteenth Conference on Coastal Engineering*, New York: American Society of Civil Engineers, pp. 1428–38.

Oltman-Shay, J., Howd, P.A. and Birkmeier, W.A. 1989. Shear instabilities of the mean longshore current. 2: field observations. *Journal of Geophysical Research*, **94**, 18031–42.

Orzech, M.D., Thornton, E.B., MacMahan, J.H., O'Reilly, W.C. and Stanton, T.P. 2010. Alongshore rip channel migration and sediment transport. *Marine Geology*, **271**, 278–91.

Osborne, P.D. and Greenwood, B. 1992. Frequency dependent cross-shore sediment transport, 2: a barred shoreface. *Marine Geology*, **106**, 25–51.

Phillips, O. 1977. *The Dynamics of the Upper Ocean*. Cambridge: Cambridge University Press.

Puleo, J.A., Blenkinsopp, C., Conley, D. *et al.* 2013. Comprehensive field study of swash-zone processes. I: experimental design with examples of hydrodynamic and sediment transport measurements. *Journal of Waterways, Port, Coastal, and Ocean Engineering*, **140**, 14–28.

Reniers, A.J.H.M., Thornton, E.B., Stanton, T.P. and Roelvink, J.A. 2004. Vertical flow structure during Sandy Duck: observations and modelling. *Coastal Engineering*, **51**, 237–60.

Ruessink, B., Miles, J., Feddersen, F., Guza, R. and Elgar, S. 2001. Modelling the alongshore current on barred beaches. *Journal of Geophysical Research*, **106**, 22451–63.

Sassi, M.G., Hoitink, A.J.F. and Vermeulen, B., 2012. Impact of sound attenuation by suspended sediment on ADCP backscatter calibrations. *Water Resources Research*, **48**(9). DOI: 10.1029/2012WR012008.

Scott, T.M., Russell, P.E. and Masselink, G. 2011. Rip current hazards on large-tidal beaches in the United Kingdom. In Leatherman, S. and Fletemeyer, J. (eds.), *Rip Currents: Beach Safety, Physical Oceanography, and Wave Modelling*. London: CRC Press, pp. 225–42.

Scott, T.M., Austin, M., Masselink, G. and Russell, P. 2016. Dynamics of rip currents associated with groynes: field measurements, modelling and implications for beach safety. *Coastal Engineering*, **107**, 53–69.

Shepard, F.P. and Inman, D.L. 1950. Nearshore water circulation related to bottom topography and wave refraction. *Eos, Transactions American Geophysical Union*, **31**(2), 196–212.

Shepard, F.P., Emery, K.O. and La Fond, E.C. 1941. Rip currents: a process of geological importance. *The Journal of Geology*, **49**(4), 337–69.

Sherman, D.J., Short, A.D. and Takeda, I. 1993. Sediment mixing depth and bedform migration in rip channels. *Journal of Coastal Research*, **15**, 39-48.

Short, A.D. 1979. Three-dimensional beach-stage model. *Journal of Geology*, **87**, 553–71.

Short, A.D. 1985. Rip-current type, spacing and persistence, Narrabeen Beach, Australia. *Marine Geology*, **65**, 47–71.

Short, A.D. 1999. Beach hazards and safety. In Short, A.D. (ed.), *Handbook of Beach and Shoreface Morphodynamics*. Chichester, UK: John Wiley & Sons, pp. 293–304.

Short, A.D. and Hogan, C.L. 1994. Rip currents and beach hazards: their impact on public safety and implications for coastal management. *Journal of Coastal Research*, **SI 12**, 197–209.

SLSA, 2010: Preventing coastal drowning deaths in Australia. National Coastal Safety Report, Surf Life Saving Australia, 22 pp.

Sonu, C.J. 1972. Field observations of nearshore current circulation and meandering currents. *Journal of Geophysical Research*, **7**, 3232–47.

Sonu, C.J., Murray, S.P., Hsu, S.A., Suhayda, J.N. and Waddell, E. 1973. Sea-breeze and coastal processes. *EOS, Transactions of the American Geophysical Union*, **54**, 820–33.

Stive, M.J.F. and Wind, H.G. 1986. Cross-shore mean flow in the surf zone. *Coastal Engineering*, **10**, 325–40.

Svendsen, I.A. 1984a. Wave height and set-up in a surf zone. *Coastal Engineering*, **8**, 303–29.

Svendsen, I.A. 1984b. Mass flux and undertow in a surf zone. *Coastal Engineering*, **8**, 347–65.

Thorne, P.D. and Hurther, D. 2014. An overview on the use of backscattered sound for measuring suspended particle size and concentration profiles in non-cohesive inorganic sediment transport studies. *Continental Shelf Research*, **73**, 97–118.

Thornton, E.B. and Guza, R.T. 1986. Surf zone longshore currents and random waves, field data and models. *Journal of Physical Oceanography*, **16**, 1165–78.

Torres-Freyermuth, A., Puleo, J.A., DiCosmo, N. *et al.* 2017. Nearshore circulation on a sea breeze dominated beach during intense wind events. *Continental Shelf Research*, **151**, 40–52.

Turner, I.L., Whyte, D., Ruessink, B.G. and Ranasinghe, R. 2007. Observations of rip spacing, persistence and mobility at a long, straight coastline. *Marine Geology*, **236**, 209–21.

USLA, 2015: National lifesaving statistics. United States Lifesaving Association. DOI: http://arc.usla.org/Statistics/view/displayAgency.asp.

Whitford, D.J. and Thornton, E.B. 1993. Comparison of wind and wave forcing of longshore currents. *Continental Shelf Research*, **103**, 1205–18.

Woodward, E., Beaumont, E., Russell, P., Wooler, A. and Macleod, R. 2013. Analysis of rip current incidents and victim demographics in the UK. *Journal of Coastal Research*, **65**(SP1), 850–5.

Wright, L.D. and Short, A.D. 1984. Morphodynamic variability of surf zones and beaches: a synthesis. *Marine Geology*, **56**, 93–118.

Wright, L.D., Guza, R.T. and Short, A.D. 1982. Dynamics of a high-energy dissipative surf zone. *Marine Geology*, **45**, 41–62.

Yuan, J. 2016. Turbulent boundary layers under irregular waves and currents: Experiments and the equivalent-wave concept. *Journal of Geophysical Research: Oceans*, **121**, 2616–40.

Yuan, J. and Madsen, O.S. 2015. Experimental and theoretical study of wave–current turbulent boundary layers. *Journal of Fluid Mechanics*, **765**, 480–523.

7

Coastal Sediment Transport

7.1 Synopsis

Modification of the coast through erosion and deposition is dependent on the divergence and convergence of sediment transport by waves and currents, with additional sediment delivered to the coast by rivers or from offshore sources. Except in a few locations in the surf zone, such as rip channels and where longshore currents are fastest, unidirectional currents in the nearshore and surf zone are generally not strong enough to erode and transport sand-sized sediment directly. Instead, sediment is set in motion by the oscillatory currents associated with waves and by the turbulent vortices generated by wave breaking, particularly where waves break as plunging breakers. The direction of net sediment transport then depends on the balance of all the forces acting on the sediment, including those due to incident and long waves, wave generated on–offshore and alongshore flows, wind-driven currents and tidal flows. It is usual to distinguish between processes that lead to the net transport of sediment onshore or offshore (shore normal transport) and those tending to move sediment alongshore (longshore sediment transport), though both processes occur simultaneously and contribute the evolution of the coast.

Fine sediments in the silt and clay size range tend to be placed in suspension and diffused uniformly through the water column by waves and currents and do not occur in appreciable amounts in the inner nearshore and surf zones. They are removed offshore or alongshore where they settle out of suspension in deep water in the outer shoreface or further offshore. Fine sediments may also be brought into estuaries, bays and lagoons, where they are deposited in areas where waves and currents weaken in the presence of vegetation such as seagrass, saltmarsh plants and mangroves. Vertical growth and seaward extension of a delta is also dependent on the deposition of sediment carried by a river that enters slow-moving or standing water. The loss of available sediment from the Mississippi River combined with sea level rise and storm activity is contributing to the rapid retraction of the Mississippi Delta. The cohesive property of these fine materials and the fact that they are often found in the presence of sheltering vegetation makes prediction of sediment entrainment quite complicated. We will therefore ignore them for now and consider them when we describe saltmarsh and mangrove coasts (Chapter 11).

7.2 Sediment Transport Mechanisms, Boundary Layers and Bedforms

7.2.1 Transport Mechanisms

Particles in the sand and gravel size fraction (~0.1–64 mm) are cohesionless and readily set in motion by waves. Particles in this size range are exchanged readily between the inner and outer shoreface and the beach, and may be transported alongshore in appreciable quantities. Alongshore variations in the amount of sediment in transport is an important control on beach morphology and whether a beach experiences erosion or deposition. Particles in this size range are characteristic of most beaches and surf zones and there is a considerable body of empirical and theoretical studies of their dynamics. We will therefore focus attention on the processes of entrainment, transport and deposition of sand, and the characteristics of the bedforms that develop across the nearshore. Larger particles (cobbles and boulders) are also cohesionless but their large size requires special treatment and transport requires very large waves (or possibly the presence of ice). Depending on the size of the sediment, the instantaneous velocities near the bed can exceed the threshold for the initiation of movement. As previously noted, unidirectional currents are generally not strong enough to erode and transport coarse sediment, with the exception of rip currents and longshore currents in extreme storms. Sediment is entrained by oscillatory currents and the turbulent vortices of wave breaking and then transported by the current. During storms or large swell waves, sediment transport can be initiated across the nearshore, and on all but the most sheltered sand beaches, or during quiet periods, sediment entrainment and transport is nearly ubiquitous in the surf zone and swash zones.

Sediment is transported in the direction of mean flow resulting from the combination of Stokes drift, undertow and wind-driven currents. The direction of net transport of sediment depends on the balance of all the forces acting on the sediment, including those due to incident and long waves, wave generated on–offshore and alongshore flows, wind-driven currents and tidal flows. The resulting net transport sediment can either be cross-shore (onshore or offshore) or alongshore, though both processes can occur simultaneously if a longshore current develops.

As is the case for unidirectional flow in rivers, sediment may be transported continuously in contact with the bed through rolling and sliding, or in intermittent contact through saltation. Most gravel sediments are transported this way, and only placed in suspension for short periods in highly energetic breaker zones. Depending on the size of the sediment or the turbulence generated by the waves and currents, sediment may also be transported in suspension. This is common throughout the surf zone during storms with large energetic waves. Suspension is also possible because of flow separation and eddy generation as the currents encounter bedforms such as ripples and dunes. The sediment is thrown into suspension during the period of maximum flows under the wave crest and trough. This sediment does not get very high in the water column – generally only a few centimetres to a few tens of centimetres above the bed, but it can significantly alter the direction of net sediment transport based on the wave period, current strength and the height to which the sediment is lifted off the bed. Under energetic conditions, there is a continuous exchange between sediments in the bed and sediments in suspension above it. In areas of breaking waves, especially plunging breakers, large vortices impinging on the bed may also throw sand into suspension throughout the water column. In practice, while reliable instrumentation now exists to measure suspended sediment concentrations some distance above the bed, it has proven difficult to measure sediment transport very close to the bed and the bedload transport itself. Consequently, there is still considerable uncertainty as to the relative contribution of the two modes of transport (Aagaard *et al.*, 2013). It is particularly difficult to predict sediment transport rates across the nearshore in the presence of bedforms and a combination of waves and currents.

7.2.2 Boundary Layers, Shear Stress and Initiation of Motion

Friction and turbulence generated by fluid motion over a boundary such as the sea bed generates a boundary layer in which the flow velocity increases away from the bed and reaches a maximum at the top of the boundary layer above which the flow is unaffected by the bed. Bedforms have a greater effect on flow in the

A.

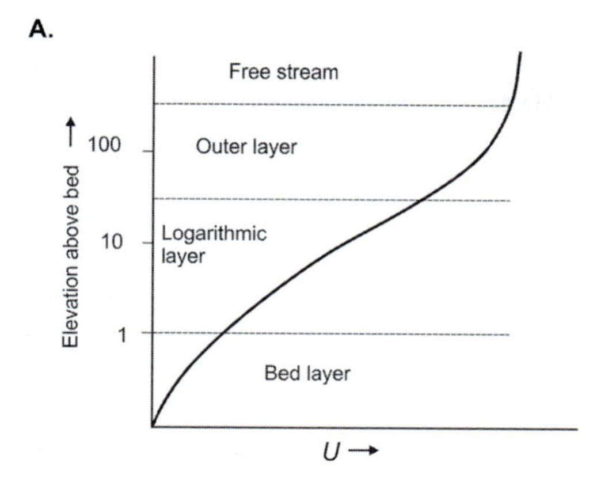

B.

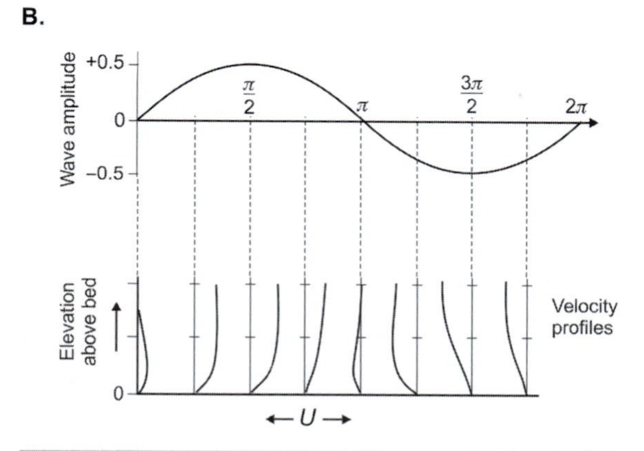

Figure 7.1 Boundary layers in coastal waters: A. the boundary layer formed under unidirectional flows such as tidal currents, rip currents and winds; B. the changing form of the oscillatory boundary layer for a single wave.

boundary layer than does sediment, particularly in the sand size range. The boundary layer under unidirectional flows consists of three principal layers – the bed layer, the logarithmic layer and the outer layer (Figure 7.1A). The bed layer may also be divided into an inner layer close to the bed and an outer buffer layer that links this to the logarithmic portion of the profile. The inner layer is termed the viscous layer when the bed is smooth (Reynolds number (Re) < 5) or the bed layer when the bed is rough and sediment on the bed projects into the turbulent flow (Re> 70). Bed roughness is produced initially by the material making up the bed (skin friction) and is generally considered to be proportional to the grain diameter D. The second contribution to bed roughness is that produced by bedforms such as ripples and dunes that are present much of the time and which have a much greater effect on the velocity profile.

Description and modelling of sediment transport under unidirectional flow in rivers and by wind action are usually based on the bed shear stress τ. This can be measured directly or, more usually, is derived from the bed shear velocity U_* which in turn is derived from the slope of the velocity profile in the logarithmic layer, using the law of the wall. In unidirectional flows the logarithmic layer is comparatively thick (on the order of tens of centimetres to several metres), and it is feasible to take the average of readings over several minutes. However, under oscillatory flow the boundary layer essentially breaks down twice during each wave period and a thick boundary layer is not able to develop. Instead the boundary layer grows and diminishes during each half of a wave period and the flow direction reverses between the crest and the trough of the wave (Figure 7.1B) – the total wave boundary layer height is usually < 0.1 m.

Measurement of the velocity profile during a wave cycle thus requires the use of instruments, such as a hot film probe, to make high-frequency measurements of the flow over very short distances close to the bed (e.g., Jonsson and Carlsen, 1976; Jensen *et al.*, 1989), and is not possible in the field. Maximum values for bed shear stress thus occur under the wave crest and trough and there are two minima in each complete wave cycle. Assuming turbulent flow conditions, the shear stress τ can be defined as:

$$\tau = (\mu + \xi)\frac{du_*}{dz}, \tag{7.1}$$

where μ is the molecular viscosity and ξ is the eddy viscosity. Molecular viscosity is a constant for given conditions and generally much smaller than the eddy viscosity, which is an effective viscosity resulting from the presence of turbulence in the water column. In rivers, an average value for ξ can be used for a given set of flow conditions but in the case of oscillatory wave motion eddy viscosity changes as the flow speeds up and then slows down with each wave passage (Wiberg, 1995).

It is possible therefore, to compute either the maximum bottom shear stress or an average value for the whole wave cycle. The maximum horizontal orbital velocity at the top of the wave boundary layer can be estimated from linear wave theory as:

$$u_0 = \frac{\pi H}{T \sin kh}. \tag{7.2}$$

This is the same as Equation 5.18 and the derivation is given in Chapter 5. The maximum bed shear stress τ_w can then be estimated from:

$$\tau_w = 0.5 \rho f_w u_0^2, \tag{7.3}$$

where ρ is the fluid density and f_w is the wave friction factor. The wave friction factor is one of those parameters that can be determined if we have measurements for all the other parameters. In predicting bed shear stress, f_w must be estimated, and a lot of work has gone into determining the best way to do this (e.g., Kamphuis, 1975; Nielsen, 1992). As would be expected from the discussion above, the friction factor should incorporate both the grain roughness and form roughness due to bedforms. Nielsen (1992) proposed the following empirical equation for obtaining f_w:

$$f_w = \exp\left[5.5\left(\frac{K_s}{d_0}\right)^{0.2} - 6.3\right] \tag{7.4}$$

where d_0 is the orbital diameter at the bed and k_s is the Nikuradse roughness length. The term k_s is made up of the grain roughness k' that is dependent on the grain size, and the form roughness k'' that is dependent on the size and type of the bedforms. According to Nielsen (1992), k' can be approximated by the diameter D of the grains making up the bed and k'' can be estimated from:

$$k'' = 8\frac{\eta^2}{\lambda}, \tag{7.5}$$

where η is the height of the bedforms and λ is the spacing between bedforms. This probably works reasonably well for relatively simple 2D ripple forms, but there is likely considerable variation for complex 3D ripples and dunes. However, the ability to approximate the maximum bed shear stress is useful for predicting the threshold of sediment motion by waves, and this approach is the start for modelling sediment transport in the nearshore.

Sand can be expected to start moving and rolling back and forth on the bed (i.e., entrained) when τ_w exceeds the threshold for the sediments making up the bed.

Similar to unidirectional flow, we can predict entrainment using the Shields diagram, which describes the relationship between the critical shear stress (expressed through the Shields parameter, θ_c) and the non-dimensional grain diameter (D_*). The θ_c is calculated as:

$$\theta_c = \frac{\tau_c}{gD(\rho_s - \rho)} \tag{7.6}$$

where g is the acceleration due to gravity, D is the grain size (mm) andτ_c is the critical shear stress. The value of D_* is given by:

$$D_* = D\left[\frac{\rho^2 g(s-1)}{\mu^2}\right]^{1/3}. \tag{7.7}$$

There is an inverse relationship between θ_c and D_* when $D_* < 10$, which means that it is more difficult to entrain small particles due to a combination of cohesion, electrostatic forces of attraction and the lack of exposure of the grain into the turbulent flow. These forces are not apparent when $D_* > 10$, such that a positive relationship develops between θ_c and D_* as initiation depends on the weight of the particle.

In simple unidirectional flows or laboratory experiments, it can be easy to identify and define incipient motion, but in the field we can expect considerable variations in grain size and in bedforms over small areas, and temporal variations in bed shear stress as a result of irregular waves (Davies, 1985). This means that we should think of the prediction of a threshold in terms of the probability (in space and time) of sediment motion. Simplified expressions are available for sediments in fine to coarse sand that provide a reasonable value, given this variability (e.g., Komar and Miller, 1973). However, during high wave events, sediment is in motion over the whole nearshore, and in practice we need not worry too much about the initiation of motion – at least for sand in the nearshore.

7.2.3 Bedform Development

As is the case for unidirectional flow in rivers and air flow, transport of sediment in bedload or close to the bed by waves or the combined action of waves and currents leads to the development of characteristic forms on the bed. Bedforms tend to be quasi-regular in morphology and spacing and their form seems to be controlled to a large extent by the diameter of the sediment, by the shear stress exerted on the bed and by the velocity asymmetry in the waves or combined wave and current flow. Bedform generation and morphology have been studied in wave tanks and oscillatory tubes (Mogridge and Kamphuis, 1972; O'Donoghue and Clubb, 2001; van der Werf *et al.*, 2006) as well as in the field (Sherman and Greenwood, 1984; Osborne and Vincent, 1993; Swales *et al.*, 2006; Miles *et al.*, 2014). We know quite a lot about ripples produced by pure oscillatory flow, but it is still very difficult to predict the type and dimensions of bedforms in the highly variable and complex flows of the breaker and surf zones. Since it takes time for bedforms to develop and evolve as the size of the waves and the strength of the current changes, the type and dimensions of the bedforms may not be in equilibrium. Combined with the spatial variation in wave heights and current strength through the nearshore, it can be difficult to predict form roughness in sediment transport models.

When waves are regular and the oscillatory flow is symmetric the sequence of bedforms observed on a sand bed as bed shear stress is increased from the critical point for initiation of motion to very high values is: plane bed (lower flow regime) → ripples → plane bed (upper flow regime). These conditions can be observed in laboratory wave tanks or in deep water some distance seaward of the breaker line as the waves begin to shoal (Li and Amos, 1999; Doucette, 2000; O'Donoghue and Clubb, 2001). The ripples that develop in these conditions are symmetric in cross section and show little evidence of migration landward or seaward. This contrasts with asymmetric ripples that develop under the highly asymmetric orbital motion under waves near the breakpoint or under unidirectional flows. Offshore of the breakpoint these ripples migrate landward with Stokes drift, while landward of the breakpoint the ripples can migrate seaward with the undertow. Consequently, there is a convergence of sediment transport at the breakpoint.

As flow speeds increase there is a transition from rolling grain ripples to vortex ripples (Bagnold, 1946). Rolling grain ripples are formed under flow conditions just above the critical shear stress, where grains are transported close to the bed and no vortex is generated in the lee of the ripple. These ripples are comparatively subdued in height. As the velocity increases, the ripples grow in height and a vortex develops on the lee side of the ripple crest during each half of the wave cycle. At even higher flow velocities, ripple height begins to decrease as large amounts of sand are transported in suspension and the subdued ripples under these conditions are termed post-vortex. Vortex ripples have well-defined crests, curved troughs and a regular spacing between the bedforms. In the field, they tend to be 2D in form, with ripple crests being continuous for distances many times longer than the spacing between ripples (Figure 7.2A). However, laboratory experiments in oscillatory wave tunnels suggest that only 3D forms are observed where there is fine sand (O'Donoghue and Clubb, 2001). As noted, ripple height η varies with flow conditions and the maximum height also increases with an increase in sediment diameter. In fine to medium sand, ripples are typically 1–5 cm in height (Figure 7.2A) and may exceed 10 cm in coarse sand and granules (Figure 7.2B). Eventually the ripples disappear entirely with sheet flow conditions over a planar bed. The transition from no sediment movement to rolling grain ripples, vortex ripples and sheet flow is described in terms of a stability diagram (Figure 7.3).

The spacing between ripples (λ) also varies with flow conditions, but the relationship is more complex than for ripple height. At low flows the spacing scales to the orbital diameter at the bed ($\lambda \sim d_o$), and it increases with increasing orbital diameter (d_o; Inman, 1957; Miller and Komar, 1980). Orbital ripples form under these conditions (Inman, 1957; Clifton, 1976; Hay, 2011). As the orbital diameter increases further, the spacing

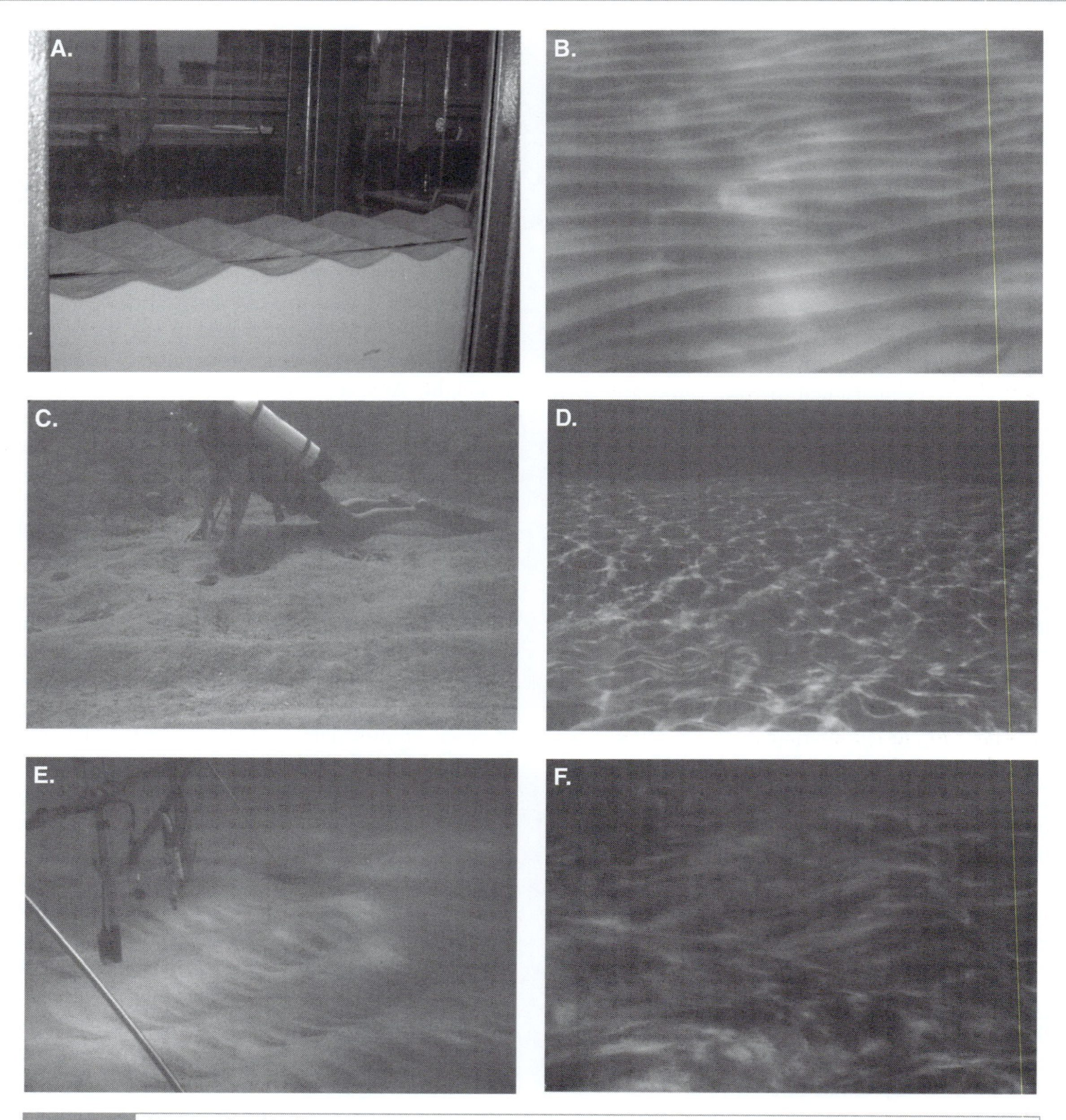

Figure 7.2 Photographs of bedforms on sandy coasts: A. Two-dimensional vortex ripples generated in a laboratory oscillatory flume, University of Aberdeen (photo Jeff Doucette). B. Two-dimensional vortex ripples characteristic of shoaling waves on a medium sand bed. Note the lateral extent of the ripple crests and the tuning fork junctions that are typical of these ripples in the field. The spacing between ripples is regular and it can be seen that this spacing adjusts to the junction (or splitting) of the crests. C. Large vortex ripples developed in coarse sand and granules, Veradero, Cuba at a depth of about 25 m. D. Three-dimensional ripples on a bar crest. E. Cross ripples or ladderback ripples (photo Jeff Doucette). F. Flow over quasi two-dimensional dunes just seaward of the breakpoint. The photograph was taken just as the wave crest passed looking obliquely offshore. Plumes of sediment are visible seaward as vortices lift off during the offshore directed portion of the flow.

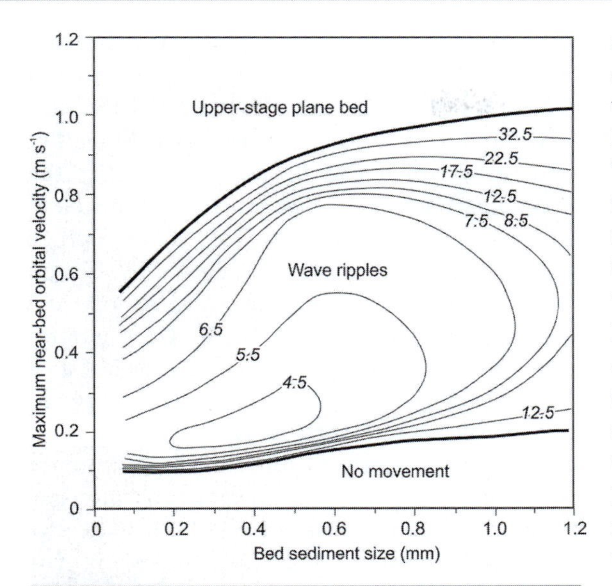

Figure 7.3 Stability diagram showing development of bed morphology (no movement, wave ripples and upper-stage plane bed) based on the combination of grain size and maximum orbital velocity near the bed. Modified from Allen (1984).

remains at some limit, or decreases slightly, and $\lambda \sim 1/d_0$. At very high flows, the wavelength of anorbital ripples become shorter than and independent of the orbital diameter. However, in the field there are rapid variations in grain size, topography, wave orbital motion and the superimposition of longshore and rip currents leading to a complex range of bedforms. In addition to 2D symmetric and asymmetric ripples, 3D ripples with short crest lengths and complex branching patterns can be observed. In troughs, where unidirectional flows associated with longshore and rip currents occur, 3D ripples characteristic of those in rivers can be observed. In the zone of shoaling waves, cross ripples may be found. These appear as 2D vortex ripples oriented diagonally to the direction of wave advance with short ripples aligned perpendicular to these in the troughs between. The whole effect looks like the rungs of a ladder and so they are sometimes called ladderback ripples (Figure 7.2E).

When the flow is strongly asymmetric in the wave shoaling zone, ripples become increasingly asymmetric and the ripple migration is in the direction of wave advance. Depending on the wave height, this asymmetric flow regime can lead to the development of lunate megaripples. These are a form of 3D dune, with an arcuate trough incised below the general bed surface and the horns of the trough pointing landward (Clifton *et al.*, 1971; Davidson-Arnott and Greenwood, 1976; Sherman *et al.*, 1993; Larsen *et al.*, 2015). Lunate megaripples are 1-4 m across and the base of the scour trough is typically 0.3–0.8 m below the flat top of the megaripple crest. In contrast to dunes in rivers, where the crest appears as a triangular feature standing above the bed level, lunate megaripple crests occupy most of the form and the troughs appear to be incised below the general bed surface. The scour trough has a slip face on both sides, and under wave action sediment avalanches into the trough and up the other side under the passage of both the wave crest and the wave trough. However, because of the strong flow asymmetry, there is a net migration in the direction of wave advance, and the resultant bedform is like that of a dune produced by unidirectional low on a river, but with a rounded trough. The megaripple troughs are spaced en echelon with distances between them on the order of 2–4 m (Clifton *et al.*, 1971; Sherman *et al.*, 1993; Gallagher, 2003). Lunate megaripples develop also in combined wave and current flows and the direction of movement is then controlled by the resultant current. Thus, lunate megaripples can migrate offshore in flows over the centre of crescentic bars or along rip current channels (Davidson-Arnott and Greenwood, 1976; Sherman *et al.*, 1993).

The transition to lunate megaripples is controlled initially by the intensity of the bed shear stress under waves. Several studies (Nielsen, 1992; Aagaard *et al.*, 2002) have found that the transition to lunate megaripples occurs when the non-dimensional shear stress (θ_{max}) is on the order of 0.8–1. However, field measurements suggest that the appearance of lunate megaripples is also dependent on the presence of either strong asymmetry in the wave orbital motion and/or relatively strong unidirectional flows superimposed on the oscillatory motion (Davidson-Arnott and Greenwood 1976; Aagaard *et al.*, 2002; Swales *et al.*, 2006). Thus, Aagaard *et al.* (2002) found that lunate megaripples disappeared at values of θ_{max} well above 1 when longshore current velocity decreased rather than as a function of increasing bed shear stress. As might be expected from unidirectional flows, the occurrence of lunate megaripples may also depend on the grain size of sediments making up the bed. They do not appear to develop in very fine sands and the sequence appears to go directly from ripples to plane bed (Clifton *et al.*, 1971; Davidson-Arnott and Pember, 1980; Doucette, 2000; Miles *et al.*, 2014).

7.2.4 Cross-Shore Variation in Bedforms

The variation in maximum orbital velocity, orbital diameter, wave skewness and Stokes drift under shoaling waves and into the breaker and surf zones produces consistent and recognisable sequences of bedforms. On a planar beach subject to large swell waves, Clifton *et al.* (1971) recognised a distinct suite of bedforms associated with a single breaker zone (Figure 7.4A). Here the large waves typically result in the formation of asymmetric ripples in quite deep water. These are succeeded by a zone of lunate megaripples just seaward of the breaker zone and these in turn wash out to form a planar bed in and around the breaker zone. The inner rough zone consists of irregular holes and humps in the inner surf zone close to the step. Flow conditions here are often complex and visibility low, so it is difficult to define the bedforms well. Finally, the swash slope is characterised by planar bedding, often in alternating laminae of light and dark minerals. Clifton *et al.* (1971) did not note the presence of cross ripples but a subsequent review by Clifton (1976) included them in a zone just seaward of the lunate megaripples and they have been observed in several areas (Ship, 1984; Osborne and Vincent, 1993).

Several breaker zones may occur on a barred nearshore, and the sequence of bedforms is more complex (Davidson-Arnott and Greenwood, 1974, 1976; Ship, 1984; Greenwood and Mittler, 1985; Swales *et al.*, 2006; see Figure 7.4B). As is the case in other sedimentary environments, migration of the bedforms and/or burial below the active surface produces distinct suites of sedimentary structures which can be used to identify the hydrodynamics of the depositional environment (Figure 7.4B). The sequence of ripples, lunate megaripples and plane bed noted by Clifton *et al.* (1971) can occur on the seaward slope and crest of each bar (Figure 7.4A, B). On the outer bars where the water depth over the crest is relatively deep the pattern down the landward slope is the reverse of the seaward slope as wave orbital motion decreases into the deeper water of the trough. In the trough both 2D asymmetric ripples aligned with waves and 3D ripples produced by longshore currents flowing in the trough may be present. This zone is often associated with the accumulation of coarse sand, shell and organic material that may give rise to larger ripples and produces a distinctive colouration and sorting of these sediments.

Cores taken from the modern environment can also provide information on sediment transport modes and directions. Thus, early field evidence to support offshore flows across outer nearshore bars during storms came from identification of seaward-dipping cross-stratification produced by the offshore migration of lunate megaripples (Davidson-Arnott and Greenwood, 1976; Greenwood and Davidson-Arnott, 1979; Greenwood and Mittler, 1979; see Figure 7.5D). Where water depth over inner nearshore bars is very shallow (e.g., at low tide) wave bores produce nearly continuous sheet flow across the bar crest leading to the formation of an avalanche slope (Figure 7.5E) and landward migration of the bar (this process is described in detail in Chapter 8). Finally, shear sorting by the swash and backwash on the beach foreshore produces distinct planar lamination, often marked by alternating bands of dark and light minerals, parallel to the foreshore slope (Figure 7.5F).

7.2.5 Bed Roughness and Sediment Suspension

In addition to the information they provide on flow conditions, especially in the interpretation

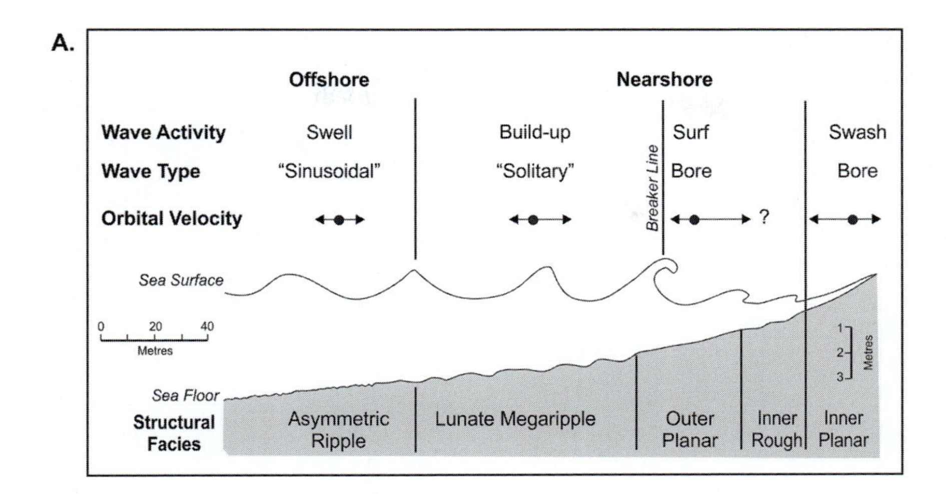

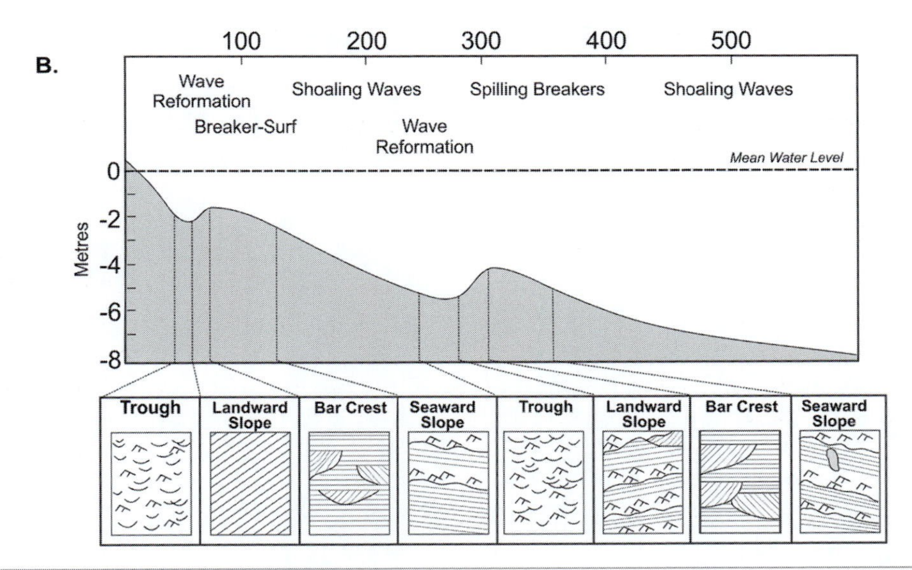

Figure 7.4 Hydrodynamics and bedform sequences in the nearshore: A. sequence of bedforms under waves on a high energy, swell coast (Clifton *et al.*, 1971); B. sequence of bedforms and sedimentary structures on a barred coastline (Davidson-Arnott and Greenwood, 1976).

of ancient sediments, the type of bedform present at any location in the zone of shoaling and breaking waves is an important determinant of the surface roughness and the vertical and temporal variation in suspended sediment concentration. The ability to predict the type of bedform present is crucial to the operation of models that are used to predict the flow field and resulting sediment transport under shoaling and breaking waves (Austin and Masselink, 2008). Where ripples are present under nearly symmetric oscillatory flow (which they nearly always are), flow over the ripple crest produces two suspension events, one associated with the onshore flow and the other with the offshore flow (Inman and Bowen, 1963; see Figure 7.6). As waves shoal and become more asymmetric, the offshore flow is suppressed and there tends to be a single peak in suspension. The height to which sand is suspended will vary with the

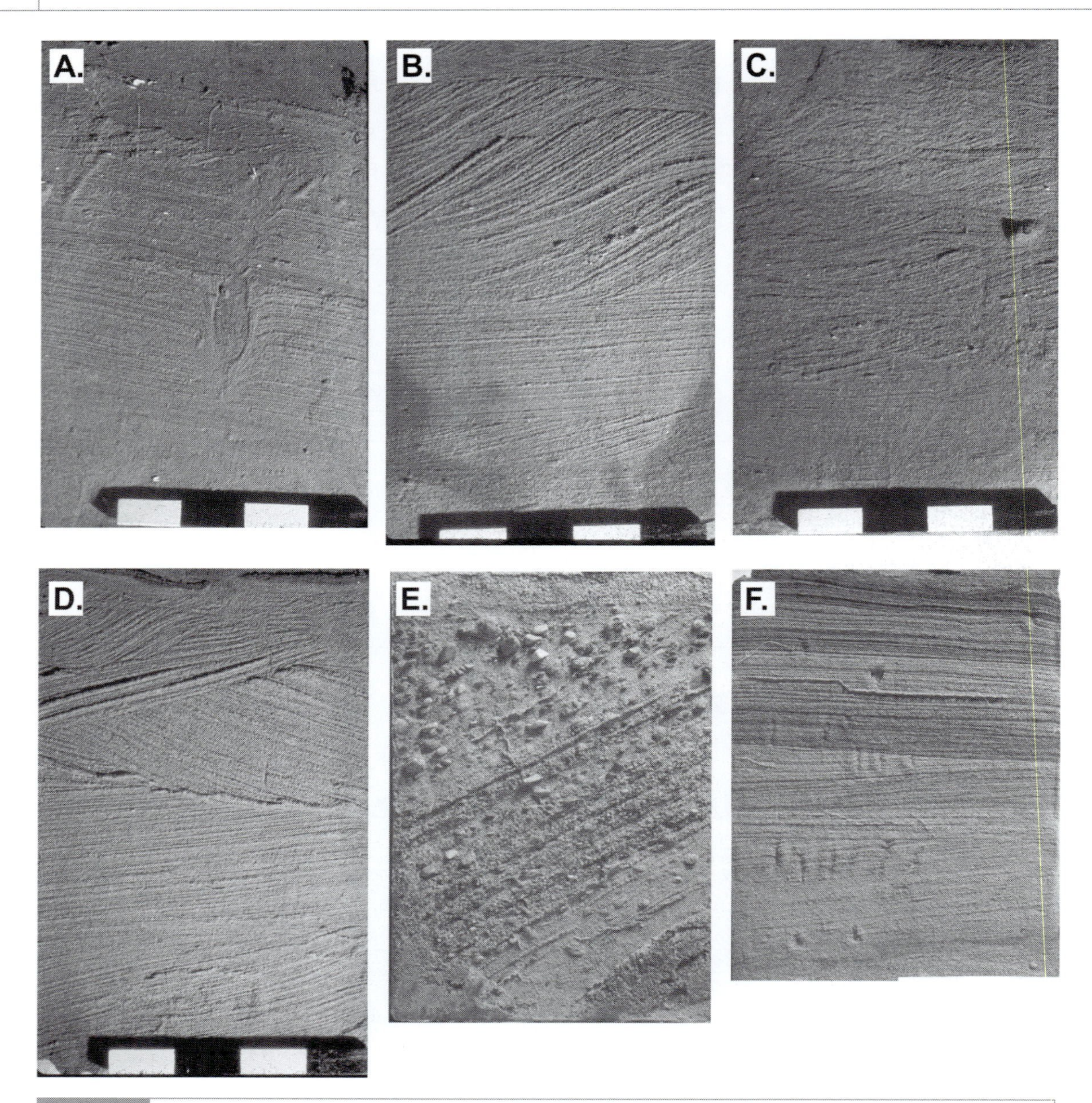

Figure 7.5 Photographs of resin peels taken from box cores showing the sedimentary structures formed by bedforms in the nearshore and foreshore. The cores were all oriented perpendicular to the shoreline with onshore to the left, and the bars on the scales are 5 cm. A. Alternating beds of ripple cross-lamination and planar bedding from the seaward slope of an outer bar in about 5 m water depth. B. Planar bedding and landward dipping cross-stratification produced by lunate megaripple migration on an outer bar crest in a water depth of 2.5 m. The black line is dyed sand placed on the surface prior to taking the core and the shape clearly marks a vortex ripple produced by low wave conditions. C. Ripple lamination in the trough landward of an outer bar. Sediments include shell hash washed over the bar crest and organic matter that accumulates in the trough under low wave conditions. D. Core taken on the crest near the centre of a crescentic outer bar showing offshore dipping cross-stratification produced by lunate megaripple migration in a zone of strong undertow or rip current activity. E. Landward-dipping cross stratification on the slip-face of an inner bar. F. Parallel lamination produced by swash and backwash on the beach foreshore.

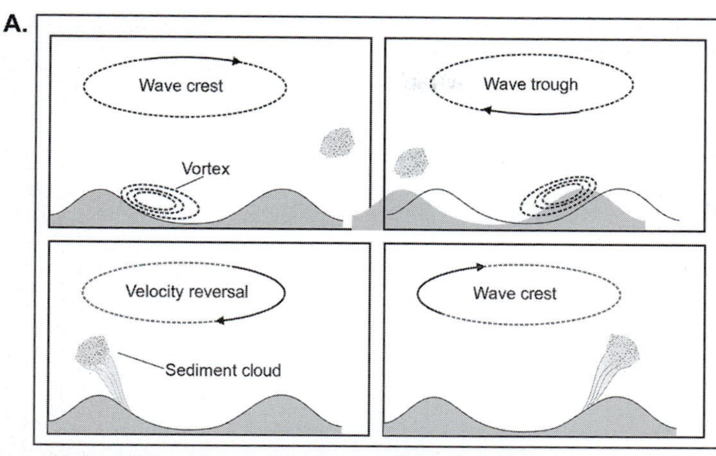

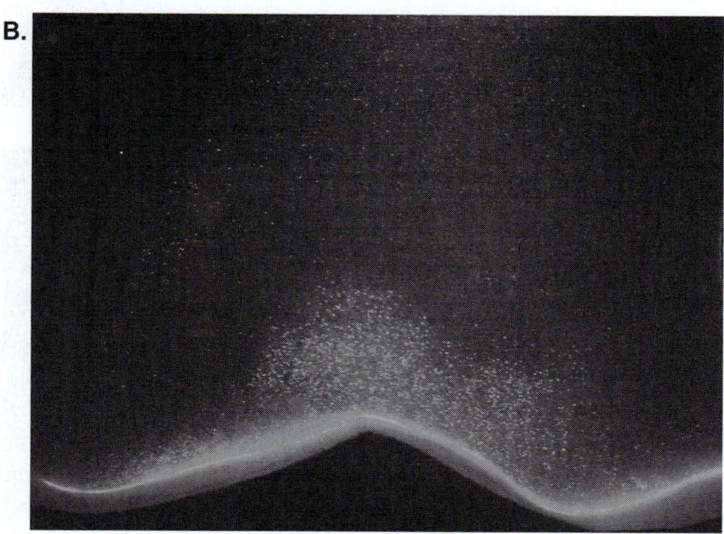

Figure 7.6 Flow over classic vortex ripples under symmetric oscillatory wave motion. A. Schematic of vortices produced on either side of the ripple crest first due to shoreward motion under the crest and then to seaward flow under the trough. As the vortices are shed during the period of flow reversal they carry sand into suspension to a height two to three times the height of the ripple (after Inman and Bowen, 1963). B. Photograph of the plume lifting off the ripple crest in an oscillatory flume (photograph Jeff Doucette).

intensity of the flows and wave period, and is much greater for flows over steep ripples with well-defined crests than for much lower, washed out ripples. In the latter case, the absence of a well-defined crest inhibits vortex formation and shedding (e.g., Osborne and Vincent, 1996; Doucette, 2000; see Figure 7.7).

Flows over the troughs of lunate megaripples can produce suspension clouds that extend much higher into the water column as a reflection of the greater turbulence across the trough. At higher flows, as the bed becomes planar, the bed roughness and bed shear stress decrease and the suspension events are much less pronounced. Under these conditions just outside the breaker zone, the water column a few tens of centimetres above the bed may be quite clear but sediment concentration within a few centimetres of the bed becomes very large as sheet flow develops. Sediment suspension in the water column then increases in the breaker and surf zones but this is largely due to turbulence advected downward from the breaking wave rather than flow across bedforms.

Ultimately simultaneous measurement of the velocity vector and suspended sediment concentration should enable determination of the net sediment transport rate. In a river, an order of magnitude estimate of suspended transport can be done relatively simply by integrating the

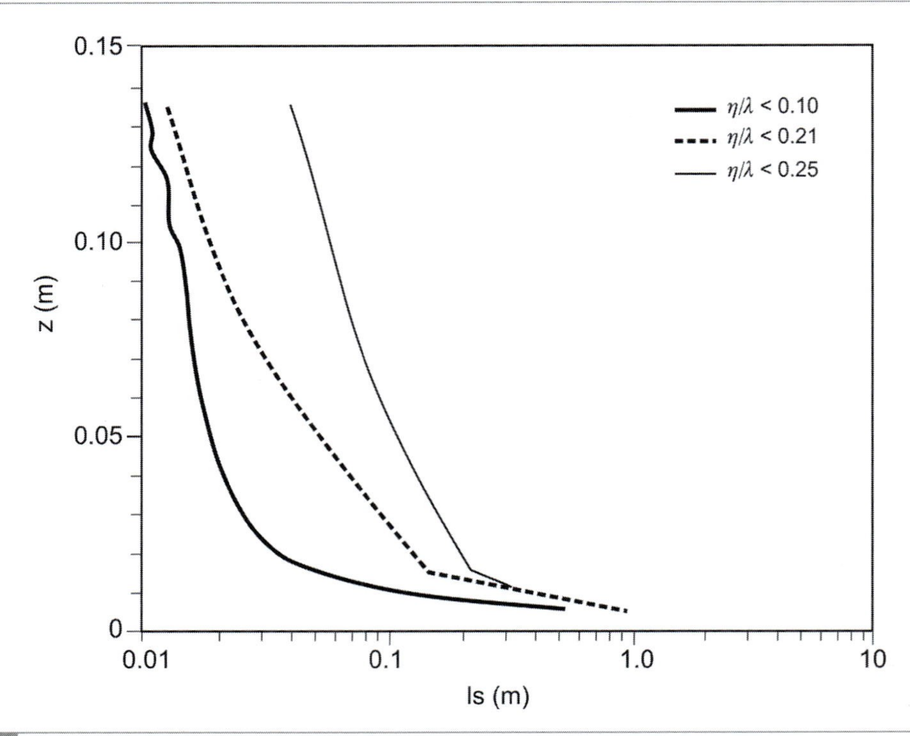

Figure 7.7 Spatially and temporally averaged sand concentration (Is) with elevation (z) above the bed over steep vortex ripples (η/λ = 0.25), flat, transitional post-vortex ripples ($\eta/\lambda < 0.1$) and steep, large-scale bedforms (η/λ = 0.21) measured with an acoustic backscatter sensor (Osborne and Vincent, 1996).

concentration over the stream depth and multiplying by the mean velocity. In the nearshore, however, it is more difficult to measure and estimate transport because of the high-frequency oscillatory motion and vertical variations in mean flow direction. Measurements are also complicated by the need to match rapidly varying concentrations with velocity asymmetries and onshore/offshore velocity fluctuations over a range of frequencies from the combination of incident waves and low-frequency motion generated by wave groups. We will examine this in more detail in the next section.

7.3 Cross-Shore Sediment Transport

As we saw in the previous section, it is possible to determine the net shore normal movement of sediment in the inner nearshore and surf zones with instruments capable of measuring instantaneous water motion and suspended sediment concentration. The net transport at any elevation can be obtained by integrating the co-variation (as measured through co-spectral analysis) of velocity and sediment concentration, which accounts for all frequencies of water motion from the relatively short-period incident waves to long infragravity waves, and the quasi-steady currents associated with the undertow and rip currents. Such an approach assumes that sediment moves at the same speed as the water. Because of the large number of instruments required to cover both vertical and horizontal variations, only a small number of such experiments have been carried out, and most of these only cover a portion of a profile normal to shore. On many beaches, onshore and offshore transport tends towards an equilibrium or balance over a period of months or years, with offshore movement dominating during periods of intense storms, and onshore movement during non-storm conditions. Depending on the magnitude and duration of the storm surge, erosion of the beach and dune occurs over hours and days, whereas recovery of the nearshore, beach and dune can take years to decades (Lee et al., 1998).

The direction of transport depends on the balance between the undertow and the skewness of the oscillatory motion. The velocity profile for undertow produced by wave breaking is onshore close to the bed, but is offshore ~10–20 cm above the bed, while above the wave troughs it is again onshore. In the presence of small waves, little sediment is suspended to a height where it can be moved offshore and the offshore flow speeds are not strong. As a result, net transport is likely to be onshore in response to the asymmetry or skewness of the oscillatory motion of the shoaling waves and the small net drift velocity onshore. However, this may be complicated a bit by the presence of ripples and vortex generation, but the effect is likely to enhance the onshore transport. This is consistent with our general observation that, under small waves, sand tends to be moved onshore, nearshore bars migrate landward and beaches accrete. As wave heights increase, the strength of the undertow will increase and more sediment will reach higher into the water column. As a result, the direction of net transport from onshore to offshore. This may be enhanced by the effect of vortex generation by plunging breakers that will dramatically increase suspended sediment concentrations in the middle of the water column (Voulgaris and Collins, 2000). This is consistent with field observations that net sand transport is offshore as wave breaking intensifies, leading to the offshore migration of nearshore bars and eventual beach erosion.

7.3.1 Suspended Sediment Concentrations

In the previous section, the direction of transport was described in a simple example in which sediment was suspended by individual waves and the direction of transport over the wave cycle was dependent on the relative strength of the undertow and wave asymmetry. Clearly the ability to measure or predict time-averaged suspended sediment concentration with height above the bed is important to developing predictions of - cross-shore sediment transport. In practice, however, the direction of net transport is also influenced by oscillatory motion in both the wind–wave frequency and infragravity frequencies as well as mean flows. It is necessary to separate the flux at each elevation due to each of these components in the frequency spectrum. This is not a trivial task, since it requires measurement of flow and concentration at a minimum of three heights above the bed at multiple locations across the nearshore profile. Results of several studies in a barred nearshore suggest that transport at wind–wave frequencies tends to be offshore during a storm, while transport at infragravity frequencies tends to be onshore (Figure 7.8A, B). The net transport then depends on the sum of transport at wind–wave and infragravity frequencies and the transport by the mean currents (Figure 7.9). The offshore-directed mean flow tends to increase in strength through the surf zone to the breakpoint (e.g., Aagaard *et al.*, 1998) and decreases seaward of the breaker zone (Ruessink *et al.*, 1998). Combined with the increase in wave skewness as waves shoal, sediment tends to converge and nearshore bars tend to migrate to the breakpoint. However, this balance is complicated by the presence of infragravity wave transport and it can be difficult to predict the direction of transport and the evolution of the nearshore (see Holman *et al.*, 2015; Inch *et al.*, 2017).

In the field, much of the data on suspended sediment concentrations in the nearshore and surf zone has come from the deployment of electromagnetic or acoustic current meters co-located with optical backscaterrance sensors (OBS) (Downing *et al.*, 1981; Beach and Sternberg, 1988). Data from the on–offshore component of a current meter and OBS deployed at Linden Beach, Nova Scotia over an intertidal bar (Dawson *et al.*, 2002) provide an example of suspension under shoaling waves near the break point (Figure 7.10A). The small-scale fluctuations in concentration are associated with the passage of individual waves over the rippled bed. Major episodes of suspension are produced by groups of larger waves which 'pump' sediment higher up

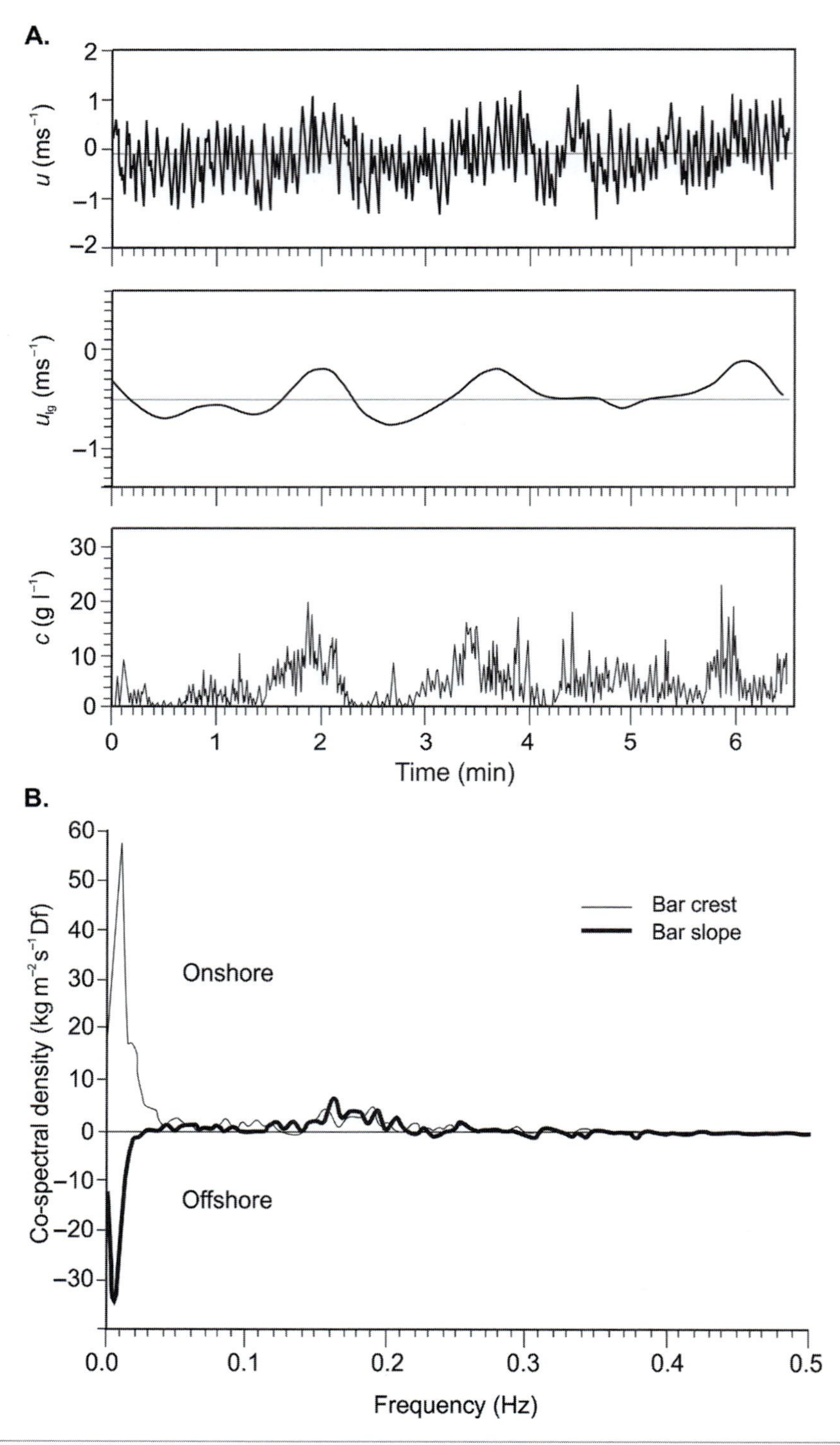

Figure 7.8 Frequency of water motion and cross-shore sediment transport. A. Time series of cross-shore velocity u, low pass velocity u_{ig} and sediment concentration (at z = 0.5 m) on a dissipative beach. Concentration maxima are in phase with the onshore stroke of the infragravity waves (after Aagaard and Greenwood, 1995). B. Reversals of sediment flux with infragravity waves on a dissipative beach – the flux is directed offshore over the lakeward bar slope and shoreward on the bar crest (Aagaard and Greenwood, 1995).

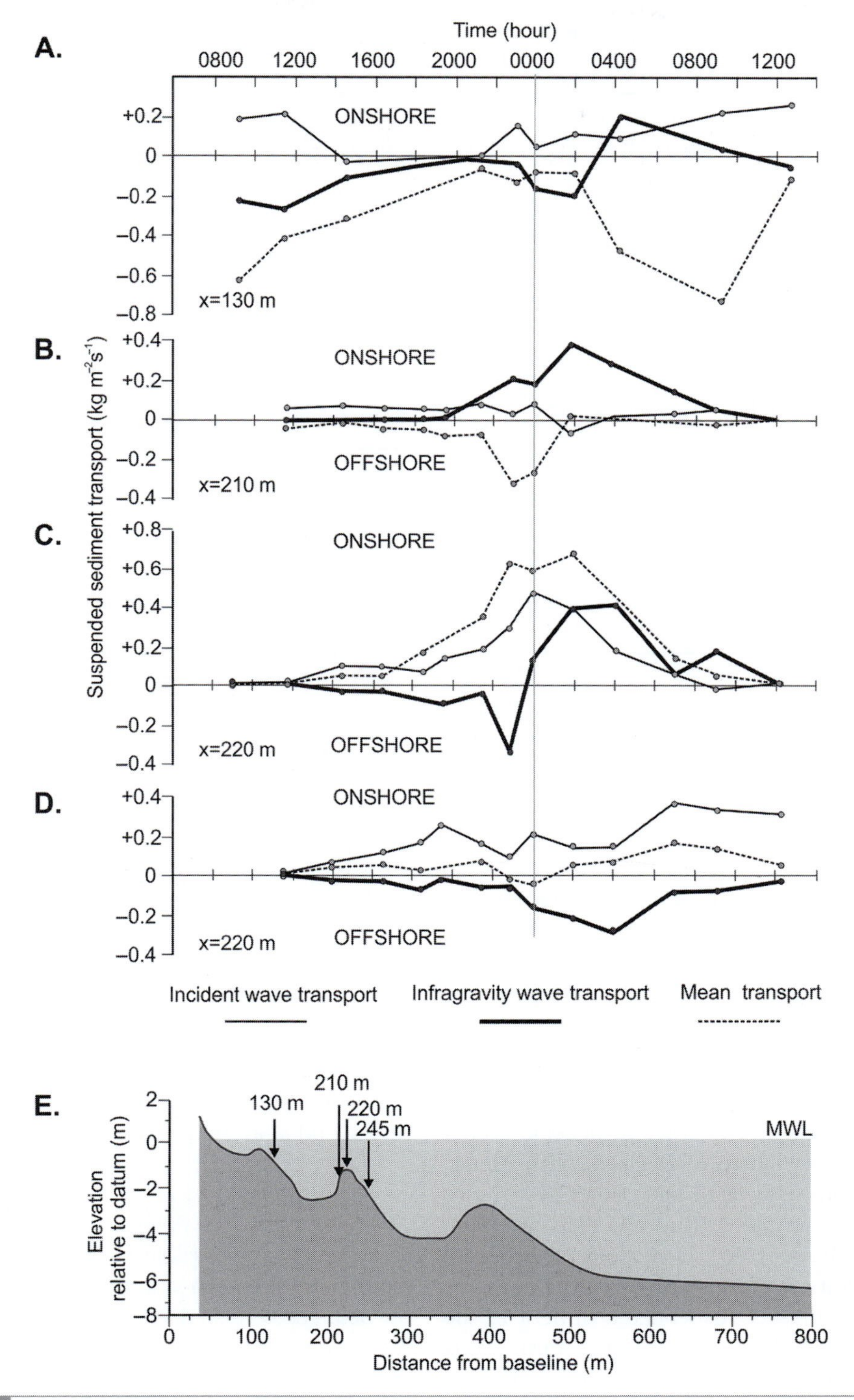

Figure 7.9 Cross-shore sediment flux measured at four locations on an intermediate beach during a moderate storm: A. 130 m offshore; B. 210 m offshore; C. 220 m offshore; and D. 240 m offshore. The profile form and location of the measurement stations is shown in E. At each station the contribution to total sediment transport is shown separately for incident waves, infragravity waves and mean flows (Aagaard and Greenwood, 1994).

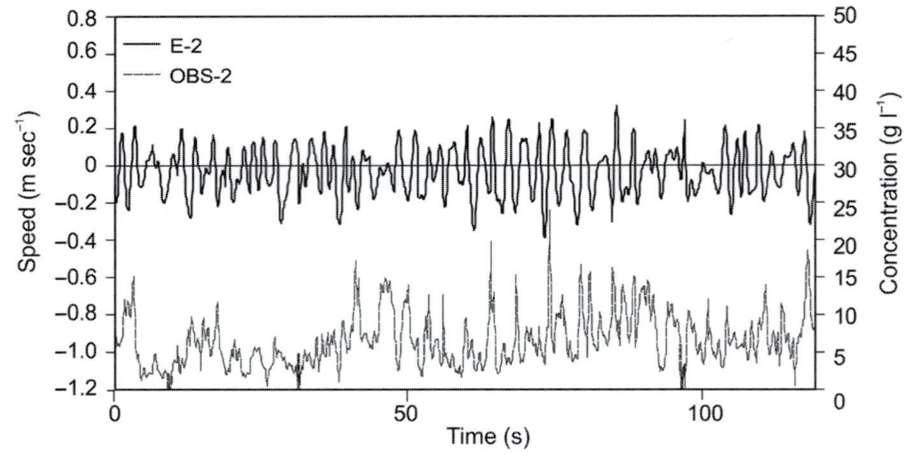

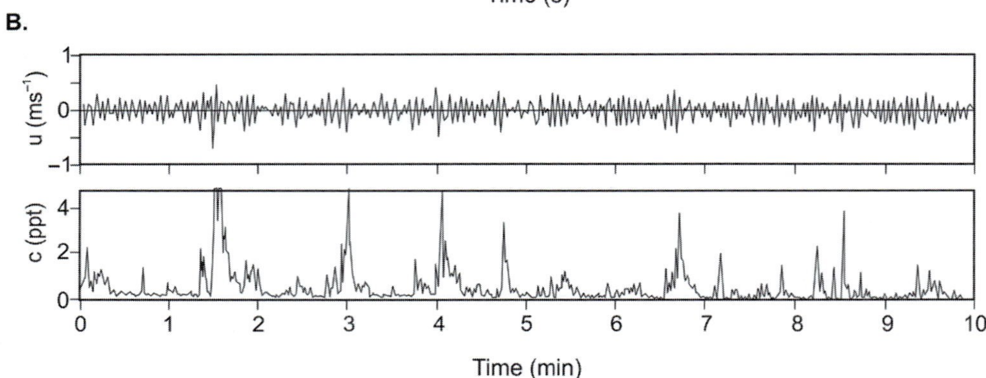

Figure 7.10 Time series of cross-shore velocity measured with an electromagnetic current meter and suspended sediment concentration measured with an OBS probe: A. shoaling waves on the crest of an intertidal bar near high tide (Dawson *et al.*, 2002); B. shoaling waves on a non-barred coast (Osborne and Greenwood, 1992). The instruments were located at an elevation of 0.1 m above the bed.

into the water column over several wave periods. Similar episodes (Figure 7.10B), can be seen for shoaling waves on a non-barred beach (Osborne and Greenwood, 1992). Larger suspension events also reflect the nature of the bed, with higher concentrations associated with flows over dunes or lunate megaripples (Hay and Bowen, 1994; Green and Black, 1999). It is important to note that OBS data are quite sensitive to the presence of small amounts of suspended sediment in the water column and to variations in grain size. It is usually necessary to calibrate the OBS in the lab against sediment taken from the study site, though it may be possible to calibrate them in situ using a laser particle size analyser (Rogers and Ravens, 2008).

The sediment transport rate (q) is calculated as the product of the suspended sediment concentration (C) and combined wave and current velocity (u) at all heights above the bed:

$$q = \int_0^h u(z)C(z)dz. \tag{7.8}$$

It is not always possible to measure suspended sediment concentration at multiple heights above the bed. The vertical distribution of sediment concentration can be predicted based on a reference concentration close to the bed. Where bedforms are present, the concentration profile is exponential and can be predicted by (Nielsen, 1986):

$$C(z) = C_{\text{ref}} \exp\left(\frac{-z}{L}\right), \tag{7.9}$$

where $C(z)$ is the mean concentration at height z above the bed; C_{ref} is the reference concentration just above the bed; and L is a mixing length. Field measurements confirm the general approach as well as the critical role of bedforms in determining the upward diffusion of sediment and the reference concentration (Green and Black, 1999; Webb and Vincent, 1999). As you can imagine, it is near impossible to measure suspended sediment transport at all heights above the bed, and everywhere across the nearshore to determine where transport is converging (deposition) and where transport is diverging (erosion). Measurements in the field provide important insight about the rates and mechanisms of sediment transport used in the development and testing of models designed to predict cross-shore sediment transport rates under waves and currents. This is discussed in the next section.

7.3.2 Sediment Transport Modelling

Sediment transport models can be used to predict morphological change in response to some modelled event or events (e.g., Schoonees and Theron (1995). Models such as SBEACH (Larson and Kraus, 1989), are used routinely in coastal engineering modelling, while others, such as the model described by Baillard (1981) have been applied extensively to test results from field experiments (Ruessink *et al.*, 1998; Masselink *et al.*, 2008). The models can be tested at two time scales: (1) against measurements of incident waves, currents and sediment transport measured at several points along a profile to determine the cross-shore gradients in transport; and (2) against measured morphological change between the beginning and end of an event. While the latter approach may appear simple, it is necessary to measure the former to determine whether the modelled mechanisms in fact agree with the measurements. In other words, it is important to determine if the model is right for the right reasons. This is not easy and it can be very expensive. It requires the deployment of large numbers of instruments in a physically demanding environment. Measurements in the inner shoreface seaward of the breaker zone are a bit easier on instrumentation because of the less energetic conditions near the bed (Ruessink *et al.*, 1998), but require more effort to install and retrieve the instruments. It is logistically much easier to install instruments in the intertidal zone (Houser *et al.*, 2006; Masselink *et al.*, 2008), and to carry out intermediate surveys of profile changes, but physical conditions are not favourable for the instrumentation. In the breaker and surf zones, especially under very energetic conditions where the surf zone may be several hundred metres wide, satisfactory testing can only be carried out where there are facilities such as the fixed pier at the Coastal Engineering Research Facility at Duck, North Carolina (Gallagher *et al.*, 1998).

The energetics approach proposed by Bagnold (1963, 1966) has provided a basis for much of the modelling of alongshore and on–offshore sand transport. In this approach, sand movement is initiated by oscillatory wave motion, with the amount of sand moving being proportional to the local rate of energy dissipation. Bedload transport is calculated as:

$$q_b = \frac{\tau_b u e_b}{\tan\phi - \tan\beta} \tag{7.10}$$

where $\tau_b u$ is the current power, e_b is the bedload efficiency factor, ϕ is the angle of internal friction for the sediment, and β is the slope of the bed. Suspended sediment transport is calculated as:

$$q_s = \frac{\tau_b u e_s}{\frac{\omega_s}{u} - \tan\beta}, \tag{7.11}$$

where e_s is the suspended load efficiency factor. In general, e_b is on the order of 0.1 in oscillatory flows, while e_s is on the order of 0.02. Both are less than 1 because some of the power is lost to friction leaving the current less efficient at transporting sediment. Waves are more efficient than steady currents at transporting sediment, but the sediment entrained by the waves is moved in the direction of the net current that is superimposed on the oscillatory motion. In applying this concept to on-offshore sediment transport, Bowen (1980) examined in detail the factors controlling both bedload and suspended load transport and considered explicitly lags between fluid flows and sediment response as well initiation of sediment motion. Bowen's model suggests that the relative importance of bedload versus suspended load transport is a function of the orbital velocity divided by the sediment fall velocity, which means that suspended transport is more important for energetic conditions and for finer sediments. Baillard (1981) adopted a similar approach, and his model has been tested in many field experiments (Gallagher *et al.*, 1998; Masselink *et al.*, 2008), and has been modified and applied to field and laboratory data or for use in numerical models (Nairn and Southgate, 1993).

There are now several numerical models designed to predict nearshore sediment transport and profile evolution, that run on desktop computers and are readily available for use by scientists and engineers. Several are used routinely in engineering applications. It is not easy to test these models adequately either in the laboratory or in the field, which means that the accuracy of the models is difficult to test. There are scaling issues in the laboratory and, even with very large and long tanks (100 m or more), the 2D nature of the processes is restricted. In the field the number of data sets that provide sufficient data to test the models thoroughly is small and they have been obtained at only a few sites (e.g., the CERF facility), so the significance of local factors (unique to the test environment) is difficult to judge. Application of the models to predict coastal change is somewhat of an art and it usually requires some sensible interpretation and verification as well as awareness of the limits and restrictions.

7.3.3 Transport Patterns in the Presence of Rips

When the nearshore morphology is 3D and rip channels are present, the patterns of sediment transport vary alongshore based on the spacing of the rip channels and shoals. If the rip channels are closely spaced, the rip flow comes at the expense of the undertow, leading to predominantly onshore-directed transport over the shoal and offshore-directed transport within the rip channel. In other words, sediment transport follows the 3D rip flow leading to the eventual landward migration of the nearshore bar. The undertow can develop and the bar migrates offshore as wave heights increase and breaking occurs on the seaward slope of the bar leading to the eventual decay of the rip channel and circulation.

Superimposed on this 3D transport pattern is the longshore current which may be continuous and relatively simple spatially on a planar beach but, as we saw in Chapter 6, where nearshore bars are present there will be more than one line of breakers in the offshore direction, and the flow pattern may be disturbed by rip cells and by rip channels and nearshore bar topography. Consequently, sand moves obliquely offshore and onshore in a complex pattern alongshore. Measurement and prediction of the transport rate are much more difficult under these conditions because of the need to determine the vertical suspended sediment concentration, the vertical variation in longshore current speed and the spatial pattern of currents. Nevertheless, conceptually, the same basic principles apply as for the simple case and we should be able to predict the total longshore transport as a product of the instantaneous velocity and sediment concentration through the breaker and surf zones (e.g. Aagaard *et al.*, 1997; Orzech *et al.*, 2010).

7.4 Longshore Sand Transport

Longshore sediment transport on the beach and within the nearshore results from the operation of three sets of processes: (1) beach drifting on the swash slope driven by oblique wave action; (2) transport by wave-generated longshore currents within and just seaward of the surf zone; and (3) transport seaward of the breaker zone by residual tidal currents and wind-driven currents. Within the breaker and surf zone, wind-driven currents and tidal currents may also contribute to longshore transport, but they are generally thought to be of lesser importance than those due to waves in this relatively shallow area.

Longshore sediment transport is extremely important on most coastlines. Locally it is important in the removal of sediment from the base of bluffs and cliffs, thus promoting further erosion and cliff retreat, and from river mouths and deltas. The sand and gravel component of this sediment is then transported alongshore to create beach and dune systems and large depositional features such as spits, baymouth barriers and barrier islands. The direction of net longshore sediment transport on a stretch of coastline may be manifested from natural features such as those noted above, and by the accumulation of sediment behind obstacles such as harbour breakwalls and groynes. Interference with the natural longshore sediment transport is the cause of many human-induced problems involving locally enhanced erosion and sedimentation, and is critical to the evaluation of many coastal engineering works. Longshore sediment transport is also a key factor in determining local beach sediment budgets and in the definition of littoral cells – these will be described later in the chapter.

We will begin with a description of the actual processes that give rise to longshore sediment transport and the relation of these to beach and nearshore form on the one hand and to the complexities of fluid motion in the nearshore and beach zones. This will be followed by a description of the techniques used to measure longshore sediment transport both instantaneously and net transport over periods of months to years and decades. Finally, modelling approaches to prediction of longshore transport on these two temporal scales will be examined as well as limitations of the data on which the models are based.

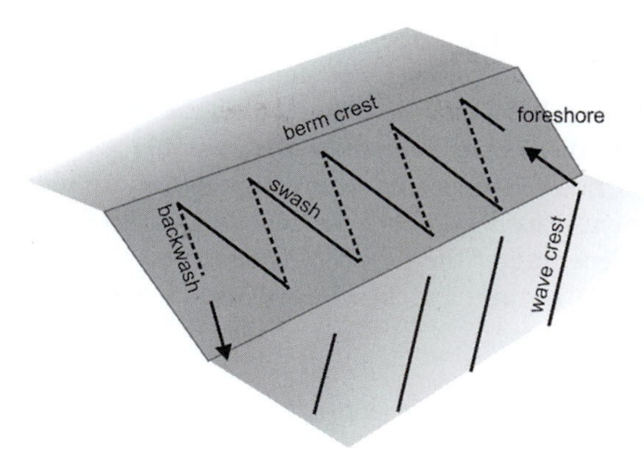

Figure 7.11 Beach drifting – schematic of the process of longshore transport on the swash slope. Swash run-up occurs perpendicular to the wave crest while return flow in the backwash occurs parallel to the beach slope producing a sawtooth alongshore motion.

7.4.1 Processes of Longshore Sediment Transport

7.4.1.1 *Beach Drifting*

Beach drifting results from wave run-up on the beach foreshore (swash slope), where waves approach at an oblique angle to the local beach. The swash (run-up) occurs at the angle of wave approach, but the return flow in the backwash is influenced primarily by gravity, and thus tends to return straight down the slope. Sediment entrained by waves is transported in the direction of wave advance in a series of sawtooth motions (Figure 7.11). Where the swash slope is steep and a substantial fraction of the swash does not infiltrate into the bed, some of the alongshore momentum in the wave uprush may carry over to the backwash so that some alongshore movement occurs during this phase as well, with the result that the transport rate is enhanced. The process is easily observed under moderate wave conditions on sandy beaches with a steep foreshore slope. The relative importance of beach drifting compared to transport in the surf zone in the total volume of sediment moved varies considerably from beach to beach, and temporally at many beaches depending on wave conditions and beach form. Gravel and cobble beaches nearly always have a steep foreshore, with relatively deep water near the base so that wave breaking occurs on the foreshore and there is rarely an extensive surf zone. As a result, beach drifting makes up a substantial fraction of the total transport and often it is the dominant process. On medium to coarse sandy beaches the relative significance will vary with the form of the foreshore and wave conditions. Where a steep foreshore and berm are present, beach drifting will be significant under low wave conditions when wave breaking takes place primarily at the beach and there is limited surf-zone development. However, under high wave conditions, waves break some distance offshore producing a wide, dissipative surf zone where most of the longshore sediment transport occurs. As we will see in Chapter 8, the swash slope is flattened by wave action under these conditions, thus reducing the effectiveness of gravity on the backwash, and the swash zone accounts for only a small portion of the profile over which longshore sediment transport takes place. On fine sand beaches steepening of the profile through berm development is rare and beach drifting is of limited significance.

7.4.1.2 *Surf-Zone and Breaker-Zone Transport*

In contrast to the swash slope, offshore motion under the wave trough is in a direction opposite to that of the wave advance and thus, ideally, there is no net longshore transport due to the wave oscillatory motion in the nearshore zone. However, as we saw in Chapter 6, longshore currents are generated in the surf zone by waves breaking at an angle to the shoreline. Sand is set in motion close to the bed by oscillatory currents moving across ripples and dunes, and may be lifted well off the bed in areas of high turbulence and vorticity, especially under plunging breakers (Figure 7.12). This suspended sediment will be carried alongshore with the current at a rate that will be some fraction of the mean speed of the current. Some longshore transport takes place in the zone of shoaling waves seaward of

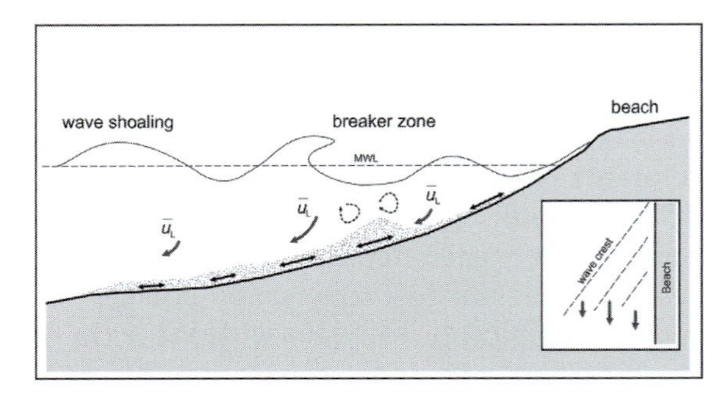

Figure 7.12 Surf-zone transport – the process of longshore transport seaward of the swash slope. Sediment is set in motion by wave oscillatory motion (bi-directional arrows) and by turbulent vortices generated by wave breaking. Where longshore currents are present due to wave approach at an angle to the shoreline (inset) sediment is transported alongshore at a rate that is a function of sediment concentration and the mean current speed $\bar{u}_L$.

the breaker zone as a result of the drift velocity near the bed, but further offshore it is likely to result from the effect of winds and tidal currents, with wave orbital motion still providing the driving force for initiation of motion.

At any instant in time, the direction and rate of longshore sediment transport (q_l) will depend on the incident wave and wind conditions, particularly the direction of wave approach relative to the shoreline orientation. However, if we are interested in shoreline evolution due to erosion and sedimentation, it is the average annual total sediment transport and the direction of the net or resultant transport that is the most crucial. The gross (or total) littoral sediment transport along a section of coast is a summation of the instantaneous volumes over the period of a year without regard to the actual direction of transport. In turn it can be subdivided into the sum of transports in one direction (to the right) and in the opposite direction (to the left). The gross littoral sediment transport Q_G over a year is thus given by:

$$Q_G = Q_L + Q_R, \tag{7.12}$$

where Q_L and Q_R are the total longshore transport to the left and right along the coast respectively.

Similarly, we can define the net sediment transport on an annual basis Q_N as the difference between the two transport volumes:

$$Q_N = Q_L - Q_R. \tag{7.13}$$

The net sediment transport is controlled by the magnitude and a frequency of waves from all directions, and is a function of the overall wave climate and the local orientation of the shoreline. It is thus possible to have very high gross transport rates with very low net rates, if the frequency and magnitude of wave approach about shore normal are nearly equal.

As we will see in the following sections, actual measurement of longshore sediment transport rates and directions is difficult and expensive, but there are several indicators that can be used to determine the direction of Q_N and these often give an order of magnitude indication of the quantity of this. These indicators can be associated with either natural or artificial features of

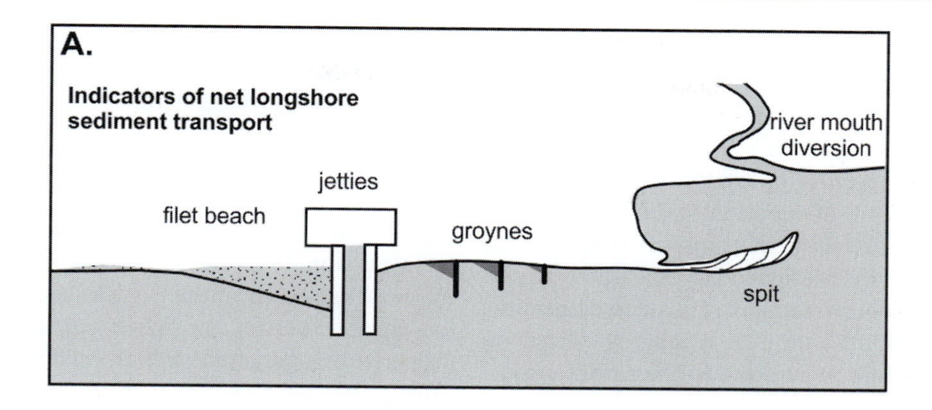

Figure 7.13 Indicators of net longshore sediment transport on the coast: A. sketch of natural and artificial indicators; and B. photograph of accumulation of sediment on the updrift side of a harbour breakwall.

the coastline such as such as groynes, jetties and harbours (Figure 7.13A). Natural features such as headlands which act to block alongshore transport and result in the impoundment of sediment on the updrift side of the structure or headland (Figure 7.13B). The result is a large increase in beach width on the updrift side of the structure and a decrease in beach width and or enhanced erosion on the downdrift side leading to a pronounced offset of the coast. Similarly, net transport in one direction can lead to diversion of the mouths of small streams where they enter a lake or ocean. On continuous sandy shorelines, the beach width may increase in the direction of net sediment transport, accompanied by increasing height and width of the dune field landward of the beach. Barrier spits, and particularly flying spits (see Chapter 10 for definition) are depositional features resulting from longshore transport and their orientation is an indication of the net drift. Similarly, on barrier island coasts, the ebb tidal delta on the seaward side of inlets separating the barrier islands tends to be skewed downdrift by the longshore current.

7.4.2 Measurement of Longshore Sediment Transport

Measurement of actual rates of sediment transport are usually carried out either to obtain values for a specific section of coastline, or to test or derive some form of equation that can be used to predict rates of sediment transport. Usually the experiments are designed to measure either the instantaneous transport rate, or to determine the net transport rate over a lengthy period of months or years. In addition to the simultaneous measurement of suspended sediment concentration and current flow by current meters and OBS probes or ADCPs (see Section 7.3), the techniques most commonly used include the injection and recovery of some suitably tagged sediment (a tracer study) or measurement of the rate of accumulation behind a barrier such as an impermeable jetty or on a depositional feature such as a spit.

7.4.2.1 *Tracers*

Most attempts to measure instantaneous rates of sediment transport in the field have utilised some form of sediment tracer. Essentially, this consists of monitoring the movement of suitably tagged particles of sediment. The two most common techniques used are: (1) to coat the grains with some form of fluorescent dye which is readily distinguishable under ultraviolet light; and (2) to tag it with a radioactive isotope which has a short half-life. Two types of experimental approach can be utilised. The Lagrangian approach involves measuring the advection rate of the tracer cloud (centroid velocity), by taking samples at suitable intervals from many grid points, determining the tracer concentrations at each point for each sample set, and then measuring the distance travelled by the tracer cloud from one sample time to the next (Silva *et al.*, 2007). The Eulerian approach involves measuring the tracer concentration at a fixed point in the system at frequent intervals, integrating this over time, and dividing this integral into the mass of tracer injected (Duane and James, 1980). The tracer can either be injected once or a continuous injection can be undertaken with the measurements downdrift focusing on determining a constant rate at the sampling point or points.

Fluorescent tracer studies usually involve coating volumes of sediment taken from the experimental site with a fluorescent dye or ink which is mixed with an adhesive to make it bond to the particles. The sand is usually mixed and dried in a cement mixer to keep the particles separate (Silva *et al.*, 2007), to ensure that the dye or ink is applied evenly and avoid changing the dynamic behaviour of the particles (e.g., by having too thick a coating, or when particles are not properly disaggregated after the coating process). During the experiment, the dyed sand is injected (placed) in the surf zone and swash slope, and its movement is determined by sampling at predetermined times and/or locations. The samples are later spread out on a surface and examined under ultraviolet light in a dark room to be identified and counted. The analysis is extremely tedious, and efforts to measure concentrations indirectly by dissolving the fluorescent coating and analysing this chemically have not proved successful. A better approach is likely to photograph the samples and to use an automated counting system to identify the fluorescing particles.

Tagging sediment with radioisotopes is a more difficult operation and the use of radioactive tracers requires more careful handling. However, the detection of sediment concentrations is more easily accomplished, since there is no need to obtain actual samples. Instead, gamma radiation emitted by the tagged grains is recorded using a scintillometer, which is traversed across the bed, with the count rate giving sediment concentration at that point. The technique can thus record the presence of buried grains, and can be used to define concentrations and the extent of the tracer cloud more precisely by carrying out several traverses. Although the technique has been used successfully in several studies (e.g., Heathershaw and Carr, 1977), getting permission to carry out a study using radioactive materials is problematic now.

Regardless of the method, there are practical problems in collecting samples in the surf zone under breaking waves, and it is obviously impractical to carry this out over the whole surf zone under conditions such as those shown in Figure 7.14. It is also necessary to take collect a

Figure 7.14 Photograph of the surf zone at Greenwich Dunes during a strong northeaster. Wave breaking is taking place over three bars and troughs with a complex inner topography that includes rip channels and oblique bars. The surf zone is about 400 m wide and under these conditions beach drifting in the swash zone is of minor significance compared to transport in the surf zone.

core sample at each location to account for the depth of mixing resulting from the migration of bedforms and to sample only from this active layer (Ciavola *et al.*, 1997). Thus, the usefulness of one of the pioneering studies of surf-zone sediment transport by Ingle (1966), who sampled only the surface sediments using Vaseline coated cards, is extremely doubtful, though the general patterns he observed are still of some value. The problems and logistics involved in carrying out a successful tracer experiment are exemplified in the work of Duane and James (1980) and Inman *et al.* (1980), and useful reviews of the techniques are to be found in Madsen (1987) and Ciavola (2004).

Radio-frequency identification (RFID) tags have also been used to track movement of gravel and cobbles in the intertidal zone (Allan et al., 2006; Dickson et al., 2011; Dolphin et al., 2016). Radio-frequency identification is an automatic identification technique, involving a reader and a transponder (the tag) which is cemented into a hole drilled into the cobble. The reader generates an electromagnetic field at multiple frequencies to interrogate the tags that can either be passive or active depending on whether there is an onboard battery. The tags can be interrogated

Figure 7.15 Photograph of a streamer trap. The fine mesh bag is attached to a rectangular frame which keeps it open. Several of these are fixed to a vertical post for deployment in the water.

both subaerially and subaqueously, but the radio frequencies are attenuated in water, and there is the potential for the tracers to be lost.

7.4.2.2 *Traps*

Bedload sediment traps, like those used in rivers, do not work well in the surf zone because of the difficulty of separating net sediment transport from the oscillatory motion under waves, and because of the effects of trap scour. Recently, attempts have been made to measure net longshore transport of suspended sediment using streamer traps (Kraus, 1987; Rosati *et al.*, 1991), and they have now been used in several studies under relatively low wave energies. The traps consist of series of polyester mesh bags a metre or more in length that are attached to a rigid rectangular frame that forms the opening to the trap. These are deployed in a vertical array on a tubular frame that is designed to be carried into the water (Figure 7.15). In the water, they are oriented into the direction of longshore current flow for a period of a few minutes. The mesh bags stream backward from the opening and permit water to flow through while retaining sand. The amount of sediment trapped decreases exponentially with distance above the bed as we would expect (Rosati *et al.*, 1991; Tonk and Masselink, 2005; Cartier and Héquette, 2015), and the amount trapped is not widely out of line with measurements using co-located OBS probes and current meters. Streamer traps provide a comforting demonstration that sand is moving and the ease of rapid deployment allows for measurement at a relatively large number of points along a profile (Cartier and Héquette, 2015). But, as noted by Tonk and Masselink (2005), it is difficult to deploy them under energetic wave conditions, they require a lot of labour and they sample for only a few minutes. Given this, they are probably best used to complement rather than replace the deployment of instruments measuring suspended sediment concentration and currents simultaneously.

7.4.2.3 *Rates of Accumulation*

Net sediment transport along a stretch of coast can frequently be estimated by measuring long-term accretion on spits, or behind barriers, such as harbour breakwalls that act as a complete sediment trap (or nearly so). Similar short-term studies in conjunction with measurement of wave data have been used to verify predictive equations based on the longshore component of wave power (Bruno *et al.*, 1980; Dean *et al.*, 1982). In this case topographic surveys upstream of the obstruction are carried out frequently to permit comparison with measured wave data over a period of weeks to months. Similarly, data from dredging of channels, inlets and for sediment bypassing can also be utilised.

Measurement of longshore transport over a period of hours to days has also been carried out by short-term impoundment (Bodge and Dean, 1987; Wang *et al.*, 1998). This involves the construction of a temporary groyne extending across the beach into the inner nearshore using, for example, a flexible mesh tube that is filled with sand or constructing a groyne with materials such as vertical steel posts and plywood. The area updrift is surveyed repeatedly simultaneously with measurements of waves and longshore currents. The approach provides a good estimate of the total transport in the swash zone and close to shore under small wave conditions. Under large wave conditions, especially on gently sloping beaches, increasing amounts of sediment are transported seaward around the temporary structure.

7.4.2.4 *Quality of Field Data*

To test the accuracy and reliability of formulae and numerical models used to predict longshore sediment transport, it is necessary to have field data collected under a range of wind and wave conditions and beach and nearshore morphologies (Schoonees and Theron, 1993; Bayram *et al.*, 2007). These comparisons also allow for the calibration of model constants and an assessment of the conditions under which the model or formula is most applicable. It is evident from the description of the measurement techniques that this is not a trivial task. There are relatively few good data sets, and they neither cover the full range of sandy beach morphologies nor the range of transport conditions. Tracer and short-term impoundment probably provide the best measure of transport rates, but most studies have been carried out under low to moderate energy wave conditions and close to the shoreline. Measurements using current meters and OBS probes can provide data on wide surf zones and under quite energetic conditions (e.g., Miller, 1999), but they often have limitations due to the small number of instruments deployed. These measurements may also be of limited usefulness because they do not provide a good measure of transport close to the bed, which may account for perhaps as much as two-thirds of the total transport (Tonk and Masselink, 2005). Until recently there were very few studies that combined a good tracer experiment and adequate determination of actual rates, with suitable measurements of the process variables, and there is so little robust field data to test theoretical and empirical predictions of instantaneous transport rates. Komar and Inman's (1970) data on Silver Strand and El Moreno beaches set the standard for this type of tracer experiment, and these results have been used extensively, despite some limitations on their applicability, particularly to beaches with relatively gentle slopes and a wide surf zone.

7.4.3 Prediction of Longshore Sediment Transport Rates

Although progress is being made in the development of numerical models that simulate all the major controls on sediment transport, the most widely used approach to the prediction of longshore sediment transport rates is a semi-empirical approach in which sediment transport rates are correlated with the longshore component of wave energy flux, P_L:

$$P_L = (ECn)_b \sin \alpha_b \cos \alpha_b, \quad (7.14)$$

where $(ECn)_b$ is the total wave energy flux evaluated at the break point and α_b is the angle the breaking wave crests make with the shoreline. Early versions of the predictive formula given in the *Shore Protection Manual* were used to predict volume transport rates, and thus had dimensional constants. Komar and Inman (1970), based on earlier work of Inman and Bagnold (1963), showed that a better approach was to predict the immersed weight transport rate I_L since this could be related to P_L by a simple dimensionless coefficient:

$$I_L = KP_L, \quad (7.15)$$

where K is an empirical constant. The immersed weight of transport can be related to a volume transport Q_L by:

$$I_L = (\rho_s - \rho)ga'Q_L, \quad (7.16)$$

where: ρ_s and ρ are the sediment and fluid densities respectively; g is the gravitational constant and a' is a correction factor for the pore space of sand (approximately 0.6 for medium sand).

Komar and Inman (1970) used the results of tracer studies carried out over several field studies to derive the relationship: I_L = 0.77 P_L (see Figure 7.16). The regression line is based solely on field data, and wave tank data plot below the line, suggesting that wave tanks are not an appropriate substitute for field data. The computations of energy used to derive this relationship is based on the root-mean-square wave height (H_{rms}), but can be modified for the significant wave height (H_s) if K is divided by 2. The coefficient of 0.77 has been used widely for predicting

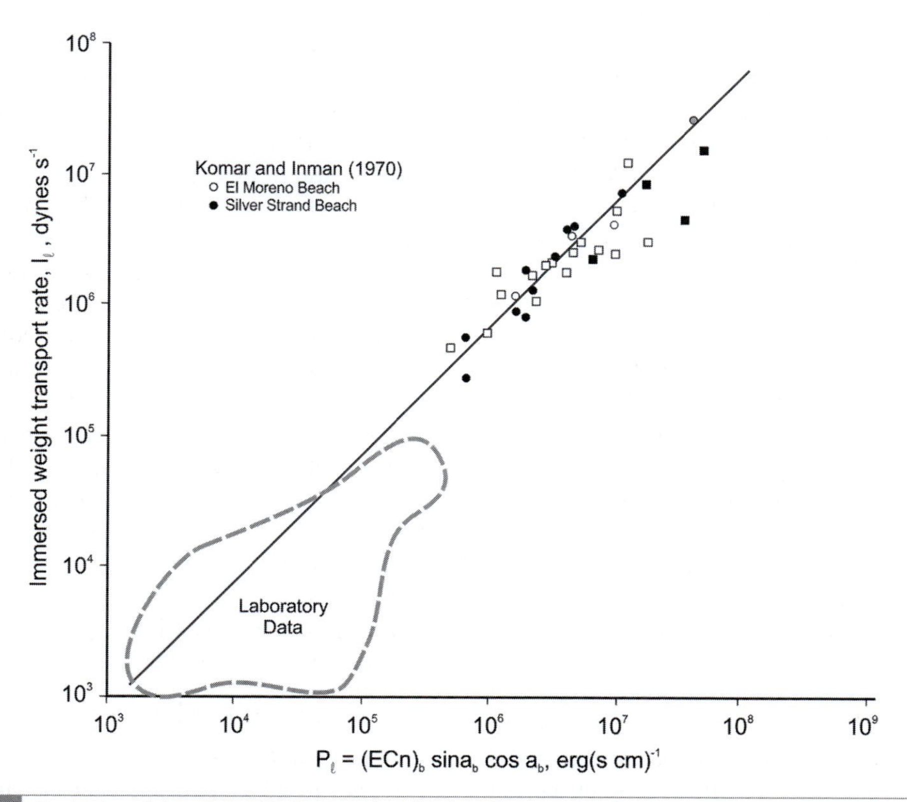

Figure 7.16 Derivation of the relationship between I_L and P_L based on available field data (Komar and Inman, 1970). Note that wave tank data plot below the line and were not used in deriving the regression.

volumes of longshore sediment transport, particularly on an annual basis, although Komar (1998) has subsequently suggested a value of 0.7 based on evaluation of additional field data. However, the revised coefficient should be treated with some caution as it is based on so little data. More recent measurements have suggested that 0.7 is probably at the low end of the scale for sand beaches, in part because values derived from accumulation behind jetties give higher values than tracer studies (Bruno *et al.*, 1980). Probably the best data on accretion behind a breakwater come from measurements at Santa Barbara, California, conducted during the National Sediment Transport Study (Dean *et al.*, 1982). They found an average value of 1.23 for K. Rosati *et al.* (1991), in a compilation of field data, suggest a value for K of 0.92.

Inman and Bagnold (1963) also relate I_L to the mean longshore current velocity and maximum wave orbital motion:

$$I_L = K'(ECn)_b \cos \alpha_b \frac{\bar{V}_L}{u_m}, \tag{7.17}$$

where $\overline{V_L}$ is the mean longshore current velocity, u_m is the maximum orbital velocity at the breaker line, and K' is a dimensionless coefficient. Equation 7.17 provides a measure of the mean stress exerted by the waves which are assumed to act to keep the sediment in motion with the transport rate then being proportional to the mean longshore current. Komar (1998) assigns a value of 0.25 to K'. There is obviously a need for further field studies to improve the data base on which these empirical relationship is based. However, simply deriving more estimates of K is not in itself particularly useful. In the first place, there is obviously not a single 'correct' value for K. Instead, K is probably a function of a complex set of variables that may include sediment size and sorting, beach slope, breaker type and substrate. Unlike transport equations for fluvial or aeolian sediment transport, there is no explicit consideration of sediment size or threshold in Equations 7.14 and 7.15. Results of Dean *et al.* (1982), and more recent work suggest that K does decrease with increasing grain size in the sand size range. However, the effect on K is relatively small for sand and it is likely only important for the transport of gravel and cobbles.

Another bulk transport expression was derived by Kamphuis (1991) based on his extensive wave tank studies:

$$Q_s = 2.27 H_{sb}^2 T_p^{1.5} m_b^{0.75} D^{-0.25} \sin^{0.6} \alpha_b, \tag{7.18}$$

where: Q_s is the longshore transport rate in kg s^{-1}; m is the beach slope; and D is the particle diameter. The expression is interesting because it includes the effects of three parameters that intuitively should influence the transport rate – wave steepness, beach slope and sediment size. Kamphuis's experiments found that transport rates were higher for plunging breakers rather than spilling breakers, hence the inclusion of wave period in the equation. The field experiments of Wang *et al.* (1998) showed better agreement with the Kamphuis equation than the CERC equation, which may reflect the relatively low wave energy conditions on the Gulf coast of Florida. However, the Kamphuis equation performed much worse in the study carried out by Tonk and Masselink (2005). One of the limitations of both Equations (7.11) and (7.13) (the CERC and Kamphuis equations) is that they only account for the contribution of waves to the longshore sediment transport. Therefore, they underestimate the transport rate when there is a substantial contribution from wind (Ciavola *et al.*, 1997; Tonk and Masselink, 2005). A formula proposed by Bayram *et al.* (2007) builds on the approach of Inman and Bagnold (1963) in that it assumes that bed sediments are put in motion by wave action and that the sediment is then transported by any type of current. It also explicitly includes sediment size through the particle fall velocity w_s,

$$Q_{lst} = \frac{\varepsilon}{(\rho_s - \rho)(1 - a) g w_s} F\bar{V} \tag{7.19}$$

where: Q_{lst} is the longshore sediment transport rate (m^3 s^{-1}); $\rho_s - \rho$ is the relative density of the sediment in water; a is the pore space of the sand; F is a measure of the suspended sediment concentration; $\bar{V}$ is the mean longshore current; and ε is a coefficient which expresses the efficiency of the waves in keeping the sand grains in suspension. A value for ε can be estimated from:

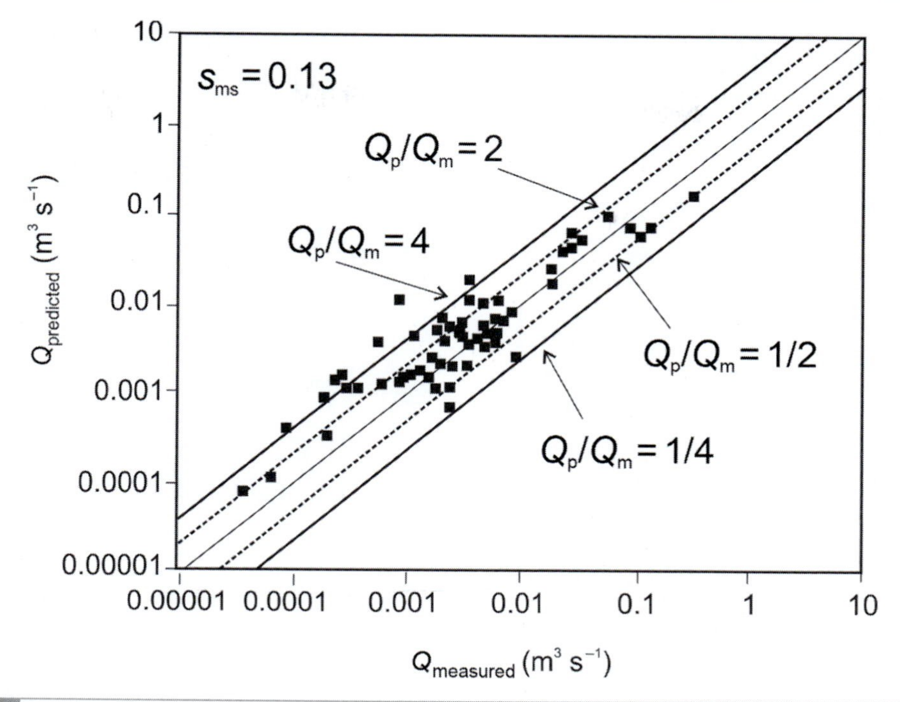

Figure 7.17 Measured longshore sediment transport rates from a verification data set versus those predicted by Equation 7.17 (from Bayram *et al.*, 2007).

$$\varepsilon = \left(9.0 + 4.0\frac{H_{sb}}{W_s T_p}\right) \cdot 10^{-5}. \tag{7.20}$$

A test of their formula against a chosen data set provided a better fit (Figure 7.17) than for Equations 7.11, 7.12 and 7.13, with the next best prediction being the formula of Kamphuis (1991). Whether the formula of Bayram *et al.* (2007) will prove to be an improvement over other less complex equations remains to be seen.

In addition to the bulk formulas discussed in this section, there are several formulas that predict bedload and suspended load transport at a point and these then can be used to predict transport across the surf zone. Bayram *et al.* (2001) tested the predictive capability of six well-known formulae against the measured cross-shore distribution of longshore sediment transport from three experiments carried out at the Coastal Research Centre at Duck, North Carolina. However, none of them provided very good estimates across the range of conditions sampled. What is clear is that improved prediction requires an increase in the amount and quality of the input data. This is particularly so because prediction of P_L and V_L are very sensitive to small changes in breaker angle and in the direction and speed of the incident winds. The potential errors tend to accumulate when we attempt to predict an average annual longshore sediment transport rate for a stretch of coast based on the wave climate, local shoreline orientation and bathymetry. Fortunately, at least some of the errors tend to cancel each other out.

Box 7.1 Longshore Sandwaves

Measurements of instantaneous longshore sediment transport and predictions using the formulae discussed in the previous section assume that sand transport takes place as individual grains and that transport is uniform along a short stretch of shoreline. However, collective movement of sediment in the form of large bedforms can also

Box 7.1 (cont.)

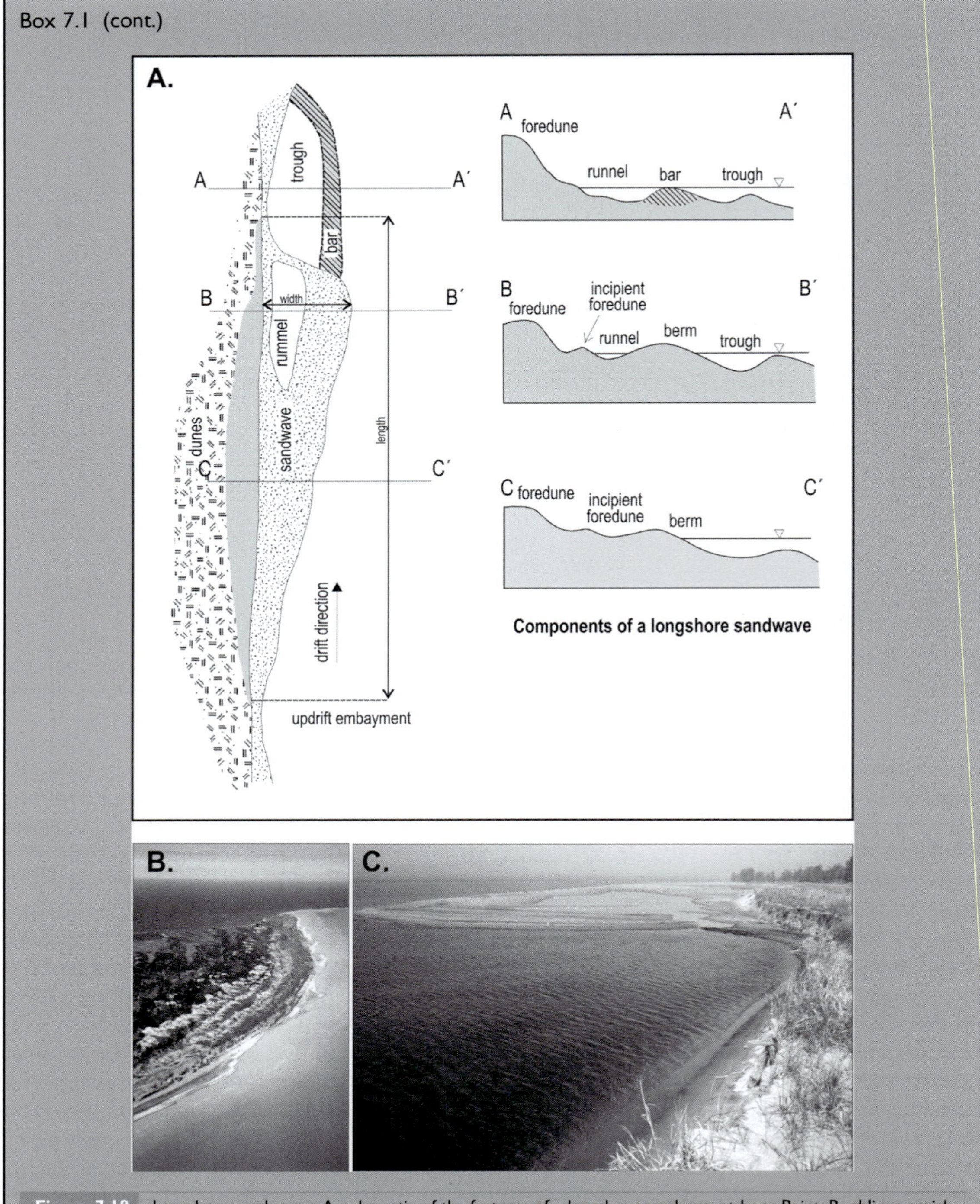

Figure 7.18 Longshore sandwaves: A. schematic of the features of a longshore sandwave at Long Point; B. oblique aerial photograph of the downdrift end of Long Point showing longshore sandwaves over a 10 km stretch of shoreline; and C. view looking updrift of the downdrift end of a longshore sandwave. The narrow beach in the foreground and evidence of dune erosion marks the erosional phase associated with passage of the sandwave immediately downdrift (Davidson-Arnott and van Heyningen, 2003).

Box 7.1 (cont.)

occur in a manner similar to the movement of large dunes and sand bars in rivers, and the onshore/offshore movement of bars in the nearshore zone (described in Chapter 8). Bruun (1954) described migrating sandwaves or sand humps on the Dutch coast, and on the Californian coast dispersal of sediment from small deltas was described in terms of accretion and erosion waves (Inman and Bagnold, 1963; Inman, 1987; Hicks and Inman, 1987). Pringle (1985) described in detail the alongshore migration of features on the Holderness coast of Yorkshire termed ords which corresponded to the erosion portion of a sandwave. More recently the term longshore sandwave has been applied to features found at Long Point (Stewart and Davidson-Arnott, 1988; Davidson-Arnott and van Heyningen, 2003) and at Southampton Beach, New York (Thevenot and Kraus, 1995).

Longshore sandwaves are best developed on sandy shorelines where there is a very strong net transport in one direction and waves approaching the coast at a high angle. Their formation appears to be an example of self-organisation in which an initial perturbation produces a positive feedback that enhances and maintains the feature and can lead to the development of a suite of similar features (Ashton *et al.*, 2001; Coco and Murray, 2007; Falqués *et al.*, 2017). Longshore sandwaves are 500–1500 m in length, 30–80 m in width at the downdrift end, and they migrate alongshore at 100–400 m a^{-1}. Because of their tapered form, alongshore migration produces a temporal alternation between a wide beach opposite the downdrift end, which protects the dune from wave erosion and supplies large amounts of sediment to the foredune. Updrift of the wave is a narrow beach characterised by erosion and scarping of the foredune (Figure 7.18). The accretion erosion cycle is easily recognised where there are prominent sandwaves (Davidson-Arnott and Stewart, 1987; Hicks and Inman, 1987; Thevenot and Kraus, 1995), but even where the sandwave form is subtler, the cycles can be distinguished in the behaviour of the toe of the foredune over several years (Verhagen, 1989; Ruessink and Jeuken, 2002) or in the form of the underwater contours (Kaergaard *et al.*, 2012).

7.5 Littoral Sediment Budget and Littoral Drift Cells

Estimation of long-term sediment transport rates is often used to determine the littoral sediment budget for a section of coast. The littoral sediment budget is often linked to the concept of a littoral drift cell, and together they provide a coastal management framework for the assessment of the potential impact of human actions on sand and gravel beaches. Many jurisdictions make use of littoral cells and littoral cell boundaries as planning units for structuring coastal zone management on clastic shorelines. The collection of data on inputs and outputs of sediment within these boundaries can be an important element of background studies carried out as part of the management exercise.

7.5.1 Littoral Sediment Budget

The littoral sediment budget is an accounting technique in which the volume (or mass) of all sediment inputs (termed sources) and outputs (sinks) to and from the beach and nearshore zone are assessed for a section or reach of the shoreline (Rosati, 2005). The most common natural sources and sinks are shown schematically in Figure 7.19. A source (and similarly a sink) can be defined either as a point source (e.g., a river mouth) where the input is confined to a local area, or a line source (such as cliff erosion) where the input takes place over a stretch of the shoreline. On a worldwide basis, inputs from rivers are probably the most important source, followed by cliff erosion and in the tropics biogenic inputs from coral reefs may dominate. In areas of recent glaciation erosion of cohesive bluffs can supply large quantities of sand and gravel, and the increasing retreat of glaciers with a warming climate may increase this contribution at high

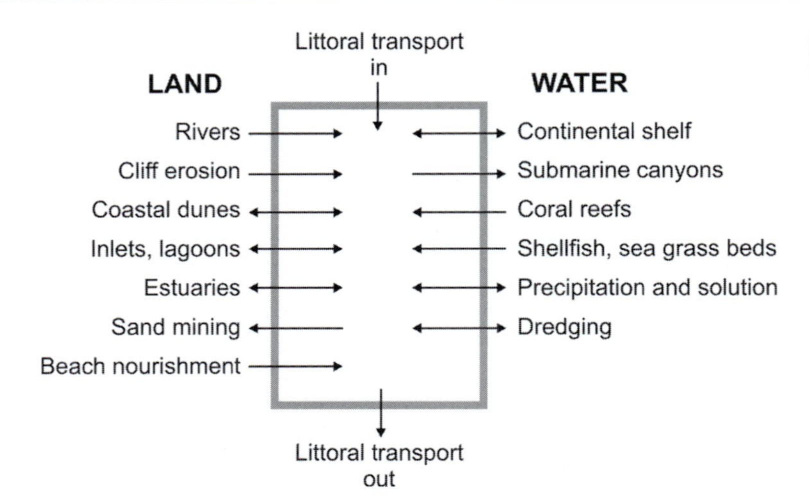

Figure 7.19 Schematic of the common natural sources and sinks for a littoral sediment budget.

latitudes. It is usually easy to isolate, from the list shown in Table 7.1, the sources and sinks that are likely to be important in a study area. Quantification of a sediment budget for a stretch of coast is not easy and is often a matter of order of magnitude estimates, informed guesses and the use of residuals where two of three components have been estimated and the remainder must constitute the third (Bowen and Inman, 1966; Rosati, 2005).

Since fine sediments are only found in small quantities in energetic littoral zones, the littoral budget tends to be quantified in terms of sand size or greater (a cut-off of 0.1 mm is generally appropriate), though exceptions need to be made on coasts such as Guyana where there is a very large supply of mud. Ideally, if the budget along a stretch of coast is positive (inputs exceed outputs) then deposition should occur and the shoreline will prograde. If it is negative, then the area will experience erosion. The shore parallel boundaries of the area within which the budget is calculated are usually defined by the landward and offshore limits of sediment transport by wave action (i.e., from the landward edge of the backshore to the outer shoreface). The shore perpendicular boundaries may be a littoral cell, or simply an arbitrary stretch of shoreline where there is a point of interest such as a spit or beach. On sandy coasts aeolian transport of sand inland represents a loss to the littoral sediment budget but it is also a gain to the complementary dune sediment budget and we will address this in Chapter 9. In addition to the natural sources, human actions can also contribute to both sediment gains and losses through activities such as jetty and seawall construction, and through dredging, beach mining and beach nourishment.

There are often good records of flows in rivers discharging at the coast and some measurements or estimates of sediment transport. Bedload transport is still difficult to measure in rivers, but there are at least several good predictive equations. Thus, an order-of-magnitude estimate of supply is often readily obtainable. Rivers draining upland areas close to the coast provide large volumes of littoral sediment directly, but much of the coarse material transported by rivers flowing across an alluvial plain tends to be deposited terrestrially. Thus, a substantial portion of the littoral sediment supplied from rivers with large deltas may come from erosion of abandoned areas of the delta, rather than directly through distributary channels. On coasts where Holocene sea level rise has drowned the lower portions of river basins, producing rias and shallow estuaries, much of the coarse sediment transported by the river may be trapped at the head of the estuary and not made available to the littoral system.

Littoral sediment supply from coastal erosion is perhaps the most easily quantified. Estimates of the volume supplied annually is determined initially by multiplying the height and length of a suitable reach by the average annual recession rate determined from mapping or comparison of historical aerial photographs and LiDAR surveys. Where there are fines present in the material supplied to the coast, the proportion of these can be subtracted to give a corrected volume.

On most coasts, it is difficult to quantify losses or gains across the offshore boundary and often a key determinant of the littoral budget is the difference between the supply of sediment by longshore transport from updrift and that is lost to longshore transport downdrift. This is especially true for sediment budget calculations determining the design criteria for the entrance to a small harbour or for assessing the potential impact of such a structure. Calculations of this nature require accurate estimates of Q_N at a point, usually through the determination of a wave climate and numerical modelling of nearshore wave transformation, with all the caveats noted in the previous section. We will return to an example of this after an examination of littoral cells.

7.5.2 Littoral Cells

An ideal littoral cell is a section of coast that contains a single or multiple sources of sediment, a well-defined and continuous zone of alongshore sediment transport, and a downdrift sink or zone where sediment is either deposited or is lost offshore (Figure 7.20A). The concept seems to have been applied first to the California coast (Bowen and Inman, 1966; Inman and Frautschy, 1966), where a series of cells with longshore sediment transport to the south was identified. In each of these cells, sediment input is primarily from rivers draining the coastal mountains, and to a lesser extent from cliff erosion. Sand is transported and generally increases southward until sand is diverted offshore and lost down the heads of submarine canyons (Figure 7.20A). The heads of the canyons are often marked by rocky headlands because of the loss of the littoral sand supply and their location within a few hundred metres of the beach is due to the relatively narrow continental shelf in California. In most littoral cells the sink is an area of net deposition marked by a feature such as a spit or dune field.

Coastal sediment compartments are used to describe the sink-to-source systems on highly embayed coasts where there are major obstacles to longshore sediment transport (Davies, 1974). These are usually large headlands that serve to restrict the angle of wave approach to the shoreline, confining sediment movement to the area between the headlands and isolating the embayments from the rest of the coast. Beaches in these embayments tend to be very stable and to develop an equilibrium form, both on–offshore and in plan view. Littoral cells can also have boundaries such as headlands and may be largely isolated from adjacent ones, but unlike coastal compartments, there can be exchange of sediment between adjacent littoral cells and the boundaries between them may be quite fuzzy. Littoral cell boundaries can be convergent, divergent or interruptive, and it is possible to recognise sub-cells within major littoral cell units (Figure 7.20B). The identification of the cells and directions of net longshore sediment transport are the starting point for collecting information on sediment inputs and outputs, and the determination of a littoral sediment budget (e.g., Sanderson and Eliot, 1999; Patsch and Griggs, 2008; Montreuil and Bullard, 2012; Thornton, 2016). This approach provides a systematic framework for research on coastal processes and coastal evolution, and for the establishment of coastal management plans. They have been adopted as the basis for coastal management in a number of jurisdictions, including Ontario and England (see Cooper and Pontee, 2006).

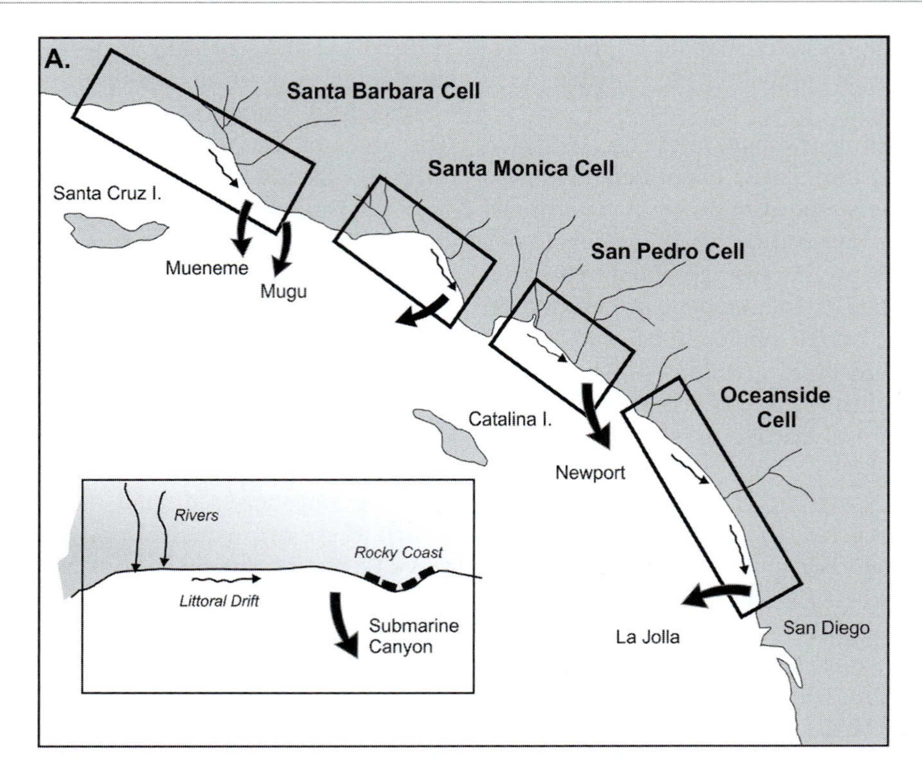

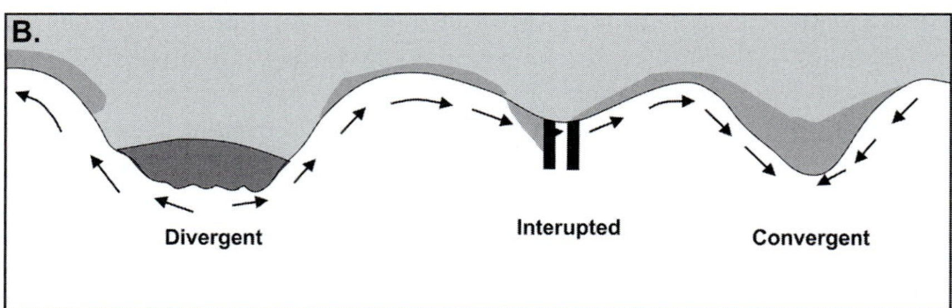

Figure 7.20 Littoral drift cells: A. Littoral cells on the California coast (after Bowen and Inman, 1970); B. types of littoral cell boundaries.

Box 7.2 Littoral Cells on the North Shore of Lake Erie

Littoral cells on the north shore of Lake Erie provide an illustration of some of the common features associated with the cells (Figure 7.21A). Limestone bedrock outcrops along the very eastern portion of the shoreline form small low headlands that are covered by till and lacustrine sediment. The rest of the shoreline along the north shore is composed of sedimentary materials, primarily till, glacio-lacustrine clay and deltaic sand and silt. Erosion of these materials, and shoreline recession, has led to the formation of bluffs that are 10–30 m high with relatively rapid retreat rates – on the order of 0.3–2 m a^{-1} (Figure 7.21B). At some locations, the retreat exceeded 5 km since the lake reached its current level ~5000 years BP. The accompanying nearshore erosion produces a steep nearshore profile in which most of the coarse sediment transport occurs through beach drifting or with the longshore current in the narrow surf zone just lakeward of the step. Bluff and nearshore erosion accounts for

Box 7.2 (cont.)

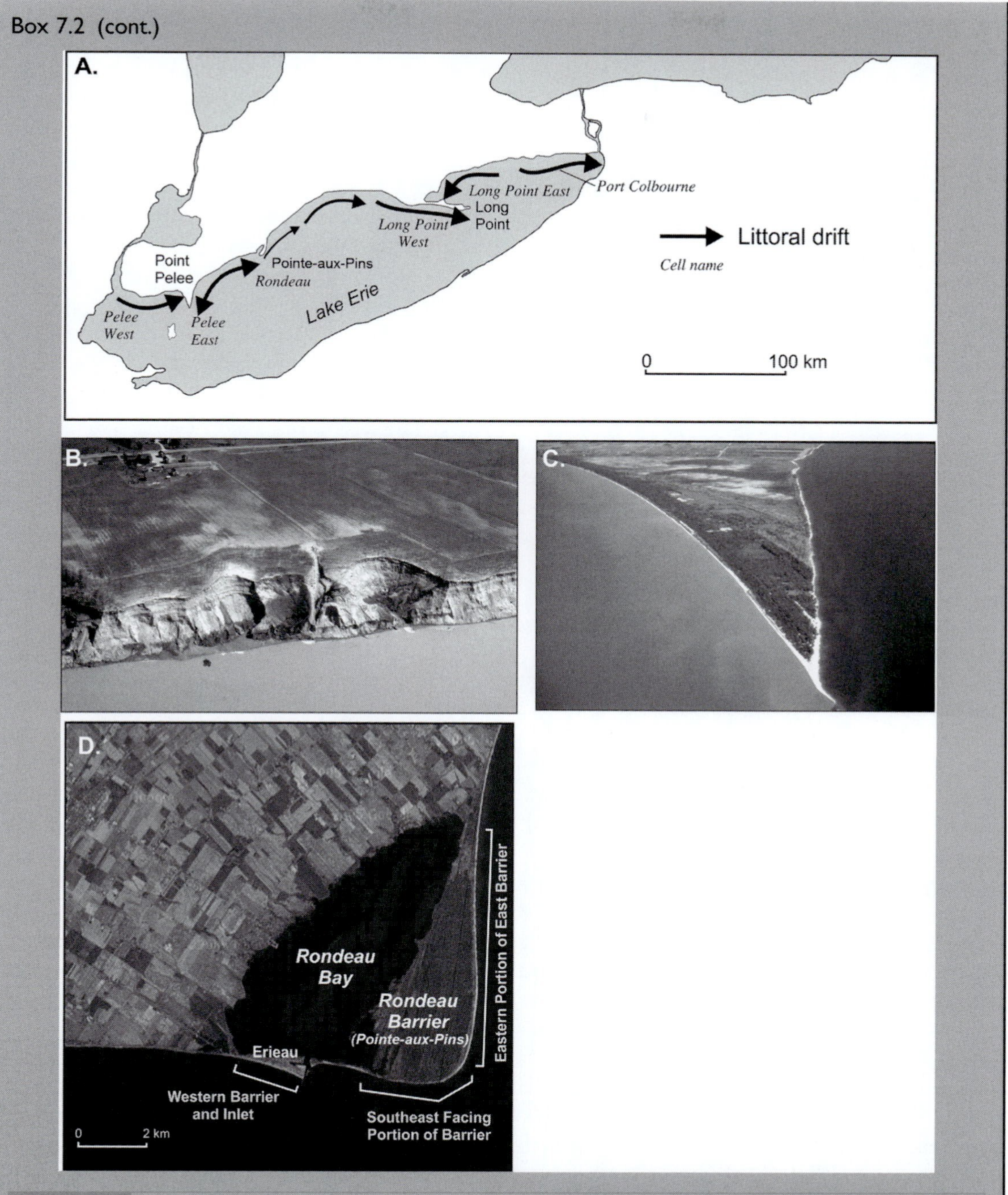

Figure 7.21 Littoral cells on the north shore of Lake Erie (after Boyd, 1981, Reinders, 1988 and Graham and Davidson-Arnott, 2005); A. map of the north shore of Lake Erie showing littoral cells, net longshore transport patterns and the major source and sink areas; B. oblique aerial photograph of a typical source area showing bluff erosion near Port Burwell, about 30 km west of Long Point. The bluffs here are about 30 m high and the recession rate ranges from 1 to 2 m a^{-1}; C. oblique aerial photograph looking north of Point Pelee. The cuspate foreland is about 8 km long and about 3 km wide at the landward margin; and D. vertical aerial photograph of the two barriers making up the central sink of the Pointe-aux-Pins system.

Box 7.2 (cont.)

almost all the littoral sediment supply, with only a very small amount coming from streams reaching the lake. The largest waves on Lake Erie are generated along the SW–NE axis of the lake, which provides the longest fetch of about 300 km. Prevailing winds are from the SW and strong westerly, northwesterly and northeaster winds can be associated with the passage of mid-latitude cyclones with direction dependent on the track of the storm. Because of the greater frequency of winds from the west, net longshore transport is towards the east over a large portion of the north shore (a similar situation exists on the south shore). Although only about 40 per cent of the eroded material contributes to the littoral sediment budget (the fines settle out in the central lake basin) the large volumes of sediment eroded and the lengthy period of stable lake level has led to the formation of three sedimentary sinks: the cuspate foreland of Point Pelee in the west, the barrier and lagoon system of Pointe-aux-Pins in the centre, and Long Point spit in the east (Figure 7.21A, C, D). The cells are given informal names here to make it easy to distinguish them.

Point Pelee is a cuspate foreland formed by the convergence of two littoral drift cells (Figure 7.21A). The coast to the west of Point Pelee is sheltered from large waves associated with NE winds by the presence of shoals and limestone outcrops around Pelee Island, and the resulting net transport from bluff erosion is to the east, even though the fetch in this direction is quite short. Large waves generated by southwest winds during storms produce a net westerly longshore transport of sediment eroded from the bluffs for tens of kilometres to the east to supply the foreland (Coakley, 1976). The fetch for westerly waves increases east of the foreland, and the fetch for northeast waves diminishes. The combined effect of this leads to a reversal of the net transport (divergent boundary) and the initiation of a littoral cell with net transport to the east. The divide between the two cells is not sharp, and there is likely a zone on the order of 10–20 km that contributes sediment to both cells, with actual transport direction varying seasonally and over several years as a function of changes in the frequency and magnitude of easterly versus westerly winds. Today, a large part of the shoreline to the east and west of Point Pelee has been armoured and the transport of sediment impeded by jetties, leading to a reduction in the amount of sediment available to the headland that is now experiencing erosion.

The second depositional complex is Pointe-aux-Pins (Figure 7.21D), which is a complex feature consisting of a short western barrier (Erieau) and an elongate eastern barrier (Rondeau) enclosing a bay that is about 10 km long and 4 km wide (Coakley, 1989). Early reports (Boyd, 1981) suggest that Pointe-aux-Pins was formed by the convergence of two cells similar to the situation at Point Pelee. However, more recent work (Graham and Davidson-Arnott, 2005) has shown that net transport is towards the east at this location and that deposition occurs because of a decrease in the transport gradient resulting from the change in shoreline orientation. The south shore of Rondeau is thus the erosional updrift end of a complex spit feature (not the downdrift end as originally proposed), and the whole feature is an example of a sink associated with an interruptive boundary.

The third major sink is Long Point, a flying spit that is fed by the littoral cell that extends over 120 km east of Rondeau (Figure 7.21A). This shoreline consists primarily of eroding bluffs that supply over 300 000 m^3 of sand and gravel annually. Long Point was initiated about 5000 years BP, and during that time the shoreline and point of attachment of the spit has migrated about 5 km inland (Coakley, 1992). The initial trigger for the diversion of sediment towards the centre of the lake may have been a cross-lake glacial moraine, but once this occurred, the deep water at the terminus of the spit acted to trap all the littoral sediment that was not deposited in the barrier itself. Because the tip of the spit is in deep water, there is little refraction around the end of the spit (Davidson-Arnott and Conliffe Reid, 1994), and most of the littoral transport that reaches the distal end goes into building the spit platform and the subaerial beach and dune complex. In contrast, Point Pelee and Rondeau have limited dune development.

The cell to the east of Long Point transports a small amount of sand westward and most of it is trapped at Turkey Point, a small spit that is building into Long Point Bay. The existence of this cell results from the growth of Long Point, which now shelters a substantial length of the coast to the east from southwest waves, thus allowing easterly winds to dominate. Further east, the sheltering effect disappears, and westward longshore transport dominates the Port Colborne cell that transports sediment towards the entrance to the Niagara River.

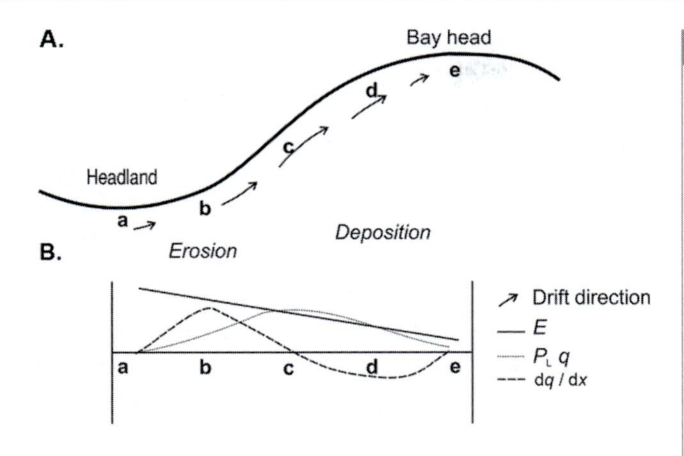

Figure 7.22 Schematic of one-half of a headland/bay beach showing the direction and magnitude of potential longshore sediment transport A. and the spatial pattern of a number of parameters resulting from shore normal wave approach and wave refraction B. E = wave energy at the break point, P_L, and q are the longshore component of wave energy flux and the volume of longshore sediment transport, and x is the alongshore distance from the headland. Points (a)–(e) mark locations of maxima and minima for key parameters. Wave energy is greatest at the headland because of convergence and least in the bay because of divergence. The rate of increase in q is a maximum at (b) and decreases to 0 at (c) which is the location of maximum longshore transport. Sediment transport decreases from a maximum at (c) to 0 at (e) where wave orthogonals are once again perpendicular to the shoreline.

7.5.3 Littoral Transport Gradients

Convergent sinks, such as those of headland/bay beaches or the cuspate foreland of Point Pelee (Figure 7.21) are easily understood. However, as shown for the origin of Pointe-aux-Pins, deposition can also occur where there is no reversal in the dominant transport direction. Instead, deposition takes place because of a change in the gradient of longshore sediment transport – in this case because the shore alignment becomes more nearly perpendicular to the southwest waves reducing the gross rate of the transport to the east. Thus, a decrease in the rate of net transport along a reach, assuming no shortage in the availability of sand, means that more sand enters the updrift boundary than leaves the downdrift boundary. The extra sand is deposited initially on the beach and nearshore, widening the beach in this area and possibly feeding more sand inland to the dune. The opposite occurs when there is an increase in the transport gradient, and more sand leaves the downdrift boundary than enters updrift. The result here is to produce beach erosion, narrowing the beach and possibly eroding material from the foredune if present.

Gradients in longshore sediment transport formed the basis for the conceptual ABC model of May and Tanner (1973). This was developed to guide the application of WAVENERG, which was one of the first computer wave refraction models made generally available to coastal geomorphologists and engineers (May, 1974). The ABC model shows one-half of a headland/bay beach, and illustrates changes in a variety of parameters for wave approach perpendicular to the shoreline that would be expected due to wave refraction processes and the change in orientation of the shoreline (Figure 7.22). Wave energy E is greatest at the headland because of convergence and least in the bay because of divergence. The longshore component of wave energy flux P_L varies with the angle of wave approach to the shoreline and the total wave energy at the break point and is zero at the headland and in the bay, where wave approach is perpendicular, and a maximum at C where the angle with respect to the shoreline is greatest. As we saw in Section 7.4.3, the immersed weight longshore sediment transport rate I_L and the volume of sediment transport rate q will be proportional to P_L and so all three of them plot on the same line (Figure 7.22). Points a and e mark the locations of maxima and minima for these parameters, and for their derivatives with distance alongshore. The rate of increase in q reaches a maximum at B and decreases to zero at C, which is the location of maximum longshore transport. The rate of sediment transport decreases rapidly between C and D, and then declines to zero at E where all longshore transport ceases.

The simple relationships between wave approach angle to the shoreline and transport gradients shown in Figure 7.22 can be applied to much more complex scenarios and form the basis for modelling of the littoral drift system using computer-based wave refraction models and local wave climate statistics. These models can be used to predict the magnitude and net direction of transport along the coast, predict areas of likely erosion and deposition and can feed into determination of the littoral sediment budget. Recent modelling efforts suggest that the primary impact of storm and wave climate changes over the next century will be associated with changes in the direction of wave advance (Slott *et al.*, 2006). As shown by Adams *et al.* (2011), a shift in waves from the northwest to the west with an increase in El Niño activity will lead to significant erosion of the southern California coast (curved section of coastline from Point Conception to San Diego). The change in wave direction and alongshore variation in sediment convergence and divergence may have a stronger influence on erosion and deposition compared to sea level rise or an increase in the magnitude and frequency of storm waves.

Output from such numerical modelling should always be compared to qualitative geomorphic indicators to ensure that all key controls have indeed been incorporated in the modelling. First-generation models such as WAVENRG use regular wave inputs, use a relatively simple topographic extrapolation algorithm, and do not allow for diffraction or wave–current interaction. The WAVENRG program was used by Lawrence and Davidson-Arnott (1997) to model sediment transport and coastal erosion on the east coast of Lake Huron where littoral cell stretched between a natural bedrock headland at Point Clark and an artificial barrier created by harbour breakwalls at Goderich. It might be expected that bluff recession rates would be associated with areas of highest wave energy flux P_B, but a plot of P_B against bluff recession rates for the littoral cell shows no significant relationship (Figure 7.23A), suggesting that the beach sediments are providing protection to some portions of the cliff shoreline. The alongshore variation in potential net longshore sediment transport S_L is plotted against the estimated cumulative supply of sediment available from bluff recession and transport from updrift (Figure 7.23B) and zones of erosion, transport and deposition are indicated based on the transport gradients. The areas with little erosion tend to correspond to zones where the cumulative supply is close to the potential transport, indicating the potential for greater protection from beach sediments while the highest erosion is associated with potential transport more than supply. The general pattern of beach width variation alongshore also fits with the predicted transport gradients (Figure 7.23C), and field evidence suggests that sheltering provided against large waves from the northwest is greater than estimated by the model. This example shows how new insights can be gained by combining with model outputs with careful field reconnaissance and that the shoreline cannot be assumed to be uniform alongshore. Similar conclusions were recently made by Slott *et al.* (2006) who used a model similar to WAVENRG to show how changes in wave direction can have a bigger impact than sea level rise.

Newer models such as MIKE21 (e.g., Johnson *et al.*, 2001) and SWAN (Booij *et al.*, 2001) combine wave prediction of random wave fields, with more sophisticated modelling of nearshore bathymetry, refraction and diffraction. Thus, they offer the potential for better modelling of waves and currents in the natural system and for the prediction of sediment transport quantity and direction in models such as Delft3D (e.g., Manson *et al.*, 2016). Nevertheless, there are still many uncertainties in the application of the models because of uncertainties in input parameters, especially the wind and wave climate and bed characteristics. These uncertainties can be reduced by comparison of modelled values with measured field values within the study area for winds, waves and currents. Thus, Manson *et al.* (2016) were able to compare

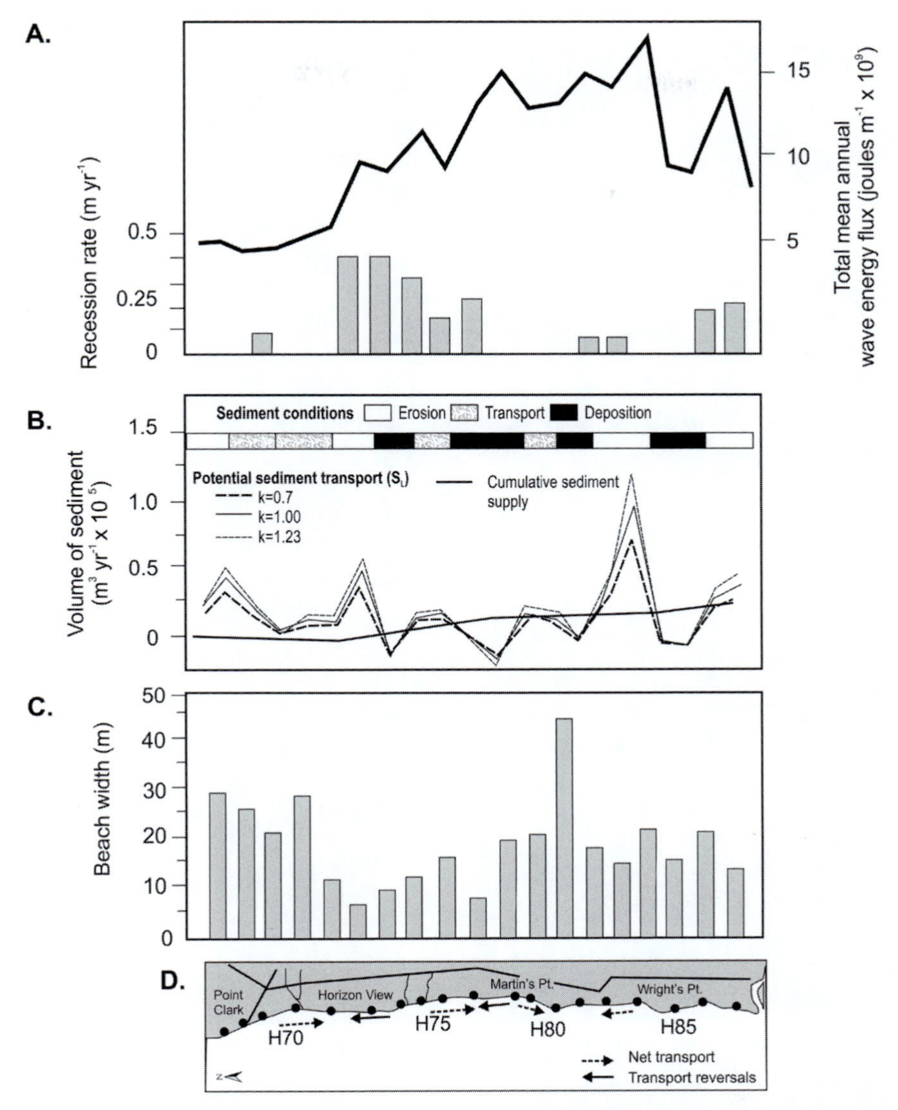

Figure 7.23 Comparison of modelling of littoral sediment transport with geomorphic indicators for a section of the Lake Huron shoreline. A. Plot of the spatial distribution of the mean annual bluff recession rate versus the mean annual total wave energy flux predicted from the model. B. The alongshore distribution of potential sediment transport based on modelling and the available littoral sediment supply based on the bluff recession rate and bluff composition. Zones of potential erosion, transport and deposition are based on the alongshore transport gradient. C. Alongshore variations in beach width measured on one day. D. Map of the coastline between Point Clark and Goderich. The dots mark locations of calculations for model values and the numbered stations are sites of long-term bluff recession monitoring (Lawrence and Davidson-Arnott, 1997).

hindcast waves for particular storms to measured wave characteristics at offshore locations (Figure 7.24A) and current speed and direction during storm and non-storm conditions to those measured by current meters. The close correspondence of measured and simulated values (Manson *et al.*, 2016) provides some confidence in the predicted values for sediment transport (Figure 7.24B, C and D). However, it is still necessary to carry out a careful assessment of the results against a range of coastal indicators and historical records.

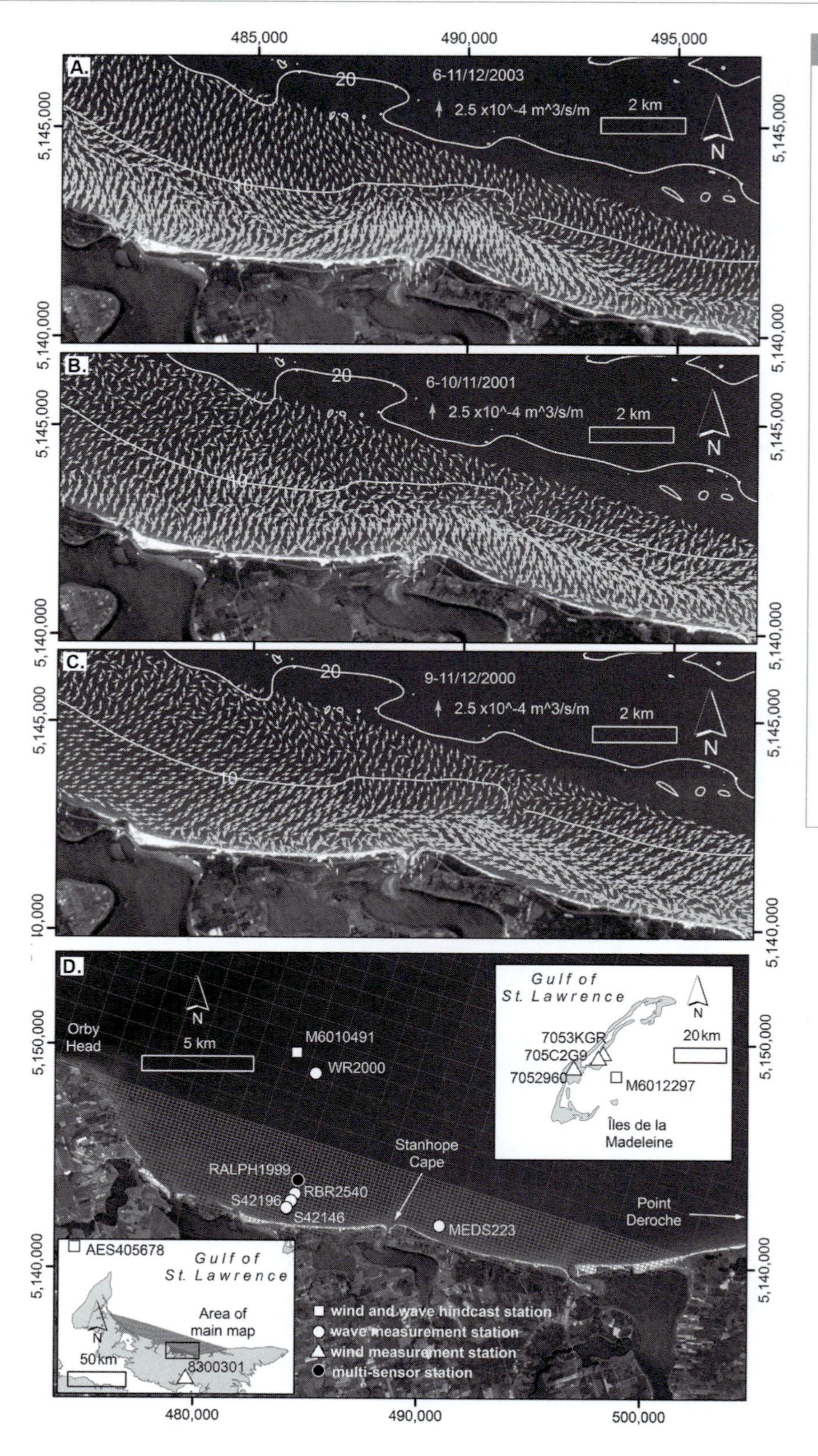

Figure 7.24 Example of output from modelling of mean sediment transport magnitudes and directions (total load) during three storms on the north coast of Prince Edward Island using the model Delft3D. A. Location of the study area within the Gulf of St Lawrence. The bottom inset shows the extent of the main model grid and the nested grid in the nearshore is shown on the main map. The locations of various instruments used to hindcast waves and to measure winds, waves and currents are shown on the main map and on the top inset of the Magdalen islands in the centre of the Gulf. B. Northeasterly storm, 6–11 December 2003. C. Northerly storm, 6–10 November 2001; D. northwesterly storm, 9–11 December 2000 (Manson *et al.*, 2016).

Further Reading

Fredsøe, J. and Deigaard, R. 1992. *Mechanics of Coastal Sediment Transport*. Singapore: World Scientific Press, 369 pp.

This book covers a lot of the basic processes related to coastal sediment transport.

References

Aagaard, T. and Greenwood, B. 1994. Suspended sediment transport and the role of infragravity waves in a barred surf zone. *Marine Geology*, **118**, 23–48.

Aagaard, T. and Greenwood, B. 1995. Suspended sediment transport and morphological response on a dissipative beach. *Continental Shelf Research*, **15**(9), 1061–86.

Aagaard, T., Greenwood, B. and Nielsen, J. 1997. Mean currents and sediment transport in a rip channel. *Marine Geology*, **140**(1–2), 25–45.

Aagaard, T., Nielsen, J. and Greenwood, B. 1998. Suspended sediment transport and nearshore bar formation on a shallow intermediate state beach. *Marine Geology*, **148**, 203–25.

Aagaard, T., Greenwood, B. and Nielsen, J. 2002. Bed level changes and megaripple migration on a barred beach. *Journal of Coastal Research*, **SI 34**, 110–16.

Aagaard, T., Greenwood, B. and Hughes, M. 2013. Sediment transport on dissipative, intermediate and reflective beaches. *Earth-Science Reviews*, **124**, 32–50.

Adams, P.N., Inman, D.L. and Lovering, J.L. 2011. Effects of climate change and wave direction on longshore sediment transport patterns in Southern California. *Climatic Change*, **109**, 211–28.

Allan, J.C., Hart, R. and Tranquili, J.V. 2006. The use of Passive Integrated Transponder (PIT) tags to trace cobble transport in a mixed sand-and-gravel beach on the high-energy Oregon coast, USA. *Marine Geology*, 232(1-2), 63–86.

Allen, J.R.L. 1984. Parallel lamination developed from upper-stage plane beds: a model based on the larger coherent structures of the turbulent boundary layer. *Sedimentary Geology*, 39, 227–42.

Ashton, A., Murray, A.B. and Arnault, O. 2001. Formation of coastal features by large-scale instabilities induced by high angle waves. *Nature*, **414**, 296–300.

Austin, M.J. and Masselink, G. 2008. The effect of bedform dynamics on computing suspended sediment fluxes using optical backscatter sensors and current meters. *Coastal Engineering*, **55**, 251–60.

Bagnold, R.A. 1946. Motion of waves in shallow water: interaction between waves and sand bottoms. *Proceedings of the Royal Society of London*, **A187**, 1–15.

Bagnold, R.A. 1963. Mechanics of marine sedimentation. In Hill, M.N. (ed.), *The Sea*. New York: Wiley-Interscience, pp. 507–28.

Bagnold, R.A. 1966. An approach to the sediment transport problem from general physics. *US Geological Survey, Professional Paper*, **422-I**, 37 pp.

Baillard, J.A. 1981. An energetics total load sediment transport model for a plane sloping beach. *Journal of Geophysical Research*, **86**, 10 938–54.

Bayram, A., Larson, M., Miller, H.C. and Kraus, N.C. 2001. Cross-shore distribution of longshore sediment transport: Comparison between predictive formulas and field measurements. *Coastal Engineering*, **44**, 79–99.

Bayram, A., Larson, M. and Hanson, H. 2007. A new formula for the total longshore sediment transport rate. *Coastal Engineering*, **54**, 700–10.

Beach, R.A. and Sternberg, R.W. 1988. Suspended sediment transport in the surf zone: response to cross-shore infragravity motion. *Marine Geology*, **80**, 61–79.

Bodge, K.R. and Dean, R.G. 1987. Short-term impoundment of longshore transport. In *The Proceedings of the Coastal Sediments '87*. Vicksburg, MS: American Society of Civil Engineers, pp. 469–83.

Booij, N., Holthuijsen, L.H. and Battjes, J.A. 2001. Ocean to near-shore wave modelling with SWAN. *Proceedings of the Fourth Conference on Coastal Dynamics*. Reston, VA: American Society of Civil Engineers, pp. 335–44.

Bowen, A.J. 1980. Simple models of nearshore sedimentation; beach profiles and longshore bars. In McCann, S.B. (ed.), *The Coastline of Canada*. Ottawa: Geological Survey of Canada, Paper 80-10, 1-11.

Bowen, A.J. and Inman, D.L. 1966. *Budget of Littoral Sands in the Vicinity of Point Arguello, California*. Washington, DC: Coastal Engineering Research Centre, Technical Memorandum **19**, 41 pp.

Boyd, G.L., 1981. *Great Lakes Erosion Monitoring Program: Final Report*. Burlington, Ontario: Ocean Science and Surveys. Bayfield Lab for Marine Science and Surveys.

Bruno, R.O., Dean, R.G. and Gable, C.G. 1980. Longshore transport evaluations at a detached breakwater. *Proceedings of the Seventeenth Coastal Engineering Conference*. New York: American Society of Civil Engineers, pp. 1453–75.

Bruun, P. 1954. Migrating sand waves or sand humps, with special reference to investigations carried out on the Danish North Sea Coast. In *Proceedings of the Fifth International Conference on Coastal Engineering*, New York: American Society of Civil Engineers, **1**(5), pp. 269–95.

Cartier, A. and Héquette, A.,2015. Vertical distribution of longshore sediment transport on barred macrotidal beaches, northern France. *Continental Shelf Research*, **93**, 1–16.

Ciavola, O. 2004. Tracers. In Schwartz, M. (ed.), *Encyclopedia of Coastal Sciences*. Dordrecht: Kluwer Academic Publishers, pp. 1253–8.

Ciavola, O., Taborda, R., Ferreira, O. and Dias, J.A. 1997. Field measurements of longshore sand transport and control processes on a steep meso-tidal beach in Portugal. *Journal of Coastal Research*, **13**, 1119–129.

Clifton, H.E. 1976. Wave-formed sedimentary structures: conceptual model. In Davis, R.A. and Ethington, R.L. (eds.), *Beach and Nearshore Sedimentation*. Tulsa, OK: Society for Economic Paleontology and Mineralogy, Special Publication **24**, pp. 126–48.

Clifton, H.E., Hunter, R.E., Phillips, R.L. 1971. Depositional structures and processes in the non-barred high-energy nearshore. *Journal of Sedimentary Petrology*, **41**, 651–70.

Coakley, J.P. 1976. The formation and evolution of Point Pelee, western Lake Erie. *Canadian Journal of Earth Sciences*, **13**, 136–44.

Coakley, J.P. 1989. The origin and evolution of a complex cuspate foreland: Pointe-aux-Pins, Lake Erie, Ontario. *Géographie Physique et Quarternaire*, **43**, 65–76.

Coakley, J.P. 1992. Holocene transgression and coastal-landform evolution in northeastern Lake Erie, Canada. In Fletcher, C.H. and Wehmiller, J.F. (eds.), *Quaternary Coasts of the United States: Marine and Lacustrine Systems*. Tulsa, OK: Society for Sedimentary Geology, Special Publication **48**, pp. 415–26.

Coco, G. and Murray, A.B. 2007. Patterns in the sand: from forcing templates to self-organisation. *Geomorphology*, **91**, 271–90.

Cooper, N.J. and Pontee, N.I. 2006. Appraisal and evolution of the littoral 'sediment cell' concept in applied coastal management: experiences from England and Wales. *Ocean and Coastal Management*, **49**, 498–510.

Davidson-Arnott, R.G.D. and Conliffe Reid, H.E. 1994. Sedimentary processes and the evolution of the distal bayside of Long Point, Lake Erie. *Canadian Journal of Earth Sciences*, **31**, 1461–73.

Davidson-Arnott, R.G.D. and Greenwood, B. 1974. Bedforms and structures associated with bar topography in the shallow-water wave environment, Kouchibouguac Bay, New Brunswick, Canada. *Journal of Sedimentary Petrology*, **44**, 698–704.

Davidson-Arnott, R.G.D. and Greenwood, B. 1976. Facies relationships on a barred coast, Kouchibouguac Bay, New Brunswick. In Davis, R.A. Jr and Ethington, R.L. (eds.), *Beach and Nearshore Sedimentation*. Tulsa, OK: Society for Economic Paleontology and Mineralogy, Special Publication **24**, pp. 149–68.

Davidson-Arnott, R.G.D. and van Heyningen, A. 2003. Migration and sedimentology of longshore sandwaves, Long Point, Lake Erie, Canada. *Sedimentology*, **50**, 1123–37.

Davidson-Arnott, R.G.D. and Pember, G.F. 1980. Morphology and sedimentology of multiple parallel bar systems, Southern Georgian Bay, Ontario. In McCann, S.B. (ed.), *The Coastline of Canada*. Ottawa: Geological Survey of Canada, Paper 80-10, pp. 417–28.

Davidson-Arnott, R.G.D. and Stewart, C.J. 1987. The effect of longshore sandwaves on dune erosion and accretion, Long Point, Ontario. *Proceedings Canadian Coastal Conference*, Ottawa: National Research Council of Canada, pp. 131–44.

Davies, A.G. 1985. Field observations of the threshold of sediment motion by wave action. *Sedimentology*, **32**, 685–704.

Davies, J.L. 1974. Coastal sediment compartment. *Australian Geographical Studies*, **12**, 139–51.

Dawson, J.C., Davidson-Arnott, R.G.D. and Ollerhead, J. 2002. Low-energy morphodynamics of a ridge and runnel system. *Journal of Coastal Research*, **SI 36**, 198–215.

Dean, R.G., Berek, E.P., Gable, C.G. and Seymour, R.J. 1982. Longshore transport determined by an efficient trap. *Proceedings of the Eighteenth Coastal Engineering Conference*, New York: American Society of Civil Engineers, pp. 954–68.

Dickson, M.E., Kench, P.S. and Kantor, M.S. 2011. Longshore transport of cobbles on a mixed sand and gravel beach, southern Hawke Bay, New Zealand. *Marine Geology*, 287(1-4), 31–42.

Dolphin, T., Lee, J., Phillips, R., Taylor, C.J. and Dyer, K. R. 2016. Velocity of RFID Tagged Gravel in a Non-uniform Longshore Transport System. *Journal of Coastal Research*, 75(sp1), 363–368.

Doucette, J.S. 2000. The distribution of nearshore bedforms and effects on sand suspension on low-energy, micro-tidal beaches in Southwestern Australia. *Marine Geology*, **165**, 41–61.

Downing, J.P., Sternberg, R.W. and Lister, C.R.B. 1981. New instrumentation for the investigation of sediment suspension processes in the shallow marine environment. *Marine Geology*, **42**, 19–34.

Duane, D.B. and James, W.R. 1980. Littoral transport in the surf zone elucidated by an Eulerian sediment tracer experiment. *Journal of Sedimentary Petrology*, **50**, 929–52.

Falqués, A., F. Ribas, F.,D. Idier, D. and Arriaga, J. 2017, Formation mechanisms for self-organized kilometer-scale shoreline sand waves, *Journal of Geophysical Research Earth Surface*, **122**, 1121–38, DOI: doi:10.1002/2016JF003964.

Gallagher, E.L. 2003. A note on megaripples in the surf zone: evidence for their relation to steady flow dunes. *Marine Geology*, **193**, 171–6.

Gallagher, E.L., Elgar, S. and Thornton, E.B. 1998. Megaripple migration in a natural surf zone. *Nature*, **394**, 165–8.

Graham, M. and Davidson-Arnott, R.G.D. 2005. Shoreline change and sediment transport patterns in the vicinity of Pointe-aux-Pins, Lake Erie: implications for past and future evolution. *Proceedings of the Canadian Coastal Conference*, Ottawa: CCSEA, 10 pp.

Green, M.O. and Black, K.P. 1999. Suspended-sediment reference concentration under waves: field observations and critical analysis of two predictive models. *Coastal Engineering*, **38**, 115–41.

Greenwood, B. and Davidson-Arnott, R.G.D. 1979. Sedimentation and equilibrium in wave-formed bars: a review and case study. *Canadian Journal of Earth Sciences*, **16**, 312–32.

Greenwood, B. and Mittler, P.R. 1979. Structural indices of sediment transport in a straight, wave-formed bar. *Marine Geology*, **32**, 191–203.

Greenwood, B. and Mittler, P.R. 1985. Vertical sequence and lateral transitions in the facies of a barred nearshore environment. *Journal of Sedimentary Petrology*, **55**, 366–75.

Hay, A., 2011. Geometric bed roughness and the bed state storm cycle. *Journal of Geophysical Research*, 116, C04017, DOI: doi:10.1029/2010JC006687.

Hay, A.E. and Bowen, A.J. 1994. Coherence scales of wave-induced suspended sand concentration fluctuations. *Journal of Geophysical Research*, **99**, 12 749–65.

Heathershaw, A.D. and Carr, A.P. 1977. Measurements of sediment transport rates using radioactive tracers. In *The Proceedings of the Coastal Sediments '77*, Reston, VA: American Society of Civil Engineers, pp. 399–416.

Hicks, D.M. and Inman, D.L. 1987. Sand dispersion from an ephemeral river delta on the Central California coast. *Marine Geology*, **77**, 305–18.

Holman, R.A., Haller, M.C., Lippmann, T.C., Holland, K.T. and Jaffe, B.E. 2015. Advances in nearshore processes research: four decades of progress. *Shore and Beach*, **83**(1), 39.

Houser, C., Greenwood, B. and Aagaard, T. 2006. Divergent response of an intertidal swash bar. *Earth Surface Processes and Landforms*, **31**, 1775–91.

Ingle, J.C. 1966. *The Movement of Beach Sand*. New York: Elsevier, 221 pp.

Inch, K., Davidson, M., Masselink, G. and Russell, P. 2017. Observations of nearshore infragravity wave dynamics under high energy swell and wind-wave conditions. *Continental Shelf Research*, **138**, 19–31.

Inman, D.L. 1987. Accretion and erosion waves on beaches. *Shore and Beach*, **55**, 61–6.

Inman, D.L. and Bagnold, R.A. 1963. Littoral processes. In Hill, M.N. (ed.), *The Sea*. New York: Wiley-Interscience, pp. 529–53.

Inman D.L. and Bowen, A.J. 1963. Flume experiments on sand transport by waves and currents. *Proceedings Eighth Conference on Coastal Engineering*. New York: American Society of Civil Engineers, pp. 137–50.

Inman, D.L. and Frautschy J.D. 1966. Littoral processes and the development of shorelines. *Proceedings Coastal Engineering Specialty Conference*. American Society of Civil Engineers, pp. 511–36.

Inman, D.L., Zampol, J.A., White, T.E. et al. 1980. Field measurements of sand motion in the surf zone. *Proceedings of the Seventeenth Coastal Engineering Conference*. American Society of Civil Engineers, pp. 1215–34.

Jensen, B.L., Summer, B.M. and Fredsøe, J. 1989. Turbulent oscillatory boundary layers at high Reynolds numbers. *Journal of Fluid Mechnics*, **206**, 265–97.

Johnson, H.K., Appendini, C.M., Soldati, M., Elfrink, B. and Sørenson, P. 2001. Numerical modelling of morphological changes due to shoreface nourishment. *Proceedings of the Fourth Coastal Dynamics Conference*. New York: American Society of Civil Engineers, pp. 878–87.

Jonsson, I.G. and Carlsen, N.A. 1976. Experimental and theoretical investigations in an oscillatory turbulent boundary layer. *Journal of Hydraulic Research*, **14**, 45–60.

Kaergaard, K., Fredsoe, J. and Knudsen, S.B., 2012. Coastline undulations on the West Coast of Denmark: Offshore extent, relation to breaker bars and transported sediment volume. *Coastal Engineering*, **60**, 109–22.

Kamphuis, J.W. 1975. Friction factor under oscillatory waves. *Journal of Waterways, Harbors and Coastal Engineering*, **101**, 135–44.

Kamphuis, J.W. 1991. Alongshore sediment transport rate. *Journal of the Waterways, Port, Coastal and Ocean Division*, **117**, 624–40.

Komar, P.D. 1998. *Beach Processes and Sedimentation*, 2nd Edition. Upper Saddle River, NJ: Prentice Hall, 544 pp.

Komar, P.D. and Inman, D.C. 1970. Longshore sand transport on beaches. *Journal of Geophysical Research*, **75**, 5914–27.

Komar, P.D. and Miller, C.M. 1973. The threshold of sediment movement under oscillatory water waves. *Journal of Sedimentary Petrology*, **43**, 1101–10.

Kraus, N.C. 1987. Application of portable traps for obtaining point measurement of sediment transport rates in the surf zone. *Journal of Coastal Research*, **2**, 139–52.

Larsen, S.M., Greenwood, B. and Aagaard, T. 2015. Observations of megaripples in the surf zone. *Marine Geology*, **364**, 1–11.

Larson, M. and Kraus, N.C. 1989. SBEACH: Numerical Model for Simulating Storm-Induced Beach Change. CERC Technical Report 89-9.

Lawrence, P.L. and Davidson-Arnott, R.G.D. 1997. Alongshore wave energy and sediment transport on south-eastern Lake Huron, Ontario, Canada. *Journal of Coastal Research*, **13**, 1004–15.

Lee, G., Nicholls, R.J. and Birkemeier, W.A. 1998. Storm-driven variability of the beach-nearshore profile at Duck, North Carolina, 1981–1991. *Marine Geology*, **148**, 163–77.

Li, M.Z. and Amos, C.L. 1999. Field observations of bedforms and sediment transport thresholds of fine sand under combined waves and currents. *Marine Geology*, **158**, 147–60.

Madsen, O.S. 1987. Use of tracers in sediment transport studies. In *The Proceedings of the Coastal Sediments '01*. New York: American Society of Civil Engineers, 424–35.

Manson, G.K., Davidson-Arnott, R.G.D., Forbes, D.L., 2016. Modelled nearshore sediment transport in open-water conditions, central north shore of Prince Edward Island, Canada. *Canadian Journal of Earth Science*, 53, 101–118.

Masselink, G., Austin, M., Tinker, J., O'Hare, T. and Russell, P. 2008. Cross-shore sediment transport and morphological response on a macrotidal beach with intertidal bar morphology, Truc Vert, France. *Marine Geology*, **251**, 141–55.

May, J.P. 1974. WAVENRG: A computer program to determine dissipation in shoaling water waves with examples from coastal Florida. In Tanner, W.F. (ed.), *Sediment Transport in the Nearshore Zone*. Tallahassee, FL: Florida State University Department of Geology: Coastal Research Notes , pp. 22–80.

May, J.P. and Tanner, W.F. 1973. The littoral power gradient and shoreline changes. In Coates, D.R. (ed.), *Coastal Geomorphology*. New York: SUNY, pp. 43–60.

Miles, J., Thorpe, A., Russell, P. and Masselink, G. 2014. Observations of bedforms on a dissipative macrotidal beach. *Ocean Dynamics*, **64**(2), 225–39.

Miller, H.C. 1999. Field measurements of longshore sediment transport during storm. *Coastal Engineering*, **36**, 301–21.

Miller M.C. and Komar, P.D. 1980. A field investigation of the relationship between oscillation ripple spacing and the near-bottom water orbital motion. *Journal of Sedimentary Petrology*, **50**, 183–91.

Mogridge, G.R. and Kamphuis, J.W. 1972. Experiments on bedform generation by wave action. In *Proceedings Thirteenth Conference on Coastal Engineering*. New York: American Society of Civil Engineers, pp. 1123–42.

Montreuil, A-L. and Bullard, J.E. 2012. A 150-year record of coastline dynamics within a sediment cell: eastern England. *Geomorphology*, **179**, 168–85.

Nairn, R.B. and Southgate, H.N. 1993. Deterministic profile modelling of nearshore processes. Part 2. Sediment transport and beach profile development. *Coastal Engineering*, **19**, 57–96.

Nielsen, P. 1986. Suspended sediment concentrations under waves. *Coastal Engineering*, **8**, 51–72.

Nielsen, P. 1992. *Coastal Bottom Boundary Layer and Sediment Transport*, Advanced Series on Ocean Engineering, 4. Singapore: World Scientific, 324 pp.

O'Donoghue, T. and Clubb, G.S. 2001. Sand ripples generated by regular oscillatory flow. *Coastal Engineering*, **44**, 101–15.

Orzech, M.D., Thornton, E.B., MacMahan, J.H., O'Reilly, W.C. and Stanton, T.P. 2010. Alongshore rip channel migration and sediment transport. *Marine Geology*, **271**(3–4), 278–91.

Osborne, P.D. and Greenwood, B. 1992. Frequency dependent cross-shore sediment transport, 2. A barred shoreface. *Marine Geology*, **106**, 25–51.

Osborne, P.D. and Vincent, C.E. 1993. Dynamics of large and small scale bedforms on a macrotidal shoreface under shoaling and breaking waves. *Marine Geology*, **115**, 207–26.

Osborne, P.D. and Vincent, C.E. 1996. Vertical and horizontal structure in suspended sand concentrations and wave-induced fluxes over bedforms. *Marine Geology*, **131**, 195–208.

Patsch, K. and Griggs, G. 2008. A sand budget for the Santa Barbara littoral cell, California. *Marine Geology*, **252**, 50–61.

Pringle, A.W. 1985. Holderness coast erosion and the significance of ords. *Earth Surface Processes and Landforms*, **10**, 107–24.

Reinders, F.J. and Associates Canada Limited 1988. Littoral Cell Definition and Sediment Budget for Ontario's Great Lakes, Final Report 1988. Ontario Ministry of Natural Resources, Conservation Authorities and Water Management Branch.

Rogers, A.L. and Ravens, T.M. 2008. Measurement of longshore sediment transport rates in the surf zone on Galveston Island, Texas. *Journal of Coastal Research*, **24**, 62–73.

Rosati, J.D. 2005. Concepts in sediment budgets. *Journal of Coastal Research*, **21**, 307–22.

Rosati, J., Gingerich, K.J., Kraus, N.C., Smith, J.M. and Beach, R.A. 1991. Longshore sand transport rate distributions measured in Lake Michigan. In *The Proceedings of the Coastal Sediments '91*. New York: American Society of Civil Engineers, pp. 156–69.

Ruessink, B.G. and Jeuken, M.C.J.L. 2002. Dunefoot dynamics along the Dutch coast. *Earth Surface Processes and Landforms*, **27**, 1043–56.

Ruessink, B.G., Houwman, K.T. and Hoekstra, P. 1998. The systematic contribution of transporting mechanisms to the cross-shore sediment transport in water depths of 3 to 9 m. *Marine Geology*, **152**, 295–324.

Sanderson, P.G. and Eliot, I. 1999. Compartmentalisation of beachface sediments along the southwestern coast of Australia. *Marine Geology*, **162**, 145–64.

Schoonees, J.S. and Theron, A.K. 1993. Review of the field-data base for longshore sediment transport. *Coastal Engineering*, 19(1-2), 1–25.

Schoonees, J.S. and Theron, A.K. 1995. Evaluation of 10 cross-shore sediment transport/morphological models. *Coastal Engineering*, **25**, 1–41.

Sherman, D.J. and Greenwood B. 1984. Boundary roughness and bedforms in the surf zone. In Greenwood B. and Davis, R.A. Jr (eds.), *Hydrodynamics and Sedimentation in Wave-dominated Coastal Environments*. Developments in Sedimentology 39, Amsterdam: Elsevier, pp. 199–218.

Sherman, D.J., Short, A.D. and Takeda, I. 1993. Sediment mixing-depth and bedform migration in rip channels. *Journal of Coastal Research*, **15**, 39–48.

Ship, R.C. 1984. Bedforms and depositional sedimentary structures of a barred nearshore system, eastern Long Island, New York. *Marine Geology*, **60**, 235–59.

Silva, A., Taborda, R., Rodrigues, A., Duarte, J. and Cascalho, J. 2007. Longshore drift estimation using fluorescent tracers: New insights from an experiment at Comporta Beach, Portugal. *Marine Geology*, **240**, 137–50.

Slott, J.M., Murray, A.B., Ashton, A.D. and Crowley, T.J. 2006. Coastline responses to changing storm patterns. *Geophysical Research Letters*, 33(18), L18404.

Stewart, C.J. and Davidson-Arnott, R.G.D. 1988. Morphology, formation and migration of longshore sandwaves; Long Point, Lake Erie, Canada. *Marine Geology*, **81**, 63–77.

Swales, A., Oldman, J.W. and Smith, K. 2006. Bedform geometry on a barred sandy shore, *Marine Geology*, **226**, 243–59.

Thevenot, M.M. and Kraus, N.C. 1995. Longshore sand waves at Southhampton Beach, New York: observation and numerical simulation of their movement. *Marine Geology*, **126**, 249–69.

Thornton, E.B., 2016. Temporal and spatial variations in sand budgets with application to southern Monterey Bay, California. *Marine Geology*, **382**, 56–67.

Tonk, A. and Masselink, G. 2005. Evaluation of longshore transport equations with OBS sensors, streamer traps, and fluorescent tracer. *Journal of Coastal Research*, **21**, 915–31.

Van der Werf, J., Ebbe J., Ribberink, J.S., O'Donoghue, T. and Doucette, J.S. 2006. Modelling and measurement of sand transport processes over full-scale ripples in oscillatory flow. *Coastal Engineering*, **53**, 657–73.

Verhagen, H.J. 1989. Sand waves along the Dutch coast. *Coastal Engineering*, **13**, 129–47.

Voulgaris, G. and Collins, M.B. 2000. Sediment resuspension on beaches: response to breaking waves. *Marine Geology*, **167**, 167–87.

Wang, P., Kraus, N.C. and Davis, R.A. Jr. 1998. Total longshore sediment transport in the surf zone: field measurement and empirical prediction. *Journal of Coastal Research*, **14**, 269–82.

Webb, M.P. and Vincent, C.E. 1999. Comparison of time-averaged acoustic backscatter concentration profile measurements with existing predictive models. *Marine Geology*, **162**, 71–90.

Wiberg, P.L. 1995. A theoretical investigation of boundary layer flow and bottom shear stress for smooth, transitional and rough flow under waves. *Journal of Geophysical Research*, **100**, 22667–79.

Part III

Coastal Systems

8

Beach and Nearshore Systems

8.1 Synopsis

The beach and nearshore are the most dynamic of coastal systems, and there is a continuous exchange of sediment between the nearshore, beach and dune based on the sequence of storm and fair-weather conditions. The equilibrium morphology of the beach and nearshore reflects a balance between the processes that act to move sediments offshore and those that tend to move them onshore. Even small waves are capable of moving sediment, but it can take a considerable amount of time for the sediment moved offshore by a storm to be moved landward and for the nearshore, beach and dune to recover. A storm can change the beach and nearshore morphology over hours and days, but it can take years to decades for the system to recover following a tropical storm or hurricane (Lee *et al.*, 1998; Houser *et al.*, 2015). In this respect, a change in the frequency and/or magnitude of storms with a changing climate will influence how a sandy coast will respond to sea level rise – this will be discussed in the next chapter.

The beach and nearshore profile is often characterised by the presence of one or more sand bars and troughs immediately landward of the bars. As will be discussed later in this chapter, the number and morphology of bars, and the dynamic movement of those bars, are controlled by factors such as the wave climate, nearshore slope and sediment size. Bars may be linear and parallel to the shore, but they also assume a variety of crescentic or rhythmic forms, with the latter usually associated with rip cell development as described in the Chapter 6. They continuously change their form and position in response to changing wave conditions, migrating offshore during storms and onshore during fair-weather conditions. On coasts with a limited fetch, which are subject primarily to storm waves, bars may be present throughout the year. However, on coasts subject to long period swell waves, prolonged periods of non-storm conditions can lead to onshore migration of the bars and their incorporation in the beach face, providing a source of sediment to the foredune if present.

Beach and nearshore morphology are superimposed on a shore-normal profile that decreases exponentially offshore. The actual steepness of the profile is primarily controlled by grain size (as measured by the settling fall velocity), with the slope decreasing with decreasing grain size. The cross-shore exchange of sediment and resulting changes in the beach and nearshore morphology can change the elevation in the bed by up to 2 m (Wright and Short, 1984). The amount of change decreases with distance offshore based on the frequency of storm waves capable of reworking sediment in deeper water. The variability and amount of change in the beach and nearshore envelope depends on the frequency and magnitude of storm events and the length of time that the beach can recover

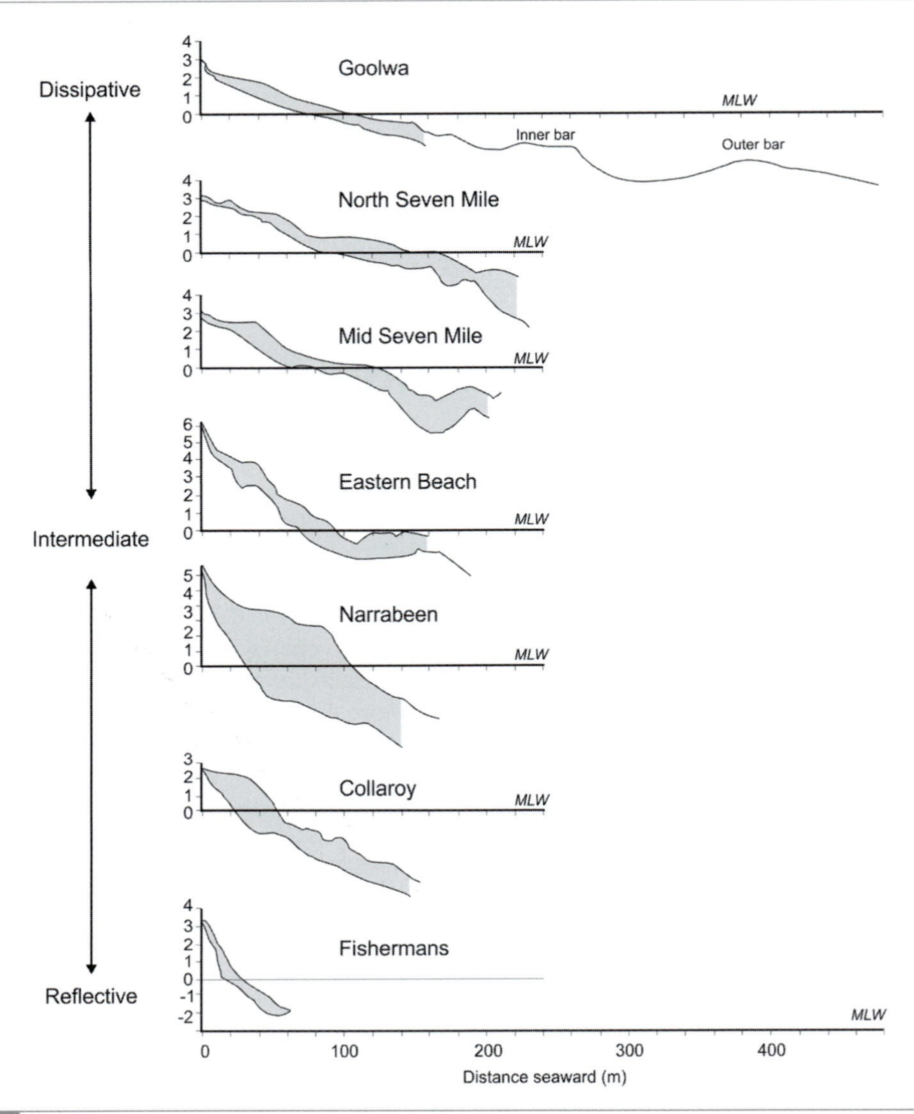

Figure 8.1 Beach and nearshore profile variability for reflective, intermediate and dissipative beaches based on measurements by Wright and Short (1984).

following each storm (Figure 8.1). In general, relatively flat (dissipative) beaches associated with frequent storm waves and/or smaller sediment exhibit a relatively small beach envelope (Wright and Short, 1984), while intermediate beaches exhibit large beach envelopes in response to the greater mobility of the beach and dune morphology. Reflective beaches dominated by relatively small, swell waves and/or larger sediment also exhibit relatively limited beach envelopes with most of the change taking place on the beach in response to the development and seaward extension of a berm (Figure 8.1).

8.2 Beach and Nearshore Sediments and Morphology

Beaches consist of an accumulation of unconsolidated sediments, ranging from fine sand to large cobbles, that have been transported and deposited

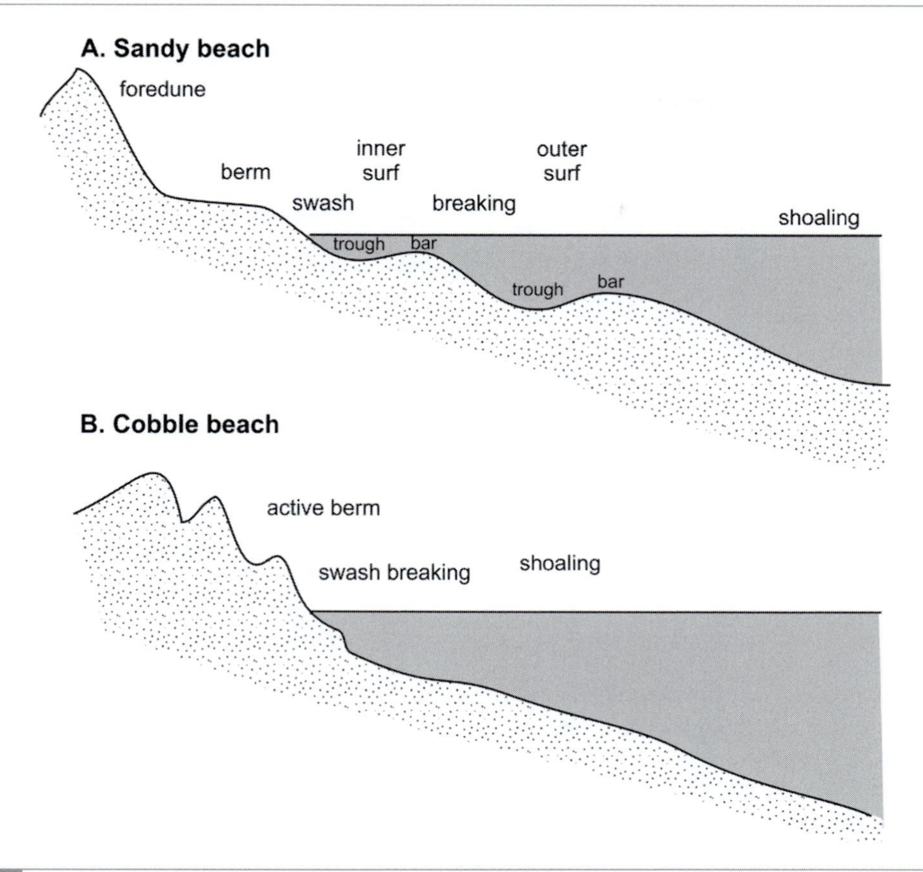

Figure 8.2 Profile of the littoral zone for A. a gently sloping sandy beach and B. a steep cobble beach showing the dynamic zones and characteristic features.

by waves and currents. Sandy beach systems (Figure 8.2A) are found along about 20 per cent of the world's coasts, and gravel or cobble beaches (Figure 8.2B) account for another 10 per cent or so, particularly in high-latitude areas. While the term beach is restricted to the zone extending from the landward limit of wave action during storms to the low tide limit (Figure 8.3), it is recognised that sediments continue offshore and that there is a significant exchange of sediments between those that are exposed on the beach and those underwater. We can characterise the whole zone over which sediment transport by waves occurs as the littoral zone (Figure 8.3), incorporating the beach and nearshore zones. The littoral system can be subdivided into three sub-systems along a profile normal to the shoreline: the outer shoreface, the inner shoreface and the beach system (Figure 8.3).

The **outer shoreface** extends offshore to the wave base. This zone is dominated by oscillatory motion generated by shoaling waves, by tidal currents and by wind-driven circulation. Large rip cell circulation extends into this zone from the breaker zone during intense storms. The slope is generally planar to curvilinear in profile, though some 3D topography may be present.

The **inner shoreface** extends roughly from MLW to seaward of the (outer) breaker zone. On sandy beaches the transition is often marked by a sharp break in slope. This zone may be absent on very steep shorelines, for example cobble beaches or on highly reflective sandy beaches. The main processes are associated with wave breaking and the generation of surf bores; mean flows such as longshore, rip currents and undertow generated by set-up; and tidal and wind-generated currents. The topography may be simple and planar or

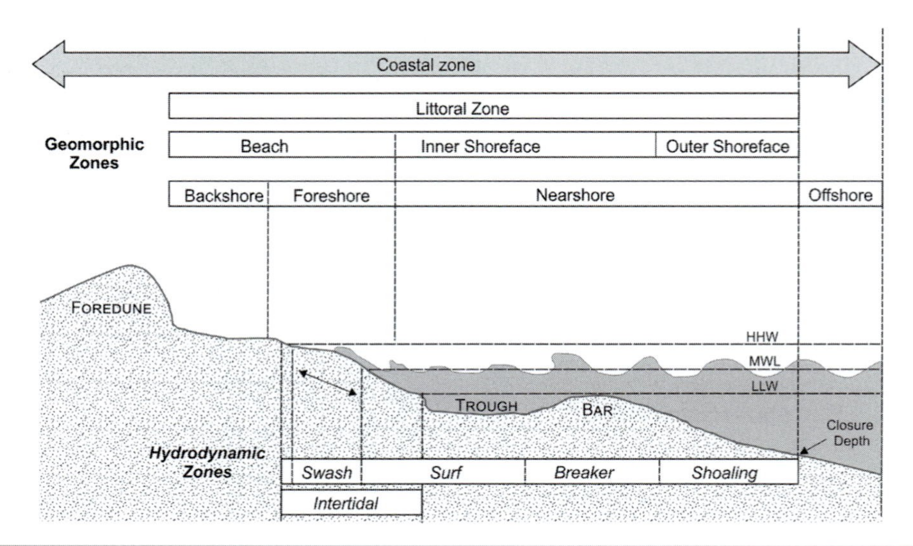

Figure 8.3 Beach nomenclature. This is a slightly modified version of Figure 2.3 to include the distinction between the inner and outer shoreface zones.

curvilinear in profile, or complex with one or more nearshore bar and trough systems present. Three-dimensional topography is common, particularly on barred shorelines.

The **beach system** extends from MLW to the landward limit of wave action. The main processes are associated with swash and backwash action. Water motion and sediment transport is influenced by infiltration into the subaerial part of the beach and by offshore flows driven by gravitational effects on the swash slope. On steep beaches and/or under low wave conditions primary wave breaking may occur here. Wave–wave interaction and wave–bed interaction may produce reflection and lead to transfers of energy to low frequencies. On low gradient beaches backwash is less effective and the beach becomes an extension of the surf zone with run-up dominated by surf beat. In the intertidal zone, ephemeral swash bars may be present, and on wide intertidal zones several permanent bars and troughs may occur.

On sandy coasts, the beach is usually backed by a dune system, and both sandy and gravel/cobble beaches can be backed by a barrier and lagoon system or by cliffs. Throughout these zones there is considerable feedback between the processes controlling sediment transport and the characteristic forms at both the micro-scale (ripples, dunes) and particularly the meso-scale (bars and troughs; swash slope forms). The interaction between coastal process and morphology is termed beach and nearshore morphodynamics, and the form of the profile is assumed to be in equilibrium with the wave, wind and tidal processes that control sediment transport, and which also reflect the influence of sediment size, shape and density. For simplicity, we will assume that the profile is entirely composed of sand or gravel/cobble, and that bedrock or mud does not outcrop along the profile. The presence of bedrock or non-erodible mud deposits leads to a cross-shore profile that is not in equilibrium with the sediment transport associated with the local waves, currents and tide (Jackson *et al.*, 2005).

8.2.1 Measurement of Morphological Change

The morphology of the beach and nearshore in sandy environments is in dynamic equilibrium with the local waves, currents and tide. Spatial variability of that morphology is associated with factors such as changes in beach orientation, wave climate or tidal range, while variability through time is associated with the sequence of storm and fair-weather waves. Understanding how and why the beach and nearshore morphology changes has received more attention than any other aspect of coastal morphology and is dependent on the ability to survey the morphology and monitor changes in that morphology.

8.2.1.1 *Nearshore Morphology*

Before the 1970s the standard technique for measuring the beach and shallow nearshore profile made use of a level and stadia rod along profiles set up normal to the shore and surveyed out to the limit of wading (Evans, 1940; King and Williams, 1949; Davis *et al.*, 1972). Since the 1960s surveying in deep-water has been carried out with echo sounders in a boat small enough to be able to permit an overlap of the surveys. Morphological change along a profile was measured through repeat surveys along closely spaced profiles and the data could be aggregated to produce a contour map (Davis and Fox, 1972; Greenwood and Davidson-Arnott, 1975; see Figure 8.4).

Today, complete mapping into quite shallow water can be carried out with multibeam echo sounding (Gorman *et al.*, 1998), and most surveys are completed using a total station or a differential GPS (DGPS) with a vertical accuracy of ±2–3 cm. The outputs from these instruments are readily incorporated into a wide range of contouring and geographic information system (GIS) software packages, which can produce digital elevation models and permit easy extraction of volume change through repetitive surveys.

A non-trivial task is that of obtaining surveys of profile change during high-energy conditions, and most morphological change has been determined from surveys carried out before and after an event. Profiles may be surveyed under moderate wave conditions using some form of buggy (e.g., the CRAB used at the field research facility at Duck, North Carolina; Plant *et al.*, 1999) or a towed sled equipped with a GPS or total station in addition to instrumentation for measuring waves currents and suspended sediment (van Maanen *et al.*, 2008). However, the sleds and buggies are only operational up to moderate wave conditions and cannot be used during storm conditions. During high wave conditions measurements can be obtained from piers such as the field research facility at Duck or from temporary structures, but these are major logistical undertakings at a few restricted sites.

Video technology has been applied for more than two decades to measure waves and swash run-up (see below) but it has also been applied to measurement of the position of nearshore bars through time exposure of wave breaking (Holman and Lippman, 1986; Konicki and Holman, 2000; Ruessink *et al.*, 2000; Holman and Stanley, 2007). This has been done at several locations worldwide using the Argus video monitoring system (Holman and Lippman, 1986), where ten-

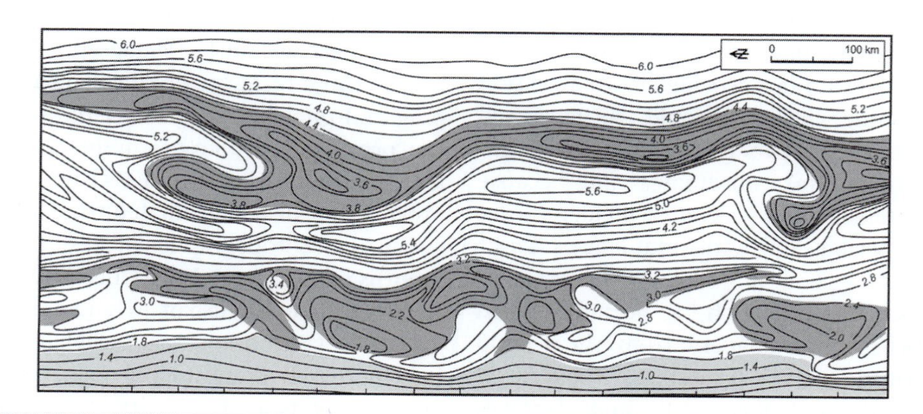

Figure 8.4 Contour map of the beach and nearshore zone associated with a crescentic two bar system in Kouchibouguac Bay, New Brunswick produced by surveying along lines at 30 m intervals and echo sounding along every second line from a small boat (from Greenwood and Davidson-Arnott, 1975).

Figure 8.5 Photograph of the nearshore bar and trough system at Greenwich Dunes, Prince Edward Island obtained using a 30 second exposure from a digital SLR camera mounted on a tower 12 m above the beach. The photograph was taken at night and the brightness has been enhanced in post processing.

minute video records are averaged to produce a single image with the location of bars and the beach face outlined by a bright band associated with the location of wave breaking and the troughs and rip channels being relatively dark. This can also be done using a digital SLR camera and taking a series of images and processing them in the same way that the Argus video scenes are processed. Alternatively, the shutter can be opened for tens of seconds to produce a single averaged image (Shand *et al.*, 2001; see Figure 8.5). Continuous monitoring over months and years provides a data set of 2D bar form and position which has greatly enhanced our knowledge of bar dynamics on this time scale. However, the images do not provide actual elevation data and some caution should be exercised because of the error bars associated with rectification to horizontal coordinates and because breaking intensity will vary with incident wave height and period, and with tidal stage (Morris *et al.*, 2001).

A new satellite approach has been developed using multi-spectral imagery from the Digital Globe WorldView3 (WV3) satellite. WorldView3 was launched in August 2014 and captures imagery every 4.5 days (or less, in some locations). These images include eight multispectral bands, with a root mean square error (RMSE) for ground position < 1 m. The yellow and coastal blue bands can be used to develop bathymetric maps at depths up to 20 m deep with high spatial accuracy (Deidda and Sanna, 2012; Miecznik and Grabowksa, 2012). Trimble (2017) used this approach to map nearshore bathymetry and identify the location and orientation of rip channels at Bondi Beach, Australia to demonstrate the utility of these images for monitoring the evolution of nearshore morphology. While the satellite provides a high spatial and spectral resolution, it has a 1-day revisit period and it cannot be used at night or when the scene is covered by clouds.

8.2.1.2 *Beach and Dune Morphology*

Mapping shoreline changes and changes in the morphology of the beach and foredune area can be completed using digital images from still and video cameras to produce digital elevation models (DEMs) through a variety of computer software packages. The technique makes use of overlapping pairs of photographs produced either in the traditional way through the movement of the camera installed in a plane, helicopter or a land-based vehicle, or using images taken from two fixed positions. In the case of aerial photography or moving vehicles the position of each digital image can be linked to real-time positional data provided by DGPS. Where fixed cameras are used on the beach control points whose position and elevation have been surveyed precisely are used to aid in rectification (Hancock and Willgoose, 2001; Delgado-Fernandez *et al.*, 2009). The advantage of these automated mapping and photogrammetric systems is that they can provide a very large number of data points for construction of the DEMs and much of the processing can be automated, thus allowing the evolution of topography over days, weeks or months to be captured (see Ollerhead *et al.*, 2013).

Since 2000, the use of airborne LiDAR to map extensive sections of beach and dune systems has increased dramatically (e.g., Zhang *et al.*, 2005; Robertson *et al.*, 2007), particularly to monitor the response and recovery of barrier islands to hurricanes and tropical storms (Houser *et al.*, 2008). It is ideal for such analyses because it provides spatially dense and accurate topographic data over a large area (Sallenger *et al.*, 2004; Zhang *et al.*, 2005). Data are collected with aircraft-mounted lasers capable of recording elevation measurements with a precision of < 0.10 m and some surveys can extend over 100 km alongshore (Wernette *et al.*, 2018). The morphology of the beach and dune can be simultaneously measured if a near-infrared laser (for ground topography) is combined with a green laser that is able to penetrate through the shallow water of the nearshore zone (Robertson *et al.*, 2007).

In recent years, unmanned aerial vehicles (UAVs), otherwise known as drones, have become increasingly popular to measure beach and dune morphology and change through a technique called structure from motion (Turner *et al.*, 2016; Moloney *et al.*, 2018). Photographs are taken of the ground surface along overlapping flight lines, and the images are combined in specialised software that relates the points in a space based on their location relative to known ground-control points. While this technique cannot be used to survey topography over as large an area as LiDAR, it can be used more frequently and is significantly less costly. The vertical precision of UAV surveys is ~0.10 m but is dependent on the accuracy of the ground control points.

8.2.1.3 *Bed Elevation Measurements*

Simple measurements of change in bed elevation at a point, and the total depth of activation can be made with rods emplaced along a profile or on a grid and which are measured before and after a storm (Greenwood *et al.*, 1979). The maximum scour depth can be resolved by placing a washer on the sand surface and then measuring the depth of burial following the storm. Results from a grid of these can be used to measure volume change in the nearshore (Greenwood and Mittler, 1984). It is possible to measure these by scuba diving during periods of moderate wave activity and tall rods placed along a line in the intertidal and shallow subtidal have been monitored through a storm (Aagaard *et al.*, 1998). Automated devices, which act in a similar fashion, have been developed using a vertical array of photo cells spaced at a small increment (Lawler, 1992), or the difference in conductivity between sediments and sea water to distinguish the bed level (Ridd, 1992; Lanckriet *et al.*, 2013). These devices both have some problems in distinguishing the bed during periods of high sediment transport and suspension because the sand–water interface becomes quite blurred. The value of these instruments is that they are relatively low cost and therefore provide the potential for deployment of sufficient sensors to give reasonable spatial coverage across the surf zone, and to provide measurements throughout a storm. Attempts have been made to use bottom mounted sonars, either fixed or mounted on a frame to permit survey of a traverse thus permitting determination of two-dimensional bedform properties and migration rates (e.g. Greenwood *et al.*, 1993).

Recently, swash zone measurements have been made using ultrasonic distance sensors, which can have vertical resolution on the same order of magnitude as the sediment (see Turner *et al.*, 2008). The distance between the sensor and either the sediment or water surface can be measured at frequencies up to 25 Hz, allowing for fine-scale and rapid measurement of uprush and backwash, percolation and the change in bed elevation following the passage of a swash bore. Houser and Barrett (2010) deployed a cross-shore transect of ultrasonic sensors over six weeks and found that erosion and deposition in the swash zone is dependent on the transformation of the incident waves by the innermost nearshore bar. An example output from their study is provided in Figure 8.6, showing the passage of individual swash bores and the resulting change in the elevation of the sediment surface.

8.2.2 Beach Sediments

The dynamics of sediment erosion, transport and deposition by waves, especially on the swash slope is greatly influenced by the density and shape of the particles making up the beach. Fine grains of heavy minerals such as magnetite are hard to erode because of their high density and they also tend to get buried below the lighter, coarser quartz grains. Irregular fragments of

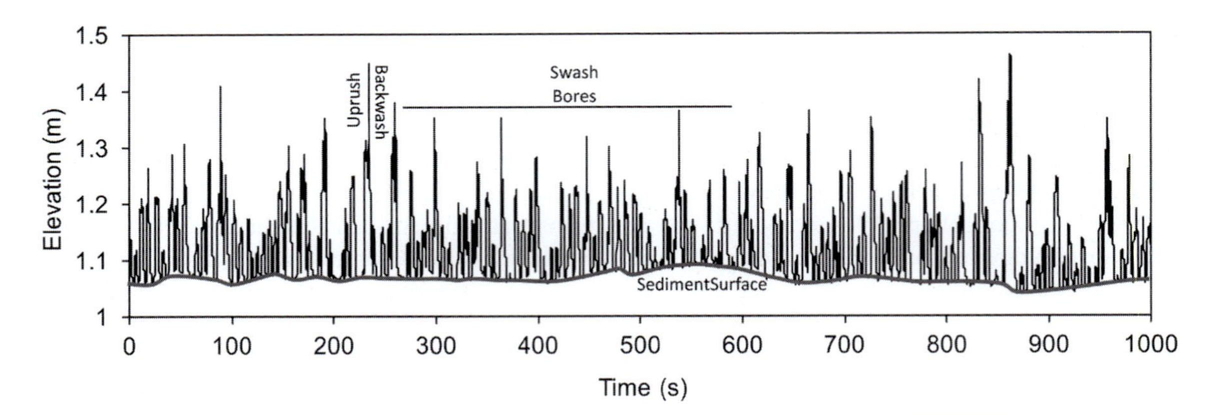

Figure 8.6 Swash bores and bed elevation in the swash zone of Pensacola Beach, Florida showing periods of erosion and deposition based on the transformation of incident waves over the innermost bar (after Houser and Barrett, 2010).

shell tend to have a much lower settling velocity than equivalent size spherical quartz grains while fragments of coral and oolites have a much lower density than the equivalent quartz spheres and are thus easier to transport by both water and wind. We can therefore characterise beaches into three types based on sediment size: (1) sand beaches (mean sediment size ranging from about 0.1–4 mm); (2) gravel beaches composed of pebbles and/or cobbles (mean grain size about 0.5–25 cm; and (3) mixed sediment beaches with a mixture of sand and pebble, or sand and cobble sediment (Figure 8.7). Two forms of mixed sediment beach can be identified (Jennings and Shulmeister, 2002), one where sediment is uniformly mixed over the profile and one where there is a distinct sorting with coarse material forming a steep upper beach and sand forming a gently sloping lower beach (Figure 8.7D). The composition and size of the beach sediments primarily reflects the source material and secondarily the processes of sorting, abrasion by wave action and chemical weathering. Since quartz crystals are highly resistant to weathering and abrasion, a large proportion of sediment transported to the beach from rivers or cliff erosion consist of quartz with secondary amounts of feldspars and perhaps some minor heavy minerals such as magnetite and garnet. In volcanic areas, black sand beaches are common, made up of fragments of andesite or basalt, or crystals of augite and hornblende. In tropical regions carbonate sediments are supplied from the abrasion of corals, coralline algae and shells (Figure 8.7A), and can be the dominant material in low-energy environments (Figure 8.7E).

It has long been recognised that particle size is an important control on the slope of the subaerial beach (Bascom, 1951; McLean and Kirk, 1969), because infiltration reduces the volume of water in the backwash (Figure 8.8). The rate of infiltration generally increases with increasing grain size, and the volume loss due to infiltration means that not all sediment brought landward by the swash uprush is returned by the succeeding backwash. The resulting sediment build-up on the upper foreshore leads to a steepening of the slope until the enhanced effect of gravity on the backwash velocity compensates for the loss of volume due to infiltration, producing an equilibrium slope (Turner and Masselink, 2009). On very fine sand beaches (0.1–0.2 mm) the slope is always gentle ($\tan\beta$ 0.03–0.005) because infiltration is low and berm development is consequently limited. At the other end of the size scale, cobble beaches always have high rates of infiltration and a steep slope ($\tan\beta$ 0.1–0.25). The slope for mixed sand and gravel beaches is quite variable, both spatially and temporally at one beach because

Figure 8.7 Photographs of beach types: A. gently sloping beach in fine carbonate sands, Sandbanks Beach, St Kitts; B. medium to coarse sandy quartz and feldspar beach with berm and steep swash slope, South Frigate Beach, St Kitts; C. cobble beach in andesite with steep swash slope, Great Salt Pond, St Kitts; D. mixed igneous and metamorphic cobbles and quartz sand beach, west coast of Ireland; E. small storm beach ridges in shells at Sabancuy on the Yucatan coast of Mexico (photo courtesy Patrick Hesp).

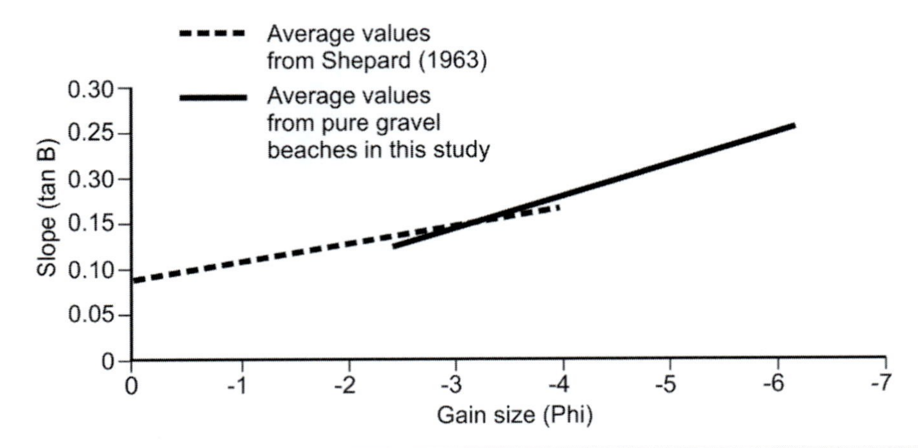

Figure 8.8 The relationship of beach foreshore slope to particle size for pure gravel beaches in New Zealand (Jennings and Shulmeister, 2002). The dashed line is the average value for sand and fine gravel beaches based on Shepard (1963).

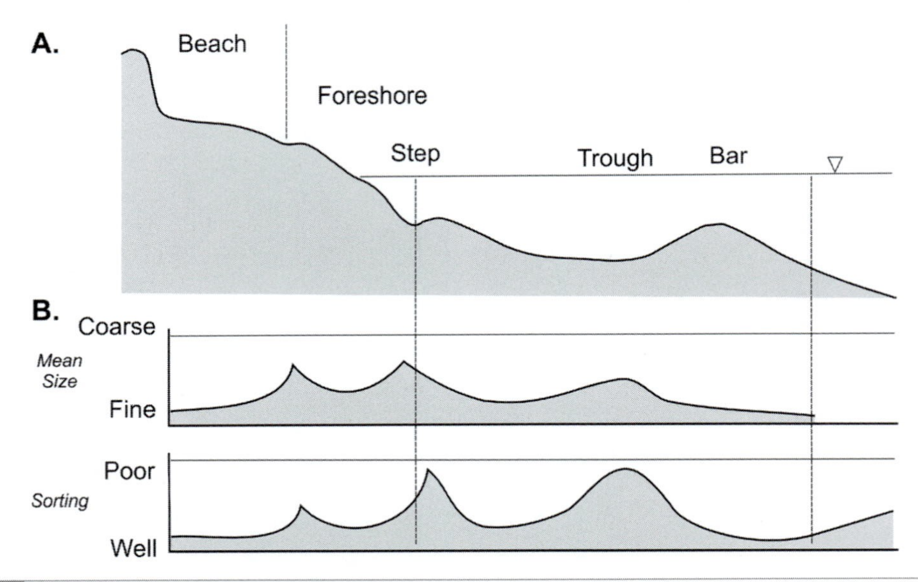

Figure 8.9 Schematic variations in sediment size and sorting on a sandy beach with one nearshore bar: A. shore-normal profile and zones; B. variations in mean grain size and sorting. The patterns are highly generalised and the *y* axis scales are arbitrary.

the finer sediments may infill the pore spaces of the larger particles, reducing infiltration and the beach gradient (McLean and Kirk, 1969). The slope of medium to coarse sand beaches does increase with increasing size but is also highly variable temporally because of changing wave and tide conditions. Consequently, modelling must consider not only grain size, but also the changes in wave and tidal forcing over time scales ranging from hours to days.

The nearshore slope on sandy beaches is comparatively gentle, and there is considerable exchange of sediment between the beach and the nearshore. Wave asymmetry and the sorting process associated with ripples tend to move coarse sediments landward towards the beach step, while finer sediments are preferentially transported offshore above the bed by the undertow or rip currents (Horn, 1992). Thus, seaward of the breaker zone grain size tends to decrease, and there is consequently a gradual decrease in the bed slope (Figure 8.9). Sorting becomes poorer towards the seaward end of the profile as the fine sediments that settle out of suspension after a storm are incorporated and, in some cases, mean sediment size increases as a result of winnowing of fines to leave a coarse lag (Liu and Zarillo, 1989). In the breaker and surf zones, grain size variations reflect the underlying bathymetry and the spatial distribution of the nearshore bars (Greenwood and Davidson-Arnott, 1972; Wang and Davis, 1998). Sediments on the crest and seaward slope of bars are generally well sorted while sediments at the base of the trough are more poorly sorted because of both the accumulation of coarse material washed over the bar and fine sediments settling out of suspension during low wave conditions (Figure 8.9). Relatively coarse sediments are also found in rip channels, where there is considerable bedload transport (Sherman *et al.*, 1993a; Thorpe *et al.*, 2014). Where the foreshore is quite steep there is a usually a distinct step present at the base characterised by the accumulation of coarse, poorly sorted sediment deposited by a vortex generated as the backwash interacts with the next wave (Larson and Sunamura, 1993; Houser, 2013). Landward of the step, mean grain size decreases and the sorting improves. Finally, on the backshore, surficial sediments are often aeolian in origin and tend to be finer and better sorted than those on the foreshore (Figure 8.9).

On most cobble beaches the profile remains steep beyond the step, and wave breaking is largely confined to the foreshore or to a point just seaward of the step. This, and the practical difficulty (and danger) of sampling cobbles in the nearshore, means that most information on particle size variation comes from the foreshore. The shape and size of the cobbles greatly influences transport and deposition, and the result is that quite distinct zones are often present (Figure 8.10). Compact and rod-shaped cobbles roll easily and tend to move up-slope with the swash and roll seaward towards the step with the backwash. They therefore, tend to be over represented near the lower part of the profile. Discs can be flipped up into the wave and present a large surface area so are often carried right to the top of the swash and thus up to and over the berm crest. Once they come to rest, they present a large surface area and thus friction tends to hold them in place leading to large concentrations of disc-shaped cobbles at the top of the profile (Bluck, 1967; Orford, 1975; Sherman *et al.*, 1993b). The spatial and temporal distribution of cobble shape and size is complex due to rapid reworking through the neap/spring tidal cycle, varying wave energy levels and alongshore variations associated with the development of rhythmic beach cusps (Sherman *et al.*, 1993b). A thorough review of cobble beach morphodynamics is presented in Orford and Anthony (2013).

8.2.3 Sandy Beach Profile

8.2.3.1 *Sweep Zone*

Repeated surveys along a profile over months and years on sandy beaches shows that there is considerable variability in shallow water and on the beach foreshore, and that the extent of the vertical change decreases into deep water to the point of closure where changes in bed elevation become imperceptible or extremely rare (Hallermeier, 1981). The envelope between the maximum bed elevation and the minimum bed elevation is termed the sweep zone, and it represents the extent of reworking by wave action and migration of topographic forms, but it does not account for a small amount of additional reworking due to bedform migration (Figure 8.11). The depth of closure conceptually marks the offshore limit of the zone of sediment circulation between the nearshore and the beach systems, although sand movement and bedform generation can occur at greater depths in response to storm waves and currents. The largest sweep zones are found on barred beaches where onshore and offshore migration of the bar and trough system lead to large vertical changes over the profile in the inner shoreface. The size and variability of the beach and nearshore morphology is an important control on the burial and emergence of previously buried oil (see Box 8.1). It is also important to the burial and emergence of archaeological sites, including shipwrecks (McNinch *et al.*, 2006).

Figure 8.10 Gravel beach sediments: A. gravel beach in mixed metamorphic rocks, Hirtle Beach, Nova Scotia; B. disc-shaped cobbles on the crest of a berm ridge; C. compact pebbles in the infill zone near the bottom of the swash slope; D. imbricate pebbles and cobbles below the berm crest; E. gravel beach in rounded pebbles and small cobbles on the east coast of Scotland. The largest cobbles are often found on the crest of the highest berm ridge, but particle size variation over the profile is complicated by the presence of smaller berms formed under varying wave and tidal conditions.

The depth of closure clearly reflects, in some way, the largest storm waves, and it is possible to predict the depth of closure using a simple empirical equation such as that provided by Hallermeier (1981):

$$h_c \approx 2H_{sig} + 11\sigma, \tag{8.1}$$

where h_c is the depth of closure, H_{sig} is the mean annual significant wave height, and σ is the standard deviation of H_{sig}. Equation 8.1 is a

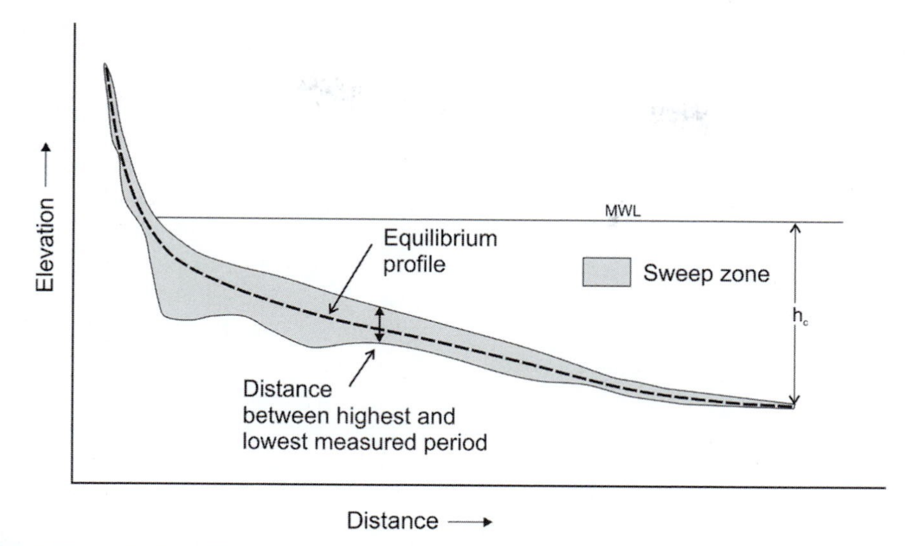

Figure 8.11 Schematic representation of the sweep zone and depth of closure for a typical barred profile in medium sand.

simplified version of an alternative equation that includes both the height and period of waves that was exceeded for only 12 hours in any one year. Since these may not be readily available, the equation does not account for sediment size, and since there is likely considerable uncertainty in the actual definition of closure depth, Equation 8.1 only provides a reasonable estimate. However, evaluation of 10 years of profile data from Duck, North Carolina gives support to the basic relationship proposed by Hallermeier (Nicholls *et al.*, 1998).

Box 8.1 Coastal Management Application

Beach and Nearshore Envelope: Implications for Oil Spills

Understanding the variability of the beach and nearshore envelope is key to determining the burial and residence time of oil spills. There are numerous oilrigs, pipelines and shipping channels along the world's coastline meaning that it is quite possible for an oil spill to contaminate beaches, degrade marine ecosystems, and/or negatively impact local economies that depend on fisheries and tourism. For example, ~257 000 barrels of oil spilled into Prince William Sound (PWS), Alaska when the Exxon Valdez ran aground in 1989. The Valdez spill occurred during strong winds and a spring tide, allowing oil to be trapped along the upper beach and seep deep into the sediment (Carls *et al.*, 2001), where it remained trapped for many years. Maximum oil concentrations were observed throughout Prince William Sound 3–6 years after the spill, but 25 years after the spill oil continues to be remobilised within PWS by storms capable of eroding the beach down to the spill layer (Xia and Boufadel, 2011). More recently, the 2010 Deepwater Horizon accident resulted in the spill of 5 million barrels of oil into the Gulf of Mexico (Boufadel *et al.*, 2014). The spill occurred in April as the nearshore and beach profile was recovering from winter storms, trapping the oil mats and tar balls at depth within the sediment. As of 2018, oil is still being remobilised along the beaches of Louisiana and northwest Florida, and remediation efforts continue at a significant cost to local agencies and at a cost to the local economy

The point of deposition on the beach is controlled by the physio-chemical composition at the time that it reaches the beach. Oil that is less dense than seawater floats on the surface and is deposited on surface sediment

Box 8.1 (cont.)

with the swash uprush, while oil that is heavier than seawater exhibits varying degrees of mixing with the sediment. Oil that is lighter than the sediment is deposited on the bottom and moves along the bed in a manner analogous to fine sand, with the sediment typically deposited at high tide. Oil from the Deepwater Horizon oil spill that made it to the shallow nearshore zone was subject to enhanced dispersion by the breaking of waves by breaking down and mixing oil parcels throughout the shallow water column (Zuijdgeest and Huettel, 2012). This natural dispersion was enhanced by the application of dispersants leading to the relatively rapid advection of the oil into permeable sands where it is buried and can persist for several years. The longer oil remains on the beach, the more it changes chemically and physically in a way that promotes burial, and the more variable that the beach and nearshore profile. In this respect, the sequence of wave activity and beach evolution that occurs over the course of an oil spill and the subsequent evolution of both oil and the beach leads to stratified layers of oil, oiled sands and clean sands.

The potential for burial depends on the beach and nearshore state at the time that the oil reaches the beach. If an oil spill happens during a low-profile state such as a 'storm' or 'winter' beach when the sediment is temporarily stored in the innermost bar(s), oil can be trapped as the beach recovers and accretes through the landward migration of the bar (Wright and Short, 1984). The oil would not be exhumed until the next storm capable of eroding the beach down to the level at which the oil was deposited or to the level at which the oil has been advected through the sediment. Exhumed oil is not necessarily removed from the beach; the oil can be moved further landward with the swash or elevated storm surge, seaward with the backwash and undertow, redeposited at the same location or buried to a greater depth that is dependent on the depth of disturbance of the swash or waves (Gonzalez *et al.*, 2009). In contrast, oil that is deposited after the beach has completely recovered will remain at or near the surface meaning that it has a relatively short residence time (at the surface) but is also easier to clean (Pontes *et al.*, 2013). However, if the oil reaches a beach and is not immediately cleaned it can end up within the backshore and dune or buried several meters (Gonzalez *et al.*, 2009), and the more it changes chemically and physically in a way that promotes burial and longer residence times.

8.2.3.2 *Equilibrium Profile Form*

The mass of sand is small compared to the forces exerted by waves, especially by the large waves that accompany storms. Observations and measurements from the field suggests that sand in the nearshore is transported as bedload and suspended load that drive the cross-shore migration of ripples and sand bars. The persistence of sand on the beach and in the nearshore, and the existence of characteristic profiles suggest that there is a dynamic equilibrium between the forces controlling sand movement, and that it should be possible to predict profile form and changes from some relatively simple combination of parameters. Much effort has gone into elucidating the nature of the controls on profile equilibrium, and on developing and justifying predictive relationships. It is useful to identify what are likely some of the important controls, but given the complexity and variability of coasts worldwide, it seems overly optimistic that their behaviour can be predicted by some simple combination of wave steepness and sediment size!

Sand size decreases offshore from the surf zone. If we ignore for the moment the small form variation due to the presence of bars and troughs in the inner nearshore on some sandy beaches, the general profile form of a sandy beach and nearshore tends to be exponential (see Figures 8.11 and 8.12). It is likely that this is related to the increase in bed shear velocity and wave asymmetry as waves shoal across the nearshore and the resulting decrease in sand size with depth and distance offshore. There have been several approaches to deriving a theoretical profile shape related to the dynamics of waves and sediment transport (e.g., Bowen, 1980) or to energy dissipation (Bruun, 1954). Dean, in various publications over several years (e.g., Dean, 1991; Dean *et al.*, 1993) promoted an equilibrium profile of the form:

$$h = Ax^m, \tag{8.2}$$

where h is the depth at an offshore distance x, A is an empirical coefficient, and the exponent m is usually assigned a value of 2/3. The coefficient A is a function of grain size and Dean (1997) derived a simple empirical relationship based on plotting A against grain size for several profiles:

$$A = 0.067w_s^{0.44} \tag{8.3}$$

where w_s is the settling velocity equivalent to the mean grain size of the sediment. An alternative approach uses an exponential model of the form (Komar and McDougal, 1994):

$$h = \frac{S_0}{k}\left(1 - e^{-kx}\right), \tag{8.4}$$

where S_0 is the slope at the shoreline, and k is an empirical constant that can be defined by evaluating the equation at a known depth and distance offshore.

While the Dean equilibrium profile is used extensively in the engineering literature, particularly as input to predicting the form of nourished beach profiles, there has been much discussion and criticism of the concept (Pilkey *et al.*, 1993; Cowell *et al.*, 1999; Are and Reimnitz, 2008; Aagaard and Hughes, 2017). It seems simplistic to use a single value for grain size, when many profiles show changes in grain size with depth. Moreover, the equilibrium profile does not directly account for the presence of nearshore bars and troughs, or the general flattening of the profile in the surf zone before the steep section of the foreshore that is characteristic of non-barred profiles. This is obviously a weakness for its application on many coasts, particularly since beach nourishment is generally applied to the shallow part of the profile, and it does not account for the tendency of profiles to evolve towards an equilibrium morphology. Finally, the equilibrium profile models assume that the profiles are entirely covered in sand and do not account for shape of the underlying bedrock or erosional surface that keep the profile out of equilibrium (Jackson *et al.*, 2005). Inman *et al.* (1993) suggested a compound profile and Stive and de Vriend (1995) modelled shoreface profile evolution based on three simple hinged panels that account for the response and evolution of the upper, middle and lower shoreface over different time scales. As a result, the entire profile does not need to follow the same equilibrium shape at any point in time.

8.2.4 Barred and Non-Barred Profiles

An examination of profiles worldwide suggests four basic profile types: (1) a smooth planar to curvilinear profile without any bars in the intertidal or nearshore; (2) a profile with one or more bars in the nearshore that exist year round; (3) a profile where bars are present some or most of the time but where the bar may disappear, usually through landward migration and 'welding' to the beach; and (4) profiles with one or more permanent bars in the intertidal zone and usually one or more in the subtidal zone as well (Figure 8.12). When bars are present, they tend to be areas of preferred wave breaking due to the shallow water over the crest, with breaking being reduced or ceasing altogether in the deeper water of the intervening troughs. Where bars are absent there is no preferential location of breaking and the width of the surf zone is a function of incident wave height and period.

Bars are usually absent on sandy beaches when some combination of steep slope (bedrock controlled) and/or small wave height (e.g., limited fetch areas within island archipelagos or in estuaries) causes the waves to break only at the shoreline. However, if the profile is only sand, and waves are on occasion large enough to break some distance offshore, the presence or absence of bars in the nearshore tends to be controlled by wave climatology and sediment size. In other words, the presence or absence of bars is a response to the dynamic controls on sediment transport. Bars are absent from low-latitude coasts dominated by swell waves only,

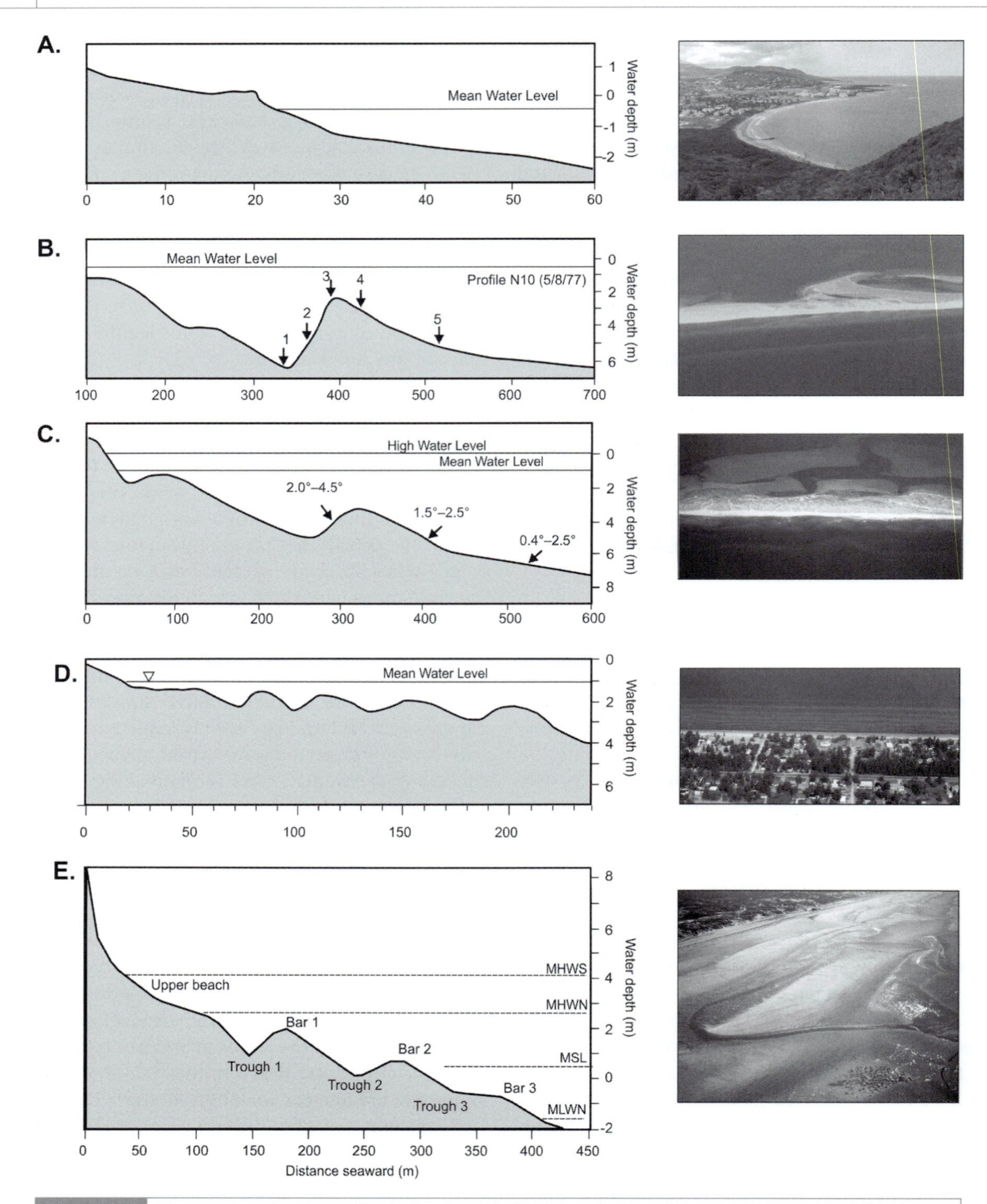

Figure 8.12 Photographs and typical profiles for beaches with differing profile form: A. non-barred beach, North Frigate Bay, windward coast St Kitts, West Indies; B. straight outer bar, Miramachi Bay, New Brunswick (Greenwood and Mittler, 1984); C. crescentic two-bar system at Kouchibouguac Bay, New Brunswick, Canada (Greenwood and Davidson-Arnott, 1975); D. multiple bars in the subtidal zone, Wasaga Beach, Georgian Bay (Davidson-Arnott and Pember, 1980); E. multiple bars in the intertidal and subtidal zone a macrotidal coast, Wisant Bay, NW France (Sedrati and Anthony, 2007).

particularly where wave refraction leads to divergence, producing low steepness waves that break at or close to the beach. Sediment transport on these coasts tends to be directed landward, building a relatively steep, reflective profile that exhibits relatively little dynamic change. Bars are also absent where relatively light winds blow over long distances such as the trade winds through the Caribbean, unless the grain size is locally small (Figure 8.12A).

In mid-latitude fetch-limited areas, such as the Great Lakes, Gulf of St Lawrence, Gulf of Mexico and the Mediterranean and Baltic Seas, bars are present on almost all sandy coasts and are a constant feature of the profile (Figure 8.12B). Here most periods of high waves and strong sediment transport are associated with the passage of mid-latitude cyclones that produce short, steep waves and are accompanied by strong onshore winds and storm surge. On the east coast of the USA bars are usually present because prevailing (westerly) winds are offshore and much of the time wave action is associated with storms generating local waves. However, on the west coast of North America (e.g., California and Oregon) and the New South Wales coast of Australia, bar generation takes place during storms with onshore migration occurring during non-storm periods driven by Pacific swell waves, and if there is a sufficiently long interval between storms the bars will weld to the beach producing a planar profile (Figure 8.12C). On the California coast the alternation of storm and non-storm waves over the year leads to a barred profile in winter and a non-barred profile in summer (Shepard, 1950a; Figure 8.13). The resulting onshore and offshore

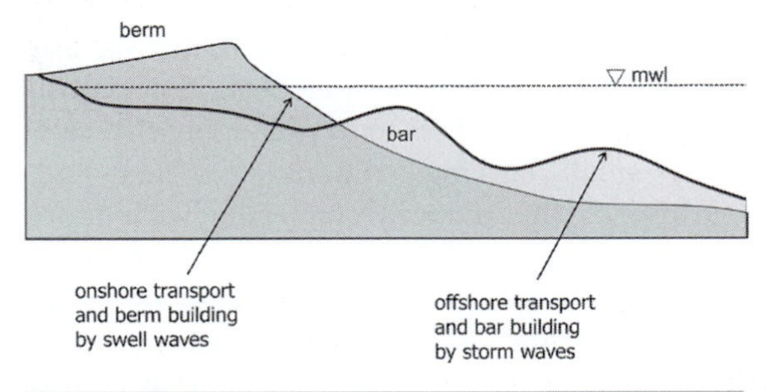

Figure 8.13 Idealised barred and non-barred profiles related to seasonal wave climate.

migration of the bar and changes in the profile shape are incorporated in the beach model developed by Short (1979) and developed further by Wright and Short (1984). These models will be described later in Section 8.3.3.

Several simple parameters have been used to discriminate between barred and non-barred states. For example, Gourlay (1968) introduced the dimensionless fall velocity parameter Ω to distinguish between barred beaches ($\Omega > 1$) and non-barred beaches ($\Omega < 1$):

$$\Omega = \frac{H_b}{w_s T}, \tag{8.5}$$

where w_s is the settling velocity of the sediment. This was used extensively by Wright and Short (1984) in Australia to distinguish different profile states from non-barred reflective beaches to fully dissipative beaches with multiple nearshore bars ($\Omega > 6$) (see Figure 8.1). The Ω parameter relates profile form to the wave steepness and profile slope (based on the known relationship between sediment size and beach slope), but physically, this parameter is the ratio of how high sediment is lifted above the bed and how far it settles back to the bed over the time it takes a wave to pass. Essentially, the larger the wave height (H_b), the higher that sediment is lifted above the bed, and the larger the grain size the faster it settles back to the bed. If sand is lifted higher under the wave crest than it can fall back to the bed under the wave trough ($H_b > w_s T$), then it tends to move offshore with the offshore flows of the wave trough and undertow to create a relatively flat dissipative profile. If sand can settle back to the bed during the passage of a single wave ($H_b < w_s T$) then sediment tends to move landward with the stronger forward flow under the crest.

Models that relate nearshore to sediment size are termed surf similarity parameters. Other surf similarity parameters that have been developed from wave tank studies including one by Dalrymple (1992):

$$P = \frac{gH_0^2}{w_s^3 T}, \tag{8.6}$$

where bars develop for $P > 9000$, and non-barred profiles occur when $P < 9000$. Alternatively, Short and Aagaard (1993) reviewed the literature on multi-barred beaches and introduced a bar parameter B_*:

$$B_* = \frac{x_s}{gT^2 \tan\beta}, \tag{8.7}$$

where x_s is the offshore distance corresponding to the depth at which the gradient becomes very small. No bars occur when $B_* < 20$ and an increasing number of bars occur for values $B_* > 20$. This parameter is based on the premise that standing infragravity waves are responsible for the formation of bars and works if waves are large enough to break offshore. However, there is mounting evidence that infragravity waves are not responsible for the formation of bars but are rather forced by the bars (Bryan *et al.*, 1998). Since the number of bars is typically associated with nearshore slope, the explanation based on infragravity waves may simply be fortuitous.

Profiles from beaches with a restricted fetch, and where bars are present all year round, offer the best opportunity to examine relationships between bar properties and controls. Some of the earliest papers on bars established that the height and spacing of bars tend to increase offshore as does the depth of water over the bar crest (Evans, 1940; King and Williams, 1949). On occasion the outermost bar is lower than the next one landward and this is likely because it is no longer active except during the largest storms. Profiles from 12 sites in the Great Lakes and Gulf of St Lawrence show the general trend related to bar height and spacing (Davidson-Arnott, 1987; Figure 8.14A). This study also found that the number of bars was highly correlated with the nearshore slope and that there was a much weaker tendency for bar number to increase with increasing fetch length for the same nearshore slope (Figure 8.14B). This is at least in part related to the distance offshore at which initial wave breaking occurs, which in turn controls the width of the zone available for wave reformation before breaking is initiated on the succeeding bar.

The height of the bars and depth of water over the crest also increases seaward and is roughly proportional to the height of the breaking waves associated with them (Pruszak *et al.*, 1997). Observations suggest that during large storms bars migrate offshore until they are in an equilibrium depth associated with the height of the breaking waves (King and Williams, 1949; Greenwood and Davidson-Arnott, 1979; Sallenger *et al.*, 1985). In other words, the bars tend to migrate towards the point of wave breaking. The position of the outermost bar is dependent on the height of the storm waves, while the position of the inner bars is dependent on whether wave breaking occurs on the next bar offshore (Houser and Greenwood, 2005). Specifically, the inner bars can only migrate offshore to a depth at which wave breaking jumps to the next bar offshore.

8.2.5 Beach Plan View

In the alongshore direction (otherwise known as plan view), sand and gravel beaches can be convex, straight or concave. In high relief areas, concave beaches tend to be found within embayments between headlands or anchored downdrift of a headland or promontory (Figure 8.15). The concave form of headland-bay beaches (Figure 8.15A) can therefore be explained in terms of wave refraction processes and the operation of the ABC model outlined in Section 7.5.3. Along the sides of the headland, waves arrive at an angle to the shoreline and any sediment supplied from cliff erosion is transported along the sides of the headland towards the head of the bay, forming a thin, drift aligned beach. At the head of the bay, refracted wave crests become more and more aligned parallel to the shoreline, thus reducing the transport gradient and leading to greater deposition and a deeper, wider beach. In the centre of the bay refraction is minimal and the beach is aligned parallel to the wave crests (i.e., swash aligned; Figure 8.15A). As the spacing between the headlands increases, the central section of the beach becomes straight with relatively small curving sections at either end. Because of the reduced sheltering with more widely spaced headlands, waves can arrive at the shoreline from an increased range of angles and the orientation of the beach tends to shift from one side to the other in response to this (Figure 8.15A). The beach may then rotate from

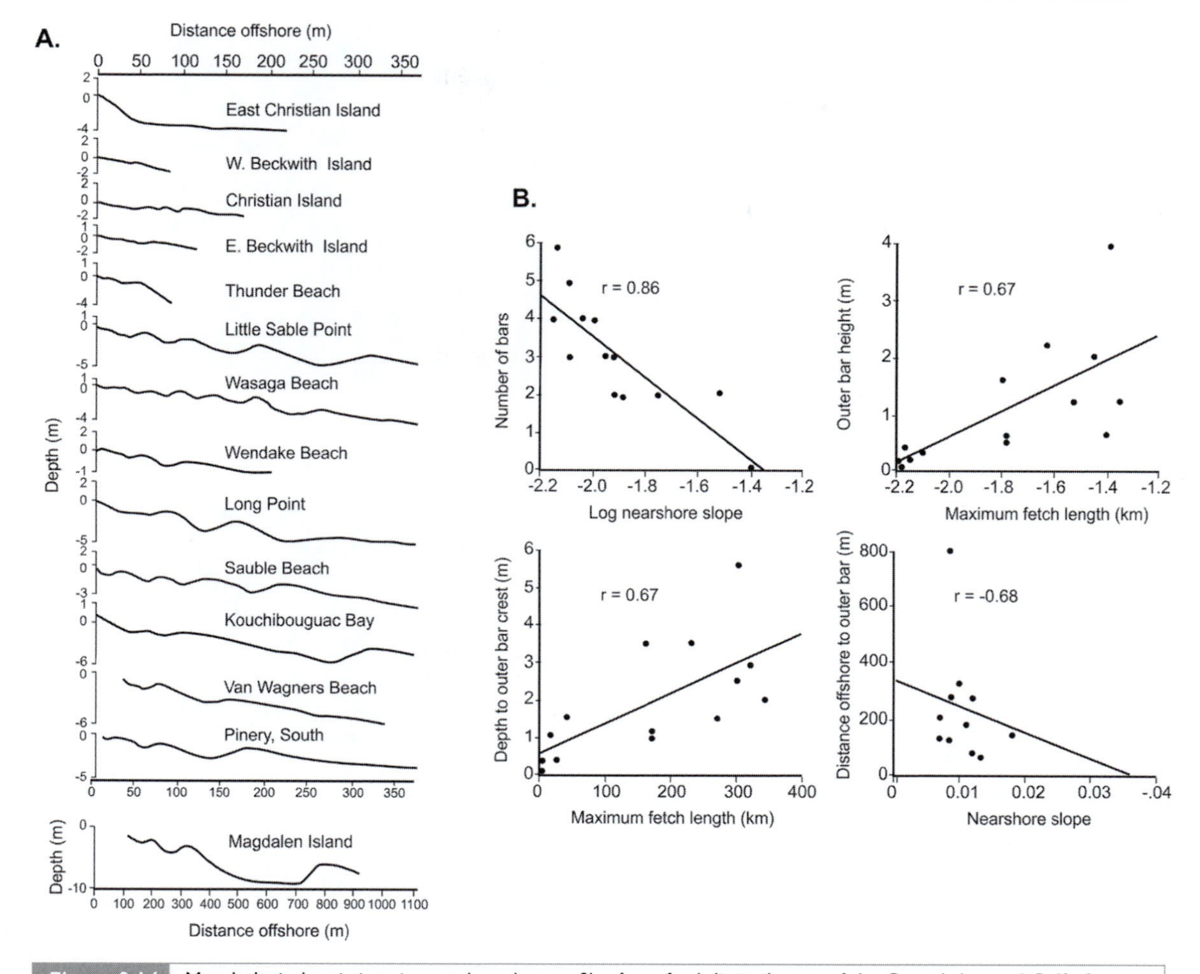

Figure 8.14 Morphological variations in nearshore bar profiles from fetch-limited areas of the Great Lakes and Gulf of St Lawrence: A. profile form; B. plots of regressions between morphological properties of the bars shown in A., and various controlling parameters (from Davidson-Arnott, 1988).

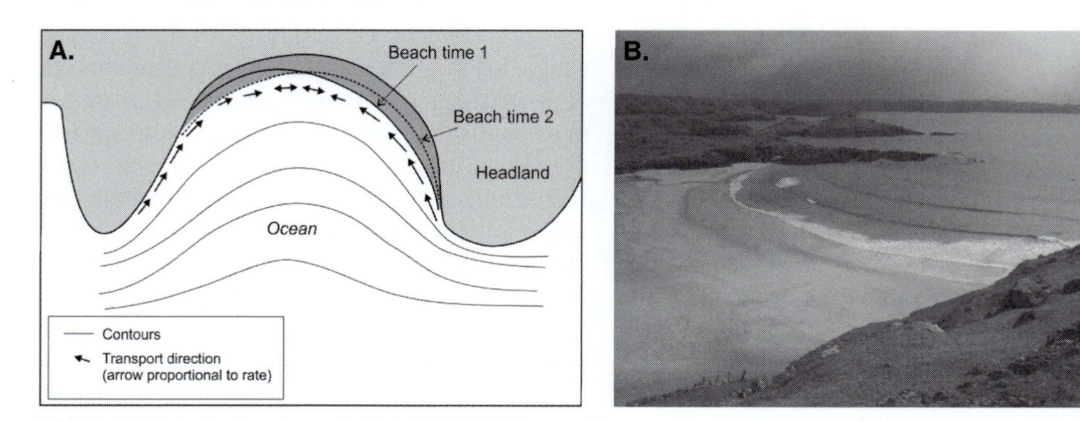

Figure 8.15 Headland–bay beaches: A. patterns of sand movement and beach alignment on an ideal headland–bay beach and rotation of the beach under differing wave direction; B. headland–bay beach, Oldshore Beg, NW Scotland.

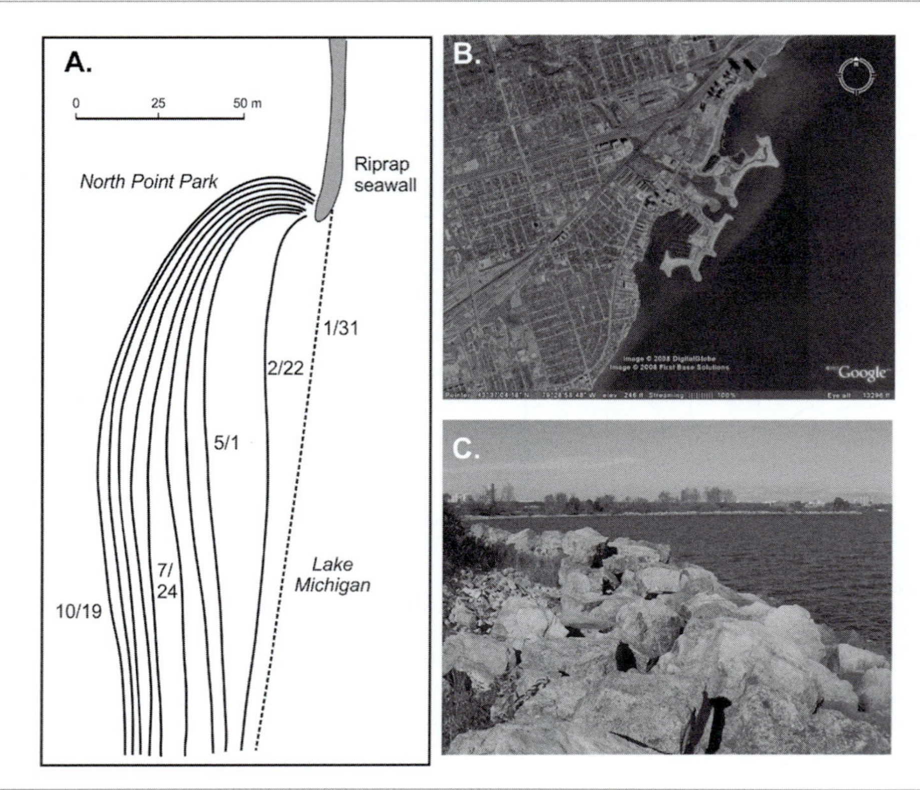

Figure 8.16 A. Evolution of a log spiral beach downdrift from a seawall on the Illinois shoreline of Lake Michigan (source: adapted from Terpstra and Chrzastowski, 1992, reproduced with permission from the Coastal Education and Research Foundation, Inc.). B. Satellite image of artificial headlands and crenulate bays on a complex landfill feature, Toronto waterfront, Ontario, Canada. C. Ground view of the armourstone structures forming one of the headland structure in B., with the Toronto skyline in the background.

one side to the other depending on individual storm events or changes in wave climate over several years (e.g., Short *et al.*, 2001; Ranasinghe *et al.*, 2004a; Short *et al.*, 2014; Harley *et al.* 2015).

On sections of exposed coast with an abundant supply of sediment and a marked longshore drift direction, the presence of widely spaced promontories or small headlands that provide some shelter from waves from the dominant direction give rise to log-spiral or zeta form beaches (Figure 8.16A). Here waves from the dominant, updrift direction refract and diffract around the headland producing a beach that locally is swash aligned in the lee of the headland with local sand transport in the direction opposite to the dominant drift direction along the coast some of the time. This produces a tight spiral form that can be described by a log-spiral function (Yasso, 1965) or by a parabolic function. The rapid evolution of a log spiral form in unconsolidated sediments following construction of a marina on Lake Michigan provides a nice example of this (Terpstra and Chrzastowski, 1992; Figure 8.16A). The concept of producing a static equilibrium beach form has been applied to shoreline stabilisation using artificial headlands (e.g. Silvester, 1974; Short and Masselink, 1999; Ojeda and Guillén, 2008). The object here is to use armourstone or similar non-erodible material to produce artificial headlands or large groynes and then to fill the area between with beach sediments. The sediments are then moved by wave action until a crenulate form is attained with the beach aligned to the waves along its length. If the beach nourishment material is quite coarse, then losses offshore and alongshore during storms are small and the beaches ideally become stable over long periods. The concept has been used extensively in the design of artificial complexes built out into the water and they serve the purpose of stabilising the shoreline and at the same time providing for several recreational activities (Figure 8.16B, C).

On exposed mainland and barrier coasts, the beach planform may be a complex response to sediment transport and the angle of wave approach (e.g., Ashton *et al.*, 2001), which may influence the shape of the coastline at a range of scales from a few tens of metres to tens of kilometres. For example, the orientation of the backbarrier shoreline tends to be controlled by the orientation of the seaward facing beach on the other side of a barrier island (see Houser, 2012). If the waves in this fetch-limited environment approach the backbarrier shoreline at a high angle an instability can develop leading to perturbations that can develop into cuspate headlands. The resulting coast is segmented and alternately swash- and drift-aligned alongshore.

8.3 Nearshore Morphodynamics

Beach and nearshore morphodynamics are generally considered at four different temporal scales:

Instantaneous (seconds to minutes) scales involving fluid processes and sediment transport processes. Morphology, incident wave field and water level are essentially constant. Local fluid and sediment transport conditions can be measured and are controlled directly by the incident wave field, winds and tides.

Short-term (hours to days) processes acting over a storm, or storm and succeeding fair weather period. Incident wave field, water levels and morphology all change.

Annual (monthly to yearly changes) and seasonal changes in morphology reflecting seasonal patterns of storm frequency and intensity and the characteristics of the incident waves and winds.

Decadal topographic change over a period of 10–50 years and longer. Seasonal patterns can still be recognised but changes reflect variations in storm intensities and in patterns of longshore migration of sandwaves as well as changes in beach and dune sediment budget.

These temporal scales are typically associated with increasing spatial scales ranging from centimetres and metres for instantaneous processes to several kilometres at decadal scales and longer. When viewed alongshore, the change in nearshore morphology and shoreline tends to follow a power-law relationship, which suggests that a single process may dominate across scales from small scale swash zone processes (< 100 m) to the scales typically associated with island curvature (> 1000 m). In other words, the changes in nearshore morphology appear to be scale invariant despite different processes and feedbacks acting across these scales (Tebbens *et al.*, 2002; Lazarus *et al.*, 2011). However, the nearshore morphology and changes in that morphology can vary alongshore in response to a variable framework geology alongshore, leading to scale-dependent variations in the nearshore morphology (Houser *et al.*, 2018).

8.3.1 Morphological Classification of Bars

Nearshore bars occur in both the subtidal and intertidal zones. They are usually oriented parallel to the shoreline, although oblique and transverse forms can occur close to the beach. The nearshore profile may have a single bar although two to four bars are common, and some very low-gradient profiles can have ten or more bars. In plan form, bars may be linear (straight), sinuous or crescentic (three-dimensional topography). While classifications in themselves do not provide an explanation for the range of bar types, they help to bring some order to the complexity of the systems observed in nature, and to the very large literature on bars which has developed over several decades. Bar morphology and dynamics have been described from a growing range of locations and as we understand more about bar processes, we can further refine the classification system to reflect similarities in bar generation and to incorporate understanding of the morphological changes that accompany bar migration on coasts with a broad range of wave conditions.

Wijnberg and Kroon (2002), in their review of nearshore bar systems, modified and updated a review and classification produced by Greenwood and Davidson-Arnott (1979). The model developed by Wijnberg and Kroon is shown in Table 8.1 along with the equivalent groups proposed by Greenwood and Davidson-Arnott – examples of these forms are shown in Figure 8.12. This model distinguishes between two different forms of 'wave-formed' bar systems: (1) outer bar systems found (and formed) some distance seaward of the shoreline and distinctly separated from processes operating on the foreshore and low tide terrace; and (2) inner bar systems found close to the shoreline, often attached to it through transverse bars, and greatly influenced by the development of rip cells and giant cusps.

The first type of bar described by Wijnberg and Kroon (2002) are two-dimensional and three-dimensional bars (Table 8.1; group VI of Greenwood and Davidson-Arnott). These are formed some distance offshore and appear to function independently from the beach. They are generated or maintained in place by steep, short-period waves during storms and dominated by shoaling waves and spilling breakers. They probably form because of sediment transport convergence with onshore transport under shoaling and breaking waves and offshore transport due to undertow (see Section 8.3.2). Straight forms reflect the dominance of 2D undertow circulation, while wavy or crescentic forms probably result from increasing influence of rip cells, either during formation, or subsequently during non-storm conditions when bars may migrate landward under swell wave conditions (see Figure 8.12B, C).

Multiple parallel bars (group III) are found in both sheltered environments and quite energetic locations (Davidson-Arnott and Pember, 1980; Davidson-Arnott and McDonald, 1989; see Figure 8.12D). The number of bars reflects primarily the existence of a very gentle slope with limited energy loss through breaking at each bar location. The gentle slope may result from the presence of very fine sand or it may be the result of a sub-horizontal underlying rock platform (Davidson-Arnott and Pember, 1980). However, processes on the outer bars are dominated by shoaling waves, spilling breakers and undertow (Davidson-Arnott and McDonald, 1989) and appear to be essentially the same as for the larger but smaller number of bars found in other locations. The outer bars are dominantly straight and continuous over long distances but the inner bars close to shore become much more complex in plan form and the longshore continuity decreases – as is case for inner bar systems in other locations. Considering their similarity to other bar systems, there no longer seems to be a good reason to have a separate class for this bar system.

Recent work on low amplitude intertidal bars (also known as ridge and runnel systems) indicates that the dominant processes on the bars, particularly the outer bars are associated with wave breaking and surf zone processes (Dawson *et al.*, 2002; Kroon and Masselink, 2002; Masselink, 2004; Masselink *et al.*, 2006; van Houwelingen *et al.*, 2006). The number of bars in the intertidal zone tends to increase with increasing tidal range but, as was noted above for multiple parallel bars, bar number appears to be more closely controlled by slope of the intertidal zone and to a lesser extent by wave steepness and absolute wave height (e.g., Reichmüth and Anthony, 2007). On intertidal ridge and runnel bars swash processes are subordinate, especially during high wave conditions, and consequently on–offshore movement of the bars reflects variations in wave height during storms and the influence of wave shoaling during non-storm conditions. At several locations where ridge and runnels bars have been described, one or more similar bars exists in the subtidal (e.g., Dawson *et al.*, 2002; Sedrati and Anthony, 2007; Price and Ruessink, 2008). There does not appear to be a good data set on the subtidal bars in locations where dynamics of the intertidal bars have been measured, but qualitative observations suggest that there is little difference between the subtidal bars and the outer intertidal ones that are exposed for short periods during spring low tides. Towards the shore, bars are increasingly modified by channels that cut through them, draining the runnels as the tide falls and their form may be modified as well by swash processes around low tide. At some sites the intertidal bars

Table 8.1 Morphological classification of nearshore bars

Bar type	Location	Slope	Wave energy	Tidal range	Bar group	Notes
Low-amplitude ridges (King and Williams, 1949)	Intertidal	0.01–0.005	Low	Meso- to macrotidal	I	Probably modified form of Group VI bars
Slip-face ridges (Davis *et al.*, 1972)	Intertidal	0.03–0.01	Low to moderate	Micro- tomesotidal	II	Migrate into intertidal zone from subtidal Group VI bars
Shore-attached bars (Sonu, 1968)	Subtidal (to intertidal)	0.03–0.005	Moderate	Micro- tomacrotidal	II	May be relatively stable on fetch limited coast
Three-dimensional longshore bars (Homma and Sonu, 1962)	Subtidal	0.03–0.005	Moderate to high	Micro- tomacrotidal	VI	3D form is probably secondary – rip cell circulation
Two-dimensional longshore bars (Evans, 1940)	Subtidal	0.03–0.005	High	Micro- tomacrotidal	VI	2D form probably reflects dominant undertow
Highly protective settings						
Transverse finger bars (Niederoda and Tanner, 1970)	Subtidal	0.01–0.0005	Very low	Microtidal	IV	Usually shore connected
Multiple-parallel (Zenkovitch, 1967)	Subtidal (and intertidal)	0.01–0.0005	Very low	Micro-(and macro)	III	Likely group VI bars formed on very gentle slopes – energy level low to moderate

Note: Wijnberg and Kroon's (2002) model, showing equivalent groups proposed by Greenwood and Davidson-Arnott (1979).

Figure 8.17 Complex inner bar and trough system exposed at spring low tide, Greenwich dues, Prince Edward Island. A. View west along the beach showing the berm crest and landward runnel with some water from the previous high tide. On the right side of the picture the low tide terrace is exposed as a flat pad with a straight inner bar and trough just seaward of it. Low waves are breaking on the bar with translatory swash bores moving across it into the trough. In the middle distance above the two figures a rip channel can be seen breaking seaward obliquely through a bar. Note, this photograph is taken with a polarising lens to enhance the contrast between sand and water. B. Photograph looking offshore at one of the rip cell systems near the top left of the previous photograph. On the left the inner bar is nearly exposed with only a few of the swash bores able to cross the bar into the trough landward of it. The trough leads into a rip channel which cuts through the bar separating it from the one on the right. Low waves can propagate through the channel and are eroding the embayment that is developing on the foreshore opposite it.

are found 2–3 wavelengths seaward of the spring high tide line and thus the influence of the shoreline is limited. However, where bars are present in the zone between MHWN and MHWS they tend to be dominated by swash rather than surf and exhibit steep slip faces on the landward side and much greater mobility (Reichmüth and Anthony, 2007), thus distinguishing them from the more stable bars of the outer system.

The inner bar system develops in quite shallow water and usually has some connection to the foreshore and low tide terrace. Bar morphology is highly varied and variable, with straight, crescentic and transverse bars all commonly formed (see Figures 8.4, 8.5). On beaches with medium to coarse sand, where a steep foreshore and berm develops under low wave conditions, the inner system is usually associated with well-developed rip cells with a quasi-regular alongshore spacing producing what is termed rhythmic topography (Falquéz *et al.*, 2008). The beach face develops embayments separated by horns which together form giant cusps (Komar, 1971, 1983; Barrett and Houser, 2012), with a spacing of less than a hundred to several hundred metres. Bars in the inner nearshore are then linked to the beach either directly to the horns or at broad pads on the low tide terrace adjacent to the horns (Figure 8.17). Water depth over the bars is shallow, especially at low tides and portions of the bars may be exposed at spring low tide. As a result, these bars are subject to shoaling and breaking waves at high tide but at low tide swash processes dominate bringing surges of water across the bar into the landward trough and often leading to the formation of a slip face. Longshore currents develop in the trough landward of the bar and then flow seaward through distinct rip channels or across the low point of crescentic bars. Where wave approach is at a high angle to the beach the whole topography, bars and rip channels, may be oriented obliquely in the direction opposite to that of wave advance, and a meandering longshore current can develop (Houser *et al.*, 2013). Waves commonly propagate through the deeper water of the rip channels and refract around the ends of transverse bars to travel up the longshore trough resulting in highly variable sediment transport directions in the trough and on the beach face (Figure 8.17).

8.3.2 Nearshore Bar Formation

The regularity of bars and their presence on so many shorelines has long stimulated interest in how they are formed, and what controls their complex morphology. Bars clearly represent the net result of sediment transport in the surf zone, and their stability over periods of days and weeks suggests that their position represents a dynamic equilibrium in which there is continuous transport of sediment within the system. There are many mechanisms that might potentially contribute to or control bar formation through their influence on sediment transport, including those associated with incident wave shoaling and breaking, infragravity waves and unidirectional rip cell circulation and undertow. One major problem has been to observe and measure the formation of a bar in the field. On coasts where bars are present all the time, onshore and offshore migration of the form has been observed, but it is only recently that there have been a couple of (rather fortuitous) measurements of bar formation close to the shore (Aagaard *et al.*, 1998; Aagaard *et al.*, 2008). On coasts such as California and New South Wales, Australia, where bar migration and welding to the shoreline can produce complete disappearance of bars, the generation of new bars takes place some distance offshore under very energetic conditions, and there does not yet appear to be a good data set documenting bar formation under these conditions (Shand *et al.*, 2001). Bars on the Dutch coast migrate offshore continuously and eventually disappear, while new ones are generated close to shore. These in turn migrate offshore, maintaining an equilibrium spacing between bars (Wijnberg and Terwindt, 1995; van Enckevort and Ruessink, 2003a). Episodes of bar switching involve the realignment of outer bars following the junction of two bars in a transition zone (Wijnberg and Wolf, 1994; Shand *et al.*, 2001). Some of the variability in the seasonal positions of the bars may reflect alongshore bar movement (van Enckevort and Ruessink, 2003b).

One obvious solution to the problem is to carry out experiments in a wave tank and, as noted earlier, there is a long history of wave tank experiments (Keulegan, 1948; King and Williams, 1949; Shepard, 1950a; Watts, 1954). As noted in the classification of Greenwood and Davidson-Arnott (1975), it was believed that the 'bars' were formed by sand moving seaward from a trough excavated by plunging breakers. However, the limited wave height (a few centimetres) in these relatively short wave tanks, and the use of monochromatic waves meant that key controls in nature were missing. While newer and larger facilities can generate full spectrum waves and the greater size permits larger waves, the significant wave heights are still quite small. For example, the Large-scale Sediment Transport Facility of the Army Corps of Engineers at Vicksburg, Mississippi has a breaking wave height (H_b) of only 0.27 m (Wang *et al.*, 2003). Finally, the laboratory experiments rarely include the effect of wind in addition to waves.

Present research seems to focus on two broad groups of models for bar formation: (1) bar formation by infragravity standing waves or edge waves; and (2) bar formation by cross-shore flows and the break-point mechanism. Other mechanisms falling outside these two principal groups may be identified as self-organisational (Wijnberg and Kroon, 2002) involving non-linear interactions in the surf zone (e.g., Damgaard Christiansen *et al.*, 1994; Boczar-Karakiewicz and Davidson-Arnott, 1987; Vittori *et al.*, 1999). However, there is some degree of self-organisation in all the models (Thornton *et al.*, 1996; Aagaard and Masselink, 1999).

8.3.2.1 *Formation Due to Standing Waves*

The models of bar formation by infragravity waves, rely on the incident waves to generate sand movement in the breaker and surf zones with the drift velocities associated with the standing waves leading to movement of sediment towards points of convergence. The standing waves associated with both leaky waves and edge waves have a cross-shore structure that produces nodal points at distances that increase exponentially offshore (see Figure 5.26), which is consistent with the increased spacing of bars (e.g., Short, 1975; Aagaard, 1990; Bauer and Greenwood, 1990; Aagaard and Greenwood, 1995). The drift velocity field in the presence of standing waves produces convergence at the nodes for bedload, and at the antinodes for suspended sediment (Carter *et al.*, 1973; Bowen, 1980). This provides some difficulties because sand is transported in both suspension and as bedload, which means that bars are likely to develop at only one or other of the nodes and it is not obvious which one it should be. O'Hare and Huntley (1994) refined the model based on the variation in short wave height and long waves for wave groups in which sediment transport is independent of height.

The alongshore shape of the infragravity waves also appears to be consistent with the alongshore variation in the morphology of nearshore bars. Leaky waves have a uniform structure alongshore and thus can be associated with the generation of long, linear bars, while edge waves have a three-dimensional form and thus the zones of convergence produce crescentic bar forms (Bowen and Inman, 1971; see Figure 8.18). The generation of edge waves with the appropriate frequency is complex, and it is not certain what mechanism (e.g., non-linear interaction or the isolation of cut-off frequencies) would provide the necessary stability over a period of many hours for the bar to generate. There is always the suspicion that where edge waves are observed they are the product of the existing bathymetry or have a form that is forced by the bars rather than the control (Huntley, 1980). As noted, the balance of opinion seems to have shifted away from models based on standing waves as the primary generator of nearshore bars, though it is still likely that they have some influence on patterns of water motion and thus on the second order form of bars, because the drift velocities can transport sand towards the bar crest.

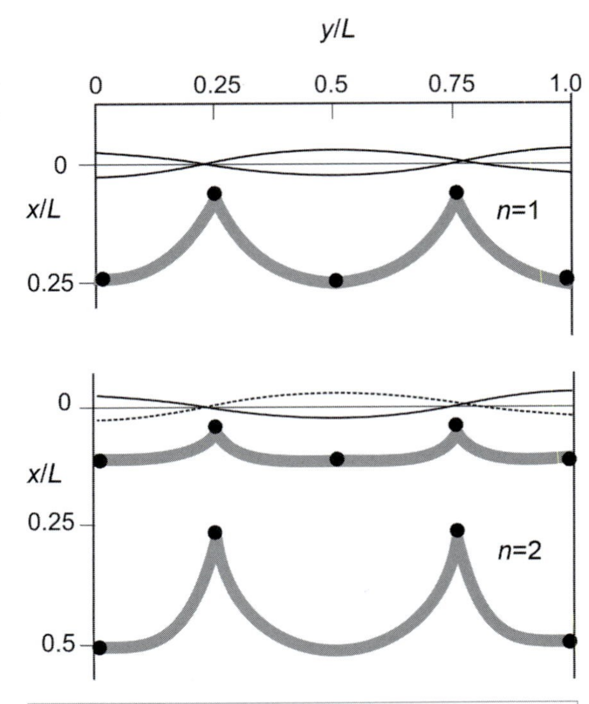

Figure 8.18 The theoretical form of crescentic bars generated by mode 1 and mode 2 edge waves (Bowen and Inman, 1971). A mode 0 edge wave has no offshore crossing point and therefore generates rhythmic topography (cusps) on the beach but not bars. Note that the wavelength of the crescent is the same for both bars produced by the mode 2 edge wave.

8.3.2.2 *Formation By Cross-Shore Flows and the Break Point Mechanism*

The second and increasingly popular bar-forming mechanism involves the convergence of sediment due to wave breaking and offshore-flowing currents. Seaward of the breaker zone sediment transport is generally landward under shoaling waves due to velocity skewness or asymmetry, and this increases towards the breaker zone. As discussed in Chapter 6, wave set-up in the surf zone generates an offshore-directed current, the undertow, that transports sediment offshore close to the bed. Because there is a distribution of the incident wave heights, the strength of the current increases seaward towards the break-point and diminishes offshore of that point (Figure 8.19A). This leads to erosion of sediment from the developing trough and deposition in the zone of convergence in the breaker zone where the landward-directed Stokes drift also reaches a maximum (Dyhr-Nielsen and Sorensen, 1970; Dally and Dean, 1984; Dally, 1987). As the bar forms, there is likely some positive feedback (Figure 8.19B), because the steepening of the

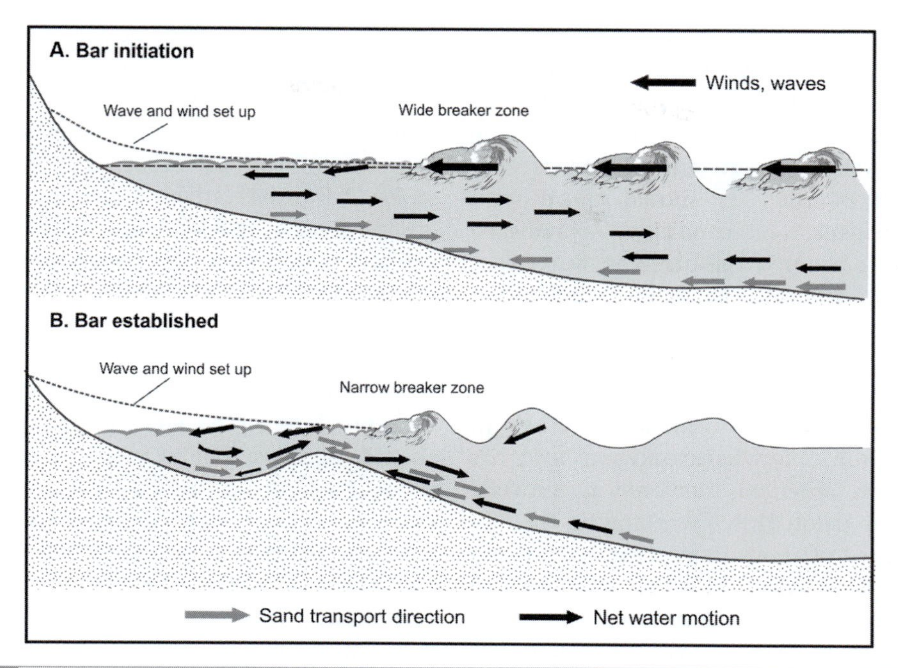

Figure 8.19 Sketch of the sediment transport patterns associated with bar formation through the convergence near the breaker zone of onshore transport under shoaling and breaking waves and offshore transport in the undertow: A. initial conditions; B. growth of the bar form leads to a narrowing of the breaker zone and increased breaker intensity (after Davidson-Arnott, 2013).

seaward slope and shallower water depth as the bar develops will lead to increased wave breaking (both the proportion of waves breaking and the intensity of breaking will increase), thus enhancing the undertow and focusing the zone of sediment transport convergence (Holman and Sallenger, 1993; Garcez Faria *et al.*, 2000). This kind of mechanism has been conjectured as far back as 1863 by Hagan writing on bar formation in the Baltic (quoted in Evans, 1940) and similar ideas are found in King and Williams (1949). Greenwood and Davidson-Arnott (1979) also put forward a conceptual model for convergence of landward transport under waves and seaward flows due to rip currents over the outer bar based on sedimentary structures in box cores (see Section 7.2.4). This model suggests that the convergence model may also explain the transition of complex three-dimensional bars to parallel two-dimensional bars during storms through weakening seaward transport of sediment through within the rip channel.

The formation of multiple bars requires the generation of distinct zones of breaking waves, with wave breaking ceasing and wave reformation in the deeper water landward of the trough. As the first bar moves further offshore and into deeper water, smaller waves can now break further landward leading to the development of another bar closer to the shore (Houser and Greenwood, 2005). Additional bars can develop as the bars migrate offshore, but eventually the outermost bar will move into a depth in which wave breaking is limited and constant Stokes drift causes the bar to either migrate landward and weld to the next bar or dissipate in height. Alternatively, on very low gradient slopes wave breaking is not concentrated at a specific location cross-shore, leading to multiple locations where the feedback between bar development and transport convergence can develop. Recent field experiments (Aagaard *et al.*, 1998; Greenwood *et al.*, 2006; Aagaard *et al.*, 2008) have provided evidence to support this model.

Key elements of the cross-shore flow and break-point mechanism are the role of morphodynamic feedback between the bar form and the wave and current hydrodynamics and the tendency for the bar form to migrate to an equilibrium position where there is a balance between the strength of the undertow and the landward transport by Stokes drift under the breaking waves. One outcome of this is that the position of the bar should move with changing incident wave conditions – landward into shallower water if the bar is in water that is too deep to generate the required undertow, and offshore if wave breaking is too intense leading to transport by the undertow onto the seaward side of the bar. Field studies in North America in the 1950s and 1960s were carried out mostly in California where only onshore bar migration was observed. However, Greenwood and Davidson-Arnott (1975) working on the east coast measured offshore movement in response to storms with greater intensity, and onshore movement with less intense storms. This oscillation in response to changing wave conditions has now been documented in many areas (e.g., Lippman *et al.*, 1993; Lee *et al.*, 1998; Plant *et al.*, 2001; Houser and Greenwood, 2005), and offshore migration over a period of months to years has been documented for coasts with a strong alongshore component of sediment transport (Ruessink *et al.*, 2000; Shand *et al.*, 2001).

Offshore migration of the outer bar takes place until it reaches a position where only a fraction of the incident waves are breaking and the mean offshore flow in the undertow will just balance the onshore transport under waves. This position and depth is estimated to be when the local relative wave height H_{rms}/h is ~ 0.3. In some cases, the storm event will not be of sufficient duration for this to be achieved and thus the response time, based on the rate of onshore or offshore migration of the bar, may be insufficient to reach the equilibrium position (Davidson *et al.*, 2013). The result is that we can envisage the bar position as being controlled by wave height and the location of wave breaking, and thus the offshore position of a bar over a year should tend toward a position determined by the largest waves (Plant *et al.*, 1999). As Houser and Greenwood (2005) note, the outer bar is subject to a wide range of wave conditions and therefore it should have a much wider range of movement as well as a greater chance, for example, of moving onshore under low waves to a position where it will be highly unstable under high waves from a rapidly developing storm. Due to the depth over the crest of the outer bar, the undertow and Stokes drift are relatively weak over most of the observed wave height distribution, which may explain the limited the responsiveness of the outer bar to changes in the incident wave field.

The innermost bars of multiple barred nearshore are subject to a much narrower range of wave conditions because of the filtering effect of wave breaking on the bars seaward of them. They should therefore exhibit much greater stability, unless the outer bar migrates offshore to a depth where it ceases to act as much of a filter. However, the relative water depth over the crest of the innermost bars and intertidal bars can vary considerably over individual tidal cycle and between neap and spring tides. This has an important determinant on the temporal and spatial pattern of the morphodynamic regime on the bar (Masselink, 1993, 2004; Kroon and Masselink, 2002; Price and Ruessink, 2008), and thus the proportion of time that conditions over the bar are dominated by shoaling, surf or swash conditions. When the water depth is small, sediment transport is predominantly onshore under swash bores and these conditions can produce rapid onshore movement and development of a steep slip face. As water levels increase, the bar can migrate offshore as wave breaking develops over the bar. Even in the absence of large changes in water level, the landward transport under swash bars can lead to what has been described as an unexpected landward migration of inner bars during large storm events.

8.3.3 Nearshore and Beach Morphodynamic Controls and Models

Based on the material covered earlier in the chapter we can identify the likely controls on the stability or dynamic range of morphological variation as being: (1) the wind and wave climate; (2) sediment size, and beach and nearshore slope; and (3) the tidal range (Table 8.2). Changes in

Table 8.2 Stability or dynamic range of morphological variation

Wind and wave climate offshore – shallow after shoaling and refraction	Sediment size beach and nearshore slope	Tidal range
Swell sea mixed	Fine, medium, coarse	Micro, meso to low, high macro macro
Low, medium, high	Low, high, low	Medium, medium low

wave conditions over periods of weeks, months and years can cause the configuration of the nearshore and beach to change. The extent of change on some coasts is quite small, and it is possible to describe a characteristic (or modal) morphology that may exhibit some variation in space and time but tends towards the same morphology throughout the year. On other coasts, there is a marked change in form with seasons driven by seasonal variations in the characteristics of incident waves. For example, the winter/summer profile described for California is a classic example of this (see Figure 8.13). Beach and nearshore form may also be driven by changes associated with longer term cycles such as ENSO that alters the dominant wind direction, wave direction, severity of storms and water level. Other beaches, such as Pensacola Beach, Florida change from a profile dominated by multiple parallel bars with winter storms to transverse bar and rip as wave heights decrease and the bars migrate landward during the summer (Houser *et al.*, 2011). This beach also exhibits a variation in the nearshore morphology alongshore in response to a transverse ridge and swale bathymetry offshore. The bars tend to be closer to the beach at the swales and are either in transverse bar and rip morphology or completely welded to the beach. Over the ridges, the bars tend to be further offshore, but can transition to a transverse bar and rip morphology through the summer. Understanding this spatial and temporal variation in the nearshore morphology is critical for beach management and safety.

While some examples of morphodynamic response had been described in work prior to 1980 (King and Williams, 1949; Davis *et al.*, 1972) it was Wright and Short (1983, 1984) working on the coast of New South Wales, Australia who put together a systematic description of beach morphodynamic stages for a high-energy microtidal coast. They attempted to define conditions that would predict the form of the beach and nearshore profile and the range of forms that could be expected on it based on the dimensionless fall velocity parameter Ω (Short, 1979; Wright *et al.*, 1982; Wright and Short, 1983, 1984; Wright *et al.*, 1985). As we have seen earlier (Equation 8.5), Ω is a form of surf similarity parameter, incorporating wave steepness and beach slope (through the dynamic relationship between sediment size and beach slope). Use of the w_s (settling velocity) parameter overcomes a tricky problem of how exactly to define beach slope on the curvilinear profile of prototype beaches and makes it easier to provide a physical explanation of this important parameter (Nielsen and Hanslow, 1991). Other difficulties arise because small waves may break on the steep, reflective portion of the foreshore while under more energetic conditions larger waves will break on the gently sloping dissipative portion of the inner nearshore.

Wright and Short (1983, 1984) distinguished between two end member beach types – dissipative and reflective – and a series of intermediate beach types. As previously noted, reflective beaches tend to occur for $\Omega < 1$ and dissipative beaches for $\Omega > 6$ while values of 2–5 are associated with intermediate beaches. Beaches that are intermediate may move between different modal stages producing a distinct suite of topographies including transverse bar and rip morphology and other cases where the innermost bar is partly or wholly attached to the shoreline. The original scheme of Wright and Short (1984) is depicted

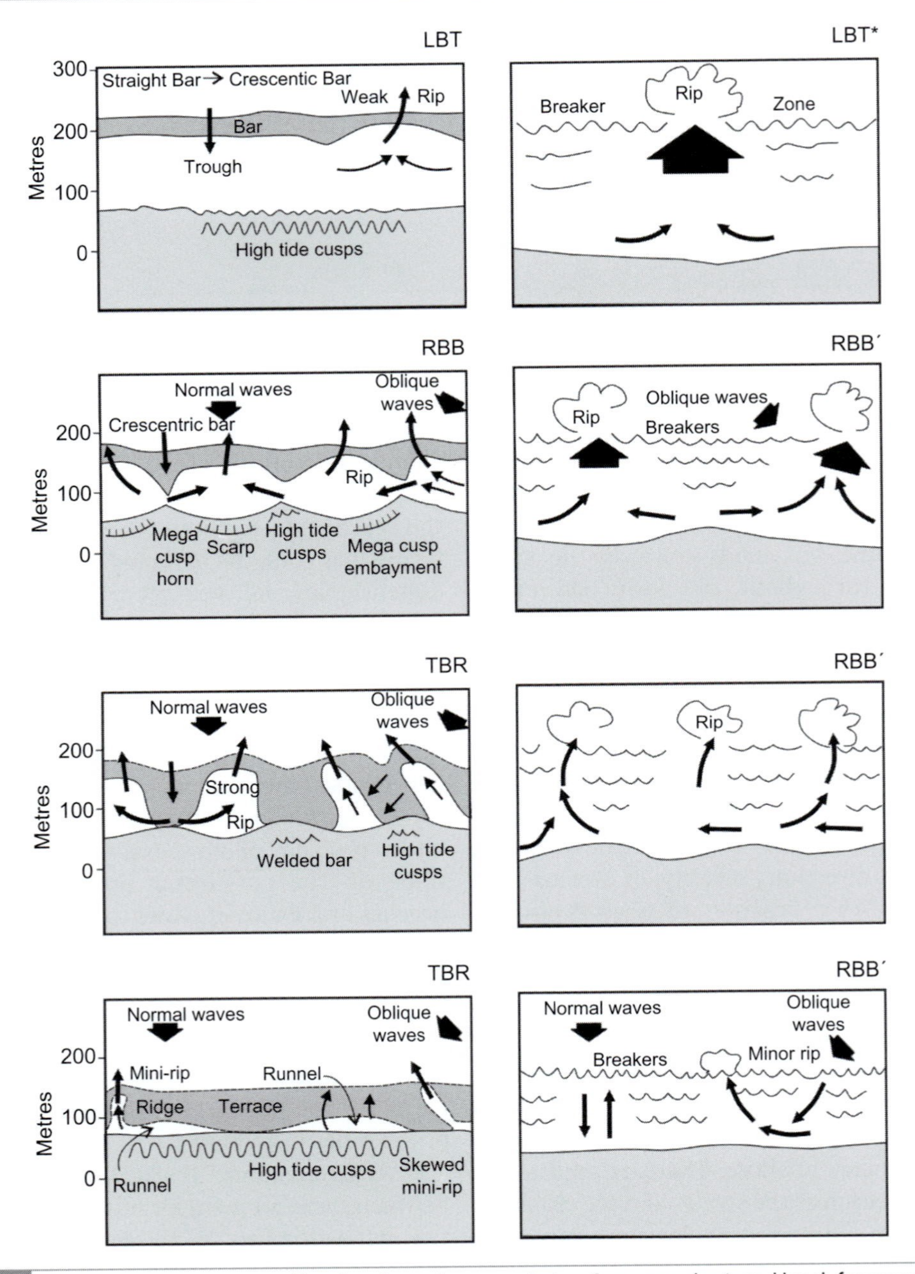

Figure 8.20 Plan views of the sequence of beach and nearshore morphology for a wave-dominated beach for accretionary (left panels) and erosional (right panels) wave conditions (Short, 2006; based on Short, 1979; Wright and Short, 1984; Sunamura, 1988 and Lippmann and Holman, 1990). LBT = longshore bar trough; RBB = rhythmic bar and beach TBR = transverse bar rip.

in Figure 8.20, but there are several variants and extension of the model including Sunamura (1988) and Lippmann and Holman (1990). The sensitivity or relative contribution among the three variables to determining the beach state is shown in Figure 8.21.

An accretionary sequence on a beach exposed to the full range of wave conditions could begin with the formation of a longshore bar and trough during an extended period (several days) of storm waves. If this is followed by an extended period (weeks) of lower, long period swell waves, the bar

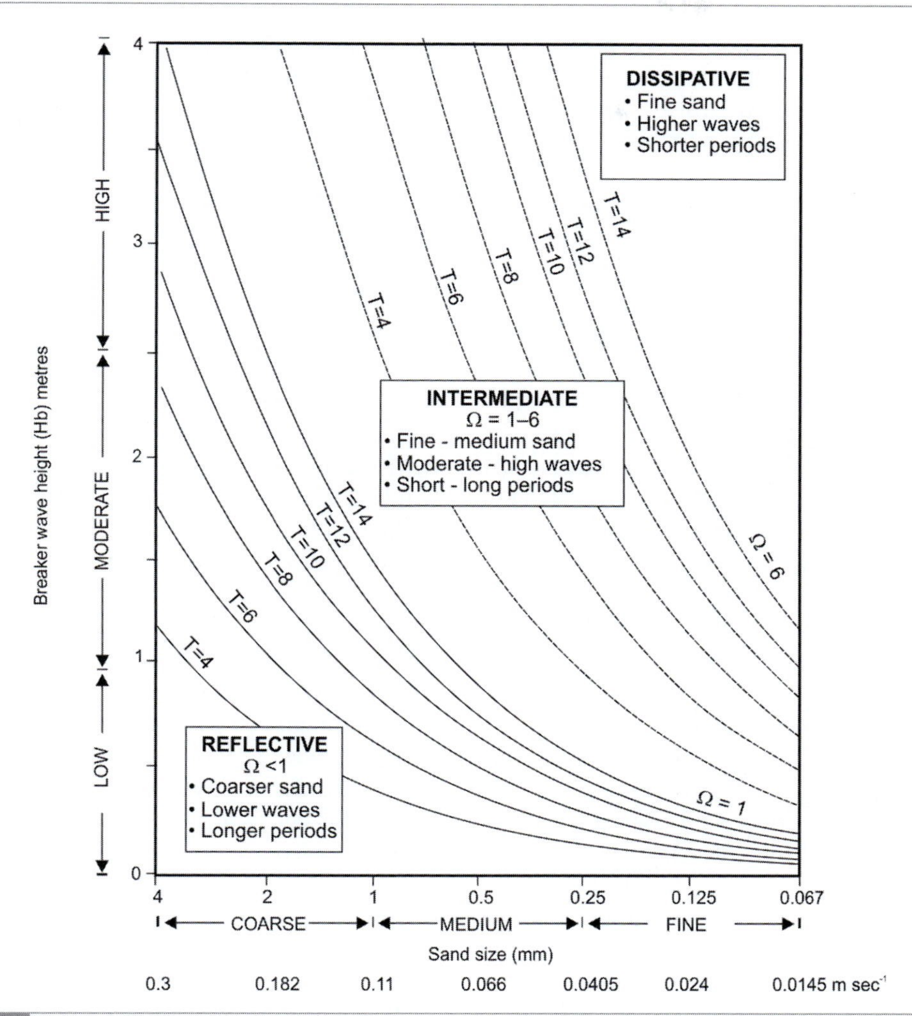

Figure 8.21 Nomogram showing the contribution of wave height, sediment size and wave period and beach type. The intersection of grain size and wave height is determined first and then the wave period is used to determine boundaries between the three domains (Short, 1999).

will migrate shoreward through a rhythmic bar and beach stage (where the bar becomes rhythmic or crescentic), through the attachment of the bar to the beach forming transverse bars with rip channels separating them, and ultimately the complete welding of the bars to form a steep, high berm (left panels of Figure 8.20). Onshore movement occurs because swell waves shoaling across the bar can move sediment landward due to the asymmetry of oscillatory motion, and there is no counterbalancing offshore transport by undertow on the seaward slope the bar. The bar becomes crescentic as rip cell circulation is established leading to predominantly onshore flow over the bar and offshore flow through the rip channel (versus undertow over the bar). This flow and transport pattern is reinforced when the bars become shore attached, leading to a deepening of the channel and a faster current Brander (1999).

The full accretionary sequence requires the right combination of wave conditions and sufficient time for the bars to slowly migrate landward. In many cases, the accretionary phase is interrupted by a storm and an erosional phase occurs with sediment and the bar(s) moving back offshore. The size of the storm required to reset the system depends on how far landward the bar migrated since the last storm. The example sequence provided in Figure 8.20 is for a coast with a single bar, but the basic beach state model can be modified to accommodate the situation where there are two bars or even three bars (Short and Aagaard, 1993). An analysis of four years of data using the Argus system (Ranasinghe *et al.*, 2004b), at Palm Beach, New South Wales provides evidence for the robustness of the stages and their sequence, and numerical modelling supports the role of high-energy wave conditions in producing the longshore bar trough (LBT) morphology and of lower wave conditions in producing the progression through the other three states. Lippmann and Holman (1990) noted a rapid transition to a linear bar under storm waves and an equally rapid (1–2 days) transformation to a rhythmic bar. However, other locations do not show this and the time scales of post-storm recovery is typically much slower.

The sequence of bar types observed on a beach ultimately depends on the distribution of wave heights and the frequency of reset storms (Castelle *et al.*, 2007). Beaches with a wide distribution of waves, such as Palm Beach, New South Wales can experience the full range of beach states, while beaches with a limited range of wave heights can exhibit a relatively stable nearshore state. For a given distribution of waves, a clustering of storms can move the bars farther offshore than a single storm alone and the bar is given an opportunity to move through several accretionary states, while storms spaced in time reset the bars frequently limiting the amount of change during and between storms. Consequently, two beaches with a similar grain size and wave climate can maintain different morphologies. The straight nearshore bar shown in Figure 8.12B is in an environment with a similar wave climate and grain size to the crescentic system in Figure 8.12C which never becomes straight. A long-term study of a double bar beach on the west coast of France shows the range of beach states possible based on the wave climate (Castelle *et al.*, 2007; Figure 8.22).

The beach stage model applies best to microtidal coasts with a broad wave climate, that is, one exposed to both periodic storms which act to move sediment offshore and generate a bar, and swell waves which are large enough to move sediment on the bar onshore thus initiating the process of bar migration. Elements of the model can be seen in field studies carried out in other locations (e.g., Lippmann and Holman, 1990), but they do not necessarily encompass the full range of nearshore and beach morphodynamics (e.g., Masselink and Pattiaratchi, 1998). In fetch-limited areas (maximum fetch < 300 km), the range of wave conditions is quite narrow and high-energy wave events are always associated with the passage of storms and strong onshore winds (Goodfellow and Stephenson, 2005). Coasts in the mid latitudes, are almost always characterised by the development of two or more bars in the nearshore (Castelle *et al.*, 2007; 2010a, b). The outer bar formed during storms can become largely inactive or relict during periods between storms because the non-storm waves are not large enough to move sediment landward, although waves with relatively low steepness may occur for a few hours after a high wind event. This behaviour is also possible in locations where wave refraction results in a large reduction in wave height. The bars in these environments are a permanent feature of the profile, with the exact morphology dependent on the frequency of storm activity.

In environments where the prevailing winds are offshore (e.g., north and east facing beaches in the Great Lakes and Gulf of St Lawrence), the inner bars may also be relatively stable during non-storm conditions because periods with waves large enough to move sediment on these bars are relatively rare or last only a few hours at a time (e.g., Greenwood and Davidson-Arnott, 1975). However, on south and west facing beaches, where there is sufficient fetch, wave action may occur on many days leading to onshore migration of the inner bars and welding to the beach (e.g., Davis and Fox, 1972; Stewart and Davidson-Arnott, 1988; see Figure 8.23A). A slip face develops on the migrating bar even before it emerges at low tide, because of wave translation across the bar crest at low tide. The crest is flattened and often the bar becomes much wider. A slip face bar may persist for weeks

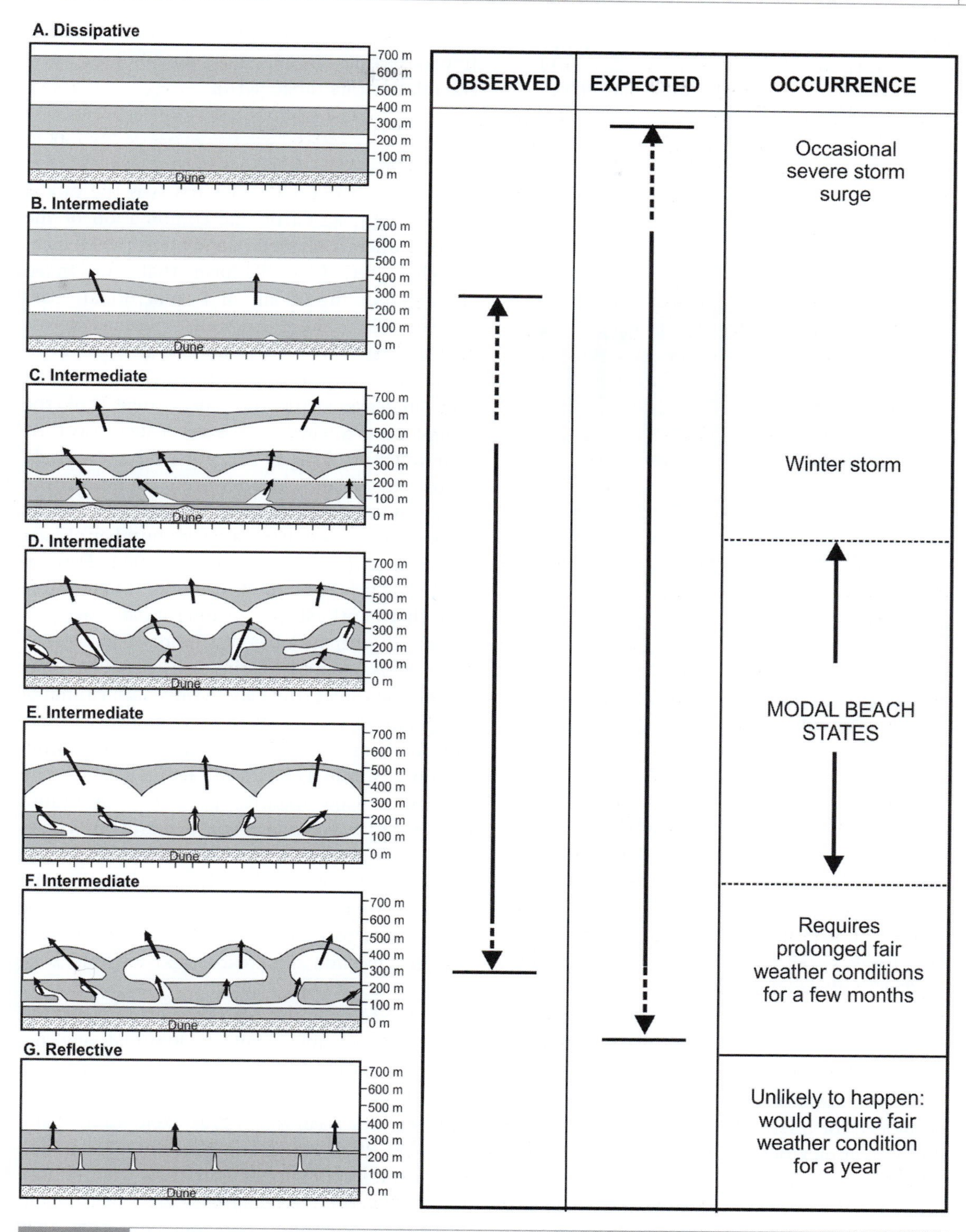

Figure 8.22 Model showing the possible range of beach states at Truc Vert beach on the Aquitaine (west) coast of France based on a long-term survey of beach states showing bar states based on the frequency of storm events. Intermediate forms labelled B.–F. have generally been observed at the site, while a fully dissipative form A. might be expected after a sequence of intense storms accompanied by storm surge (Castelle *et al.*, 2007).

A.

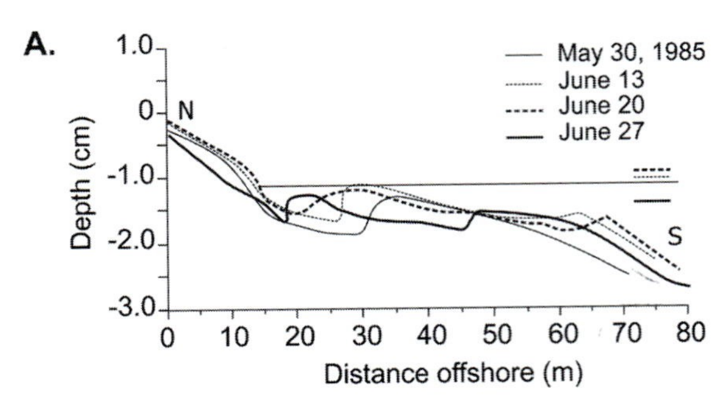

B.

Figure 8.23 Intertidal bar welding: A. onshore migration of inner sandbars as slip face ridges and welding to the beach at Long Point, Lake Erie during summer low wave conditions: B. photograph of a slip face bar in the intertidal zone, Skalingen, Denmark (Houser *et al.*, 2006). (Photo courtesy Troels Aagaard.)

in the intertidal zone of relatively fine-grained beaches (Houser *et al.*, 2006; see Figure 8.23B), and its position may shift offshore under higher wave conditions and onshore when wave energy and storm surge decline or it may merge with the beach and be reset by another period of high wave activity (Quartel *et al.*, 2007).

8.3.4 Nearshore and Beach Models for Tidal Environments

The beach stage model of Wright and Short was developed for microtidal beach settings and additional complexities are introduced as the tidal range increases to meso- and macrotidal (Masselink and Hegge, 1995). Tides can introduce large changes in water level, creation of drainage channels at low tide and strong tidal currents flowing alongshore in the trough (Sedrati and Anthony, 2007). Masselink and Short (1993) describe the impact of tides on nearshore morphology using the relative tide range (RTR):

$$RTR = \frac{TR}{H} \tag{8.8}$$

where TR is the tidal range and H is the modal wave height. Based on the combination of the two parameters Masselink and Short (1993) identify characteristic profile forms that is expanded to include mesotidal and macrotidal bars (Figure 8.24). The breakpoint migrates across the leading wave to the development of nearshore bars at multiple locations. The number of bars increases with the tidal range, but the height of the bars decreases because the time that wave breaking is concentrated at the individual bar crests is also reduced. On quite steep slopes, two to four bars may be present and as the slope becomes less steep a larger number of bars are formed (Hale and McCann, 1982; Dawson *et al.*, 2002; Masselink *et al.*, 2006), in the same way that gentle slopes on microtidal coast promote a greater number of subtidal bars. Masselink *et al.* (2006) distinguish between low-amplitude ridges (the classic ridge and runnel of King and Williams, 1949) and sandwaves (Hale and McCann, 1982), but in view of the consolidation of subtidal classes in Section 8.3.1, it may be best to view them as part of a continuum where the wave and hydrodynamic processes are similar and the prime difference is a response to decreasing intertidal gradient. When the tidal range gets very large (high macrotidal) the profile form tends to become rectilinear with a steep upper slope and flat, featureless low-tide zone (Levoy *et al.*, 2000).

Bars are common features within the intertidal zone of meso- and macrotidal environments, and there has been considerable work on these bars on both sides of the English Channel (Michel and Howa, 1999; Reichmüth and Anthony, 2002; Kroon and Masselink, 2002; Masselink *et al.*, 2008). In general, these studies have found that the intertidal bars are a permanent feature of the intertidal zone and that they respond in a similar way as subtidal bars to changes in wave energy (Masselink *et al.*, 2005).

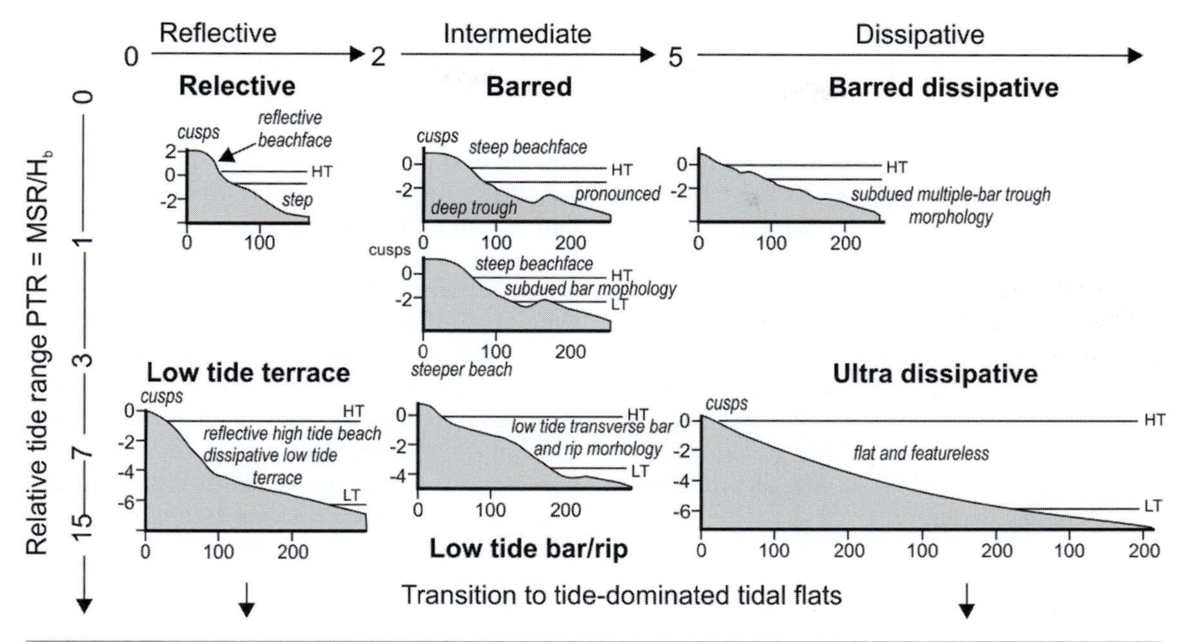

Figure 8.24 Beach states as a function of the dimensionless fall velocity and the relative tidal range. Adapted from Masselink and Short (1993). Reproduced with permission from the Coastal Education and Research Foundation, Inc.

The landward migration of intertidal bars can be an important supply of sediment to the foredune (Aagaard *et al.*, 2004). The (swash) bar migrates landward with the asymmetry of the swash bores and the offshore flow through drainage channels as a weak rip current. If the bar is able to migrate to the top of the foreshore during spring tides, or during a storm, the sediment dries and can be transported landward by an onshore wind.

8.4 Beach Morphodynamics

As we would expect from the discussion in Section 8.2.2, the extent to which the beach responds to changes in wave and tidal forcing is dependent on beach sediment size. Change in the beach profile tends to be muted for beaches in fine sand, which maintain a gently sloping planar profile. However, beaches composed of medium and coarse sand experience quite large changes in form that reflect a tendency towards a dynamic equilibrium driven by the interaction of swash and the beach water table. On gentle slopes, the swash zone merges into the nearshore zone without any notable break, but with steeper slopes, and especially on beaches with coarse sand and gravel, the transition to the nearshore is marked by a distinct step at the base of the foreshore (Austin and Buscombe, 2008).

8.4.1 Sandy Beach Morphodynamics

Beach form changes readily in response to changes in wave conditions between storm events and periods of swell waves or fair-weather conditions (Ortega-Sánchez *et al.*, 2008; Quartel *et al.*, 2008). This response is modulated by changes in water level due to wave and wind set-up, which affects the landward limit of wave action and to diurnal and neap/spring tidal cycles. The general tendency for equilibrium beach slope to increase with increasing grain size was noted in Section 8.2.2. Grain size also influences the dynamic response of beaches to variations in wave characteristics, overall wave energy, and to water level variations, especially through interaction of the swash with the beach water table (Buscombe and Masselink, 2006). In general, periods with high waves, especially steep, short period waves during a storm, leads to erosion of the beach and flattening of the beach profile producing the storm profile of Figure 8.13. High water levels associated with storm surge and high tide may permit waves to reach the backshore and lead to erosion of aeolian sediments at the back of the beach and perhaps to erosion of the foredune (see Section 9.3.1). Extended periods of relatively low waves lead to onshore transport of sediment from the inner nearshore and the build-up of the beach to form a berm, commonly known as a 'fair-weather' profile (Figure 8.13). The beach water table plays an important role in modulating both the erosion of the berm during periods of high waves and the build-up of the beach during periods of low waves.

8.4.1.1 *Beach Water Table*

The beach water table represents the upper level of groundwater that underlies the beach (Figure 8.25). At the landward margin of the beach the groundwater continues under features such as sand dunes or bluffs and there is often a hydraulic gradient towards the beach that may, for example, lead to a zone of seepage near the dune toe. The groundwater table outcrops at the seaward margin of the beach, usually on the middle to lower portion of the foreshore. The beach groundwater system interacts with the atmosphere, receiving inputs of precipitation through infiltration and losing mass through capillary rise and evaporation (Horn, 2002; Schmutz and Namikas, 2013). On occasion, the groundwater table may be at the surface over a portion of the beach profile (Schmutz and Namikis, 2018), for example where the berm crest is built up above the beach surface, leading to the formation of a runnel which may be filled with water for part of the tidal cycle (Figure 8.26; see Figure 8.17A). On a fine sand beach, the surface slope may not be much greater than the slope of the groundwater table with the result that the groundwater table is close to the surface over most of the profile. Under these circumstances, capillary rise maintains a damp beach surface and evaporation leads to deposition of salts.

There is considerable interaction between the groundwater and the ocean at the seaward margin of the beach, due to movement of the mean water level with tide and wind effects, and to the effects of swash and backwash action. On the rising tide, the increase in the mean water level results in swash reaching further up the foreshore and the dry portion of the foreshore slope is inundated more and more frequently. This produces a net infiltration of water into the beach surface which infiltrates to the groundwater and leads to a rise in the groundwater table at that location (Figure 8.27A). Conversely, the swash zone migrates offshore and downward on

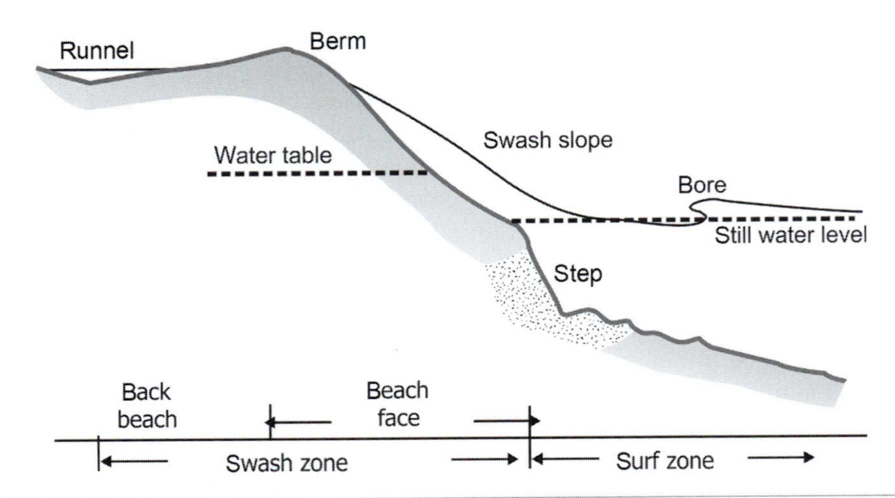

Figure 8.25 Schematic profile through the lower beach and foreshore showing the main topographic features and the location of the groundwater table. The profile is shown here for a beach in medium sand with a well-developed berm.

the falling tide, and there is exfiltration from a seepage zone which results in a fall in the groundwater table close to the foreshore slope (Figure 8.27B). The mean water level can drop more quickly than water can be discharged through exfiltration, with the result that the groundwater table can become decoupled from sea level (Horn, 2002). This happens particularly during spring tides when the low tide terrace seaward of the bottom of the foreshore may be exposed and seepage occurs over quite a wide zone (Figure 8.17A).

8.4.1.2 *Berm Building*

The migration of the swash zone up and down the foreshore with the diurnal tidal cycle produces a distinctive cycle of 'cut and fill' on the swash slope that was described in a classic paper by Duncan (1964; see Figure 8.26). As the tide rises, a wedge of sediment is moved up the swash slope by the waves, with a zone of erosion occurring on the lower portion where entrainment is made easier by the positive pore water pressure generated in the exfiltration zone. At the top of the swash, water infiltrates into the unsaturated zone above the outcrop of the water table, thus reducing the volume of the backwash and allowing sediment to be deposited there. Sediment is stranded near the top of the swash on the falling tide, but erosion occurs on the lower portion as the effluent zone migrates downward. During low wave conditions, this process results in a steepening of the foreshore slope and the building of a berm crest over a period of days, while the foreshore may prograde due to the landward movement of sediment from the inner nearshore (Hine, 1979).

Berm building under low wave conditions, due to the interaction between waves and the beach groundwater table, is further enhanced by the spring/neap tidal cycle. As the tidal range increases during the transition from neap to ebb tides, the high tide limit of the swash migrates further up the foreshore and leads to overtopping of the berm for a period of tens of minutes to an hour or more. Sediment tends to be deposited at

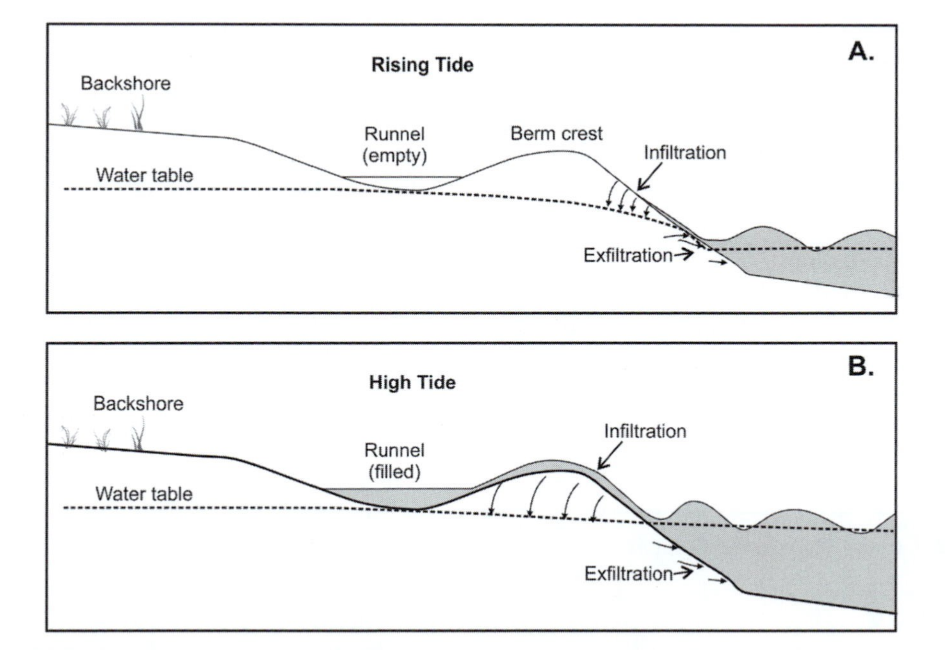

Figure 8.26 Dynamics of the groundwater table under: A. rising tide; and B. falling tide. During the rising tide increasing swash excursion leads to infiltration of water to the groundwater and a local increase in the level of the groundwater table. During the falling tide swash does not reach as far up the swash slope and there is net exfiltration of water from the groundwater which is evident as seepage. This produces a net draw down of the beach groundwater table near the berm.

the berm crest and down the landward slope as water infiltrates into the beach. At spring tides, water may inundate the whole zone, flooding the runnel and extending right up the beach (Figure 8.17A). The result of this is to build up the height of the berm. On the return to neap tides, waves no longer overtop the berm at high tide and changes take place only on the swash slope, often leading to the seaward building of the berm (Weir *et al.*, 2006). During this period, the berm crest dries out and there is the potential for considerable aeolian transport of sediment which may complete the infilling of the runnel and transfer sediments to the backshore and embryo dune area (see also Anthony, 2013).

Early studies (Bagnold, 1940; Bascom, 1953) have suggested that berm height increases with increasing wave height, although this may be more evident in wave tank studies than in the field where tidal fluctuations and varying wave conditions make it more difficult to extricate this influence from among several others. Some work has also suggested that berm height increases with increasing grain size and certainly the height of gravel and cobble berms appears to be higher for given wave conditions (Bascom, 1953). Field studies are also complicated by the fact that an increase in wave height is often accompanied by increased wave and wind set-up, which leads to erosion and flattening of the berm.

8.4.1.3 *Bar Welding*

The emergence and onshore migration of swash bars, which may be the final phase of onshore migration of bars from the nearshore zone, is the second major mechanism of beach restoration and berm building during the interval between storms. The emergence and transformation to a berm occurs in a process termed bar welding and, as shown in Figure 8.24, may include the welding of more than one bar as new bars are formed in the inner surf zone and migrate onshore. Bar welding and berm formation are the final phase of the accretional cycle of the beach stage model but, as we have seen, swash bar formation and bar welding can also occur where bars continue to exist in the nearshore (Aagaard *et al.*, 2005). Like the landward migration of swash bars, bar welding can be an important supply of sediment to the beach and ultimately the foredune (Aagaard *et al.*, 2004).

How sediment moves from the welded bar to the beach remains unclear and can occur through either continued landward migration of the bar and incorporation into the beach, or through the advection of sediment from the bar to the swash without the migration of the entire form (Houser and Barrett, 2010). Regardless of the mechanism, the welding of the innermost bar creates a gentle slope within the inner-nearshore that causes final wave breaking over a broad area, that in turn creates swash bores that tend to be depositional. If wave breaking occurs closer to the seaward margin of the beach, the swash bores become asymmetric and offshore transport and swash steepening is promoted. In this respect, bar welding not only provides a supply of sediment to the base of the foreshore, but also transforms the wave field to promote accretion. It also means that the behavior of the swash zone and all zones further landward are controlled by the behaviour of the nearshore bars and their ability to transform the incident wave field.

Box 8.2 Coastal Management Application

Beach Nourishment and Mega-Nourishment

Beach nourishment involves the importation of sand (occasionally gravel) from an onshore or offshore source and its placement on a beach. Nourishment is carried out in order to mitigate hazards such as flooding and erosion and to restore and enhance the attractiveness of the beach for recreation and tourism (Figure 8.27). Its use has grown rapidly since the 1950s, reflecting the increased demand for properties on the coast and beach tourism, especially

Box 8.2 (cont.)

Figure 8.27 Photographs the beach along the Gold Coast, Queensland, Australia taken in: A. 2013, showing the narrow beach following several years of erosion during storms; and B. 2017, showing the wide beach following a major beach nourishment project.

in countries such as the USA, Australia and those bordering the Mediterranean. In the past three decades its use has been encouraged as a 'soft' alternative to shore protection structures, which have been perceived to be destructive of sandy beach habitats (Williams *et al.*, 2016). In the US, beach nourishment has been used particularly to restore beaches where sand has been lost as a result of reduced sediment supply (e.g., Valverde *et al.*, 1999). Often erosion results from human actions (e.g., interception of longshore drift through construction of harbour walls, construction of dams on rivers supplying sand to the coast and sand mining) and frequently the replenishment is initiated after major storm events when the remaining sand is removed offshore and alongshore.

There has been much written about individual nourishment projects (e.g., Castelle *et al.*, 2009; Ludka *et al.*, 2018; Marinho et al., 2018), and about the practice of nourishment (Dean, 2002). However, many concerns have been raised about poor post-construction monitoring of many nourishment projects which has made it difficult to assess critical factors such as the stability and longevity of the fill material, the protection afforded to development along the coast, whether the economic benefits outweigh the costs, and whether there were any significant environmental impacts (Ludka *et al.*, 2018). In many cases projects are carried out in the aftermath of a major erosional event and there is insufficient time to carry out effective benefit–cost analysis and to evaluate alternative long-term plans that might involve relocation of structures within the hazard zone. Following Hurricane Sandy in October 2012 there was extensive beach scraping to provide temporary protection to development along much of the New Jersey coast (Figure 8.28). A study of 40 km of shoreline between Manasquan Inlet and Sandy Hook (Messaros *et al.*, 2018) documented the placement of about 11×10^6 m^3 of beach nourishment during the period 2008–16. Nearly half of this was placed following the impact of Hurricane Sandy. The study by Messaros *et al.* (2018) shows a net negative littoral sediment budget along this portion of the New Jersey coast and points to the need for continuing nourishment in order to compensate for the losses. It is doubtful if this offers a long-term economic solution, particularly in the face of ongoing sea level rise here, and in many other parts of the world (Parkinson and Ogurcak, 2018).

Except for sand bypassing schemes, where there is no fixed limit on operations, almost all beach nourishment schemes can be described as short term with a design life of < 20 years. They are typically applied to a few kilometres of shoreline with the volume of nourishment on the order of 50 000 m^3 to 1–2 $\times$ 10^6 m^3

Box 8.2 (cont.)

Figure 8.28 Beach scraping along the northern New Jersey shoreline following the impact of Hurricane Sandy in October 2012. The photograph was taken one year later in October 2013 prior to nourishment. There is no room onshore for the development of a natural foredune and, in the absence of vegetation, the beach scraping provides limited protection against another storm. Meanwhile, it is vulnerable to erosion due to onshore aeolian transport (see, e.g., Smyth and Hesp, 2015).

Figure 8.29 The Sandmotor nourishment in 2014 two years after placement. (Google Earth image.)

(e.g., Valverde *et al.*, 1999; Cooke *et al.*, 2012). The Dutch 'De Zandmotor' or Sandmotor has been called a mega nourishment scheme because it is designed to supply sediment to beaches and dunes along a 10–20 km stretch of coast over a period of 20 years from a single large nourishment source of about 21×10^6 m^3 of sand (Stive *et al.*, 2013; Figure 8.29). The nourished material was dredged from about 10 km offshore during 2011 and 2012 (Stive *et al.*, 2013). It extends about 2 km alongshore and up to 1 km offshore, and has a purpose designed shape that includes a lagoon and a pond which can be used for recreational purposes (Figure 8.29). The nourished material was placed at a drift divide on the coast and it has been estimated that about 60 per cent of the material is transported to the north and 40 per cent to the south (Stive *et al.*, 2013). The size of the scheme and concentration in one relatively small area has economic advantages in reduced cost for the sediment used for nourishment and for the operation itself. It also has a much smaller impact on the coast (Stive *et al.*, 2013; Brown *et al.*, 2016).

The Sandmotor project appears to be working largely as designed in terms of the spread of the nourished material along the coast (Stive *et al.*, 2013; Brown *et al.*, 2016; Stronkhorst *et al.*, 2018). However, at the site itself sand supply to the foredunes has been much less than anticipated because of the high concentration of shells within the dredged sediments. Winnowing of the fine sand has resulted in a concentration of shells to produce an armoured surface which acts to restrict aeolian sand transport from the subaerial part of the dredged sediments (Figure 8.30). The development of surface armouring, together with the effects of surface moisture and fetch

Box 8.2 (cont.)

Figure 8.30 Photograph taken in November, 2018 of the development of surface armouring by shells on the Sandmotor nourishment site. (Photo courtesy Irene Delgado-Fernandez.)

limitations, introduces many complexities in the modelling of sand supply to the foredune over time (Hoonhout and de Vries, 2019).

The Sandmotor is designed to be part of the Dutch 'working with nature' approach adopted to allow as far as possible the development of natural coastal dunes to provide protection against rising sea level and severe storm events, as well as to prevent salt water intrusion into the terrestrial groundwater. While beach nourishment has a lower environmental impact than hard structures, the dredging of 21×10^6 m^3 of sediment from offshore is hardly trivial, and this has to be considered in light of the annual total nourishment for all of the Dutch coast which is on the order of $8–10 \times 10^6$ m^3. It works here on the Dutch coast because of the relatively simple morphology, an abundant supply of borrow material offshore, and the history and resources of several hundred years of holding the sea back. The larger issue for coastal communities around the world is how to incorporate beach nourishment within a broader strategy of adaptation to sea level rise over the next 50–100 years. While it may be possible in some areas to use beach nourishment to 'hold the line' (Houston, 2017; Mesaros *et al.*, 2018) this is a very expensive proposition and one that seems improbable for most countries (Parkinson and Ogurcak, 2018).

8.4.1.4 *Beach Cusps*

As we saw in Section 8.3.1, the development of rip current cells and associated rhythmic inner bars leads to the development of cusps opposite the rip channels and horns opposite the feeder current divides. These 'giant cusps' (Komar, 1971, 1983) are best developed when a steep, reflective foreshore slope and berm have developed. They have an alongshore wavelength that is controlled by the dimensions of the rip cells and the rhythmic inner bars (usually 100–300 m), and they may persist for weeks or even a few months. It follows that the presence of these cusps can be used to determine whether there are rip channels present within the inner-nearshore.

In contrast, beach cusps are much smaller features with length scales on the order of a few metres to a few tens of metres (Figure 8.31). They consist of an embayment between two horns

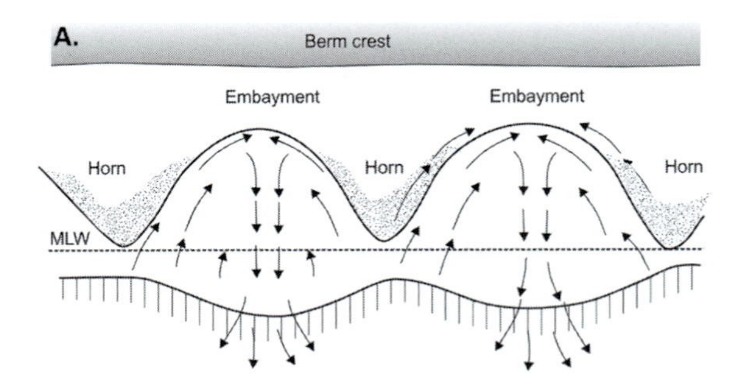

Figure 8.31 Beach cusps: A. sketch of swash and backwash flows in developing cusps. The swash uprush tends to be diverted away from the high points of the cusp horns towards the centre of the embayment, especially near the end of the swash and beginning of the backwash; B. photo of cusps on a sand beach (courtesy Patrick Hesp).

which point seaward. They form best when there is some a combination of a steep foreshore and relatively low waves that permits the backwash cycle from one wave to be completed or nearly completed before the next wave breaks. Cusps also appear to develop more often in medium to coarse sand or gravel, which permits some infiltration of the swash during uprush, particularly on the cusp horns. Cusps have attracted a lot of attention – probably much more than is deserved given their limited significance for beach dynamics – but we all have a fascination for trying to understand any natural phenomenon that gives rise to a regular pattern. While cusps do have a regular pattern and a characteristic spacing Figure 8.31A, B), actual field measurements can show quite a large coefficient of variation for their wavelength (Nolan *et al.*, 1999). There are useful reviews in Seymour and Aubrey (1985), Masselink *et al.* (1997), Komar (1998) and Almar *et al.* (2008).

Observations show that near the top of the swash, run-up flow near the horns tends to be divided and directed towards the middle of the embayment (Figure 8.31A). There is some infiltration of water near the top of the cusp horns, and this, together with the diversion towards the centre of the embayment, results in most of the backwash being concentrated in the embayment. The concentration of backwash in the embayment produces a deep flow that tends to scour the embayment, thus enhancing the alongshore relief and at the end of the backwash this jet can flow seaward beyond the beach in a form of miniature rip (Figure 8.31A). Once initiated, cusps can develop through erosion of the embayment, deposition on the ridges that form the cusp horns or some combination of the two. There are conflicting reports of whether cusps are erosional or depositional features – whether their formation is accompanied by net accretion or erosion of the foreshore slope is still not clear. It seems likely that either can be true, but the conditions under which they develop usually favour stability or slight accretion of the foreshore. In addition, a change to steep, higher frequency waves usually results in erosion of the foreshore and destruction of any cusps present.

The low waves and steep foreshore slope are favourable to the generation of edge waves, and so it is not surprising that one of the proposed mechanisms for cusp formation is the presence of a mode 0 edge wave (Guza and Inman, 1975, Huntley and Bowen, 1978). The edge wave that leads to cusp formation may be either synchronous (period equal to the incident waves), or more often subharmonic (period equal to one-half the incident wave frequency). The beach cusp spacing based on the edge wave mechanism can be predicted from:

$$\lambda_c = m\frac{g}{\pi}T_i^2 \sin\beta, \tag{8.9}$$

where λ_c is the cusp spacing, T_i is the incident wave frequency and m is 1 for synchronous edge waves and 0.5 for subharmonic edge waves. While several studies have suggested that edge waves are responsible for the initiation of cusps (Huntley and Bowen, 1978; Sallenger, 1979) many recent studies have failed to establish this link (Holland and Holman, 1996; Masselink *et al.*, 1997; Almar *et al.*, 2008). A limitation of the edge wave model is that once the cusps develop, they tend to suppress edge wave motion.

An alternate model that seems to be gaining favour is that cusps simply represent a form of self-organisation (Werner and Fink, 1993; Wang and Sherman, 2016). Cusps arise because of non-linear interactions between flow, sediment transport and morphology, provided that suitable conditions exist to reinforce the original perturbation that initiates the interaction. The results of field studies by Holland and Holman (1996) and Masselink *et al.* (1997) appear to favour this mechanism, as do the 3 years of beach video monitoring by Almar *et al.* (2008). Under this model of formation, the key control on cusp spacing is simply the horizontal distance of the swash excursion (Werner and Fink, 1993; Coco *et al.*, 1999). A field study by Wang and Sherman (2016) found that the cusp spacing (of 7.6 m) at Portmore Beach, Ireland is best explained by the swash excursion length (4.7 m) and not a synchronous standing edge wave, which should have generated a cusp spacing of 130 m.

References

Aagaard, T. 1990. Infragravity waves and nearshore bars in protected, storm-dominated coastal environments. *Marine Geology*, **94**, 181–203.

Aagaard, T. and Greenwood, B. 1995. Longshore and cross-shore suspended sediment transport at far infragravity frequencies in a barred environment. *Continental Shelf Research*, **15**, 1235–49.

Aagaard, T. and Hughes, M.G. 2017. Equilibrium shoreface profiles: a sediment transport approach. *Marine Geology*, **390**, 321–30.

Aagaard, T. and Masselink, G. 1999. The surf zone. In Short, A.D. (ed.), *Handbook of Beach and Shoreface Morphodynamics*. Chichester, UK: John Wiley & Sons, pp. 72–113.

Aagaard, T., Nielsen J. and Greenwood, B. 1998. Suspended sediment transport and nearshore bar formation on a shallow intermediate state beach. *Marine Geology*, **148**, 203–25.

Aagaard, T., Davidson-Arnott, R., Greenwood, B. and Nielsen, J. 2004. Sediment supply from shoreface to dunes: linking sediment transport measurements and long-term morphological evolution. *Geomorphology*, **60**(1–2), 205–24.

Aagaard, T., Kroon, A., Andersen, S. *et al.* 2005. Intertidal beach change during storm conditions; Egmond, The Netherlands. *Marine Geology*, **218**, 65–80.

Aagaard, T., Kroon, A., Hughes, M.G. and Greenwood, B. 2008. Field observations of nearshore bar formation. *Earth Surface Processes and Landforms*, **33**, 1021–32.

Almar, R., Coco, G., Bryan, K.R. *et al.* 2008. Video observations of beach cusp morphodynamics. *Marine Geology*, **254**, 215–23.

Anthony, E.J. 2013. Storms, shoreface morphodynamics, sand supply, and the accretion and erosion of coastal dune barriers in the southern North Sea. *Geomorphology*, **199**, 8–21.

Are, F. and Reimnitz, E. 2008. The A and m coefficients in the Bruun/Dean equilibrium profile equation seen from the Arctic. *Journal of Coastal Research*, **24 SP2**, 243–9.

Ashton, A., Murray, A.B. and Arnault, O. 2001. Formation of coastal features by large-scale instabilities induced by high angle waves. *Nature*, **414**, 296–300.

Austin, M.J. and Buscombe, D. 2008. Morphological change and sediment dynamics of the beach step on a macrotidal gravel beach. *Marine Geology*, **249**, 167–83.

Bagnold, R.A. 1940. Beach formation by waves: some model experiments in a wave tank. *Journal of the Institute of Civil Engineers*, **15**, 27–52.

Barrett, G. and Houser, C., 2012. Identifying hotspots of rip current activity using wavelet analysis at Pensacola Beach, Florida. *Physical Geography*, **33**(1), 32–49.

Bascom, W.N. 1951. The relationship between sand size and beach face slope. *Transactions of the American Geophysical Union*, **52**, 866–74.

Bascom, W.N. 1953. Characteristics of natural beaches. *Proceedings of the Fourth Coastal Engineering Conference*. New York: American Society of Civil Engineers, pp. 163–80.

Bauer, B.O. and Greenwood, B. 1990. Modification of a linear bar–trough system by a standing edge wave. *Marine Geology*, **92**, 177–204.

Bluck, Bj. 1967. Sedimentation of beach gravels: examples from south Wales. *Journal of Sedimentary Petrology*, **37**, 128–56.

Boczar-Karakiewicz, B. and Davidson-Arnott, R.G.D. 1987. Nearshore bar formation by non-linear wave processes: A comparison of model results and field data. *Marine Geology*, **77**, 287–304.

Boufadel, M.C., Abdollahi-Nasab, A., Geng, X., Galt, J. and Torlapati, J. 2014. Simulation of the landfall of the deepwater horizon oil on the shorelines of the Gulf of Mexico. *Environmental science & technology*, **48** (16), 9496–505.

Bowen, A.J. 1980. Simple models of nearshore sedimentation; beach profiles and longshore bars. In McCann, S.B. (ed.), *The Coastline of Canada*. Ottawa: Geological Survey of Canada Paper 80-10, pp. 1–11.

Bowen, A.J. and Inman, D.L. 1971. Edge waves and crescentic bars. *Journal of Geophysical Research*, **76**, 8862–71.

Brander, R.W. 1999. Field observations on the morphodynamic evolution of a low-energy rip current system. *Marine Geology*, **157**, 199–217.

Brown, J.M., Phelps, J.J.C., Barkwith, A. *et al.* 2016. The effectiveness of beach mega-nourishment, assessed over three management epochs. *Journal of Environmental Management*, **184**, 400–8.

Bruun, P. 1954. *Coast Erosion and the Development of Beach Profiles*. Washington, DC: US Beach Erosion Board, Technical Memorandum, **44**, 66 pp.

Bryan, K.R., Howd, P.A. and Bowen, A.J. 1998. Field observations of bar-trapped edge waves. *Journal of Geophysical Research: Oceans*, **103**(C1), 1285–305.

Buscombe, D. and Masselink, G. 2006. Concepts in gravel beach dynamics. *Earth-Science Reviews*, **79**, 33–52.

Carls, M.G., Babcock, M.M., Harris, P.M. *et al.* 2001. Persistence of oiling in mussel beds after the Exxon Valdez oil spill. *Marine Environmental Research*, **51**(2), 167–90.

Carter, T.G., Liu, P.L. and Mei, C.C. 1973. Mass transport by waves and offshore sand bedforms. *Journal of Waterways, Harbors and Coastal Engineering*, **99**(2), 165–84.

Castelle, B., Bonneton, P., Dupuis, H. and Sénéchal, N. 2007. Double bar beach dynamics on the high-energy meso-macrotidal French Aquitanian Coast: a review. *Marine Geology*, **245**(1–4), 141–59.

Castelle, B., Turner, I.L., Bertin, X. and Tomlinson, R., 2009. Beach nourishments at Coolangatta Bay over the period 1987–2005: impacts and lessons. *Coastal Engineering*, **56**, 940–50.

Castelle, B., Ruessink, B.G., Bonneton, P. *et al.* 2010a. Coupling mechanisms in double sandbar systems. Part 2: impact on alongshore variability of inner-bar rip channels. *Earth Surface Processes and Landforms: The Journal of the British Geomorphological Research Group*, **35**(7), 771–81.

Castelle, B., Ruessink, B.G., Bonneton, P. *et al.* 2010b. Coupling mechanisms in double sandbar systems. Part 1: patterns and physical explanation. *Earth Surface Processes and Landforms*, **35**(4), 476–86.

Coco, G., O'Hare, T.J. and Huntley, D.A. 1999. Beach cusps: a comparison of data and theories for their formation. *Journal of Coastal Research*, **15**, 741–9.

Cooke, B.C., Jones, A.R., Goodwin, I.D. and Bishop, M.J. 2012. Nourishment practices on Australian sandy beaches: a review. *Journal of Environmental Management*, **113**, 319–27.

Cowell, P.J., Hanslow, D.J. and Meleo, J.F. 1999. The shoreface. In Short, A.D. (ed.), *Handbook of Beach and Shoreface Morphodynamics*. Chichester, UK: John Wiley & Sons, pp. 39–71.

Dally, W.R. 1987. Longshore bar formation: surf beat or undertow? In *The Proceedings of the Coastal Sediments '87*. New York: American Society of Civil Engineers, pp. 71–86.

Dally, W.R. and Dean, R.G. 1984. Suspended sediment transport and beach profile evolution. *Journal of Waterways, Port, Coastal and Ocean Engineering*, **110**, 15–33.

Dalrymple, R.A. 1992. Prediction of storm/normal beach profiles. *Journal of Waterways, Port, Coastal and Ocean Engineering*, **118**, 193–200.

Damgaard Christensen, E., Deigaard, R., Fredsøe, J. 1994. Sea bed stability on a long straight coast. *Proceedings of Twenty-fourth Conference on Coastal Engineering*. New York: American Society of Civil Engineers, pp. 1865–79.

Davidson, M.A., Splinter, K.D. and Turner, I.L. 2013. A simple equilibrium model for predicting shoreline change. *Coastal Engineering*, **73**, 191–202.

Davidson-Arnott, R.G.D. 1987. Controls on the formation and form of barred nearshore profiles, *Geographical Review*, **78**, 185–93.

Davidson-Arnott, R.G. 1988. Temporal and spatial controls on beach/dune interaction, Long Point, Lake Erie. *Journal of Coastal Research*, **SI 3**, 131–6.

Davidson-Arnott, R.G.D. 2013. Nearshore bars. In Shroder, J.F. (ed.), *Treatise on Geomorphology*, vol. 10. San Diego: Academic Press, pp. 130–48.

Davidson-Arnott, R.G.D. and McDonald, R.A. 1989. Nearshore water motion and mean flows in a multiple parallel bar system. *Marine Geology*, **86**, 321–38.

Davidson-Arnott, R.G.D. and Pember, G.F. 1980. Morphology and sedimentology of multiple parallel bar systems, Southern Georgian Bay, Ontario. In McCann, S.B. (ed.), *The Coastline of Canada*. Ottawa: Geological Survey of Canada Paper **80-10**, pp. 417–28.

Davis, R.A. Jr and Fox, W.T. 1972. Coastal Processes and nearshore sand bars. *Journal of Sedimentary Petrology*, **42** 401–12.

Davis, R.A. Jr, Fox, W.T., Hayes, M.O. and Boothroyd, J.C. 1972. Comparison of ridge and runnel systems in tidal and non-tidal environments. *Journal of Sedimentary Petrology*, **42**, 413–21.

Dawson, J.C., Davidson-Arnott, R.G.D. and Ollerhead, J. 2002. Low-energy morphodynamics of a ridge and runnel system. *Journal of Coastal Research*, **SI 36**, 198–215.

Dean, R.G. 1991. Equilibrium beach profiles: characteristics and applications. *Journal of Coastal Research*, **7**, 53–84.

Dean, R.G. 1997. Models for barrier island restoration. *Journal of Coastal Research*, **13**, 694–703.

Dean, R.G., 2002. *Beach Nourishment: Theory and Practice*. Singapore: World Scientific Press, 399 pp.

Dean, R.G., Healy, T.R. and Dommerholt, A.P. 1993. A 'blind-folded' test of equilibrium beach profile concepts with New Zealand data. *Marine Geology*, **109**, 253–66.

Deidda, M. and Sanna, G. 2012. Bathymetric extraction using WorldView-2 high resolution images. *International Archives of the Photogrammetry, Remote Sensing and Spatial Information Sciences*, **39**(B8), 153–7.

Delgado-Fernandez, I., Davidson-Arnott, R. and Ollerhead, J. 2009. Application of a remote sensing technique to the study of coastal dunes. *Journal of Coastal Research*, **25**(5), 1160–7.

Duncan, R. 1964. The effects of water table and tidal cycle on swash-backwash sediment distribution and beach profile development. *Marine Geology*, **2**, 186–97.

Dyhr-Nielsen, M. and Sørensen, T. 1970. Some sand transport phenomena on coasts with bars. *Proceedings of the Twelfth Coastal Engineering Conference*. New York: American Society of Civil Engineers, pp. 855–66.

Evans, O.F. 1940. The low and ball of the east-shore of Lake Michigan. *Journal of Geology*, **48**, 467–511.

Falqués, A., Dodd, N., Garnier, R. *et al.* 2008. Rhythmic surf zone bars and morphodynamic self-organization. *Coastal Engineering*, **55**, 622–41.

Garcez-Faria, A.F., Tornton, E.B., Lippmann, T.C. and Stanton, T.P. 2000. Undertow over a barred beach. *Journal of Geophysical Research*, **105**(C7), 16999–7010.

González, J., Figueiras, F.G., Aranguren-Gassis, M. *et al.* 2009. Effect of a simulated oil spill on natural assemblages of marine phytoplankton enclosed in microcosms. *Estuarine, Coastal and Shelf Science*, **83**(3), 265–76.

Goodfellow, B.W. and Stephenson, W.J. 2005. Beach morphodynamics in a strong-wind bay: a low energy environment? *Marine Geology*, **214**, 101–16.

Gorman, L., Morang, A. and Larson, R. 1998. Monitoring the coastal environment; Part IV: mapping, shoreline change and bathymetric analysis. *Journal of Coastal Research*, **14**, 61–92.

Gourlay, M.R. 1968. *Beach and Dune Erosion Tests*. Delft Hydraulics Laboratory Report No. M935/M936.

Greenwood, B. and Davidson-Arnott, R.G.D. 1972. Textural variation in the sub-environments of the shallow-water wave zone, Kouchibouguac Bay, New Brunswick. *Canadian Journal of Earth Sciences*, **9**, 679–88.

Greenwood, B. and Davidson-Arnott, R.G.D. 1975. Marine bars and nearshore sedimentary processes, Kouchibouguac Bay, New Brunswick, Canada. In Hails, J. and Carr, A. (eds.), *Nearshore Sediment Dynamics and Sedimentation: An Interdisciplinary Review*. Chichester, UK: John Wiley & Sons, pp. 123–50.

Greenwood, B. and Davidson-Arnott, R.G.D. 1979. Sedimentation and equilibrium in wave-formed bars: a review and case study. *Canadian Journal of Earth Sciences*, **16**, 312–32.

Greenwood, B. and Mittler P.R. 1984. Sediment flux and equilibrium slopes in a barred nearshore. *Marine Geology*, **60**, 79–98.

Greenwood, B., Hale, P.B. and Mittler, P.R. 1979. Sediment flux determination in the nearshore zone. *Proceedings Workshop on Instrumentation for Currents and Sediments in the Nearshore Zone*. Ottawa: National Research Council of Canada, pp. 99–115.

Greenwood, B., Richards, R.G. and Brander, R.W. 1993. Acoustic imaging of sea-bed geometry: a high resolution remote tracking sonar (HERTSII). *Marine Geology*, **112**, 207–18.

Greenwood, B., Permanand-Schwatrz, A. and Houser, C.A. 2006. Emergence and migration of a nearshore bar: sediment flux and morphological change on a multi-barred beach in the Great Lakes. *Géographie Physique et Quaternaire*, **60**, 31–47.

Guza, R.T. and Inman, D.L. 1975. Edge waves and beach cusps. *Journal of Geophysical Research*, **80**, 2997–3012.

Hale, P. B. and McCann, S.B. 1982. Rhythmic topography in a mesotidal, low-wave energy environment. *Journal of Sedimentary Petrology*, **52**, 415–29.

Hallermeier, R.J. 1981. A profile zonation for seasonal sand beaches from wave climate. *Coastal Engineering*, **4**, 253–77.

Hancock, G. and Willgoose, G. 2001. The production of digital elevation models for experimental model landscapes. *Earth Surface Processes and Landforms*, **26**, 475–90.

Harley, M.D., Turner, I.L. and Short, A.D. 2015. New insights into embayed beach rotation: The importance of wave exposure and cross-shore processes. *Journal of Geophysical Research: Earth Surface*, **120**(8), 1470–84.

Hine, A.C. 1979. Mechanisms of berm development and resulting beach growth along a barrier spit complex. *Sedimentology*, **26**, 333–51.

Holland, K.T. and Holman, R.A. 1996. Field observations of beach cusps and swash motions. *Marine Geology*, **134**, 77–93.

Holman, R.A. and Lippman, T.C. 1986. Remote sensing of nearshore bar systems: making morphology visible. In *The Proceedings of the Coastal Sediments '87*. New York: American Society of Civil Engineers, pp. 929–44.

Holman, R.A. and Sallenger, A.H. 1993. Sand bar generation: a discussion of the Duck experiment series. *Journal of Coastal Research*, **SI 15**, 76–92.

Holman, R.A. and Stanley, J. 2007. The history and technical capabilities of Argus. *Coastal Engineering*, **54**, 477–91.

Hoonhout, B. and de Vries, S. 2019. Simulating spatiotemporal aeolian sediment supply at a mega nourishment. *Coastal Engineering*, **145**, 21–35.

Horn, D.P. 1992. A numerical model for shore-normal sediment size variation on a macro-tidal beach. *Earth Surface Processes and Landforms*, **17**, 755–73.

Horn, D.P., 2002. Beach groundwater dynamics. *Geomorphology*, **48**, 121–46.

Houser, C. 2012. Feedback between ridge and swale bathymetry and barrier island storm response and transgression. *Geomorphology*, **173**, 1–16.

Houser, C. 2013. Flow Separation over a Prograding Beach Step at Pensacola Beach, Florida. *Journal of Coastal Research*, **29**(6), 1247–56.

Houser, C. and Barrett, G. 2010. Divergent behavior of the swash zone in response to different foreshore slopes and nearshore states. *Marine Geology*, **271**(1–2), 106–18.

Houser, C. and Greenwood, B. 2005. Profile response of a lacustrine multiple barred nearshore to a sequence of storm events. *Geomorphology*, **69**, 118–37.

Houser, C., Greenwood, B. and Aagaard, T. 2006. Divergent response of an intertidal swash bar, *Earth Surface Processes and Landforms*, **31**, 1775–91.

Houser, C., Hapke, C. and Hamilton, S. 2008. Controls on coastal dune morphology, shoreline erosion and barrier island response to extreme storms. *Geomorphology*, **100**(3–4), 223–40.

Houser, C., Barrett, G. and Labude, D. 2011. Alongshore variation in the rip current hazard at Pensacola Beach, Florida. *Natural Hazards*, **57**(2), 501–23.

Houser, C., Arnott, R., Ulzhöfer, S. and Barrett, G. 2013. Nearshore circulation over transverse bar and rip morphology with oblique wave forcing. *Earth Surface Processes and Landforms*, **38**(11), 1269–79.

Houser, C., Wernette, P., Rentschlar, E. *et al.* 2015. Post-storm beach and dune recovery: Implications for barrier island resilience. *Geomorphology*, **234**, 54–63.

Houser, C., Wernette, P. and Weymer, B.A. 2018. Scale-dependent behaviour of the foredune: Implications for barrier island response to storms and sea-level rise. *Geomorphology*, **303**, 362–74.

Houston, J.R. 2017. Shoreline change in response to sea-level rise on Florida's west coast. *Journal of Coastal Research*, **336**, 1243–60.

Huntley, D.A. 1980. Edge waves in a crescentic bar system. In McCann, S.B. (ed.), *The Coastline of Canada*. Ottawa: Geological Survey of Canada Paper **80-10**, pp. 111–21.

Huntley, D.A. and Bowen, A.J. 1978. Beach cusps and edge waves. *Proceedings Sixteenth Conference on Coastal Engineering*. New York: American Society of Civil Engineers, pp. 1378–93.

Inman, D.L., Elwany, M.H. and Jenkins, S.A. 1993. Shore rise and bar berm on ocean beaches. *Journal of Geophysical Research*, **98**, 18181–99.

Jackson, D.W.T., Cooper, J.A.G. and Del Rio, L. 2005. Geological control of beach morphodynamic state. *Marine Geology*, **216**(4), 297–314.

Jennings, R. and Shulmeister, J. 2002. A field based classification scheme for gravel beaches. *Marine Geology*, **182**, 211–28.

Keulegan, G.H. 1948. An Experimental Study of Submarine Sandbars. Washington, DC: US Beach Erosion Board, Technical Report **3**, 40 pp.

King, C.A.M. and Williams, W.W. 1949. The formation and movement of sand bars by wave action. *Geographical Journal*, **112**, 70–85.

Komar, P.D. 1971. Nearshore cell circulation and the formation of giant cusps. *Geological Society of America Bulletin*, **82**, 2643–50.

Komar, P.D. 1983. Rhythmic shoreline features and their origin. In Gardner, R. and Scoging, H. (eds.), *Mega-Geomorphology*. Oxford: Clarendon Press, pp. 92–112.

Komar, P.D. 1998. *Beach Processes and Sedimentation*, 2nd Edition. Upper Saddle River, NJ: Prentice Hall, 544 pp.

Komar, P.D. and McDougal, W.G. 1994. The analysis of beach profiles and nearshore processes using the exponential beach profile form. *Journal of Coastal Research*, **10**, 59–69.

Konicki, K.M. and Holman, R.A. 2000. The statistics and kinematics of transverse sand bars on an open coast. *Marine Geology*, **169**, 69–101.

Kroon A. and Masselink, G. 2002. Morphodynamics of intertidal bar morphology on a macrotidal beach under low-energy wave conditions. North Lincolnshire, England. *Marine Geology*, **190**, 591–608.

Lanckriet, T., Puleo, J.A. and Waite, N. 2013. A conductivity concentration profiler for sheet flow sediment transport. *IEEE Journal of Oceanic Engineering*, **38**(1), 55–70.

Larson, M. and Sunamura, T. 1993. Laboratory experiment on flow characteristics at a beach step. *Journal of Sedimentary Research*, **63**(3), 495–500.

Lawler, D.M. 1992. Design and installation of a novel automatic erosion monitoring system. *Earth Surface Processes and Landforms*, **17**, 455–63.

Lazarus, E., Ashton, A., Murray, A.B., Tebbens, S. and Burroughs, S. 2011. Cumulative versus transient shoreline change: dependencies on temporal and spatial scale. *Journal of Geophysical Research: Earth Surface*, **116**(F2), FO2014.

Lee, G., Nicholls, R.J. and Birkemeier, W.A. 1998. Storm-driven variability of the beach-nearshore profile at Duck, North Carolina, 1981–1991. *Marine Geology*, **148**, 163–77.

Levoy, F., Anthony, E.J., Monfort, O. and Larsonneur, C. 2000. The morphodynamics of megatidal beaches in Normandy, France. *Marine Geology*, **171**, 39–59.

Lippmann, T.C. and Holman, R.A. 1990. The spatial and temporal variability of sand bar morphology. *Journal of Geophysical Research*, **106**, 973–89.

Lippmann, T.C., Holman, R.A. and Hathaway, K.K. 1993. Episodic, non-stationary behaviour of a double bar system at Duck, N.C., USA, 1986–1991. *Journal of Coastal Research*, **SI 15**, 49–75.

Liu, J.T. and Zarillo, G.A. 1989. Distribution of grain size across a transgressive shoreface. *Marine Geology*, **87**, 121–36.

Ludka, B.C, Guza, R.T. and O'Reilly, W.C. 2018. Nourishment evolution and impacts at four southern California beaches: A sand volume analysis. *Coastal Engineering*, **136**, 96–105.

Marinho, B., Coelho, C., Larson, M. and Hanson, H. 2018. Monitoring the evolution of nearshore nourishments along Barra-Vagueira coastal stretch, Portugal. *Ocean and Coastal Management*, **157**, 23–39.

Masselink G. 1993. Simulating the effects of tides on beach morphodynamics. *Journal of Coastal Research*, **SI 15**, 180–97.

Masselink G. 2004. Formation and evolution of multiple intertidal bars on macrotidal beaches: application of a morphodynamic model. *Coastal Engineering*, **51**, 713–30.

Masselink, G. and Hegge, B. 1995. Morphodynamics of meso- and macrotidal beaches: examples from central Queensland, Australia. *Marine Geology*, **129**, 1–23.

Masselink, G. and Pattiaratchi, C.B. 1998. The effects of sea breeze on beach morphology, surf zone hydrodynamics and sediment resuspension. *Marine Geology*, **146**, 115–35.

Masselink, G. and Short, A.D. 1993. The effect of tide range on beach morphodynamics, a conceptual model. *Journal of Coastal Research*, **9**, 785–800.

Masselink G. Hegge, B and Pattiaratchi, C.B. 1997. Beach cusp morphodynamics. *Earth Surface Processes and Landforms*, **22**, 1139–55.

Masselink, G., Evans, D., Hughes, M.G. and Russell, P. 2005. Suspended sediment transport in the swash zone of a dissipative beach. *Marine Geology*, **216**(3), 169–89.

Masselink, G., Kroon, A. and Davidson-Arnott, R.G.D. 2006. Intertidal bar morphodynamics: a review. *Geomorphology*, **73**, 33–49.

Masselink, G., Austin, M., Tinker, J., O'Hare, T. and Russell, P. 2008. Cross-shore sediment transport and morphological response on a macrotidal beach with intertidal bar morphology, Truc Vert, France. *Marine Geology*, **251**, 141–55.

McLean, R.F. and Kirk, R.M. 1969. Relationship between grain size, size-sorting, and foreshore slope on mixed sand-shingle beaches. *New Zealand Journal of Geology and Geophysics*, **12**, 138–55.

McNinch, J.E., Wells, J.T. and Trembanis, A.C. 2006. Predicting the fate of artefacts in energetic, shallow marine environments: an approach to site management. *International Journal of Nautical Archaeology*, **35**(2), 290–309.

Messaros, R.C., Rosati, J.D., Buonaiuto, F. *et al.* 2018. Assessing the coastal resilience of Manasquan Inlet to Sea Bright, New Jersey: regional sediment budget 1992–2003. *Journal of Coastal Research*, **34**, 955–67.

Michel, D. and Howa, H.L. 1999. Short-term morphodynamic response on a ridge and runnel system on a mesotidal sandy beach. *Journal of Coastal Research*, **15**, 428–37.

Miecznik, G. and Grabowska, D. 2012. Worldview-2 bathymetric capabilities. In *Algorithms and Technologies for Multispectral, Hyperspectral, and Ultraspectral Imagery XVIII*, vol. 8390. International Society for Optics and Photonics, p. 83901J.

Moloney, J. G., Hilton, M. J., Sirguey, P. and Simons-Smith, T. 2018. Coastal dune surveying using a low-cost remotely piloted aerial system (RPAS). *Journal of Coastal Research*, **34**, 1244–55.

Morris, B. A., Davidson, M. A. and Huntley, D. A. 2001. Measurements of the response of a coastal inlet using video monitoring techniques. *Marine Geology*, **175**, 251–72.

Nicholls, R.J., Birkemeier, W.A. and Lee, G-H. 1998. Evaluation of depth of closure using data from Duck, NC, USA. *Marine Geology*, **148**, 179–201.

Niedoroda, A.W. and Tanner, W.F. 1970. Preliminary study of transverse bars. *Marine Geology*, **9**(1), 41–62.

Nielsen, P. and Hanslow, D.J. 1991. Wave runup distributions on natural beaches. *Journal of Coastal Research*, **7**, 1139–52.

Nolan, T.J., Kirk, R.M. and Shulmeister, J. 1999. Beach cusp morphology on sand and mixed sand and gravel beaches, South Island, New Zealand. *Marine Geology*, **157**, 185–98.

O'Hare, T.J. and Huntley, D.A. 1994. Bar formation due to wave groups and associated long waves. *Marine Geology*, **116**, 313–25.

Ojeda, E. and Guillén, J. 2008. Shoreline dynamics and beach rotation of artificial embayed beaches. *Marine Geology*, **253**, 51–62.

Ollerhead, J., Davidson-Arnott, R., Walker, I.J. and Mathew, S. 2013. Annual to decadal morphodynamics of the foredune system at Greenwich Dunes, Prince Edward Island, Canada. *Earth Surface Processes and Landforms*, **38**(3), 284–98.

Orford, J.D. 1975. Discrimination of particle zonation on a pebble beach. *Sedimentology*, **22**, 441–63.

Orford, J.D. and Anthony, E. 2013. Coastal gravel systems. In Shroder, J.F. (ed.), *Treatise on Geomorphology*. San Diego, CA: Academic Press, pp. 245–66.

Ortega-Sánchez, M., Fachin, S., Sancho, F. and Losada, M.A. 2008. Relation between beachface morphology and wave climate at Trafalgar beach (Cadiz, Spain). *Geomorphology*, **99**, 171–85.

Parkinson, R.W. and Ogurcak, D.E. 2018. Beach nourishment is not a sustainable strategy to mitigate climate change. *Estuarine, Coastal and Shelf Science*, **212**, 203–9.

Pilkey, O.H., Young, R.S., Riggs, S.R. *et al.* 1993. The concept of shoreface profile of equilibrium: a critical review. *Journal of Coastal Research*, **9**, 255–78.

Plant, N.G., Holman, R.A., Freilich, M.H. and Birkemeir, W.A. 1999. A simple model for interannual bar behaviour. *Journal of Geophysical Research*, **104**, **C7**, 15755–76.

Plant, N.G., Freilich, M.H. and Holman, R.A. 2001. Role of morphologic feedback in surf zone sand bar response. *Journal of Geophysical Research*, **106**, 973–89.

Pontes, J., Mucha, A.P., Santos, H. *et al.* 2013. Potential of bioremediation for buried oil removal in beaches after an oil spill. *Marine Pollution Bulletin*, **76**(1-2), 258–65.

Price, T.D. and Ruessink, B.G. 2008. Morphodynamic zone variability on a microtidal barred beach. *Marine Geology*, **251**, 98–109.

Pruszak, Z., Rozynski, G. and Zeidler, R.B. 1997. Statistical properties of multiple bars. *Coastal Engineering*, **31**, 263–80.

Quartel, S., Ruessink, B.G. and Kroon, A. 2007. Daily to seasonal cross-shore behaviour of quasi-persistent intertidal beach morphology. *Earth Surface Processes and Landforms*, **32**, 1293–307.

Quartel, S., Kroon, A. and Ruessink, B.G. 2008. Seasonal accretion and erosion patterns of a microtidal sandy beach. *Marine Geology*, **250**, 19–33.

Ranasinghe, R., McLoughlin, R., Short, A. and Symonds, G. 2004a. The Southern Oscillation Index, wave climate and beach rotation. *Marine Geology*, **204**, 273–87.

Ranasinghe, R., Symonds, G., Black, K. and Holman, R. 2004b. Morphodynamics of intermediate beaches: a video imaging and numerical modelling study. *Coastal Engineering*, 51, 629–55.

Reichmüth, B. and Anthony, E.J. 2002. The variability of ridge and runnel beach morphology: examples from northern France. *Journal of Coastal Research*, **36** (sp1), 612–21.

Reichmüth, B. and Anthony, E.J. 2007. Tidal influence on the intertidal bar morphology of two contrasting macrotidal beaches. *Geomorphology*, **90**, 101–14.

Ridd, P.V. 1992. A sediment level sensor for erosion and siltation detection. *Estuarine, Coastal and Shelf Science*, **35**, 355–62.

Robertson, W., Zhang, K. and Whitman, D. 2007. Hurricane-induced beach change derived from airborne laser measurements near Panama City, Florida. *Marine Geology*, **237**, 191–205

Ruessink, B.G., van Enckvort, I.M.J., Kingston, K.S. and Davidson, M.A. 2000. Analysis of two- and three-dimensional nearshore bar behaviour. *Marine Geology*, **169**, 161–83.

Sallenger, A.H. 1979. Beach cusp formation. *Marine Geology*, **29**, 23–37.

Sallenger, A.H. Jr, Holman, R.A. and Birkemeier, W.A. 1985. Storm-induced response of a nearshore bar system. *Marine Geology*, **64**, 237–58.

Sallenger, A.H., Wright, C.W., Guy, K. and Morgan, K. 2004. Assessing storm-induced damage and dune erosion using airborne lidar: examples from Hurricane Isabel. *Shore & Beach*, **72**(2), 3–7.

Schmutz, P.P. and Namikas, S.L. 2013. Measurement and modeling of moisture content above an oscillating water table: implications for beach surface moisture dynamics. *Earth Surface Processes and Landforms*, **38**(11), 1317–25.

Schmutz, P.P. and Namikas, S.L. 2018. Measurement and modeling of the spatiotemporal dynamics of beach surface moisture content. *Aeolian Research*, **34**, 35–48.

Sedrati, M. and Anthony, E. 2007. Storm-generated morphological change and longshore sand transport in the intertidal zone of a multi-barred macrotidal beach. *Marine Geology*, **244**, 209–29.

Seymour, R.J. and Aubrey, D.G. 1985. Rhythmic beach cusp formation: a conceptual synthesis. *Marine Geology*, **65**, 23–37.

Shand, R.D., Bailey, D.G. and Shepherd, M.J. 2001. Longshore realignment of shore-parallel sand-bars at Wanganui, New Zealand. *Marine Geology*, **179**, 147–61.

Shepard, F.P. 1950a. *Beach cycles in southern California*. Washington, DC: US Beach Erosion Board, Technical Memorandum **15**, 32 pp.

Shepard, F.P. 1950b. *Longshore bars longshore troughs*. Washington, DC: US Beach Erosion Board, Technical Memorandum **20**, 38 pp.

Shepard, F.P. 1963. *Submarine Geology*. 2nd Edition. New York: Harper and Rowe, 412 pp.

Sherman, D.J., Short, A.D. and Takeda, I. 1993a. Sediment mixing-depth and bedform migration in rip channels. *Journal of Coastal Research*, **SI 15**, 39–48.

Sherman, D.J., Orford, J.D. and Carter, R.W.G. 1993b. Development of cusp-related, gravel size and shape facies at Malin Head, Ireland. *Sedimentology*, **40**, 1139–52.

Short, A.D. 1975. Offshore bars along the Alaskan Arctic coast. *Journal of Geology*, **83**, 209–21.

Short, A.D. 1979. Three dimensional beach stage model. *Journal of Geology*, **87**, 553–71.

Short, A.D. 2006. Australian beach systems: nature and distribution. *Journal of Coastal Research*, **221**, 11–27.

Short, A.D. and Aagaard, T. 1993. Single and multi-bar beach change models. *Journal of Coastal Research*, **SI 15**, 141–57.

Short, A.D. and Masselink, G. 1999. Embayed and structurally controlled beaches. In Short, A.D. (ed.), *Handbook of Beach and Shoreface Morphodynamics*. Chichester, UK: John Wiley & Sons, pp. 230–50.

Short, A.D., Trembanis, A.C. and Turner, I.L. 2001. Beach oscillation, rotation and the Southern oscillation, Narrabeen Beach, Australia. *Coastal Engineering*, **276**, 2439–52.

Short, A.D., Bracs, M.A. and Turner, I.L. 2014. Beach oscillation and rotation: local and regional response at three beaches in southeast Australia. *Journal of Coastal Research*, **70**(1), 712–17.

Silvester, R. 1974. *Coastal Engineering*, 2 volumes. Amsterdam: Elsevier.

Smyth, A.G. and Hesp, P.A. 2015. Aeolian dynamics of beach scraped ridge and dyke structures. *Coastal Engineering*, **99**, 38–45.

Sonu, C.J. 1968. Collective movement of sediments in nearshore environments. *Proceedings of the Eleventh Coastal Engineering Conference*. New York: American Society of Civil Engineers, pp. 373–400.

Sonu, C.J. 1972. Field observation of nearshore circulation and meandering currents. *Journal of Geophysical Research*, **77**(18), 3232–47.

Stewart, C.J. and Davidson-Arnott, R.G.D. 1988. Morphology, formation and migration of longshore sandwaves: Long Point, Lake Erie, Canada. *Marine Geology*, **81**, 63–77.

Stive, M.J.F. and de Vriend, H.J.. 1995. Modelling shoreface profile evolution. *Marine Geology*, **126**, 235–48.

Stive, M.J.F., De Schipper, M.A., Luijendijk, A.P. *et al.* 2013. A new alternative to saving our beaches from sea-level rise: the sand engine. *Journal of Coastal Research*, **29**, 1001–8.

Stronkhorst, J., Huismana, B., Giardinoa, A., Santinellia, G. and Duarte Santos, F. 2018. Sand nourishment strategies to mitigate coastal erosion and sea level rise at the coasts of Holland (The Netherlands) and

Aveiro (Portugal) in the 21st century. *Ocean and Coastal Management*, **156**, 266–76.

Sunamura, T. 1988. Beach morphologies and their change. In Horikawa, K. (ed.), *Nearshore Dynamics and Coastal Processes*. Tokyo: University of Tokyo Press, pp. 136–66.

Tebbens, S.F., Burroughs, S.M. and Nelson, E.E. 2002. Wavelet analysis of shoreline change on the Outer Banks of North Carolina: an example of complexity in the marine sciences. *Proceedings of the National Academy of Sciences of the USA*, **99**(suppl 1), 2554–60.

Terpstra, P.D. and Chrzastowski, M.J. 1992. Geometric trends in the evolution of a small log-spiral embayment on the Illinois shore of Lake Michigan. *Journal of Coastal Research*, **8**, 603–17.

Thornton, E.B., Hurmiston, R.T. and Birkmeir, W. 1996. Bar/trough generation on a natural beach. *Journal of Geophysical Research*, **101 C5**, 12097–110.

Thorpe, A., Miles, J., Masselink, G. and Russell, P. 2014. Bedform dynamics in a rip current. *Journal of Coastal Research*, **70**(1), 700–05.

Trimble, S.M. 2017. Addressing the International Rip Current Health Hazard. Doctoral dissertation, Texas A & M University. Available online at http://hdl.handle.net/1969.1/169643.

Turner, I.L. and Masselink, G. 1998. Swash infiltration-exfiltration and sediment transport. *Journal of Geophysical Research: Oceans*, **103**(C13), 30813–24.

Turner, I.L., Russell, P.E. and Butt, T. 2008. Measurement of wave-by-wave bed-levels in the swash zone. *Coastal Engineering*, **55**(12), 1237–42.

Turner, I.L., Harley, M.D. and Drummond, C.D. 2016. UAVs for coastal surveying. *Coastal Engineering*, **114**, 19–24.

Valverde, H.R., Trembanis, A.C. and Pilkey, O.H. 1999. Summary of beach nourishment episodes on the US east coast barrier islands. *Journal of Coastal Research*, **15**, 1100–18.

van Enckevort, I.M.J. and Ruessink B.G. 2003a. Video observations of nearshore bar behaviour. Part 1: alongshore uniform variability. *Continental Shelf Research*, **23**, 501–12.

van Enckevort, I.M.J. and Ruessink B.G. 2003b. Video observations of nearshore bar behaviour. Part 2: alongshore non-uniform variability. *Continental Shelf Research*, **23**, 513–32.

van Houwelingen, S., Masselink, G. and Bullard, J. 2006. Characteristics and dynamics of multiple intertidal bars, north Lincolnshire, England. *Earth Surface Processes and Landforms*, **31**, 428–43.

van Maanen, B, Ruiter, P.J. de, Coco, G., Bryan K.R. and Ruessink, B.G. 2008. Onshore sandbar migration at Tairua Beach (New Zealand): numerical simulations and field measurements. *Marine Geology*, **253**, 99–105.

Vittori, G., De Swart, H.E. and Blondeaux, P. 1999. Crescentic bedforms in the nearshore region. *Journal of Fluid Mechanics*, **381**, 271–303.

Wang, J. and Sherman, D.J. 2016. Cusp Development on a Gravel Beach. *Journal of Coastal Research*, **75**(1), 937–41.

Wang, P. and Davis, R.A. Jr. 1998. A beach profile model for a barred coast: case study from Sand Keys, west-central Florida. *Journal of Coastal Research*, **14**, 981–91.

Wang, P., Ebersole, B.A. and Smith, E.R. 2003. Beach-profile evolution under spilling and plunging breakers. *Journal of Waterways, Port, Coastal and Ocean Engineering*, **129**, 41–6.

Watts, G.M. 1954. Laboratory study of the effects of varying wave period on beach profiles. Washington, DC: US Beach Erosion Board, Technical Memorandum, **53**, 19 pp.

Weir, F.M., Hughes, M.G. and Baldock, T.E. 2006. Beach face and berm morphodynamics fronting a coastal lagoon, *Geomorphology*, **82**, 331–46.

Werner, B.T. and Fink, T.M. 1993. Beach cusps as self-organised patterns. *Science*, **260**, 968–71.

Wernette, P., Houser, C., Weymer, B.A. *et al.* 2018. Influence of a spatially complex framework geology on barrier island geomorphology. *Marine Geology*, **398**, 151–62.

Wijnberg, K.M. and Kroon, A. 2002. Barred beaches. *Geomorphology*, **48**, 103–20.

Wijnberg, K.M. and Terwindt, J.H.J. 1995. Quantification of decadal morphological behaviour of the central Dutch coast. *Marine Geology*, **126**, 301–30.

Wijnberg, K.M. and Wolf, F.C.J. 1994. Three-dimensional behaviour of a multiple bar system. *Proceedings of Coastal Dynamics '94*. New York: American Society of Civil Engineers, pp. 59–73.

Williams, A.T., Giardino, A. and Pranzini, E. 2016. Canons of coastal engineering in the United Kingdom: seawalls/groynes, a century of change? *Journal of Coastal Research*, **32**(5), 1196–211.

Wright, L.D. and Short, A.D. 1983. Morphodynamic of beaches and surf zones in Australia. In Komar, P.D. (ed.), *Handbook of Coastal Processes and Erosion*. Boca Raton, FL: CRC Press, pp. 35–64.

Wright, L.D. and Short, A.D. 1984. Morphodynamic variability of surf zones and beaches: a synthesis. *Marine Geology*, **56**, 93–118.

Wright, L.D., Guza, R.T. and Short, A.D. 1982. Dynamics of a high energy surf zone. *Marine Geology*, **45**, 41–62.

Wright, L.D., Short, A.D. and Green, M.O. 1985. Short-term changes in the morphodynamic states of beaches and surf zones: an empirical predictive model. *Marine Geology*, **62**, 339–64.

Xia, Y. and Boufadel, M.C. 2011. Beach geomorphic factors for the persistence of subsurface oil from the Exxon Valdez spill in Alaska. *Environmental monitoring and assessment*, **183**(1–4), 5–21.

Yasso, W.E. 1965. Plan geometry of headland bay beaches. *Journal of Geology*, **73**, 702–14.

Zhang, K., Whitman, D., Leatherman, S. and Robertson, W. 2005. Quantification of beach changes caused by Hurricane Floyd along Florida's Atlantic coast using airborne laser surveys. *Journal of Coastal Research*, **21**, 123–34.

Zenkovich, V.P. 1967. *Processes of Coastal Development*. New York: Interscience Publishers, 738 pp.

Zuijdgeest, A. and Huettel, M. 2012. Dispersants as used in response to the MC252-spill lead to higher mobility of polycyclic aromatic hydrocarbons in oil-contaminated Gulf of Mexico sand. *PloS One*, **7**(11), e50549.

9

Coastal Sand Dunes and Aeolian Processes

9.1 Synopsis

Waves and currents in the nearshore zone are essential to the evolution and maintenance of sandy beaches, which are often fringed by aeolian (wind-blown) sand dunes. Coastal dunes vary in geometry and range in size across a broad spectrum of scales. At one end are very small incipient forms less than 0.1 m in height and 2–3 m in length that may change their form on a daily basis. At the other end of the spectrum are massive features that are 100 m or more in height and 1000 m in length. These larger forms may appear as multiple groups that extend several kilometres alongshore and inland as part of major sandy barrier systems situated on low coastal plains that have evolved over centuries or longer. The most seaward dune is referred to as the foredune because it is the first (or only) dune in a sequence of dunes extending inland (Figure 9.1). Unlike their arid counterparts, sandy dunes along coasts have a steep stoss slope that is typically partially or completely vegetated. Colonising vegetation plays a fundamental role in the evolution, stabilisation, and maintenance of coastal sand dunes and dune ridge complexes.

Figure 9.1 Single foredune with recent sedimentation on the crest after a major wind event. Photo taken in early spring before vegetation has had time to green up. (Greenwich Dunes, Prince Edward Island, Canada.)

On some beaches with limited sediment supply or restricted accommodation space (e.g., a narrow beach backed by a sea wall or a cliff) only a single coastal dune may develop with little aeolian activity farther inland. Such foredunes exist in a state of dynamic equilibrium because aeolian sand transport to the dunes (resulting in net dune growth) is countered by erosion of the dune toe by wave scarping during large storms, especially during spring high tides. During extreme storm events, the foredune can be breached and eroded completely leaving behind a relatively featureless overwash flat, which is typical of small barrier islands and baymouth bars (see Chapter 10). The complex and infrequent interplay of sediment supply

Figure 9.2 Complex dune field (Lanphere Dunes, Humboldt Bay, California, USA) consisting of a scarped foredune with multiple blowouts transitioning landward to parabolic dunes and then to an extensive open sand sheet with transgressive dunes en echelon in the background. Note incipient dune hummocks on the backshore developing seaward of the foredune. (Photo courtesy of David Kenworthy.)

and removal via nearshore and aeolian processes collectively referred to as beach–dune interaction, will dictate the long-term fate of the coastal foredune.

Along coasts where there is an abundant supply of sediment and sufficiently strong onshore winds to mobilise beach sands frequently, a transgressive dune complex may evolve behind the foredune, usually as a consequence of the development of blowouts along the foredune ridge and subsequent migration of parabolic dunes inland (Figure 9.2). Alternatively, on coastlines with rapid shoreline progradation, the foredune ridge may become stranded because a new incipient foredune that grows in front of the old foredune on the widening backshore will trap most of the onshore-moving sand. Figure 9.3 is a schematic representation of both types of shoreline morphologies. If the sequence of foredune stranding and incipient foredune growth repeats itself over many decades, a succession of parallel beach ridges evolves that is especially visible on air photos and satellite imagery (Figure 9.4).

Coastal foredunes and aeolian processes are an integral part of the overall morphodynamics of sandy beaches, serving as a storage mechanism for sediment that is frequently exchanged with the nearshore (Houser *et al.*, 2018a). Thus, coastal dune systems serve to protect the shore from progressive erosion and inundation. In addition to their natural beauty, coastal dunes provide critical ecological habitat for plants and animals that are uniquely adapted for life in this dynamic sandy environment (Maun, 2009; Zarnetske *et al.*, 2010). Moreover, dunes are a significant resource for human activities, including a host of recreational activities, their role in preventing saltwater intrusion into coastal aquifers, the protection they offer against storm surge and extreme wave action, and the opportunities they provide for sand mining given the well-sorted and clean nature of dune sands.

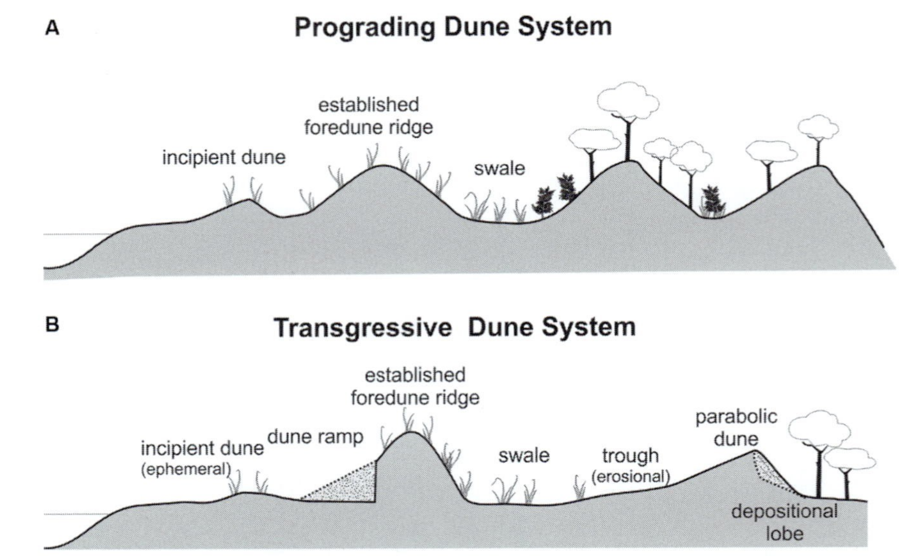

Figure 9.3 Dune profiles across: A. a prograding system with incipient dunes forming in front of an established foredune, with a series of relict foredune ridges in the landward direction where mature vegetation has grown; and B. a transgressive (erosional) system with a scarped foredune and recently infilled dune ramp, transitioning to swales and erosional troughs due to parabolic dune migration in the landward direction.

Figure 9.4 Beach ridges formed by a process of progressive shoreline progradation that strands or isolates prior foredunes in inland locations. Often, these ridges consist of coarse sediment laid down by waves during storms, on top of which an aeolian cap of finer sand is deposited. (Piaçabuçu, São Francisco River Delta, Brazil; Google Earth.)

9.2 Morphology and Structure of Coastal Dunes and Dune Fields

Coastal sand dunes are formed by the transport of sediment from the foreshore and backshore by wind action to a new position farther inland where deposition occurs. Using the terminology of a sediment budget, the beach is the 'source' or supply of sand whereas the dunes represent the sediment 'sink' or reservoir. During intense storms, this relationship may be reversed because dune erosion releases sediment back to the nearshore zone.

The grain-size distributions of dune sands tend to be very well sorted, with modal sizes between 0.2 mm and 0.6 mm, which represents the size fractions that are most easily mobilised. The density ratio of quartz sand (2650 kg m^{-3}) to salt water (1030 kg m^{-3}) is about 2.65, whereas for air (1.15 kg m^{-3}) it is about 2300, or three orders of magnitude greater than for water. As a consequence, wind speed has to be quite fast (usually in excess of 5–6 m s^{-1}) to move small sand-sized particles. When sand particles are mobilised by wind, typically the fine fraction is removed from the surface first because the small particles are easier to entrain and transport. This process of winnowing out the fine sizes leaves the coarse fraction (>1 mm) behind on the beach, which explains why coastal dunes rarely contain sand coarser than 1 mm. However, a coarse-grained lag deposit made of pebbles and cobbles, as found commonly in dry-land environments (referred to as desert pavement), rarely develops on beaches because the surface materials are frequently reworked by waves. Size fractions less than about 0.15 mm are equally scarce in coastal dunes because fine sediments are generally not prevalent in active nearshore environments. The combination of wave and current action in the surf zone easily entrains and transports such fine-grained sediments away in suspension (see Chapter 7).

9.2.1 Dune Types

Coastal dunes are classified according to their orientation, shape and stability. They can have a shore-parallel (transverse) orientation, which is commonly the case with foredunes (Figure 9.1), or they can have morphologies that are oblique or perpendicular to the shoreline, as with the trailing arms of parabolic dunes (Figure 9.5A). There are two-dimensional and three-dimensional shapes depending on how laterally extensive and complex the dune shape is (Figure 9.5B). A distinction is often made (Table 9.1) between transgressive dunes that are mobile and migrate inland versus stabilised (impeded dunes) that are largely fixed in place by vegetation or anchored by some obstruction such as a rock outcrop or cliff (Pye, 1983). In the former case, sand is readily mobilised from a bare sand surface by wind action, and the geometry and orientation of the dunes closely reflect the pattern of wind flow, especially the dominant competent winds. In the latter, the vegetation cover serves to trap sand and often controls the shape and geometry of the dune (Hesp, 1983; 1989).

There is a continuum of coastal dune forms, with varying degrees of vegetation cover and stabilisation depending on stage of plant growth and succession as well as on seasonal changes in temperature, moisture availability and human impact. At one end of the continuum, the temperatures are too cold or there is too little moisture or nutrient supply to support vegetation of any kind, except perhaps micro-biota living in between and on individual sand grains. Alternatively, the sediment supply and deposition rates may be so large that vegetation is unable to colonise the unstable sand surface. Similarly, human disturbance from vehicular and pedestrian traffic or from weed-spraying efforts to control invasive species may kill vegetation and expose bare sand surfaces to wind. These unvegetated sand surfaces are very easily mobilised by wind, and the resulting transgressive dunes are free to evolve and migrate inland. The south coast of Namibia provides good examples, where there is virtually no vegetation cover to prevent aeolian sand transport (Figure 9.6). In some instances, the mobility of free dunes can be constrained by the presence of the water table in interdune depressions (Figure 9.7A; see also Figure 9.5B), or by hardpan layers such as clay

Figure 9.5 Dune orientation relative to shore. A. Shore-perpendicular trailing arms of transgressive parabolic dunes at Guadalupe, California, USA. Note almost complete absence of foredunes due to wave erosion (foreground), and large, relict parabolics in background (Google Earth). B. Complex dune field with large, shore-parallel foredune with blowout hollows (foreground) and hummocky dune remnants transitioning to parabolics in background. Note swale with small pond in the middle of the photo. (Mason Bay, Stewart Island, New Zealand.)

Table 9.1 Spectrum of dune types based on stability characteristics, ranging from free dunes to impeded dunes (after Pye, 1983)

	Vegetation Cover	Dune Form	Comment
Free Dunes			
⇑	absent or scarce	barchans, barchanoid ridges; transverse and oblique ridges	wide range of dune forms similar to desert dunes; found on beaches and dune fields
	fringes and borders only	precipitation ridges,	found at downwind border of transgressive dune fields;
		blowouts, parabolic dunes	eroded foredunes and stabilised dune fields
	extensive coverage	stabilised ridges, dune fields	wide range of formerly active dune forms on transgressive dune fields
	extensive to complete coverage	incipient dunes, nebkhas,	small, often laterally discontinuous forms developed in and around vegetation; mostly stable;
		foredune ridges	large stable, laterally continuous features formed parallel to the beach
⇓	complete coverage	relic foredune ridges, transverse ridges	formed on prograding or regressive shorelines; cut off from beach by development of new foredune(s)
Impeded Dunes			

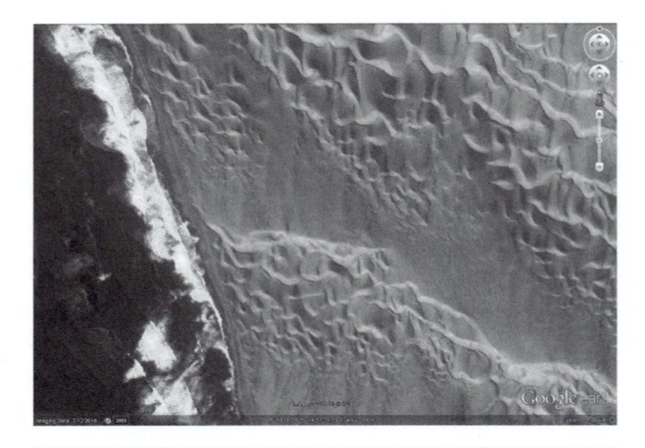

Figure 9.6 Coastal dune development along the Namibian coast, north of Luderitz. Note absence of proper foredune and no vegetation (Google Earth).

pans, ferricretes (limonites) and calcretes on deflation surfaces between dunes (Figure 9.7B). However, the growth of vegetation is clearly the most important control on dune stability, especially on foredunes. This was understood by early European settlers in North America and New Zealand, for example, who introduced various types of beach grasses to stabilise the dunes to protect adjacent agricultural land from sand incursions. This practice had significant consequences for the eco-morphodynamics of many coastlines across the world, some of which are now being reverse engineered to their more natural dynamic state by removal of the invasive species and re-planting of native plants (see Box 9.1).

Figure 9.7 Controls on sand transport and dune migration. A. Deflation basin (swale) behind a large foredune with exposed (perched) water table, which restricts sand transport across the swale area. Ocean is toward rear of photo, with prevailing winds from top right to lower left. (Lanphere Dunes, Humboldt Bay, California, USA). B. Indurated paleo-layer with iron-bearing and other heavy minerals that has been exposed by aeolian erosion on a large blowout (Mason Bay, Stewart Island, New Zealand).

Box 9.1 Coastal Management Application

Managing Invasive Plants and Restoring Dune Dynamics

The history of human colonisation of new territory is full of examples where traditions and practices that suited a homeland context were transplanted, either purposefully or inadvertently, to new environments, often with unintended consequences. The introduction of non-native plant species to recently occupied coastal landscapes demonstrates how human action can influence the geomorphology of natural systems in ways that are just beginning to be understood. One example of such a coastal landscape-altering species is European beachgrass (*Ammophila arenaria*) and its relative the American beachgrass (*Ammophila breviligulata*), which is native to the Atlantic coast and Great Lakes region of North America. Both species (commonly referred to as marram grass) have spread to temperate coasts around the world, where they out-compete native sand-colonising plant species. The outcome in most settings is a monoculture of marram grass, a loss of dune dynamism, a loss of native biodiversity and alteration of the characteristic dune landscape. Because the morphology of coastal dunes is intricately related to the type of vegetation cover, a shift from native to non-native species often leads to a change in the aeolian sediment transport regime as well as the broader dune ecosystem conditions. Locations with (naturally) sparse or patchy native vegetation tend to have very active transport regimes and low or irregular foredunes, transitioning to transgressive dune forms in the hinterland (e.g., open sand sheets, semi-vegetated parabolic dunes and transverse dunes). In contrast, a dense cover of vegetation will stabilise dunes in ways that lead to vertical accretion and very steep foredune profiles (sometimes with blowouts backed by parabolic dunes). The latter situation is typical of marram grass (Ruggiero *et al.*, 2018).

Programmes are underway in several countries to eradicate, or at least manage, invasive plant species and to restore the dynamics of the dune regime associated with native plants. Two examples are provided below.

Lanphere and Ma-le'l Dunes, Humboldt Bay National Wildlife Refuge, Northern California

(Courtesy of Andrea Pickart, US Fish and Wildlife Service)

The Lanphere and Ma-le'l Dunes were the site of the first experimental project intended to restore natural coastal dune processes on the west coast of North America. The dune system is characterised by a foredune–blowout–

Box 9.1 (cont.)

parabolic dune complex fronting deflation basins and plains upwind of a larger active parabolic dune and transgressive dune field (Figure 9.2). The naturally occurring vegetation is a dune mat consisting of a variety of low and slowly growing plants with variable spatial cover (i.e., some dense patches and some open sand areas). In the 1970s, non-native European beachgrass (*A. arenaria*) invaded the dunes displacing native vegetation and creating a dense monoculture that stabilised the foredunes and disrupted sediment exchange between the beach, foredune and backdune zones. Beginning in 1992 and with an expanded effort in 2005, much of the marram was removed and native dune-building American dunegrass (*Elymus mollis*) was planted (Figure 9.8). Removal of marram is an arduous process involving labour crews that dig out the rhizomes to a depth of 0.3 m, thus killing off the biomass.

Figure 9.8 Before and after photos of a section of the Ma-le'l Dunes (Humboldt Bay, California) showing original dense cover of *A. arenaria* (2005) and transformation to a more open and active surface of native plants (2011 and 2014) after a restoration project. (Photos taken by Andrea Pickart.)

Box 9.1 (cont.)

This was done over two consecutive growing seasons to ensure 100 per cent mortality. Populations of marram continue to exist to the north and south of the treatment sites, and rhizomes are dispersed by wave action following foredune erosion events (with lesser spread from seed dispersal). This requires ongoing maintenance of the treatment sites with annual '*Ammophila* sweeps'. Monitoring and follow-up research have demonstrated that native communities of plants, invertebrates and vertebrates have recovered to reference conditions, and foredune morphodynamics are being restored.

Mason Bay, Stewart Island, New Zealand

(Courtesy of Mike Hilton, University of Otago, Dunedin, New Zealand)

Mason Bay (Oneroa) hosts one of the largest and least modified transgressive dune systems remaining in New Zealand. It contains a diversity of dune and dune-related landforms, shaped by exposure of the west coast of Stewart Island (Rakiura) to the prevailing 'roaring forties', and it is a key site for the conservation of native biodiversity. European beachgrass (*A. arenaria*) is the main threat to this biodiversity, due to its ability to thrive and colonise dunes in high rates of sedimentation and a dynamic aeolian environment. Marram disperses between sites by floating rhizomes and within sites by wind-blown seed and vegetative growth. How and exactly when marram grass was introduced to the dune systems of Stewart Island is uncertain. It was likely planted at the southern end of Mason Bay at Kilbride in the 1940s by farmers. It may have also established from rhizomes that drifted from the South Island of New Zealand. The spread of marram grass to the northern sections of Mason Bay most certainly occurred by this process – the first aerial photograph in1958 records multiple scattered marram grass nebkha in the foredune zone in the northern part of the bay. Marram invasion resulted in the formation of a large and very stable foredune, and simultaneous progradation (approximately 70 m), between the early 1950s and 2000. Shoreline advance ceased about this time, and marram progressively invaded the dune hinterland displacing native flora and fauna. The development of a large foredune had effects far inland. The new foredune prevented sand exchange between the beach and the wider dune system (kilometres inland of the foredune), which resulted in areas of sand deficit, the reorganisation of landforms and the loss of habitat for native species.

The Department of Conservation commenced a dune restoration programme at Mason Bay in 2000. The programme targeted marram grass (using a grass-selective herbicide) and also tree lupin (*Lupinus arboreus*). The herbicide was applied by knapsack and vehicle-mounted pumps across isolated patches of marram in the hinterland, and is being applied across the foredune using helicopters. There is no concern that the foredune will be adversely affected by this process, since the foredune is entirely an artefact of marram invasion. Indeed, the

Figure 9.9 Marram grass has been reduced to zero density across most of the central hinterland of Mason Bay dunes (Stewart Island, New Zealand) and on some sections of the foredune. This image compares a section of fully vegetated, stable foredune (left) and the devegetated foredune (right) after treatment. The sand released from the eroding foredune in the treated area will benefit native plant communities downwind. The position of the shoreline prior to the introduction of marram grass was just behind the foredune, and was buried due to dune progradation as a result of the establishment of the marram grass. (Photo taken by Mike Hilton.)

Box 9.1 (cont.)

expectation is that the foredune will eventually erode, thereby restoring sand supply to the dune system in the hinterland, to the benefit of native plant communities. Fortunately there are large numbers of native plants remaining in the wider dune system to provide a source of seed to recolonise the areas cleared of marram grass, although the expectation is the shoreline will retreat to its pre-marram location. The project has been successful with removal of marram throughout the hinterland, and the gradual erosion of the foredune (Figure 9.9). The management effort needs to be sustained for many more decades because marram grass is able to regrow from dormant buds on rhizomes and from buried seed.

9.2.2 Primary Dunes: the Foredune and Incipient Dunes

The argument has been made that the only distinctly coastal dune is the foredune (Bauer and Sherman, 1999), because its overall morphology is very different from other dunes that are sculpted by the wind alone and therefore have an aerodynamic shape. Specifically, the foredune has a steep windward (stoss) slope due to frequent wave scarping at the dune toe and a gentle lee slope due to the growth of vegetation (Figure 9.10). In contrast, most other aeolian dunes such as those found in deserts or in the inland portions of coastal dune complexes have shallow to moderate stoss slopes that serve as transport surfaces leading toward a depositional zone near the dune crest (Figure 9.11). Aerodynamically shaped dunes also have very steep lee (downwind) slopes that are prone to periodic avalanching (Figure 9.12), which is not typically the case for foredunes unless their crests are rapidly accreting. Moreover, since the foredune is always the most seaward dune along the coastline, and the only one that is influenced directly by nearshore processes, it is considered the 'primary' coastal dune even when it is smaller in size and volume than secondary dunes farther inland.

A key characteristic of the foredune is that it forms a continuous or semi-continuous shore-

Figure 9.10 Classic foredune geometry with steep stoss slope that was scarped by waves and is being repaired by deposition of a sand ramp at the toe that is being rapidly re-vegetated. In contrast, the lee slope is gentle and fully vegetated. Aluminum frame is for measuring topographic changes (Greenwich Dunes, Prince Edward Island, Canada).

Figure 9.11 Large, barchanoid-shaped dunes on a sand plain, Jerricoacarra, Brazil. Sand is supplied from beaches on the upwind side of a peninsula and forms a dune field that migrates into the sea on the lee side of the peninsula. Note the gentle stoss slope leading toward a sharp crest and then to a steep avalanche lee slope on these unvegetated dunes. Air flow is from left to right.

parallel ridge that backs the main beach irrespective of the prevailing or dominant wind direction (e.g., Figure 9.2). In part, this is due to the role of vegetation in anchoring the dune location on the backshore, but also because of the trimming action of large waves that erode the foredune toe in a fashion that follows the trend of the shoreline. Hesp (1983) proposed a conceptual model of foredune evolution that begins with the emplacement of a wrack line on the upper backshore during a storm (Figure 9.13A). Vegetative debris in the wrack usually includes bits of viable rhizomes, plant stems and seeds, and if the wrack becomes partly buried by sand, the seeds may germinate or the rhizomes may sprout new growth thereby initiating the first phase of plant colonisation on the backshore. Sand moving onshore or alongshore by wind encounters the emergent seedlings, and the enhanced roughness induces deposition. A linear, shore-parallel ridge of sand begins to accumulate along the line of vegetation, and this is the incipient (or embryo) foredune (Figure 9.13B). Additional sand brought onshore by aeolian action encounters the vegetation as well as an adverse slope, both of which force further deposition. As the incipient foredune grows in size, it begins to have a pronounced effect on the near-surface boundary layer flow, and this flow–form feedback further enhances the tendency for sediment deposition (Walker and Hesp, 2013) on the incipient foredune and in its lee. In turn, the sand partially buries the plants and provides nutrients that are essential for plant growth, initiating a positive feedback cycle of dune and vegetation growth.

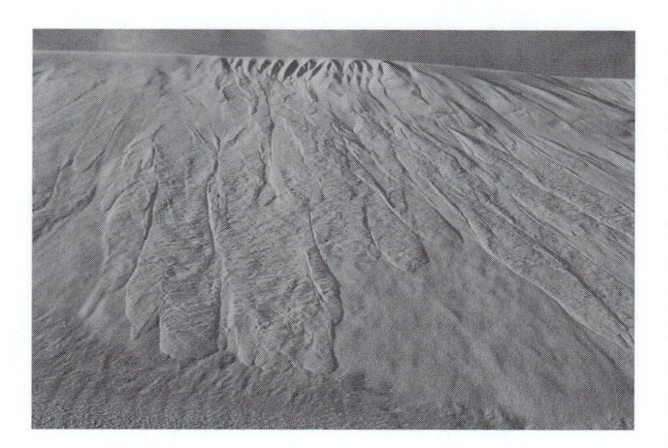

Figure 9.12 Sand avalanche lobes on the lee slope of a transgressive dune. The upper slope is over-steepened by sediment accumulation during active transport events, leading to instabilities that initiate avalanching. The sand eventually comes to rest at the dynamic angle of repose (approximately 30 degrees).

The Hesp (1983) model of incipient dune formation requires wrack or debris to be emplaced on the backshore in front of existing dunes or on

Figure 9.13 Incipient foredune evolution. A. Wrack line emplaced during a storm slows wind and causes sand deposition in front of an existing, older foredune (Doughboy Bay, Stewart Island, New Zealand). B. Incipient foredune evolving on the backshore of a wide beach in front of an established foredune that is much larger and fully vegetated. The plants on the incipient foredune are able to induce sediment deposition thereby growing the incipient foredune and preventing sediment from reaching the older foredune (Humboldt Bay, California, USA).

an overwash flat without dunes (Carter *et al.*, 1992), which serve as the nucleus for aeolian deposition. These nascent aeolian deposits are in a precarious state because the next storm may bring large waves that completely erode whatever sand has been emplaced by aeolian processes. However, if conditions allow progressive aeolian deposition over months and years, the incipient dunes will become large enough to survive the erosional effects of even an extreme storm. The toe region may become severely scarped during such large storms, but the dune core retains its structural integrity. At this stage, these dunes are properly referred to as established foredunes, or simply, a foredune.

On many stable and eroding shorelines, incipient dunes may be ephemeral features that develop in front of an established foredune over a period of months to several years and are then removed by wave action during large storms. Thus, incipient dunes and dune hummocks become a recurring, characteristic feature of the shore when monitored over long time frames (i.e., decades).

9.2.3 Secondary Dunes: Blowouts, Parabolic Dunes and Dune Fields

Dunes and dune-like features found inland of the foredune are referred to as secondary dunes that collectively make up dune fields, which are usually morphologically very complex (Figure 9.14). Under what conditions might beach sand be transported past the foredune in volumes large enough to create extensive dune fields?

In many coastal settings, the vegetation cover on foredunes is disturbed from time to time due to natural (e.g., fires, blights, droughts, storms) or anthropogenic (e.g., trampling, trail riding, coastal development, weed spraying, overgrazing) causes resulting in the reactivation of aeolian activity. Although this can occur anywhere on a stabilised dune field, including relict foredunes, remobilisation frequently occurs on laterally continuous foredune system because humans and animals make trails that lead to the beach or because a section of the foredune is particularly susceptible to disturbance. Notch-like erosional features will develop in the foredune through which wind is funnelled and accelerated (Figure 9.15A). The sand that is eroded from the foredune notch is transported a short distance landward and deposited forming a lip. With time, more and more sand is eroded from the throat, and the depositional lip migrates inland.

Figure 9.14 Complex dune field with blowouts, shadow dunes and trailing arms of parabolic dunes climbing a steep slope with rocky outcrops visible in the foreground (Mason Bay, Stewart Island, New Zealand).

9.2.3.1 *Blowouts*

If the notch in the foredune deepens and widens it may create a feature referred to as a blowout (Figure 9.15B). Often, these features are circular or oblong, with depth and elongation increasing from an initially shallow depression (Hesp, 2002). Because they have steep erosional walls with a distinct lip or circular ridge, they are called saucer or cup blowouts. Wind and sand from the foredune notch is funnelled against the erosional wall and steered sideways, spiralling its way around the blowout and eventually out the far end (Hesp and Hyde, 1996; Hesp and Pringle, 2000; Hansen *et al.*, 2009). At the outlet of the blowout, a sand ramp develops that often has an avalanche slope to the lee. Well-established saucer blowouts can be tens of metres in diameter and several metres deep, and they progressively elongate yielding a trough shaped depression. Processes of blowout development and healing are described by Gares and Nordstrom (1995) and Hesp *et al.* (2016), while wind

Figure 9.15 Blowouts in foredunes. A. Erosional notch on the seaward-facing (stoss) side of a large foredune. Wind is preferentially funnelled from the beach (lower part of photo) and accelerated through the notch and into the backdune zone (Mason Bay, Steward Island, New Zealand). B. Fully formed saucer blowout in the lee of a large foredune. Note erosional notch in the foredune to upper right leading to an erosional wall on the distal side of the blowout (Greenwich Dunes, Prince Edward Island, Canada).

and sand circulation through blowouts has been modelled using computational fluid dynamics (e.g., Smyth et al., 2014).

Blowouts in vegetated dunes may stabilise over time depending on the particular combination of wind and weather conditions, sand supply and vegetation propagation (Schwarz *et al.*, 2019). Stabilisation is promoted when vegetation becomes established on the bottom of the depression and then spreads upwards to colonise the sides. This can occur where the bottom of the blowout intersects the groundwater table, but often there is sufficient moisture near the surface to retard sand movement and permit plant establishment. Plant colonisation may also be aided by the slumping of intact blocks carrying plants. Thus, many blowouts stabilise over time, or they may be in-filled gradually as sand blows in from the beach (Gares and Nordstrom, 1995). The result of these blowout processes is that the simple, two-dimensional topography of many foredunes becomes increasingly complex over time

9.2.3.2 *Parabolic Dunes*

If the blowout notch does not plug up, the saucer shape will elongate in the landward direction and the erosional form will begin to migrate inland. When this occurs there is a transition from a blowout to a parabolic dune. Parabolic dunes are U-shaped migrating dunes with trailing arms that are very common in secondary dune fields (Figure 9.16A). They can be hundreds of metres across and more than 1 kilometre in length (Figure 9.16B). The dune head is a zone of active migration, and the landward (downwind) slope of the dune head is usually a slip face dominated by avalanche processes, as is typical of aerodynamically shaped dunes. The migration process leaves behind trailing arms on either side of the parabolic dune thereby indicating the pathway that the dune took while migrating inland (Figure 9.16C). With adequate sediment supply from the coast, the parabolic arms can be kilometres long, but if the sand supply is limited the arms may be scavenged and progressively eroded, with sediment being transported to the parabolic head.

Sometimes parabolic dunes migrate inland in isolation, but more commonly there can be multiple parabolic dunes transgressing the landscape in groups (Figure 9.16A). At some point the active migration ceases, either because the head of the parabolic dune encounters an obstacle such as a river or cliff, which poses a physical barrier to further migration. Parabolics sometimes advance into forested areas where the trees cause a reduction in wind speed at the forest edge. The rate of landward migration of the dune is then much

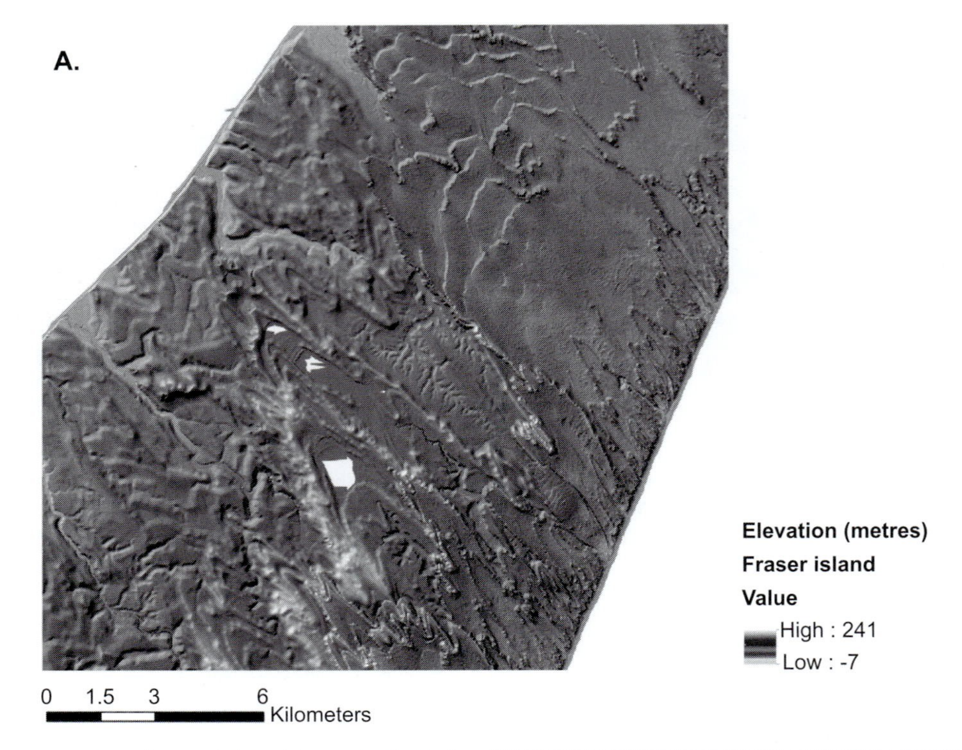

Figure 9.16 Parabolic dunes. A. Inactive and active parabolic dunes shown in a DEM of Fraser Island (near Brisbane, Queensland, Australia). The sequence of parabolics (lower portion of DEM) and transverse dunes (upper portion of DEM) range in age from currently active dunes near the southeast coast up to, perhaps, 400 000 years toward the northwest coast. Slow uplift has preserved old shorelines, and the growth of forest vegetation with decomposing organic material has yielded buried impervious layers that have allowed perched lakes to develop in the dune depressions (blocked out areas in DEM) (DEM courtesy of Dr Jamie Shulmeister, Dan Ellerton and Munkzoolboo Purev, University of Queensland). B. Large sequence of parabolic dunes migrating inland en echelon over a forested lowland area. The parabolic head (left of photo) is approximately 2 km from the coast (to right of photo) and is separated from newer, more active dune heads by a vegetated bowl. Another parabolic is entering from the right of the photo, whereas remnants of very old parabolics have been stabilised by trees (in the distant right of photo) (Mason Bay, Stewart Island, New Zealand). C. Partly vegetated trailing arm of a parabolic dune, looking seaward to source of wind and sediment. Newer dunes are establishing in the erosion hollow of the parabolic (Doughboy Bay, Stewart Island, New Zealand).

Figure 9.17 A. Depositional avalanche face of a migrating coastal dune, advancing into a forested area and gradually inundating and killing the trees (La Grande Dune de Pilat, France). B. Vegetated, relict parabolic dunes extending (left to right) several kilometres inland from the coast. Note signs of partial reactivation in blowouts to right of photo (Mason Bay, Stewart Island, New Zealand). (DEM in A. kindly provided by Munkzoolboo Purev and Dan Ellerton, University of Queensland. It was processed using LiDAR data acquired on behalf of the Queensland State Government in 2009.)

slower but can take place over many decades and reach several kilometres inland (Clemmensen *et al.*, 2001; Hesp and Martínez, 2007). Trees that are buried during the advancing phase die off and are eventually unearthed when the erosional bowl of the parabolic migrates through, leaving a sandy landscape punctuated by barren tree branches (Figure 9.17A). In other instances, there can be a change in the environmental or climatic conditions, which may lead to vegetation growth on the parabolic dune and thus stabilisation (Figure 9.17B).

9.2.3.3 *Dune Fields*

The reworking of a dune field during a new aeolian re-activation phase may create a range of free dune forms such as barchanoid ridges, transverse ridges or large sand sheets, which may occupy hundreds to thousands of square kilometres. The orientation of these secondary features is no longer controlled by proximity to ocean waves, and thus, tend to be much more reflective of the dominant wind direction as well as the nature of vegetation cover, the depth to the water table, and any controls imparted by bedrock outcrops. In many cases periodic disturbance on time scales of centuries produces a succession of overlapping dune forms with sediments from the older episodes being buried beneath younger deposits (Hesp and Thom, 1990; Clemmensen *et al.*, 2001). In New South Wales, Australia, transgressive dune fields show multiple phases of development with periods of stability and foredune growth followed by disturbance and inland migration of sand leading to burial of older transgressions (Lees, 2006). The episodic transgressive dune field development may result from disturbance caused by large storms, periods of drought, fluctuating lake levels (Olson, 1958a; Loope and Arbogast, 2000) or human activities (Clemmensen *et al.*, 2007), and may be moderated by the need to develop a critical mass of sand in the foredune so that the migrating dune front is able to bury vegetation landward of it.

Transgressive dune fields, sand sheets and large-scale sand seas have downwind borders consisting of precipitation ridges, while deflation basins and plains develop upwind (Cooper, 1958; Hesp and Martínez, 2007). Some of the largest transgressive dune fields can be found along the coast of Brazil (Hesp *et al.*, 2007a; Giannini *et al.*, 2007; Martinho *et al.*, 2008), the east coast of Australia (Hesp and Thom, 1990), Denmark (Clemmensen *et al*, 2007; Aagaard *et al.*, 2007), Argentina (Tripaldi and Forman, 2007), and

South Africa (Illenberger and Rust, 1988). In many cases these dune fields were established during the Holocene sea-level transgression when large volumes of sediment were brought onshore across the continental shelf. Similar processes were involved in transgressive dune fields formed in the lower Great Lakes of the US and Canada (Martini, 1981; Davidson-Arnott and Pyskir, 1988; Arbogast and Loope, 1999).

9.2.4 Shadow Dunes and Nebkha (Coppice Dunes)

Objects on the sand surface such as driftwood, seaweed or isolated plants serve to alter the near surface flow conditions in complicated ways. Depending on the size, geometry and orientation of the object, the modified flow around and over-top of the object can lead to scour or to sand deposition (Figure 9.18), which is thought to depend on the nature of coherent flow structures such as corkscrew vortices that develop around the object (McKenna Neuman and Bédard, 2015). Under the proper moisture and nutrient conditions, plants may grow and progressively modify the wind and sand transport patterns. In the simplest case, the plant slows the wind sufficiently so as to induce deposition downwind of it. This is especially true for relatively porous plants such as isolated grasses and sedges that do not entirely block the wind in front of the plant but reduce the flow velocity through the stems. A type of sand deposit is created in the lee of the plant that is referred to as a shadow dune because deposition occurs in the downwind 'shadow' of the flow obstruction (Hesp, 1981; Figure 9.19). Shadow dunes can be a few centimetres to several metres in height and 1–20 m in length. The specific geometry depends in large part on the nature of the flow obstruction and prevailing wind direction. Shadow dunes often appear in clusters, and as they grow in size they will interact and produce hybrid en echelon forms (Figure 9.19D).

Depending on the porosity and growth characteristics of the plant, the locus of sand deposition may be at the base of the plant rather than in its lee. This occurs with plants that are very efficient at stopping sand movement through the stem and leaf system, leading to partial burial of the plant basal structure. If nutrient and moisture conditions are favourable, the growth of the plant is able keep pace with the rate of deposition, producing a sand hummock that has a partial or complete vegetative cover. These features are known as nebkha (derived from the Arabic word 'nabkha' meaning small sand hillock) or coppice dunes (Figure 9.20). They are usually found in isolation, and are common in

Figure 9.18 Flow obstructions on sand surfaces creating complex deposition and erosion patterns. A. Piece of driftwood inducing scour in the lee with depositional trailing arms on either side. Note how isolated clumps of vegetation in the background have deposition in the lee rather than scour. B. Complex scour patterns in front of and around a broken plastic bin with deposition in the interior section of the bin (flow from right to left; kiwi footprints for scale).

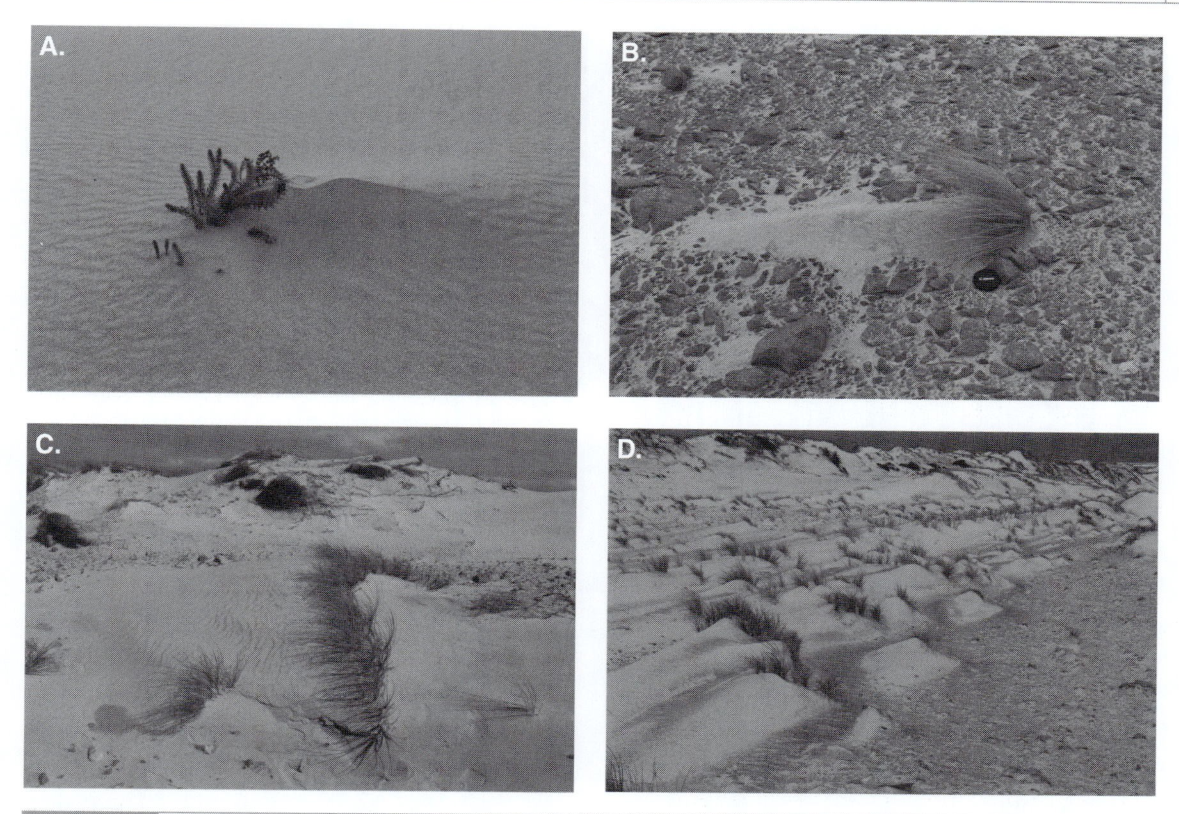

Figure 9.19 Shadow dunes. A. Small deposit in the lee of plant stems on a flat, rippled sand surface (air flow from left to right). B. Small deposit in the lee of a plant clump with long bendable stems on a coarse lag deflation layer (from right to left). C. Large shadow dune with a fringe of porous sedge on the crest that has grown with the dune. Air flow is from right to left, and sand ramp that has developed in front of the shadow dune is due to blockage of air flow by the sand deposit rather than the vegetation. D. En echelon shadow dunes coalescing.

arid environments as well as coastal dune complexes. Similar to shadow dunes, they range in size from a few centimetres to several metres in height, but their shape is usually not aerodynamic as with shadow dunes but hummocky and rounded. Depending on the rate of sediment supply and the structural properties of the plant, nebkha may show asymmetry in both morphology and plant health (Figure 9.20C).

Although the classic nebkha shape is more rounded than oblong, the nebkha itself is a barrier to air flow and sand transport across the dune. In cases where sand supply is limited but wind is ample, there can be very deep scour hollows around the base of the dune, and these maintain the rounded or tear-drop shape of the nebkha. In other instances, a shadow dune may form in the lee of the nebkha, thereby creating a hybrid form with a windward side that is fully vegetated and a leeward side that is an open sand surface extending downwind (Figure 9.21).

9.3 Foredune Morphodynamics and Maintenance

The conceptual model of foredune evolution proposed by Hesp (1983) describes situations for which a beach–dune system evolves from an essentially flat sandy surface to the growth of incipient foredunes and dune hummocks, and

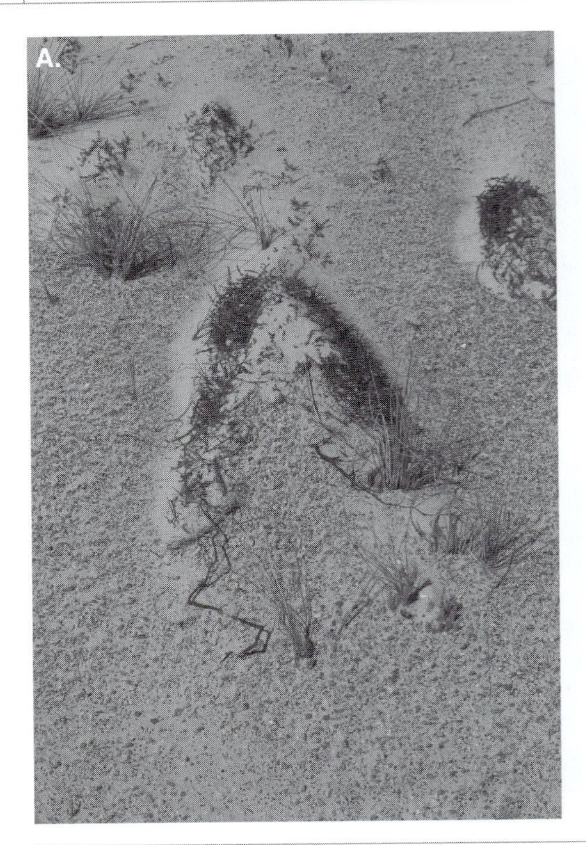

Figure 9.20 Nebkha or coppice dunes. A. Small, incipient nebkha forming with a v-shaped plant that has trapped sand on the windward and leeward side as well as within the structure of the plant. Note difference between this example and the incipient shadow dune shown in Figure 9.19B. B. Typical nebkha with avalanche slope in lee. Scour hollows around base are indicated by coarse lag deposits. C. Internal structure of nebkha showing how sediment is trapped within the plant community, indicating co-evolution (air flow from left to right).

Figure 9.21 Example of hybrid nebkha-shadow dune that displays elements of both classic dune forms. This is a medium-sized dune (approximately 1 m high) with dense cover of grasses on windward side, scour hollows around the front and side and shadow deposits in the lee.

ultimately, to the development of a mature beach–foredune morphology. The model describes progressive genesis or developmental growth rather than equilibrium behaviour or foredune maintenance. Although recurring phases of complete foredune destruction and re-formation do occur (e.g., on barrier islands prone to overwash during intense storms; see Chapter 10, Section 10.3), more often the foredune system is only partially eroded and therefore exists in a long-term state of equilibrium. The beach–dune system moves through cyclical phases of partial foredune erosion during extreme storms followed by dune rebuilding via aeolian processes during inter-storm periods. The foredune may change its morphology and position, but over the long term (i.e., several decades) it simply oscillates back and forth adopting several interim states that are all within the range of normal steady-state behaviour.

9.3.1 Wave-Cut Scarps and Sand Ramps (Aprons)

Major storm events cause beach erosion (see Chapter 8), which often manifests as an overall lowering of the foreshore and backshore. During intense storms with a large storm surge, especially during spring tides, the base of the foredune can also be eroded thereby creating a scarp (Figure 9.22). The sediment that is eroded from the beach and foredune is moved offshore or alongshore within the littoral cell. However, during fair weather conditions, sediment will move back onshore to rebuild the beach, and then it is becomes available for aeolian transport. The foredune plays an important role in this exchange of sediment because it serves as a sand supply (buffer) during the erosional phases and as a storage reservoir during the rebuilding phases. Because the scarp face that is carved into the foredune toe is usually very steep, there are often slope failures in the form of slump blocks and sand avalanches on the seaward margin of the dune. During the storm, these materials are easily removed by waves because they have little cohesive strength. After the storm has receded, slope instabilities above the scarp may lead to more slumping as well as sand avalanching, and these deposits accumulating at the foredune toe will initiate the process of dune ramp development.

Erosional scarps are significant barriers for aeolian sediment transport because they prevent sand from being delivered from the lower beach to the foredune stoss slope and crest. A zone of flow stagnation develops at the base of a scarp in which the onshore flow slows down because of the physical barrier that the scarp presents to the wind field. The incoming air is forced overtop the stagnation zone while the near-surface flow slows down or stalls completely. This reduction in near-surface air flow causes sand deposition at the base of the scarp. Under some situations, such as with very large scarps and perpendicular wind, a reverse eddy or vortex forms in the toe region, which leads to the development of echo dunes in front of the scarp (Figure 9.23). However, with oblique wind approach angles, the wind and sand that move across the backshore are deflected alongshore at the base of the scarp (Bauer *et al.*, 2012; Hesp *et al.*, 2015). These conditions promote the accumulation of sand at the base of the scarp in the form of a sand ramp or sand apron (Figure 9.24).

Small scarps are filled rapidly because the sand ramp will not require a large volume of sand to completely rebuild, especially if there are slump blocks or avalanche lobes already occupying the toe region. In situations where a wave-cut scarp is carved into an irregular foredune with blowout notches, the sediment

Figure 9.22 Wave cut scarps on foredunes. A. Foredune scarp on North Stradbroke Island (Queensland, Australia) showing viable exposed rhizomes and angle-of-repose avalanche cones beginning to fill the toe region with dry sand from above. B. Large scarp cut into a foredune that removed the entire stoss slope up to the crest. Note major slump blocks with intact vegetated surfaces on top (Greenwich Dunes, Prince Edward Island, Canada).

Figure 9.23 Echo dunes. A. Small echo dunes with partial vegetation regrowth. B. Large, isolated echo dunes in front of a cliff with avalanche deposits. C. Continuous sand sheet separated from the scarp by a scour hollow. Note lee-side slope that is maintained at a steepness greater than the angle of repose because of the vortex action in the scour hollow.

conveyor from the beach to the dunes is usually established first in the low-lying blowout notches (Figure 9.24B). The overall slopes are gentler in the notches, and therefore they are less susceptible to scarp formation by wave action as well as being easier to heal by sand ramp development. After the sand ramp has re-established in front of the notch, sediment can move up the sand ramp, through the notch, and onto the lee surface of the foredune, thereby feeding sand to the hinterland. On a laterally continuous dune without notch blowouts, the re-establishment of an extensive sand ramp is a prerequisite for delivery of sand to the stoss slope and dune crest (Figure 9.24C; see also Figure 9.1), thereby leading to foredune growth and progradation.

The long-term maintenance and growth of foredunes is therefore critically dependent on the sand ramp healing process, which is an essential prerequisite to re-establishing a continuous sediment transport pathway between the beach and the stoss slope of the foredune (Christiansen and Davidson-Arnott, 2004). Sediment delivery to the foredune crest and beyond is most effective when there is an extensive, smoothly continuous, and gently sloping sand ramp fronting the dune. However, such ramps develop only after a significant period of aeolian activity that has not been interrupted by beach erosion due to storm waves. The foredune is therefore in a state of dynamic equilibrium that is sensitive to the balance between phases of storm-induced wave erosion that scarp the foredune toe and aeolian sand delivery that rebuild the ramp. The morphology of the foredune will vary from month to month and year to year depending on the sequencing of intense storms

Figure 9.24 Dune ramp building is essential to re-connecting the continuous sand transport pathway from the beach to the foredune. A. Steep, partially complete sand ramp in front of a large dune. B. Sand ramp building into a notched section of a scarped foredune, which has allowed sand to be plastered on the marram grass downwind. Note scour hollows to right and left where the scarp is steep and the marram grass is not covered in sand. C. A smoothly continues sand ramp at the base of this vegetated dune is allowing sediment to move up the stoss slope of the dune and increase its volume.

and intervening periods of beach rebuilding (Davidson-Arnott *et al.*, 2018).

9.3.2 The Role of Vegetation in Dune Rebuilding Processes

After a foredune erosion event has occurred, it is very common to find remnant and exposed rhizomes hanging from the scarp wall or extending out on the backbeach (Figure 9.25A). Rhizomes are typically buried, and by growing horizontally and vertically they allow the plant to expand and propagate new nodes into previously unoccupied areas. Depending on the plant species, new sprouts (consisting of a main stem, several tillers with leaves, and a root wad) may appear at nodes that are spaced approximately 0.05 to 0.15 m or so (Figure 9.25B). *Ammophila* and *Spinifex* rhizomes may extend tens of metres horizontally and can be buried by more than a metre of sand. If exposed during an erosion event, they will sit on the sand surface, grow horizontally (usually downslope) and sprout new nodes.

The presence of vegetation on freshly eroded sand surfaces can aid the foredune healing process through the promotion of sand deposition (Figure 9.26). There has been much research on the effect of plants on wind flow and turbulence, and to modelling their effect on the erosion potential from semi-arid areas and agricultural fields (Wolfe and Nickling, 1993; Lancaster and Baas, 1998; Finnegan, 2000; Leenders *et al.*, 2007; de Langre, 2008). The effectiveness of vegetation in reducing transport is primarily controlled by

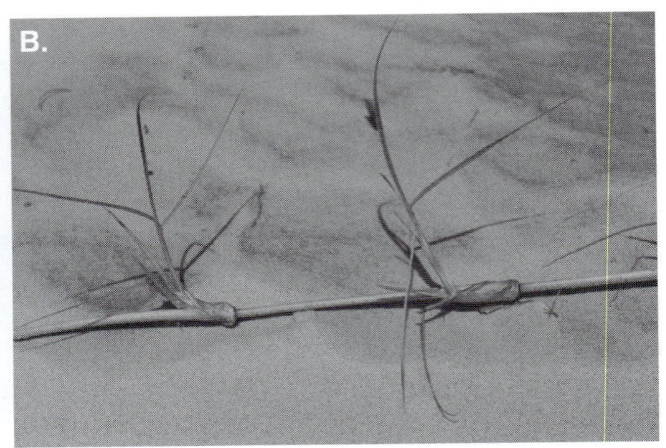

Figure 9.25 Marram grass rhizomes (West Beach, Calvert Island, British Columbia, Canada). A. Rhizomes that were previously buried in about 0.25 cm of sand are now exposed after a dune-scarping event and trying to re-establish on the avalanche surface at the toe of the foredune. B. Close up of rhizome showing plant structure. A single rhizome or runner can be many metres in length and host several main stems each of which may sprout roots.

Figure 9.26 A. Example of rhizome partly buried by aeolian transport showing how the plant can re-establish itself and extend seaward across the beach in front of the previously scarped foredune (background of photo). B. Plant runners growing seaward across the back beach.

per-cent surface cover, but it is also a complex function of plant height, stem density, form and flexibility, as well as the spatial pattern of plant clusters and bare sand patches (Olson, 1958b; Hesp, 1989; Arens *et al.*, 2001). Vegetation cover is typically sparse at the seaward edge of the foredune but increases towards the stoss slope and crest. Sand is easily transported through the bare patches between individual plants and clumps of plants, but as the vegetation cover increases, more sand is deposited. Field studies show that the sand transport rate is reduced to negligible amounts compared to a bare surface for a vegetation cover of 20–25 per cent (Lancaster and Baas, 1998; Kuriyama *et al.*, 2005).

9.3.3 Environmental Gradients and Plant Zonation on Foredunes

Pioneering beach plants such as *Ammophila* and *Spinifex* have evolved coping strategies to live in environments where they are progressively buried – they actually require such burial to thrive (Moreno-Casasola, 1986; Maun, 2009). *Ammophila breviligulata*, for example, is able to survive burial of 1 m or more (Maun and Lapierre, 1984; Maun, 2004), and propagates rapidly under such conditions. However, the backshore is a zone of considerable stress for plants (Hesp and Martinez, 2007; Miyanishi and Johnson, 2007; Miot da Silva *et al.*, 2008; Lane *et al.*, 2008), because sandy substrates do not provide an organic-rich soil from which roots can draw minerals, nutrients and moisture very effectively. In addition, temperatures can be very hot during the day and there is continuous exposure to salt spray that coats plant surfaces. As a result, the plants that grow in these harsh environments are limited to a few pioneering species such as grasses and sedges that are adapted to the stress (Ranwell, 1972; Packham and Willis, 1997; Maun, 2009). The primary outcome of this biological adaptation is that the pioneering plants are able to trap more and more sediment, which helps them to out-compete other species while enabling the formation of incipient dunes (Figure 9.27).

As described earlier, the growth of incipient dunes in front of an established foredune prevents sediment from being transported to the old foredune, thereby stranding it in its landward position. This process of foredune stranding implies that the pioneering plants that originally co-evolved with the old foredune will have to cope with slightly different moisture, nutrient and sediment conditions. Other types of plants will begin to invade the stabilised dune surface and potentially out-compete the pioneering plants that originally helped the foredune to grow. The classic model of Cowles (1899) emphasised the role of a succession of plant species in such an evolving foredune, which leads to stabilisation of the surface and improved moisture retention and nutrient availability as grasses are replaced by bush and then forest. Although originally proposed as a temporal (evolutionary) model, the successional sequence at a single location has a spatial analogue in prograding dune systems, such that the pioneering plants are most prominent on the incipient dunes and the successional plants (broad-leafed plants, shrubs, trees) are most prevalent farther inland (Doing, 1985; Hesp, 1991; Hesp and Martínez, 2007; Dech and Maun, 2007).

Figure 9.28 shows a foredune system that reflects the spatial gradient in decreasing environmental stress away from the beach. The influence of limiting growth factors is so common to foredunes across the world that the vegetation assemblages often look very similar regardless of location even though the plant species themselves may be different. Thus, the form and evolution of foredunes, and especially the incipient

Figure 9.27 Pioneering plants on sand ramps trap sediment thereby leading to incipient dune growth seaward of the established foredune. Whether the incipient dune will continue to grow depends on the frequency and timing of wave-erosional events relative to aeolian deposition. (Near Port Lincoln, South Australia.)

Figure 9.28 Example of the spatial zonation of plants on a prograding foredune system with pioneering species on the incipient dunes and woody shrubs and trees on the older foredunes. (Near Port Lincoln, South Australia.)

dunes, on a scale of years to decades is remarkably similar from place to place because of the structure and growth habits of the colonising vegetation and the successional plants that follow. Where the pioneering species are less tolerant of burial, transgressive dune fields may be more common. The classic Coos Bay transgressive dune field described by Cooper (1958) developed because of the inability of native species to withstand burial, and thus, foredunes were nearly absent. Similarly, in New Zealand the native pingao grass (*Desmoschoenus spiralis*) is relatively slow growing and unable to withstand rapid rates of burial leading to generally greater degree of foredune instability and development of transgressive dune fields.

The co-evolution of foredunes and plants as described in the model of Cowles (1899) is largely a reflection of the reduction in the frequency and magnitude of disturbances from waves, as well as a shift in the intensity of the limiting growth factors such as exposure to salt spray and moisture stress. Thus, just as the plants have an influence on the evolution of the foredune, so does the morphological context influence the nature of plant growth. For example, trees that are adapted to salt spray (such as coconut and manchineel) are common very close to the shoreline on low-energy tropical beaches where dune accretion rates are small. Where the beach sediment budget is nearly neutral and the dune budget is positive, there may be little change in the gradient of salt spray and sand burial over decades, and consequently little change in the plant communities. On the other hand, disturbances due to erosion by waves, blowouts and parabolic dune development can quickly lead to a change in the dominant vegetation assembly locally, or over large parts of the dune field depending on the severity of the disturbance (Saunders and Davidson-Arnott, 1990; Maun, 2004). Thus, primary dune colonisers such as *Ammophila* or *Ipomea* can be found far inland in locations where there has been local disturbance and reactivation of sand movement. Similarly, large cottonwood trees (*Populus deltoides*) are a common coloniser on coastal dunes around the Great Lakes, even in areas with a large sediment supply because there is no salt spray to cope with. Progressive foredune growth may lead to an increase in the water table in the dune slack landward of the dune lee slope and thus to replacement of plants associated with dry conditions by plants such as rushes and sedges.

9.4 Aeolian Processes on Beaches and Dunes

Coastal sand dunes are created and modified by aeolian sediment transport processes, so a general introduction to the nature of fluid flow and the movement of sediment in air is necessary to fully understand the development and maintenance of dunes. The seminal contributions of Bagnold (1941) to aeolian sediment transport are still widely cited and referred to, but excellent summaries related to coastal dunes can be found in Pye (1983), Horikawa *et al.* (1986), Nickling and Davidson-Arnott (1990), as well as in the compendium of aeolian chapters in the *Treatise on Geomorphology* (Shroder, 2013, vol. 11). Here, we begin with the classic approach to wind flow in a boundary layer, working toward a predictive relationship between bed shear velocity, u_*, and sediment transport rate, q. To large degree these are idealised conceptualisations of aeolian transport mechanics that are applicable in wind tunnels and over extensive, flat sand sheets in desert-like environments. Natural beaches, in contrast, are exceedingly complex, and there are a host of factors that need to be taken into consideration that affect both the capacity of the wind to transport sediment (referred to as transport limitations) and the supply of sediment that is available for transport at any specific location and time (referred to as supply limitations). Thus, the use of most sediment transport equations that were developed for simple situations under equilibrium conditions do not always provide reliable predictions of sediment transport rate on beaches. Much more research is needed in order to predict the future evolution of beach–dune systems under uncertain scenarios of climate change and human development of the coastline.

Air Versus Water

Aeolian geomorphologists are concerned with how the flow of air moves sediment and yields aeolian sediment features. Although both air and water are Newtonian fluids and subject to the same physics, they differ significantly in their properties (e.g., density, viscosity) and their flow characteristics (i.e., turbulence), with significant implications for the potential of air to entrain, transport and deposit sediment. Liquid water is about 1000 times more dense than air at normal temperatures and pressures, so the relative density (or submerged weight) of sand particles in air is very much greater than in water. Whereas even a relatively weak water current (e.g., 0.3 m s^{-1}) or small ocean wave can move sand, wind is only able to mobilise fine sand particles when wind speeds exceed about 6 m s^{-1}. In many coastal locations, wind speeds in excess of 10 m s^{-1} are quite common, whereas speeds greater than 30 m s^{-1} are relatively unusual except during extreme storms. The situation on Mars, Venus or the Moon is different because the atmospheres are made of different gases (and have different densities) and because the gravitational fields are different from Earth.

9.4.1 Aeolian Boundary Layers on Beaches

Wind blowing across a beach is affected by frictional resistance at the bed, which leads to the development of a boundary layer in the near-surface region, as is typical of any fluid flowing over a fixed surface. The boundary layer is a zone of reduced flow speed, which manifests as a wind speed profile with the slowest moving air closest to the bed and fastest air high above the surface (in the free stream). Figure 9.29 shows an anemometer tower deployed to measure the wind speed profile on a beach (Figure 9.29A), as well as the resulting time series from each of the rotating-cup anemometers on the tower for a short time interval (Figure 9.29B). Averaging of the time series yields a mean wind speed for each anemometer, and the data pairs (height, mean speed) can be plotted on a graph to yield a wind speed profile. Two characteristic profiles for weak and strong incident winds are shown in Figure 9.29C. The shape of the profile, which is classically parabolic upwards, depends in part on the nature of the incident wind as well as the roughness properties of the surface and the topographic slope of the beach.

Although not immediately apparent from the smooth shape of the wind speed profiles shown in Figure 9.29C, the boundary layer consists of an inner layer very close to the bed (of the order of a few mm to cm thick; dominated by viscous effects) and an outer layer that extends from about 1–2 metres above the bed to several tens of metres or more over rough terrain (dominated by turbulent transfers of energy). In between the inner and outer layers is a transitional layer for which the increase in wind speed away from the surface is proportional to the logarithm of height above the bed (e.g., Bauer, 2013). The wind speed profile for air blowing over a smooth sand surface (without sediment transport) is described by the logarithmic velocity profile equation:

$$U_z = \frac{u_*}{\kappa} \ln\left(\frac{z}{z_o}\right), \tag{9.1}$$

where U_z is average wind speed at height, z, above the surface, z_o is the aerodynamic surface roughness length (indicative of the relative surface roughness), κ is the von Karman constant (approximately equal to 0.4), and u_* is the bed shear velocity. Figure 9.30A shows three different wind speed profiles plotted on a graph with arithmetic scaling of the axes. All have the classic parabolic shape whereas a transformation of the height axis into logarithmic units demonstrates that the profiles are now linear (Figure 9.30B). This is the basis of Equation 9.1, which leads to the immediate conclusion that the bed shear velocity (u_*) is proportional to the slope of the wind velocity profile when plotted on a log-linear graph. Thus, u_* is not a true velocity but rather a scaling parameter that is related to the shear stress (τ) imparted to the bed by the fluid flow above the bed. The precise definition of shear velocity is

$$u_* \equiv \sqrt{\frac{\tau}{\rho}}, \tag{9.2}$$

A.

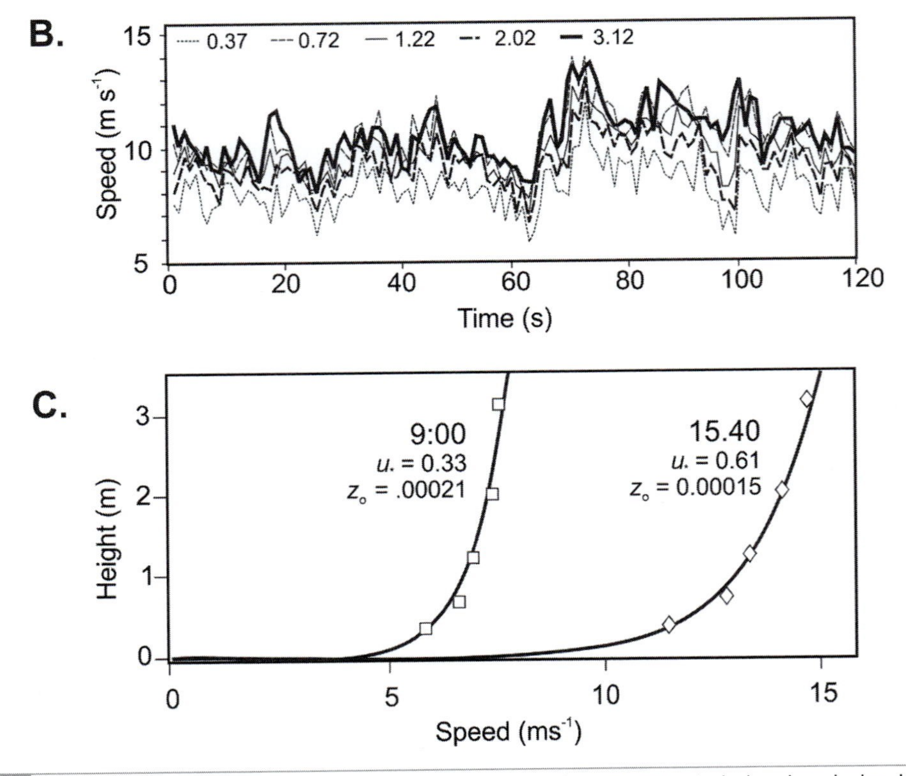

Figure 9.29 In most aeolian experiments, a vertical tower holding several anemometers is deployed on the beach to measure the wind conditions A., which can be represented as a time series plot B. that shows the data from each of the anemometers as a function of time. Usually, the highest anemometer (at 3.12 m, in this example) measures the fastest wind speed whereas the lowest, near-surface anemometer (at 0.37 m) measures the slowest wind speed. When the data are averaged across time, each of the average wind speeds can be plotted against the height of the anemometer, yielding a vertical wind speed profile C. that has a characteristic parabolic shape. Here, u_* is shear velocity (m s^{-1}), and z_o is roughness length (m).

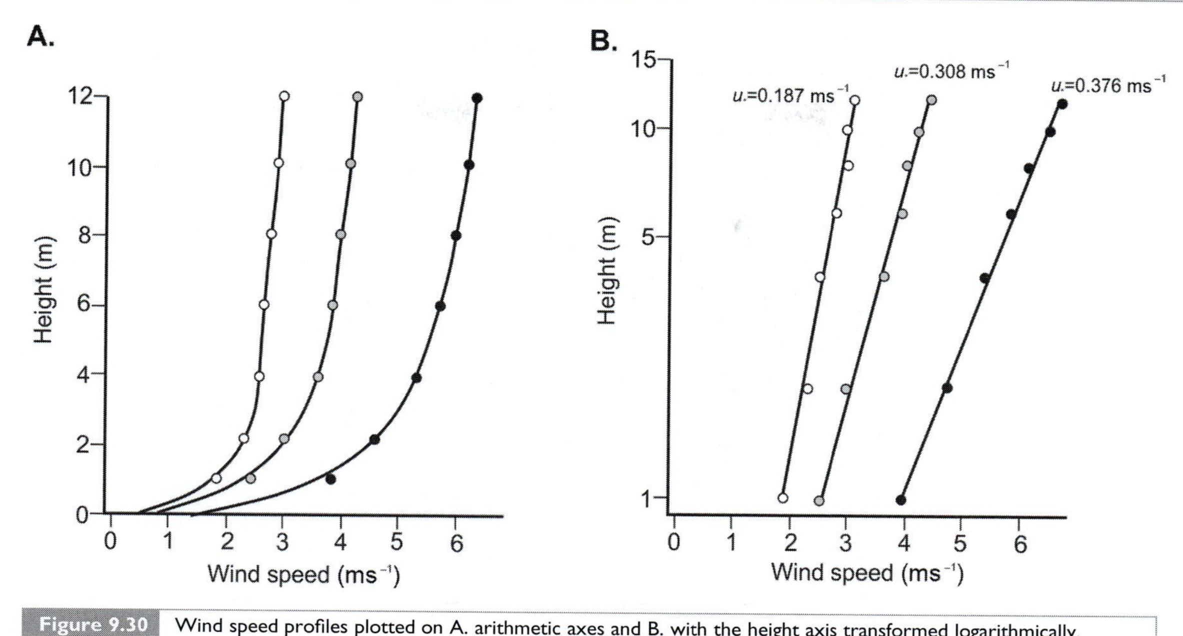

Figure 9.30 Wind speed profiles plotted on A. arithmetic axes and B. with the height axis transformed logarithmically.

where ρ is the fluid density. However, shear stress, τ, is generally unknown and difficult to measure, so Equation 9.1 is used to estimate the shear stress via the shape of the wind speed profile because it is a reflection of the strain within the boundary layer. It is standard practice in both wind tunnel and field experiments to estimate u_* from measurements of wind speed at several heights within the logarithmic boundary layer through the use of a regression of wind speed (the dependent variable) against the log of height. This regression technique also provides direct estimates of the roughness length, z_o, as discussed in Bauer *et al.*, 1992.

Bagnold (1941) proposed a value of $D/30$ for z_o, where D is the grain diameter. This approximation applies to smooth, static sand beds because the sand grains are the main roughness elements on the surface (larger grains imply a rougher surface). If the wind speed increases to the point where sediment is entrained and transported in the flow field, the wind speed profile is altered by the presence of sand in the flow field (Bauer *et al.*, 2004). Momentum is extracted from the air flow by the sand, which leads to a reduction in wind speed in the near-surface layer, a steepening of the speed profile, and an increase in shear velocity. As a consequence, a new roughness length, z'_o, develops, which is usually larger than z_o. This appears to be related both to the actual effect of sand in transport in the lower layers and to the development of ripples on the sand surface, which happens very quickly after the initiation of sediment transport. The presence of ripples and dunes, as well as vegetation and other obstacles such as driftwood, on the beach surface produces additional frictional roughness that further alters flow close to the bed, and thus, the shape of the velocity profile. Where the surface is covered by tall grasses or shrubs the wind speed profile is often displaced upwards from the surface (Olson, 1958c), to a new reference plane, which is a function of the height, density, porosity and flexibility of the roughness elements. This phenomenon is well known in flow over forest canopies and agricultural fields. The upward displacement is termed the zero plane displacement height, z_d. For these situations the wind profile equation becomes:

$$U_z = \frac{u_*}{\kappa} \ln\left(\frac{z - z_d}{z_o}\right). \tag{9.3}$$

The displacement height can be thought of as the level above the measurement surface, but within the plant canopy, that the mean wind speed tends to zero. However, this concept of a displacement height is of limited value for estimating wind transport on vegetated dunes where most erosion takes place in bare patches between vegetation elements and where gusts and variations in wind direction change the relative exposure of sand to transport over very short time periods and distances. Thus, Equation 9.1 is used more frequently by aeolian geomorphologists working on coasts.

In wind tunnel experiments the wind speed can be held constant and the instantaneous measurements of wind speed (from second-to-second) will not deviate significantly from the mean wind speed obtained after averaging over several minutes. In the field, the flow is highly turbulent and unsteady, with wind gusts well above the mean that are separated by lulls when the flow speed is much below the average. The time series presented in Figure 9.29B show examples of this variability on a natural beach. From a statistical perspective, the instantaneous wind measurements are uncertain estimates of the broader processes characterising the wind field, so the most sensible approach is to use the mean wind speeds from each anemometer to define the wind speed profile. The mean profile is then used to estimate u_* and z_o, which highlights the fact that these parameters are mean properties of the flow field (not instantaneous measures). Typically, a record of 10 or 15 minutes is used – long enough to get a robust mean but short enough to assume temporal stationarity in the record. Shorter averaging intervals are often used in sediment transport studies because there is more information on the temporal changes in wind strength, but the statistical uncertainty increases proportionally as averaging intervals are shortened. Namikas *et al.*, (2003) demonstrated that the error bars for estimating u_* decrease rapidly for averaging times of < 3 minutes. This is of critical importance because sediment transport is predicted on the basis of shear velocity, so if the estimate of shear velocity is uncertain, so too will be the prediction of sediment transport rate. There is a trade-off between getting more information in the time domain and the overall statistical reliability of that information.

The discussion above pertains to changes in surface roughness that may evolve at one location over time due to an increase in wind speed. But there are also spatial changes in surface roughness when the wind flows across an ocean or lake surface and encounters the beach–foredune system. Whether the surface becomes aerodynamically smoother or rougher dictates how the wind profile will respond to the new frictional drag (Bauer, 2013). The flow adjustments are almost immediate in the inner boundary layer (close the roughness elements) and the effects eventually diffuse upward and downwind via turbulent mixing processes. But on beaches and dunes, where there can be rapid changes in the surface form (e.g., from waves to foreshore, backshore, and foredune) and considerable variation in vegetation height and density (grasses, shrubs, trees), there can be a continuous sequence of boundary layer adjustments that are quite complex. As a result, the logarithmic portion of the boundary layer (also referred to as the constant stress layer) from which u_* can be estimated may only be a few tens of centimetres thick, thus making it difficult to place a minimum of three anemometers within this layer. This remains a fundamental challenge for aeolian geomorphologists working on beaches because wind speed profiles that do not conform to the logarithmic model are quite common. There is little guidance in the literature on how to overcome this challenge, so by convention, most aeolian geomorphologists try to use only wind speed profiles that have large R^2 values (i.e., conform well to the logarithmic model) when predicting sediment transport. Increasingly, aeolian geomorphology is becoming reliant on computational models to determine stress distributions across beach–dune systems, but field verification of the model results remains a difficult undertaking.

9.4.2 Initiation of Sediment Movement (Entrainment) by Wind

Wind flow over a sandy beach or dune surface has the capacity to move sediment if the fluid forces (lift and drag) acting on individual grains are sufficiently large to overcome their tendency to stay in place (Figure 9.31). Particles on the bed are controlled by the force of gravity, specifically their submerged weight in air. But the lodgement position of the particle relative to its neighbours is critically important because some grains are partly buried (i.e., difficult to mobilise) whereas others are perched and relatively exposed (i.e., easily mobilised). The forces acting on a spherical grain sitting on a bed of spherical particles of similar size can be resolved using force vectors and a moment analysis. Ultimately, such an analysis leads to the conclusion that sediment entrainment is related to the Shields parameter (τ_*), which is a non-dimensional shear stress that is defined by

$$\tau_* = \frac{\tau}{(\rho - \rho_s)\ gD}, \tag{9.4}$$

where τ is the bed shear stress, g is gravitational acceleration, D is particle diameter, and ρ and ρ_s are the density of the fluid and sediment, respectively. The relationship is non-linear and depends critically on the turbulent character of the boundary layer flow, as shown on a Shields curve. In practice, it is difficult to measure the bed shear stress directly, even in a wind tunnel, so the near-bed shear velocity, u_* , is substituted into the numerator of Equation 9.4 on the basis of the relationship in Equation 9.2 (i.e., $\tau = \rho\, u_*^2$).

Bagnold (1941) performed extensive wind tunnel experiments and discovered that the threshold shear velocity, u_{*t}, at which particle motion is initiated depends on particle size and relative density. He proposed the following predictive relation:

$$u_{*t} = A\left(\frac{\rho_s - \rho}{\rho} gD\right)^{0.5}, \tag{9.5}$$

where A is an empirical coefficient that is approximately equal to 0.1 for particles with a Grain Reynolds number, $\mathrm{Re}_G > 3.5$. Close examination of this relation indicates that it is very similar to the Shields' parameterisation, with the main difference being that Bagnold used A while Shields used τ_*, both of which are complex empirical parameters. Bagnold's keen insight into sediment entrainment in air, specifically, led him to propose that there were actually two different thresholds to consider (Figure 9.32). When immobile particles are sitting on the bed, the wind speed needs to be increased to a certain critical level to initiate sediment motion, which Bagnold referred to as the fluid threshold because the initial motion is triggered by fluid forces alone. However, once particles are in motion, they will make contact with other

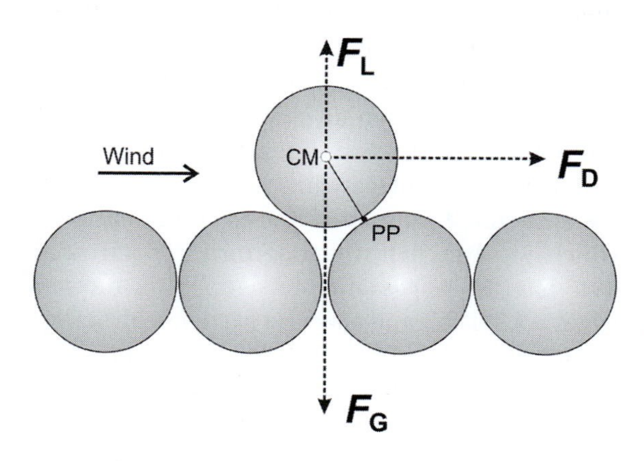

F_L	Lift Force
F_D	Drag Force
F_G	Gravity Force
CM	Centre of Mass
PP	Pivot Point

Figure 9.31 Forces acting on a spherical particle sitting on a horizontal bed of similar particles. Gravity, F_G, acts vertically downward whereas the lift force, F_L, acts vertically upward. The fluid drag force, F_D, is horizontal and in the direction of fluid flow, which is perpendicular to the lift force. In this idealised example, all forces act through the centre of mass of the particle, which is typically not the case for non-spherical particles. In order for the particle to be dislodged, it needs to be lifted or rotated around its contact pivot point, PP, which involves moment arms that are levered according to the distance between CM and PP.

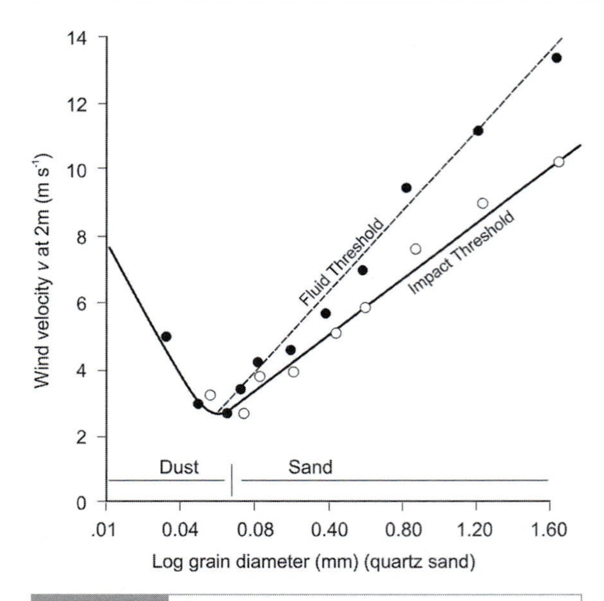

Figure 9.32 The wind velocity threshold for initiation of motion as a function of grain size assuming a particle density equivalent to quartz. Note that the wind speed required to initiate motion of resting particles by fluid forces alone (fluid threshold) is greater than the wind speed required to keep grains in motion once saltation has been induced (impact threshold) (Baas, 2019).

particles sitting on the bed on their return trajectories. These impacts serve to move sediment downwind, and as a consequence it is possible to reduce the wind speed by about 20 per cent from the initial fluid threshold and still keep sediment in motion. This is referred to as the impact threshold, and it seems peculiar to sand transport in air but less so in water. Figure 9.32 shows that quartz particles around 0.09 mm in diameter are the easiest to entrain, at a shear velocity of about 0.16 m s^{-1}. Such a shear velocity requires an actual wind speed of about 4 m s^{-1} or greater (measured at a height of 2–3 metres above the sand surface). The threshold shear velocity increases rapidly with increasing grain size and density, as one might expect. The upper limit for sediment transport under natural conditions in air is about 2.5 mm, although the mean grain size of most dune sands is < 0.5 mm. The threshold shear velocity increases into the silt and clay range because of cohesive bonding between fine particles as well as the plate-like geometry of clay particles in particular, which makes dust transport more complex than sand transport. On natural beaches, sediment entrainment is complicated by the presence of a range of grain sizes, variations in shape and mineralogy, the development of ripples, changes in surface moisture both spatially and temporally, and the unsteady and turbulent character of wind in nature. Thus, quantifying the threshold speed (or shear stress) for sediment entrainment on a natural beach is a difficult and imprecise undertaking (e.g., Davidson-Arnott and Bauer, 2009).

9.4.3 Modes of Sand Transport

Once particles have been entrained from the bed, they are generally transported in the direction of the wind with some lateral spread. As is the case with sediment transport in water, aeolian sediment transport occurs in several modes, including (1) bedload, which involves rolling, sliding or creeping along the bed; (2) suspension load, in which particles infrequently touch the bed and are suspended in the air; and (3) saltation load, which involves ballistic trajectories of particles that has the appearance of a hopping motion (Figure 9.33). Silt and clay size particles, although harder to entrain from the bed, are generally carried via suspension due to turbulent eddies that can keep fine-grained particles in the air over long distances and can take them to extreme heights. However, because such fine-grained particles are rarely found in beach materials, aeolian suspension is generally unimportant in the context of coastal dunes. Fine to medium sand can be launched into suspension for short periods of time, for example, in the lee of large dune crests with avalanche slopes, but sand-sized particles are generally too large to be maintained in suspension for more than a few seconds.

The dominant mode of sand transport across beaches is saltation, which consists of grains being launched to heights of a few centimetres or so from the bed, although trajectories as high as 0.5 m or more are not uncommon in strong winds. Saltating particles rise very quickly (almost vertically) in the initial phase of the trajectory, and then are accelerated forward by the wind. As they are carried downwind, they reach

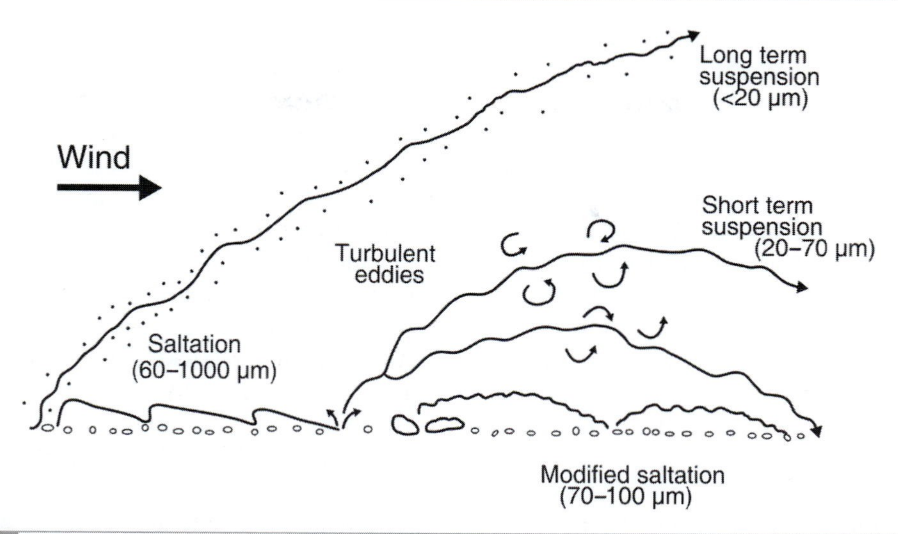

Figure 9.33 Modes of sediment transport in aeolian systems. In general, the finest particles stay in suspension whereas the heavier particles are in semi-continuous contact with the bed, moving either in the mode of classic saltation (ballistic trajectories), modified saltation (shorter, infrequent hops), or by rolling or sliding along the bottom (creep) in consequence of being hit by saltating particles on their return trajectories to the bed (after Nickling and Davidson-Arnott, 1990).

their maximum height and begin to fall back to the bed under the influence of gravity. Downwind travel distances range from several centimetres to many metres. The final phase of the trajectory is a shallow angle return to the bed with the particle impacting the sand surface with considerable velocity and momentum. It is thought that these quickly moving particles actually contribute positive momentum to the air very close to the bed. In part, this may explain why the impact threshold for particle entrainment during active transport conditions is slightly less than the fluid threshold necessary to entrain stationary particles (Figure 9.32).

The classic saltation trajectory is associated with high-energy saltons, which are distinguished from low-energy saltons that only travel a few millimetres upward and a few centimetres downwind in a process known as modified saltation or reptation. Reptation results primarily from the crater-like impact or splash of high-energy saltons with the bed, thereby ejecting several low-energy reptating grains in multiple directions, including laterally and upwind as well as the predominant downwind direction. Bagnold (1941) estimated that about 75 per cent of the sand being transported at any instant in any location is via saltation. The remaining 25 per cent moves as bedload. The proportion of sediment moving as bedload decreases with decreasing particle size, and bedload may be as small as 10–15 per cent of the total transport for surfaces that are dominantly fine sand.

Observations in wind tunnels and from careful field experiments show that when saltation is fully developed, the concentration of grains (as well as the transport rate) is greatest in the zone within about 2–4 cm of the bed (e.g., Farrell and Sherman, 2013; Bauer and Davidson-Arnott, 2014; O'Brien and McKenna Neuman, 2018). Grain concentrations drop off rapidly above 10–15 cm, and there are relatively few grains travelling above about 30–40 cm (although particles at face height have been experienced by many field researchers during strong winds!). The decrease in transport rate with height above the bed is referred to as the vertical flux profile. It is usually modelled as a negative exponential function or a power function, but there can be great variability in the range of regression parameters derived from such models, and therefore universal prediction remains elusive. Nevertheless, because of the rapid increase in wind speed with distance above the bed, grains that are ejected higher into the air stream are accelerated more (O'Brien and McKenna Neuman, 2018), and have much longer hop lengths than ones that only reach a few centimetres in height.

Field measurements based on trapping of sand in transport across the backshore of a beach in California (Namikas, 2003), show that the majority of vertical hops were less than 10 cm (Figure 9.34A) while horizontal hop lengths were typically on the order of 10–50 cm (Figure 9.34B). On natural beaches there is a much greater range of saltation trajectories than is commonly observed in carefully controlled wind tunnel studies. Where a thick layer of dry sand exists on the backshore and the beach is relatively smooth, transport conditions may closely resemble those observed in the wind tunnel. However, if sand is transported across a damp, much harder surface (e.g., where dry sand has been stripped to expose damp sand below) or across an armoured surface of pebbles, the loss of energy on impact is reduced and rebounds are much more elastic so particles are transported faster and higher (McKenna Neuman and Muljaars Scott, 1998; Namikas *et al.*, 2009). On softer beds, more of the incoming energy from saltons is diffused through the bed by a process of grain rearrangement within the bed rather than contributing to the splashing of new particles into the airstream.

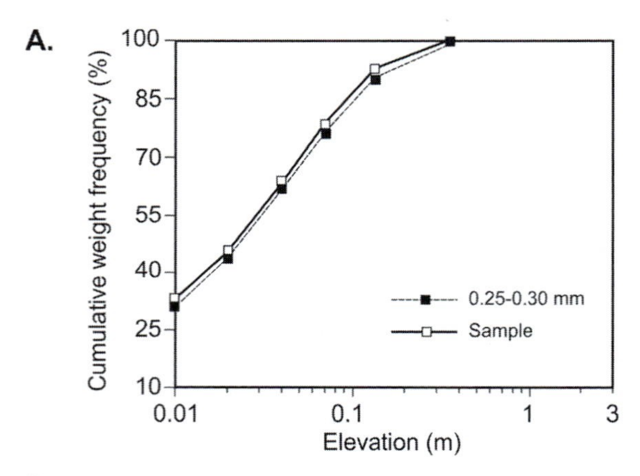

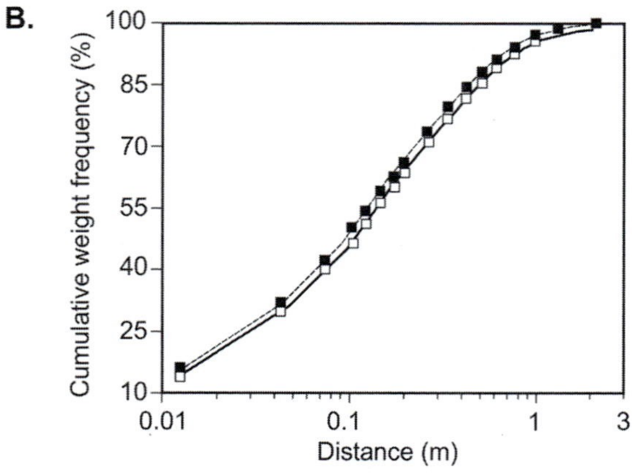

Figure 9.34 Cumulative frequency distributions of the weight of trapped sand as a function of A. height above the bed and B. horizontal distance downwind. Segmented traps were used to collect the data. 'Sample' refers to the entire sample of trapped sand in each trap segment, and the same trend occurs for the sand fraction in the range 0.25 mm to 0.3 mm (after Namikas, 2003).

9.4.4 Predicting the Instantaneous Sediment Transport Rate

A number of relatively simple semi-empirical formulae have been proposed to predict the mean sand transport rate under aeolian action, and almost all of these relations incorporate a cubic function of shear velocity, u_*^3, or mean wind speed, U^3. The coefficients in these equations are usually derived from wind tunnel studies and therefore the use of these formulae must be treated with appropriate discretion. One of the earliest to develop an expression for predicting the sand transport rate was Bagnold (1941) and his formula is still used extensively:

$$q = C_B\sqrt{\frac{D}{D_r}}\frac{\rho}{g}u_*^3, \tag{9.6}$$

where D is grain diameter (mm); D_r is a reference grain diameter usually taken as 0.25 mm, and C_B is an empirical coefficient (dimensionless) that depends on the sorting of the sediment (1.5 nearly uniform sand, 1.8 naturally graded sand, 2.8 for a wide range of sizes).

Bagnold's equation does not include a threshold shear velocity and has been criticised on that basis. However, a threshold condition is implicit to the derivation of his formula and the way Bagnold (1941) calculated shear velocity for use in Equation 9.6. Several subsequent expressions have included an explicit threshold term, for example that of Kawamura (1951):

$$q = C_K \frac{\rho}{g}(u_* - u_{*t})(u_* + u_{*t})^2, \quad (9.7)$$

where C_K is another empirical coefficient (dimensionless) equal to 2.78, and u_{*t} is the threshold shear velocity that can be estimated from Equation 9.5 or by another means. There are several other transport formulae that use this basic approach (Zingg, 1953; Lettau and Lettau, 1977), while the model of Kadib (1964) incorporates a transport intensity function based on Einstein's bedload transport model. Small differences in these formulations actually result in quite large absolute differences in the predicted transport rate (Nickling and Davidson-Arnott, 1990; Sherman *et al.*, 1998).

9.4.5 Equilibrium Transport and the Saturated Flux Assumption

The overall performance of relations such as Equations 9.6 and 9.7 to predict sediment transport rate in natural systems is disappointingly poor (Sherman and Li, 2012). The best results are obtained when the field site conforms closely to the idealistic conditions in a wind tunnel – a steady, uniform wind blowing over a horizontal bed of loose, dry particles of uniform size with no upwind supply restrictions or downwind flow controls. Under these circumstances it can be assumed that the number of grains being ejected from any position on the bed is equal to the number of grains falling back to the bed at that same location. In other words, there is no net erosion or deposition locally, and the transport system is in equilibrium. Moreover, if the wind is carrying the greatest number of particles that it has the capacity to carry (i.e., no supply restrictions), the system is said to be at saturation. The saturated flux condition therefore requires that there is adequate sand supply from upwind and that there is nothing to prevent more particles from leaving the surface except the limiting capacity of the wind to carry sediment. Most natural beach–dune systems do not conform well to these idealised assumptions. Indeed, there are a host of challenges not considered in the equilibrium approach that are arguably of greater importance when trying to quantify and model aeolian sediment transport.

9.4.5.1 *Complications Leading to Non-Equilibrium Conditions*

On natural beaches and coastal dune systems, measurements of sand transport and subsequent deposition over periods of months to years reveal that actual values are generally less than the predicted values based on standard equilibrium equations such as those of Bagnold or Kawamura. Aeolian geomorphologists working in the field have shown that the idealised assumptions underlying the equilibrium approach and the saturated flux assumption are unrealistic for natural beaches (Sarre, 1989; Kroon and Hoekstra, 1990; Davidson-Arnott and Law, 1990, 1996; Bauer and Sherman, 1999; Meur-Férec and Ruz, 2002; Sherman and Li, 2012). A number of factors act to limit the entrainment of sand from the beach surface, and thus, the transport rate will be less than predicted on the basis of wind strength alone.

9.4.6 Supply Limiting Factors

Supply limiting factors (Nickling and Davidson-Arnott, 1990; de Vries *et al.*, 2014) include those that act to reduce the amount of sand that is available to be entrained and transported by the wind. In the simplest instance, if there is no sand on the bed (e.g., a concrete parking lot or a grassy lawn) there can be no sand transport even if there is ample wind. Similarly, there may be no sand source upwind of the point of interest, as is the case on beaches with onshore wind because the air flow will transition from the water surface to the sandy beach surface. Thus, a range of supply limiting factors may preclude the transport rate from achieving the theoretical maximum, and these are typically due to the influence of: (1) surface moisture; (2) binding salts and other chemicals; (3) frozen pore water; (4) pebble and shell lags; (5) algal mats and vegetation cover; (6) roughness elements and obstructions; and (7) snow or ice cover.

In general, a supply limiting factor implies that there is some additional force (beyond gravitational attraction) that prevents sand grains from being entrained from the bed or transported as easily as they might if the sediment was dry and non-cohesive. In effect, the cohesion that results from surface moisture or binding salts raises the threshold shear velocity required to lift particles off the bed. If the value of u_{*t} increases in Equation 9.5, the difference between u_* and u_{*t} decreases in Equation 9.7, which implies that the transport rate, q, will be less when cohesive forces are at play. In this sense, u_{*t} may be viewed as an integral property of the surface that depends on a range of physical factors controlling how much sediment can be released from the surface.

With onshore winds, the point at which entrainment is initiated is a complex function of wind speed, wind approach angle, and surface moisture content (Oblinger and Anthony, 2008). It has long been known that moisture acts to bond particles as a consequence of adhesion of water molecules to soil particles. A tension force is produced at the points of contact between sand particles where capillary water forms a meniscus that binds the adjacent particles together, producing an apparent cohesion that increases the force required to dislodge particles from the bed. There are several models of these forces that predict the effect of moisture on the threshold condition (McKenna Neuman and Nickling, 1989; Cornelis *et al.*, 2004; Darke and McKenna Neuman, 2008) and two useful reviews are provided by Namikas and Sherman (1995) and Cornelis and Gabriels (2003).

Wind tunnel studies (e.g., Belly, 1964) and theoretical models (McKenna Neuman and Nickling, 1989; Cornelis *et al.*, 2004) indicate that surface moisture content of 2–3 per cent is sufficient to severely restrict aeolian sand transport. However, field studies have documented active transport conditions occurring with surface moisture content >5 per cent and in some cases >10 per cent (Sarre, 1989; Gares *et al.*, 1996; Davidson-Arnott and Dawson, 2001; Wiggs *et al.*, 2004a; Davidson-Arnott *et al.*, 2008; Bauer *et al.*, 2009). As McKenna Neuman and Langston (2006) point out, the wind tunnel experiments use a flat, uniformly moist surface, small test area and short run lengths, all of which tend to reduce the natural complexity inherent to field situations. As a result, while wind tunnel studies provide some insights into the physics of the entrainment process with moist sand surfaces, the empirical and theoretical models based on them produce predictions that do not accord well with field measurements. There are three major factors accounting for this discrepancy: (1) instantaneous wind speeds may exceed the threshold for motion even though the mean wind speed lies below the threshold; (2) the moisture content of a natural beach varies spatially (both across the surface and vertically within the beach) as well as temporally so that the threshold may be exceeded in some areas even though the majority of the area lies below the threshold; and (3) once small amounts of sand begin to move, the impact of saltating grains results in the ejection of grains from the downwind surface even though the bonding forces are too large for entrainment by fluid forces alone.

On natural beaches, sources of moisture include: precipitation; condensation (dew); spray from breaking waves; wave run-up (which in turn is controlled by wave characteristics, tidal stage, storm surge and the beach topography); capillary rise from water table or moist subsurface; and seepage from the base of the foredune. These, together with grain size and packing, determine the gradient in moisture with depth from the surface. Moisture will also vary within the beach sediments as a function of packing, small-scale bedding features and beach topography. What is increasingly recognised is the highly dynamic nature of beach surface moisture (Jackson and Nordstrom, 1997; 1998; Wiggs *et al.*, 2004a; Yang and Davidson-Arnott, 2005; McKenna Neuman and Langston, 2006; Bauer *et al.*, 2009; Schmutz and Namikas, 2013). The greatest controls on beach moisture are precipitation and groundwater fluctuations due to tidal cycles and storm surge. The tidal cycles are critically important on many beaches in determining the extent of dry sand available for transport (Atherton *et al.*, 2001). The rising tide inundates the beach whereas on the ebb phase there is a time lag while beach groundwater drains through the pore spaces of the beach sediments. Spring tides, in particular, have the capacity to flood the backshore especially when coupled with storm surge and wave set-up (Bauer *et al.*, 2009). Excellent summaries of beach groundwater dynamics and implications for sediment transport are given in papers by Turner and Masselink (1998), Horn (2002), Namikas *et al.* (2010), and Schmutz and Namikas (2013).

The rate of surface evaporation is also important. Moisture content tends to be quite uniform over small areas when the sand is either very dry (< 1%) or when it is quite damp (> 10%), but at intermediate moisture contents there can be considerable spatial variability (Yang and Davidson-Arnott, 2005). At scales of metres to tens of metres there is a general trend for moisture to decrease from the foreshore to the upper beach, although there are complex patterns due to surface features such as berms, cusps, ridge-runnel systems, aeolian sandwaves and sand ramps (Figure 9.35A). Surface moisture content will also vary temporally with high temperatures and sunshine leading to rapid drying (Figure 9.35B), while cool temperatures during the night can lead to condensation and a rise in moisture content. The rate of surface drying increases with increasing wind speed so that a thin moist layer can dry out very quickly leading to entrainment and rapid stripping of that layer. Near the foreshore where the sediments are consistently wet, stripping off such a surface layer of sand exposes the damp sand below, and therefore the sand supply is shut down temporarily while more drying occurs. The result is that where the sand is generally quite moist (as on the lower beach), erosion takes place intermittently with sand transport initiated during gusts after the surface has had a chance to dry. Evaporation during subsequent periods of gentle winds prepares the sand surface for further entrainment during the next gust sequence. Once a sand grain is

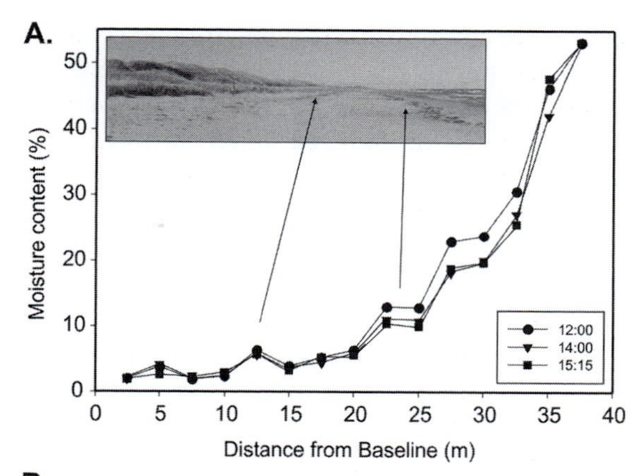

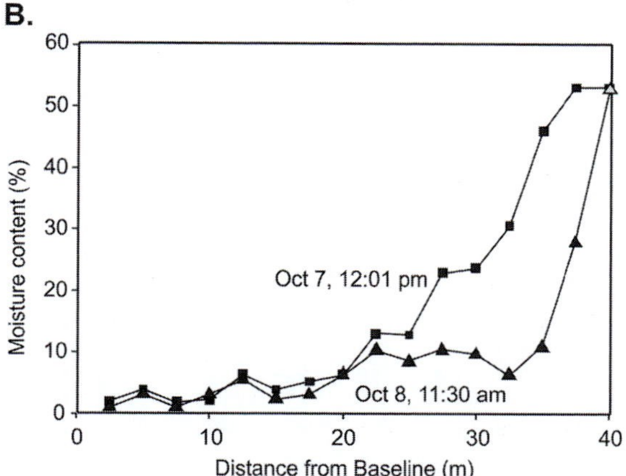

Figure 9.35 A. Spatial and temporal variations in beach surface moisture at Greenwich Dunes, Prince Edward Island, Canada. Typical spatial patterns of beach moisture from moist conditions on the foreshore (right) to dry conditions at the base of the foredune (left). B. Moisture profiles from same location as A., showing that surface drying after one day is most pronounced on the berm above the foreshore. C. Alternating bands of moist and dry sand due to migration of aeolian sandwaves alongshore (Skallingen, Denmark).

entrained, any remaining moisture on the grain surface is quickly evaporated, and thus there is a build-up of dry sand downwind that can be transported readily, leading to beach surfaces that have complex patterns of wet and dry patches (Figure 9.35C).

The effect of binding salts in beach sediments is similar to that of surface moisture, which is to increase the threshold wind speed required to entrain sediments and thereby reduce the equilibrium transport rate. Despite the widespread presence of salts in the sediments of marine beaches, few studies have been conducted on the topic and relatively little is known about how important they are to aeolian transport. Salts can accumulate in surficial sediments either through aerosol deposition (from spray derived from the wave breaking zone) or via beach groundwater fluxes driven by intense evaporation during sunny, dry days. Sometimes a crust develops on the surface that is very resistant to aeolian entrainment, but it is often difficult to determine whether the crusts are due to moisture or salts without careful inspection (Figure 9.36). In either case, these crusts are difficult for aeolian forces to break down, but they are easily interrupted by human and animal trampling or by vehicular traffic. They are usually ephemeral features because the salts can be dissolved during moist periods (e.g., early morning dew) or flushed away during rain storms, and in the case of moisture, it will be evaporated during the day.

The break-up of surface crusts usually leads to exposure of loose sediments that are available for aeolian transport. This is not possible, however, when the surface sediments are frozen. Warmer temperatures and/or solar radiation leading to

Figure 9.36 Surface crusts due to moisture A., B. and binding salts C.

sublimation or melting of the ice are required before the wind is able to mobilise frozen sediments. Clearly, this situation is difficult to predict and it requires a great deal of additional information beyond wind speed and grain size. Similarly, beaches that are frequently covered in snow (Figure 9.37A) will have virtually no aeolian sediment transport during the winter season despite encountering some of the most severe storms and fastest winds (Delgado-Fernandez and Davidson-Arnott, 2011). Algal mats and accumulation of seaweed (Figure 9.37B) or other flotsam on the beach also acts to protect the surface underneath from wind action and may trap sand that is being transported from bare areas. Branches and logs perform a similar function and on many beaches (Figure 9.37C), such as those on the coast of Oregon, Washington, Alaska and British Columbia where logging is prevalent. Complex log matrices have been shown to serve as a depositional reservoir in front the foredune and therefore large accumulations of logs can control sand supply to the foredune and potentially regulate foredune development (Anderson and Walker, 2006; Grilliot *et al.*, 2018).

Pebbles and cobbles deposited on the beach also serve to protect the sand surface from aeolian entrainment in a similar fashion to the development of pebble lag surfaces in arid and semi-arid regions (Figure 9.38A). Laboratory and field experiments show that the effects of pebble lag development are complex (Nickling and McKenna Neuman, 1995; Davidson-Arnott *et al.*, 1997). Sand transport rates are initially enhanced over surfaces with pebbles relative to bare sand

Figure 9.37 Surface covers on sand due to A. snow, B. seaweed and C. large woody debris, all of which control how aeolian sediment moves across or is trapped on the beach.

surfaces, partly because sand rebounding from large pebbles will be ejected higher into the boundary layer and therefore gain more momentum on their return to the sand surface. In addition, a sparse scattering of isolated pebbles increases scour due to flow acceleration around and over the pebbles with resultant turbulent wakes that are very effective in entraining sand-sized particles in the vicinity of the pebbles. But transport rate decreases quite rapidly if the degree of surface cover by pebbles increases beyond about 30 per cent, until transport is almost completely shut down at a surface coverage of about 50 per cent (Figure 9.38B).

The topographic slope of the bed is also known to affect the threshold of movement and

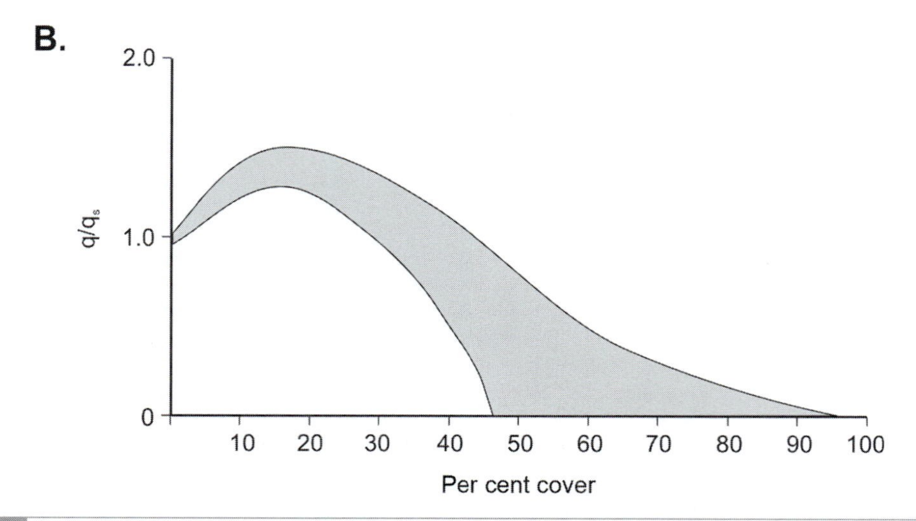

Figure 9.38 A. Pebble lag (pavement) surface developed in an inter-dune depression (Mason Bay, Stewart Island, New Zealand). B. Generalised schematic showing normalised sediment transport on the vertical axis and percent of the surface covered by pebbles on horizontal axis (after Logie, 1982; Nickling and McKenna Neuman, 1995; Davidson-Arnott *et al.*, 1997).

the rate of transport (Iversen and Rasmussen, 1994). Various algorithms have been proposed to adjust predicted sand transport rates on sloping surfaces (Sherman *et al.*, 1998) but relatively little is known about how the direction of transport changes. The effects are small for slopes < 10–$15°$, and therefore beach slope is typically ignored when predicting sand transport. This is not the case for dunes, however, where slope is an important factor, in part because the influence of gravity becomes more pronounced. In essence, it is more difficult for the wind to move a grain of sand uphill than it is across a flat surface. However, an equally important consideration is the influence of topography on the flow field, which serves to alter the potential of the wind to entrainment and transport sediment. These are the focus of the next few sections.

9.5 Flow Modification and Wind Steering by Topography

The wind blowing over an open ocean or lake surface is generally simpler to model than the complex boundary layer adjustments that occur over land because of the large variability in the type and configuration of surfaces that the wind encounters (e.g., beach, dunes, vegetated fields, forests, urban areas, etc.). When the wind blows directly onshore, the flow first traverses the foreshore zone, which has a gentle slope on dissipative beaches but may be very steep on reflective beaches. Flow will be accelerated up the foreshore toward the berm crest, often promoting erosion on the upper foreshore depending on the slope angle, moisture conditions and incident wind speed. A zone of flow separation may develop landward of the berm crest, but regardless there will be gradual transitions in the inner boundary layer dynamics from the water to the land (Figure 9.39). Not only is the mean wind speed (and possibly direction) affected by the subtle topographical changes encountered on the beach, but so are the turbulence intensity and available shear stress, which are critical parameters for controlling sediment transport potential.

When the wind encounters the foredune, the boundary layer will be modified in pronounced ways (Rasmussen, 1989; Arens *et al.*, 1995; Nickling and McKenna Neuman, 1999). On foredunes with wave-cut scarps, the flow actually stalls at the base of the dune (see discussion in Section 9.3.1). In contrast, on foredunes with gentle stoss slopes, the flow slows down slightly at the toe region and then speeds up due to flow compression as the wind moves up the stoss slope of the dune (Hesp *et al.*, 2005; Walker et al., 2009a). The increase in wind speed (referred to as wind speed-up) is often parameterised by taking the measured wind speed at any position on the dune (the mid-stoss slope, for example) and dividing it by the wind speed at the same height above the sand surface at a location on the beach far in front of the dune. The normalised wind speed is equal to 1 (or 100%) on the beach, and it becomes slightly smaller (or even zero or negative) at the dune toe, followed by increases to values above 1 toward the crest. In extreme cases, a low-level jet with values up to 2 (200%) can be found over dune crests (Hesp *et al.*, 2013). The effect is most pronounced with winds that are directly onshore, but is reduced as the angle of wind approach becomes more oblique because the apparent slope (the slope that the wind encounters) decreases. The topographic speed-up effect is counterbalanced by the drag effect of vegetation on the dune slope, and where this consists of shrubs, the effect of the vegetation may outweigh that of the topography, resulting in near-surface speed-down. Clearly, all these modifications to the flow field will have important implications for the capacity of the wind to entrain and transport sand on the foredune.

Although the fluid dynamics for wind flow over foredunes can be very complex, simplified expressions such as Bernoulli's law can be used to provide insight into the fundamental nature of flow adjustments and wind steering due to topographic modulation (Hesp *et al.*, 2015). In general, winds that are oriented directly onshore or directly offshore are easiest to predict, and the main issue is whether the flow field will remain attached to the surface or whether there will be flow separation at the dune crest. The importance of flow separation is critical to understanding the evolving shape of the dune

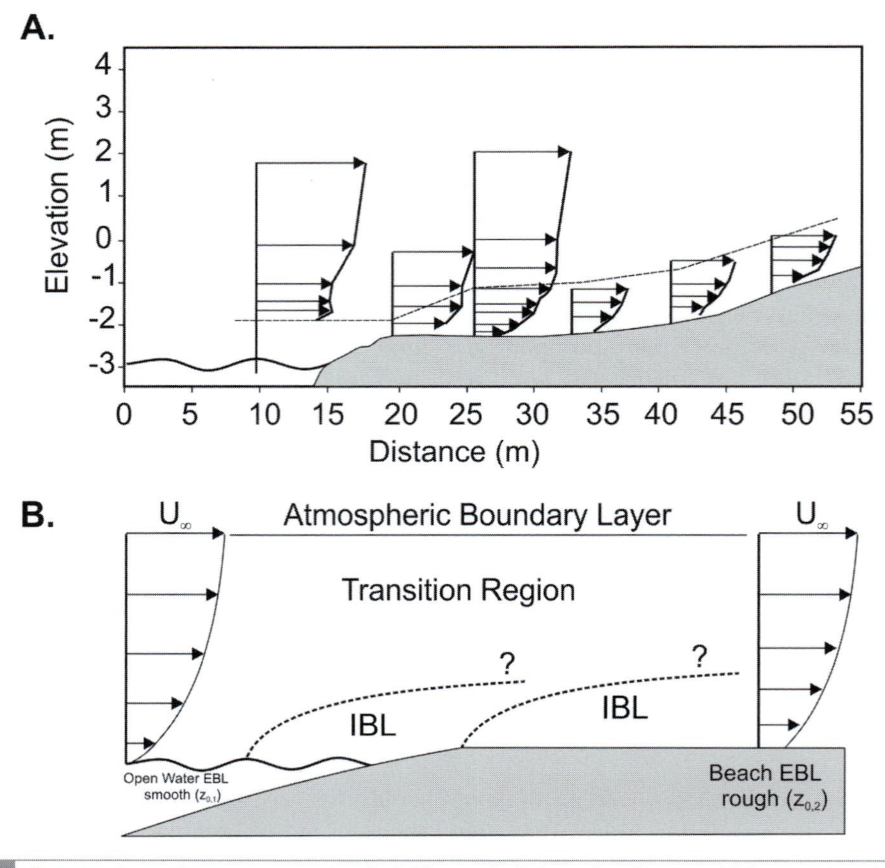

Figure 9.39 Boundary layer adjustments over a beach. A. Measured wind speed profiles at multiple cross-shore locations from the surf zone to the backshore showing vertical growth of the inner boundary layer (indicated by the dashed line) and zones of near-surface flow speed-up due to compression. B. Conceptual version of the empirical data indicating the likelihood that there are multiple internal boundary layers (IBL) that merge into one another as a consequence of roughness changes across the surf zone, up the foreshore, and across the backshore. 'EBL' indicates an equilibrium boundary layer. Adapted from Bauer *et al.*, (1996). Reproduced with permission from the Coastal Education and Research Foundation, Inc.

and the potential downwind distance of transport. Until recently, there was relatively little interest in offshore winds, but it has been demonstrated with field measurements and supported by computational fluid dynamical models that flow separation and subsequent eddy recirculation in the lee of the dune can lead to onshore sand transport on the beach (Lynch *et al.*, 2008; Walker *et al.*, 2009a; Jackson *et al.*, 2011; 2103; Bauer *et al.*, 2015). This may prove to be an important process on coasts with persistent offshore winds because onshore sand transport contributes to re-healing wave-cut scarps and re-establishing sand ramps at the base of the foredune.

When the wind approaches the shoreline at oblique angles, considerable flow steering can take place, which serves to re-orient the primary direction of the wind as well as the sediment transport pathways across the beach and foredune. The degree of steering depends on several factors, including: (1) free-stream wind speed; (2) incident wind angle relative to the foredune orientation; (3) slope angle of the foredune; (4) height of the dune; and (5) surface roughness and degree of vegetation cover. A conceptual model based on extensive field measurements is provided in Figure 9.40, which shows that onshore and offshore winds that are oriented

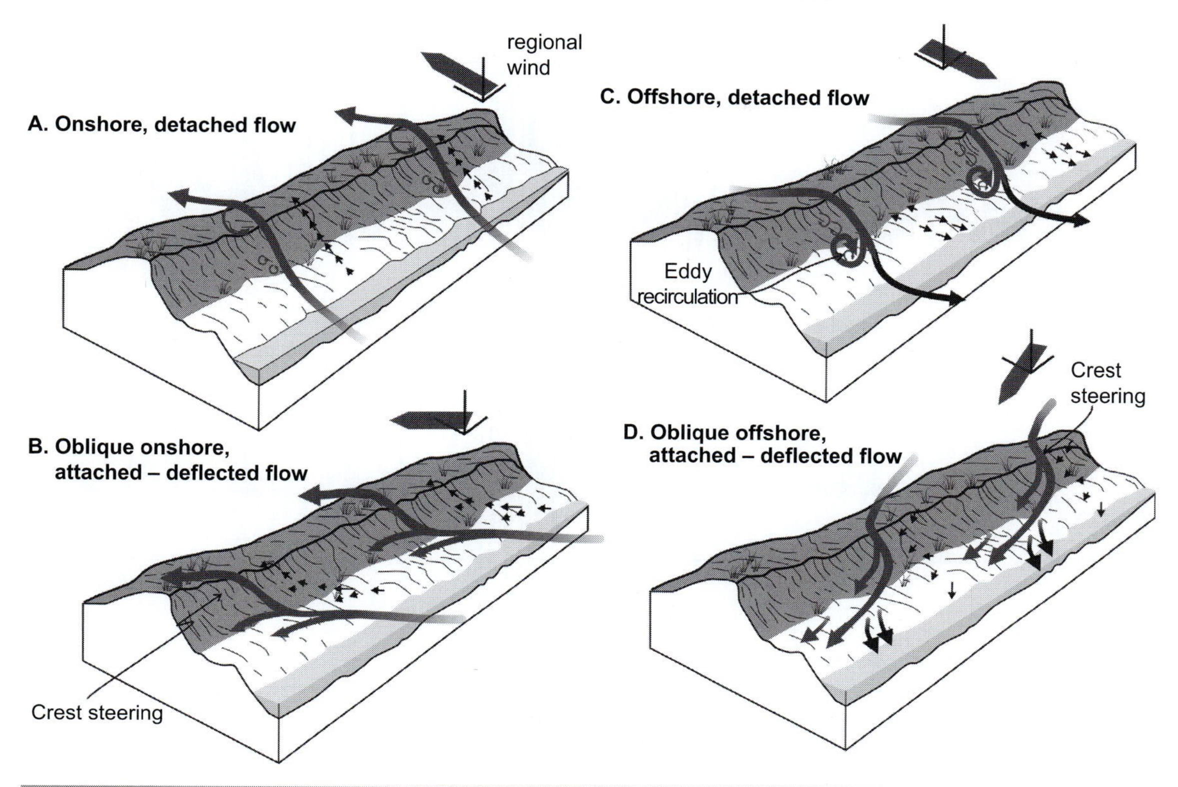

Figure 9.40 Conceptual model of wind steering across a beach–dune system for onshore, offshore and oblique wind angle approaches (after Bauer *et al.*, 2012). The regional (incident) wind at free-stream height is indicated by a large bold arrow and Cartesian coordinates in the upper right corner of each panel. The bold flow lines show the near-surface, topographically modified wind directions, whereas the short arrows indicate the likely sediment transport pathways.

perpendicular to the dune typically lead to flow separation in the lee of the dune crest. With oblique approach angles, the situation can be quite complicated. For example, with oblique onshore wind, the wind is often steered in a direction parallel to the shoreline or the beach, leading to alongshore sediment transport, whereas on the stoss slope of the dune, the wind is steered up the foredune slope (i.e., more perpendicular than the incident wind angle). This leads to a situation referred to as sediment pathway decoupling because the sand that is mobile on the beach moves alongshore and is not delivered up into the stoss slope. Thus, sediment delivery to the foredune crest can only come from local entrainment on the stoss slope rather than the backshore. This appears to be the situation on beach–dune systems that have a flat beach backed by a steep foredune or scarped foredune. If there is a well-developed sand ramp that provides a smooth transition from the beach to the stoss slope, a continuous pathway for sediment is available to deliver sand from the beach to the dune.

9.5.1 Wind Unsteadiness and Transport Intermittency

On natural beaches, wind speed and direction is usually highly variable, with important implications for predicting the rate of sand transport. Figure 9.41 shows actual field measurements from a dune crest of wind speed and direction as well as the time series of sand particle counts from a vertical array of laser counters (Bauer and Davidson-Arnott, 2014). The 5-minute mean speed (Figure 9.41A) was slightly below 10 m s^{-1}

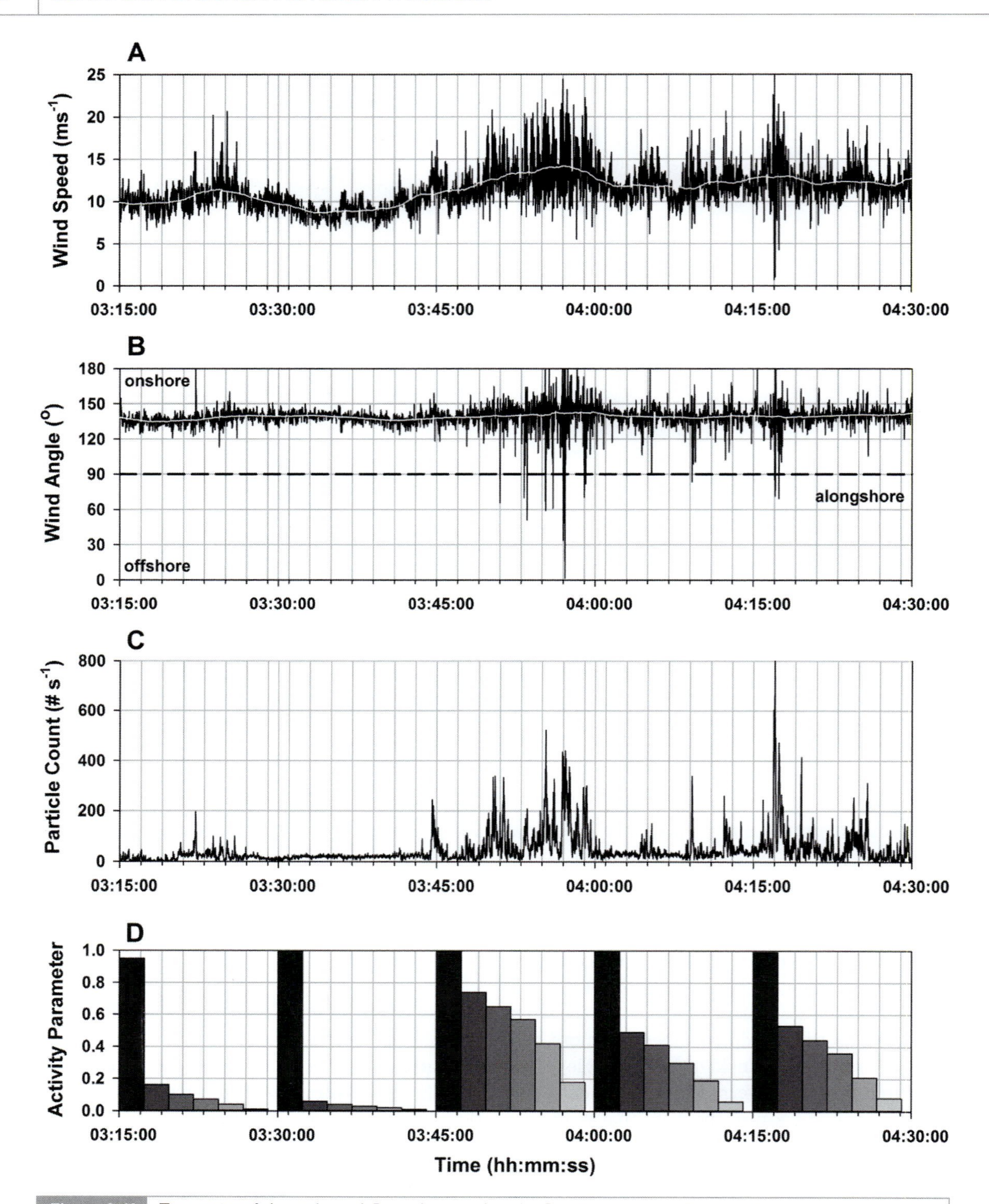

Figure 9.41 Time series of: A. wind speed; B. wind approach angle; C. sediment particle flux; and D. activity parameter (intermittency) recorded at 1 Hz for a 75-minute measurement period. Wind speed and direction are from 3D sonic anemometer at 0.2 m above the bed. Smoothed trend lines are 5-minute moving averages. Particle flux is the vertically integrated instantaneous (1 Hz) count summed over six laser particle counters (LPC) in a vertical array. Activity parameter (AP) refers to proportion of time that sand particles were recorded. Every 15-minute segment has six vertical bars that indicates AP for each of the six sensors in the vertical array (left-most bar is the lowermost LPC and right-most bar is the highest LPC in the array). (After Bauer and Davidson-Arnott, 2014.)

at the start of the time series with most of the fluctuations in the range of +/– 3 m s^{-1} about the mean. The wind direction (Figure 9.41B) was very constant at an obliquely onshore direction during this period. However, after about 03:45 am, the mean wind speed increased to about 13 m s^{-1} with wind gusts reaching almost 25 m s^{-1} when the frontal disturbance passed over the experimental site. The mean wind direction remained obliquely onshore, but individual gusts were associated with radical shifts in wind orientation, in some cases yielding offshore phases (e.g., immediately prior to 04:00 am). The impact on sand transport, as shown in Figure 9.41C) was pronounced, with an early period of relatively marginal transport (less than 50 counts per second) followed by periods of intense transport reaching above 200 counts per second. These intense periods of transport during pronounced gusts are referred to as sediment flurries (Bauer and Davidson-Arnott, 2014), and they are commonly found in sediment transport time series. Aeolian researchers have therefore shifted their focus away from using only mean quantities (and equilibrium relationships such as Equation 9.7) and toward the investigation of wind unsteadiness and its impact on transport potential. Unfortunately, there is as yet no practically applicable equation that can be used across a broad range of situations where wind unsteadiness is prevalent.

Wind unsteadiness associated with pronounced gusts produces four scenarios for the initiation of sediment motion and for sustaining aeolian saltation for extended periods of time. First, the mean wind speed can be very slow with fluctuations that never exceed the entrainment threshold. In this trivial case, there can be no sediment transport. Second, the mean wind speed can be moderately fast but still below the threshold of motion. The equations would predict no transport based on the mean wind speed. However, during some of the more extreme gusts, the wind speed can exceed the threshold and then isolated sediment flurries might develop. This is particularly true when the mean wind speed is only incrementally below the entrainment threshold, which implies that most gusts will exceed the threshold. In this instance, the transport time series will have many flurries with either extended or brief lulls depending on how close the mean is to the threshold speed. The flurries will be of all different sizes and shapes, some of quite long duration. Third, the mean wind speed could be slightly greater than the entrainment threshold, in which case the transport equations would predict continuous sand transport. However, given the gusty nature of the wind, it is likely that lulls will reduce the instantaneous wind speed to values below the threshold, thereby producing periods in the transport record when saltation ceases. This third situation is very similar to the second situation with the exception that the predicted equilibrium transport rate would differ. Moreover, the discrete nature of individual flurries under the second situation will transition into a series of amalgamated flurries with only a few isolated periods of transport cessation. Thus, the differences are more in the degree of transport rather than in kind. Finally, there can be situations when the mean wind speed is well above the threshold for sediment motion and even the lulls are too short and moderate to cause transport to cease entirely. There will be periods of reduced transport interspersed among transport peaks, as seen in the spiky nature of Figure 9.41C, but some marginal transport is sustained due to the inherent momentum within the saltation curtain.

The qualitative descriptions of these four transport scenarios are complemented by ongoing research on transport intermittency, which refers to how (dis)continuous the transport time series is (Sherman *et al.*, 2017). An intermittency parameter, as proposed by Stout and Zobeck (1997), can be used to quantify the degree to which sediment transport is continuous in time, and it is calculated by dividing the number of observations with active transport by the total number of observations in the record. For example, if the time series was 10 minutes long (i.e., 6000 seconds) with a sampling rate of 1 Hz, and if only 1800 seconds had transport readings, then the intermittency parameter would have a value of 0.3 (i.e., 1800/6000). In other words, transport was active for 30 per cent of the time during the measurement interval, which is why

Davidson-Arnott and Bauer (2009) proposed the term activity parameter (AP) in lieu of intermittency parameter. Figure 9.41D shows values of the activity parameter for every 15-minute segment of the total time series, with each vertical bar representing values for laser counters located progressively higher in the vertical array. What is immediately apparent is that it matters very much where the instrument is located. Instruments close to the surface generally measure much more (continuous) sediment transport than instruments farther away from the sand surface. The laser counter located within the lower 2 cm measured almost continuous transport during this event, whereas the highest sensor (at about 50 cm) had activity parameter values less than 0.1 and often only around 0.02. Of course, this is consistent with what is known about the geometric decrease in transport flux with height during aeolian saltation, but it is also indicative of the challenges associated with quantifying sediment transport in the context of unsteady wind.

To add to the complexity, various studies have measured aeolian saltation when the instantaneous wind is substantially below the threshold for entrainment (likely due to a 1–2 second response lag between the wind and sand transport) and also no transport when the wind speed is substantially above the threshold, perhaps due to supply limitations (Wiggs *et al.*, 2004b; Davidson-Arnott *et al.*, 2005; 2008; Davidson-Arnott and Bauer, 2009). In the field, the concept of a precisely defined threshold of motion based on mean grain size and mean wind speed is not nearly as straightforward as one might anticipate from wind tunnel studies or by invoking a time-fraction-equivalence scheme (Stout, 2004). There is a need for formulating a threshold condition based on probabilistic foundations that include information on wind unsteadiness, grain size distributions, surface moisture content, and a range of other supply-limiting factors.

9.5.2 Boundary Layer Turbulence and Coherent Flow Structures

The recognition that short-term variability (or unsteadiness) in wind speed was coupled to flurries in the sediment transport rate (or intermittency), led aeolian geomorphologists to investigate the nature of fluid turbulence in near-surface boundary layers. Turbulence refers to the range of semi-random fluctuations in a flow field that occur across a broad range of frequencies from very high (e.g., 50 Hz or greater) to very low (e.g., several minutes). These fluctuations are classically associated with eddies of different spatial scales that are embedded in the flow field and therefore carried along (i.e., advected) in the direction of flow at the mean wind speed. Because these eddies are oriented in all different directions (and likely have different shapes), their bulk interaction causes the instantaneous velocity field to fluctuate about the mean condition in almost random fashion. An anemometer deployed within the boundary layer will therefore produce a time-series trace that is not perfectly flat (aligned with the mean speed) but rather will have deviations from the mean (see Figure 9.41A) that are typically in the range of +/– 10–25 per cent of the mean values, with peaks that may exceed 50 per cent or more.

Turbulent flow is usually parameterised using a non-dimensional relationship known as the Reynolds number:

$$\mathrm{Re} = \frac{U\ \rho\ L}{\mu}, \tag{9.8}$$

where U is the mean flow velocity, ρ is the fluid density, L is a characteristic length scale (e.g., height above the bed), and μ is the dynamic viscosity of the fluid. The flow in a given direction can be separated into mean and fluctuating components, as follows:

$$u = U + u' \tag{9.9}$$

where u is the instantaneous velocity (as measured by a rapid response anemometer, for example), U is the mean (time-averaged) flow velocity, and u' is the fluctuating component of the velocity field due to turbulent eddies. Recognising that a velocity field has three cardinal coordinate directions (x, y, z), each of which will have associated mean velocity vectors (U, V, W) and fluctuating components (u', v', w') leads directly to a statistical interpretation of turbulence. For example, the standard deviation, SD_u, or root-mean-square, u_{rms}, of a wind-speed time series is simply,

$$SD_u = u_{rms} = \sqrt{\frac{1}{N}\sum(u-U)^2} = \sqrt{\overline{u'^2}}. \quad (9.10)$$

It follows that terms such as $\overline{u'^2}$ are proportional to the variance of the time series and that terms such as $\overline{u'v'}$ are proportional to the covariance between the wind vectors in the two cardinal directions indicated. It can be shown (e.g., Bauer, 2013) that this provides a means by which to estimate the local shear stress within the boundary layer, as follows:

$$\tau_z = -\rho\overline{u'w'}, \quad (9.11)$$

where τ_z is the local or Reynolds shear stress associated with the velocity measurements taken at height, z, above the sand surface. Recalling that Equation 9.2 defines the relationship between bottom shear stress and the shear velocity, u_*, which was derived from the vertical velocity profile, leads to the following identity:

$$u_* = -\overline{u'w'}. \quad (9.12)$$

In theory, therefore, it should be possible to estimate the shear stress on the bed, which is proportional to the sediment transport rate (Equations 9.6 and 9.7), using either a vertical array of cup anemometers or a single, fast-response anemometer that can measure the velocity vectors in the horizontal and vertical directions simultaneously (cf., Bauer *et al.*, 1996).

The advent of fast-response sonic anemometry (van Boxel *et al.*, 2004; Walker, 2005; Walker et al., 2009b) and a range of new technologies that allow rapid measurement of sand grains in saltation (Bauer *et al.*, 2018) have yielded new insights into sediment transport across beach–dune systems in relation to turbulence. The following generalisations can be made:

1. Over very wide, flat beaches, equilibrium transport conditions are likely to be attained, especially on the backbeach zone. The foreshore, in contrast, is problematic because of moisture and slope effects as well as the influence of a surface roughness transition from water to sand. Similarly, steep narrow beaches with complex topography are difficult to model, and the rate of sediment transport is typically not in local equilibrium with the wind field.
2. As the flow field approaches the foredune, the mean flow slows down, especially in front of large steep dunes or those with wave-cut scarps. This results in the rate of sediment transport decreasing and subsequent deposition. However, if there is a smooth, gently sloping sand ramp leading to the stoss slope, sediment transport can be sustained by enhanced levels of turbulence despite a decrease in mean wind speed (Wiggs *et al.*, 1996). Elevated levels of Reynolds shear and normal stresses and turbulence intensity are routinely found at the toe region of large dunes (e.g., Chapman *et al.*, 2012; 2013), especially those with substantial collections of logs and large woody debris in front of the dune (Grilliot *et al.*, 2018).
3. The stoss slopes of dunes are very complicated in regard to wind steering and to changes in the turbulent boundary layer, in part because there are many parameters to consider (e.g., vegetation cover, slope angle, dune height, wind approach angle, etc.) but also because there have not been many studies of turbulence over coastal dunes. However, it does appear that Reynolds stress is an unreliable predictor of the actual stress exerted on the sediment surface of the dune, and that there is only a weak relationship with sediment transport rate. This possibility was recognised in the late 1990s by studies (e.g., Sterk *et al.*, 1998) that showed that there was a better relationship between sediment transport flurries and the Reynolds normal stress ($\rho\ \overline{u'^2}$) than the Reynolds shear stress ($\rho\ \overline{u'w'}$). This is supported by many recent observations on flow and sediment transport over coastal foredunes (e.g., Walker *et al.*, 2009a; 2017; Chapman *et al.*, 2013).

One of the intriguing aspects of turbulence that continues to attract considerable attention is the possibility that semi-ordered vortices within the flow field (referred to as coherent flow

Figure 9.42 Aeolian sediment streamers traveling across a moist, hard beach at low tide. Transport direction is obliquely offshore (right to left in photo).

structures) may be closely related to sediment transport events. Baas and Sherman (2005) showed that the character of streamers or sand snakes (Figure 9.42) evolves with increasing wind speed, from isolated, singular features to multiple, interwoven streamers to full-blown saltation clouds that have complex patterns of transport intensity. Dupont *et al.* (2013) simulated streamer-like patterns using a numerical flow model coupled to a sediment transport model. Nevertheless, a critical review of the relevance of coherent flow structures to aeolian saltation (Bauer *et al.*, 2013), indicates that there is, as yet, little direct evidence that the structures embedded in the turbulent boundary layer scale with size and spacing of streamers. Nevertheless, the importance of turbulence within the flow field, more generally, cannot be understated.

9.6 Geometric Controls on Sand Delivery to Foredunes

As the wind field transitions from the nearshore zone onto the beach (see Figure 9.39), the boundary layer encounters a step change in the available supply of sediment, from zero over water to some finite value on the foreshore depending on grain size and moisture content. Clearly, the transport system is unable to achieve its full potential (i.e., maximum transport rate) instantaneously at the water–shore interface because there is no sediment contribution from upwind. The wind may pick up a single grain or two of loose sand on the lower foreshore, and these grains will move downwind a short distance before returning to the surface. At this downwind location (the mid foreshore, perhaps) conditions may be slightly drier, and the impact of the incoming grains may eject several other grains into the air. These grains, in turn, will accelerate downwind before colliding with the sand surface on the upper foreshore or berm, and eject even more loose grains. Once transport is initiated, the downwind increase in transport rate is usually rapid and geometric.

The saltation cascade has been documented extensively in wind tunnel experiments (e.g., Nickling, 1988), and it is well known that a series of complex adjustments takes place involving momentum transfers between the grains and the wind, often with periods of disequilibrium (e.g., Owen, 1964; Shao and Raupach, 1992). Eventually, however, at some downwind location, the saltation system comes into equilibrium with the new boundary layer, such that there is a predictable relationship between the fluid shear stress and the transport rate (as embodied in Equations 9.6 and 9.7 or some variant thereof). The distance over which the system has achieved its full potential to transport sediment is known as the critical fetch distance. At this point and beyond, the fluid threshold is no longer relevant, but the impact threshold dominates (see Figure 9.32), which implies that most of the interesting physics of equilibrium saltation are embedded in the cratering and ejection processes rather than simple notions of fluid drag and lift.

The fetch effect, more generally, is defined as the progressive increase in sand transport rate downwind from the boundary of a no transport zone such as the upwind end of the sand bed in a wind tunnel, the edge of an agricultural field, or the swash zone on a beach (Gillette *et al.*, 1996; Bauer and Davidson-Arnott, 2002; Delgado-Fernandez, 2010). The significance of this effect is that if the available fetch distance, which is governed by beach width, wind speed, and bed characteristics, is shorter than the critical fetch distance required for equilibrium transport to be achieved, then the actual transport rate at any point on the foreshore and beach will be less than that predicted by any of the equilibrium transport equations. Field studies of wind erosion on agricultural fields have shown that the fetch effect can extend for distances of tens to hundreds of metres, in part because the process involves the gradual abrasion of silt and clay clods by saltating sand (Gillette *et al.*, 1996).

The fetch effect was first noted on beaches by Svasek and Terwindt (1974), and a geometric model was proposed by Bauer and Davidson-Arnott (2002; Figure 9.43). Since then there have been a number of field studies that have reported critical fetch distances of tens of metres (Davidson-Arnott and Law, 1990; Nordstrom and Jackson 1992, 1993; van der Wal, 1998; Davidson-Arnott and Dawson, 2001; Davidson-Arnott *et al.*, 2008; Bauer *et al.*, 2009). Several of the studies on beaches have taken place with sand transport over a moist beach surface and thus the distance to reach a transport limited situation can be attributed to a reduction in the rate of grain ejection due to the apparent cohesion and probably also to increased intermittency close to the swash limit (Davidson-Arnott *et al.*, 2005, 2008; Bauer *et al.*, 2009). However, the fetch effect has also been observed for a relatively dry beach (Davidson-Arnott and Law, 1990; van der Wal, 1998; Davidson-Arnott *et al.*, 2008) and Dong *et al.* (2004) reported on fetch effects observed with dry sand in a long (16 m) wind tunnel. They measured sediment transport at 1 m intervals down the tunnel, and they observed that transport continued to increase up to the end of the working section. The increase over the first few metres was exponential but the increase with distance beyond this was much slower (Figure 9.43C), which may reflect gradual increases in the number of grains travelling at higher elevations and hence at greater speeds (Dong *et al.*, 2004; Wang *et al.*, 2008).

Figure 9.43A indicates that when the beach width is greater than the critical fetch, $w > F_c$, the fetch effect does not have any impact on sand transport across the dune line and the actual transport rate, q, per unit width is equal to the maximum transport rate, q_m. However, when $w << F_c$, q will be $<< q_m$ for shore normal winds and this brings into play a complex interaction between wind angle, fetch length and the cosine effect (Davidson-Arnott and Dawson, 2001; Bauer and Davidson-Arnott, 2002). The sand transport rate is usually expressed in terms of the mass of sediment transported per unit time per unit width along the direction of transport (e.g., $kgm^{-1}\ s^{-1}$). In Figure 9.43A it can be seen that sand transported between two parallel streamlines across a unit width, b, is delivered across a length of dune front, l, which lies oblique to the wind. The geometric relationship is $b = l\ (\cos\alpha)$. As the wind angle varies from shore perpendicular to an oblique angle, the length l increases such that a unit mass of sand across b is deposited over an increasing length of dune front. Therefore, the predicted sand delivery to the foredune must be adjusted to account for this cosine effect, ultimately yielding zero transport into the foredune when the wind is parallel to the beach. For very narrow beaches the wind angle at which maximum transport into the dune occurs may be as much as 60° to 70° from shore perpendicular, though the absolute values of transport into the dune will be considerably less than q_m.

Although the geometric model of Bauer and Davidson-Arnott (2002) provides a robust framework for predicting long-term sand supply to the foredune, successful implementation requires considerable data on beach conditions that are not easily obtained over long periods of time (Lynch *et al.*, 2006). For example, one might believe that strong onshore winds would yield ideal conditions for sediment delivery to the foredune because the cosine effect is minimal. However, these conditions are also accompanied by large waves and higher storm surge. As a consequence, the width of the beach is effectively reduced because of the wet conditions on the foreshore and lower beach (Bauer *et al.*, 2009), sometimes to the point where waves inundate the entire beach surface thereby shutting down the aeolian transport system. The outcome of this is to enhance the relative importance of

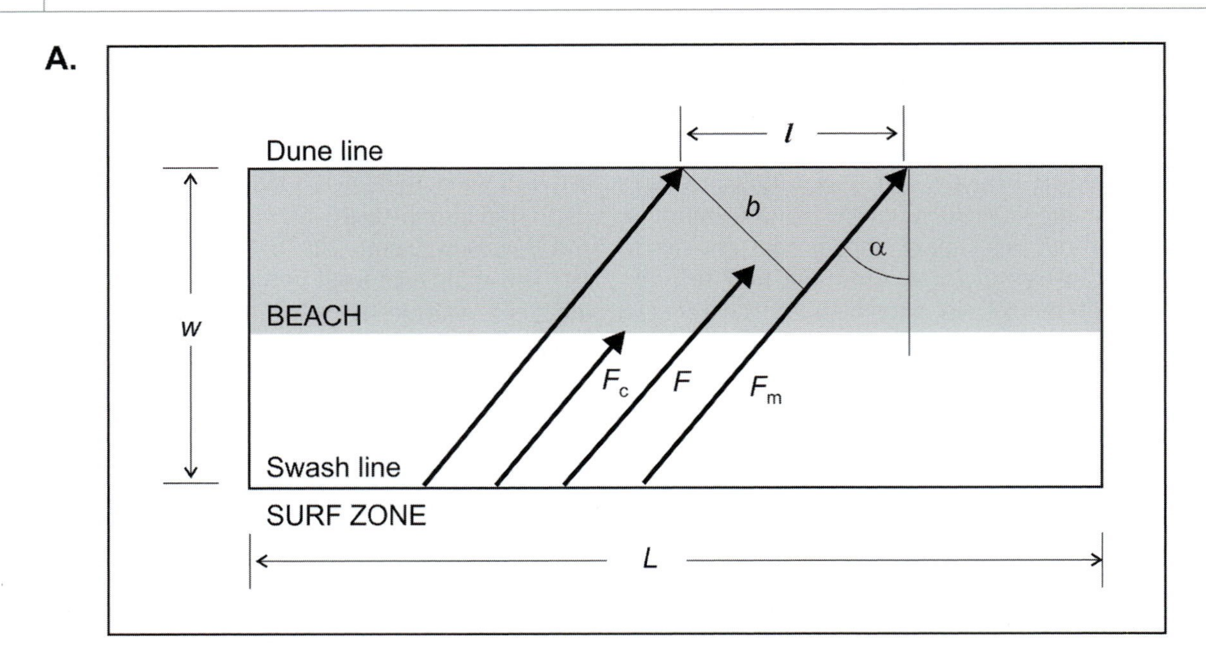

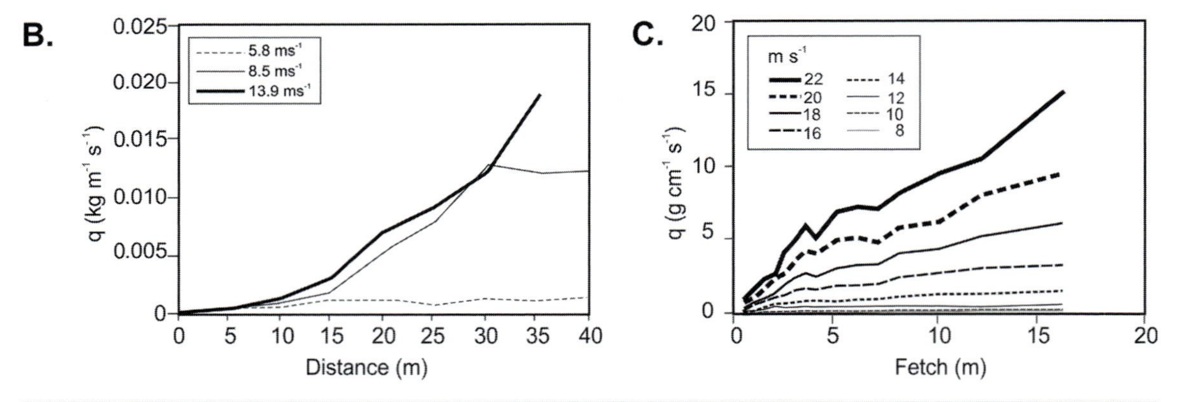

Figure 9.43 The fetch effect. A. Basic terminology defining the fetch effect on a rectangular beach of length *L* and width *w*. The beach is defined here as the zone between the limit of wave uprush on the beach and the dune line (limit of dune vegetation or break in slope), F_c is the critical fetch length required to achieve equilibrium transport under the existing conditions (wind speed, grain size, moisture content, etc.) and the shaded area is the region landward of F_c where sediment transport is at the maximum; F_m is the maximum fetch which is determined by the beach width and wind angle α relative to shore perpendicular. Distance, *l*. represents a unit alongshore length at the dune line mapped out by two parallel streamlines of the wind field separated by the perpendicular distance *b* such that $b = l(\cos\alpha)$ (after Bauer and Davidson-Arnott (2002). B. Increase in sand transport rate, measured with vertical sand traps, with distance from the top of the swash at Long Point, Lake Erie for three different wind speeds and relatively dry sand (after Davidson-Arnott and Law, 1990). The traps were set up along a line parallel to the wind direction and distances are measured with respect to the swash limit. Note that equilibrium (constant) transport is achieved for winds of 5.8 m s^{-1} (F_c = 15 m) and 8.5 m s^{-}1 (F_c = 30 m) but not for winds of 13.8 m s^{-1}; C. Increase in sand transport rate measured at 1 m intervals along a wind tunnel for a range of wind speeds, which also show constant transport in the downwind direction for slow speeds but not fast speeds (original data kindly supplied by Dr Z. Dong).

oblique winds in delivering sand to foredunes, especially where beach width is narrow ($w <$ 40–60 m) under non-storm conditions (Cloutier and Héquette, 1998). Similarly, there can be seasonal controls on sediment delivery to foredunes that confound our ability to predict foredune growth (e.g., Delgado-Fernandez and Davidson-Arnott, 2011).

9.7 Prediction of Long-Term Sediment Delivery to Foredunes

If an unlimited supply of dry sand is available on the beach at all times of the year, and assuming no other transport restrictions are active, it should be possible to use the average annual wind climate and any of several sediment transport equations (e.g., Equations 9.6 or 9.7) to predict how much sand could be supplied to the foredune from the beach every year. Such a calculation leads to a value of *potential* sediment delivery to the foredune, and it is based on a method outlined by Fryberger and Dean (1979) to predict the sand drift potential in desert regions. Standard meteorological wind data, measured at a height of 10 m, are categorised into speed and direction classes to drive a sediment transport relation such as a modified Lettau and Lettau (1977) equation:

$$Q \propto U^2(U - U_t)t \tag{9.13}$$

where Q is the amount of sand drift (expressed in vector units), U is mean wind speed at 10 m, U_t is the threshold wind speed, and t is the frequency of occurrence of the class expressed as a per cent of the total period. The approach has been applied to coastal dunes and dune fields (e.g., Wal and McManus, 1993; Olivier and Garland, 2003; Hesp *et al.*, 2007b; Lynch *et al.*, 2006). The calculations can be programmed easily on a personal computer (Saqqa and Saqqa, 2007), although it is critical to use wind direction and speed classes rather than continuous time series (see Pearce and Walker, 2005). This approach is useful for assessing the relative magnitude of potential sediment transport to foredunes at different beaches or to provide a gross estimate of the likely maximum sediment delivery on a mean annual basis. But caution needs to be exercised when comparing these values to actual long-term measurements of foredune growth because of a range of seasonally mediated controls on aeolian sediment delivery from the beach to the dunes. In the end, the measured (actual) transport across a beach is almost always less than the predicted transport based on the method proposed by Fryberger and Dean (1979).

In the mid and high latitudes there are marked seasonal patterns in wind regime, plant growth and sand transport that are important considerations that need to be taken into account when predicting foredune growth (see Box 9.2), particularly where winters are cold enough for the ground to freeze and for snow to blanket the beach and foredune. This affects both the timing of sand supply to the foredune and the locations where transport and deposition are most active. Delgado-Fernandez and Davidson-Arnott (2011) monitored a beach on the north shore of Prince Edward Island for a period of nine months (September 2007 to May 2008) using automated, tower-based photogrammetry with ground control points, erosion pins and continuously measuring instrumentation that characterised the wind conditions and the sediment transport intensity. A total of 184 wind events were detected during the measurement period, each nominally capable of transporting sediment because the mean wind speed was above the threshold of motion for the beach sediments. Approximately 60 per cent (109) of these were classified as having the potential to induce sediment transport at very small magnitude whereas only 15 were classified as having the potential to induce large or very large magnitude transport events, based on wind speed alone. The winter season contained most of the large magnitude events, followed by the fall, with the smallest transport potential occurring in summer. Thus, from the standpoint of the capacity of the wind to carry sediment toward the foredune, the winter season would seem to have been the most important on this beach. However, the sediment transport sensors showed that only 66 (35%) wind events produced measurable sand transport, of which 40 were characterised by only trace amounts of transport. Over the nine-month period, only 26 of 184 wind events were able to transport significant quantities of sediment, most of which occurred in the fall. Winter events were ineffective in mobilising sediment because the beach and dune were covered in snow and ice.

9.8 Long-Term Foredune Evolution and Beach–Dune Interaction

The transfer and exchange of sediment between the nearshore, beach and foredunes occurs on very short-term scales, driven by the repetitive action of individual waves and wind gusts that are sustained over the longer duration of storms and wind events. Yet, the overall morphology of

Box 9.2 A Conceptual Model of Seasonal Foredune Growth

Long-term measurements at Long Point, Lake Erie, by Law and Davidson-Arnott (1990) yielded a conceptual model of how foredune development is influenced by weather conditions and vegetation growth patterns during different seasons (Figure 9.44). Recent measurements at Greenwich Dunes in the Gulf of St Lawrence have provided support for the seasonal effects described in Figure 9.44 (Ollerhead *et al.*, 2012; Walker *et al.*, 2017). The model is generally applicable to much of eastern Canada, the northeast United States and other mid-latitude shorelines that experience winter snow and ice.

Figure 9.44 Cartoon summarising the seasonal pattern of erosion and deposition of sand on the beach and foredune on a mid-latitude coast with winter ice (modified after Law and Davidson-Arnott, 1990).

Box 9.2 (cont.)

In the late spring and summer, sand transport rates are generally small because the wind events are benign. Plants grow vigorously during summer but sand supply from the beach is small and deposition is confined to the incipient dune zone. In the early autumn (fall), transport increases considerably as the wind events become stronger, and much of the sand delivered from the beach is trapped primarily in the thick vegetation at the base of the stoss slope of the foredune and on the incipient dune if one is present. In late autumn and into the early winter, sand transport is at a maximum and a combination of burial and plant die-back permits sand to move up the stoss slope and onto and over the foredune crest. In the winter, sand supply from the beach is reduced because of freezing of interstitial water and because of the snow and ice cover. However, towards the end of the winter, sublimation and melting expose the beach and permit some sand supply from the beach to the foredune. There is also considerable reworking of sediment near the dune crest because of the reduced vegetation cover, and reduced snow cover. At lower latitudes or warmer locations where the beach seldom freezes, e.g., western Europe and the west coast of the US and Canada, winter months tend to have transport activity that is similar to that of late fall.

As deposition within the plants occurs, it is necessary to consider the accommodation space provided by the plant cover – that is, the volume of sediment that can be trapped by the plants and the area sheltered by them. With grasses such as *Ammophila* and *Uniola* the height and surface area is reduced with deposition, thus reducing the overall impact on wind flow and the size of the sheltered area. In mid and high latitudes, the grasses reach their maximum height in the early autumn at a time when the delivery of sand from the beach is likely to be at its greatest. As the grasses die back there is some loss of foliage and at the same time there is increasing burial of the vegetated area closest to the beach in the incipient dune zone and at the base of the foredune (Figure 9.45) so that sediment can pass through the incipient dune zone and up the stoss slope of the foredune towards the dune crest. The maximum depth of sediment that can be trapped and retained over the winter period depends on the vegetation height and is generally on the order of 0.2–0.5 m. Where sediment is trapped by annuals such as *Cakile edentula* on the foreshore, a mound of sand may be preserved over the winter but new plants will have to germinate and establish quickly in the spring in order to preserve the deposit. In low latitudes, the plants are active all year round but the pioneering vegetation typically has smaller accommodation space than the taller grasses found in the mid latitudes.

Shrubs such as sea grape and trees obviously have a much greater accommodation space than grasses and stoloniferous plants, but at the same time they grow much more slowly and are less able to tolerate substantial amounts of burial. Thus the controls on the timing of sediment supply from the beach and the accommodation space within vegetation on the incipient dune and foredune will vary depending on the climatic zone.

Figure 9.45 Accretion of sand in marram grass in the embryo dune and foredune zones at Long Point, Lake Erie: A. photograph looking alongshore at the beginning of September when the vegetation is at its highest; B. photograph taken at the end of November showing formation of a small ridge in the incipient dune zone, burial of vegetation and accumulation towards the base of the foredune.

any beach–dune system (e.g., the beach width and steepness, the size and stability of the foredune) closely reflects the nature of a range of broader controls, such as the wind and wave climatologies, the littoral sediment budget, the recent and past trends in relative sea-level fluctuations and the impact of anthropogenic disturbances. The term beach–dune interaction (Psuty, 1988) is particularly useful in this context because it emphasises the role played by the littoral cell (i.e., the waves, currents, tides, and sediment supply) in influencing the development of the foredune system. The many interactions inherent to beach–dune interaction are summarised schematically in Figure 9.46. The diagram is useful in distinguishing between the internal workings of the beach–dune system and the external controls that influence the system. In addition, it highlights the large number of alternative pathways that lead to positive and negative feedback relationships, with the implication that beach–dune interaction is complex and richly nuanced. The reader is referred to a special issue of the *Journal of Coastal Research* (1988, SI 3) dedicated to the topic of beach–dune interaction as well as a series of summary papers by Sherman and Bauer (1993), Bauer and Sherman (1999), and by Houser and Ellis (2013), as well as an in-depth discussion of how the various controls and responses are scale dependent by Walker *et al.* (2017).

9.8.1 Conceptual Models

It is important to appreciate that the foredune that is visible on any given beach today has both a history and a future. The long-term trajectory of the foredune is only partly dependent on the ability of the wind to move sediment across the beach and into the foredune (de Vries *et al.*, 2012). It also depends on whether there is an ample supply of sediment available to the aeolian system when it is active and whether stored sediment in the foredune is periodically removed. In both cases, the evolution of the foredune, past and future, is closely associated with what happens in the nearshore zone and likely farther offshore as well. This association between the size and geometry of the foredune relative to

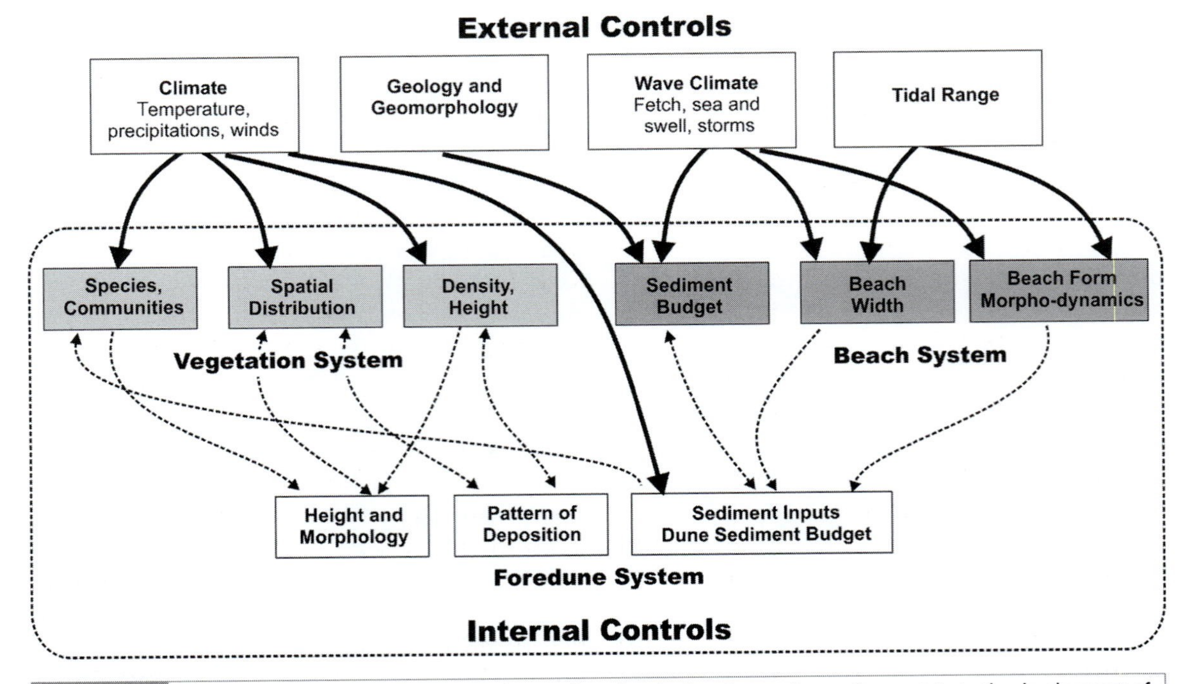

Figure 9.46 Schematic of the external and internal controls on beach–dune interaction as they pertain to the development of coastal foredunes (Reed *et al.*, 2009)

the energetics of the nearshore zone has been recognised for a long time but was formalised in several conceptual models (Short and Hesp, 1982; Psuty, 1988) that are being continually updated (e.g., Hesp, 2002) to account for greater complexity than was originally envisioned. Brodie *et al.*, (2017) and Palmsten *et al.* (2017) provide a short summary of the technical challenges associated with modelling foredune evolution in a way that properly accounts for beach–dune interaction.

9.8.1.1 *Beach stage Model of Short and Hesp (1992)*

As discussed in Chapter 8, the beach stage model for nearshore morphodynamics differentiates between dissipative, intermediate, and reflective beach types. At about the same time that this idea was being widely accepted for the nearshore environment (e.g., Short, 1979; Wright and Short, 1983; 1984), Short and Hesp (1982) suggested that the character and evolution of foredunes were broadly related to beach stage, based on their observations along beaches of southeast Australia. Dissipative beaches are characteristically wide and flat with small grain size, and therefore, the potential for aeolian sediment transport is large given the extensive fetch distances. Moreover, shallow sloping beaches are susceptible to storm surge and therefore the foredunes are periodically destabilised, which may initiate phases of dune transgression. At the other end of the continuum, reflective beaches are typically narrow and steep with large grain sizes. The potential for sediment transport is reduced relative to dissipative beaches, and few episodes of foredune destabilisation occur. This leads to stable foredune development, reduced vegetation disturbance and limited transport of sand beyond the foredune. Intermediate beach types have greater probabilities of foredune destruction and often the foredunes have blowouts and are backed by large-scale parabolic dunes. The general aspects of the model regarding the aeolian sediment transport potential on reflective versus dissipative beaches were verified by Sherman and Bauer (1993) and incorporated in a computer simulation model by Sherman and Lyons (1994).

9.8.1.2 *Sediment budget Model of Psuty (1988)*

An alternative conceptual model of beach–dune interaction was proposed by Psuty (1988) and elaborated in several publications over a number of years (e.g., Psuty, 2004). This model focuses directly on the nature of sediment supply rather than on beach form. In retrospect, it is evident that the model of Short and Hesp (1982) had several underlying assumptions that preclude universal applicability, including relatively stable sea level, a microtidal coast and a positive littoral sediment budget, which are all inherent to the southeast coast of Australia. The Psuty model, in contrast, related the evolution of the foredune system to the nature of the sediment budget of the beach (nearshore) in relation to the sediment budget of the dune (Figure 9.47). The beach budget can be positive, in which case the shoreline will likely prograde, or it can be negative, in which case there will be shoreline transgression. The latter may occur because of reductions in sediment supply to the littoral cell (e.g., construction of a major dam on the primary river feeding the littoral cell) or due to relative sea level rise (RSLR). A neutral beach budget leads to a stable shoreline, which is the state that is most conducive to foredune stability and growth, although it is often argued that a slightly negative beach budget favours foredune growth because it leads to periodic destabilisation of the foredune stoss slope. The foredune budget dictates how much sand is available for dune building, either in the foredune or in the back-dune region. It is difficult to conceive of a situation with a large positive littoral sediment budget and a negative dune budget so the dashed curve on the right side of the diagram should more realistically flatten off around the neutral line.

Based on the scheme of Psuty (1988), supplemented by Nickling and Davidson-Arnott (1990), the response of the beach–dune system to different sediment budget combinations is shown schematically in Figure 9.47B. A negative littoral sediment budget leads to transgression of the shoreline. If the beach and dune sediment budget

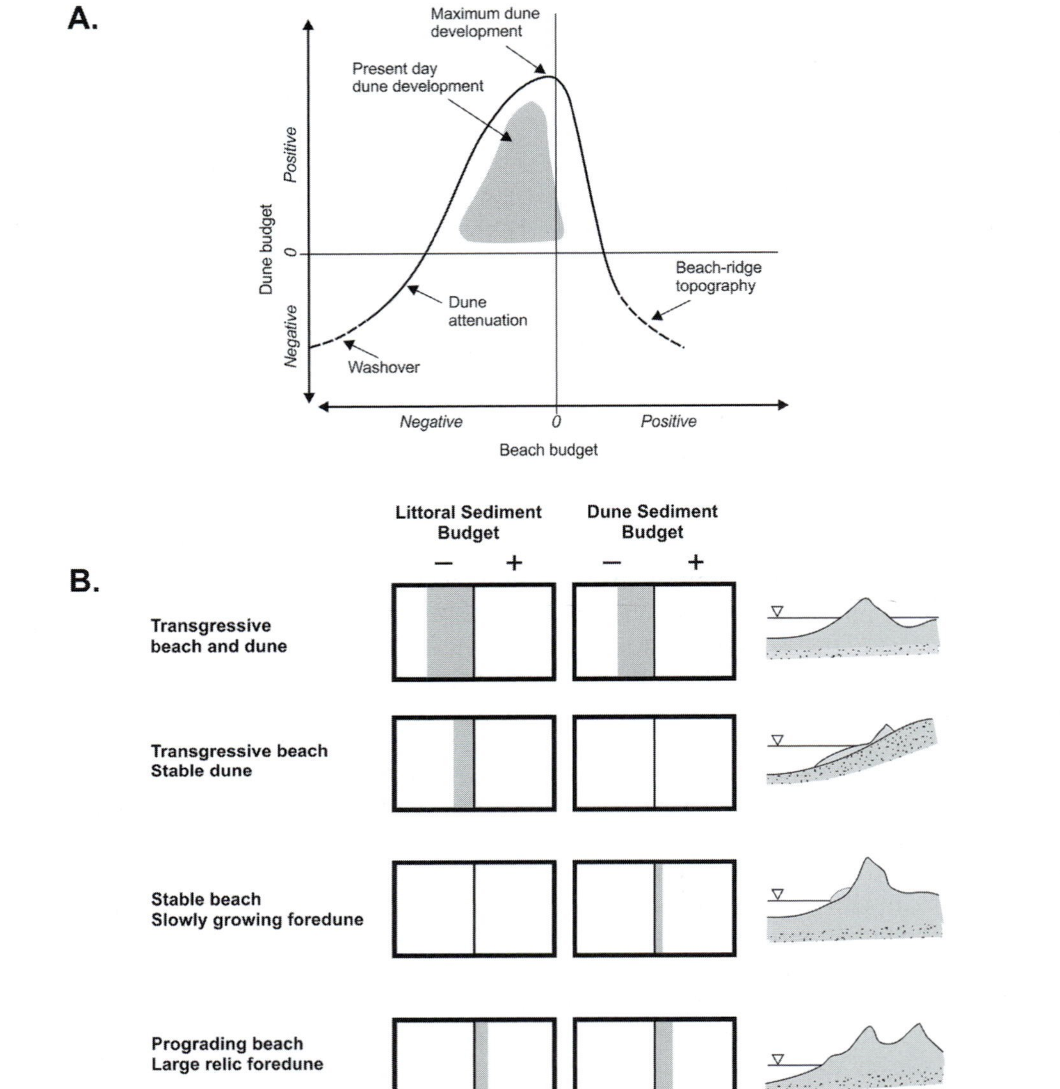

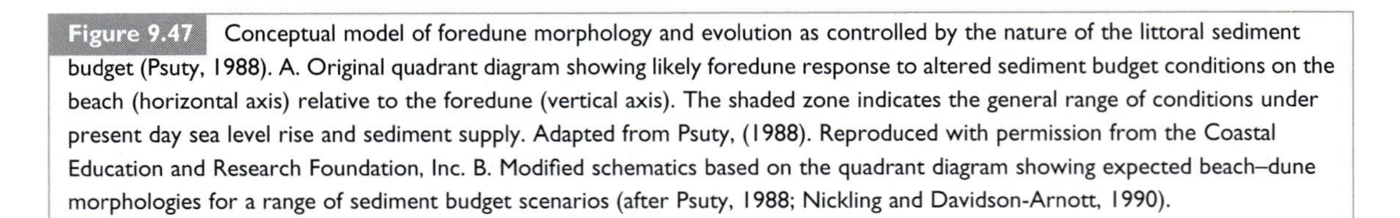

Figure 9.47 Conceptual model of foredune morphology and evolution as controlled by the nature of the littoral sediment budget (Psuty, 1988). A. Original quadrant diagram showing likely foredune response to altered sediment budget conditions on the beach (horizontal axis) relative to the foredune (vertical axis). The shaded zone indicates the general range of conditions under present day sea level rise and sediment supply. Adapted from Psuty, (1988). Reproduced with permission from the Coastal Education and Research Foundation, Inc. B. Modified schematics based on the quadrant diagram showing expected beach–dune morphologies for a range of sediment budget scenarios (after Psuty, 1988; Nickling and Davidson-Arnott, 1990).

are both greatly negative, the foredune system may eventually be destroyed by wave scarping, breaching of blowouts and overwash. If the deficit of the littoral sediment budget is moderate, then shoreline transgression still occurs but the dune sediment budget may be maintained. Under these conditions the foredune will migrate inland at the same rate as shoreline transgression. A single, tall foredune ridge with steep stoss and lee slopes is the expected form. Where the littoral budget is near neutral, the shoreline position remains fixed and the foredune is maintained or grows slowly. Finally, where both the littoral and foredune budgets have a surplus, progradation takes place, with the rate of progradation depending on the relative size of the littoral and dune sediment budget surpluses. As the rate of progradation increases, the time available for building each foredune ridge decreases, resulting in the construction of more, but smaller dune ridges.

9.8.1.3 *Foredune evolution model of Hesp (2002)*

The beach stage model and the sediment budget model are the basis for a more recent foredune evolution model proposed by Hesp (2002) that includes the effects of large storm events in scarping the foredune as well as the erosional or depositional trend of the coastline within which the dunes are situated (Figure 9.48). The range of scenarios reflects the large number of factors that can influence foredune development, and the model makes explicit that the morphology of the foredune, as viewed today, depends on the sequence of events leading up to today as well as the short-term influences and large-scale contexts that control future evolution. For example, a stable foredune in Stage 1 may experience progressive erosion through Stage 5 and therefore change its form and morphology from a continuous foredune to a sequence of blowouts and hummocks. Stages 1, 2 and 3 are typically found on stable or prograding coasts, whereas future growth sequences for these stages are shown in Figure 9.48A, depending on the sediment supply available from the littoral zone. In contrast, Stages 3, 4 and 5 are commonly found on eroding coasts where the littoral sediment budget is negative, and typical sequences of landward transgression are shown in Figure 9.48B. Figure 9.48C shows the situation when Stage 5 transitions to a rapidly transgressing sand sheet or parabolic dune field. Figure 9.48D indicates that there can be periodic storms that yield high-water events with large waves that are sufficient to scarp the dune toe or lead to overwash, as is common on barrier islands (see Chapter 10). Under these circumstances, the long-term evolution of the foredune through the various stages may be interrupted. There can be acceleration of the erosive trend (e.g., Stage 5 to overwash to sand sheet) or, on accreting coasts with positive beach and dune budgets, there can be a re-healing process as diagrammed in Stages 3b, 4b and 5b. The generalised model of Hesp (2002) for long-term foredune evolution therefore highlights the fact that the morphology of a dune is a complex manifestation of recent events (e.g., wave scarping followed by ramp building and incipient foredune growth) as mediated by the long-term tendencies of the overall coast (erosion versus deposition), especially the littoral sediment budget.

9.8.2 Beach–Dune Interaction in Practice

The conceptual models of beach–dune interaction described above are somewhat abstract in that they provide only the broad framework in which to think about long-term foredune evolution as a consequence of beach–dune interaction. Several case studies demonstrate the practical nature of these conceptual models in understanding the evolution of foredunes (e.g., Aagaard *et al.*, 2004; Cohn *et al.*, 2018). On very long beaches (tens of kilometres or longer), for example, it is not unusual for the character of the foredunes to change from one end of the beach to the other. Miot da Silva *et al.* (2010) describe a situation on Moçambique Beach in southern Brazil where the dominantly southerly waves are responsible for an alongshore gradient in wave energy, grain size and surf-zone type. The combination of a wide beach, maximum sediment supply, and strong wave-energy dissipation in the northern regions of the beach is reflected in the morphology of the dunes, in contrast to

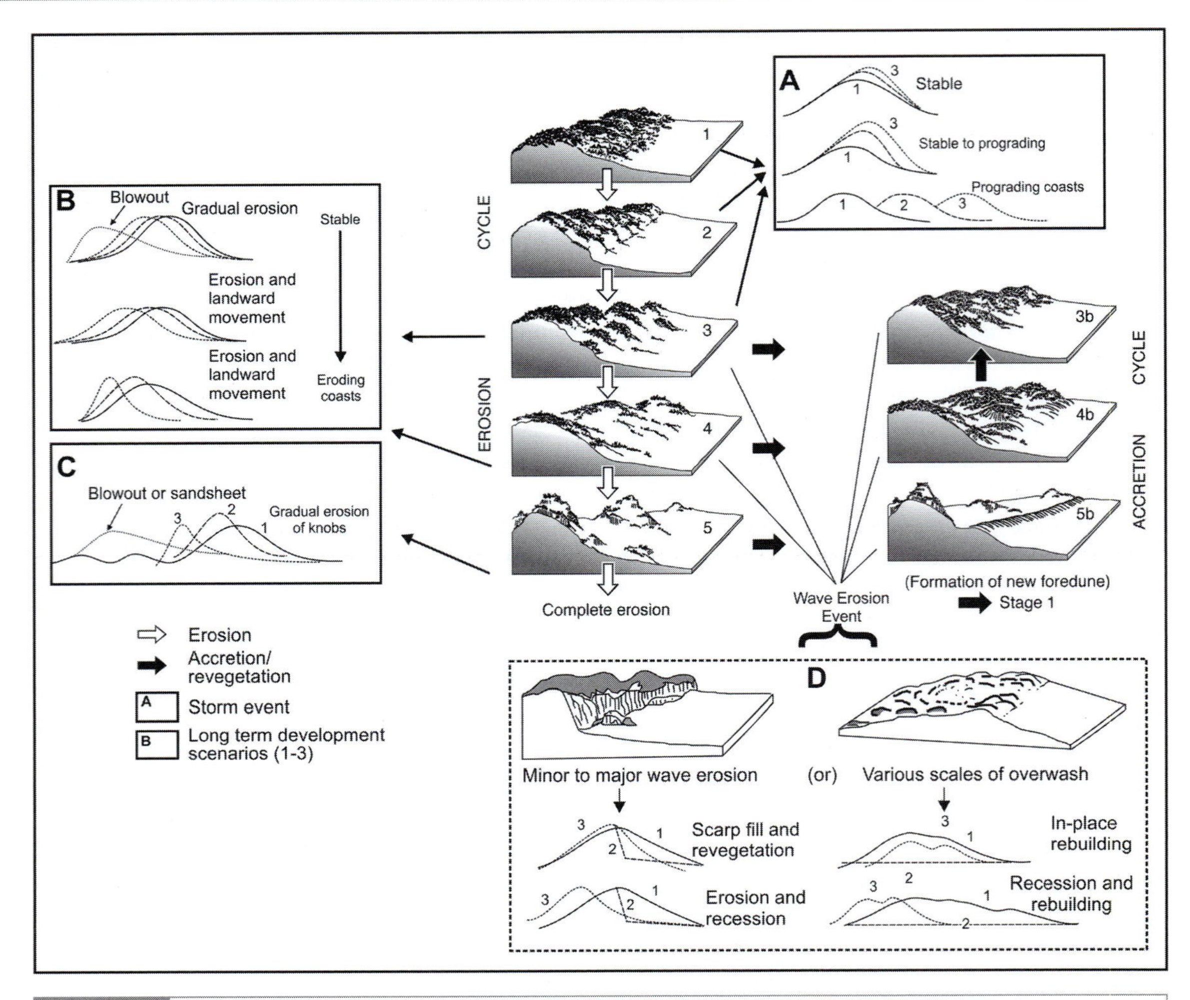

Figure 9.48 Generalised model proposed by Hesp (2002) of foredune morphology and evolution as dependent on whether the coast is stable, prograding (Stages 1–3 and sequences in A., or eroding (Stages 1–5 and sequences in B. and C.). The expected impact of wave erosion and overwash on dune form are shown in D., with several alternative pathways of accretion leading to profile Stages 1, 2 or 3.

conditions in the southern regions where the beach and dune sediment budgets are negative or neutral. Psuty (2004) suggested that a gradient of decreasing sediment supply away from the source (e.g., a river mouth) will be reflected in a similar gradient in the dune sediment budget and therefore also in the foredune form. There are obvious implications for understanding the impact of engineering projects that modify sediment supply to the littoral drift system.

Some nearshore systems are strongly three-dimensional with features such as rhythmic cusps, crescentic bars, rip channels and alongshore migrating sandwaves (see Chapter 8). These features alter the width of the beach, and hence the fetch distance, in complex ways alongshore, and therefore they are likely to have an influence on aeolian sediment delivery to the foredunes backing various beach segments. In the case of transitory features such as swash cusps, the impact is minimal, but if the features are fixed in place, such as large rip channels in embayed beaches that are constrained by the geometry of headlands or groins, the impact may be pronounced. Stewart and Davidson-Arnott (1988) describe a situation where large sandwaves were documented along the north shore of Lake Erie (Ontario) (see Box 7.1). Because these large-scale features migrate very slowly, the beach behind the widest portions of the sandwaves has a greater fetch distance for periods of months to years. Thus, there is more aeolian transport to the foredunes, which may assist in the regeneration of dune vegetation and the growth of incipient dunes. In a slightly different context, Houser *et al.* (2018b) demonstrated that alongshore variations in beach width and dune height on barrier islands can be controlled by the geological framework of the nearshore and offshore zones. Specifically, they showed that in the case of Santa Rosa Island (northwest Florida), there was a series of transverse ridges offshore that controlled the sediment supply to different sections of the barrier island thereby controlling the width of the island and the size of the dunes, ultimately influencing where overwash occurred during Hurricane Ivan. Similar findings were reported by Wernette *et al.*, (2018) for Padre Island National Seashore (Texas) where the dune morphology was linked to a sequence of paleochannels in the offshore zone.

Beach–dune interaction in its most basic form involves the imbalance/balance between sand transport by aeolian processes from the beach to the dune and removal of sand from the dune by wave action during storms, returning it to the foreshore or nearshore zone. The long-term influence of aeolian transport into the dunes is to extract sand from the littoral sediment budget, which may lead to a lowering and narrowing of the beach if there is insufficient replacement from offshore or updrift. In this case of inadequate sediment supply, the beach and foredune become progressively more susceptible to wave action, as suggested in Figure 9.47. However, scarping of the foredune by waves will return sediment to the littoral system, which is then available to rebuild the beach by the action of swell waves. Thus, there can be a delicate balance between the foredune sediment budget and the littoral sediment budget that occurs in a sequence of cyclical exchanges, as suggested by Figure 9.48. At any point in time, the dune morphology will reflect the most recent event that influenced the exchange (e.g., wave scarp versus dune ramp). Over the longer term, the progressive erosion or accretion of the foredune system will depend on the erosional or depositional state of the nearshore system, and hence the historical trend in shoreline transgression or progradation. Foredunes, therefore, are almost always complex, compound features incorporating varying degrees of cut and fill across a range of spatial scales.

References

Aagaard, T., Davidson-Arnott, R., Greenwood B. and Nielsen, J. 2004. Sediment supply from shoreface to dunes: linking sediment transport measurements and long term morphological evolution. *Geomorphology*, **60**, 205–24.

Aagaard, T., Orford, J. and Murray, A.S. 2007. Environmental controls on coastal dune formation; Skallingen Spit, Denmark. *Geomorphology*, **83**, 29–47.

Anderson, J.L. and Walker, I.J. 2006. Airflow and sand transport variations within a backshore-parabolic dune plain complex: NE Graham Island, British Columbia, Canada. *Geomorphology*, **77**, 17–34.

Arbogast, A.F. and Loope, W.L. 1999. Maximum-limiting ages of Lake-Michigan coastal dunes: their correlation with Holocene lake level history. *Journal of Great Lakes Research*, **29**, 72–382.

Arens, S.M., van Kaam, P. and van Boxel, J.H. 1995. Airflow over foredunes and implications for sand transport. *Earth Surface Processes and Landforms*, **20**, 315–25.

Arens, S.M., Baas, A.C.W., van Boxel, J.H. and Kalkman, C. 2001. Influence of reed stem density on foredune development. *Earth Surface Processes and Landforms*, **26**, 1161–76.

Atherton, R.J., Baird, A.J. and Wiggs, G.F.S. 2001. Intertidal dynamics of surface moisture content on a meso-tidal beach. *Journal of Coastal Research*, **17**, 482–9.

Baas, A.C.W., 2019. Grains in motion. In Livingston, I. and Warren, A., *Aeolian Geomorphology: A New Introduction*, John Wiley & Sons Ltd, 27–60.

Baas, A.C.W. and Sherman, D.J. 2005. Formation and behaviour of aeolian streamers. *Journal of Geophysical Research*, **110**, F03011. DOI:10.1029/2004JF000270.

Bagnold, R.A. 1941. *The Physics of Blown Sand and Desert Dunes*. London: Marrow, 265 pp.

Bauer, B.O. 2013. Fundamentals of aeolian sediment transport boundary-layer processes. In Shroder, J.F. (ed.), Lancaster, N., Sherman, D.J., Baas, A.C.W. (vol. eds.), *Treatise on Geomorphology*. San Diego, CA: Academic Press. vol 11, *Aeolian Geomorphology*, pp. 7–22.

Bauer, B.O. and Davidson-Arnott, R.G.D. 2002. A general framework for modelling sediment supply to coastal dunes including wind angle, beach geometry and fetch effects. *Geomorphology*, **49**, 89–108.

Bauer, B.O. and Davidson-Arnott, R.G.D. 2014. Aeolian particle flux profiles and transport unsteadiness. *Journal of Geophysical Research*, **119(7)**, 1542–63.

Bauer, B.O. and Sherman, D.J. 1999. Coastal dune dynamics: Problems and prospects. In Goudie, A.S., Livingstone I. and Stokes S. (eds.), *Aeolian Environments, Sediments and Landforms*. Chichester, UK: John Wiley and Sons, pp. 71–104.

Bauer, B.O., Sherman, D.I. and Wolcott, I.F. 1992. Sources of uncertainty in shear stress and roughness length estimates derived from velocity profiles. *Professional Geographer*, **44**, 453–64.

Bauer, B.O., Davidson-Arnott, R.G.D., Sherman, D.J. *et al.* 1996. Indeterminacy in eolian sediment transport across beaches. *Journal of Coastal Research*, **12**, 641–53.

Bauer, B.O., Houser, C.A. and Nickling, W.G. 2004. Analysis of velocity profile measurements from wind tunnel experiments with saltation. *Geomorphology*, **59**, 81–98.

Bauer, B.O., Davidson-Arnott, R.G.D., Hesp P.A. *et al.* 2009. Aeolian sediment transport conditions on a beach: Surface moisture, wind fetch, and mean transport rate. *Geomorphology*, **105**, 106–16.

Bauer, B.O., Davidson-Arnott, R.G.D., Walker, I.J., Hesp, P.A. and Ollerhead, J. 2012. Wind Direction and Complex Sediment Transport Response Across a Beach-Dune System. *Earth Surface Processes & Landforms*, **37**(15), 1661–77.

Bauer, B.O., Walker, I.J., Baas, A.C.W. *et al.* 2013. Critical reflections on the coherent flow structures paradigm in aeolian geomorphology. In Venditti, J.G., Best, J.L., Church, M. and Hardy, R.J. (eds.), *Coherent Flow Structures at Earth's Surface*. Chichester, UK: John Wiley and Sons, Ltd., pp. 111–34.

Bauer, B.O., Hesp, P.A., Walker, I.J. and Davidson-Arnott, R.G.D. 2015. Sediment transport (dis)continuity across a beach-dune profile during an offshore wind event. *Geomorphology*, **245**, 135–48.

Bauer, B.O., Davidson-Arnott, R.G.D., Hilton, M.J. and Fraser, D. 2018. On the frequency response of a Wenglor particle-counting system for aeolian transport measurements. *Aeolian Research*, **32**, 133–40. DOI: https://doi.org/10.1016/j.aeolia.2018.02.008.

Belly, P.-Y. 1964. *Sand Movement by Wind*. US Army Corps of Engineers CERC, Technical Memorandum, 1, 38 pp.

Brodie, K. L., Palmsten, M.L. and Spore, N. J. 2017. *Coastal foredune evolution, Part 1: Environmental factors and forcing processes affecting morphological evolution*. ERDC/CHL CHETN-II-56. Vicksburg, MS: US Army Engineer Research and Development Center. DOI: http://dx.doi.org/10.21079/11681/21468.

Carter, R.W.G., Bauer, B.O., Sherman, D.J. *et al.* 1992. Dune development in the aftermath of stream outlet closure: examples from Ireland and California. In Carter, R.W.G. *et al.* (eds.), *Coastal Dunes: Geomorphology, Ecology, and Management for Conservation: Proceedings of the Third European Dune Congress, Galway Ireland. June 17-21*. Rotterdam: A.A. Balkema, pp. 57–69.

Chapman, C., Walker, I.J., Hesp, P.A., Bauer, B.O. and Davidson-Arnott, R.G.D. 2012. Turbulent Reynolds stress and quadrant event activity in wind flow over a coastal foredune. *Geomorphology*, **151–2**, 1–12.

Chapman, C., Walker, I.J., Hesp, P.A. *et al.* 2013. Reynolds stress and sand transport over a foredune. *Earth Surface Processes and Landforms*, **38**(14), 1735–47.

Christiansen, M. and Davidson-Arnott, R.G.D. 2004. The effects of dune ramps on sediment supply to coastal foredunes, Skallingen Denmark. *Geografisk Tidsskrift* (Danish Journal of Geography), **104**, 29–41.

Clemmensen, L. B., Pye, K., Murray, A. and Heinemeier, J. 2001. Sedimentology, stratigraphy and landscape evolution of a Holocene coastal dune system, Lodbjerg, NW Jutland, Denmark. *Sedimentology*, **48**, 3–27.

Clemmensen, L.B., Bjørnsen, M., Murray, A. and Pedersen, K. 2007. Formation of aeolian dunes on Anholt, Denmark since AD 1560: a record of deforestation and increased storminess. *Sedimentary Geology*, **199**, 171–87.

Cloutier, M. and Héquette, A. 1998. Aeolian and overwash sediment transport across a low barrier spit, southeastern Canadian Beaufort Sea. *Zeitschrift fur Geomorphologie*, 42, 349–65.

Cohn, N., Ruggerio, P., de Vries, S. and Kaminsky, G.M. 2018. New insights on coastal foredune growth: the relative contributions of marine and aeolian processes. *Geophysical Research Letters*, **45**(10), 4965–73.

Cooper, W.S. 1958. *Coastal Sand Dunes of Oregon and Washington*. Memoir (Geological Society of America), vol. 72. New York: Geological Society of America, 169 pp.

Cornelis, W.M. and Gabriels, D. 2003. The effect of surface moisture on the entrainment of dune sand by wind: an evaluation of selected models. *Sedimentology*, **50**, 771–90.

Cornelis, W.M., Gabriels, D. and Hartmann R. 2004. A parameterisation for the threshold shear velocity to initiate deflation of dry and wet sediment. *Geomorphology*, **59**, 43–51.

Cowles, H.C. (1899). *The Ecological Relations of the Vegetation of the Sand Dunes of Lake Michigan*. Chicago: University of Chicago Press, 119 pp.

Darke, I. and McKenna Neuman, C. 2008. Field study of beach water content as a guide to wind erosion potential. *Journal of Coastal Research*, **24**, 1200–8.

Davidson-Arnott, R.G.D. and Bauer, B.O. 2009. Aeolian sediment transport on a beach: Thresholds, intermittency and high frequency variability. *Geomorphology*, **105**, 117–26.

Davidson-Arnott, R.G.D. and Dawson, J.D. 2001, Moisture and fetch effects on rates of aeolian sediment transport, Skallingen, Denmark. *Proceedings Canadian Coastal Conference*, Ottawa: Canadian Coastal Science and Engineering Association, pp. 309–21.

Davidson-Arnott, R.G.D. and Law, M.N. 1990. Seasonal patterns and controls on sediment supply to coastal foredunes, Long Point, Lake Erie. In Nordstrom, K.F., Psuty, N.P. and Carter R.W.G. (eds.), *Coastal Dunes: Form and Process*. Chichester, UK: John Wiley and Sons, pp. 177–200.

Davidson-Arnott, R.G.D. and Law, M.N. 1996. Measurement and prediction of long-term sediment supply to coastal foredunes. *Journal of Coastal Research*, **12**, 654–63.

Davidson-Arnott, R.G.D. and Pyskir, N.M. 1988. Morphology and formation of an Holocene coastal dune field, Bruce Peninsula, Ontario. *Géographie Physique et Quaternaire*, **42**, 163–70.

Davidson-Arnott, R.G.D., White, D.C. and Ollerhead, J. 1997. The effect of pebble lag concentration on aeolian sediment transport on a beach. *Canadian Journal of Earth Sciences*, **34**, 1499–508.

Davidson-Arnott, R.G.D., MacQuarrie, K. and Aagaard, T. 2005. The effect of wind gusts, moisture content and fetch length on sand transport on a beach. *Geomorphology*, **68**, 115–29.

Davidson-Arnott, R.G.D., Yang, Y, Ollerhead, J., Hesp, P.A. and Walker, I.J. 2008. The effects of surface moisture on aeolian sediment transport threshold and mass flux on a beach. *Earth Surface Processes and Landforms*, **33**, 55–74.

Davidson-Arnott, R.G.D., Hesp, P., Ollerhead, J. *et al.* 2018. Sediment budget controls on foredune height: Comparing simulation model results with field data. *Earth Surface Processes and Landforms* **43**, 1798–810.

Dech, J.P. and Maun, M.A. 2007. Zonation of vegetation along a burial gradient on lee slopes of Lake Huron sand dunes. *Canadian Journal of Botany*, **83**, 227–36.

de Langre, E. 2008. Effects of wind on plants. *Annual Review of Fluid Mechanics*, **40**, 141–68.

Delgado-Fernandez, I. 2010. A review of the application of the fetch effect to modelling sand supply to coastal foredunes. *Aeolian Research*, **2**, 61–7.

Delgado-Fernandez, I. and Davidson-Arnott, R.G.D. 2011. Meso-scale modelling of aeolian sediment input to coastal dunes: the nature of aeolian transport events. *Geomorphology*, **126**, 217–32.

de Vries S., Southgate, H.N., Kanning, W. and Ranasinghe, R. 2012. Dune behavior and aeolian transport on decadal timescales. *Coastal Engineering* **67**, 41–53.

de Vries, S., Van Thiel de Vries, J., van Rijn, L.C., Ranasinghe, R. and Arens, S.M. 2014. Aeolian sediment transport in supply limited situations. *Aeolian Research*, **12**, 75–85.

Doing, H. 1985. Coastal fore-dune zonation and succession in various parts of the world. *Vegetatio*, **61**, 65–75.

Dong, Z., Wang, H., Liu, X., Li, F. and Zhao, A. 2004. Experimental investigation of the velocity of a sand cloud blowing over a sandy surface. *Earth Surface Processes and Landforms*, **29**, 343–58.

Dupont, S., Bergametti, G.B.M. and Simoens, S. 2013. Modelling saltation intermittency. *Journal of Geophysical Research Atmospheres* **118**, 7109–28.

Farrell, E.J. and Sherman, D.J. 2013. Estimates of the Schmidt number for vertical flux distributions of wind-blown sand. *Journal of Coastal Research*, **SI 65**, 1289–94. DOI: 10.2112/SI65-218.1.

Finnegan, J. 2000. Turbulence in plant canopies. *Annual Review of Fluid Mechanics*, **32**, 519–71.

Fryberger, S.G. and Dean, G. 1979. Dune forms and wind regime. In McKee, E.D. (ed.), Global Sand Seas, US Geological Survey, Professional Paper, 1052, pp. 137–69.

Gares, P.A. and Nordstrom, K.F. 1995. A cyclic model of foredune blowout evolution for a leeward coast:

Island Beach, New Jersey. *Annals of the Association of American Geographers*, **85**, 1–20.

Gares, P.A., Davidson-Arnott, R.G.D., Bauer, B.O. *et al.* 1996. Alongshore variations in aeolian sediment transport, Carrick Finn. *Journal of Coastal Research*, **12**(3), 673–82.

Giannini, P.C.F., Sawakuchi, A.O., Martinho, C.T. and Tatumi, S.H. 2007. Eolian depositional episodes controlled by late Quaternary relative sea level changes on the Imbituba–Laguna coast (southern Brazil). *Marine Geology*, **237**, 143–68.

Gillette, D.A., Herbert, G., Stockton, P.H. and Owen, P.R. 1996. Causes of the fetch effect in wind erosion. *Earth Surface Processes and Landforms*, **21**, 641–59.

Grilliot, M., Walker, I.J. and Bauer, B.O. 2018. Airflow dynamics over a beach and foredune system with large woody debris. *Geosciences* **8**(5), 147. DOI: https://doi.org/10.3390/geosciences8050147.

Hansen, E., DeVries-Zimmerman, S., van Dijk, D and Yurk, B. 2009. Patterns of wind flow and aeolian deposition on a parabolic dune on the southeastern shore of Lake Michigan. *Geomorphology*,**105**, 147–57.

Hesp, P.A. 1981. The formation of shadow dunes. *Journal of Sedimentary Petrology*, **51** 101–12.

Hesp, P.A. 1983. Morphodynamics of incipient foredunes New South Wales, Australia. In Brookfield, M.E and Ahlbrandt, T.S. (eds.), *Eolian Sediments and Processes*. Amsterdam: Elsevier, pp. 325–42.

Hesp, P.A. 1989. A review of the biological and geomorphological processes involved in the initiation and development of incipient foredunes. *Proceedings of the Royal Society of Edinburgh*, **96B**, 181–201.

Hesp, P. A. 1991. Ecological processes and plant adaptations on coastal dunes. *Journal of Arid Environments*, **21**, 165–91.

Hesp, P.A. 2002. Foredunes and blowouts: initiation, geomorphology and dynamics. *Geomorphology*, **48**, 245–68.

Hesp, P.A. and Hyde, R. 1996. Geomorphology and dynamics of a trough blowout. *Sedimentology*, **43**, 505–25.

Hesp, P.A. and Martínez, M.L. 2007. Disturbance processes and dynamics in coastal dunes. In Johnson, E.A. and Miyanishi, K. (eds.), *Plant Disturbance Ecology*. San Diego, CA: Academic Press, pp. 215–47.

Hesp, P.A. and Pringle, A. 2000. Wind flow and topographic steering within a trough blowout. *Journal of Coastal Research*, **SI 34**, 597–601.

Hesp, P.A. and Thom, B.G. 1990. Geomorphology and evolution of active transgressive dune fields. In Nordstrom, K.F., Psuty, N.P. and Carter, R.W.G. (eds.), *Coastal Dunes: Form and Process*. Chichester, UK:, John Wiley and Sons, pp. 253–88.

Hesp, P.A., Davidson-Arnott, R.G.D., Walker, I.J. and Ollerhead, J. 2005. Flow dynamics over a foredune at Prince Edward Island, Canada. *Geomorphology*, **65**, 71–84.

Hesp, P.A., Dillenburg, S. R., Barboza, E. G. *et al.* 2007a. Morphology of the Itapeva to Tramandai transgressive dune field barrier system and mid- to late-Holocene sea level change. *Earth Surface Processes and Landforms*, **32**, 407–14.

Hesp, P.A, Abreu de Castilhos, J., Miot da Silva, G. *et al.* 2007b. Regional wind fields and dune field migration, Southern Brazil. *Earth Surface Processes and Landforms*, **32**, 561–73.

Hesp, P.A., Walker, I.J., Chapman, C., Davidson-Arnott, R.G.D. and Bauer, B.O. 2013. Aeolian dynamics over a coastal foredune, Prince Edward Island, Canada. *Earth Surface Processes and Landforms*, **38**, 1566–75.

Hesp, P.A., Smyth, T.A.G., Nielsen, P. *et al.* 2015. Flow deflection over a foredune. *Geomorphology* **230**, 64–74.

Hesp, P.A., Smyth, T.A.G., Walker, I.J., Gares, P.A. and Wasklewisz, T. 2016. Flow within a trough blowout at Cape Cod. *Journal of Coastal Research*, **SI 75**, 288–92.

Horikawa, K., Hotta, S., and Kraus, N.C. 1986, Literature review of sand transport by wind on a dry sand surface. *Coastal Engineering*, **9**, 503–26.

Horn, D.P. 2002. Beach groundwater dynamics. *Geomorphology* **48**(1), 121–46.

Houser, C. and Ellis, J. 2013. Beach and dune interaction. In Shroder, J.F. (ed. in Chief), Sherman, D.J. (vol. ed.), *Treatise on Geomorphology* vol. 10. *Coastal Geomorphology*. San Diego, CA: Academic Press, pp. 267–88.

Houser, C., Barrineau, P., Hammond, B. *et al.* 2018a. Role of the foredune in controlling barrier island response to sea level rise. In Moore, L.J. and Murray, A.B. (eds.), *Barrier Dynamics and Response to Climate Change*. New York: Springer.

Houser, C., Wernette, P. and Weymer, B.A. 2018b. Scale-dependent behaviour of the foredune: Implications for barrier island response to storms and sea-level rise. *Geomorphology*, **303**, 362–74.

Illenberger, W.K. and Rust, I. C. 1988. A sand budget for the Alexandria coastal dune field, South Africa. *Sedimentology*, **35**, 513–22.

Iversen, J.D. and Rasmussen, K.R. 1994. The effect of surface slope on saltation threshold. *Sedimentology*, **41**, 721–28.

Jackson, D.W.T., Beyers, J.H.M, Lynch, K. *et al.* 2011. Investigation of three-dimensional wind flow behaviour over coastal dune morphology under offshore winds using computational fluid dynamics (CFD) and ultrasonic anemometry. *Earth Surface Processes and Landforms*, **36**, 1113–24.

Jackson, D.W.T., Beyers, M., Delgado-Fernandez, I. *et al.* 2013. Airflow reversals and alternating corkscrew vortices in foredune wake zones during perpendicular and oblique offshore winds. *Geomorphology*, **187**, 86–93.

Jackson, N.L. and Nordstrom, K.L. 1997. Effects of time-dependent moisture content of surface sediments on aeolian transport rates across a beach, Wildwood, New Jersey, USA. *Earth Surface Processes and Landforms*, **22**, 611–21.

Jackson, N.L. and Nordstrom, K.L. 1998, Aeolian transport of sediment on a beach during and after rainfall, Wildwood, NJ, USA. *Geomorphology*, 22, 151–57.

Kadib, A.L. 1964. Calculation procedure for sand transport by wind on natural beaches. US Army Corps of Engineers, CERC Miscellaneous Paper, 264.

Kawamura, R. 1951. Study of sand movement by wind. University of Tokyo, Report of the Institute of Science and Technology, 5(3/4), 95–112.

Kroon, A. and Hoekstra, P. 1990. Eolian sediment transport on a natural beach. *Journal of Coastal Research*, **6**, 367–80.

Kuriyama, Y., Mochizuki, N. and Nakashima, T. 2005. Influence of vegetation on aeolian sand transport rate from a backshore to a foredune at Hasaki, Japan. *Sedimentology*, **52**, 1123–32.

Lancaster, N. and Baas A. 1998. Influence of vegetation cover on sand transport by wind: Field studies at Owens Lake, California. *Earth Surface Processes and Landforms*, **23**, 69–82.

Lane, C., Wright, S.J., Roncal, J. and Mashinski, J. 2008. Characterizing environmental gradients and their influence on vegetation zonation in a sub-tropical coastal sand dune system. *Journal of Coastal Research*, **24**, 213–24.

Law, M.N. and Davidson-Arnott, R.G.D. 1990. Seasonal controls on aeolian processes on the beach and foredune. In Davidson-Arnott, R.G.D. (ed.), *Proceedings of the Symposium on Coastal Sand Dunes*. Ottawa: National Research Council of Canada, pp. 49–68.

Leenders, J.K., van Boxel, J.H. and Sterk, G. 2007. The effect of single vegetation elements on wind speed and sediment transport in the Sahelian zone of Burkina Faso. *Earth Surface Processes and Landforms*, **32**, 1454–74.

Lees, B. 2006. Timing and formation of coastal dunes in northern and eastern Australia. *Journal of Coastal Research*, **22**, 78–89.

Lettau, K. and Lettau, H. 1977. Experimental and micrometeorological field studies of dune migration. In Lettau, K. and Lettau, H. (eds.), *Exploring the World's Driest Climate*. Madison, WI: University of Wisconsin Press, IES Report **101**, pp. 110–47.

Logie, M. 1982. Influence of roughness elements and soil moisture on the resistance of sand to wind erosion. In Yaalon, D.H. (ed.), *Aridic Soils and Geomorphic Processes*. Catena Supplement 1. Braunschweig: Catena-Verlag, pp. 161–73.

Loope, W.L. and Arbogast, A.F. 2000. Dominance of an ~150-year cycle of sand supply in late Holocene dune-building along the eastern shore of Lake Michigan. *Quaternary Research*, **54**, 414–22.

Lynch, K., Jackson, D.W.T. and Cooper, A. 2006. A remote-sensing technique for the identification of aeolian fetch distance. *Sedimentology*, **53**, 1381–90.

Lynch, K., Jackson, D.W.T. and Cooper, A. 2008. Aeolian fetch distance and secondary airflow effects: the influence of micro-scale variables on meso-scale foredune development. *Earth Surface Processes and Landforms*, **33**, 991–1005.

Martinho, C.T., Dillenburg, S.R. and Hesp, P.A. 2008. Mid to late Holocene evolution of transgressive dune fields from Rio Grande do Sul coast, southern Brazil. *Marine Geology*, **256**, 49–64.

Martini, I. P. 1981, Coastal dunes of Ontario: Distribution and geomorphology. *Géographie Physique et Quaternaire*, **35**, 219–29.

Maun, M. A. 2004. Burial of plants as a selective force in sand dunes. In Martínez, M.L. and Psuty, N.P. (eds.) *Coastal Dunes: Ecology and Conservation*. Ecological Studies vol. 171. Berlin: Springer-Verlag, 119–35.

Maun, M.A. 2009. *The Biology of Coastal Sand Dunes*. Oxford, UK: Oxford University Press, 265 pp.

Maun, M.A. and Lapierre, J. 1984. The effects of burial by sand on *Ammophila breviligulata*. *The Journal of Ecology*, **72**, 827–39.

McKenna Neuman, C. and Bédard, O. 2015. A wind tunnel study of flow structure adjustment on deformable sand beds containing a surface mounted obstacle. *Journal of Geophysical Research: Earth Surface*, **120**, 1824–40. DOI: 10.1002/2015JF003475.

McKenna Neuman C. and Langston G. 2006. Measurement of water content as a control of particle entrainment by wind. *Earth Surface Processes and Landforms*, **31**, 303–17.

McKenna Neuman, C. and Muljaars Scott, M. 1998. A wind tunnel study of the influence of pore water on aeolian sediment transport. *Journal of Arid Environments*, **39**, 403–19.

McKenna Neuman, C. and Nickling, W.G. 1989. A theoretical and wind tunnel investigation of the effects of capillary water on the entrainment of sediment by wind. *Canadian Journal of Soil Science*, **69**, 79–96.

Meur-Férec, C. and Ruz, M-H. 2002. Transports éoliens réels et théoreiques en haut de plage et sommet de dune (Wisant, pas de Calais, France). (Observed and predicted rates of aeolian sand transport on the upper beach and dune top, Wissant beach, Northern France.) *Géomorphologie: Relief, Processus, Environnement*, **4**, 321–34.

Miot da Silva, G. and Hesp, P.A. 2010. Coastline orientation, aeolian sediment transport and foredune and dune field dynamics of Moçambique Beach, Southern Brazil. *Geomorphology*, **120**(3–4), 258–78.

Miot da Silva, G., Hesp, P.A., Peixoto, J. and Dillenbeurg, S.R. 2008. Foredune vegetation patterns and alongshore environmental gradients: Moçambique Beach, Santa Caterina Island, Brazil. *Earth Surface Processes and Landforms*, **33**(10), 1557–73. DOI: 10.1002/esp1633.

Miyanishi, K. and Johnson E.A. 2007. Coastal dune succession and the reality of dune processes. In Johnson, E.A. and Miyanishi, K. (eds.), *Plant Disturbance Ecology*. San Diego, CA: Academic Press, pp. 215–47.

Moreno-Casasola, P. 1986. Sand movement as a factor in the distribution of plant communities in a coastal dune system. *Vegetatio*, **65**, 67–76.

Namikas, S.L. 2003. Field measurement and numerical modelling of aeolian mass flux distribution on a sandy beach. *Sedimentology*, **50**, 303–26.

Namikas, S.L. and Sherman, D.J. 1995. A review of the effects of surface moisture content on aeolian sand transport. In Tchakerian, V.P. (ed.), *Desert Aeolian Process*. London: Chapman and Hall, pp. 269–93.

Namikas, S.L., Bauer, B.O. and Sherman D.J. 2003. Influence of averaging interval on shear velocity estimates for aeolian transport modelling. *Geomorphology*, **53**, 235–46.

Namikas, S.L., Bauer, B.O., Edwards, B.L., Hesp, P.A. and Zhu, Y. 2009. Measurements of aeolian mass flux distributions on a fine-grained beach: Implications for grain-bed collision mechanics. *Journal of Coastal Research*, **SI 56**, 337–41.

Namikas, S.L., Edwards, B.L., Bitton, M.C.A., Booth, J.L. and Zhu, Y. 2010. Temporal and spatial variability in the surface moisture content of a fine-grained beach. *Geomorphology*, **114**, 303–10.

Nickling, W.G. 1988. The initiation of particle movement by wind. *Sedimentology*, **35**, 499–511.

Nickling, W.G. and Davidson-Arnott, R.G.D. 1990. Aeolian sediment transport on beaches and coastal sand dunes. In Davidson-Arnott, R.G.D. (ed.), *Proceedings of the Symposium on Coastal Sand Dunes*. Ottawa: National Research Council of Canada, pp. 1–35.

Nickling, W.G. and McKenna Neuman, C. 1995. Development of deflation lag surfaces. *Sedimentology*, **42**, 403–14.

Nickling, W.G. and McKenna Neuman, C. 1999. Recent investigations of air flow and sediment transport over desert dunes. In Goudie, A.S., Livingstone, I. and Stokes, S. (eds.), *Aeolian Environments: Sediments and Landforms*. Chichester, UK: John Wiley & Sons, pp. 15–47.

Nordstrom, K.F. and Jackson, N.L. 1992. Effect of source width and tidal elevation changes on aeolian transport on an estuarine beach. *Sedimentology*, **39**, 769–78.

Nordstrom, K.F. and Jackson, N.L. 1993. The role of wind direction in eolian transport on a narrow sand beach. *Earth Surface Processes and Landforms*, **18**, 675–85.

Oblinger, A. and Anthony, E. 2008. Surface moisture variations on a multibarred macrotidal beach: Implications foe aeolian sand transport. *Journal of Coastal Research*, **24**, 1194–9.

O'Brien, P. and McKenna Neuman, C. 2018. An experimental study of the dynamics of saltation within a three-dimensional framework. *Aeolian Research*, **32**, 62–71.

Olivier, M.J. and Garland, G.G. 2003. Short-term monitoring of foredune formation on the east coast of South Africa. *Earth Surface Processes and Landforms*, **28**, 1143–55.

Ollerhead, J., Davidson-Arnott, R., Walker, I.J. and Mathew, S. 2012. Annual to decadal morphodynamics of the foredune system at Greenwich Dunes, Prince Edward Island, Canada. *Earth Surface Processes and Landforms*, **38**, 284–98.

Olson, J.S. 1958a. Lake Michigan dune development 3. Lake level, beach and dune oscillations. *Journal of Geology*, **66**, 473–83.

Olson, J.S. 1958b. Lake Michigan dune development 2. Plants as agents and tools in geomorphology. *Journal of Geology*, **66**, 345–51.

Olson, J.S. 1958c. Lake Michigan dune development 1. Wind velocity profiles. *Journal of Geology*, **66**, 254–63.

Owen, P.R. 1964. Saltation of uniform grains in air. *Journal of Fluid Mechanics*, **20**, 225–42.

Packham, J.R. and Willis, A.J. 1997. *Ecology of Dunes, Salt Marsh and Shingle*. London: Chapman and Hall, 334 pp.

Palmsten, M.L., Brodie, K.L. and Spore, N.J. 2017. *Coastal Foredune Evolution, Part 2: Modelling Approaches for Meso-Scale Morphologic Evolution.* ERDC/CHL CHETN-II-57. Vicksburg, MS: US Army Engineer Research and Development Center. DOI: http://dx.doi.org/10.21079/11681/21627.

Pearce, K.I. and Walker, I.J. 2005. Frequency and magnitude biases in the Fryberger model with implications for characterising geomorphically effective winds. *Geomorphology*, **68**, 39–55.

Psuty, N.P. 1988, Sediment budget and dune/beach interaction. *Journal of Coastal Research, Special Issue*, **3**, 1-4.

Psuty, N.P. 2004. The coastal foredune: a morphological basis for regional coastal dune development. In Martínez, M.L. and Psuty, N.P. (eds.) *Coastal Dunes: Ecology and Conservation*. Ecological Studies vol. 171. Berlin: Springer-Verlag, pp. 11–28.

Pye, K. 1983. Coastal dunes. *Progress in Physical Geography*, **7**, 531–97.

Ranwell, D.S. 1972. *Ecology of Salt Marshes and Sand Dunes*. London: Chapman and Hall, 258 pp.

Rasmussen, K.R. 1989. Some aspects of flow over coastal dunes. *Proceedings of the Royal Society of Edinburgh*, **96B**, 129–47.

Reed, D.J., Davidson-Arnott, R.G.D. and Perillo, G.M.E. 2009. Estuaries, coastal marshes, tidal flats, and coastal dunes. In Slaymaker, O. (ed.). *Landscape Changes in the 21st Century*. Cambridge: Cambridge University Press, pp. 130–67.

Ruggerio, P., Hacker, S., Seabloom, E. and Zarnetske, P. 2018. The role of vegetation in determining dune morphology, exposure to sea-level rise, and storm-induced coastal hazards: a Pacific northwest perspective. In Moore L. and Murray A. (eds.), *Barrier Dynamics and Responses to Changing Climate*. Cham: Springer International Publishing, pp. 337–361. DOI: doi.org10.1007 978-3-319-68086-6_11.

Saqqa, W.A. and Saqqa, A.W. 2007. A computer program (WDTSRP) designed for computation of sand drift potential (DP) and plotting sand roses. *Earth Surface Processes and Landforms*, **32**, 832–40.

Sarre, R.D. 1989. Aeolian sand drift from the intertidal zone on a temperate beach: Potential and actual rates. *Earth Surface Processes and Landforms*, **14**, 247–58.

Saunders, K.E. and Davidson-Arnott, R.G.D. 1990. Coastal dune response to natural disturbances. In Davidson-Arnott, R.G.D. (ed.), *Proceedings of the Symposium on Coastal Sand Dunes*. Ottawa: National Research Council of Canada, pp. 321–46.

Schmutz, P.P. and Namikas, S.L. 2013. Measurement and modelling of moisture content above an oscillating water table: implications for beach surface moisture dynamics. *Earth Surface Processes and Landforms*, **38**, 1317–25.

Schwarz, C., Brinkkemper, J. and Ruessink, G. 2019. Feedbacks between biotic and abiotic processes governing the development of foredune blowouts: a review. *Journal of Marine Science and Engineering*, **7**, 2. DOI: doi:10.3390/jmse7010002.

Shao, Y. and Raupach, M.R. 1992. The overshoot and equilibrium of saltation. *Journal of Geophysical Research*, **97**, 559–64.

Sherman, D.J. and Bauer, B.O. 1993. Dynamics of beach-dune systems. *Progress in Physical Geography*, **17**, 413–47.

Sherman, D.J. and Li, B. 2012. Predicting aeolian sand transport rates: a reevaluation of models. *Aeolian Research*, **3**, 371–78.

Sherman, D.J. and Lyons, W. 1994. Beach state controls on aeolian sand delivery to coastal dunes. *Physical Geography*, **15**, 381–95.

Sherman, D.J., Namikas, S.L., Jackson, D.W.T. and Wang, S.L. 1998. Wind blown sand on beaches: An evaluation of models. *Geomorphology*, **22**, 113-133.

Sherman, D.J., Li, B., Ellis, J.T. and Swann, C. 2017. Intermittent aeolian saltation: A protocol for quantification. *Geographical Review*, **108**, 296–314.

Short, A.D. 1979. Three dimensional beach stage model. *Journal of Geology*, **87**, 553-571.

Short, A.D. and Hesp, P.A. 1982. Wave, beach and dune interactions in southeastern Australia. *Marine Geology*, **48**, 259-284.

Shroder, J.F. (Editor-in-Chief), 2013. *Treatise on Geomorphology*. San Diego, CA: Academic Press, 14 volumes.

Smyth, T.A.G., Jackson, D. and Cooper, A. 2104. Airflow and aeolian sediment transport patterns within a coastal trough blowout during lateral wind conditions. *Earth Surface Processes and Landforms*, **39**, 1847–54.

Sterk, G., Jacobs, A.F.G. and van Boxel, J.H. 1998. The effect of turbulent flow structures on saltation sand transport in the atmospheric boundary layer. *Earth Surface Processes and Landforms*, **28**, 877–87.

Stewart, C.J. and Davidson-Arnott, R.G.D. 1988. Morphology, formation and migration of longshore sandwaves: Long Point, Lake Erie, Canada, *Marine Geology*, **81**, 63–77.

Stout, J.E. 2004. A method for establishing the critical threshold for Aeolian transport in the field. *Earth Surface Processes and Landforms*, **29**, 1195–207.

Stout, J.E. and Zobeck, T.M. 1997. Intermittent saltation. *Sedimentology*, **44**, 959–70.

Svasek, J.N., *and* Terwindt, J.H.J. 1974. Measurement of sand transport by wind on a natural beach. *Sedimentology*, **21**, 311–22.

Tripaldi, A. and Forman, S.L. 2007. Geomorphology and chronology of Late Quaternary dune fields of western Argentina. *Palaeogeography, Palaeoclimatology, Palaeoecology*, **251**, 300–20.

Turner, I.L. and Masselink, G. 1998. Swash infiltration-exfiltration and sediment transport. *Journal of Geophysical Research*, **102**(C13), 30813–24.

van Boxel, J.H., Sterk, G. and Arens, S.M. 2004, Sonic anemometers in aeolian transport research. *Geomorphology*, **59**, 131–47.

van der Wal, D. 1998. Effects of fetch and surface texture on aeolian sand transport on two nourished beaches. *Journal of Arid Environments*, **39**, 533–47.

Wal, A. and McManus, J. 1993. Wind regime and sand transport on a coastal beach-dune complex, Tentsmuir, eastern Scotland. In Pye, K. (ed.), *The Dynamics and Environmental Context of Aeolian Sedimentary Systems* London: Geological Society Special Publication, **72**, pp. 159–72.

Walker, I.J. 2005. Physical and logistical considerations of using ultrasonic anemometers in aeolian sediment transport research. *Geomorphology*, **68**, 57–76.

Walker, I.J. and Hesp, P.A. 2013. Fundamentals of aeolian sediment transport: airflow over dunes. In Lancaster, N., Sherman, D.J. and Baas, A.C.W. (eds.), vol 11: *Aeolian Geomorphology*. In Shroder, J.F. (ed. in chief) *Treatise on Geomorphology*. Oxford: Elsevier, pp. 109–33.

Walker, I.J., Hesp. P.A., Davidson-Arnott, R.G.D. Bauer, B.O. and Ollerhead, J. 2009a. Response of three-dimensional flow to variations in the angle of incident flow and profile form of dunes: Greenwich Dunes, Prince Edward Island, Canada. *Geomorphology*, **105**, 127–38.

Walker, I.J., Davidson-Arnott, R.G.D., Hesp, P.A., Bauer, B.O. and Ollerhead, J. 2009b. Mean flow and turbulence response in airflow over foredunes: New insights from recent research. *Journal of Coastal Research*, **SI 56**, 366–70.

Walker, I.J., Davidson-Arnott, R.G.D., Bauer, B.O. *et al.* 2017. Scale-dependent perspectives on the geomorphology and evolution of beach-dune systems. *Earth-Science Reviews*, **171**, 220–53.

Wang, D., Wang, Y., Yang, B. and Zhang, W. 2008. Statistical analysis of sand grain/bed collision process recorded by high-speed digital camera. *Sedimentology*, **55**, 461-470.

Wernette, P., Houser, C., Weymer, B.A. *et al.* 2018. Influence of a spatially complex framework geology on barrier island geomorphology. *Marine Geology*, **398**, 151–62.

Wiggs, G.F.S., Livingstone I. and Warren, A. 1996. The role of streamline curvature in sand dune dynamics: evidence from field and wind tunnel measurements. *Geomorphology*, **17**, 29–46.

Wiggs, G.F.S., Baird, A.J. and Atherton, R.J. 2004a. The dynamic effect of moisture on the entrainment and transport of sand by wind. *Geomorphology*, **59**, 15–30.

Wiggs, G.F.S., Atherton, R.J. and Baird, A.J. 2004b. Thresholds of aeolian sand transport: establishing suitable values. *Sedimentology*, **51**, 95–108.

Wolfe, S.A. and Nickling, W.G. 1993. The protective role of sparse vegetation in wind erosion. *Progress in Physical Geography*, **17**, 50–68.

Wright, L.D. and Short, A.D. 1983. Morphodynamic of beaches and surf zones in Australia. In Komar, P.D. (ed.), *Handbook of Coastal Processes and Erosion*. Boca Raton, FL: CRC Press, pp. 35–64.

Wright, L.D. and Short, A.D. 1984. Morphodynamic variability of surf zones and beaches: a synthesis. *Marine Geology*, **56**, 93–118.

Yang, Y. and Davidson-Arnott, R.G.D. 2005. Rapid measurement of surface moisture content on a beach. *Journal of Coastal Research*, **21**, 447–52.

Zarnetske, P.L., Seabloom, E.W. and Hacker, S.D. 2010. Non-target effects of invasive species management: beach grass, birds, and bulldozers in coastal dunes. *Ecosphere* **1**, 13.

Zingg, A.W. 1953. Wind tunnel studies of the movement of sedimentary material. *Proceedings of the Fifth Hydraulic Conference*, Bulletin **34**, Iowa City: University of Iowa Institute of Hydraulics, pp. 111–35.

10

Barrier Systems

10.1 Synopsis

The term barrier is used to describe a range of emergent depositional landforms that are separated from the mainland coast by a lagoon, bay or marsh. Barriers make up more than 10 per cent of the world's coastline and are particularly well developed on microtidal trailing edge coasts such as the east coast of North America, South America, the east coast of Australia, southern Africa and western Europe from Holland to Denmark and parts of the Baltic. Barriers are also formed in large lakes such as the North American Great Lakes, and one of the earliest descriptions of barrier spits was written by G.K. Gilbert (1890) on barriers preserved along the shoreline of Lake Bonneville. They can range in size from a few tens of metres in length for small barriers across a stream mouth, to spits that may be tens of kilometres long and hold many millions of cubic metres of sediment, to barrier island and spit chains that extend hundreds of kilometres.

While barriers may be composed entirely of cobbles or gravels, they are more commonly composed of sand-sized sediment. There are sufficient differences in dynamics to warrant a distinction between sandy barriers and gravel or cobble barriers. Gravel and cobble barriers can only be built to the limit of wave action during storms and are always subject to overtopping unless there has been a fall in sea level. In contrast, sandy barriers can be built well above the limit of wave action through the transport of sand inland by wind and the development of dune systems. As a result, the subaerial component of sandy barriers consists of the beach and backshore, dunes and a variety of back barrier sediments that are washed over during storms or transported by wind both from the seaward side and from the lagoon or bay. Because of this cross-shore exchange of sediment and alongshore transport, the subaerial component overlies a subaqueous platform. The lagoon or bay, intertidal flats and marshes that are protected by the barrier are usually highly productive ecosystems composed of marshes and seagrass beds that provide the base for a food web that includes a wide range of shellfish, crustaceans and invertebrates, nurseries for fish and habitats for a wide range of birds and mammals. It is the unique ecological make-up of the lagoon, bay or marsh that provides evidence that barriers migrate landward in response to sea level rise.

Barriers are the most dynamic of depositional coastal landforms and they respond rapidly to changes in littoral sediment supply and sea level, as well as to the dynamic processes associated with severe storms. During storms, high waves and storm surge may lead to breaching of the high point of gravel barriers or the dunes of sandy barriers, forming overwash fans on the landward side and occasionally to the formation of an inlet connecting the lagoon or bay to the ocean. Permanent inlets may form a connection between the lagoon, or bay, and the open ocean with tides driving water into and out of the bay through the inlet. Erosion and sedimentation in and around the inlets lead to inlet migration, infilling and sometimes closure. The elevation of the dunes on sandy barriers or the berm on gravel barriers is an important control on the rate and extent of washover during storms, and thereby control the rate of island migration. Most modern barrier systems are only a few thousand years old because of rapid shoreline change during the Holocene transgression. In areas with large volumes of available sediment and coastal progradation such as the southeast coast of Australia there may be one or more older barriers preserved inland of the modern coast. However, on the east coast of North America many early barriers were drowned and reworked during the transgression.

10.2 Barrier Types and Morphology

10.2.1 Definition and Morphological Classification of Barriers

There have been several attempts to produce a classification of barrier types including extensive descriptions in Zenkovitch (1967). A simple one based on barrier geometry and number of free ends is shown in Figure 10.1 (Ollerhead, 1993). The grouping of features at this level is quite useful, and it is relatively simple to assign barriers to one of the classes. In general, there is a tendency for stability of the barrier to increase from barrier islands with two free ends, to spits with one free end, to barriers that are attached at both ends.

10.2.1.1 *No Free Ends*

Barriers with no free ends include tombolos, and a variety of barriers built from one shore to the other across bays and estuaries (Figure 10.1). Natural tombolos are barriers that link islands to each other or to the mainland coast and they generally form because of wave refraction and diffraction around the island and the development of a littoral transport gradient that leads to deposition in the lee of the island or islands (Zenkovitch, 1967; Sanderson and Elliot, 1996; Flinn, 1997; Mariner *et al.*, 2008). Depending on the orientation of the barrier with respect to prevailing winds and waves, one side of the tombolo may be more sheltered than the other but in

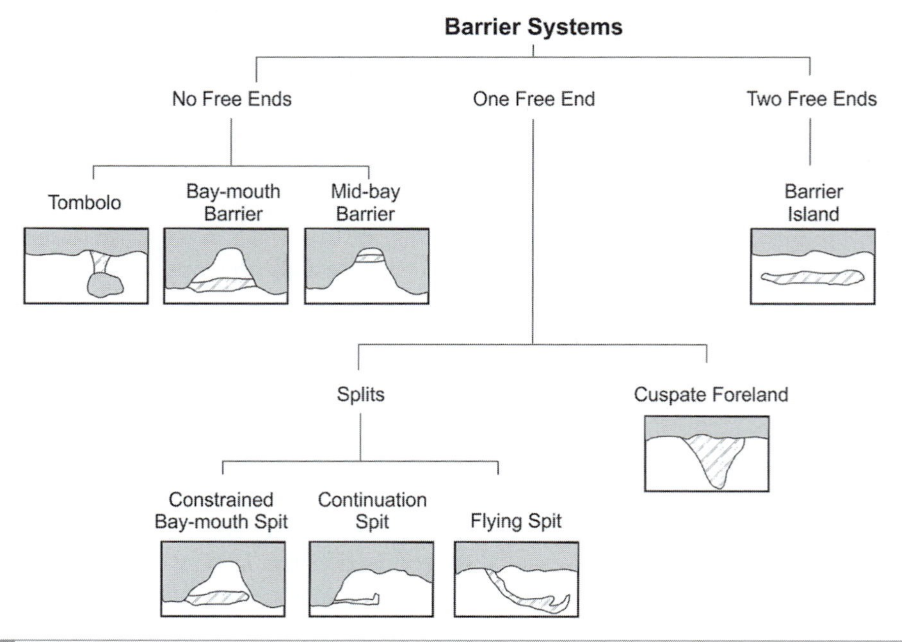

Figure 10.1 Classification of barrier systems based on morphology and geometry (after Ollerhead, 1993).

Figure 10.2 Examples of barrier with no free ends. A. View south over the Friar's Bay tombolo connecting volcanic outcrops, SE Peninsula, St Kitts. The distant cloud covered peak is the volcanic peak on the island of Nevis; B. Baymouth barriers formed in gravel, Cape Breton, Nova Scotia. Note the small channel connecting the open ocean to the bay. C. Mid-bay barrier, Gaspe Peninsula, Quebec, Canada. The barrier provides road access across the bay as well as a small, sheltered harbour for fishing boats. Note the saltmarsh in the protected bay to the right.

most cases one is subject to greater wave activity than the other, and this may include breaching and overwash. In many cases beaches on both sides are separated by a pond or marsh (Figure 10.2A), though this may be filled in over time by dune formation. On a much smaller scale, construction of an offshore detached breakwater may lead to the formation first of a beach salient opposite the breakwater, and this in turn may grow towards the breakwater to form a small tombolo (Silvester and Hsu, 1993; Bowman and Pranzini, 2004).

Baymouth and mid-bay barriers form in response to littoral sediment transport at the entrance to a bay or along the sides within the bay (Figure 10.2B, C). In some cases, where there is littoral transport from both sides, the barrier may build from both ends. However, there is often a net littoral transport from one side resulting in the building of a spit across the bay from the updrift end. Where a baymouth barrier forms early, it reduces wave action within the bay and a single barrier is formed. Where the entrance to the bay is comparatively deep, a mid-bay or baymouth barrier may develop first and subsequently a baymouth barrier forms, or there may never be sufficient sediment to form an outer barrier. If the area enclosed by the barrier is quite large there may be sufficient tidal flow to maintain an open entrance or inlet through the barrier, either permanently or episodically (Figure 10.2B).

10.2.1.2 *One Free End*

This class includes a range of spit forms in which there is a pronounced littoral drift from one

Figure 10.3 Examples of barriers with one free end: A. oblique aerial view of a cuspate foreland on the southwest shore of Anguilla. Note the carbonate sands building over coral reef; B. Fish Point, a cuspate foreland, formed primarily in sand and gravel, at the south end of Pelee Island, western Lake Erie; C. view looking south across the inlet entrance to Omaha spit, on the east coast north of Auckland, New Zealand – an example of a constrained spit; D. Cape Dundas spit, Georgian Bay, Ontario. This is a continuation spit built primarily in shingle supplied by erosion of sandy dolomite from a lower unit of the Niagara escarpment (see Figure 13.7). Ongoing isostatic uplift has resulted in preservation of the older shingle ridges at elevations up to 5 m above the present lake level.

direction, and cuspate forelands (or headlands) that form where there is a convergence of littoral drift from both directions. Cuspate forelands and headlands require that littoral drift from both directions be of the same order of magnitude (otherwise a spit would likely form) and thus they occur most often where there are features of the shoreline shape and orientation, or fetch, that are conducive to this (Sanderson *et al.*, 2000). They may form on the leeward coast of an island or reef because of wave refraction around the island (Figure 10.3A), or as an extension of a point where there is wave convergence (Figure 10.3B). They may also form on straight shorelines where the shore perpendicular fetch is restricted and there is substantial littoral drift from oblique waves from both the right and left. Point Pelee at the western end of Lake Erie is an example of this (see Box 7.2 and Figure 7.21C). The large, complex sand and gravel barrier forming the Dungeness foreland on the south coast of England is of comparable size and has also evolved over the past 5000 years (Long *et al.*, 2006). The resilience of these features can in part be traced to their location at zones of sediment transport convergence, though there may be considerable change in form and appearance through reworking of the beach, dune and marsh complexes that make up the barrier (Coakley, 1976; Long *et al.*, 2006).

Spits are much more numerous than cuspate forelands, and worldwide may be more common than barrier islands. They are formed at the downdrift end of a littoral cell system where there is an abrupt change in shoreline orientation, for example where sediment builds across a bay. They also form where there is an obstruction such as a rock reef or a glacial moraine that initiates deposition and deflection of the littoral transport. It is useful to distinguish between continuation spits, where the direction of spit progradation is parallel to the updrift shoreline, and flying spits that leave the coast at an acute angle. Where continuation spits built almost across a bay, further progradation and evolution may be prevented by removal of sediment by strong tidal currents flowing into and out of the remaining opening. This prevents closure to form a baymouth barrier and the feature is termed a constrained spit (Figure 10.3C).

Spits can range in length from < 100 m for small spits in estuaries and bays (Figure 10.3D), to features such as Long Point in Lake Erie which is about 40 km long and up to 5 km wide at the distal (downdrift) end. Wave refraction around the distal end of the spit leads to a pronounced curvature of the shoreline, and consequently of the foredune behind it. Migrating swash bars and sandwaves attach to the downdrift end and extend the spit both downdrift as well as normal to the shoreline. The result is to form a series of curved relict foredunes, termed recurves, with interdune swales or marshes between them. Where water at the distal end is quite shallow, refraction is pronounced and the curvature can approach 90°. Where the spit builds into deep water, the refraction becomes less pronounced and so do the recurves (see Box 10.3).

10.2.1.3 *Two Free Ends*

Barrier islands occur on gently sloping coastal plains in many parts of the world and tend to form chains of islands, separated by tidal inlets, that run parallel to the mainland coast and enclose a lagoon or bay 1–5 km wide. By strict definition, they should have two free ends but some barriers in a chain may be attached to the mainland at one end or at a point where transgression has bridged the lagoon (Figure 10.4), and it is probably best to consider the whole chain as barrier islands rather than trying to separate some into the spit category. The barrier islands in Kouchibouguac Bay, shown in Figure 10.4, form part of a larger chain of barriers on the southeast coast of New Brunswick. In turn, this chain is one of four chains of barriers in the Gulf of St Lawrence that occur along the coasts of New Brunswick, Prince Edward Island and the Isles de la Madeleine (Owens, 1974; McCann, 1979). They exist in microtidal environments subject to storm-wave activity, and where ice is present for several months of the year. The Gulf of St Lawrence barriers can be regarded as the northernmost extension of the barrier systems that extends southward along much of the east coast of the US and into the Gulf of Mexico. Padre Island, on the Texas coast, is the longest barrier island in the world (at ~180 km), and is believed to have formed through longshore extension of smaller islands that had been previously separated by fluvial systems (Wernette *et al.*, 2018).

10.2.2 Structure and Components of Barrier Systems

While there are some differences between the three major types of barrier recognised above, it is possible to recognise six major components or sub-environments (McCubbin, 1982; Oertel, 1985), which are common to almost all barriers (Figure 10.5). These include: (1) the mainland coast which is protected by the barrier and which may include bays and estuaries extending some distance inland; (2) the lagoon, bay or extensive marsh that separates the barrier from the mainland; (3) the subaerial barrier, including the beach, dune and backbarrier deposits; (4) the subaqueous platform of sediments which underpins the barrier and on which the subaerial portion of the barrier is built; (5) the shoreface extending offshore from the exposed beach; and (6) inlets and associated tidal deltas that are typically associated with barrier islands, but may also be found, at least temporarily, on all of the other barrier types.

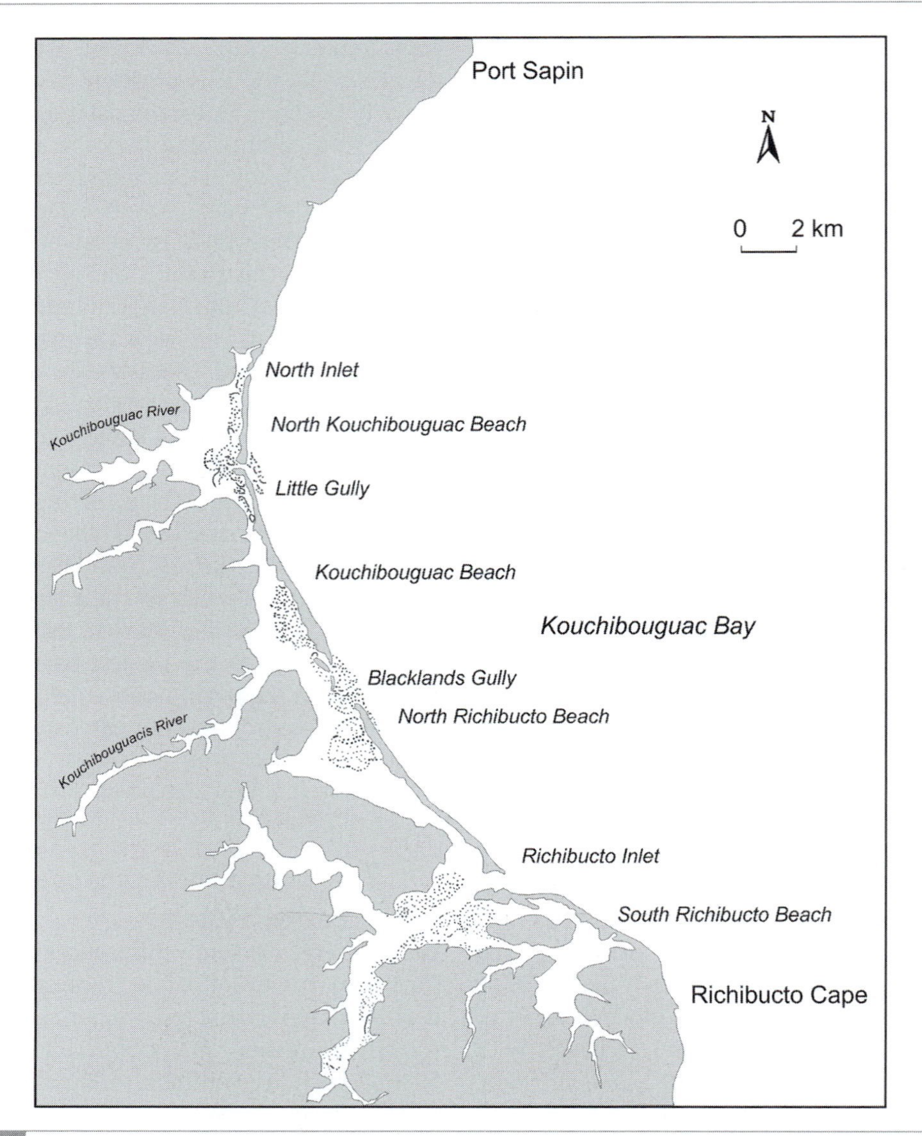

Figure 10.4 Map of the Kouchibouguac Bay barrier island system, New Brunswick. As a result of transgression the northern part of the barrier system has become fixed to the mainland forming a series of dune ridges backed by ponds and marsh. There are three permanent inlets opposite the major estuaries and occasionally a temporary inlet such as North Inlet is opened during a storm.

The mainland associated with barrier islands is generally a low-lying delta or coastal plain, often characterised by shallow bays and estuaries that have been drowned near the end of the Holocene transgression. Spits and baymouth barriers may form on much steeper coasts and often are attached to headlands at the entrance to a bay or along the sides. However, the bay itself must be shallow enough relative to sediment supply, to permit the barrier platform to develop and the subsequent growth of the subaerial component. The mainland may have a range of protected sandy beaches, tidal flats or marshes. The other five components are depositional environments characterised by a wide range of sediment types and sedimentary processes. There are frequent transfers of sediment from one component to the other and the evolution of the barrier results

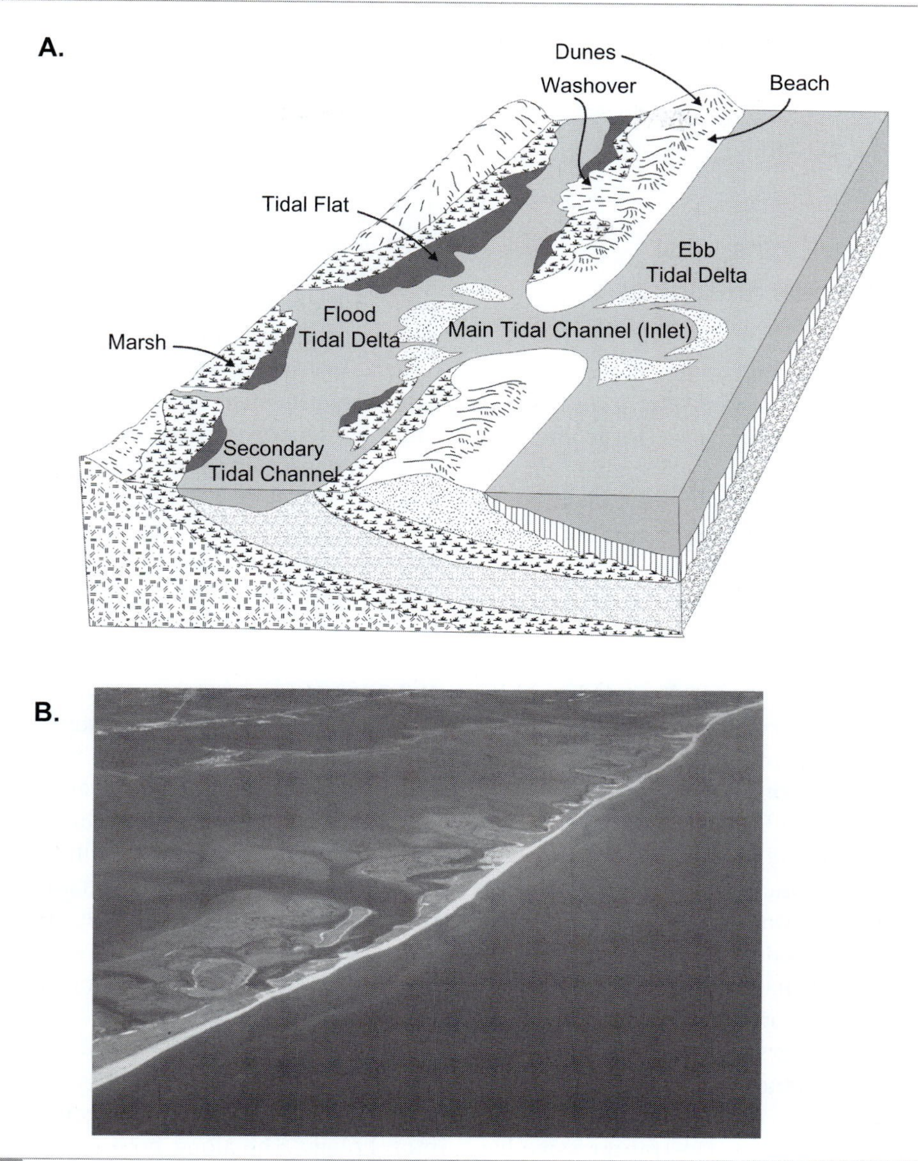

Figure 10.5 Components of a barrier island system: A. plan view; B. oblique aerial photograph of a portion of the Neguac barrier, northern New Brunswick, Gulf of St Lawrence.

in changes to the direction and intensity of processes controlling erosion and sedimentation. As a result, most barrier systems are highly dynamic on a range of time scales from decades to centuries. Some features are common to more than one of the depositional environments (Figure 10.5). Beaches extend along the length of the seaward side of the barrier and are a major part of the attraction for recreational purposes. However, beaches are also found in places along the lagoon and on the mainland and estuaries, especially where the fetch is long enough to permit waves to remove fines. Tidal flats are formed on the ebb and flood tidal deltas of inlets as well as in the lagoon or bay. Marshes develop along protected shorelines on the lagoon coast of the barrier as well as on the mainland. Finally, dunes develop along both the exposed coast and onshore of the lagoon and mainland beaches, though the narrow beaches of the lagoon generally restrict the size of dunes that develop along this low-energy shoreline.

10.3 Barrier Dynamics: Overwash and Inlets

The impact of storms on barriers depends on the elevation of the total water level (tide + storm surge + wave run-up) relative to the geometry of the coast, which is largely dependent on the height and alongshore extent of the foredune (Thieler and Young, 1991; Sallenger, 2000; Morton, 2002; Nott, 2006; Houser *et al.*, 2008a; Houser and Hamilton, 2009). The impact of a storm can range from minor scarping at the base of the dune to overwash through gaps in the dune line or complete inundation and breaching to form inlets. Overwash and inlets formed during hurricanes, tropical storms and mid-latitude storms are common features of the barrier islands along the Eastern Seaboard and Gulf of Mexico of the United States. Overwash is particularly characteristic of transgressive barriers (or transgressive portions of barriers, such as the proximal (updrift) end of barrier spits), because they are relatively narrow and the volume of sand in dunes is more limited than stable or progradational barriers.

Both overwash and inlet formation during storms are important controls on the dynamics and evolution of barriers. On transgressive barriers, overwash moves sediment from the seaward side of the island to the backbarrier shoreline causing the island to migrate landward, a process termed a rollover. Without overwash, and to a lesser degree dune blowouts, barrier islands would not be able to maintain form and ecological function with sea level rise – they would simply drown in place. Alternatively, a reduction in the amount of sediment available from alongshore and/or offshore sources can limit dune development (or recovery following storms), thereby allowing the dunes to be over washed by relatively small waves and surge in the absence of sea level rise (e.g., Psuty, 2008). In other words, overwash acts to preserve the barrier as a depositional landform during periods of sea level rise, and the dune is an important control on the frequency and extent of overwash. If an island is inundated during a storm due to an extreme surge or small dunes, the return (or seaward) flow of water from the lagoon or bay can cause deep scouring, leading to the formation of an inlet. Along developed barriers, the return flow of the storm surge tends to follow gaps between structures, roads and breaks in shore protection structures, leading to scours that can persist for years after a storm (Sherman *et al.*, 2013).

Modern barrier islands are a product of sea level rise over the late Holocene in which processes associated with marine transgression modified older and irregularly preserved stratigraphic units of variable ages, compositions and geometries, thereby providing the subsurface platform and sediment supply for barrier island transgression. Whether the result of glacial activity during the Pleistocene (Schupp *et al.*, 2006) or the transgression of the island (Houser, 2012), the framework geology influences barrier island evolution through variations in bathymetry, and sediment texture, thickness and supply (Belknap and Kraft, 1985; Houser and Mathew, 2011). The presence of Pleistocene fluvial channels beneath Padre Island (Weymar *et al.*, 2018) and the Outer Banks in North Carolina (Lazarus *et al.*, 2011) are associated with corresponding variations in the nearshore and dune morphology. The alongshore variation in framework geology in responsible for an alongshore variation in dune height that influences how the island responds to and recovers from storms (Houser *et al.*, 2018).

10.3.1 Storm Surge and Overwash

10.3.1.1 *Storm Surge, Erosion and Overtopping*

Overwash occurs when storm waves overtop the foredunes or when erosion and scarping of the base of the foredune are so severe that a breach is formed, and water and sediments are carried inland. The resulting sedimentary deposit is termed a washover fan or washover terrace depending on the lateral extent. As noted, the general requirements for overtopping or breaching to occur are elevated water levels due to storm surge and high waves that erode and flatten the beach profile, thus allowing wave action to reach the dune. Storm duration is also important because it determines the number of high tides over which erosion can occur, and coincidence with spring tides increases the potential for the combined surge and tidal elevation to exceed the threshold for generating overwash. Much of our basic understanding of the overwash process on barriers comes from studies in the 1960s and 1970s of overwash processes, documentation of volumes of erosion and deposition on the beach and on the barrier, and measurement of the sedimentary characteristics of washover deposits (e.g., Hayes, 1967; Pierce, 1970; Schwartz, 1975; Leatherman, 1976; Leatherman *et al.*, 1977; Cleary and Hosier, 1979; Orford and Carter, 1984).

Recent papers have distinguished between two major overwash regimes (Morton *et al.*, 2000; Sallenger, 2000; Orford *et al.*, 2003; Donnelly *et al.*, 2006): (1) run-up overwash where the storm surge level is below the level of the backshore ridge or dunes but wave run-up results in some swash overtopping the crest; and (2) inundation overwash where the elevation of the storm surge is above the backshore crest or foredune and water flows landward nearly continuously (Figure 10.6). It should be recognised that beach width and backshore ridge or dune elevation vary alongshore, and at the same time the storm surge elevation will vary temporally with meteorological conditions and with the tidal cycle, as well as with the nearshore bathymetry (Figure 10.7). Increasing storm surge, and wave run-up duration and elevation, produces an increase in the extent and severity of overwash. On the other hand, increasing dune height and width, the extent of vegetation cover and the presence of shrubs, and the time since previous overwash events tend to decrease the potential for overwash. The result is that there is a continuum of overwash conditions (Claudino-Sales

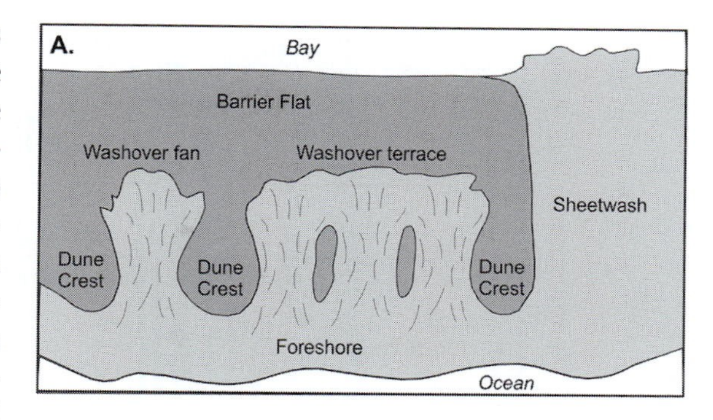

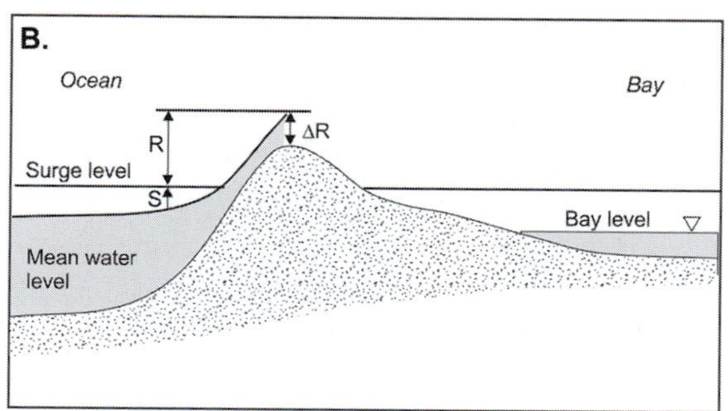

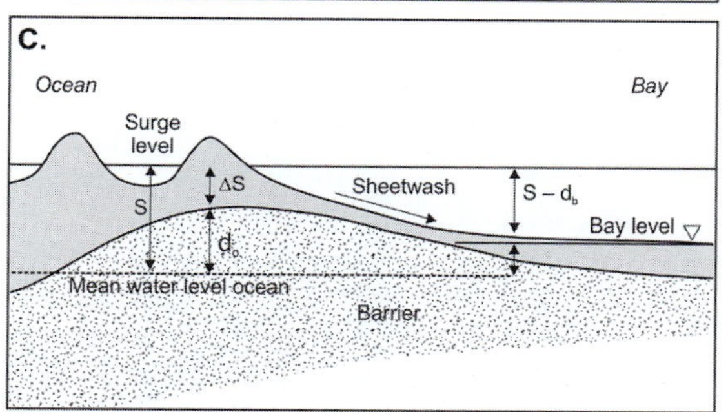

Figure 10.6 Overwash regimes: A. definition sketch of the three principal forms of overwash morphology and sediments; B. definition sketch showing overwash by wave run-up. S is the storm surge elevation and R the run-up elevation superimposed on this. The quantity DR is the amount by which the elevation R exceeds the dune or beach crest height d_c; C. definition sketch showing overwash by inundation. The extent of inundation is indicated by the quantity DS, which is the amount by which the storm surge elevation exceeds d_c (after Donnelly *et al.*, 2006).

et al., 2008; Matias *et al.*, 2008), beginning with occasional overtopping at low points and the reactivation of recent overwash channels, through nearly continuous overtopping and washover fan formation at many places (Figure 10.8), to complete inundation along a

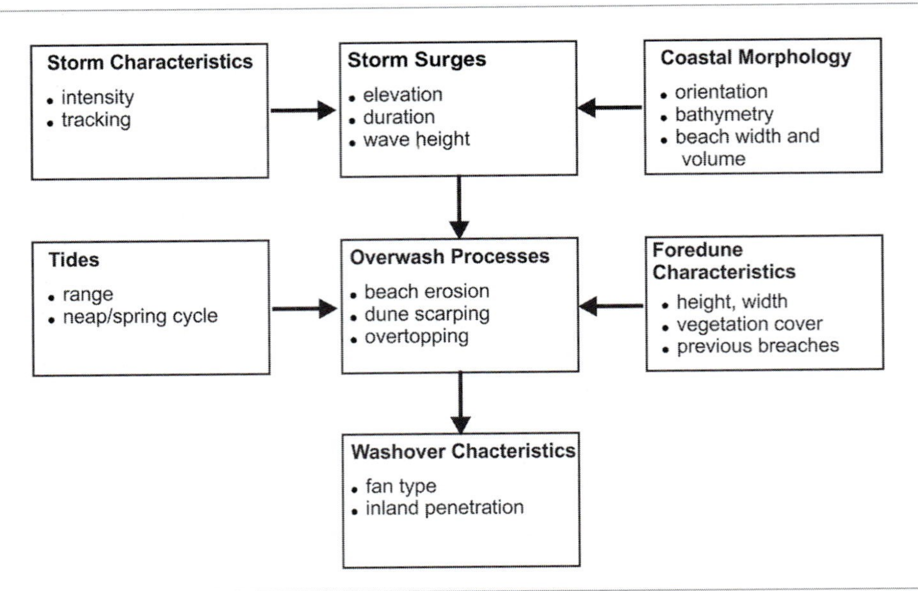

Figure 10.7 Schematic model of factors controlling the proportion of the shoreline overwashed and the severity and inland penetration of the overwash itself.

Figure 10.8 Extensive overwash, Cavendish Spit, Prince Edward Island generated by a storm on 28 December 2004. The spatial pattern of overtopping and erosion appears to reflect the alongshore spacing of rip cell/giant cusps with the dune remaining intact opposite the locations with a wider beach.

substantial length of the barrier (Houser *et al.*, 2008a). Where there are relatively high, wide foredunes, overwash is confined to locations where the dune is vulnerable, usually pre-existing washover channels, points where the foredune is quite narrow due to erosion from earlier storms (Houser, 2012), and at 'hot spots' where locally the beach is quite narrow.

Routine mapping of barrier systems using LiDAR and extraction of DEMs from historical aerial photogrammetry has permitted mapping of pre- and post-storm topography along large sections of barrier coastline and documentation of the response of the coast to intense storms (Stockdon *et al.*, 2002; Leatherman, 2003; Morton and Sallenger, 2003; Robertson *et al.*, 2007; Houser *et al.*, 2008a). Example LiDAR and oblique aerial photographs are provided in Figure 10.9 from Santa Rosa Island in northwest Florida before and after Hurricane Ivan in September 2004. This technology is enabling quantification of the factors controlling the extent and severity of overwash and the development of predictive models of dune survival (Morton, 2002; Stockdon *et al.*, 2007; Claudino-Sales *et al.*, 2008; Houser *et al.*, 2008a; Price *et al.*, 2008). Stockdon *et al.* (2007) tested the model developed by Sallenger (2000) on the impact of Hurricane Bonnie (27 August 1998) and Hurricane Floyd (16 September 1999) on a 50 km stretch of the North Carolina coast. The model was successful in predicting the overwash regime with an accuracy of 85–90 per cent but tended to

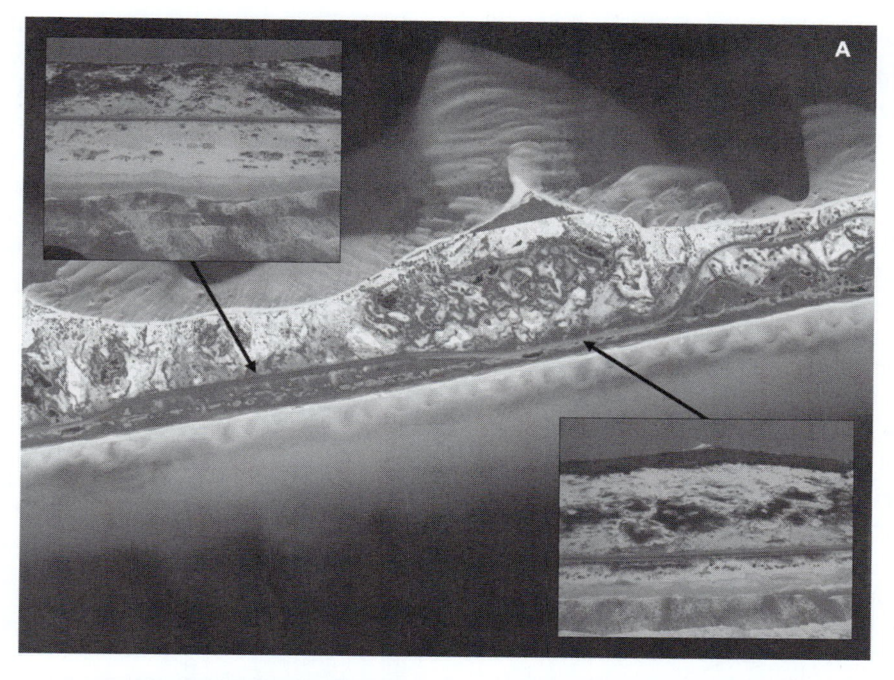

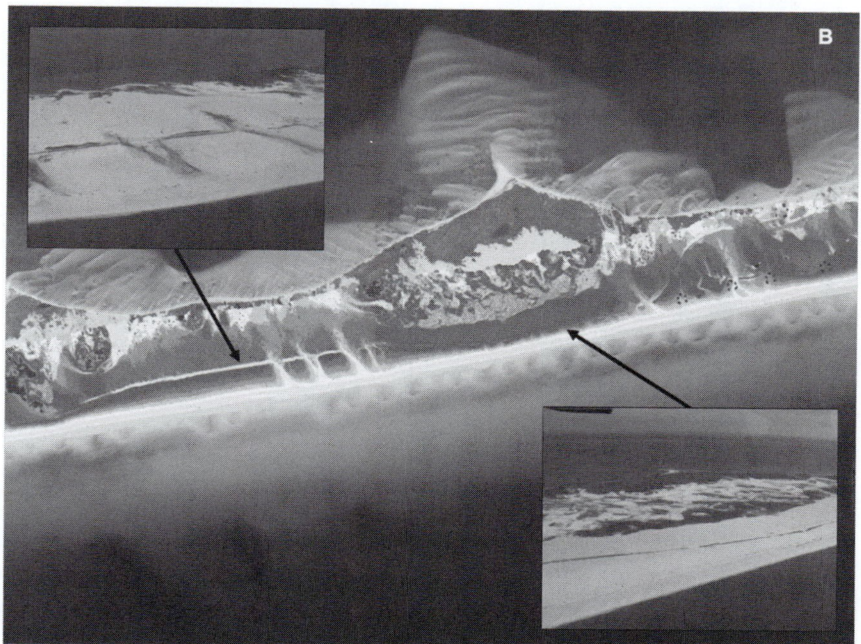

Figure 10.9 LiDAR maps and oblique imagery of Santa Rosa Island near Pensacola, Florida A. before and B. after Hurricane Ivan in September 2004.

overestimate the impact and predicted inundation in many areas where it did not occur. There have been several attempts to develop models to predict the evolution of dune erosion and ultimately overwash (Kriebel and Dean, 1985; Larson *et al.*, 2004), and a variety of routines are now included in SBEACH (Donnelly *et al.*, 2006).

10.3.1.2 *Overwash Processes and Washover Fan Characteristics*

In the early stages of run-up, overwash waves may erode the top of the beach and cliff the embryo dune (Figure 10.10A), and occasional wave overtopping will generally result in the transport of some sediment over the crest with little impact on crest height. At higher storm surge levels or later in a storm, wave overtopping is more frequent and the erosion occurs through the dune ridge, carving a channel and allowing surges of water to run through to the backbarrier flats (Figure 10.10B). Deepening and widening of the channel and gap within the dune line produces a positive feedback, because more surges flow through the channel and the flow is deeper (Houser, 2013). The greater volume and speed of these surges permit the water to flow much further across the backbarrier, perhaps aided by a raised water table or inundation from the lagoon side, and deposition rapidly increases the length and breadth of the fan (Figure 10.10C). An early example of pre- and post-storm surveys across a dune and washover channel (Leatherman *et al.*, 1977) illustrates nicely erosion of the beach and throat of the overwash channel (Figure 10.11). The strength of the overwash depends in part on the width of the barrier, with the hydraulic gradient across the barrier being larger for narrow barriers. If the overwash deposits sediment into the backbarrier lagoon it leads either to extension of the backbarrier shoreline

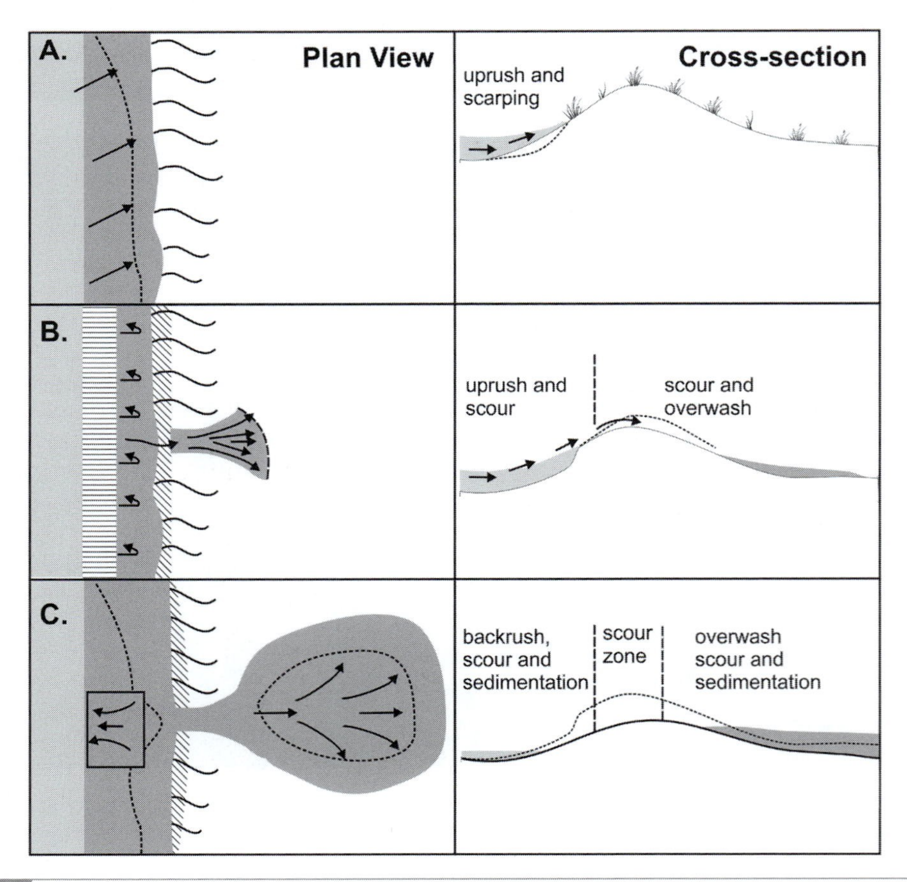

Figure 10.10 Plan and cross-sectional sketches of stages in the development of an isolated overwash fan: A. initial erosion of the backshore and foredune with minor overtopping; B. frequent overtopping leading to scour of the dune forming the inlet throat and initiating fan development at the edge of the backbarrier; C. continued erosion of the overwash throat leading to deepening and widening with deposition on the backbarrier leading to landward and lateral extension of the fan.

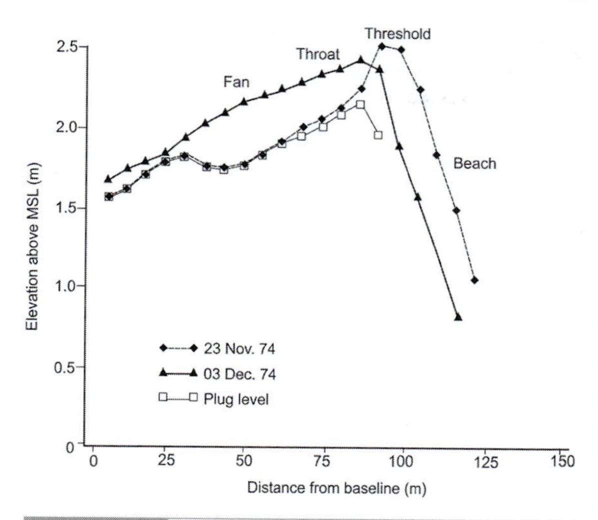

Figure 10.11 Pre- and post-storm survey of a washover throat showing beach erosion and deposition of sediment on the fan (after Leatherman *et al.*, 1977).

Figure 10.12 Overwash deposition past the backbarrier shoreline along Santa Rosa Island near Pensacola, Florida after Hurricane Ivan.

Figure 10.13 A small pit in an overwash fan, Long Point, Lake Erie showing about 40 cm of deposition from the overwash event of 2 December 1985. The sediments near the base initially buried grasses covering the backbarrier surface and are somewhat disturbed, but overlying sediments are characterised by planar bedding on a nearly horizontal surface.

(Houser, 2012; Figure 10.12) or to a net loss of sediment from the island.

Sediment deposited in the overwash fan comes both from erosion of the foredune and upper beach, and in the later stages of the storm it may also come from the lower beach. As a result, extensive overwash can lead to a net loss of sediment from the foredune, thereby limiting recovery of the dune without a new supply of sediment from offshore or alongshore sources (Houser *et al.*, 2015). Much of the material is deposited as thin units of planar bedding dipping gently towards the lagoon. In the throat where flow speeds are higher and erosion dominates sediments are usually inversely graded, often with fine, heavy minerals at the base. On the overwash fans the beds are normally graded (Switzer and Jones, 2008), with coarse sediment at the bottom of each unit and fining upwards (Figure 10.13). The lower units are often disrupted by the effects of flow around or thorough vegetation and hardened structures on the backbarrier surface, but this is reduced as these obstacles become buried. Foreset beds are formed where the fan progrades into ponds or standing water in marshes at the fringes of the barrier (Schwartz, 1975). Overwash fans are often reactivated by subsequent storms because of the absence of a protective dune and thus fans may contain depositional units from more than one overwash event (Davidson-Arnott and Fisher,

1992; Sedgwick and Davis, 2003; Wang and Horowitz, 2007; Switzer and Jones, 2008). The bare sediments on the washover fan provide a source area for aeolian transport, and thus the fan surface is often subject to deflation in the period between overwash events and before the regeneration of vegetation. This exposes and concentrates pebbles from the upper beach that may be dispersed within the general sandy matrix, and acts to armour the surface (Figure 10.14A). The sediment eroded from the fan surface may be trapped by vegetation around the margin of the fan, producing low dunes. On developed barriers, the erosion of structures and roads can lead to extensive debris fields that create a lag that limits aeolian transport and the development of dunes (Houser, 2009a; Figure 10.15).

10.3.1.3 *Washover Healing and the Role of Vegetation*

Although some earlier work on barrier overwash and response to storms had touched on the evolution of overwash fans and barrier island recovery from storms (e.g., Godfrey and Godfrey, 1973) it was the work of Cleary and Hosier (1979) that first clearly identified the concept of inlet and overwash healing. Based on extensive work on North Carolina barriers, both in the field and from the study of historical aerial photographs, they documented the history of overwash produced by tropical and extratropical storms, the form of overwash fans and the subsequent regeneration of vegetation and growth of new dunes. Cleary and Hosier (1979) recognised that the evolution of breaches through the dune line and the washover fans behind them followed recognisable patterns that were controlled by both physical factors, such as waves, aeolian sediment transport and sediment size, as well as biological factors determining the rate of colonisation of the bare surfaces and the distribution of species doing so. Cleary and Hosier (1979) identified two models of cyclic washover and recovery or healing – one related to fine grained washovers and the other to coarse grained washovers. While the models identified an ideal cycle from overwash to stabilisation and dune regeneration, it was recognised that the process of washover

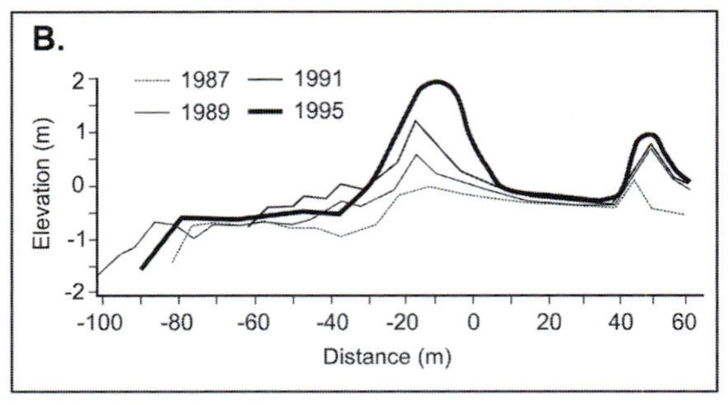

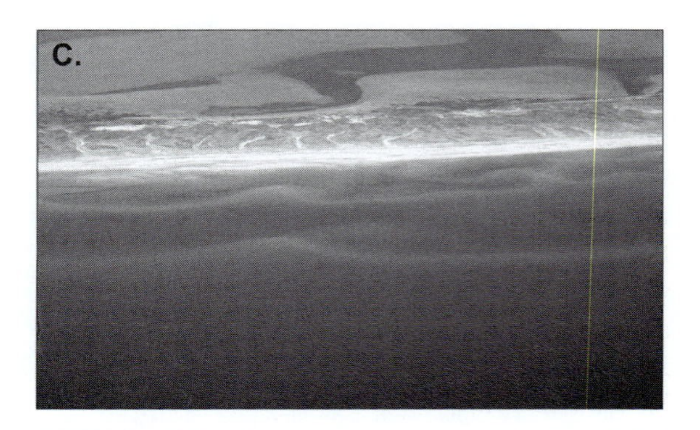

Figure 10.14 Washover healing. A. View looking offshore of the throat of a small washover (the washover on the right side of the photo in Figure 10.17D) showing vegetation establishment at the top of the beach and growth of a foredune across the entrance. There has been some deflation of the fan surface leading to the development of a pebble lag. B. Profiles along the centre line of the washover fan showing growth of the foredune and of a small dune marking the landward edge of the fan. C. Oblique aerial photograph of a former inlet and washover fan surface, North Richibucto Beach, New Brunswick (Figure 10.4). Healing has progressed through the coalescence of small nebkha dunes on the fan surface with washover channels remaining active over much of the healing period because of the low backbeach elevation.

Figure 10.15 A. Oblique photograph of extensive overwash and scouring along Santa Rosa Island after Hurricane Ivan and B. ground photograph of the resulting road debris that has created a lagged surface that limits aeolian transport and dune recovery.

healing could be interrupted at any time by a new storm or storms that would then reset the recovery process. Subsequent work has shown that the process of washover healing is highly varied because of differences in the frequency and magnitude of overwash events, differences in barrier morphology, sediment budget and stability, and differences in the species composition of vegetation on the barrier. Where barriers, or sections of barriers, have a negative budget unvegetated washovers and active washover channels may always be present, and the barrier may be characterised by washovers at various stages of healing (Matias *et al.*, 2008). Following high-magnitude storm, large sections of the barrier may be subject to inundation overwash because the dunes are smaller (Fritz *et al.*, 2007; Froede, 2008; Weymer *et al.*, 2013; Houser *et al.*, 2015), and washover healing will likely take many years with all sections of the barrier passing through the healing stages at roughly the same time (Houser *et al.*, 2018).

The width of the washover throat, the area and landward extent of the fan surface, elevation of the backshore and of the fan, both with respect to the potential for further overwash and proximity to the water table, will all influence the extent of aeolian sand transport and the development of dunes in the backbarrier region. Where the washover is quite narrow and the fan size is small, most aeolian activity will be limited to reworking of sediments on the washover fan and only small amounts of sand will be transferred through the throat from the beach (Figure 10.14A). Pioneer vegetation is quickly re-established from established dunes on either side and trapping of sediment in this quickly leads to the building of a foredune across the entrance to the washover (Figure 10.14A, B) cutting off further sand transport inland. Vegetation gradually becomes established on the washover fan surface and the process of healing may be completed within 6–10 years. Where the washover is much wider and a broad terrace is formed, the degree of aeolian activity as well as further overwash may restrict foredune development. Instead, isolated patches of vegetation become established on the surface forming small vegetated islands or nebkha dunes. As these coalesce, overwash becomes restricted to a few well-defined channels and later these too become infilled (Figure 10.14C). This may take 10–15 years for healing to be completed. At Greenwich Dunes, Prince Edward Island a catastrophic storm in 1923 completely removed the foredune along a 5 km stretch of coast, forming a washover terrace that extended inland 100–400 m (Mathew *et al.*, 2010; Walker *et al.*, 2017). The absence of vegetation permitted large amounts of sand to be transported continuously from the beach across the washover terrace, leading to the formation of large transgressive dunes inland and hindering vegetation establishment. As a result, it took several decades to establish a continuous foredune once more and about 70 years for complete stabilisation of the inland dunes.

The type of vegetation that recolonises a beach following a storm depends on the frequency of overwash (Wolner *et al.*, 2013). Dunes that experience frequent overwash are colonised by maintainer species that have adjusted to frequent disturbance through a shorter lifespan and the production of numerous offspring. The low-profile of maintainer species does not promote the deposition of sediment transferred from the beach to the backshore as rapidly as dune-builder vegetation. As a result, the washover channel does not recover and the frequency of washover increases thereby reinforcing the maintainer species and low-elevation dunes and gaps (Stallins and Parker, 2003; Wolner *et al.*, 2013). Dune building vegetation can develop when and where disturbance is minimised, because they require time for expansion through rhizomes, and for the new seedlings to grow to a size capable of promoting the deposition of sediment and rapid dune growth. The different growth strategies and tolerance to burial creates an eco-geomorphic feedback that reinforces the height, extent and volume of the foredune, with different species giving rise to different dune shapes (Ruggiero *et al.*, 2018). It follows that an increase in the frequency and/or magnitude of storm surge has the potential to promote burial-tolerant species and the development of low dunes and a new equilibrium state (Hosier and Cleary, 1977; Stallins and Parker, 2003; Wolner *et al.*, 2013), though this low-dune state may also be achieved by feedbacks between the growth rate of dunes and storm frequency alone, without the aid of maintainer species (Vinent and Moore, 2015; Moore *et al.*, 2018).

10.3.1.4 *Post-Storm Beach and Dune Recovery*

The time required for the dune to recover to its pre-storm height and volume, or to a height that prevents significant erosion and breaching during the next storm event, depends on the level of impact and can range anywhere from a couple of weeks and months for minor scarping to almost a decade in areas where the dune was completely overwashed (Sallenger, 2000). Results from Galveston Island, Texas (Morton *et al.*, 1994) and Santa Rosa Island, Florida (Houser *et al.*, 2015), suggest that post-storm dune recovery can take up to 10 years following a storm that causes extensive washover. Post-storm recovery of dune to its pre-storm state or some long-term equilibrium height follows a sigmoidal curve first described in the growth model of Verhulst (1838). The sigmoidal curve reflects the slow recovery of the beach and backshore to provide a supply of sediment to be captured by the vegetation that needs to recolonise the eroded area before the dune can recover. This is consistent with the four stages of dune recovery described by Morton *et al.* (1994):

Stage 1 Immediately following a storm the berm begins to recover and the beach face steepens as sediment is returned to the beach face and the beach undergoes gradual accretion as the innermost bar migrates landward and welds to the beach face. As described in Chapter 8, this leads to a steep beach ridge in reflective environments or a low-gradient berm in more dissipative environments. Depending on the severity of the storm, this can take several weeks for minor storms up to an entire year following a major storm that moves the nearshore bars far offshore.

Stage 2 Continued recovery of the beach leads to deposition in the backshore and the lengthening of the available fetch for aeolian transport to the developing dune. This can either occur between storms, when winds tend to be below the transport threshold, or during storms, through the landward migration of subtidal and intertidal bars (Houser and Greenwood, 2005; 2007), the alongshore migration of sandwaves (Davidson-Arnott and van Heyningen, 2003) or in response to lake levels (Saunders and Davidson-Arnott, 1990). As the beach widens, the amount of available dry sediment increases, allowing for transport from the beach to the dune system. Widening of the backshore can take several years following a storm.

Stage 3 The wider fetch allows for aeolian transport across the beach and backshore to the embryo dune when either dune building or maintainer species have been able to colonise.

Since storm winds capable of entraining sediment are usually accompanied by elevated water levels (Ruz and Meur-Ferec, 2004; Delgado-Fernandez and Davidson-Arnott, 2011) and precipitation (Keijsers *et al.*, 2012), it is reasonable to assume that sediment only becomes available to the dunes when the backshore expands (Davidson-Arnott, 1988; Davidson-Arnott and Law, 1990, 1996; Bauer and Davidson-Arnott, 2003; Houser, 2009b). This part of the recovery can take several years.

Stage 4 The final stage of recovery depends on the establishment of vegetation and involves the growth of 'taller, wider, continuous, and more densely vegetated' dunes (Morton *et al.* 1994). Depending on the extent to which the roots and rhizomes are impacted, this stage of recovery can take 2–8 years if the roots and rhizomes are not destroyed during the disturbance (Brodhead and Godfrey, 1979; Houser *et al.*, 2013). The faster the vegetation emerges, the more likely it will remain viable (Maun, 2009) and the dunes can recover faster.

Due to the disparate time scales of erosion by elevated storm surge and post-storm recovery, the resiliency of barrier islands is dependent on changes in the frequency of storm events. A rapid succession of (even relatively weak) storms may lead to widespread erosion and washover that can leave an island especially vulnerable to further overtopping for decades (Houser and Hamilton, 2009). This is because the dune remains small and potentially discontinuous and the threshold for storm surge to exceed the dune crest elevation is lower, which increases the potential for washover during subsequent storms and a loss of the recovered sediment to be moved landward and made unavailable for further recovery without a new supply of sediment from offshore or alongshore. If, however, most of the sediment remains in the nearshore or in a recently recovered beach profile, the recovery will be delayed and it is possible for the dune to recover to its pre-storm height and volume at or near its original position (Houser *et al.*, 2015, 2018). Understanding the response and recovery of barrier islands to storm activity is important to our ability to predict the impact of sea level rise on them.

10.3.2 Tidal Inlets

Tidal inlets are openings that separate barrier islands, portions of a barrier spit, baymouth or mid-bay barrier, or lie between one end of a barrier and the mainland coast. Inlets provide a connection between the sea and the lagoon, bay or estuary landward of the barrier and therefore allow for the exchange of water between the two water bodies (Figure 10.19). This exchange is driven primarily by the rise and fall of the tides, with water flowing into the lagoon during the flood tide and out during the ebb (Seabergh, 2006). Where there is substantial freshwater input into the lagoon, the volume of the ebb flow may be substantially enhanced over the flood. Inlets located opposite estuaries where the flow is directed by a channel incised in bedrock may be highly stable in location (Figure 10.19A), while others on barriers with a longshore parallel lagoon may migrate rapidly downdrift or close off when a competing inlet opens. On some microtidal coasts, where barriers close off a river mouth, inlets may only open seasonally when flooding from the river breaches the barrier or sporadically because of an intense storm (Ranasinghe and Pattiaratchi, 1999). Many inlets provide an entrance to a safe harbour for recreational and fishing boats and in some cases for commercial shipping. Often this requires dredging the inlet channel and the entrance to it and installing jetties on one or both sides of the entrance to keep the dredged channel from silting up (Seabergh, 2006).

10.3.2.1 *Tidal Prism and Flows Through Inlets*

There is a continuous movement of sand into the inlet entrance by waves and littoral currents, and the influx of sediment has the potential to choke the inlet over time and close it off. This is opposed by the tidal currents into and out of the inlet opening, and the ability of these currents to keep the inlet open then depends on the volume and associated velocity of flow through

Box 10.1 Overwash and Lake Level Cycles

Long-term water level fluctuations in the Great Lakes over years to decades impose a cyclical pattern on the timing of dune development and overwash activity (Olson, 1958; Davidson-Arnott and Fisher, 1992; Bray and Carter, 1992). In Lake Erie, the range of lake level over several decades is on the order of 1.5 m, with high lake levels being experienced in 1952–3, 1972–3, 1985–6 and 1997–8 (Figure 10.16A). When intense storms occur around the periods of high lake level they produce extensive overwash on barriers around the lake and in the other Great Lakes. These washovers are frequently reactivated by less intense storms over the 2–4-year period that is characteristic of the high-water phase. The susceptibility to overwash is also influenced by the local littoral sediment budget through its control on beach width and dune volume.

A study of overwash on Long Point using historical aerial photographs and field measurements during and after the 1985–6 high water period (Davidson-Arnott and Fisher, 1992) illustrates the temporal and spatial patterns of overwash and washover healing. The long-term littoral sediment budget is negative in the proximal and central zones but positive at the distal end which is characterised by wide, progradational beaches (Figure 10.17). One or more intense storms occurred during each of the first three phases noted above and this is reflected in the high proportion of the spit occupied by active washover fans visible on historical aerial photographs and oblique photography taken each year from 1985 to 1988. While overwash occurred over 30–50 per cent of the shoreline in the proximal end in each of the three high water phases, it occupied less than 5 per cent of the 10 km stretch of the distal end. In 1985, overwash at the distal end was confined to a single location at the downdrift end of a longshore sandwave where the beach was very narrow (Figure 10.17D). Almost all the hover formed in the proximal and central zones were reactivated during moderate storms in 1986 and 1987 but the wider beach resulting from longshore sandwave migration protected the two washovers at the distal end (Figure 10.17E).

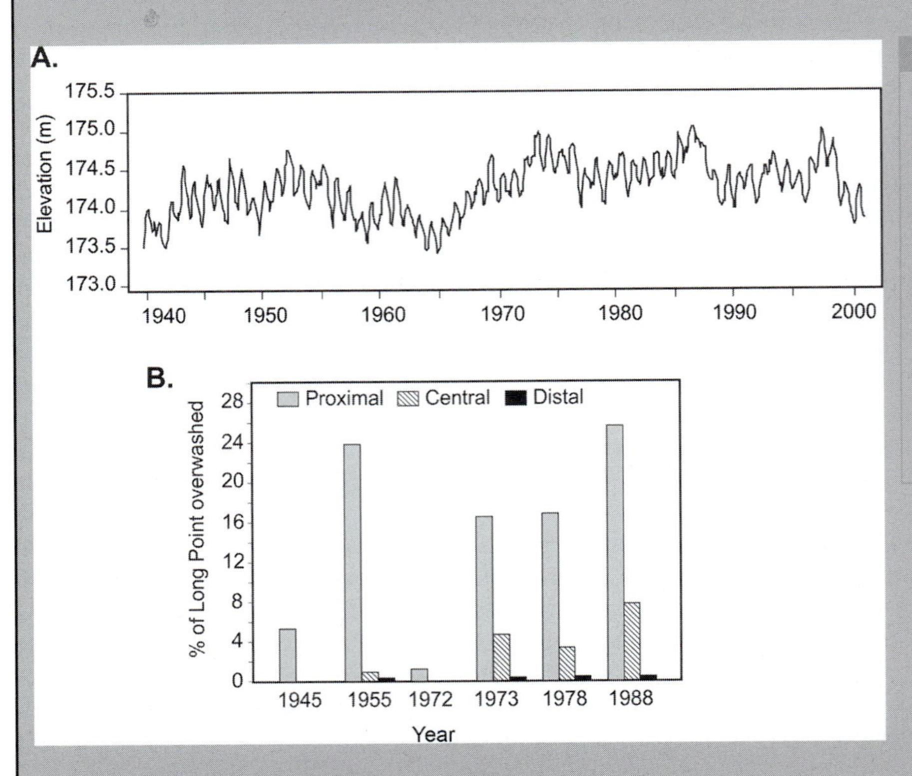

Figure 10.16 Washovers on Long Point in relation to lake level fluctuations in Lake Erie: A. mean monthly water level in Lake Erie 1950–2000; B. proportion of each zone occupied by washovers based on analysis of historical aerial photographs (after Davidson-Arnott and Fisher, 1992).

Box 10.1 (cont.)

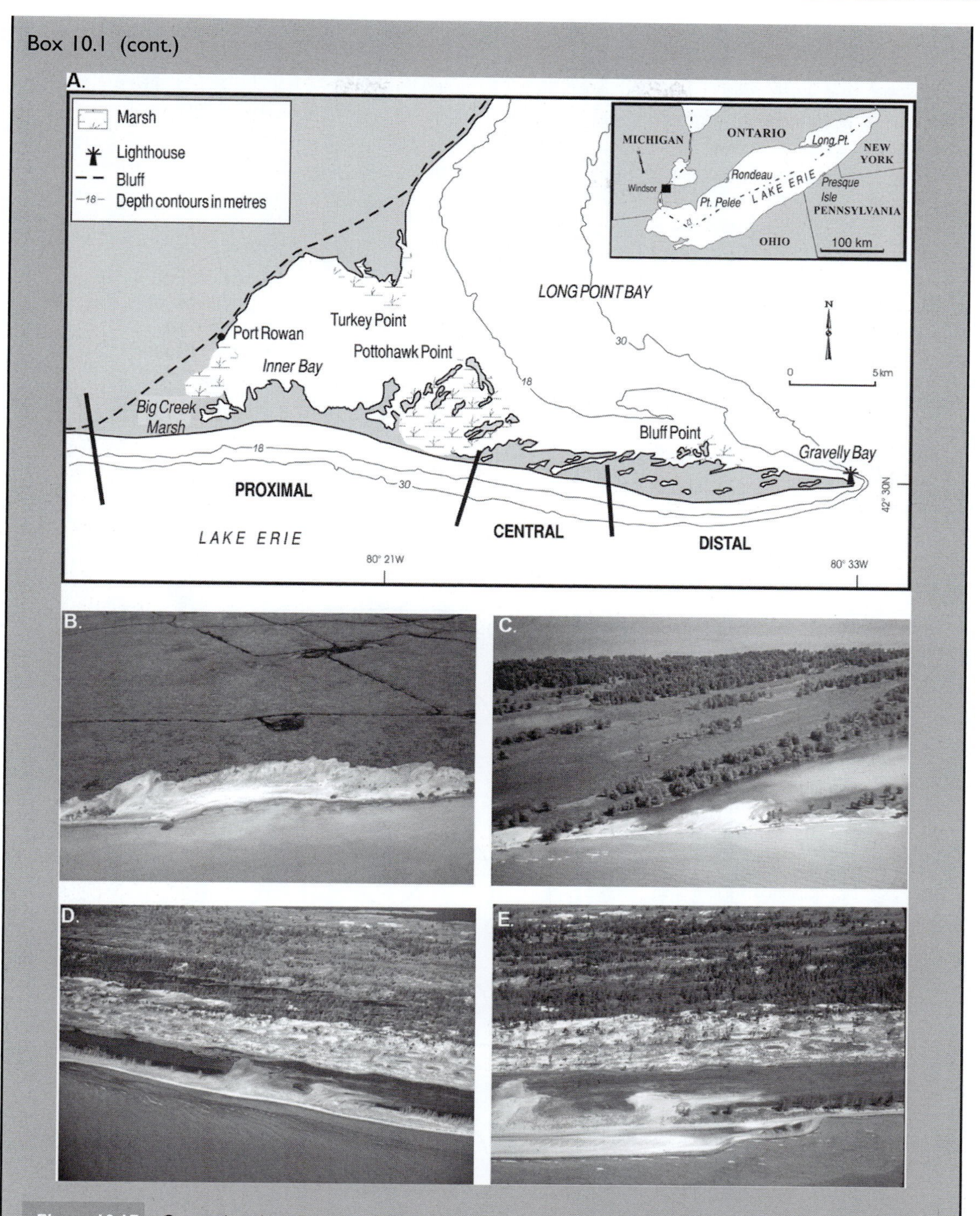

Figure 10.17 Overwash on Long Point, Lake Erie resulting from a storm on 2 December 1985. A. A map of Long Point spit showing the division into proximal, central and distal zones. B. Photograph of inundation overwash into marshes at the proximal end of Long Point spit following the 2 December 1985 storm. C. Overwash into ponds between old dune recurves in the central section. D. Oblique aerial photograph taken in June 1986 showing the formation of two small washover fans at the distal end. Note the narrow beach and the presence of a nearshore bar which is attached to the downdrift end of a longshore sandwave just off the picture to the left. E. Photograph taken 5 June 1987 showing migration of a longshore sandwave in front of the overwash areas. The wide beach associated with the sandwave prevented reactivation of these two fans during a severe storm on 15 December 1987.

Box 10.1 (cont.)

Following the approach of Cleary and Hosier (1979), Davidson-Arnott and Fisher (1992) produced a schematic model relating of overwash occurrence and washover healing to long-term lake level cycles (Figure 10.18). There were no major storm events during the succeeding high water phase of 1998 and as a result the healing cycle was completed over all sections of the spit. Similar temporal patterns can be associated with El Niño events on the west coast of the US, and with shifts in the tracks of major hurricanes and extra tropical storms affecting the East and Gulf coast barriers of North America (Forbes *et al.*, 2004).

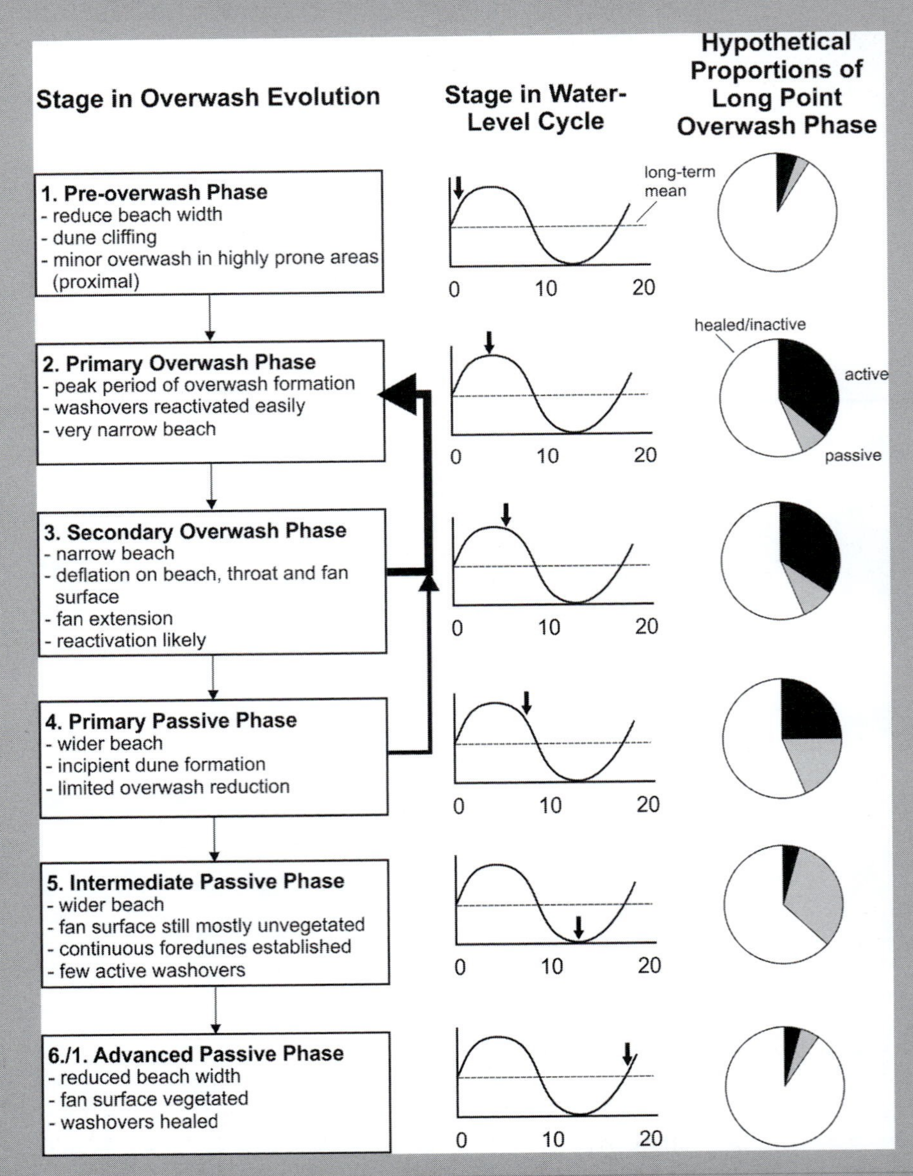

Figure 10.18 Schematic model of overwash occurrence and washover healing in relation to long-term lake level cycles on the Great Lakes (Davidson-Arnott and Fisher, 1992). Overwash potential is highest during the high water phase and the potential for washover reactivation decreases during the healing phases and water level declines.

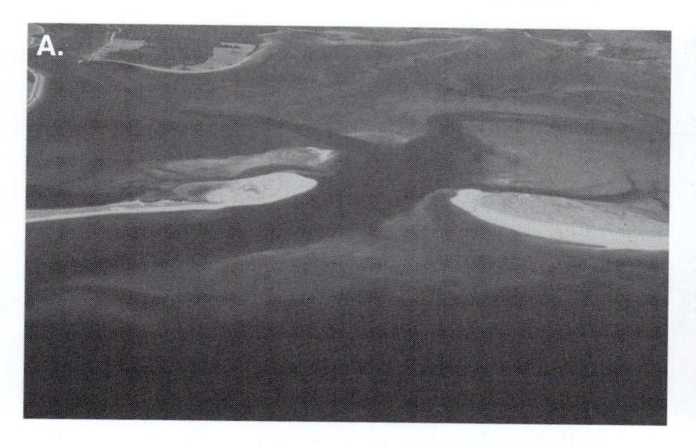

Figure 10.19 Tidal inlets. A. Little Gulley inlet separating the North and South Kouchibouguac barrier islands (see Figure 10.4). The inlet location is highly stable because of its location opposite the estuary with the drowned river channel cut in sandstone bedrock. B. North Rustico inlet on the north shore of Prince Edward Island. Like many of the small inlets through barriers on the shorelines of the Gulf of St Lawrence, the inlet entrance is stabilised by jetties to protect small harbours for fishing boats. C. North Inlet, a temporary inlet through the North Kouchibouguac barrier which was breached in 1970 and closed by 1976.

the channel. Ignoring any freshwater inflow from rivers, the volume of water flowing into the inlet on the flood tide generally equals the volume flowing out on the ebb and is termed the tidal prism. The tidal prism is determined by the area of the lagoon, estuary or bay draining through the inlet and the tidal range:

$$Q_p = AR, \tag{10.1}$$

where: Q_p is the tidal prism, A is the area of the lagoon draining to the inlet and R is the tidal range. Determining a value for A is complicated because the basin sides are often very gently sloping, though an approximation can be obtained using the perimeter outlined by the mid-tide line. It will likely be complicated further where tidal conditions extend inland for some distance along estuaries, because the tidal range will decrease with distance from the inlet entrance. Finally, where there is more than one inlet entrance opposite the backbarrier lagoon, the ebb and flood flows may not be equal, and demarcating the drainage divide between inlets may be difficult. An examination of the map in Figure 10.4 will show that these complexities apply to the tidal prism of Little Gulley (Figure 10.19A).

Inlet morphology reflects the interaction between flows through the inlet entrance, sand transport by waves and the evolution of the two ends of the barriers on either side of the entrance. The stability of an individual inlet depends on the relative magnitude of flows through the inlet to wave energy and littoral

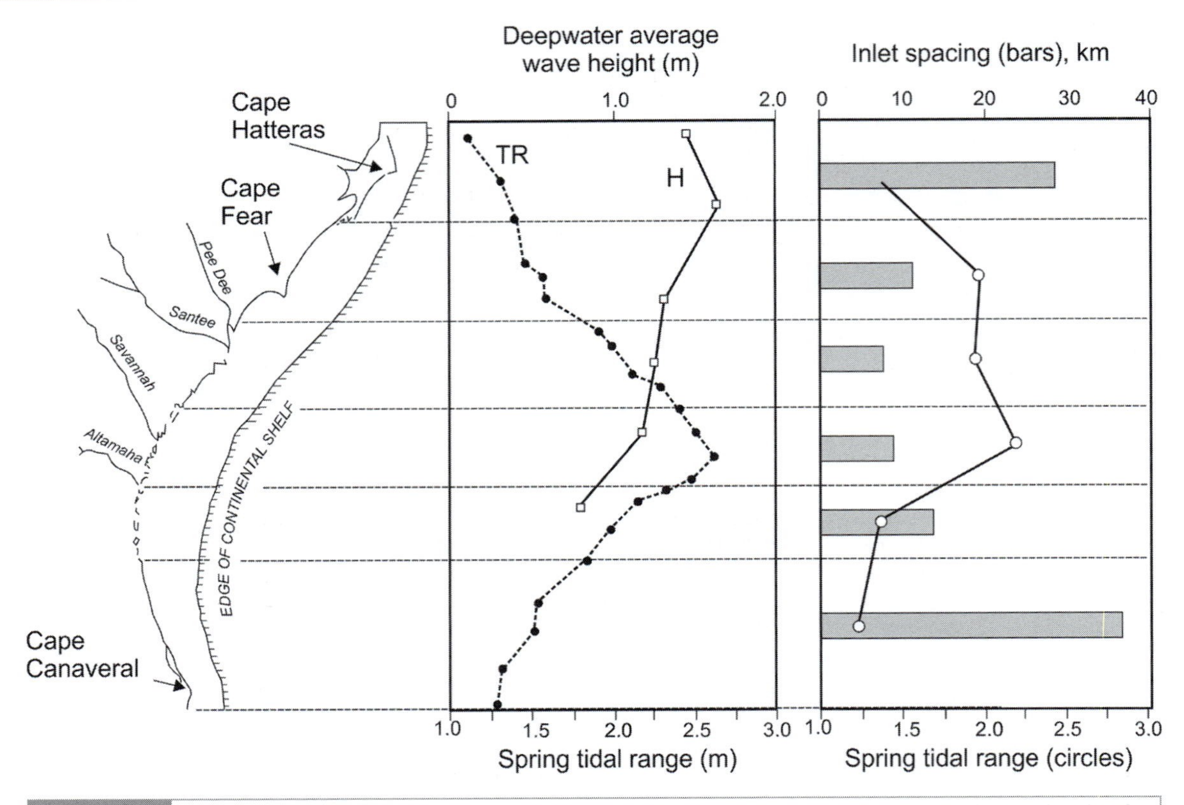

Figure 10.20 Patterns in the number and spacing of tidal inlets as a function of wave versus tidal energy along six lengths of shoreline in the Georgia Bight, USA (from FitzGerald, 1996).

sediment transport. Where wave energy and littoral transport are large, more sediment is brought to the inlet entrance and greater flows, and hence a larger tidal prism, are required to flush sediment through and to keep it open. We should expect, therefore, to find that the spacing between inlets increases with increasing wave energy, and decreases with increasing tidal range. This has been demonstrated for barrier island chains in several areas (Hayes, 1979; FitzGerald, 1996). Where chains of barrier islands extend over hundreds of kilometres it is possible to demonstrate the relationship between inlet spacing and the two controlling variables, as has been shown for the Georgia Bight (FitzGerald, 1996; see Figure 10.20). The spacing of inlets and their form in relation to tidal versus wave energy also influences the form of the barrier islands (Hayes, 1979). Barriers on microtidal coasts tend to be long and thin, with little variation in width along the length of the barrier (Figure 10.21A). Padre Island along the southern Texas coast is in a microtidal low-wave energy environment and is the largest barrier island in the world. It used to extend for several hundred kilometres before the Mansfield Channel was dredged to allow for navigation. In contrast, inlets on mesotidal coasts tend to be relatively short, narrow in the centre, with wider sections near the inlets due to progradation (Figure 10.21B). Being fond of food, Hayes (1979) labelled microtidal barriers hot dogs and mesotidal barriers drumsticks, and these terms seem to have stuck. While barrier form is also influenced by other factors, including sea level history, sediment abundance and the area of the lagoon behind the barrier, tidal range and wave energy are the most important.

The major feature of an inlet is a rectangular channel in which the cross section is adjusted to the tidal prism that flows through it each tidal

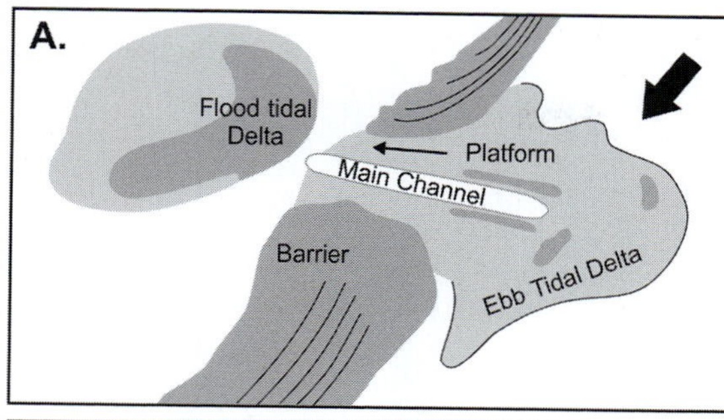

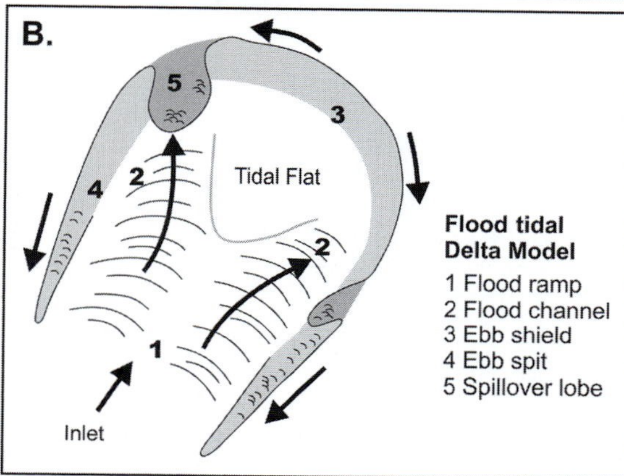

Figure 10.21 Schematic models of ebb and flood tidal deltas. Arrows indicate dominant direction of tidal currents: A. ebb tidal delta with terminal lobe and swash bars that are exposed at low tide; B. flood tidal delta showing bifurcation of inlet channels across the flood ramp (after Hayes, 1980).

cycle. Flow speeds are highest in the channel between the islands and decreases as the flow begins to spread out, both on the lagoon side and on the ocean side, leading to sediment deposition in those regions. The channel is often scoured of fine sand, leaving a lag of coarse sand and gravels, often mixed with clam and mussel shells. Sand is deposited at either end of the inlet entrance because of the increase in channel cross section, decreased flow speed and, on the ocean side, the interaction of the ebb flow with waves. The result is the creation of two large depositional sand bodies – an ebb tidal delta on the seaward side of the inlet and a flood tidal delta on the landward side (Figure 10.21). Details of the morphology, channel patterns and size of the ebb and flood tidal deltas vary from inlet to inlet but there are several characteristics that are common to most or all (Hayes, 1980; Fitzgerald, 1996).

The primary feature of the ebb tidal delta (Figure 10.22A) is the terminal lobe that extends seaward from the end of the main ebb channel and ideally has a curved form with a broad shallow platform, and several shoals or swash bars that are often emergent at spring low tide. The terminal lobe is built out over the barrier platform and extends some distance offshore beyond the beach, and it has one or more channels across it. The speed of the ebb flow tends to be greatest close to low tide and continues while the tide is already rising on the ocean coast. As a result, the flood tidal waters are initially confined to marginal flood channels on either side of the main channel until water levels on the ocean side exceed those in the lagoon and flow into the lagoon is established over the entire inlet. The model shown in Figure 10.22A is typical of the mesotidal New England coast characterised by comparatively high wave energy conditions. Hayes (1980) suggests that the ebb tidal deltas become more elongate and more ebb-dominated in South Carolina because of reduced wave energy and perhaps larger tidal prism while ebb tidal deltas on the microtidal Gulf coast are much less developed. Where the tidal prism is large, because of an extensive estuary behind the inlet for example, both the ebb and flood tidal deltas may be large, though most of the area will be subtidal (Reinson, 1977).

Where the net littoral sediment transport is modest relative to tidal current, discharge through the inlet, the terminal lobe is symmetric (Figure 10.22A), but where there is a relatively large transport in one direction the ebb flow is deflected downdrift. This causes the swash platform to extend parallel to the coast, with the channel between it and the end of the downdrift barrier. Sediment bypassing along the extended ebb tidal delta lobe results in the downdrift barrier becoming thinner at this location and increases the potential for overwash. At the same time, the extended alongshore flow of the ebb tide becomes more and more inefficient and there is a tendency for sediment deposition to block the end of the channel. When conditions

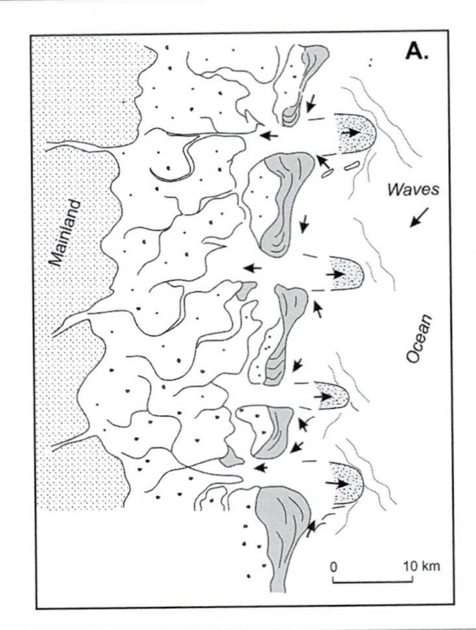

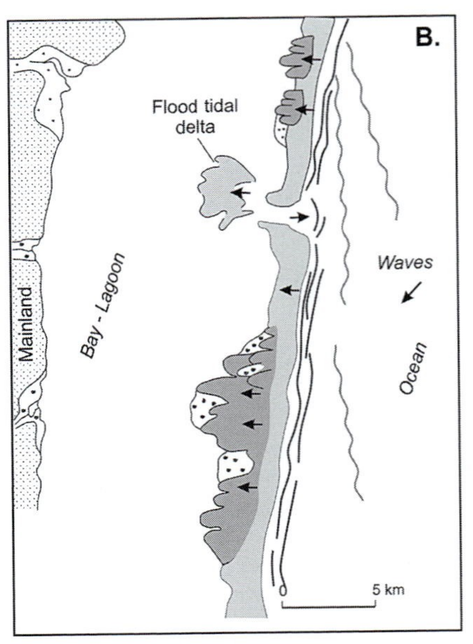

Figure 10.22 The general form of barrier islands as a function of tidal range: A. microtidal coasts characterised by long, narrow hotdog-shaped barriers; B. mesotidal coasts characterised by short, bulbous drumstick-shaped barriers (after Hayes, 1979).

are favourable, such as spring tides with offshore winds and low waves for example, the ebb tide flow may break through the ebb tidal delta opposite the inlet entrance, thus restoring the shortest route seaward. The extended downdrift portion of the delta lobe may then be driven onshore, widening the beach and barrier at the updrift end and helping to stabilise it. These breaches of the ebb tidal delta occur episodically every few years to decades.

The flood tidal delta is like the ebb tidal delta in that it forms because of flow expansion and decrease in speed as the flood tidal currents spread out into the lagoon (see Figures 10.18A, 10.21B). The flood ramp consists of lobes of sediment divided by bifurcation of the flood channels often with quite complex lobate shoals. Ebb flows tend to be concentrated in channels around the margin and along the landward sides of the barrier. The effects of wave action are generally much smaller here than on the ebb tidal delta.

10.3.2.2 *Inlet Formation, Stability and Evolution*

Inlets and the adjacent ends of barriers are dynamic features, and they respond relatively rapidly to changing sediment supply, storm events and wave regime as well as to factors that control the tidal prism. Inlets may form as downdrift extension of a spit building across a bay narrows the gap between the distal end of the spit and the downdrift coast. However, many inlets form because of the breaching of a barrier system during a storm. As we have seen in the previous section, overwash and breaching of the foredune is common on many barriers during intense storms. However, on the backbarrier the pulses of swash bore transport result in deposition and the building of a fan rather than erosion of a channel. Instead, inlet formation seems to occur primarily because of a coincidence of conditions that favour the generation of unidirectional flow from the lagoon side near the end of a storm event (Pierce, 1970; Greenwood and Keay, 1979). These conditions likely involve the generation of a large storm surge, which elevates the water levels in the lagoon. A switch of winds from onshore to offshore, coincident with a falling tide on the ocean coast, then produces a situation where the water level is much higher on the lagoon side and currents flow across a low-lying area driven by winds and the hydraulic

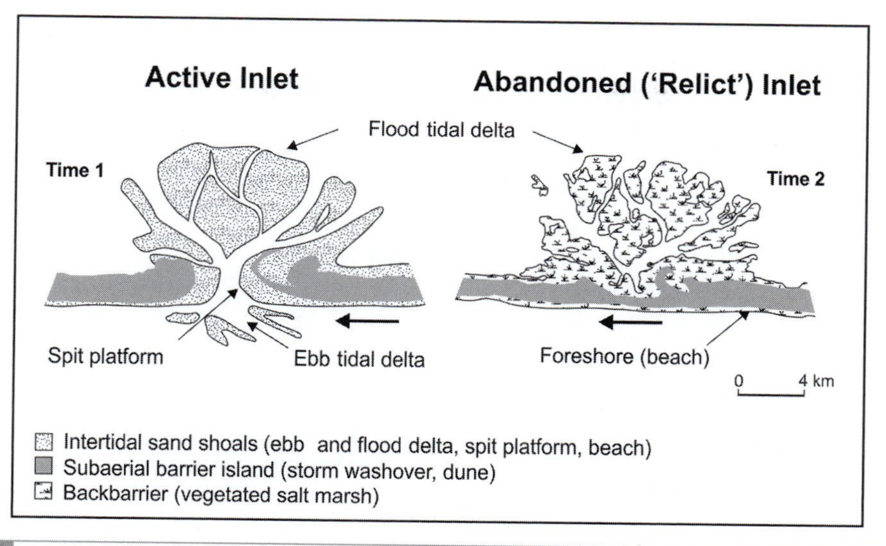

Figure 10.23 Morphologic evolution of a wave-dominated inlet following inlet closure. The inlet is closed through downdrift migration of the updrift barrier and gradual choking of the inlet entrance by swash bars. Flood tidal delta shoals become vegetated by saltmarsh leaving them as islands surrounded by old tidal channels. Arrows indicate the direction of longshore sediment transport (after Moslow and Tye, 1985).

head. These quickly scour a channel through the sand below mean sea level, and the subsequent tidal current flows erode a deeper, wider channel. The long-term stability of these inlets then depends on the drainage area of the lagoon that the inlet can capture. If the drainage area is too small to be stable, then the inlet will close over a few years to decades (Greenwood and Keay, 1979; Ashley, 1987). Once the inlet is closed, the ebb tidal delta is reworked by wave action and disappears but the flood tidal delta is preserved in the lagoon, with tidal channels meandering between saltmarshes occupying the former shoals (Moslow and Tye, 1985; see Figure 10.23). Examples of former ebb tidal deltas can be seen in the lagoon in on the left of the photo in Figure 10.5.

There have been extensive studies of tidal inlet geometry and stability, related to design considerations for stabilisation of harbour entrances (O'Brien, 1969; Fitzgerald and Fitzgerald, 1977; Bruun, 1978; Pacheco *et al.*, 2008), and for the transport of sediment (Fiechter *et al.*, 2006). Inlets adjust their cross-sectional area over time in response to changes in tidal prism. The intertidal area in the lagoon may be reduced by sediment deposition, both naturally over a long time and through human intervention in the form of infilling and dyking (Kragtwijk *et al.*, 2004; Bertin *et al.*, 2005). The resulting reduction in flow volume and speed through the inlet leads to infilling of the channel and reduction in the cross-sectional area until a new equilibrium is restored. Changes in the tidal prism may also result from the opening and closing of inlets in a multi-inlet system. These changes may occur naturally because of storm impacts, as was the case in the opening of North Inlet in 1970 (Greenwood and Keay, 1979). They may be the result of human intervention such as the large-scale effects associated with the closure of the Zuiderzee in 1932 (Kragtwijk *et al.*, 2004), or the more limited impacts of artificial opening of small inlets such as one through the Ria Formosa barrier system (Vila-Concejo *et al.*, 2003). Where there is a loss of some of the tidal prism the inlet may be reduced in size or may silt up completely. Where there is an increase in the tidal prism the inlet may become completely unstable and migration of the inlet position may be initiated because of changes to flows within the lagoon and location of the inlet watershed. Forbes and Solomon (1999) describe the effects of artificial closure of an inlet through a barrier island on Rustico Bay on the north shore of Prince Edward Island that provides a compelling cautionary tale of unexpected consequences on the stability and operation of the remaining inlet.

Inlets for estuaries that extend inland for some distance and control the flow of a significant portion of the tidal prism, tend to be relatively stable in position through time. The same is often true for inlets located between a headland or side of a large bay and the downdrift end of a spit. However, inlets may migrate downdrift on barriers where there is a strong net littoral drift in one direction, and where there are no features that tend to fix their position. Sediment entering the updrift side of the inlet entrance tends to build outward into the channel, constricting flow and leading to erosion of the end of the adjacent barrier. The result is to produce a depositional unit that is as thick as the depth of the inlet channel with the major bedding planes dipping in the direction of inlet migration (Kumar and Sanders, 1974; Hayes, 1980). The base of the unit is formed in the lag gravels and shells that make up the floor of the inlet channel fining upward through cross-bedded sand units and later capped by aeolian dune deposits.

10.3.3 Lagoons

Coastal lagoons are an important component of the barrier system. They are particularly rich ecologically because of nutrients brought into the system from both the land and the ocean and the production of organic matter in the lagoon (e.g., in beds of eel grass – *Zostera marinera* – and fringing saltmarshes). Lagoon beaches, sand flats and mudflats provide food for migrating birds and the shallow waters are nurseries for many fish and shellfish. We can distinguish lagoons from estuaries based on form and alignment, though there is some overlap between the two in all definitions. Estuaries extend inland for a considerable distance along a drowned river valley and are often oriented perpendicular to the coast, while lagoons generally have their long axis parallel to the coast and to the barrier. Estuaries are also distinguished by having a substantial freshwater input with mixing occurring downstream from the tidal limit. Lagoons too can have a considerable freshwater input, but it is not a condition of the definition. In many cases the tidal water bodies behind major barrier chains include both lagoons and estuaries, as is evident from Figure 10.4. It should also be recognised that similar lagoons exist behind coral reefs.

Kjerfve (1986) distinguished between three types of lagoons based on the extent of water exchange between the lagoon and the ocean:

1. **Choked lagoons** are found along microtidal coasts where there is moderate to high wave energy and considerable sediment availability. They usually have only a single inlet and the inlet may be closed seasonally (Ranasinghe and Pattiaratchi, 1999). As a result, wind forcing is dominant and residence times may be weeks or months.
2. **Restricted lagoons** usually have two or more entrances. Circulation is controlled both by tidal currents and wind forcing and they are vertically mixed.
3. **Leaky lagoons** may stretch along the coast for tens of kilometres and have multiple inlets with relatively unimpeded exchange of water with the ocean.

The waters of restricted and leaky lagoons usually range from brackish close to freshwater inputs to salinity close to that of the ocean, but hypersaline conditions may occur in choked lagoons in arid and semi-arid regions (Kjerfve and Magill, 1989).

Lagoons may expand in size and grow deeper under rising sea level, but they are primarily areas of net deposition and therefore, their evolution is controlled by both rates of sediment infilling and sea level changes. Whether lagoons expand or contract depends on the balance between the rate of net sediment input and the rate of sea level rise (Nichols, 1989; Cooper, 1994; Morton *et al.*, 2000). The main sources and pathways of sediment input to coastal lagoons are:

1. **Littoral sediments** – sand and gravel supplied from offshore and alongshore. These can reach the lagoon directly through transport by tidal

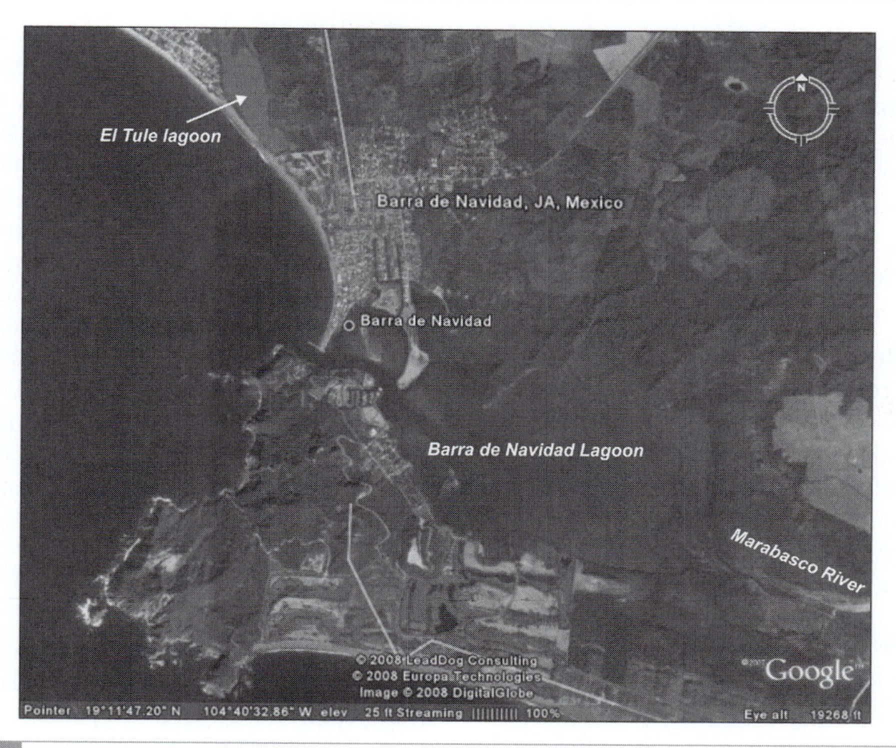

Figure 10.24 Portion of a Landsat image of Jalisco, Mexico showing the Barra de Navidad lagoon and surrounding area. The delta of the Marabasco River has expanded rapidly as a result of increased sediment input from the surrounding hills. The El Tule lagoon near the top left of the picture is an example of a choked lagoon that is only open seasonally during the wet season.

currents into the inlet and deposition on the flood tidal delta or transport further into the lagoon by waves and currents. Littoral sediments may also reach the ocean side of the lagoon through overwash, aeolian processes and through temporary inlets.

2. **Fluvial sediments** – fine to coarse sediments brought into the landward side of the lagoon primarily by rivers draining the uplands surrounding the lagoon and ephemeral streams acting on coastal bluffs. On relatively steep coasts the coarse sediment may be deposited directly in the lagoon, with fines being dispersed throughout the lagoon and possibly some through the inlet to the ocean. On gently sloping coasts, much of the coarse material may be trapped some distance inland at the tidal limit on shallow estuaries and thus only silts and clays reach the lagoon proper.
3. **Chemical and organic precipitates** – carbonates primarily from shellfish growing within the subtidal and intertidal zones of the lagoon can be a significant input to the system and organic matter accumulation in saltmarshes and seagrass beds may also be significant in some lagoons.

Field studies of lagoons of the coastal plain of the USA suggest that rates of infilling are either roughly in equilibrium with sea level rise or not keeping pace with it (Nichols, 1989; Morton *et al.*, 2000), but sedimentation in areas with steeper coasts, such as the west coast of Mexico (Moore and Slinn, 1984) and the east coast of South Africa, often exceeds the rate of sea level rise (e.g., Cooper, 1994). An example of this is provided by the Barra de Navidad, a choked lagoon in Jalisco, Mexico (Holland, 2005). This has an area of 3.6 km^2 with a single outlet and a large sediment input from drainage of the highlands behind it (Figure 10.24). Rates of fluvial sedimentation have been increased due to agricultural and logging activities in the surrounding river catchments and by clearing for urban and tourism development so that the lagoon now appears to have a net positive sediment budget (Holland, 2005; Méndez Linares *et al.*, 2007).

10.4 Barrier Spit Morphodynamics

Barrier systems are dynamic features that evolve rapidly in response to the processes controlling sediment supply, and to change in controlling variables such as sea level, wind climate and evolution of the mainland coast (Figure 10.25). As we saw in Section 10.3, morphodynamic processes, such as overwash, inlet formation, inlet migration and washover healing, can result in significant changes to barrier systems over periods of years and decades. In response, many barrier systems, and particularly barrier spits and barrier islands, evolve rapidly and have a life span that may be only a few decades to a few thousand years. There is now considerable information on the sedimentary architecture and evolution of spits and other barriers from excavation of features preserved in former glacial and pluvial lakes (Gilbert, 1890; Jewel, 2007), or in areas of rapid isostatic uplift (Nielsen *et al.*, 1988; Mäkinen and Räsänen (2003). Extensive coring through modern features worldwide has provided many insights into barrier evolution and particularly the effects of the Holocene transgression (Héquette and Ruz, 1991; Soons *et al.*, 1997; Anthony and Blivi, 1999; Otvos and Giardino, 2004; Harvey, 2006; Long *et al.*, 2006; Tomazelli and Dillenberg, 2007; Storms *et al.*, 2008), and the vertical stratigraphy of the core data may be extended through the use of seismic profiling and ground penetrating radar (van Heteren *et al.*, 1996; Jol *et al.*, 2003; Novak and Pedersen, 2006; Weymer *et al.*, 2013; Wernette *et al.*, 2018).

10.4.1 Barrier Spit Progradation and Recurve Development

A clear feature of all spits is the existence of a significant volume of net longshore sediment transport from updrift of the point of attachment to the mainland, to the downdrift or distal end of the spit. The transport direction reflects local coastal orientation with respect to the wind and wave climate of the region, and there must be a

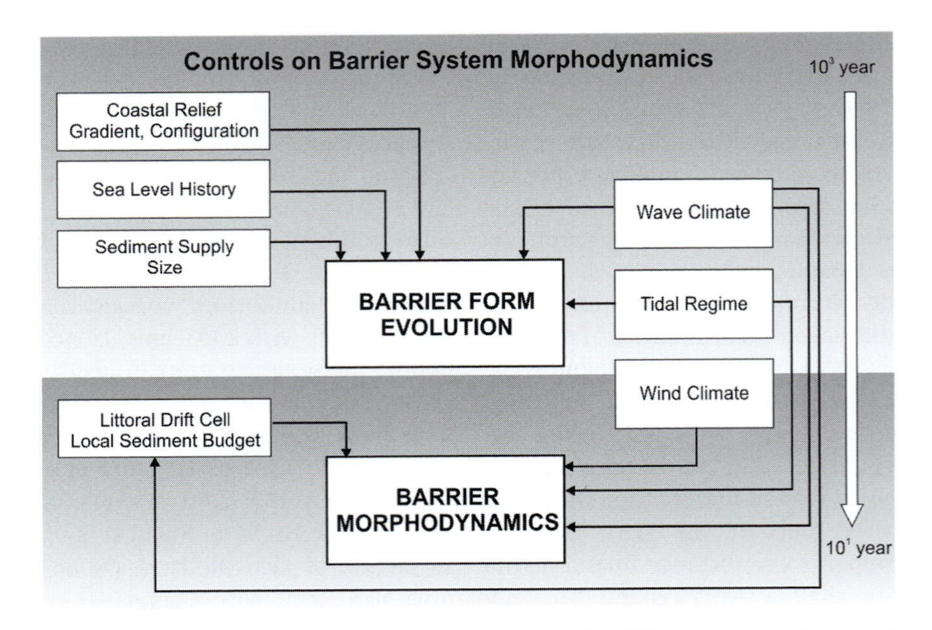

Figure 10.25 Schematic model of the factors controlling barrier system morphodynamics and evolution and the interactions between them.

substantial source of littoral sediments from coastal erosion or fluvial sediment inputs updrift to provide for the initiation and continued evolution of the spit. Sand and gravel are transported along the ocean (or lake) shore of the spit by waves and currents within the nearshore and swash zones. There is usually a decreasing gradient of sediment transport towards the distal end (discussed further in the next section), and this is reflected by deposition leading to wide beaches, seaward progradation, and the development of a succession of fore dune ridges and intervening swales or ponds (Figure 10.26A, B). The distal end of the spit is the zone where most or all the active increase in spit size occurs through progradation normal to the shoreline, and through extension of the spit and the spit platform (Meistrell, 1972) in the direction of net sediment transport.

At the terminus, sand in the nearshore is transported into increasingly deeper water leading to deposition and extension of the spit platform (Figure 10.26B). Refraction of the dominant waves over and around the platform results in the development of a curvature of the shoreline and transport of sediment towards the bay. Littoral sediment transport may be slowed further, and the curvature of the shoreline enhanced by the effects of waves from the direction opposite to the direction of net littoral drift, sometimes producing an exaggerated 'hook' shape to the end of the spit (Figure 10.26B). The shallow platform, and periodic downdrift extension by sandwaves and swash bars (Davidson-Arnott and van Heyningen, 2003; Park and Wells, 2007), often results in the enclosure of a large pond or runnel and the development of a new foredune ridge some distance from the next oldest one (Figure 10.26C). The result is to produce a series of elongate ponds facing the bay which over time become infilled through marsh development and the reworking of material by waves generated within the protected bay (Davidson-Arnott and Conliffe Reid, 1994).

10.4.2 Sediment Budget and Spit Evolution

The overall growth of the depositional body forming the spit and its platform requires a net input of sediment from updrift and thus the overall sediment budget exercises an important control on both the size of the spit and its rate of growth. As the spit develops, sediment budgets can be constructed for sub-units of the spit based either on spatial location (e.g., the proximal and distal ends of the spit) or on sub-environments such as the platform, beach, dunes and backbarrier marshes. The evolution of a simple spit can be modelled theoretically by simulating wave refraction and the longshore sediment transport gradient (Petersen *et al.*, 2008). Similar modelling can be carried out for examples based on bathymetry and wave climate (Davidson-Arnott and Conliffe Reid, 1994; Park and Wells, 2007; Allard *et al.*, 2008). Semi-quantitative estimates of historical sediment budgets can be quantified from reconstruction of the evolution of the plan form of the spit from historic maps and aerial photographs along with some data on the thickness of the sedimentary unit from borehole data or seismic lines (Coakley, 1992; Soons *et al.*, 1997; Park and Wells, 2007; Allard *et al.*, 2008). The combination of sediment budget modelling, and estimates of volumes of sediment accumulation at different locations from borehole data with littoral transport modelling, can provide significant insights into the pattern of evolution determined from maps and aerial photographs.

However, there are several factors that also influence the local sediment budget both spatially and temporally, and which therefore have an impact on how the spit evolves. These include: changes to the volume of sediment supply from the updrift portion of the littoral cell; local changes resulting from the impact of storms such as overwash and the opening or closing of inlets; changes to the wave climate because of climatic factors or evolution of other coastal features; and sea level change. Changes to the sediment input from updrift may result from natural events, including the evolution of other barriers or changes to the locations of distributaries on a river delta. Changes may also result from human intervention such as the building of jetties at harbours updrift, or reduction in the sediment supply to the coast from rivers or cliff erosion because of dams or shore protection

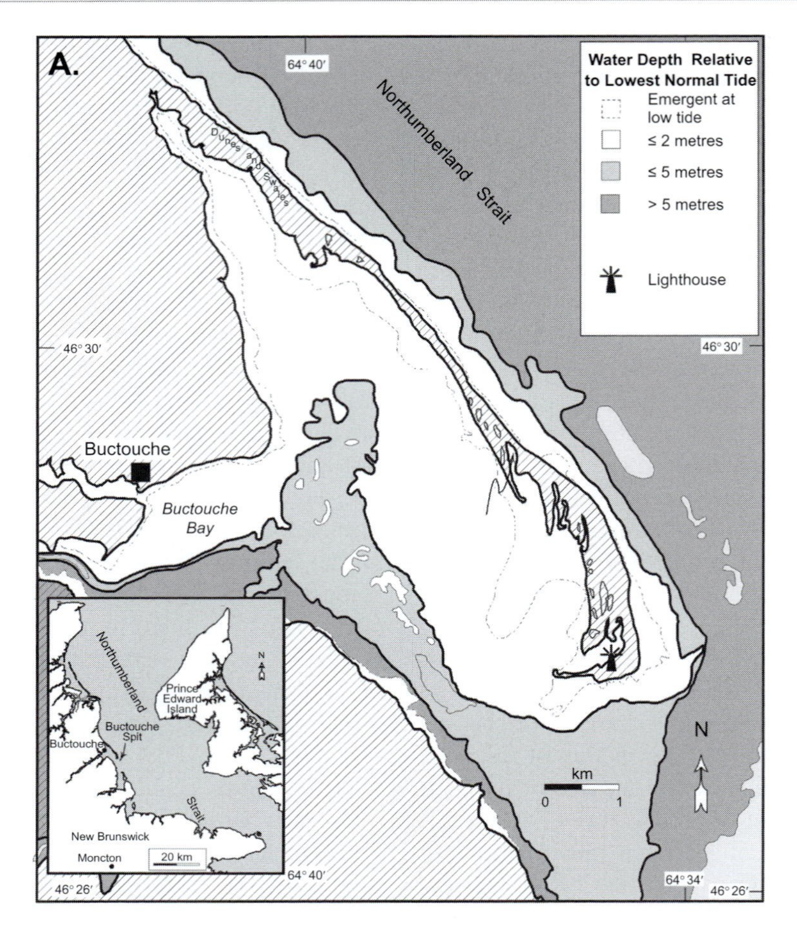

Figure 10.26 Examples of sandy spit development. A. Map of Buctouche Spit, New Brunswick, Canada. B. Oblique aerial photograph of Buctouche Spit. The spit is about 14 km long from the downdrift end to the point of attachment near the top of the photo. The extensive shoals of the spit platform are evident at the downdrift end. The main tidal channel of the constricted entrance to Buctouche Bay can be seen at the lower left. C. Oblique aerial photo of the distal end of Long Point, Lake Erie. The progradational portion of the spit is about 10 km long.

Box 10.2 Comparison of Buctouche and Long Point Spits

The role of refraction over the shallow platform in controlling shoreline and dune ridge curvature is illustrated by comparison of distal end of Buctouche spit with that of Long Point (Figure 10.27). Buctouche has built across the deepest part of Buctouche Bay and the platform is presently

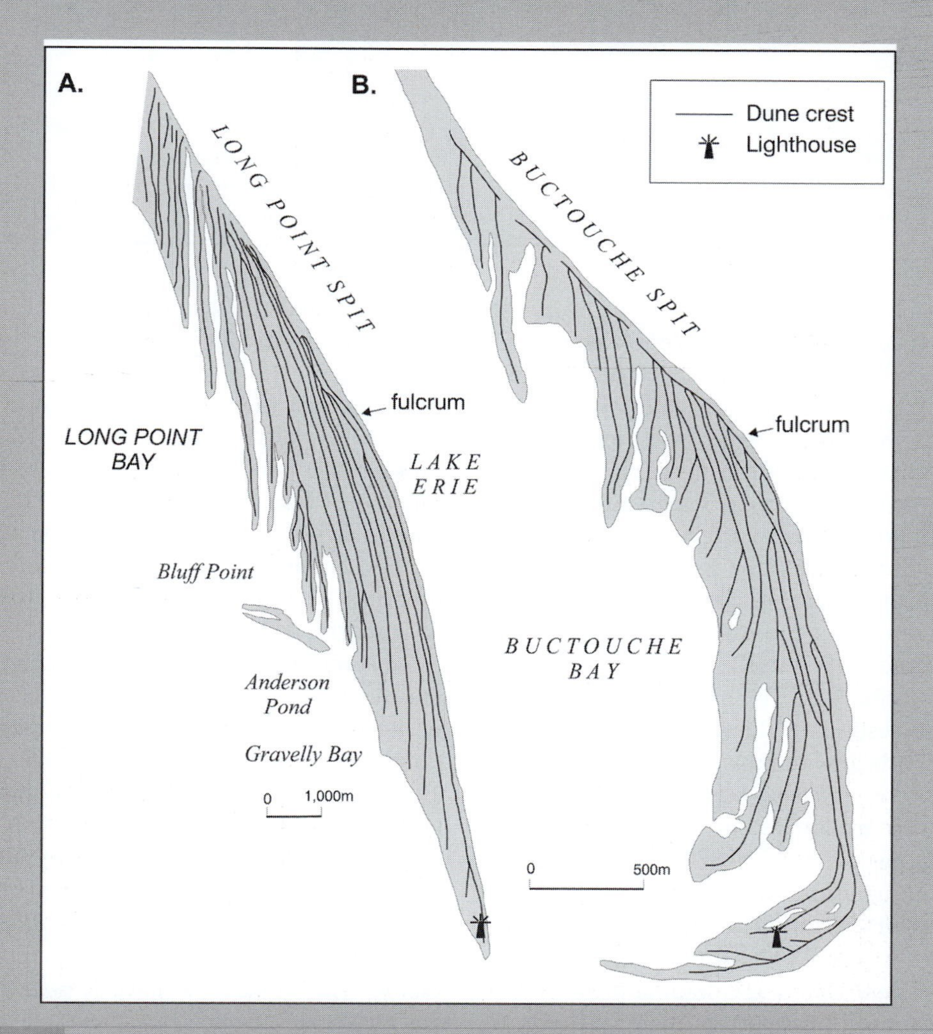

Figure 10.27 Comparison of the evolution of dune ridge curvature at the distal end of A. Long Point and B. Buctouche spits. The map of Long Point has been flipped so that the orientation of the two spits is the same. Note the differences in the scale of the two maps reflecting the much greater absolute size of Long Point.

extending into water depths < 6 m whereas Long Point is extending into a basin that is 30 m deep in the centre of Lake Erie. The result is that wave refraction occurs for almost all waves affecting Buctouche (Ollerhead, 1993), but there is almost no refraction around the end of Long Point (Davidson-Arnott and Conliffe Reid, 1994). As a result, the curvature of dune ridges on Buctouche has been increasing over the past few hundred years, while the distal end of Long Point has become less and less curved.

Box 10.2 (cont.)

Downdrift extension of the spit requires deposition of sufficient sediment to build the spit platform up to the surface and thus the rate of spit extension is a function of the littoral sediment transport rate and the depth of water into which the spit is building. There will be rapid extension when the sediment budget is large and the depth of the basin is small and vice versa (Rucińska-Zjadacz and Wróblewski 2018). Sediment supply to Long Point has been estimated to be on the order of 1.0×10^6 m^3 a^{-1}, though the actual supply of sand and gravel may be about half this. However, because the platform is building into about 30 m of water, a large volume of sediment is required to bring it to the lake surface and the rate of subaerial spit extension is only about 4 m a^{-1} (Coakley, 1992). This is similar to the Hel barrier spit in the Baltic which is currently building into even deeper water (about 50 m) Sediment supply to Buctouche spit is about 15 000 m^3 a^{-1} and spit extension over the past 150 years has also been about 4 m a^{-1} (Ollerhead and Davidson-Arnott, 1995), because the platform is only building into about 8 m of water.

structures. Locally, on the spit itself, overwash can lead to sand transport through washovers onto the backbarrier or into the lagoon, rather than being stored in the embryo dune and foredune system. Thus, large volumes can be extracted from the beach until the foredune system is restored. Decadal scale shifts in wind patterns, storm tracks or ice cover can also produce significant variations in the littoral transport rate over a period of several years.

There is also a significant effect of the evolution of the spit itself on the sediment budget alongshore that is characteristic of almost all spits, and especially flying spits. Initially when the spit first begins to grow the entire spit likely has a positive sediment budget. However, as it extends, the proximal end of the spit often develops a negative sediment budget because of a change in the littoral transport gradient. This is brought about because the distal end begins to shelter the proximal end from waves from directions opposite to that of the dominant waves. The result is that the transport gradient at the proximal end increases – in effect more sand is moved towards the end of the spit that comes in from updrift. The result is that the spit becomes segmented into a transgressive proximal end characterised by overwash and temporary inlet formation (a process sometimes termed necking) and a progradational downdrift end characterised by wide beaches and the development of a relict foredune dune field. The combination of transgression at the proximal end, and progradation at the distal end, results in the spit evolving into a distinct S shape. As transgression takes place in the proximal end the original spit sediments are reworked until none of the original subaerial barrier remains. Within the transgressive portion of the spit there may be a transitional zone, where remnants of the original dune recurves are preserved as dune ridges at a distinct angle to the modern shoreline and foredune.

Transgression of the proximal end, overwash and instability are heightened when the spit is attached to an eroding shoreline. This is the case for both Long Point and Buctouche spits. Erosion around the point of attachment results in its downdrift migration. A breach of the narrow neck of the spit during a severe storm or series of storms can cut off sediment supply to the downdrift portion of the spit leading to the formation of a barrier island and potentially to the complete reworking and disappearance of that section. This appears to have happened on several occasions at Spurn Head, a small spit which is supplied with sediment from erosion of the bluffs along the Holderness coast and builds southward across the Humber estuary on the northeast coast of England (de Boer, 1964). The existence of historical maps and information going back several centuries enabled de Boer to reconstruct not only the shape and position of the spit over time but also some of the events that led to breaching. He was thus able to make a convincing case for a cycle of formation,

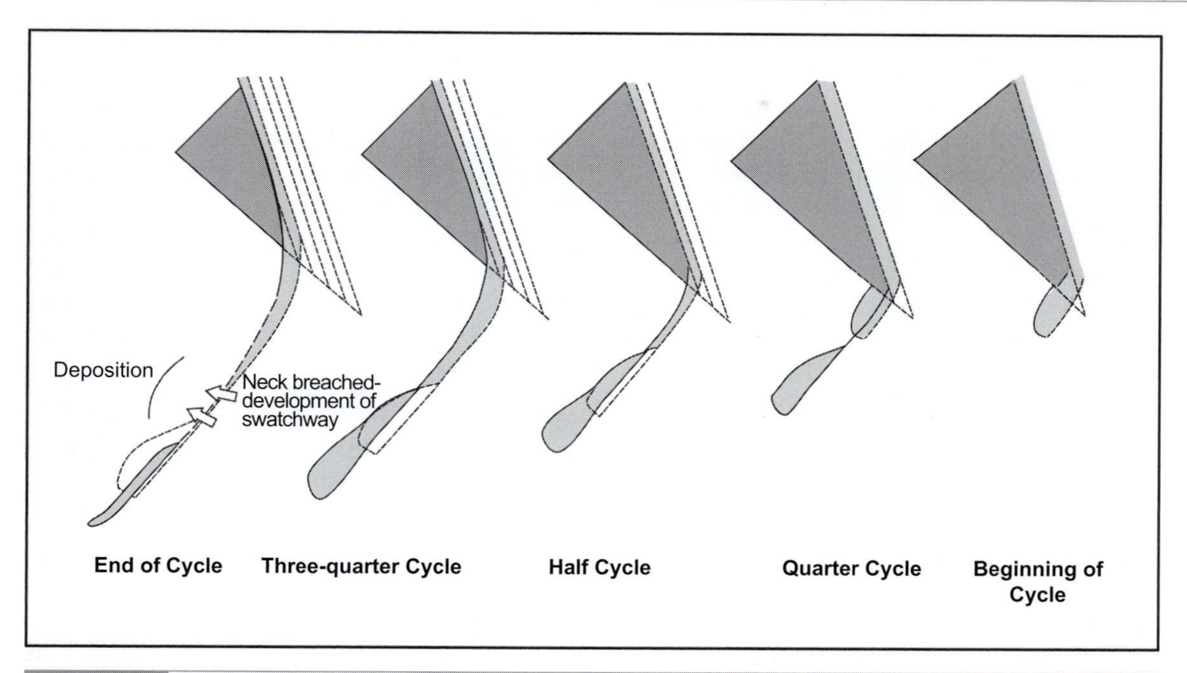

Figure 10.28 Schematic diagram of a cycle of accretion and erosion in the evolution of Spurn Head (after de Boer, 1964).

extension and destruction (de Boer, 1964; see Figure 10.28). Similar cycles of barrier overstepping have been postulated for several other areas including Buctouche spit (Ollerhead and Davidson-Arnott, 1995), the gravel barriers of Story Head, Nova Scotia (Forbes *et al.*, 1990) and Duxbury Beach spit, Massachusetts (Hill and Fitzgerald, 1992).

10.5 Barrier Islands

Barrier islands are generally restricted to comparatively wide, gently shelving coasts that permit the barrier to establish some distance offshore, enclosing a lagoon between it and the mainland shoreline. They are therefore found most commonly on trailing edge coasts in areas where there is an abundant supply of sand such as the east coast of North America, western Europe between Holland and the Jutland peninsula, and on the southeast coast of Africa from South Africa to Mozambique. Because they are not attached to the mainland, barrier islands can migrate relatively rapidly in response to changing sea level and sediment supply, with the landward migration of the island accomplished during large storms capable of overwashing the dune. Most barrier islands have formed under rising sea levels near the end of the Holocene transgression and few exist in the location where they were originally formed, although they have undergone significant transformation as water levels stabilised through the late Holocene. Eustatic and isostatic sea level rise in response to a changing climate has the potential to start a new period of island evolution, which will put the economic and ecological value of these islands at risk.

10.5.1 Origin and Formation of Barrier Islands

Competing models of barrier island genesis led to some heated discussions in the 1960s and 1970s, and a plea by Schwartz (1971) to accept that there is evidence for barrier island formation through several different mechanisms. While there may still be disagreement over the origin of a barrier island or barrier island system, there is now general recognition that there are at least three valid mechanisms for barrier island formation and, in some cases, barrier formation and evolution involves more than one mechanism. The three primary mechanisms are: (1) Spit detachment or segmentation by inlet formation (Fisher, 1968); (2) nearshore aggradation of shoals (Otvos, 1981); and (3) detachment of a dune ridge complex through submergence and flooding of low-lying areas landward of the dune ridge (Hoyt, 1967). The applicability of a model for barrier formation is clearly dependent on the general coastal morphology, shelf width, sediment sources and abundance, and the late Holocene sea level history. These factors likely account for differences in the amount of emphasis given to a theory in different coastal regions. Given the rapidity with which barrier systems evolve, the mode of formation probably is of limited significance in controlling barrier dynamics compared to the factors noted above and so less attention has been paid to it in the past couple of decades.

10.5.1.1 *Spit Detachment*

A simple scenario for barrier island formation through spit detachment assumes that there is a shallow headland bay complex in an area with a well-defined net longshore transport leading to the formation of a drift-aligned spit. Extension of the spit across the bay leads to thinning of the neck of the spit which makes it susceptible to overwash and inlet formation during severs storms. If the spit has extended far enough, the inlet can capture a sufficiently large tidal prism to be self-sustaining and the detached downdrift portion of the spit now evolves into a barrier island (Fisher, 1968). The inlet and the barrier migrate downdrift, while the spit begins a new phase of extension and progradation. This model of formation requires a relatively shallow embayment so that the newly formed barrier island continues to receive sediments through inlet bypassing. If the bay is too deep or the tidal currents are too strong, then the detached island quickly disappears, producing the cyclic model of spit development shown in Figure 10.28. Barrier islands on the Mozambique coast appear to have formed by this mechanism (Armitage *et al.*, 2006).

Barriers may also evolve in a similar way through the erosion of delta lobes. This is the case for some transgressive barriers in the Gulf of Mexico, such as Isles Derniers and the Chandleur Islands associated with abandoned lobes of the Mississippi delta (Penland *et al.*, 1988), and similar spit and barrier island development have been observed at deltas such as the Rhone in France and the Po in Italy (Simeoni *et al.*, 2007). The model of Penland *et al.* (1988) involves the following stages:

Stage 1 Initiation of erosion and transgression, as active sediment supply through the distributary channels is reduced (Figure 10.29).

Stage 2 Development of an erosional headland as the distributary channel recedes with accompanying spit and barrier island development.

Stage 3 Disappearance of the barrier as they are cut off from further sediment supply by transgression and submergence of the delta lobe and thinning of the barriers as they spread, leaving a relict shoal to mark the location of the barrier platform.

The delta lobe barriers of the Mississippi Island received considerable attention following Hurricane Katrina. While the loss of the islands was inevitable given the lack of a modern sediment supply and continued subsidence of the delta in that region, the islands protected New Orleans and oil and gas installations on the delta.

10.5.1.2 *Shoal Aggradation*

An understanding of the dynamic nature of nearshore bars in relation to waves and water levels led to the discounting of early hypotheses of barrier island formation through the emergence of nearshore bars. However, where there is a large shoal area some distance offshore, it is possible for emergent intertidal features resembling swash bars to form and grow to the limit of wave action forming a nucleus on which flotsam can accumulate. Colonisation by plants results in the trapping of sediment to form low dunes and the island may grow through further accretion in much the same way that coral cays develop on barrier reefs and carbonate platforms. Such an origin has been postulated for many of the

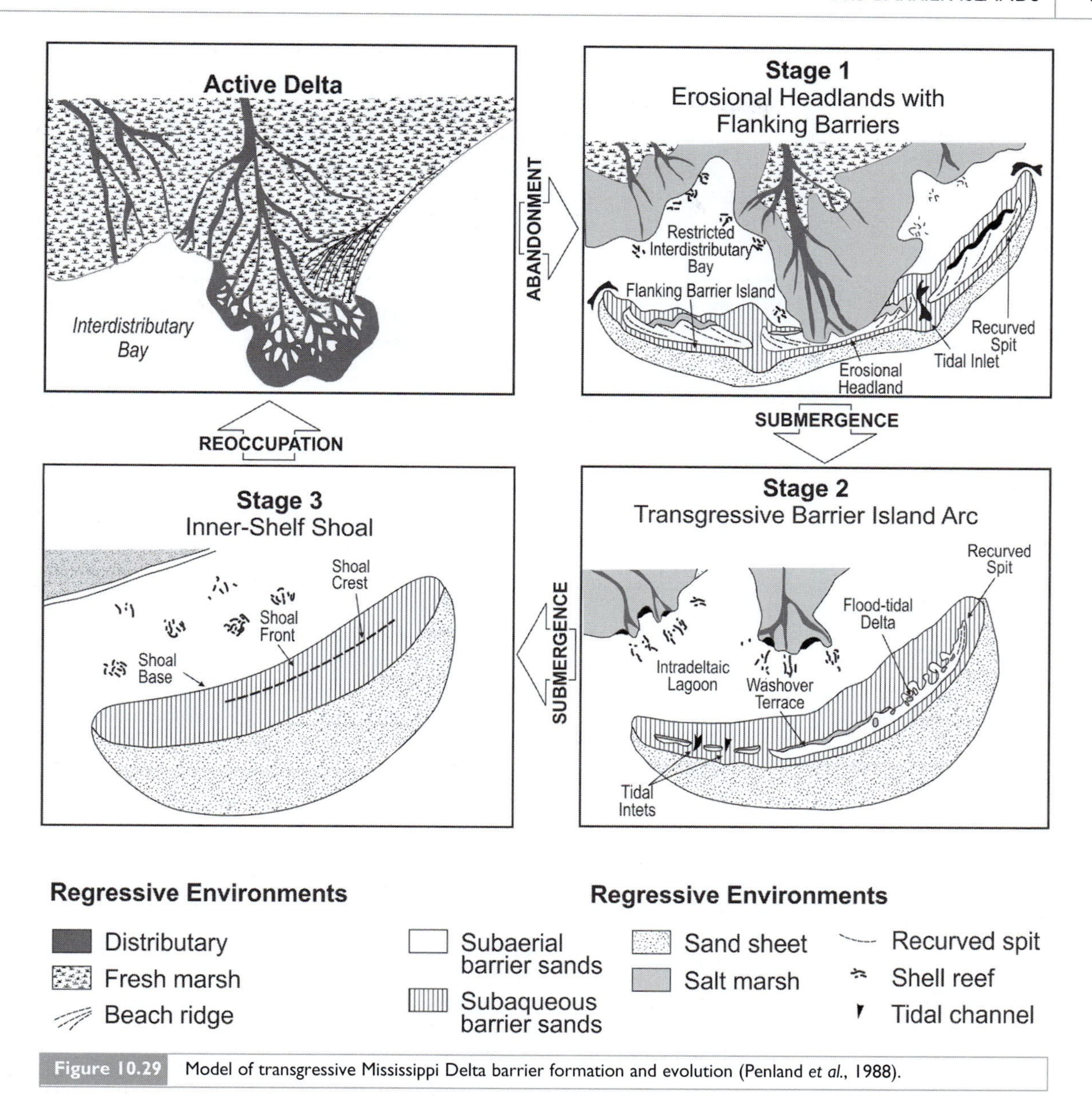

Figure 10.29 Model of transgressive Mississippi Delta barrier formation and evolution (Penland *et al.*, 1988).

Mississippi barrier islands to the east of the delta (Otvos, 1981; Otvos and Giardino, 2004), including Petit Bois and Horn Islands. The emergent model is partly supported by the emergence of shoals following Hurricane Katrina, and other barrier shorelines on the middle Atlantic coast of North America (Oertel and Overman, 2004).

Davis *et al.* (2003) suggest a variant of barrier emergence model for barrier islands along the west coast of the Florida peninsula. The Gulf coast in this region is a low-gradient, low-energy environment and they envisage the formation of a muddy mangrove or saltmarsh environment as the coastal plain was drowned. As the coastline position stabilised about 3000 years ago reworking by waves separated increasing amounts of sand from the Pleistocene sediments, leading to the formation of offshore shoals that gradually accreted into the intertidal and finally supratidal with aeolian sediment transport and

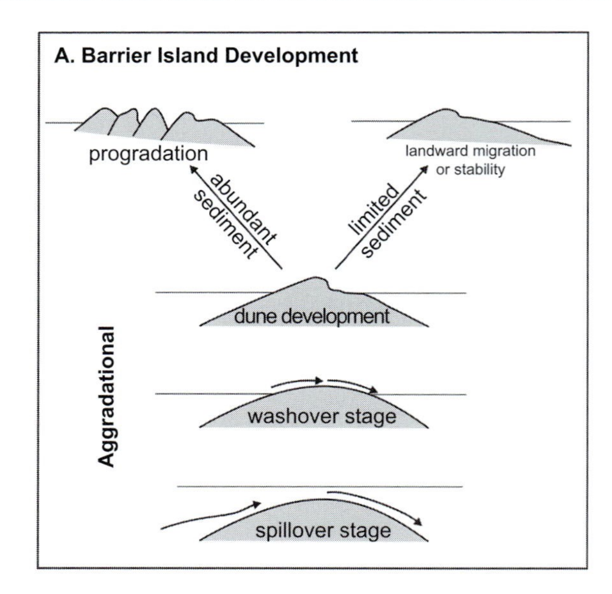

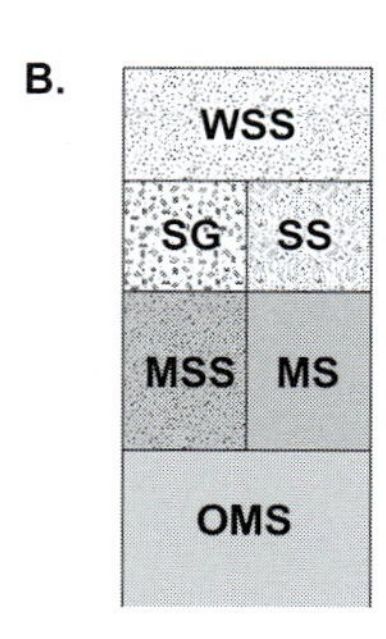

Figure 10.30 Stratigraphic model showing development of barriers on the west coast of the Florida panhandle by shoal aggradation: A. stages in the evolution of the barriers; B. simplified stratigraphy of the barriers based on the facies identified by Davis *et al.* (2003). WSS well sorted sand – modern dune and beach sand. SG shell gravel – washover channels and inlets. SS shelly sand – beach and nearshore. MS muddy sand – low-energy seagrass beds on either side of barrier. MSS muddy shelly sand – low-energy seagrass beds on either side of barrier. OMS organic muddy sands – basal Holocene unit with high organic matter content indicative of saltmarsh or mangrove.

dune development (Figure 10.30A). Coring along nine transects shows a relatively simple stratigraphy for all the barriers (Figure 10.30B), with four major sedimentary units reflecting the transition from the organic rich mangrove or saltmarsh sediments through low-energy seagrass stabilised beds on the seaward or landward side of the barrier and then sand and shell deposits associated with shallow marine sediments, beach and dune sediments and coarser sediments associated with overwash and inlet formation. Where there is abundant sediment available progradation occurs, otherwise the barrier may be transgressive.

10.5.1.3 *Mainland Beach Detachment*

Hoyt (1967) developed a theory of barrier island formation based primarily on sea level rise and transgression leading to the flooding of low-lying areas landward of a dune ridge on a gently sloping coastal plain. The theory was based primarily on work on barrier systems of the US east coast where transgression continued through much of the Holocene. The theory envisages the existence of substantial foredune ridges backed by a low coastal plain. Rapid sea level rise could lead to flooding of the area landward of the ridge to produce a lagoon while ongoing wave and aeolian processes act to maintain the ridge and associated beach. This was supported indirectly by the work of Swift (1975), though he noted that most contemporary barriers on this coast have migrated landward with the transgression and that the theory was likely more applicable to the early transgression. On a coastal plain, this requires the rate of submergence to exceed the rate of dune transgression. Much of the logic for the theory is based on the apparent absence of open marine sediments landward of the barrier ridge – these should be present if the barrier was formed through shoal aggradation or spit detachment. This model is consistent with geologic data collected at Santa Rosa Island in northwest Florida (Houser, 2012), as well as Padre Island along the Texas coast (Wernette *et al.*, 2018). In the latter case, there is evidence that the island started as a chain of islands separated by rivers that coalesced in the late Holocene as the coastal plain was inundated and the river outlets retreated landward (Wernette, 2018). However, theories for the formation of barriers through Holocene can be controversial as we are left with only geologic records and there are no modern cases of island formation. For example, Otvos (1982) presents evidence that Santa Rosa Island

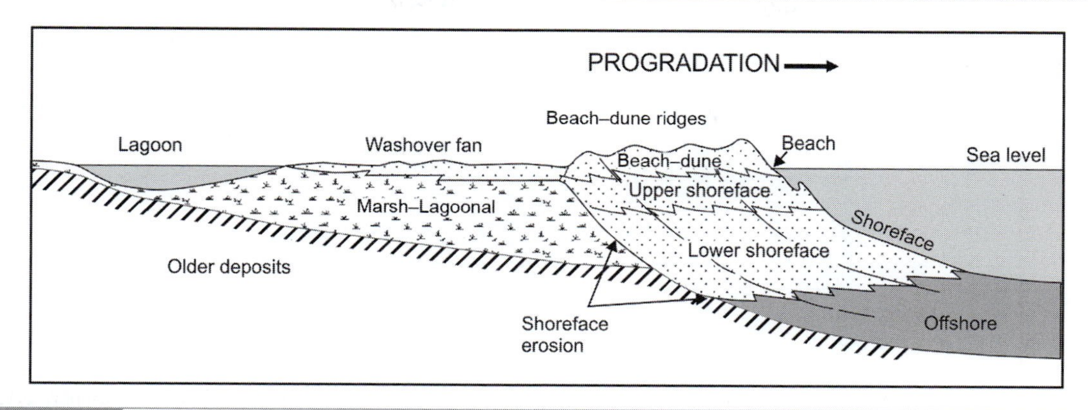

Figure 10.31 Progradation of a barrier island under stable sea level. Beach and dune sands build out across upper and lower shoreface sediments. Backbarrier marsh and washover fan sediments overly lagoonal sediments, but development of new foredune ridges produces a wider dune field and reduces the possibility of new overwash occurrence.

developed through spit elongation in the late Holocene, but bathymetric and geologic data presented by Houser (2012) suggests island transgression, which would support the detachment model. It is most likely that islands have experienced a far more complicated development and evolution than is suggested by these simple conceptual models and that more studies are required.

10.5.2 Barrier Response to Sediment Budget and Sea Level Change

Modern barrier islands and barrier systems generally continue to evolve in response to the factors that have controlled development towards the end of the Holocene. The major controls on this evolution are the local littoral and dune sediment budgets, and relative sea level change. Both the sediment budget and sea level rise influence the response and recovery of dunes to storms that move sediment landward and forces the island to migrate. In this respect, an understanding of barrier island evolution over longer spatial and temporal scales is dependent on beach and dune interaction on much shorter time and space scales.

10.5.2.1 *Sediment Budget*

When there is abundant sediment supply, the beach and nearshore will build seaward (or prograde). This leads to the development of new foredune ridges and increases the width of the barrier system (Figure 10.31). While backbarrier marsh and overwash deposits tend to be deposited in the early phase of progradation, the potential for further overwash is diminished as the island widens. However, disturbance to the relict dune system may lead to the development of migrating parabolic dunes that can cover the older backbarrier sediments and infill the lagoon. This is a characteristic feature of many of the Holocene barrier systems on the coast of New South Wales, Australia (Roy *et al.*, 1994; Hesp and Short, 1999; Harvey, 2006; Goodwin *et al.*, 2006), where evidence of several episodes of dune destabilisation and transgression can often be found. Progradation can occur relatively uniformly along the whole length of a barrier, and this is often the case for baymouth or midbay barriers anchored to headlands or the sides of large embayments. It may also be limited to the updrift and downdrift portions of barrier islands that are adjacent to inlets or the distal portion of barrier spits.

Most modern barrier systems are only a few thousand years old because of rapid shoreline change during the Holocene transgression. The Holocene transgression resulted in the reworking of sediments deposited on the inner continental shelf by glaciers, rivers and by wave action during falling sea levels at the end of the Wisconsinan glaciation. Much of this is now stored in

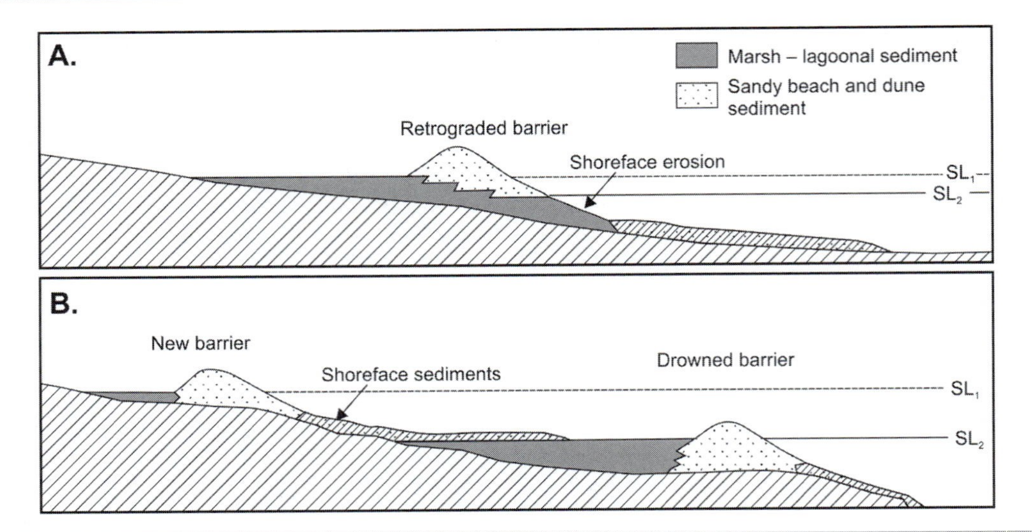

Figure 10.32 Transgression of a barrier island due to sea level rise. A. Rising sea level leads to landward migration of the beach and dune system and reworking of the inner shoreface. The barrier overrides marsh and lagoon sediments which are later eroded as they become exposed at the seaward side. B. If sea level rises too quickly or there is insufficient sediment in the system then the barrier may be drowned leaving portions of the barrier and lagoon sediments stranded on the continental shelf.

the form of beach and dune systems, including those associated with barriers, and large amounts are now stranded estuaries, in relict foredune plains and transgressive dune fields. As a result, the sediment supply to many barrier systems is now much lower than it was several hundreds to thousands of years ago (not including the effect of shoreline protection measures such as groynes and jetties), and this has the effect of increasing the proportion of barriers that have a negative sediment budget over some or all their length. The effect of a negative budget locally is to reduce beach width and ultimately to make that location more susceptible to overwash (Jiménez and Sánchez-Arcilla, 2004). Barrier islands such as the Isles Derniers and the Chandleur Islands off the Mississippi delta that are now largely detached from their sediment supply are rapidly thinning through loss of sediment at the margins of chain and divided into smaller and smaller islands through overwash and inlet formation during hurricanes (Penland *et al.*, 1988). Where barriers are supplied primarily through longshore sediment transport, reduced supply may lead to transgression across the back-barrier and lagoon, ultimately resulting in welding of the barrier to the mainland.

10.5.2.2 *Sea Level Rise*

Regardless of the formation model, most modern barriers were initiated during the Holocene transgression on the continental shelf. Ideally, rising sea level leads to a continual transfer of sediments from the seaward side of the barrier to the landward side, building a base upon which the dune system can migrate inland and upward (Figure 10.32A). Sediment transfers can take place through three primary mechanisms: (1) transfers through the tidal inlets into the lagoon system through the building of flood tidal delta shoals and sedimentary deposits within the lagoon itself; (2) overwash of sediments onto the back barrier and into the lagoon (Claudino-Sales *et al.*, 2008; Houser *et al.*, 2008a); and (3) aeolian sediment transport either directly over the foredune crest onto the lee slope and beyond or through the development of blowouts and parabolic dunes (Jewell *et al.*, 2014, 2017). All three mechanisms may occur within one barrier system, but the relative contribution of each is highly variable from one barrier system to another. Infilling of the lagoon is counterbalanced by rising sea level, leading to flooding of the coastal plain landward and so all segments of the barrier migrate landward synchronously. There is some room here for dynamic equilibrium with the rate of sea level rise because a higher rate of sea level rise triggers greater erosion of the foredune ridge system and therefore increases the rate of landward transfers. There may be limited recovery from overwash before another series of events occurs, creating an upper limit for the ability of the system to adjust.

If the rate of sea level rise is faster than landward transfer of sediment can occur, the barrier will become completely overwashed and the subaerial and inner shoreface sediments will be smeared across the marsh and lagoon sediments. These sediments are then preserved on the continental shelf (Figure 10.32B). There is evidence for drowned barrier sediments, rather than preserved barriers, in several areas including Australia (Roy *et al.*, 1994) and the east coast of the USA (Swift, 1975). Actual barrier overstepping as proposed by Rampino and Sanders (1980) among others, probably requires a very rapid sea level rise or significant loss of sediment to preserve the subaerial components of the system intact but examples have been suggested, for example, for the northern Adriatic shelf off Italy (Storms *et al.*, 2008). The loss of sediment from alongshore has resulted in Follets Island, Texas transitioning from an island with well-developed dunes and limited landward retreat to an island with little elevation experiencing rapid retreat (Harter *et al.*, 2015).

There is a growing body of literature on the potential acceleration of barrier erosion and migration because of increased rates of sea level rise and attempts to model this on a decadal scale (Dean and Maurmeyer, 1983; Dubois, 1995; Eitner, 1996; List *et al.*, 1997; Moore *et al.*, 2007; FitzGerald *et al.*, 2008). Dean and Maurmeyer adapted the Bruun rule for the special case of a barrier island and their model allowed for overwash and inlet transfers. While some authors still favour this simple approach (Zhang *et al.*, 2004), others have suggested that response to sea level rise is more complex (Cooper and Pilkey, 2004). On a longer time-scale, the morphological behaviour model of Cowell *et al.* (1995) provides an alternative approach and this approach has been utilised in several studies (Stolper *et al.*, 2005; Moore *et al.*, 2007).

Much field and modelling research is devoted to understanding how barrier islands will respond to future sea level rise, particularly those islands that are considered ecologically, economically and/or socially important. Resiliency is the ability of a landform to return to its previous disturbance state (Woodroffe, 2007). Barrier islands are considered resilient with a rise in sea level if they can maintain elevation, width and volume (see Godfrey and Godfrey, 1973; Hosier and Cleary, 1977), and ecological function, which in turn depends on the frequency and extent of washover, the ability of the dune and vegetation to recover. As noted, our understanding of large-scale island transgression to sea level rise (in the past and in the future) depends on the exchange of sediment as the islands respond to and recover from storms. It is difficult, however, to predict the future of barrier island based on their response to the Holocene transgression since most modern barriers have been significantly altered over the last century.

10.6 Anthropogenic Impacts and Natural Hazards

Human activity has become an important part of many barrier islands, particularly in the United States where development often includes infrastructure to support recreation and navigation. Tourism and development began in the second half of the nineteenth century (Ehlers and Kunz, 1993; Meyer-Arendt, 1990; Nordstrom, 2004), and a large part of this expansion was associated with private vehicles and the need for roads and community services (Meyer-Arendt, 1991). To protect against erosion during tropical storms and hurricanes, roads are often stabilised through hard structures and, in rare cases, beach and dune nourishment, and they are underlain by gravel and shell material. As previously noted, the base material and hard structures are redistributed by washover, creating a non-erodible lag

Figure 10.33 Driving on the beach is permitted on sections of Padre Island on the coast of Texas. Driving A. reduces the elevation of the beach and dune making it more susceptible to scarping and washover compared to B. areas where driving is restricted and dune-building vegetation can establish on the backshore.

that limits aeolian transport and dune development (Houser *et al.*, 2008b; Houser, 2009a), and roads perpendicular to the beach create preferential conduits for the return flow leading to pronounced scours and even island breaches (Sherman *et al.*, 2013). On some islands, including Galveston and Padre Island in Texas, driving is permitted on the beach leading to a reduction in the elevation of the beach and dune toe and the pulverisation of wrack and vegetation (Houser *et al.*, 2013; Figure 10.33), which promotes dune scarping by even small storms leading to the development of blowouts that move sediment towards the backbarrier (Jewell *et al.*, 2014, 2017).

Dune recovery can also be impacted by the emplacement of jetties and groynes that limit or eliminate the transport of sediment alongshore. We have touched on the problems posed by jetty construction and dredging at inlets in Section 7.4.1. Jetties tend to trap sediment on the updrift side and thus pose a problem for erosion of the downdrift side. For example, the jetty at Ocean City, Maryland resulted in a significant reduction in the amount of sediment available for dune development on Assateague Island just downdrift. The result has been a rapid transgression of the northern part of the island (Leatherman, 1976), with recession rates ranging from 11 m a^{-1} to 12.2 m a^{-1} (Thornberry-Ehrlich, 2005). Similarly, the emplacement of jetties and groynes on Galveston Island has limited the alongshore transport of sediment leading to limited dune development and enhanced shoreline recession that prompted the construction of the Galveston seawall following a hurricane in 1915.

Building of recreational homes, permanent residences, apartments and service facilities on barrier islands and barrier spits that are transgressive under rising sea level poses major problems for coastal managers. Buildings and supporting infrastructure restrict the space available for dunes to develop, such that dunes on developed coasts tend to be lower and narrower compared to undeveloped sections of coast (Nordstrom *et al.*, 2006). Buildings also act as large-scale roughness elements that reduce wind speeds and limit aeolian transport by onshore winds, but can increase stability by blocking offshore winds. Nordstrom and McCluskey (1985) argue that even a moderate level of development can strongly affect sediment transport and dune development. Dune heights are further limited by pedestrian traffic across the dune, leading to the loss of vegetation and the development of low points within the dune that can lead to blowouts. Nordstrom *et al.* (2000, 2002) suggest that the impact of pedestrians can be limited by restricting or rerouting paths along with the placement of sand fencing (Miller *et al.*, 2001), altered beach-cleaning activities (Mendelssohn *et al.*, 1991; Williams *et al.*, 2008), and development planning (Nordstrom and Jackson, 1998). Without adopting these strategies, anthropogenic activities can lead to a reduction in dune height and extent, which can reinforce and even amplify, to varying degrees, the vulnerability of a barrier island, and commercial and residential development, to storm waves and surge (Nordstrom, 2000).

Stable and progradational baymouth and mid-bay barriers, in areas where there is sufficient sediment supply to counteract present rates of sea level rise, pose relatively few problems for the management of human activities. The stability of these barriers means that there should be no need for shore protection, and the relatively simple setback and access restrictions designed for the management of coastal dunes should be sufficient to prevent large-scale destabilisation and parabolic dune development. However, transgressive barrier islands and barrier spits pose much more of a problem for human activities, especially for the control of inlets for navigation purposes and development of housing, roads and other infrastructure. There are now numerous examples of the folly of building on transgressive barrier systems on the US east and Gulf coasts in the face of severe northeasters or hurricanes. Pictures of a section of barrier north of Galveston before and after Hurricane Ike struck the region in September 2008 provide dramatic evidence of the extent of devastation caused by overwash, even to buildings constructed on stilts to allow water to flow beneath them (Figure 10.34). Similar destruction of houses on a barrier can occur with strong mid-latitude storms. The storm on 2 December 1985 destroyed about 80 houses and cottages along a stretch of shoreline on the proximal end of Long Point (Figure 10.35).

Figure 10.34 Pair of oblique aerial photographs of a portion of the barrier north east of Galveston, Texas showing destruction in the wake of landfall by Hurricane Ike on 13 September 2008. (Photo credit: US Geological Survey).

The US barrier coasts have been hit by a series of severe hurricanes over the last two decades and the extent of the losses has resulted in insurance companies refusing to insure against damage due to wave action and flooding associated with severe storms (H. John Heinz III Center for Science, Economics and the Environment, 2000). It has also resulted in changes to the operation of the Federal Emergency Measures Agency (FEMA) to attempt to force states to prohibit rebuilding within designated hazard zones by withdrawing funding to support this initiative.

Figure 10.35 Destruction in the wake of an intense storm on 2 December 1985 along the proximal end of Long Point. A. Overwash resulted in the destruction of the foredune. The cottage behind it was lifted off its foundations and carried it 75 m into the backbarrier marsh. B. Destruction of cottage along Hastings Drive. The circles indicate the location of houses that were transported into the marsh.

Box 10.3 Coastal Management Application

Planning, Setbacks and Managed Retreat

There are a wide range of approaches to addressing flooding and erosion hazards posed to coastal communities by severe storms accompanied by large waves and storm surge, as well as shoreline recession resulting from long-term erosion and/or sea level rise. These include: (1) a range of 'hard' engineering structures such as groynes, seawalls and revetments; (2) soft engineering approaches such as beach nourishment (Box 8.2); and (3) the application of planning restrictions, setbacks and managed retreat within the framework of an integrated coastal management plan. In many areas this latter approach is likely to be far less costly than engineering approaches and it reduces the need for widespread evacuations in the face of major storms and the destruction of properties that is often a consequence (Figure 10.36). Some form of coastal management involving planning restrictions on development, setbacks and the implementation of managed retreat in the face of long-term coastal erosion have been proposed and implemented in many countries in the past few decades. However, they have met with varying degrees of success and they have often produced fierce resistance from private property owners, developers and real estate agencies, among others.

Coastal management of sandy coasts is generally carried out within the framework of littoral cells or coastal compartments (e.g., Nicholls *et al.*, 2013; Thom *et al.*, 2018), because of the need to determine long-term shoreline

Figure 10.36 Buildings occupying the backshore and foredune such as those at Santa Rosa Island, Florida are exposed to damage and destruction as a result of large waves and high water levels associated with severe storms (Photo courtesy of Patrick Hesp).

Box 10.3 (cont.)

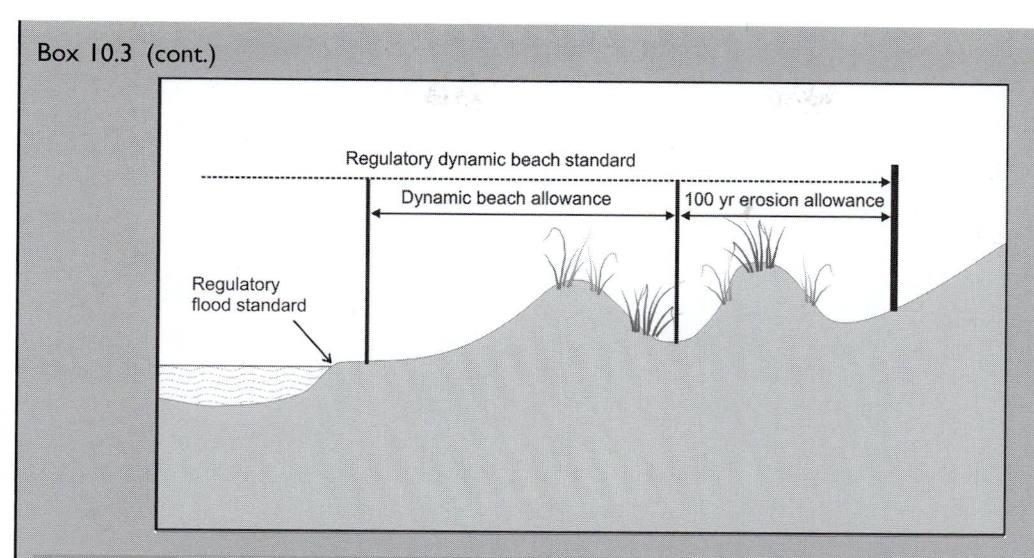

Figure 10.37 An example of setback limits for a sandy beach and dune system – the dynamic beach regulations of the Ontario policy for management of Great Lakes shoreline. The setback begins with determination of the flood limits associated with the 1/100 year flood event, the dynamic beach allowance determines the setback account for erosion of foredune during the 1/100 year storm event. On shorelines subject to erosion an additional setback is determined based on 100 times the average annual recession rate.

stability and whether there is a need to incorporate shoreline recession in the planning process. In Ontario, setback for sandy beaches on the Great Lakes shoreline come under the dynamic beach section of the management policy (Ontario Ministry of Natural Resources, 2001). This gives explicit recognition to the functioning of the foredune as a reservoir of sand that is activated through erosion during severe storms with large waves and storm surge (see Section 9.3). As presented in Figure 10.37, the setback is determined by three components: (1) the flood standard, which is the 100-year flood elevation based on the combined probability of fluctuating water level and the magnitude of storm surge events plus a 15 m allowance for wave uprush; (2) the dynamic beach allowance which is designed to accommodate erosion of the incipient dune and foredune during the 100-year event; and (3) the 100-year erosion allowance (100 times the average annual recession rate). In Ontario conservation authorities implement the setback policy, within management boundaries determined by littoral cells, and they maintain a GIS which maps the 100-year flood elevation and setback lines which are updated as new modelling and data become available. This utilisation of this type of data base is widespread among shoreline management agencies worldwide (Williams *et al.*, 2018).

On stable coasts (i.e., those with no long-term shoreline recession), simple planning setbacks may be implemented to restrict new development from areas subject to flooding due to high water levels, including storm surge (e.g., Chaumillon *et al.*, 2017), and to short-term erosion of coastal dunes as part of beach/dune interaction. The 100-year flood elevation (or alternative return period) can be determined from a combination of historical records and the modelling of storm surge elevation associated with storm events of a particular magnitude and direction. On marine coasts, the 100-year flood elevation will incorporate the tidal regime and seasonal fluctuations as well as storm surge elevation. In the Ontario case, an additional 15 m is added as a precaution, recognising the spatial and temporal variation in beach width and backshore elevation which will influence the extent to which the base of the foredune is subject to erosion (Figure 10.37). Once the position of the regulatory flood standard has been determined the dynamic beach allowance is determined. In the absence of detailed geomorphic studies it is set at a minimum of 30 m landward of the flood standard. The dynamic beach setback is applied to all new development along the coast and has been implemented with relative success (Figure 10.38), and only moderate opposition from private property owners and developers.

Box 10.3 (cont.)

Figure 10.38 Photo taken in September 2015 of new 'cottage' development on the Lake Huron shoreline south of Grand Bend. The buildings are set back landward of the toe of the foredune. The stoss slope of the foredune is vegetated primarily by *Ammophila breviligualta* in the foreground with another pioneering grass, *Calamovilfa longifolia*, occupying the upper slope and crest. The bench visible as a dark band from right to left across the centre of the photo marks the limit of dune erosion during the last major storm event in the high water phase in 1985–6.

On sandy beach and dune coasts the implementation of planning setbacks becomes much more difficult when the coast is subject to long-term recession due to erosion resulting from a negative littoral sediment budget and/or sea level rise (e.g., Abbott, 2013; Leatherman, 2018). In the past few decades determination of coastal recession rates has become routine and the predicted recession for a chosen time interval can be easily incorporated into a GIS. In Ontario the chosen limit is 100 years and this is added to the setback for flooding and for dynamic beach (Figure 10.37). In recognition of the continuous nature of recession, determination of the setback limit is carried out for individual development applications based on the date of the application rather than on a fixed limit which might only be updated on a decadal basis. Implementing the setback for new development on an eroding coast is also relatively straightforward. However, such an approach is usually implemented within a framework of managed retreat, whether that is spelled out explicitly or is implicit within the regulations. This envisions the removal or destruction of properties once shoreline recession poses an immediate danger to the property and to human life during a storm event. The result is that property values decline as the distance to the shoreline is reduced and property owners, faced with this and the prospect of imminent property loss, often demand that the recession process be halted using hard or soft engineering approaches. This then leads to conflict which pits management policies based on science and cost/benefit analysis against the political realities of unhappy and vocal shoreline residents (Abbot, 2013; Jongejan *et al.*, 2017; Neal *et al.*, 2018; Williams *et al.* 2018; Leatherman, 2018).

References

Abbott, T. 2013. Shifting shorelines and political winds: the complexities of implementing the simple idea of shoreline setbacks for oceanfront developments in Maui, Hawaii. *Ocean and Coastal Management*, **73**, 13–21.

Allard, J., Bertin, X., Chaumillon, E. and Pouget, F. 2008. Sand spit rhythmic development: A potential record of wave climate variations? Arçay spit, western coast of France. *Marine Geology*, **253**, 107–31.

Anthony, E.J. and Blivi, A.B. 1999. Morphosedimentary evolution of a delta-sourced, drift-aligned sand barrier-lagoon complex, western Bight of Benin. *Marine Geology*, **158**, 161–76.

Armitage, S.J., Botha, G.A., Duller, G.A.T. *et al.* 2006. The formation and evolution of the barrier islands of Inhaca and Bazaruto, Mozambique. *Geomorphology*, **82**, 295–308.

Ashley, G.M. 1987. Assessment of the hydraulics and longevity of Wood End Cut (inlet), Cape Cod, Massachussetts, USA. *Journal of Coastal Research*, **3**, 281–95.

Bauer, B.O. and Davidson-Arnott, R.G. 2003. A general framework for modelling sediment supply to coastal dunes including wind angle, beach geometry, and fetch effects. *Geomorphology*, **49**(1–2), 89–108.

Belknap, D.F. and Kraft, J.C. 1985. Influence of antecedent geology on stratigraphic preservation potential and evolution of Delaware's barrier systems. *Marine geology*, **63**(1–4), 235–62.

Bertin, X., Chaumillon, E., Sottolichio, A. and Pederos, R. 2005. Tidal inlet response to sediment infilling of the associated bay and possible implications of human activities: the Marennes-Oléron Bay and the Maumusson Inlet, France. *Continental Shelf Research*, **25**, 1115–31.

Bowman, D. and Pranzini, E. 2004. Reversed responses within a segmented detached breakwater, the Tuscany coast, Italy: a case study. *Coastal Engineering*, **49**, 263–74.

Bray, T.F. and Carter, C.H. 1992. Physical processes and sedimentary record of a modern, transgressive, lacustrine barrier island. *Marine Geology*, **105**, 155–68.

Brodhead, J.M. and Godfrey, P.J. 1979. *The Effects of Off-road Vehicles on Coastal Dune Vegetation in the Province Lands, Cape Cod National Seashore, Massachusetts* (No. 32). Amherst, MA: Environmental Institute.

Bruun, P. 1978. *Stability of Tidal Inlets: Theory and Engineering* Amsterdam: Elsevier, 506 pp.

Chaumillon, E., Bertina, X., Fortunato, A.B. *et al.* 2017. Storm-induced marine flooding: lessons from a multidisciplinary approach. *Earth-Science Reviews*, **165**, 151–84.

Claudino-Sales, V., Wang, P. and Horwitz, M.H. 2008. Factors controlling the survival of coastal dunes during multiple hurricane impacts in 2004 and 2005: Santa Rosa barrier island, Florida. *Geomorphology*, **95**, 295–315.

Cleary, W.J. and Hosier, P.E. 1979. Geomorphology, washover history, and inlet zonation: Cape Lookout, NC to Bird Island, N.C. In Leatherman, S.P. (ed.), *Barrier Islands: From the Gulf of St. Lawrence to the Gulf of Mexico*. New York: Academic Press, 237–71.

Coakley, J.P. 1976. The formation and evolution of Pt. Pelee, western Lake Erie. *Canadian Journal of Earth Sciences*, **13**, 136–44.

Coakley, J.P. 1992. Holocene transgression and coastal landform evolution in northeastern Lake Erie, Canada. In Fletcher, C.H. Ill and Wehmiller, J.F. (eds.), *Quaternary Coasts of the United States: Marine and Lacustrine Systems*. Tulsa, OK: Society of Economic Paleontologists and Mineralogists, Special Publication **48**, pp. 415–26.

Cooper, J.A.G. 1994. Lagoons and microtidal coasts. In Carter, R.W.G. and Woodroffe, C.D. (eds.), *Coastal Evolution*. Cambridge: Cambridge University Press, pp. 219–65.

Cooper J.A.G. and Pilkey, O.H. 2004. Sea level rise and shoreline retreat: time to abandon the Bruun rule. *Global and Plantetary Change*, **43**, 157–71.

Cowell P.J., Roy P.S. and Jones R.A. 1995. Simulation of large-scale coastal change using a morphological behaviour model. *Marine Geology*, **126**, 46–61.

Davidson-Arnott, R.G. 1988. Temporal and spatial controls on beach/dune interaction, Long Point, Lake Erie. *Journal of Coastal Research*, SI 3, 131–6.

Davidson-Arnott, R.G.D. and Conliffe Reid, H.E. 1994. Sedimentary processes and the evolution of the distal bayside of Long Point, Lake Erie. *Canadian Journal of Earth Sciences*, **31**, 1461–73.

Davidson-Arnott, R.G.D. and Fisher, J.D. 1992. Spatial and temporal controls on overwash occurrence on a Great Lakes barrier spit. *Canadian Journal of Earth Sciences*, **29**, 102–17.

Davidson-Arnott, R.G. and Law, M.N. 1990. Seasonal patterns and controls on sediment supply to coastal foredunes, Long Point, Lake Erie. In Nordstrom, K.F., Psuty, N.P. and Carter R.W.G. (eds.), *Coastal Dunes: Form and Process*. Chichester, UK: John Wiley and Sons, pp. 177–200.

Davidson-Arnott, R.G. and Law, M.N. 1996. Measurement and prediction of long-term sediment supply to coastal foredunes. *Journal of Coastal Research*, **12**, 654–63.

Davidson-Arnott, R.G.D. and van Heyningen, A. 2003. Migration and sedimentology of longshore sandwaves, Long Point, Lake Erie, Canada. *Sedimentology*, **50**, 1123–37.

Davis, R.A. Jr, Yale, K.E., Pekala, J.M. and Hamilton, M.V. 2003. Barrier island stratigraphy and Holocene history west-central Florida. *Marine Geology*, **200**, 103–23.

Dean, R.G. and Maurmeyer, E.M. 1983. Models for beach profile responses. In Komar, P.D. (ed.), *Handbook of Coastal Processes and Erosion*. Boca Raton, FL: CRC Press, pp. 151–66.

de Boer, G. 1964. Spurn Head: its history and evolution. *Transactions of the Institute of British Geographers*, **34**, 71–89.

Delgado-Fernandez, I. and Davidson-Arnott, R. 2011. Meso-scale aeolian sediment input to coastal dunes: the nature of aeolian transport events. *Geomorphology*, **126**(1-2), 217–32.

Donnelly, C., Kraus, N. and Larson, M. 2006. State of knowledge of modelling of coastal overwash. *Journal of Coastal Research*, **22**, 965–91.

Dubois R.N. 1995. The transgressive barrier model: an alternative to two-dimensional volume balanced models. *Journal of Coastal Research*, **11**, 272–86.

Ehlers, J. and Kunz, H. 1993, July. Morphology of the Wadden Sea natural processes and human interference. In *Coastlines of the Southern North Sea*. New York: American Society of Civil Engineers, pp. 65–84.

Eitner, V. 1996. Geomorphological response of the East Friesian barrier islands to sea level rise: an investigation of past and future evolution. *Geomorphology*, **15**, 57–65.

Fiechter, J., Steffen, K.L., Mooers, C.N.K. and Haus, B.K. 2006. Hydrodynamics and sediment transport in a southeast Florida tidal inlet. *Estuarine, Coastal and Shelf Science*, **70**, 297–306.

Fisher, J.J. 1968. Barrier island formation: discussion. *Geological Society of America Bulletin*, **79**, 1421–26.

FitzGerald, D.M. 1996. Geomorphic variability and morphologic and sedimentologic controls on tidal inlets. *Journal of Coastal Research*, **SI 23**, 47–71.

FitzGerald, D.M. and FitzGerald, S.A. 1977. Factors influencing tidal inlet throat geometry. In *The Proceedings of the Coastal Sediments '77*. New York: American Society of Civil Engineers, pp. 563–71.

FitzGerald, D.M., Fenster, M.S., Argow, B.A. and Buynevich, I.V. 2008. Coastal impacts due to sea-level rise. *Annual Review of Earth and Planetary Sciences*, **36**, 601–47.

Flinn, D. 1997. The role of wave diffraction in the formation of St. Ninian's ayre in Shetland, Scotland. *Journal of Coastal Research*, **13**, 202–8.

Forbes, D.L. and Solomon S.M. 1999. Inlet division and coastal instability following tidal prism diversion. In *The Proceedings of the Coastal Sediments '99*. Reston, VA: American Society of Civil Engineers, pp. 1418–33.

Forbes, D.L., Taylor, R.B., Orford, J.D., Carter, R.W.G. and Shaw, J. 1990. Gravel-barrier migration and overstepping. *Marine Geology*, **97**, 305–13.

Forbes, D.L., Parkes, G.S., Manson, G.K. and Ketch, L.A. 2004. Storms and shoreline retreat in the southern Gulf of St Lawrence. *Marine Geology*, **210**, 169–204.

Fritz, H.M., Blount, C., Sokoloski, R. *et al.* 2007. Hurricane Katrina storm surge distribution and field observations on the Mississippi Barrier Islands. *Estuarine, Coastal and Shelf Science*, **74**, 12–20.

Froede, C.R. Jr. 2008. Changes to Dauphin Island, Alabama, brought about by Hurricane Katrina (August 29, 2005). *Journal of Coastal Research*, **24**, 110–17.

Gilbert, G.K. 1890. *Lake Bonneville*. Washington, DC: US Geological Survey Monograph vol. 1, 438 pp.

Godfrey, P.J. and Godfrey, M.M. 1973. Comparison of ecological and geomorphic interactions between altered and unaltered barrier island systems in North Carolina. In Coates, D.R. (ed.), *Coastal Geomoprphology*. New York: Publications in Geomorphology, SUNY, pp. 239–58.

Goodwin, I.D., Stables, M.A. and Olley, J.M. 2006. Wave climate, sand budget and shoreline alignment evolution of the Iluka-Woody Bay sand barrier, northern New South Wales, Australia, since 3000 yr BP. *Marine Geology*, **226**, 127–44.

Greenwood, B. and Keay, P.A. 1979. Morphology and dynamics of a barrier breach: a study in stability. *Canadian Journal of Earth Sciences*, **16** 1533–46.

H. John Heinz III Center for Science, Economics and the Environement 2000. *The Hidden Costs of Coastal Hazards*. Washington, DC: Island Press, 220 pp.

Harter, C., Figlus, J. and Dellapenna, T.I.M. 2015. The role of hurricanes on the morphological evolution of a sediment-starved barrier island along the upper Texas coast: Follets Island. In *The Proceedings of the Coastal Sediments 2015*. Singapore: World Scientific, DOI: https://doi.org/10.1142/9789814689977_0005.

Harvey, N. 2006. Holocene coastal evolution: barriers, beach ridges, and tidal flats of South Australia. *Journal of Coastal Research*, **22**, 90–9.

Hayes, M.O. 1967. Hurricanes as geological agents, south Texas coast. *American Association of Petroleum Geologists Bulletin*, **51**, 937–42.

Hayes, M.O. 1979. Barrier island morphology as a function of tidal and wave regime. In Leatherman S.P. (ed.), *Barrier Islands: From the Gulf of St. Lawrence to the Gulf of Mexico*. New York: Academic Press, pp. 1–27.

Hayes, M.O. 1980. General morphology and sediment patterns in tidal inlets. *Sedimentary Geology*, **28**, 139–56.

Héquette, A.and Ruz, M.H. 1991. Spit and barrier island migration in the Southeastern Canadian Beaufort Sea. *Journal of Coastal Research*, **7**, 677–98.

Hesp, P.A. and Short, A.D. 1999. Barrier morphodynamics. In Short, A.D. (ed.), *Handbook of Beach and Shoreface Morphodynamics*. Chichester, UK: John Wiley and Sons, pp. 307–33.

Hill, M.C. and Fitzgerald, D.M. 1992. Evolution and Holocene stratigraphy of Plymouth, Kingston and Duxbury Bays, Massachusetts. In Fletcher, C.H. lll and Wehmiller, J.F. (eds.), *Quaternary Coasts of the United States: Marine and Lacustrine Systems*. Tulsa, OK: Society of Economic Paleontologists and Mineralogists, Special Publication, **48**, pp. 45–56.

Holland, T.L. 2005. Landscape changes in a coastal lagoon system, Jalisco, Mexico: implications for Barra de Navidad Lagoon. MSc Thesis, University of Guelph, 114 pp.

Hosier, P.E. and Cleary, W.J. 1977. Cyclic geomorphic patterns of washover on a barrier island in southeastern North Carolina. *Environmental Geology*, **2**(1), 23–31.

Houser, C. 2009a. Synchronisation of transport and supply in beach–dune interaction. *Progress in Physical Geography*, **33**(6), 733–46.

Houser, C. 2009b. Geomorphological controls on road damage during Hurricanes Ivan and Dennis. *Journal of Coastal Research*, **25**(3), 558–68.

Houser, C. 2012. Feedback between ridge and swale bathymetry and barrier island storm response and transgression. *Geomorphology*, **173**, 1–16.

Houser, C. 2013. Alongshore variation in the morphology of coastal dunes: implications for storm response. *Geomorphology*, **199**, 48–61.

Houser, C. and Greenwood, B. 2005. Hydrodynamics and sediment transport within the inner surf zone of a lacustrine multiple-barred nearshore. *Marine Geology*, **218**(1–4), 37–63.

Houser, C. and Greenwood, B. 2007. Onshore migration of a swash bar during a storm. *Journal of Coastal Research*, 23(1), 1–14.

Houser, C. and Hamilton, S. 2009. Sensitivity of post-hurricane beach and dune recovery to event frequency. *Earth Surface Processes and Landforms*, **34**(5), 613–28.

Houser, C. and Mathew, S. 2011. Alongshore variation in foredune height in response to transport potential and sediment supply: South Padre Island, Texas. *Geomorphology*, **125**(1), 62–72.

Houser, C., Hapke, C. and Hamilton, S. 2008a. Controls on coastal dune morphology, shoreline erosion and barrier island response to extreme storms. *Geomorphology*, **100**(3–4), 223–40.

Houser, C., Hobbs, C. and Saari, B. 2008b. Posthurricane airflow and sediment transport over a recovering dune. *Journal of Coastal Research*, **24**(4), 944–53.

Houser, C., Labude, B., Haider, L. and Weymer, B. 2013. Impacts of driving on the beach: case studies from Assateague Island and Padre Island National Seashores. *Ocean and Coastal Management*, **71**, 33–45.

Houser, C., Wernette, P., Rentschlar, E. *et al.* 2015. Post-storm beach and dune recovery: implications for barrier island resilience. *Geomorphology*, **234**, 54–63.

Houser, C., Wernette, P. and Weymer, B.A. 2018. Scale-dependent behaviour of the foredune: Implications for barrier island response to storms and sea-level rise. *Geomorphology*, **303**, 362–74.

Hoyt, J.H. 1967. Barrier island formation. *Geological Society of America Bulletin*, **79**, 1125–36.

Jewell, M., Houser, C. and Trimble, S. 2014. Initiation and evolution of blowouts within Padre Island National Seashore, Texas. *Ocean and Coastal Management*, **95**, 156–64.

Jewell, M., Houser, C. and Trimble, S. 2017. Phases of blowout initiation and stabilization on Padre Island revealed through ground-penetrating radar and remotely sensed imagery. *Physical Geography*, **38**(6), 556–77.

Jewell, P. 2007. Morphology and paleoclimatic significance of Pleistocene Lake Bonneville spits. *Quaternary Research*, **68**, 421-430.

Jiménez, J.A. and Sánchez-Arcilla, A 2004. A long-term (decadal scale) evolution model for microtidal barrier systems. *Coastal Engineering*, **51**, 749–64.

Jol, H.M., Lawton, D.C. and Smith, D.G. 2003. Ground penetrating radar: 2-D and 3-D subsurface imaging of a coastal barrier spit, Long Beach, WA, USA. *Geomorphology*, **53**, 165–81.

Jongejan, R., Ranasinghe, R., Wainwright, D., Callaghan, D.P. and Reyns, J. 2017. Drawing the line on coastline recession risk. *Ocean and Coastal Management*, **122**, 87-94.

Keijsers, J.G.S., Poortinga, A., Riksen, M.J.P.M. and de Groot, A.V. 2012. Connecting aeolian sediment transport with foredune development. In Kranenburg, W., Horstman, E., & Wijnberg, K. M. (eds.), *NCK-days 2012 Crossing Borders in Coastal Research. Jubilee Conference Proceedings 20th NCK-Days*. Enchede: University of Twente, pp. 153–6.

Kjerfve, B. 1986. Comparative oceanography of coastal lagoons. In Wolfe, D.A. (ed.), *Estuarine Variability*. New York: Academic Press, pp. 63–81.

Kjerfve, B. and Magill, K.E. 1989. Geographic and hydrodynamic characteristics of shallow coastal lagoons. *Marine Geology*, **88**, 187–99.

Kragtwijk, N.G., Zitman, T.J., Stive, M.J.F. and Wang, Z.B. 2004. Morphological response of tidal basins to human interventions. *Coastal Engineering*, **51**, 207–21.

Kriebel, D.L. and Dean, R.G. 1985. Numerical simulation of time-dependent beach and dune erosion. *Coastal Engineering*, **9**, 221–45.

Kumar, N. and Sanders, J.E. 1974. Inlet sequence: a vertical succession of sedimentary structures and textures created by the lateral migration of tidal inlets. *Sedimentology*, **21**, 491–532.

Larson, M., Erikson, L. and Hanson, H. 2004. An analytical model to predict dune erosion due to wave impact. *Coastal Engineering*, **51**, 675–96.

Lazarus, E., Ashton, A., Murray, A.B., Tebbens, S. and Burroughs, S. 2011. Cumulative versus transient shoreline change: dependencies on temporal and spatial scale. *Journal of Geophysical Research: Earth Surface*, **116**(F2).

Leatherman, S.P. 1976. Barrier island dynamics: overwash processes and aeolian transport. *Proceedings of the Fifteenth Coastal Engineering Conference*, New York: American Society of Civil Engineers, pp. 1958–74.

Leatherman, S.P. 2003. Shoreline change mapping and management along the US East Coast. *Journal of Coastal Research*, **SI 38**, 5–13.

Leatherman, S.P. 2018. Coastal Erosion and the United States National Flood Insurance Program. *Ocean and Coastal Management*, **156**, 35–42.

Leatherman, S.P., Williams, A.T. and Fisher, J.S. 1977. Overwash sedimentation associated with a large-scale northeaster. *Marine Geology*, **24**, 109–21.

List, J.H., Sallenger, A.H. Jr, Hansen, M.E. and Jaffe, B.E. 1997. Accelerated sea-level rise and rapid coastal erosion: testing a causal relationship for the Louisiana barrier islands. *Marine Geology*, **140**, 347–65.

Long, A.J., Waller, M.P. and Plater, A.J. 2006. Coastal resilience and late Holocene tidal inlet history: the evolution of Dungeness Foreland and the Romney Marsh depositional complex (UK). *Geomorphology*, **82**, 309–36.

Mäkinen, J. and Räsänen, M. 2003. Early Holocene regressive spit-platform and nearshore sedimentation on a glaciofluvial complex during the Yoldia Sea and the Ancylus Lake phases of the Baltic Basin, SW Finland. *Sedimentary Geology*, **158**, 25–56.

Mariner, N., Goiran, J.P. and Morhange, C. 2008. Alexander the Great's tombolos at Tyre and Alexandria, eastern Mediteranean. *Geomorphology*, **100**, 377–400.

Mathew, S., Davidson-Arnott, R.G.D. and Ollerhead, J. 2010. Evolution of a beach/dune system following overwash during a catastrophic storm: Greenwich Dunes, Prince Edward Island, 1936–2005. *Canadian Journal of Earth Sciences*, **47**, 273–90.

Matias, A., Ferreira, Ó, Vila-Concejo, A., Garcia, T., and Dias, J.A. 2008. Classification of washover dynamics in barrier islands. *Geomorphology*, **97**, 655–74.

Maun, M.A. 2009. *The Biology of Coastal Sand Dunes*. Oxford: Oxford University Press, 280 pp.

McCann, S.B. 1979. Barrier islands in the southern Gulf of St. Lawrence. In Leatherman S.P. (ed.), *Barrier Islands: From the Gulf of St. Lawrence to the Gulf of Mexico*. New York: Academic Press, pp. 29–63.

McGubbin, D.G. 1982. Barrier-island and strand plain facies. In Scholle, P.A. and Spearing, D. (eds.), *Sandstone Depositional Environments*. Tulsa, OK: American Association of Petroleum Geologists, pp. 247–79.

Meistrell, F.J., 1972. The spit-platform concept: laboratory observation of spit development. In Schwartz, M.L. (ed.), *Spits and Bars*. Stroudsburg, PA: Dowden, Hutchinson and Ross, pp. 225–83.

Mendelssohn, I.A., Hester, M.W., Monteferrante, F.J. and Talbot, F. 1991. Experimental dune building and vegetative stabilization in a sand-deficient barrier island setting on the Louisiana coast, USA. *Journal of Coastal Research*, **7**(1), 137–49.

Méndez Linares, A.P., López-Portillo,J., Hernández-Santana, J.R., Ortiz Pérez, M.A. and Oropeza Orozco, O. 2007. The mangrove communities in the Arroyo Seco deltaic fan, Jalisco, Mexico, and their relation with the geomorphic and physical–geographic zonation. *CATENA*, **127**, 127–42.

Meyer-Arendt, K.J. 1990. Recreational business districts in Gulf of Mexico seaside resorts. *Journal of Cultural Geography*, **11**(1), 39–55.

Meyer-Arendt, K.J. 1991. Tourism development on the north Yucatan coast: human response to shoreline erosion and hurricanes. *GeoJournal*, **23**(4), 327–36.

Miller, D.L., Thetford, M. and Yager, L. 2001. Evaluation of sand fence and vegetation for dune building following overwash by Hurricane Opal on Santa Rosa Island, Florida. *Journal of Coastal Research*, **17**(4), 936–48.

Moore L.J., List J.H., Williams, S.J. and Stolper D. 2007. Modelling barrier island response to sea-level rise in the Outer Banks, North Carolina. In *The Proceedings of the Coastal Sediments '07*. Reston, VA: American Society of Civil Engineers, pp. 1153–64.

Moore, L.J., Goldstein, E.B., Vinent, O.D. *et al.* 2018. The role of ecomorphodynamic feedbacks and landscape couplings in influencing the response of barriers to changing climate. In Moore, L. and Murray, A. (eds.), *Barrier Dynamics and Response to Changing Climate*. Cham: Springer, pp. 305–36.

Moore, N.H. and Slinn, D.J. 1984. The physical hydrology of a lagoon system on the Pacific coast of Mexico. *Estuarine, Coastal and Shelf Science*, **19**, 27–42.

Morton, R.A. 2002. Factors controlling storm impacts on coastal barriers and beaches: a preliminary basis for real-time forecasting. *Journal of Coastal Research*, **18**, 486–501.

Morton, R.A. and Sallenger, A.H. 2003. Morphological impact of extreme storms on sand beaches and barriers. *Journal of Coastal Research*, **19**, 560–74.

Morton, R.A., Paine, J.G. and Gibeaut, J.C. 1994. Stages and durations of post-storm beach recovery, southeastern Texas coast, USA. *Journal of Coastal Research*, **10**(4), 884–908.

Morton, R.A., Ward, G.H. and White, W.A. 2000. Rates of sediment supply and sea-level rise in a large coastal lagoon. *Marine Geology*, **167**, 261–84.

Moslow, T.F. and Tye, R.S. 1985. Recognition and characteristic of Holocene tidal inlet sequences. *Marine Geology*, **63**, 129–51.

Neal, W.J., Pilkey, O.H., Cooper, J.A.G. and Longo, N.J. 2018. Why coastal regulations fail. *Ocean and Coastal Management*, **156**, 21–34.

Nichols, M.N. 1989. Sediment accumulation rates and relative sea-level rise in lagoons. *Marine Geology*, **88**, 201–19.

Nicholls, R.J., Townend, I.H., Bradbury, A.P., Ramsbottom, D. and Day, S.A. 2013. Planning for long-term coastal change: experiences from England and Wales. *Ocean Engineering*. **71**, 3–16.

Nielsen, L.H., Johannessen, P.N. and Surlyk, F. 1988. A late Pleistocene coarse-grained spit-platform sequence in northern Jylland, Denmark. *Sedimentology*, **35**, 915–37.

Nordstrom, K.F. 2004. *Beaches and Dunes of Developed Coasts*. Cambridge Cambridge University Press, 356 pp.

Nordstrom, K.F. and Jackson, N.L. 1998. Effects of a high rise building on wind flow and beach characteristics at Atlantic City, NJ, USA. *Ocean and Coastal Management*, **39**(3), 245–63.

Nordstrom, K.F. and McCluskey, J.M. 1985. The effects of houses and sand fences on the eolian sediment budget at Fire Island, New York. *Journal of Coastal Research*, **1**(1), 39–46.

Nordstrom, K.F., Lampe, R. and Vandemark, L.M. 2000. Reestablishing naturally functioning dunes on developed coasts. *Environmental Management*, **25**(1), 37–51.

Nordstrom, K.F., Jackson, N.L., Bruno, M.S. and de Butts, H.A. 2002. Municipal initiatives for managing dunes in coastal residential areas: a case study of Avalon, New Jersey, USA. *Geomorphology*, **47**(2–4), 137–52.

Nordstrom, K.F., Jackson, N.L., Klein, A.H., Sherman, D.J. and Hesp, P.A. 2006. Offshore aeolian transport across a low foredune on a developed barrier island. *Journal of Coastal Research*, **22**(5), 1260–7.

Nott, J., 2006. Tropical cyclones and the evolution of the sedimentary coast of northern Australia. *Journal of Coastal Research*, **22**(1), 49–62.

Novak, B. and Pedersen, G.K. 2006. Sedimentology, seismic facies and stratigraphy of a Holocene spit-platform complex interpreted from high-resolution shallow seismics, Lysegrund, southern Kattegat, Denmark. *Marine Geology*, **162**, 317–35.

O'Brien, M.P. 1969. Equilibrium flow areas and inlets on sandy coasts. *Journal of Waterways, Harbors and Coastal Engineering*, **15**, 43–52.

Oertel, G.F. 1985. The barrier island system. *Marine Geology*, **63**, 1–18.

Oertel, G.F. and Overman, K. 2004. Sequence morphodynamics at an emergent barrier island, middle Atlantic coast of North America. *Geomorphology*, **58**, 67–83.

Ollerhead, J. 1993. The evolution of Buctouche Spit, New Brunswick, Canada. PhD Thesis, University of Guelph, 156 pp.

Ollerhead, J. and Davidson-Arnott, R.G.D. 1995. The evolution of Buctouche Spit, New Brunswick, Canada. *Marine Geology*, **124**, 215–36.

Olson, J.S. 1958. Lake Michigan dune development 3. Lake level, beach and dune oscillations. *Journal of Geology*, **66**, 473–83.

Ontario Ministry of Natural Resources. 2001. Great Lakes–St. Lawrence River System and Large Inland Lakes Technical Guides for flooding, erosion and dynamic beaches in support of Natural Hazards Policies 3.1 of the Provincial Policy Statement (1997) of the Planning Act. Published by Watershed Science Centre, Trent University Peterborough, Ontario, Canada Copyright © 2001 The Queen's Printer for Ontario.

Orford, J.D. and Carter, R.W.G. 1984. Mechanisms to account for the longshore spacing of overwash throats on a coarse clastic barrier in southeast Ireland. *Marine Geology*, **56**, 207–26.

Orford, J.D., Jennings, S. and Pethick, J. 2003. Extreme storm effect on gravel dominated barriers. In *The Proceedings of the Coastal Sediments '03*. Reston, VA: American Society of Civil Engineers, 14 pp.

Otvos, E.G. 1981. Barrier island formation through nearshore aggradation: stratigraphic and field evidence. *Marine Geology*, **43**, 195–243.

Otvos, E.G. 1982. Santa Rosa island, Florida panhandle, origins of a composite barrier island. *Southeastern Geology*, **23**(1), 15–23.

Otvos, E.G. and Giardino, M.J. 2004. Interlinked barrier chain and delta lobe development, northern Gulf of Mexico. *Sedimentary Geology*, **169**, 47–71.

Owens, E.H. 1974. Barrier beaches and sediment transport in the southern Gulf of St. Lawrence. In *Proceedings of the Fourteenth Coastal Engineering Conference*. New York: American Society of Civil Engineers, pp. 1177–93.

Pacheco, A., Vila-Concejo, A., Ferreira, Ó. and Dias, J.A. 2008. Assessment of tidal inlet evolution and stability using sediment budget computations and hydraulic parameter analysis. *Marine Geology*, **247**, 104–27.

Park, J.Y. and Wells, J.T. 2007. Spit growth and downdrift erosion: results of longshore transport modelling and morphologic analysis at the Cape Lookout cuspate foreland. *Journal of Coastal Research*, **23**, 553–68.

Penland, S., Boyd, R. and Suter, J.R. 1988. Transgressive depositional systems of the Mississippi Delta plain: a model for barrier shoreline and shelf sand development. *Journal of Sedimentary Petrology*, **58**, 932–49.

Petersen, D., Deigaard, R. and Fredsøe, J. 2008. Modelling the morphology of sandy spits, *Coastal Engineering*, **55**, 671–84.

Pierce, J.W. 1970. Tidal inlets and washover fans. *The Journal of Geology*, **78**, 230–4.

Price, A.J., Miller, D.L. and Branch, L.C. 2008. Identification of structural and spatial features that influence storm-related dune erosion along a barrier-island-island ecosystem in the Gulf of Mexico. *Journal of Coastal Research*, **24**, 168–75.

Psuty, N.P. 2008. The coastal foredune: a morphological basis for regional coastal dune development. In Martínez, M.L. and Psuty, N.P. (eds.), *Coastal Dunes* Berlin, Heidelberg: Springer, pp. 11–27.

Rampino, M.R. and Sanders, J.E. 1980. Holocene transgression in south-central Long Island, New York. *Journal of Sedimentary Petrology*, **50**, 1063–80.

Ranasinghe, R. and Pattiaratchi, C. 1999. The seasonal closure of tidal inlets: Wilson Inlet – a case study. *Marine Geology*, **37**, 37–56.

Reinson, G.E. 1977. Tidal-current control of submarine morphology at the mouth of the Miramichi estuary, New Brunswick. *Canadian Journal of Earth Sciences*, **14**, 2524–32.

Robertson, W.V., Zhang, K. and Whitman, D. 2007. Hurricane-induced beach change derived from airborne laser measurements near Panama City, Florida. *Marine Geology*, **237**, 191–205.

Roy, P.S., Cowell, P.J., Ferland, M.A. and Thom, B.G. 1994. Wave-dominated coasts. In Carter, R.W.G. and Woodroffe, C.D. (eds.), *Coastal Evolution*. Cambridge: Cambridge University Press, 517 pp.

Rucińska-Zjadacz, M. and Wróblewski, R. 2018. The complex geomorphology of a barrier spit prograding into deep water, Hel Peninsula, Poland. *Geo-Marine Letters*, **38**(6), 513–25. DOI 10.1007/s00367–018-0554-6.

Ruggiero, P., Hacker, S., Seabloom, E. and Zarnetske, P. 2018. The role of vegetation in determining dune morphology, exposure to sea-level rise, and storm-induced coastal hazards: a US Pacific northwest perspective. In Moore L. and Murray A. (eds.), *Barrier Dynamics and Response to Changing Climate*. Cham: Springer, pp. 337–61.

Ruz, M.H. and Meur-Ferec, C. 2004. Influence of high water levels on aeolian sand transport: upper beach/

dune evolution on a macrotidal coast, Wissant Bay, northern France. *Geomorphology*, **60**(1–2), 73–87.

Sallenger, A.H. 2000. Storm impact scale for barrier islands. *Journal of Coastal Research*, **16**, 890–95.

Sanderson, P.G. and Elliot, I. 1996. Shoreline salients, cuspate forelands and tombolos on the coast of western Australia. *Journal of Coastal Research*, **12**, 761–73.

Sanderson, P.G., Eliot, I., Hegge, B. and Maxwell, S. 2000. Regional variation of coastal morphology in southwestern Australia: a synthesis. *Geomorphology*, **34**, 73–88.

Saunders, K.E. and Davidson-Arnott, R.G.D. 1990. Coastal dune response to natural disturbances. In Davidson-Arnott, R.G.D. (ed.), *Proceedings of the Symposium on Coastal Sand Dunes*. Ottawa: National Research Council of Canada, pp. 321–46.

Schupp, C.A., McNinch, J.E. and List, J.H. 2006. Nearshore shore-oblique bars, gravel outcrops, and their correlation to shoreline change. *Marine Geology*, **233** (1–4), 63–79.

Schwartz, M.L. 1971. The multiple causality of barrier islands. *Journal of Geology*, **79**, 91–4.

Schwartz, R.K. 1975. Nature and Genesis of some Storm Washover Deposits. US Army Corps of Engineers, Coastal Engineering Research Centre, Technical Memorandum 61, 67 pp.

Seabergh, W.C. 2006. Hydrodynamics of tidal inlets. Coastal Engineering Manual 1110-2-1100, US Army Corps of Engineers, Chapter II-6, 79 pp.

Sedgwick, P.E. and Davis, R.A. Jr. 2003. Stratigraphy of washover deposits in Florida: implications for recognition in the stratigraphic record. *Marine Geology*, **200**, 31–48.

Sherman, D.J., Hales, B.U., Potts, M.K. *et al.* 2013. Impacts of Hurricane Ike on the beaches of the Bolivar Peninsula, TX, USA. *Geomorphology*, **199**, 62–81.

Silvester, R. and Hsu, J. 1993. *Coastal Stabilisation: Innovative Concepts*. Englewood Cliffs, NJ: Prentice Hall, 578 pp.

Simeoni, U., Fontolan, G., Tessari, U. and Corbau, C. 2007. Domains of spit evolution in the Goro area, Po Delta, Italy. *Geomorphology*, **86**, 332–48.

Soons, J.M., Schulmeister, J. and Holt, S. 1997. The Holocene evolution of a well nourished gravelly barrier and lagoon complex, Kaitorete 'Spit', Canterbury, New Zealand. *Marine Geology*, **138**, 69–90.

Stallins, J.A. and Parker, A.J., 2003. The influence of complex systems interactions on barrier island dune vegetation pattern and process. *Annals of the Association of American Geographers*, **93**(1), 13–29.

Stockdon, H.F., Sallenger, A.H., Holman, R. and Howd, P.A. 2007. A simple model for the spatially-variable coastal response to hurricanes. *Marine Geology*, **238**, 1–20.

Stolper D., List J.H. and Thieler, E.R. 2005. Simulating the evolution of coastal morphology and stratigraphy with a new morphological-behaviour model (GEOMBEST). *Marine Geology*, **218**, 17–36.

Storms, J.E.A., Weltje, G.J., Terra, G.J., Cattaneo, A., and Trincardi, F. 2008. Coastal dynamics under conditions of rapid sea-level rise: late Pleistocene to early Holocene evolution of barrier–lagoon systems on the northern Adriatic shelf (Italy). *Quaternary Science Reviews*, **27**, 1107–23.

Swift, D.J.P. 1975. Barrier island genesis: evidence from the central Atlantic Shelf, eastern USA. *Sedimentary Geology*, **14**, 1–43.

Switzer, A.D. and Jones, B.G. 2008. Setup, deposition and sedimentary characteristics of two storm overwash deposits, Abrahams Bosom Beach, southeastern Australia. *Journal of Coastal Research*, **24**, 189–200.

Thieler, E.R. and Young, R.S., 1991. Quantitative evaluation of coastal geomorphological changes in South Carolina after Hurricane Hugo. *Journal of Coastal Research*, SI 8, 187–200.

Thom, B.G., Eliot, I., Eliot, M. *et al.* 2018. National sediment compartment framework for Australian coastal management. *Ocean and Coastal Management*, **154**, 103–20.

Thornberry-Ehrlich, T.L. 2005. Assateague Island National Seashore geologic resource management issues scoping summary. US Department of the Interior, National Park Service, available online at https://irma.nps.gov/DataStore/DownloadFile/596388.

Tomazelli, L. and Dillenburg, S. 2007. Sedimentary facies and stratigraphy of a last interglacial coastal barrier in south Brazil. *Marine Geology*, **244**, 33–45.

Van Heteren, S., FitzGerald, D.M., Barber, D.C., Kelley, J.T. and Belknap, D.F. 1996. Volumetric analysis of a New England barrier system using Ground-Penetrating-Radar and coring techniques. *The Journal of Geology*, **104**, 471–83.

Verhulst, P.F. 1838. Notice on the law which the population follows in its growth. *Correspondence Mathematique et Physique*, **10**, 113–26.

Vila-Concejo, A., Ferreira, Ó., Matias, A. and Dias, J.M.A. 2003. The first two years of an inlet: sedimentary dynamics. *Continental Shelf Research*, **23**, 1425–45.

Vinent, O.D. and Moore, L.J. 2015. Barrier island bistability induced by biophysical interactions. *Nature Climate Change*, **5**(2), 158–62.

Walker, I.J., Davidson-Arnott, R.G.D., Bauer, B.O. *et al.* 2017. Scale-dependent perspectives on the geomorphology of beach-dune systems. *Earth Science Reviews*, **171**, 220–53. DOI:10.1016/j.earscirev.2017.04.011.

Wang P. and Horowitz, M.H. 2007. Erosional and depositional characteristics of regional overwash deposits caused by multiple hurricanes. *Sedimentology*, **54**, 545–64.

Wernette, P., Houser, C., Weymer, B. *et al.* 2018. Long-range dependence in coastal framework geology: asymmetries and implications for barrier island resiliency. *Earth Surface Dynamics Discussions*, 1–23.

Weymer, B.A., Houser, C. and Giardino, J.R. 2013. Post-storm evolution of beach-dune morphology: Padre Island National Seashore, Texas. *Journal of Coastal Research*, **31**(3), 634–44.

Weymer, B.A., Wernette, P., Everett, M.E. and Houser, C. 2018. Statistical modeling of the long-range-dependent structure of barrier island framework geology and surface geomorphology. *Earth Surface Dynamics*, **6**(2), 431–50.

Williams, A., Feagin, R. and Stafford, A.W. 2008. Environmental impacts of beach raking of Sargassum spp. on Galveston Island, TX. *Shore and Beach*, **76**, 63–9.

Williams, A.T., Rangel-Buitrago, N., Pranzini, E. and Anfuso, G. 2018. The management of coastal erosion. *Ocean and Coastal Management*, **156**, 4–20.

Wolner, C.W., Moore, L.J., Young, D.R. *et al.* 2013. Ecomorphodynamic feedbacks and barrier island response to disturbance: insights from the Virginia Barrier Islands, Mid-Atlantic Bight, USA. *Geomorphology*, **199**, 115–28.

Woodroffe, C.D. 2007. The natural resilience of coastal systems: primary concepts. In McFadden, L. Penning-Rowsell, E. and Nicholls, R.J. (eds.), *Managing Coastal Vulnerability*. Amsterdam: Elsevier, pp. 45–60.

Zenkovitch, V.P. 1967. *Processes of Coastal Development*. Edinburgh: Oliver and Boyd, 738 pp.

Zhang K, Douglas B.C. and Leatherman, S.P. 2004. Global warming and coastal erosion. *Climate Change*, **64**, 41–58.

11

Saltmarshes and Mangroves

11.1 Synopsis

Saltmarshes and mangroves are wetlands formed in the intertidal zone of sheltered coasts, notably in bays, lagoons, delta distributaries and estuaries. The vegetation of the two environments is quite different: mangroves are composed of a number of tree species, while saltmarshes are dominated by grasses and herbs. Mangroves are confined to the tropics and the lower latitudes of the subtropics, and are the dominant intertidal vegetation in these regions. Saltmarshes are found throughout the mid and high latitudes and may also occur in some areas in the tropics. Plants of both ecosystems are salt-tolerant because they are frequently inundated by salt water and they are also adapted to growing under anaerobic conditions. The stress imposed by frequent inundation and waterlogged soils generally produces a gradient with one or two species dominating at lower levels and species richness increasing towards the high intertidal and beyond as the frequency of inundation decreases.

The movement of tidal waters across both mangroves and saltmarshes results in the development of tidal creek systems which act to move water, sediments and organic matter into and out of the systems. The vegetation of mangrove swamps and saltmarshes act to reduce wave action and the speed of tidal currents, thus promoting deposition of fine material. Vertical accretion in both environments ranges from a few millimetres to several centimetres a year and includes both organic material and mineral grains. However, net vertical accretion over decades is generally much less because of compaction of sediments and loss of some of the carbon from roots and leaf litter.

Both ecosystems are highly productive and act to support a wide range of fish, shellfish, migratory birds and animals, both within the environment and in the surrounding waters. While both ecosystems may be under stress from sea level rise, the greatest impact on them has come from a wide range of human activities, including dyking and drainage for agriculture, harbours and urban development. The building of dykes and embankments may enhance the potential adverse effects of sea level rise through inhibiting landward movement – termed coastal squeeze.

11.2 Saltmarsh and Mangrove Ecosystems

Saltmarshes and mangroves are types of halophytic wetland ecosystem that are found along marine coasts (Adam, 1990; Robertson and Alongi, 1992; Hogarth, 1999; Mitsch and Gosselink, 2000; Scott *et al.*, 2014). They are found in the upper intertidal zone in areas that are sheltered from high wave action, which in turn permits the establishment of vegetation communities. Because they occupy a zone of transition between terrestrial and marine ecosystems, marshes and mangroves play a critical role in sediment exchange with adjacent mudflats, estuaries and open coastal waters and with sediments transported from uplands to the coast. Fine sediment accretion is not necessary per se for the establishment of mangrove or saltmarsh communities, but standing biomass leads to the rapid attenuation of waves and currents capable of importing coarse sediment (sand and gravel) and thus it promotes the deposition, and particularly the retention, of fine sediment and organic matter. Saltmarshes and mangroves thus act as sinks for fine sediments that are brought in by the tidal waters and for the accumulation of organic matter (Gordon *et al.*, 1985; Woodroffe, 1992; Boyd *et al.*, 2017). As a result, they also act as a sink for contaminants (Scrimshaw *et al.*, 1996).

Rapid vertical accretion can occur in locations where there is an abundant supply of sediment (e.g., in estuaries and river deltas), producing a dominantly minerogenic deposit. However, where there is much less sediment available, accretion is predominantly from organic material such as roots, leaves and stems, and the resulting deposit is dominantly organogenic. Both saltmarshes and mangroves develop on nearly horizontal or gently sloping platforms with tidal creeks providing a conduit for the flow of water into and out of the zone (Figure 11.1). A significant fraction of inundation may occur from water coming across the marsh or mangrove margin, and in some cases this may be more important than that supplied from tidal creeks. Drainage may be restricted over portions of the surface while some areas near the landward margin may be inundated only a few times a year during extreme astronomical tides and/or storm surge events.

Saltmarshes are vegetated by grasses, herbs and small shrubs and are found throughout the middle and high latitudes (Figure 11.1A). Mangroves, in contrast, are primarily woody trees and shrubs that are almost exclusively tropical (Figure 11.1B) though their range does extend into the subtropics (Hogarth, 1999; Stokes *et al.*, 2010). Mangrove communities are the predominant intertidal plant association in tropical intertidal zones, though small areas of saltmarsh may also be found. Areas of saltmarsh and mangrove development include river mouths, estuaries, drowned river valleys and deltas, back-barrier lagoons and bays, natural embayments and sheltered areas behind islands and coral reefs (Allen and Pye, 1992; Woodroffe, 1992; Hogarth, 1999). Mangroves in particular may develop on open coasts where wave energy is dissipated over a wide shallow nearshore along deltas and muddy shorelines, such as those of Guyana, and major river deltas in Asia, such as those of the Ganges and Mekong.

The vegetation of both saltmarshes and mangroves is adapted to three major forms of stress: high levels of salinity; inundation for periods of hours (and sometimes days or weeks in the case of mangroves in some locations); and waterlogged and anaerobic soils. In general, these stress levels are highest at the seaward margins of marsh or mangrove communities where inundation periods are longest and most frequent, and consequently there may be more waterlogging. This tends to produce a zonation of plant species parallel to the shoreline, with dominance of a few species close to the margin where stress levels are generally highest, and increasing diversity towards higher ground. However, evaporation of sea water from areas such as pannes on saltmarshes or enclosed lagoons can produce salinities that are much higher than the 30–35‰ of sea water and local reversal of the zonation trend or the complete die back of vegetation can occur. At a small scale greater diversity is produced by small topographic irregularities that lead to improved or reduced drainage, by proximity to tidal creeks and by changes in local substrate or slope (Figure 11.1). Where saltmarshes and mangroves occur in the upper reaches of estuaries, along deltaic distributaries and in some closed lagoons, they may also be subject to periods where the water has a very low salinity. Saltmarshes can grade into tidal freshwater marshes inland or upriver and mangrove communities give way to other forest or herb communities that can survive inundation by fresh or brackish water.

Figure 11.1 Vegetation of saltmarshes and mangroves. A. View eastward with the sea to the left of the upper marsh at Stiffkey, north Norfolk, England. The elevation of the marsh platform here is about 2.7 m OD, which is close to MHWS. The lower areas are revegetated pans dominated by *Salicornia europaea*, with a few patches of *Spartina anglica*. The slightly higher areas around the pan are covered mainly by *Limonium vulgare* (purple flowers) and *Atriplex portulaciodes*, with some *Puccinellia maritima* (photo Helene Burningham). B. Oblique aerial view of mangroves fringing the landward side of the lagoon behind the Belize barrier reef. Mangroves occupy the mainland coast at the bottom of the picture and the margins of a small coral cay. The centre of the cay is largely bare of mangroves, possibly because of insufficient nutrient supply or excessive salinity.

Saltmarshes and mangroves are generally areas of high primary productivity. Decaying organic matter provides the beginning of a complex food chain that exists both within the marsh or mangrove community, and in the adjacent tidal flats and lagoon or estuarine waters to which organic matter is exported during ebb tides (Gordon *et al.*, 1985; Alongi, 1998). Most of the net primary productivity of mangroves comes from above the ground whereas the below ground contribution of saltmarsh grasses may be 1.5 times that of the above ground component (Alongi, 1998, p. 48). Mangroves and saltmarshes provide shelter for juvenile fish, and tidal mudflats (and some mangroves) support large populations of shellfish. Tidal creeks and mudflats are staging and wintering habitats for a wide variety of shorebirds and waterfowl, and mangrove forests support a rich fauna. Both marshes and mangroves have supported a wide range of subsistence or harvesting activities and both have suffered destruction in the face of human actions such as drainage for intensive agriculture (including shrimp farming in tropical areas), construction of housing, port and navigation facilities, as well as utilities such as power plants. Recognition of the valuable role they play ecologically as well as their potential role as a buffer against storms and tsunamis, sea level rise and siltation of coral reefs has stimulated increased research into the biology and geomorphology of saltmarshes and mangroves and to the incorporation of their preservation as a significant component of integrated coastal management.

11.3 Saltmarshes

Coastal saltmarshes form in the upper intertidal zone in latitudes ranging from the Arctic to the subtropics, and consequently they develop under a wide range of climatic conditions (Scott *et al.*, 2014). There are extensive saltmarshes in the Arctic through Alaska, Canada, Russia and Scandinavia, with the most extensive areas being in southern Hudson Bay and James Bay. They are found in the mid latitudes throughout eastern and western North America, western Europe and the Mediterranean, the east Asian coasts of Japan and Korea, as well as on the southern coasts of Australia, New Zealand, Argentina and Chile. Extensive saltmarsh systems on the east coast of North America and western Europe are developed within shallow drowned river mouths and in the lagoons behind barrier islands and spits. A large proportion of the research on saltmarsh development has taken place in these environments.

11.3.1 Saltmarsh Form, Flooding and Vegetation Zonation

Morphologically saltmarshes consist of a gently sloping vegetated platform, dissected by a network of tidal creeks that increase in width and depth seaward (Pethick, 1992). The seaward margin (strictly speaking the margin bordering the ocean, bay, estuary or tidal creek from which tidal waters flood the marsh) may be gently sloping with no visible change in slope from the vegetated surface to the fronting mud or sand flat; or it may consist of a low cliff with an abrupt transition. An idealised cross section shows a transition from the upland area with an elevation above that of inundation by marine waters, through the high marsh zone, low marsh zone and onto the tidal mudflats that often occupy the intertidal zone below the level to which saltmarsh vegetation can become established. Examples of this zonation are shown for: (1) Allen Creek marsh, a hypertidal marsh on the Cumberland Basin, in the upper Bay of Fundy, Canada (Figures 11.2; 11.3); and (2) for Skallingen, a microtidal back barrier marsh on the west coast of Denmark at the north end of the Wadden Sea (Figures 11.4; 11.5). In the low marsh zone along the marsh edge and the banks of tidal creeks, vegetation is submerged on almost every tide and can be submerged for 6 hours or more depending on the tidal regime. Towards the high marsh zone the frequency and depth of inundation decrease and within the high marsh zone vegetation may be submerged only briefly during spring high tides. Close to the upland boundary, inundation only occurs during exceptional spring tides and as a result of high water levels associated with storm surge. In the first case the period

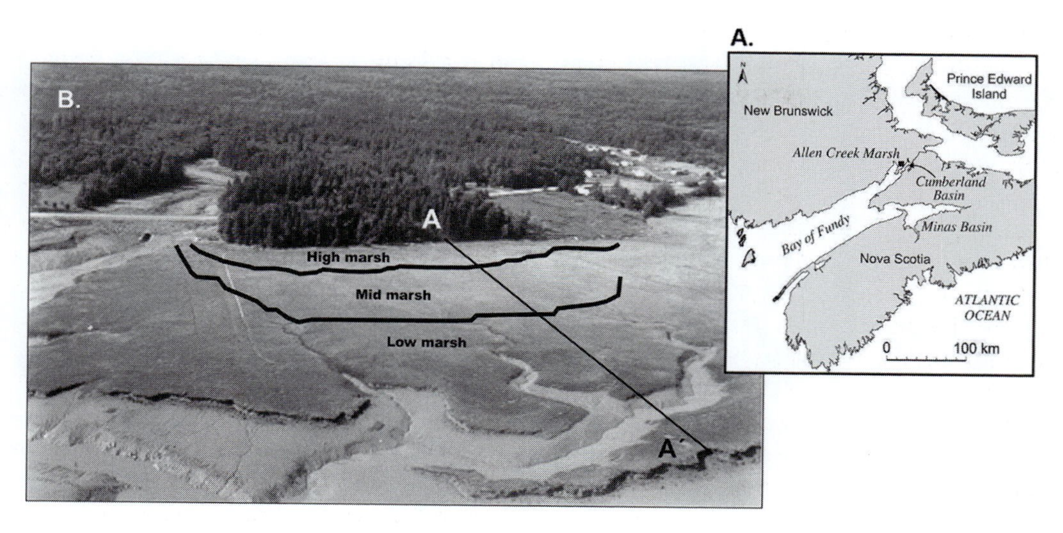

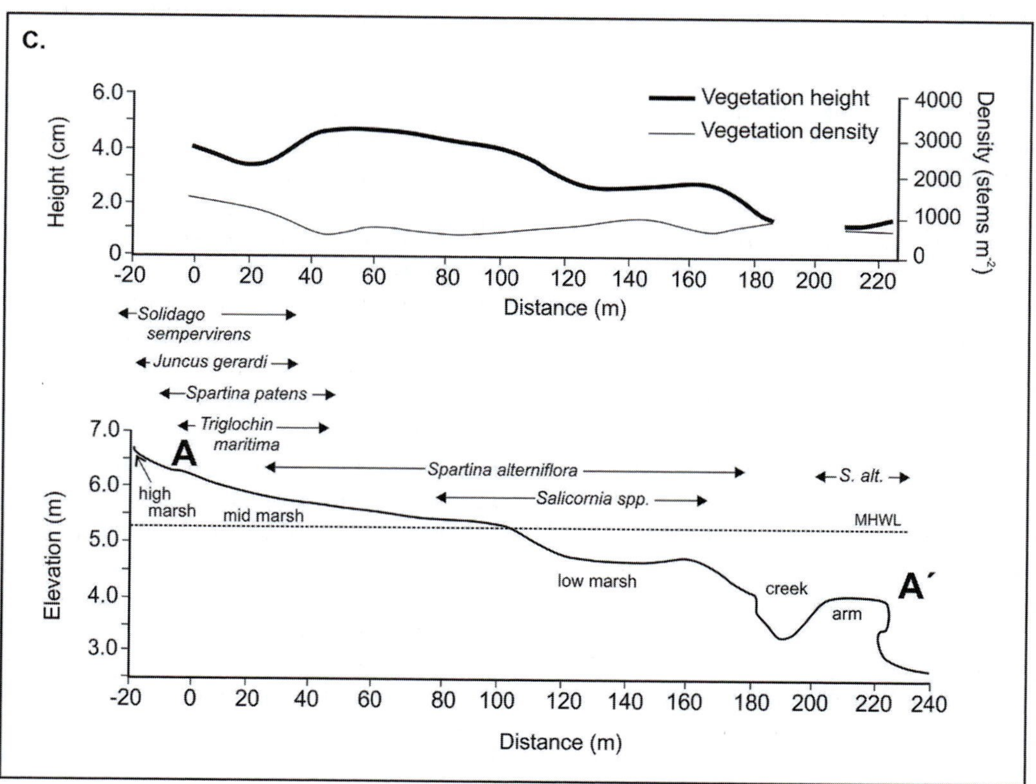

Figure 11.2 Allen Creek marsh, Cumberland Basin, Bay of Fundy. A. Location of Allen Creek marsh and the Cumberland Basin within the Bay of Fundy. B. Oblique aerial photograph showing the main topographic features and approximate limits of vegetation zonation (van Proosdij *et al.*, 1999). The marsh is presently erosional, so the margins bordering on the Cumberland Basin (bottom of photograph) are marked by a low cliff. Allen Creek is a small perennial stream which is tidal in its lower reach, while the marsh surface has a number of tidal creeks which only drain the marsh surface and which continue onto the fronting mudflats. C. Topographic profile, distribution of dominant plant species and vegetation characteristics along line AA′ shown in B. (After van Proosdij *et al.*,1999.)

Figure 11.3 Photographs looking landward from a platform on Allen Creek marsh in late June at: A. low tide; B. spring high tide. The vegetation in the foreground in A. is *Spartina alterniflora* which is nearing its maximum height. This portion of the low marsh is submerged by water up to a metre deep on spring high tides.

of inundation is very short, but storm surge events may lead to submergence for hours and occasionally one or two days in the case of the Skallingen backbarrier marsh (Bartholdy and Aagaard, 2001).

The depth and duration of submergence limit the amount of time that respiration and photosynthesis can take place, with the result that saltmarsh vegetation does not grow well much below the mean high tide level. This varies spatially with location on the profile and daily with the neap/spring tidal cycle, and it increases with increasing tidal range. Some portion of the leaves of grasses on the low marsh may be above the water for much of the time on microtidal marshes, but on macrotidal marshes the whole plant may be submerged for several hours (Figure 11.3B). Growth of vegetation on the low marsh may also be affected by cold temperatures in the substrate that may take some time to warm up after flood waters recede. The result is that biomass productivity on the marsh is strongly influenced by location on the profile and thus the frequency and duration of submergence (Gordon *et al.*, 1985). Net primary productivity of saltmarshes can also be expected to decrease towards higher latitudes because of the decrease in growing season.

Saltmarsh productivity and the zonation and abundance of species are also strongly influenced by salinity of the water and the substrate (Adam, 1990; Mitsch and Gosselink, 2000). In the low marsh, where plants are flooded once or twice daily, salinity is similar to that of the ocean waters. Towards the upper marsh, salinity may increase as a result of a combination of moderate flooding frequency, poor drainage and greater evaporation, and this may limit both the species and their growth. In some very poorly drained areas salt pannes develop; salinity in them may exceed 80‰ and completely inhibit vegetation growth (Figure 11.6). Generally, the low marsh where stresses are highest is dominated by one or two species and species abundance increases in the high marsh zone where conditions are more favourable for a wider range of plants and competition becomes more important (Gray, 1992). Individual plant species commonly occupy quite a wide range here, but are abundant over a much narrower range (Figure 11.2B and Table 11.1). However, while elevation and distance from the creek margin may exercise some control on the occurrence of particular plant species, the resulting distribution is usually quite complex and reflects small topographic irregularities as well as historical and random factors (Silvestri *et al.*, 2005).

11.3.2 Controls on Saltmarsh and Tidal Creek Development

11.3.2.1 *Saltmarshes*

The characteristics and evolution of saltmarshes (Figure 11.6) are determined by a wide range of

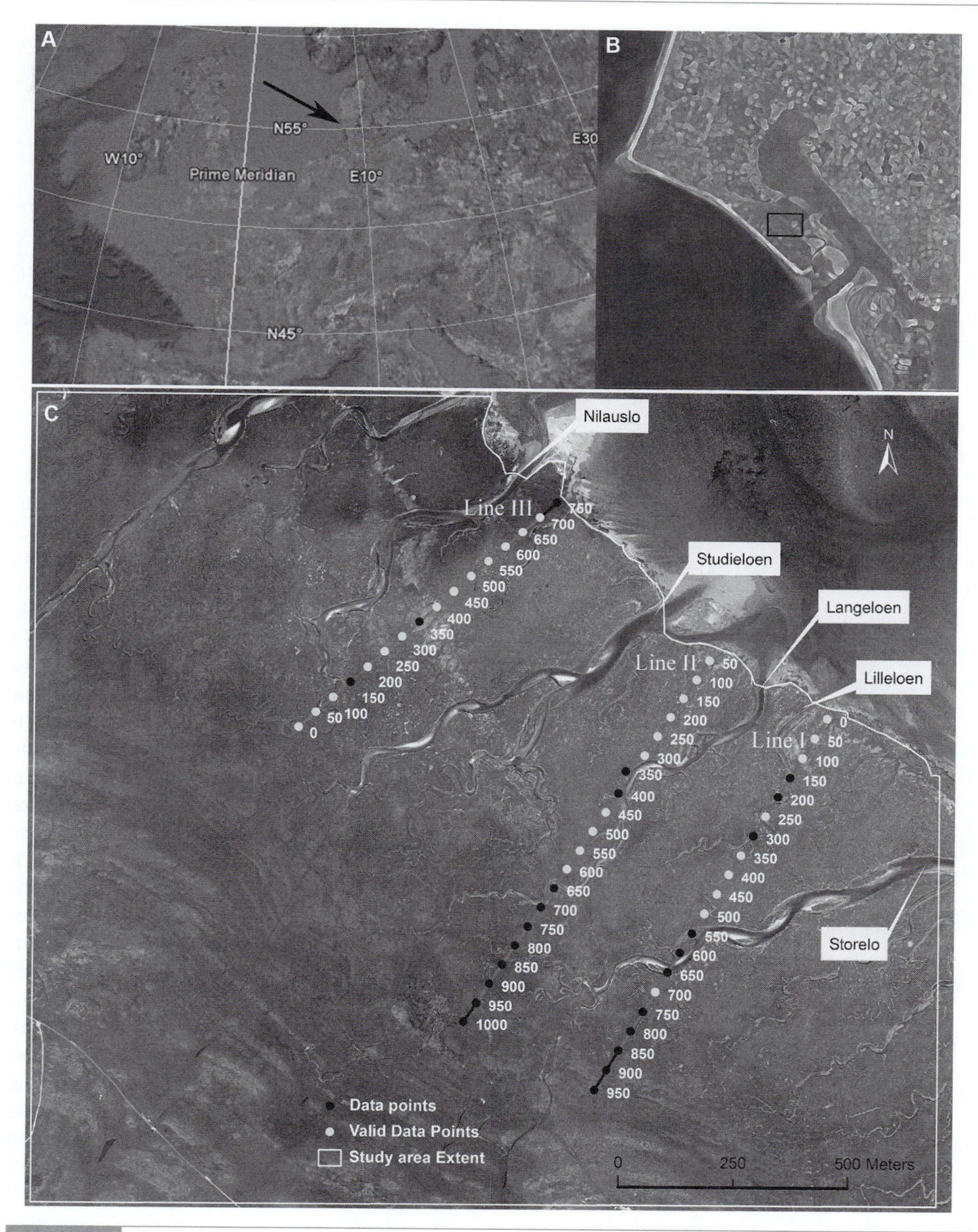

Figure 11.4 Skallingen marsh, west coast of Denmark. A. Location of Skallingen barrier spit in relation to the North Sea and the barrier islands of the Wadden Sea. B. Satellite image of the Skallingen barrier spit. C. Satellite image of a portion of the backbarrier marsh showing the division between the low marsh and high marsh and profiles. (After Bartholdy *et al.*, 2010a; B. and C. courtesy Jesper Bartholdy.)

Table 11.1 Distribution of plant species on the back barrier marsh at Skallingen spit (after Bertholdy *et al.*, 2010a and Kim *et al.*, 2016)

Location	High marsh	Mid marsh	Low marsh
Elevation above datum (m)	5.8–6.2	5.3–5.8	4.8–5.3
Dominant species	*Spartina patens*	*Spartina alterniflora*	*Spartina alterniflora*
Subordinate species	*Atriplex sp* *Dischilia spicata* *Elymus arenarius* *Glaux maritima* *Hierochloe odorata* *Juncus gerardi* *Limonium nashi* *Plantago maritima* *Puchinella maritima* *Solidago sepmiverens* *Triglochin maritima*	*Triglochin maritima* *Spartina patens* *Limonium nashi* *Salicornia spp* *Suaeada maritima*	*Salicornia spp* *Suaeada maritima* *Puchinella maritima* *Limonium nashi*

Figure 11.5 View landward across the lower saltmarsh at Skallingen, Denmark showing the low surface gradient on the saltmarsh platform and a small tidal creek draining the marsh in the foreground. The marsh is about 1500 m wide at this location and the dunes marking the seaward margin of the barrier can be seen in the background. Herbaceous vegetation that characterises the low marsh is dominantly *Pucinellia maritima* and *Suaedia maritima*.

physical and biological controls and processes, including: climate, shoreline configuration and wave climate, tidal range, sediment sources and volume of sediment input, sea level history, and vegetation characteristics and dynamics (Luternauer *et al.*, 1995; Mitch and Gosselink, 2000; Scott *et al.*, 2014). Climatic factors in middle and high latitudes control the occurrence and distribution of the species of vegetation that colonise the marsh as well as influencing physical processes such as marsh hydrology and salinity, wave climate and wave-induced sediment transport, and the occurrence of ice and factors associated with below freezing temperatures (Gordon and Desplanque, 1983; Argow *et al.*, 2011; Coulombier *et al.*, 2012). The shoreline configuration influences the local wave climate, the slope and nature of the intertidal substrate, and the circulation pattern of tidal currents.

Tidal range and the tidal regime (semi-diurnal, mixed or diurnal) influence the hydrodynamics of flow in tidal creeks and over the marsh surface as well as the extent and duration of inundation (Healey *et al.*, 1981; French and Stoddart, 1992). In turn, all of these influence the vertical and horizontal extent over which saltmarsh development takes place. The tidal range classes established in Chapter 3 can be used to divide saltmarshes into microtidal (<2 m), mesotidal (2–4 m); macrotidal (4–6 m) and hypertidal (>6m).

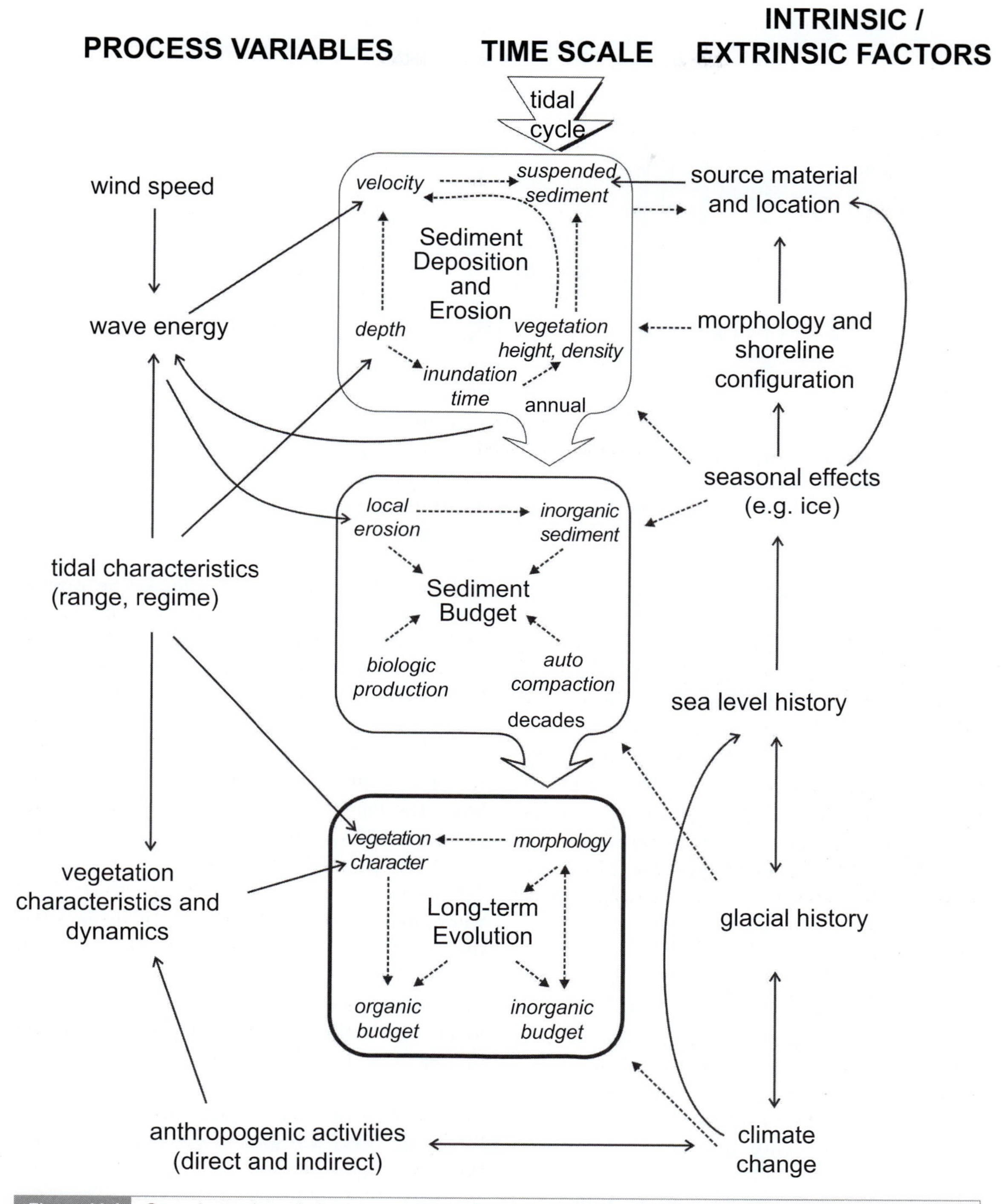

Figure 11.6 Controls on saltmarsh development at varying time scales (after Davidson-Arnott *et al.*, 2002).

Figure 11.7 Colonisation of a tidal mudflat, Peck's Cove, Bay of Fundy by *Spartina alterniflora*. The marsh is about 15 years old and the vegetation front has advanced at 8–10 m a^{-1}. The solid line marks the junction of the marsh with the tidal flats. The mudflats in the middle distance (between the solid and dashed lines) are up to 600 m wide at low tide. Note the fine sediments trapped by the vegetation which contrast with the coarse sand and gravel of the beach which is exposed to wave action only during spring high tides.

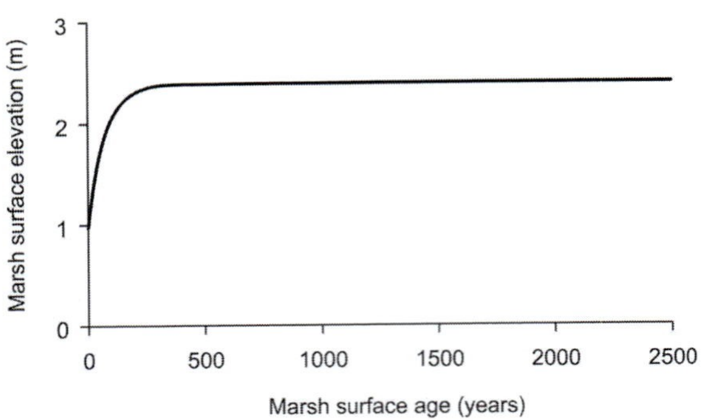

Figure 11.8 Conceptual model of the relationship between marsh height and age (after Pethick, 1981).

The nature of sediment sources and the volume of sediment supply will influence the characteristics of the marsh substrate, sedimentation patterns and the potential rate of sediment accumulation, vertical accretion and vegetation development. In estuaries, river mouths and deltas sediments may be derived primarily from fluvial inputs, while in embayments and in barrier/lagoon systems they may be derived largely from shoreline erosion and alongshore and onshore sediment transport by waves. Sediment-laden waters may reach the marsh directly in the case of marshes located on open coast or embayments, or indirectly through estuaries and inlets. In some areas such as rocky coasts, or enclosed lagoons subject to infrequent overwash, the supply of mineral sediment may be very low and marsh accretion is largely by organic accumulation. In other areas such as the Bay of Fundy in Canada or the Wash in Great Britain, large volumes of sediment are supplied through coastal erosion of fine-grained rocks.

Saltmarsh development is commonly pictured as beginning with the colonisation of intertidal sand or mudflats by vascular plants that are both halophytic and tolerant of repeated submergence for periods of up to several hours (Figure 11.7). The establishment of plants then encourages the deposition of fine sediments and the accumulation of organic matter, leading to the vertical growth of the developing marsh surface and the integration of the tidal creek network into the saltmarsh tidal creek network (Redfield, 1972). Under a stable sea level, vertical growth of saltmarshes is limited by tidal range. The rate of upward growth of saltmarshes, and of minerogenic saltmarshes in particular, has been shown to be asymptotic based on sampling of 14 saltmarshes in the UK (Pethick, 1981; see Figure 11.8). There is rapid growth in the early phase and very low growth rates once a mature marsh surface is established near or above the mean high tide level. The height of the marsh surface can thus be expressed as an asymptotic curve:

$$h = a - be^{-ct}, \qquad (11.1)$$

where h is the height of the marsh surface, t is the age of the marsh and a, b and c are constants.

The actual rate of growth will vary from marsh to marsh depending on the supply of sediment and rate of accumulation of organic matter. The model is based on the assumption that the rate of sediment deposition decreases rapidly once the marsh surface is above mean high tide because there is a rapid drop-off in flooding frequency, and thus the potential for sediment deposition, above this elevation. However, under rising sea level, thick deposits can accumulate as long as the vertical growth can keep pace with the rate of sea level rise (Allen, 1990, French, 1993; Reed *et al.*, 2018).

Figure 11.9 Poorly developed drainage and pannes on the marsh surface of a backbarrier marsh, Culbin, Scotland.

Figure 11.10 Lower portion of a tidal creek at Allen Creek marsh near the marsh margin where the depth and width are largest. The creek is just to the left of profile AA′ in Figure 11.2B.

11.3.2.2 *Tidal Creek Networks*

A characteristic feature of most saltmarshes is the development of a network of tidal creeks which resemble drainage basins of fluvial systems, with small tributaries forming near spring high tide and the depth and width of the creeks increasing towards the marsh margin (Steel and Pye, 1997). These creeks often continue out onto the intertidal mudflats in front of the marsh (Figure 11.2B; 11.4B). The pattern of the creek networks is generally dendritic, but a range of other patterns can be identified (Steel and Pye, 1997), including reticulate, and complex patterns on poorly drained areas with numerous ponds (Figure 11.9). A major difference between the tidal creek system and fluvial systems is that there are generally no topographic high points marking the drainage divides. Marsh platforms have little relief and the total elevation difference on the platform will usually be less than half that of the spring tidal range. The tidal range limits the maximum channel depth and thus channels on microtidal coasts will be very much shallower than those on macrotidal coasts (Figures 11.5 and 11.10).

Flow over the marsh surface and within the channels of the tidal creeks is not driven by topographic slope, except at the end of the ebb, but rather by the hydraulic head resulting from the lag between tidal elevation in the bay or estuary and that on the marsh surface, and by the total tidal prism over the marsh surface (Allen, 2000; O'Laughlin and van Proosdij, 2013). As a result maximum flow speeds will be greater in tidal creeks on wide marshes such as Skallingen (Figure 11.4B), than on narrow marshes such as Allen Creek (Figure 11.2B). On the rising tide, flow in the channels and across the vegetated marsh surface is retarded by friction, aided by the shallower flow and so the water surface slopes from the seaward margin towards the landward edge. Likewise, on the ebb tide water levels seaward of the marsh fall more rapidly than on the marsh generating an hydraulic head in the opposite direction. When water covers the marsh surface, flow will be influenced by residual tidal currents, flows in the basin and also by winds.

Where saltmarshes develop through the growth of vegetation out onto tidal mudflats it is usually thought that the creek system on the marsh is inherited with very little change from the pre-existing tidal creek network on the mudflats. As vegetation becomes established, trapping of fine sediment leads to the vertical growth of the marsh and the channels become fixed in place because of bank stability. Under this condition, evolution of the channel network occurs very slowly (Allen, 2000). The bed and bottom portion of the channel banks on the lower portion of the tidal creek within the marsh may be free of vegetation (Figure 11.9), but the upper reaches are usually shallow and may be completely vegetated. Once the landward portion of the marsh nears the upper limit of vertical growth and inundation is infrequent, the headwaters of the creek channels become silted up. This may also account for the infrequent occurrence of tidal creek networks in some Australian saltmarshes that may occupy the upper intertidal zone landward of a zone of fringing mangroves (Adam, 1997).

Where saltmarshes develop on backbarrier sands there may be no pre-existing channel network and drainage is often very poor, with many small ponds remaining on the surface after the tide has receded (Figure 11.10). In this case, the drainage system develops on the marsh surface after formation of the marsh and may result from the gradual connection of ponds on the surface (Allen, 2000; Perillo and Iribarne, 2003; Bartholdy *et al.*, 2004). At Skallingen the vegetation colonisation of the backbarrier began a little over 100 years ago and the creek network seen in Figure 11.4C evolved rapidly over the first half century (Bartholdy *et al.*, 2018). Expansion of the network occurred both laterally through new tributary growth and through headward extension of the creeks which intercepted and drained many of the ponds on the high marsh (Bartholdy *et al.*, 2018). Spontaneous formation of a tidal creek network has been observed on the floor of a newly created marsh within the Venice lagoon (D'Alpaos *et al.*, 2007).

11.3.3 Saltmarsh Hydrodynamics

The rise and fall of the tides generate flows first in the tidal creeks and then across the marsh surface as the level exceeds the creek banks and/or the marsh margin (Figure 11.11). During neap tides only a portion of the low marsh may be flooded, but at spring high tides flooding may extend to and across much of the high marsh. Where the tidal creek channels are marked by levees, portions of the marsh surface may remain dry for some period during the rising tide until the levee is overtopped. At the same time, on wide marsh platforms, portions of the surface towards the landward end of the marsh may be flooded before others that are closer to the margin if there is a protective chenier, beach ridge or dune marking the transition to the tidal flats. In many cases, however, flooding of the marsh surface occurs nearly simultaneously from the margin and from overtopping of creek banks. We can thus distinguish between conditions during the early and late part of the tidal cycle when the waters are largely confined to the tidal creek network, and those near high tide when the marsh surface itself is covered and water movement may be influenced by tidal currents in the larger basin and by wind action in addition to hydraulic gradients due to the tidal rise and fall. These flows are critical to the import of siliclastic material to the marsh and the export of both siliclastic and organic materials from the marsh.

Figure 11.11 The transition from flows in tidal creeks to overmarsh flow on a microtidal marsh in a small basin on the lower LaHave estuary, Nova Scotia. The tide has filled the tidal creek channel and is now advancing up the small creeks on the marsh surface as well as over the creek margins. The low marsh is dominated by *Spartina alterniflora* with the high marsh in the foreground marked by the presence of *Spartina patens*.

11.3.3.1 *Tidal Creeks*

Where the marsh surface provides an extensive flat surface close to the creek channels there may be quite sudden large increases in the supply of water out of the channels on the rising tide and into the channel on the falling tide around the bankfull stage (French and Stoddart, 1992). An example of velocity-stage curves for gauging stations near the mouth and in the middle section of a tidal creek developed in marshes behind Scolt Head Island, Norfolk, England is shown in Figure 11.12A and B (French and Stoddart, 1992). These data show that flow speeds can reach nearly 1 m s^{-1} on the ebb tide when the stage

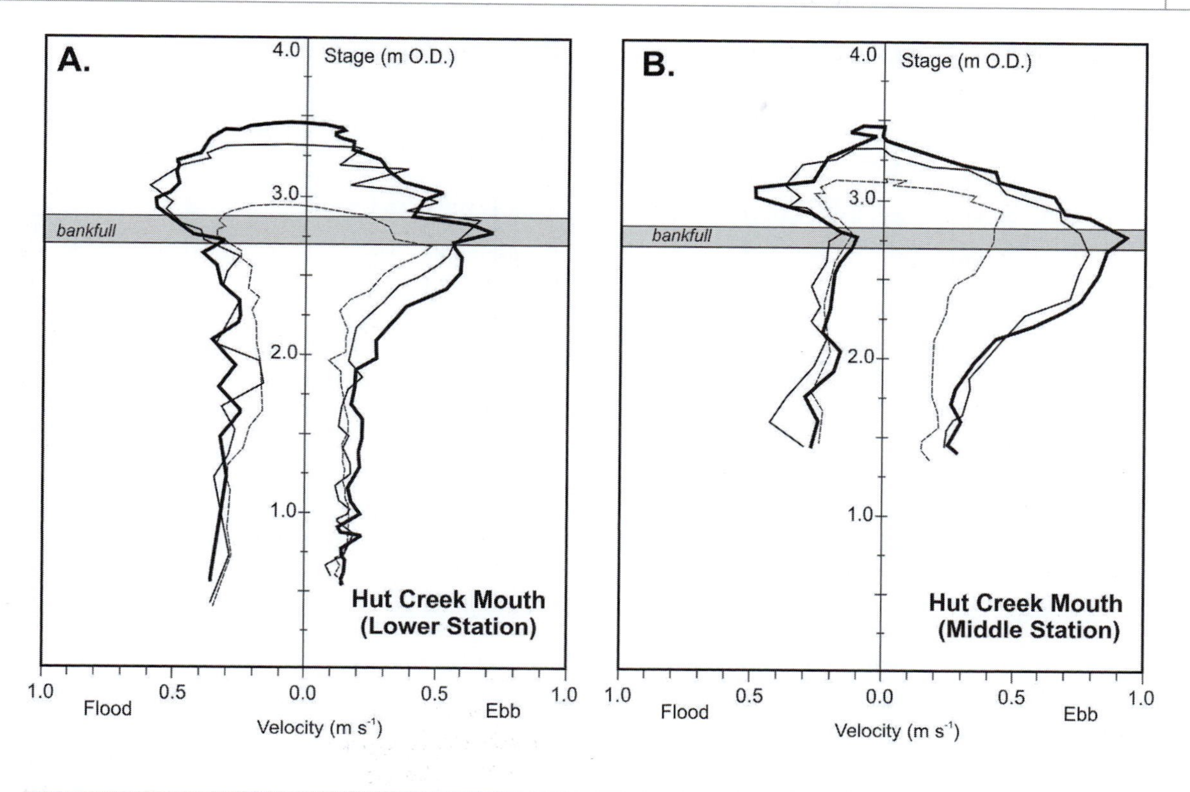

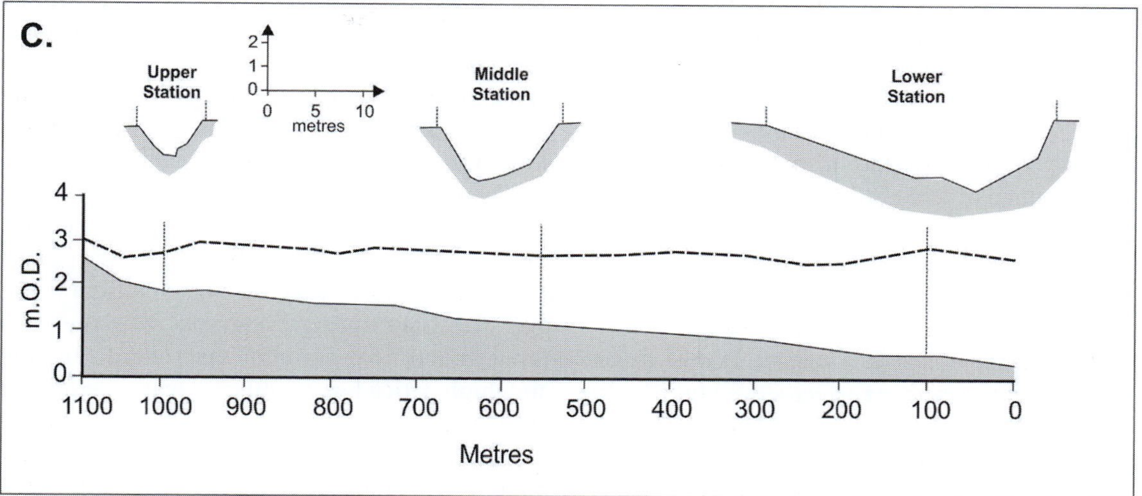

Figure 11.12 Flows speeds versus stage elevation for selected tides at two locations on Hut Creek, Scolt Head Island: A. lower station about 100 m from the mouth; B. middle station about 600 m from the mouth of the creek. Cross profiles at the stations and long profile of the creek are shown in C. (After French and Stoddart, 1992.)

crosses the bankfull elevation on the middle station (about 600 m from the creek mouth) and that speeds here are higher than at the lower station (about 100 m from the mouth). These speeds are high enough to erode the bank and to flush out sediments that might be deposited in the channel during neap tides. Both stations also show a distinct ebb asymmetry – higher speeds on the ebb

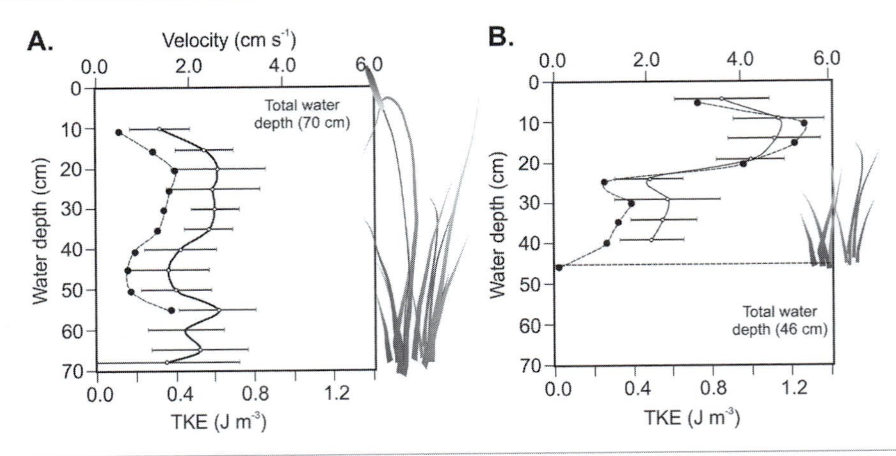

Figure 11.13 Velocity and total kinetic energy (TKE) profiles collected in emergent and submerged *Spartina alterniflora* canopies: A. velocity and TKE profiles for the natural vegetation with leaves of the higher plants reaching the water surface; B. the profiles after vegetation had been clipped to produce an average height of about 25 cm. Error bars show vertical turbulence intensity associated with each measurement. (After Leonard and Croft, 2006.)

than on the flow – which is commonly found in many tidal creeks (Bayliss-Smith *et al.*, 1979; Allen, 2000). These findings contrast with measurements of flows in tidal creeks on Allen Creek marsh (Schostak *et al.*, 2000) where maximum flow speeds were generally < 0.1 m s^{-1}. Despite the very large tidal range and the erosional appearance of the tidal creek channels (Figure 11.2B), the short channel length and absence of levees produce a small tidal prism for an incremental change in stage elevation and serve to highlight the control of water surface slope and tidal prism on flow speeds.

11.3.3.2 *Marsh Surface*

Flow over the saltmarsh surface is still driven primarily by the water surface slope but it is also influenced by winds and potentially by wave action. It is also greatly influenced by the presence of vegetation on the marsh surface which retards flow speeds and alters the velocity profile. Flow speeds are greatly dampened within the vegetation canopy and both vertical and horizontal eddies are reduced. The effects on mean flow speeds and turbulence intensity are greatly affected by the stem density, height and stiffness of the vegetation (Leonard and Luther, 1995; Christiansen *et al.*, 2000; Shi *et al.*, 2000; Nepf, 2012; Houser *et al.*, 2015). As might be expected, flow and turbulence reduction are greater as the vegetation height increases (Figure 11.13), and the same kind of reduction occurs as the total vegetation biomass increases. Where the whole plant is submerged there is a rapid dampening of turbulence below the top of the canopy and this likely occurs because the ends of the leaves tend to curve and present a floating mat near the surface, particularly late in the growing season when the stems are longest and most flexible.

The effect of vegetation on flow varies with the depth of submergence of the plants and with the absolute and relative height of the plants. This will therefore vary through the tidal cycle, over the neap–spring cycle and with the tidal range (microtidal through macrotidal). In mid and high latitudes it will also vary seasonally with the growth stage of the plants. Because the stems of most saltmarsh grasses are tightly bunched there are gaps around each plant close to the bed and flow speeds around each plant are generally faster at this level than a bit higher in the canopy where the leaves tend to spread out and occupy most of the space (Figure 11.14). Thus, some of the highest speeds over the marsh are measured soon after, and near the end of inundation. This may account for some of the rapid increases in flow in the tidal creek channels around these points. Flow speeds in the

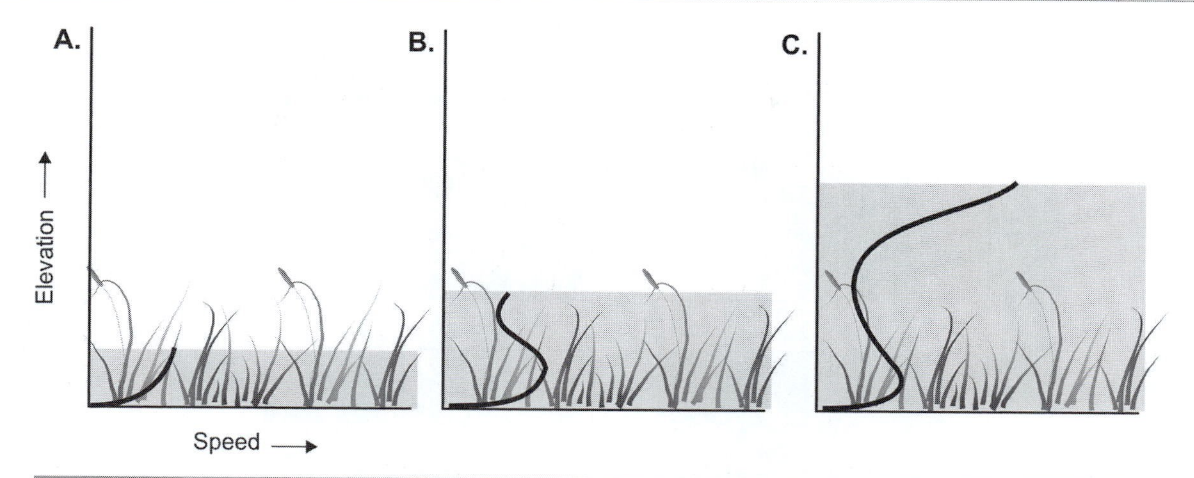

Figure 11.14 Schematic of the effect of depth of submergence on typical velocity profiles within saltmarsh grass canopies where: A. water level is just above the marsh surface; B. water level is near the top of the canopy; C. water level is substantially higher than the vegetation canopy. Flow around the stems of individual plants within a few centimetres of the bed tends to be faster than that a bit higher because of the spread of the plant leaves. Once most plants are submerged flow speeds above the canopy are much greater.

canopy are relatively slow until most of the plants are submerged and then flow speed increases dramatically in the area above the canopy and may be enhanced by skimming flow similar to wind flow over beach vegetation (Figure 11.15).

It is now possible to simulate flows in the tidal creek channels and over vegetated marsh surface using numerical models (Temmerman *et al.*, 2005; Ashall *et al.*, 2016; Wu *et al.*, 2017). Ashall *et al.* (2016) used Delft3D together with a vegetation module to model flow over the tidal mudflats and vegetated marsh surface, and in tidal creeks at Kingsport saltmarsh. The marsh is located in the Cornwallis Estuary in the hypertidal upper Minas Basin, Bay of Fundy. Simulated flows for four times around high tide are shown in Figure 11.15, together with measured flows at four points in a transect across the marsh surface and into a second order tidal creek. It can be seen that flows here are primarily parallel to the tidal creek 1.5 hours before and after high tide. Flow speeds on the marsh surface 0.5 hours before high tide are much slower than in the tidal creek channel because of the drag by vegetation when flow depth is small but flow direction is similar to that in the creek channel. At the start of the ebb, 0.5 hours after high tide, flow speeds on the marsh are much larger because of reduced drag from the vegetation due to greater depth of water over the surface and the flow direction is determined in part by flows in the larger basin. At this time flows are still onshore, illustrating the hydrodynamic lag between the initiation of a fall in tidal elevation and the start of ebb flows in the channel.

Saltmarsh vegetation also plays an important role in dampening wave action in marshes exposed to wave action (Allen, 2000; Yang *et al.*, 2012; Leonardi *et al.*, 2018). Waves may propagate from a bay or estuary across the mudflats in front of the marsh or they may be locally generated at high tide, especially during storm surge events where the whole marsh surface is inundated. Measurements of attenuation across macrotidal saltmarshes show a 50 per cent reduction in wave energy within 10–20 m of the edge of vegetation (Moller and Spencer, 2002; Moller, 2006; see Figure 11.16). As is the case for unidirectional flows, the effectiveness of vegetation in dampening wave energy varies with the stiffness of plant stems and with the height and density of the standing biomass (Moller, 2006; Moller *et al.*, 2014). It is likely more effective for short-period waves than for long-period ones.

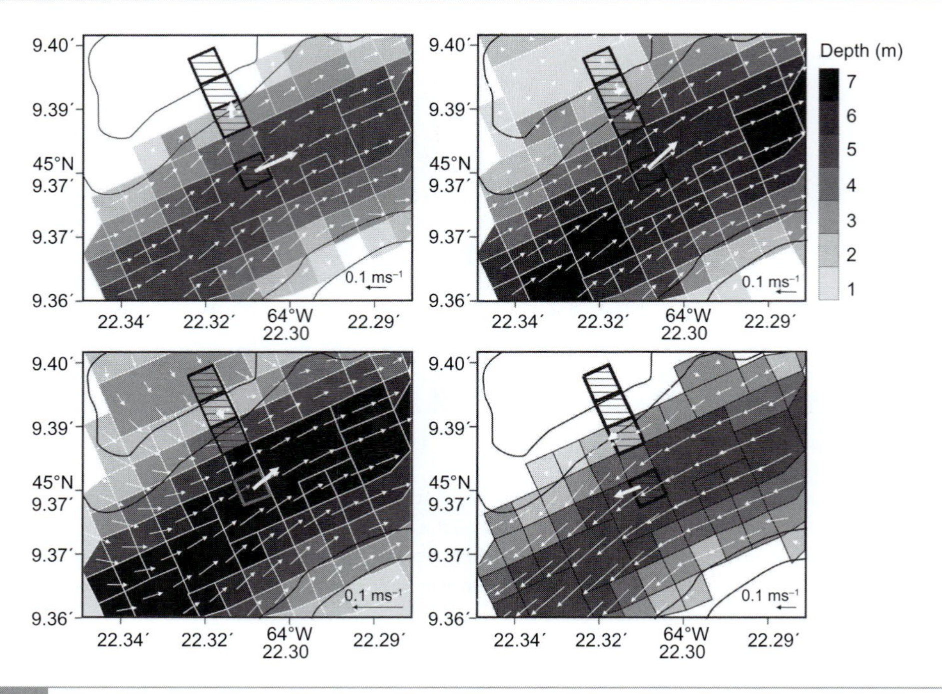

Figure 11.15 Simulated velocity vectors (white) with water depth in metres (grey scale) in model grid cells in a macrotidal channel, Minas Basin, Bay of Fundy. The four squares with heavy borders indicate grid cells where instruments are located, the large white arrows indicate velocity observations and the lines indicate vegetation polygon borders between high marsh vegetation, low marsh vegetation, and unvegetated mud. Vectors are shown at hourly intervals before and after high tide on YD 145 (25 May): 1.5 h before high tide (top left); 0.5 h before high tide (top right); 0.5 h after high tide (bottom left); and 1.5 h after high tide (bottom right). (After Ashall *et al.*, 2016, figure 9).

11.3.4 Saltmarsh Sedimentation and Dynamics

An understanding of the sedimentary dynamics of coastal saltmarshes, particularly within the context of a sediment budget can be useful for examining the response of the saltmarsh system to changes in one or more controlling variables such as sediment supply and sea level. These studies can also be seen as part of a broader effort to understand the complex interaction between the biotic and abiotic components of saltmarsh ecosystems and of the patterns of energy flow, nutrient and sediment cycling within the marsh and between the marsh and coastal waters (Adam, 1990; Mitsch and Gosselink, 2000). Sediment deposition on the marsh surface can be seen as being a function of the availability of sediment and the opportunity for this sediment to be brought onto the marsh and to settle out on it (Reed, 1989; van Proosdij *et al.*, 2000). Both of these vary spatially and temporally. Sediment availability may reflect the passage of storms or cold fronts, storm surges, heavy rainfall events and river flooding all of which may act to increase the concentration of total suspended solids (TSS) in the water column. The opportunity for deposition depends on the frequency and duration of flooding of the marsh surface, which is controlled by the tidal regime, marsh elevation and storm surge effects. The potential deposition rate will also depend on the effects of vegetation on reducing current and wave energy as well as sediment size and flocculation. All of these factors will vary between marshes and spatially within a marsh.

Several different aspects related to the sediment budget and mass balance of coastal saltmarshes can be recognised in recent studies, including: (1) measurement of sediment dynamics (erosion, transport and deposition), and sediment budgets associated with individual tidal

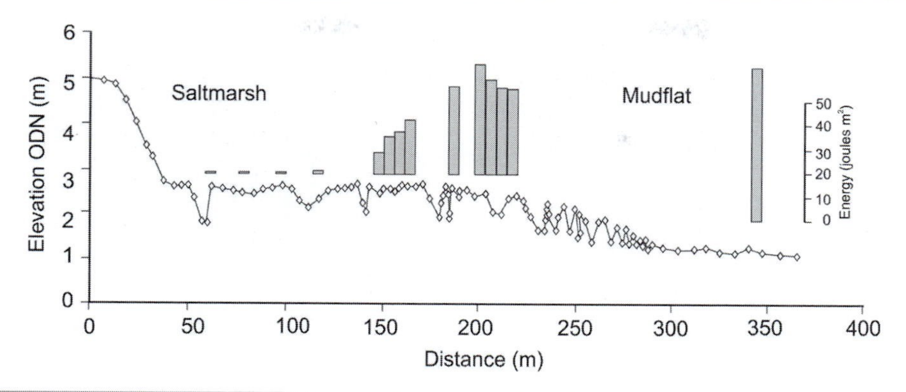

Figure 11.16 Change in significant wave height measured across the tidal flat and saltmarsh at Tillingham, UK. Vegetation height is shown for five locations on the marsh surface. (Modified from Moller and Spencer (2002). Reproduced with permission from the Coastal Education and Research Foundation, Inc.

cycles and/or the biweekly spring–neap cycle in tidal creeks and over the marsh surface; (2) measurement of sediment accumulation and rates of vertical growth as well as changes in the areal extent of marshes and tidal creek networks over periods of 10^{-1}–10^{3} years; and (3) use of simulation models to explore one or more aspects of the controls on saltmarsh dynamics and evolution. In this section the focus is on deposition over individual tidal cycles and on seasonal patterns. The long-term mass balance and response to sea level change are considered in the following section.

11.3.4.1 *Organic Accretion*

Much of the above ground organic matter production is broken down on the marsh surface and exported to the surrounding tidal flats, estuary or bay (Alongi, 1998; Mitsch and Gosselink, 2000; Mueller *et al.*, 2016). Small amounts of material from the above ground production can be incorporated into the saltmarsh sediments but the most significant contribution to increasing marsh elevation comes from root production at and below the ground surface (Nyman *et al.*, 2006). Because the saltmarsh soils are anaerobic beginning a few millimetres below the surface, the rate of breakdown of the organic matter tends to decrease rapidly after one or two years (Callaway *et al.*, 1997) and substantial beds of peat can form where inorganic sediment input is low. Within the root zone, if inorganic sedimentation is low, pore space may account for 80 per cent or more of the volume and this means that over time we can expect considerable compaction of the soil to take place (Bartholdy *et al.*, 2010b). In many estuarine and deltaic marshes the inorganic input is much higher than the organic component, and the final contribution of organic matter to saltmarsh accretion is usually modelled as being quite small (e.g. French, 1993; Temmerman *et al.*, 2004). However, in areas such as coastal Louisiana, the supply of inorganic sediments is low and thus most of the accretion may come from organic production (Nyman *et al.*, 2006). In modelling saltmarsh evolution on a decadal time scale it is important to account for the organic matter contribution to the saltmarsh budget but modelling it directly is still difficult and is usually accomplished by making use of empirical data on compaction rates.

Box 11.1 Measurement of Accretion and Surface Elevation Change

Sedimentary deposits are all subject to the process of autocompaction whereby the volume of the deposit decreases over time as a result of a set of physical, chemical and biological processes. A portion of this comes about from physical compaction leading to a rearrangement of the packing of mineral grains and concurrent reduction in the pore space. A similar process can take place in plant organic matter, but in addition plant matter

Box 11.1 (cont.)

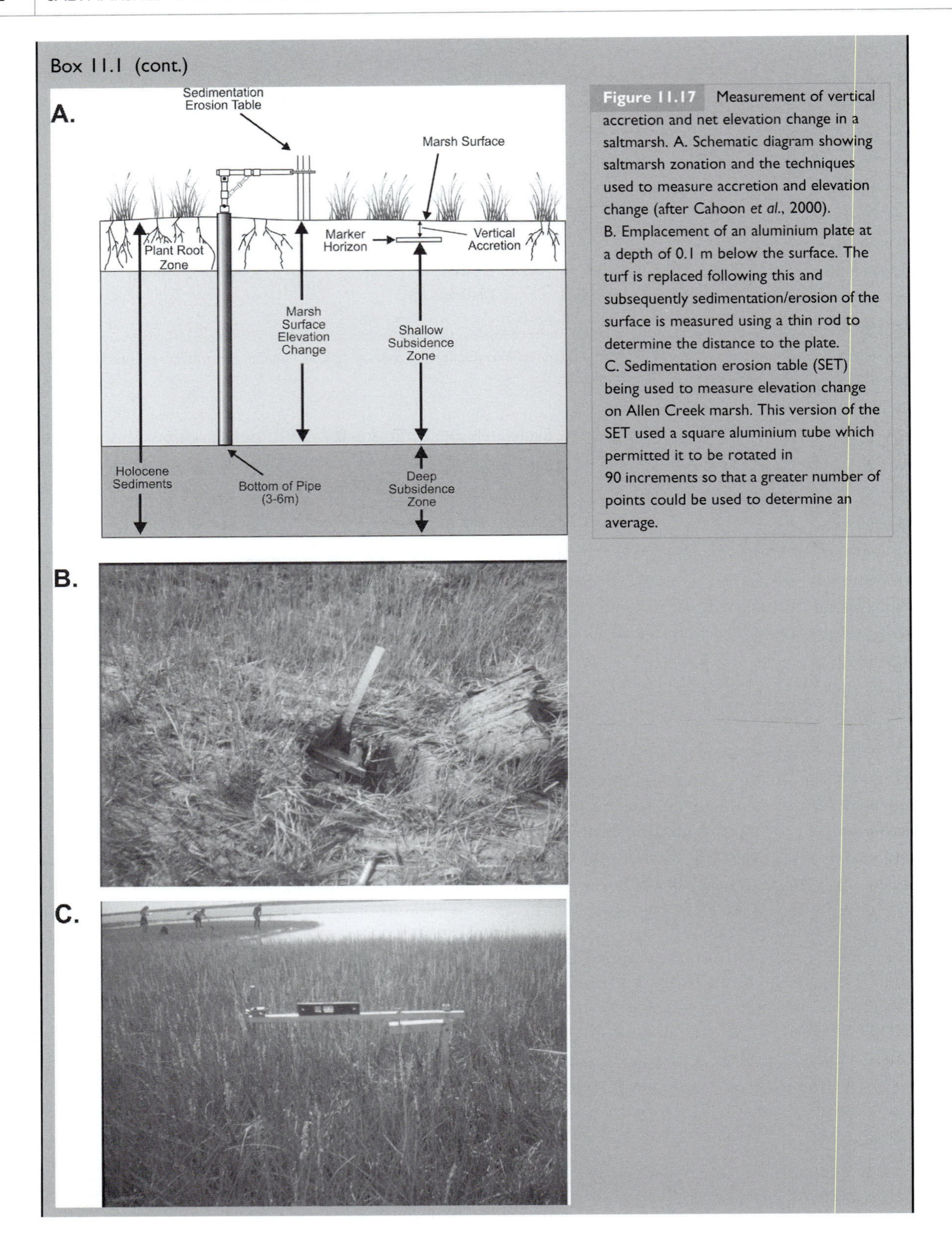

Figure 11.17 Measurement of vertical accretion and net elevation change in a saltmarsh. A. Schematic diagram showing saltmarsh zonation and the techniques used to measure accretion and elevation change (after Cahoon *et al.*, 2000).
B. Emplacement of an aluminium plate at a depth of 0.1 m below the surface. The turf is replaced following this and subsequently sedimentation/erosion of the surface is measured using a thin rod to determine the distance to the plate.
C. Sedimentation erosion table (SET) being used to measure elevation change on Allen Creek marsh. This version of the SET used a square aluminium tube which permitted it to be rotated in 90 increments so that a greater number of points could be used to determine an average.

Box 11.1 (cont.)

may decay through a number of biological processes. Chemical processes can lead to the dissolution of both plant material and carbonates, though some of these processes may add material through cementation (Allen, 2000). Autocompaction is especially important in saltmarshes because of the fine-grained nature of many saltmarsh sediments, the significance of flocculation in inorganic sedimentation, and the initially high amount of organic matter in the root zone near the surface. One manifestation of this is a tendency for the bulk density of cores in saltmarshes to increase with depth and to reach a nearly constant figure at some depth on the order of 3–5 m. Alternately, the rate of autocompaction can be determined from the difference between the change in elevation of the marsh surface over time and the accretion on the surface (Cahoon *et al.*, 1995, 2000; Bartholdy *et al.*, 2010b).

Measurement of surface accretion on the marsh surface on an annual to decadal scale is most often determined using some form of subsurface marker horizon (Figure 11.17A). This may be a layer of sand, clay, brick dust or other readily identifiable sediment, or an aluminium plate which is buried a few centimetres below the surface (Steers, 1948; Wood *et al.*, 1989; Nolte *et al.*, 2013). Distance to the marker horizon is commonly determined by taking short cores and measuring the depth below the surface. If an aluminium plate is used (Figure 11.17B), the distance to the plate can be measured by probing through the marsh surface with a thin rod. Seasonal or annual sedimentation is measured as a greater distance to the plate and vice versa for erosion.

The sedimentation erosion table (SET) is used to determine the net elevation change of the marsh surface by allowing for autocompaction (Boumans and Day, 1993). A pipe is driven into the marsh surface until it reaches a basal unit where compaction is minimal (Figure 11.17C). At Allen Creek this was a till or mudflat surface below the base of the marsh. Measurements were made by slotting the SET into the pipe and taking measurements of the distance to the bed at five points along three mutually orthogonal lines for a total of 15 points. The assumption is made that the pipe in the marsh is fixed and the elevation of the top of the pipe can be checked every two or three years using a total station or DGPS to ensure that this is indeed the case. The rate of autocompaction can then be determined by subtracting the surface elevation change measured by the SET from the accretion/erosion change determined from the aluminium plate or marker horizon.

11.3.4.2 *Sedimentation over Tidal Cycles*

Sediments can reach the marsh surface through a variety of pathways, including runoff from upland areas, overwash of barriers and deposition of dust from atmospheric fallout (including radioactive elements from airborne testing of nuclear weapons which now provides a convenient dating horizon). In many marshes most of the mineral sediments are brought to the marsh as suspended sediment from the bay or estuary during periodic tidal flooding of the surface and episodically during storm surge events. This sediment is dominantly in the silt and clay size range though wave action may transport sand from adjacent intertidal sand flats if these are present. The focus here is on transport of the fine fraction. The amount and pattern of deposition over a tidal cycle will depend initially on the suspended sediment concentration of waters reaching the marsh through tidal creeks and over margin flows and then on factors that determine flow speeds and the settling of particles from suspension.

The concentration of TSS reaching the marsh surface averaged over a period of years will vary with the setting of the marsh and the available sediment sources. In relatively narrow estuaries measurements show a strong positive relationship between increasing tidal height (neap to spring tides) and increasing TSS (Allen and Duffy, 1998; Temmerman *et al.*, 2003; Murphy and Voulgaris, 2006), likely as a result of increased flow speeds and turbulence of tidal currents associated with the larger tidal prism. In other areas the correlation with tidal range is lower, with wind and wave action playing a much stronger role (Leonard *et al.*, 1995a; van Proosdij *et al.*,

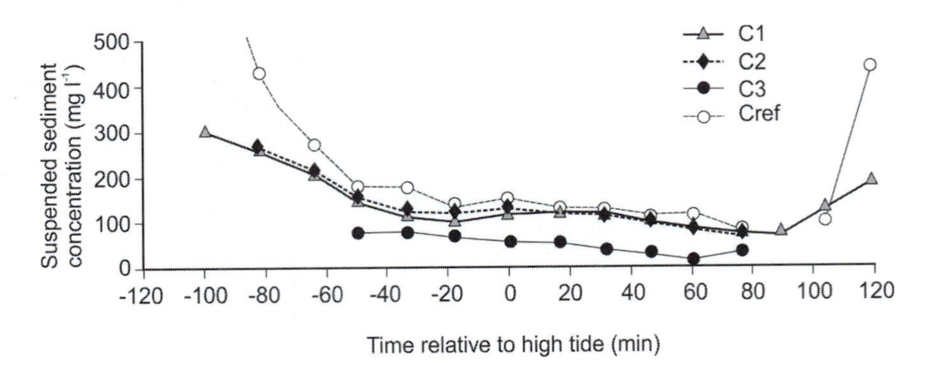

Figure 11.18 Suspended sediment concentration measured by optical backscatterance sensors (OBSs) along a profile from the main tidal creek and on the marsh surface at Allen Creek marsh over an individual tidal cycle. The locations of the sensors are: Cref, tidal creek; C1, top of creek bank; C2 and C3 on the marsh surface at an increasing distance from the creek. (After van Proosdij, 2001.)

2006a; Bartholdy, 2012), and rainfall events during low tide may also mobilise sediment on the surface of the marsh, tidal creeks and tidal flats (Mwamba and Torres, 2002; Murphy and Voulgaris, 2006).

Water flowing through the tidal creek network and over the marsh margin brings sediment in suspension onto the marsh surface. The TSS concentration tends to be highest initially because of remobilisation of sediment on tidal mudflats, in the tidal creeks and on the marsh surface, especially if there is some wave activity (Figure 11.18). Concentration decreases as flow depth increases and remains steady over much of the flood over the surface. It then decreases during the ebb as a result of deposition on the marsh surface (van Proosdij *et al.*, 2000; Christiansen *et al.*, 2000). There may be a short period of increased TSS near the end of the ebb, again because of sediment remobilisation in shallow flows (Figure 11.18).

It is not really feasible to calculate net deposition based on changes in the TSS concentration either in a tidal creek or on the marsh surface over a tidal cycle. However, on marshes where wave action is restricted there may be a strong positive correlation between net deposition over a tidal cycle and the magnitude of TSS concentration in an adjacent tidal creek. Sediment brought onto the marsh surface in suspension is deposited on the surface of the marsh, on leaf litter and on the stems and leaves of the plants. The settling velocity of individual silt and clay particles is very small and much of the time, even with reduced flow speed and turbulence within the marsh vegetation, there is insufficient time for sediments in all but the lowest few cm to settle. However, sediment can reach the marsh surface in a variety of ways other than settling of individual particles. A considerable proportion of the fine sediment making up the TSS is in the form of flocs that have a settling velocity less than a quartz particle of equivalent size but still much greater than that of the individual particles which make up the floc (Christiansen *et al.*, 2000; O'Laughlin *et al.*, 2014). Thus, a much greater mass of sediment can be deposited over a tidal cycle than would be predicted from the settling velocity of individual particles. Sediments may be deposited on, or adhere to, litter on the marsh surface and to the stems and leaves of the vegetation and then reach the surface as a result of rainfall during low tide. Finally, some sediments settle out of suspension within small pools remaining on the marsh surface at the end of the ebb tidal cycle. This tendency to fill in hollows is an important element in producing and maintaining the relatively flat marsh surface.

Sediment deposition on the marsh over a tidal cycle is commonly measured by collecting sediment on filter paper placed in a suitable container on the marsh surface or by some form of open cylinder or flat trap (Reed, 1989; Nolte *et al.*, 2013). These work well in low-energy

Figure 11.19 Measurement of sediment deposition on the marsh surface: A. sediment deposition on a Petri dish trap (circled; after Reed, 1989) and the modified Guelph trap; B. series of Guelph traps used for measuring deposition over various stages of the tidal cycle. The traps were fitted with lids and installed near a platform from which the lids could be lifted off at various stages of the tide. The trap on the left was exposed for the whole tidal cycle and the other three (from left to right) were exposed halfway through the flood, at the end of the flood, and halfway through the ebb.

environments but, where the traps may be subject to some wave action at high tide, a more robust frame is required (Figure 11.19A). These traps can be used to give an indication of relative rates of deposition over tidal cycles and used to test simple relationships between deposition rate and tidal range or spatial patterns related to distance from the marsh margin or a tidal creek (Allen and Duffy, 1998; Temmerman *et al.*, 2003, 2005; van Proosdij *et al.*, 2006a). However, they do not provide reliable measures of the true depositional mass.

Some early literature suggested that deposition occurred primarily at slack tide but more recent work suggests that it likely occurs over the whole tidal cycle (French and Spencer, 1993; Christiansen *et al.*, 2000). Measurements at Allen Creek made use of traps with covers that could be removed during the period of inundation to determine the relative mass of sediment deposited at various tidal stages (van Proosdij *et al.*, 2000; see Figure 11.19B). These showed that deposition tended to be highest near the beginning and end of the period of inundation, when water depths are smallest, and relatively steady during inundation. Under moderate wave conditions, sediment was deposited only at the beginning and end of inundation when water depth was too shallow to permit wave activity to reach the location.

Studies of deposition at a number of marsh sites show that spatial and temporal patterns of deposition over tidal cycles are controlled by several interrelated variables including: tidal elevation, TSS concentration in waters of the tidal basin, relative roughness of vegetation, distance from the margin or tidal creek and marsh topography. Recent modelling efforts have stressed the contribution of vegetation to saltmarsh sedimentation through the reduction of wave energy and current speeds on the marsh surface as well as the direct capture of particles in suspension in the water column (Mudd *et al.*, 2010; Fagherazzi *et al.*, 2012; Kakeh *et al.*, 2016). This is in addition to the contribution of below ground organic matter. Because of the complexities involved there are considerable differences between models in their conceptual approach and in the assumptions of the models. An example is provided by the approach of Mudd *et al.* (2010) who model the rate of change at a point on the marsh surface by:

$$\frac{\partial \xi_s}{\partial t} = Q_c/\rho_s + Q_s/\rho_s + O - E - \mathrm{Cmp}, \qquad (11.2)$$

where ξ_s is the elevation of the marsh surface, t is time, Q_c is the rate of sediment mass captured by plant stems per unit area of marsh, ρ_s is the density of marsh sediments, Q_s is the rate of mass settling out of suspension on the marsh surface, E is the erosion rate and Cmp is the rate of compaction. Getting all this right is not a trivial task. Aside from the sediment density, all the other components require a separate modelling endeavour based on limited field data.

On Allen Creek marsh, where flow depth over the surface is often much greater than vegetation height and levees are poorly developed, distance from the marsh or tidal creek margin is an important control on the spatial pattern of deposition, especially under calm or low wave conditions. Results of measured deposition for single tides (Figure 11.20), show that deposition is

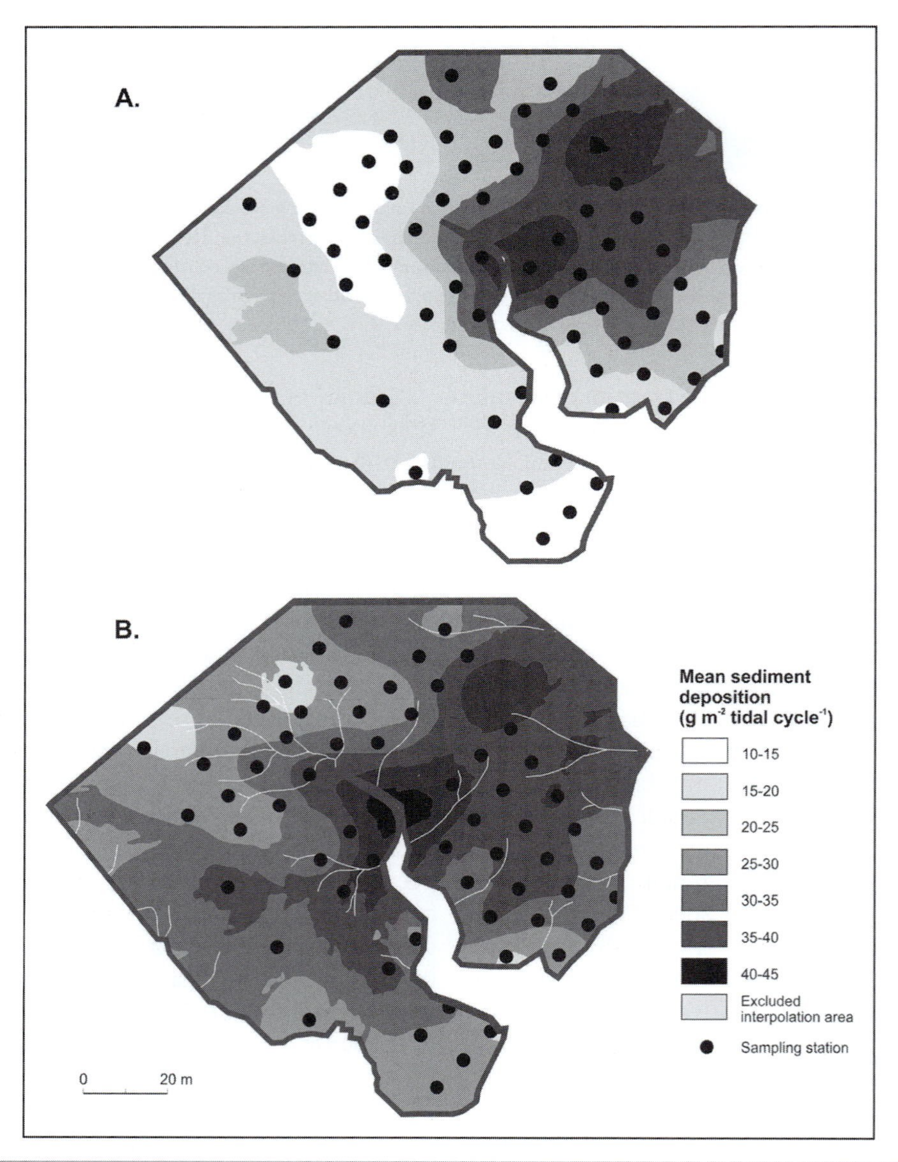

Figure 11.20 Sediment deposition over single tides for a portion of Allen Creek marsh based on measurements made with sediment traps similar to the one shown in Figure 11.19A. The black dots show the location of the traps and the maps of deposition are produced by 2D kriging of the data using a spherical model with a lag distance of 5 m: A. mean of sediment deposition per tidal cycle in June and July 1997 for four tides where waves were < 0.2 m in height at the marsh margin; and B. sediment deposition during a single tide with calm conditions on 24 June 1997. The creek in the centre of the map is shown in Figure 11.10. (After van Proosdij *et al.*, 2006a.)

highest at the upper end of the low marsh and it tends to decrease with distance from the main tidal creek channel and landward towards the upper marsh where duration of inundation is much lower (van Proosdij *et al.*, 2006a). However, under conditions with waves >0.2 m there is little deposition in the middle of the marsh because of the effects of wave orbital motion. Instead, the largest values for deposition are found on the upper marsh where wave action is reduced by shallower water depth and increased dampening by vegetation (van Proosdij *et al.*, 2006a). On other marshes the pattern of decreasing deposition away from tidal creeks or the marsh margin is much stronger (e.g., Reed *et al.*, 1999; Temmerman *et al.*, 2003; Alexander *et al.*, 2017). Sediment deposition toward the high marsh, that is with increasing elevation, on most saltmarshes tends to decrease because of the reduced frequency and duration of inundation. There may also be a zone of low deposition near the marsh margin because of resuspension by wave activity.

11.3.4.3 *Seasonal and Annual Variations in Deposition and Erosion on the Marsh Surface*

Measurements of sedimentation and net accretion on the marsh surface over periods of months to a decade or more can be made with some version of the techniques described in Box 11.1. The advent of differential GPS has made relocation of marker horizons or plates and the pipes for SET stations much simpler and the techniques have been applied in a number of locations. Seasonal variations in accretion of the surface have also been evaluated indirectly by comparing levels of TSS, measured continuously at a station in a suitable location on a tidal creek or in the bay or estuary, in conjunction with measurement of deposition over individual tidal cycles at intervals during the same period (Murphy and Voulgaris, 2006).

Seasonal variations in sedimentation can be expected to reflect either variations in the TSS concentration of waters reaching the marsh (i.e., in the available supply of sediment) or variations in the factors controlling erosion and deposition on the marsh surface itself. Higher values for TSS usually reflect either periods of higher rainfall and runoff in rivers supplying sediment to estuaries or the coast, or strong winds and wave activity which mobilise sediments in the nearshore or on intertidal flats (Reed, 1989; Temmerman *et al.*, 2003). Reduced vegetation cover due to dieback in winter months might be expected to reduce deposition on the marsh surface, or at least result in increased remobilisation by currents and wave action. However, in the absence of significant wave action it seems that considerable deposition can occur on the marsh surface even with greatly reduced vegetation cover.

Whether there are strong seasonal signals in accretion on the marsh surface will therefore vary with the physical marsh setting (estuary, bay, backbarrier) and with the climatic factors controlling rainfall, and wind and wave action. Measurements over 2.5 years at two sites in North Inlet, South Carolina (Childers *et al.*, 1993), which are influenced by freshwater sediment inputs, illustrate the spatial variability in response both between marshes in different locations and between upper and lower marsh stations on the same marsh (Figure 11.21). At Oyster Landing deposition occurs in all seasons except winter. At South Town Creek erosion occurred in the autumn. Deposition was often greatest in the winter, with the high marsh site showing much greater variability. Temmerman *et al.* (2003) also found higher deposition in the winter months, but a number of studies in the US have found higher deposition in the summer (Leonard *et al.*, 1995b; Murphy and Voulgaris, 2006). Neubauer *et al.* (2002) found highest rates of deposition during the summer months along creek margins in a tidal freshwater marsh but no seasonal pattern in the interior away from the streams. At Allen Creek, marsh deposition is highly variable seasonally, but the pattern is different between high marsh and low marsh sites (Figure 11.22). Here, deposition on the mid and high marsh is high in the fall when TSS concentration is highest because of stronger wind and wave action and wave energy is dampened by vegetation on the low marsh. It is also high in the

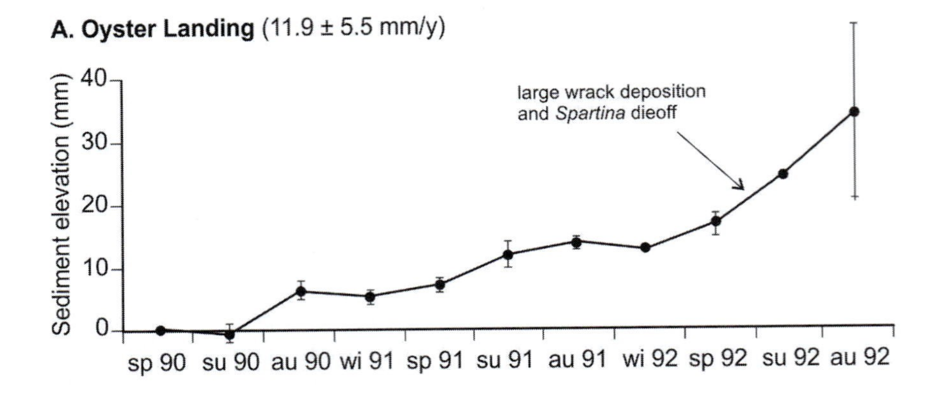

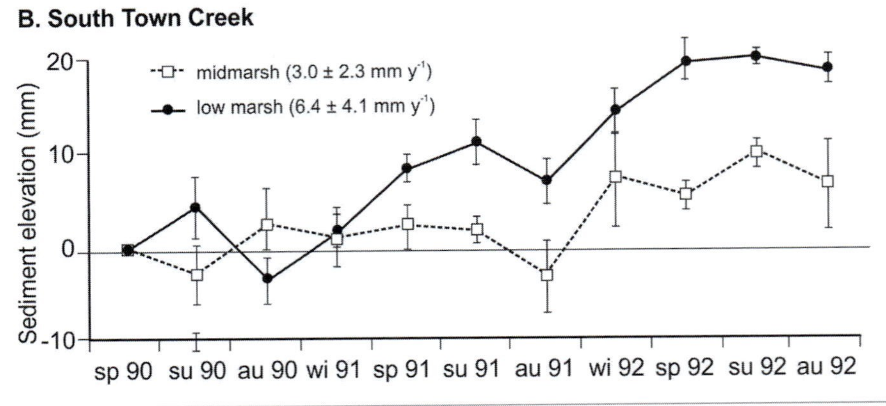

Figure 11.21 Changes in the elevation of the marsh surface over time at two marshes in North Carolina (after Childers *et al.*, 1993). Sediment input is chiefly from river sources, with Oyster Landing being closer to a source and therefore having higher inputs. Reproduced with permission from the Coastal Education and Research Foundation, Inc.

winter, in part for the same reason, but also because of the stranding of blocks of ice near the spring high tide limit and subsequent deposition of sediment from the ice due to melting in late winter (van Proosdij *et al.*, 2006b). Argow *et al.* (2011) report similar patterns related to ice on marshes in New England, and provide a detailed description of the processes associated with the tidal regime that controls the incorporation of sediment into the ice and the subsequent rafting of the ice blocks onto the marsh surface.

Deposition on the low marsh tends to be lower during this time because of hindered deposition and possibly erosion of the surface by waves coming over the marsh margin. In the late spring and summer, deposition is higher here because of the reduced wave energy levels and greater vegetation height and density. Most seasonal studies that are carried out over just a few years have found considerable variability from year to year, for example, because of the effects of one storm or a mild winter.

While deposition over tidal cycles and seasonally is highly variable, measurements made over a period of several years tend to show a more consistent spatial pattern. Deposition rates decrease with elevation on the high marsh reflecting the lower frequency of inundation and this is consistent with the asymptotic model of saltmarsh evolution (Figure 11.7). Measurement of deposition on Hut marsh by Steers between 1935 and 1947 (Steers, 1948; Figure 11.23A), and subsequent measurements by Stoddart *et al.* (1989) show a consistent trend and it is notable that the scatter decreases with increasing period of measurement (Figure 11.23B). This pattern is best developed on the high marsh where inundation occurs only on extreme spring tides or during storm surges.

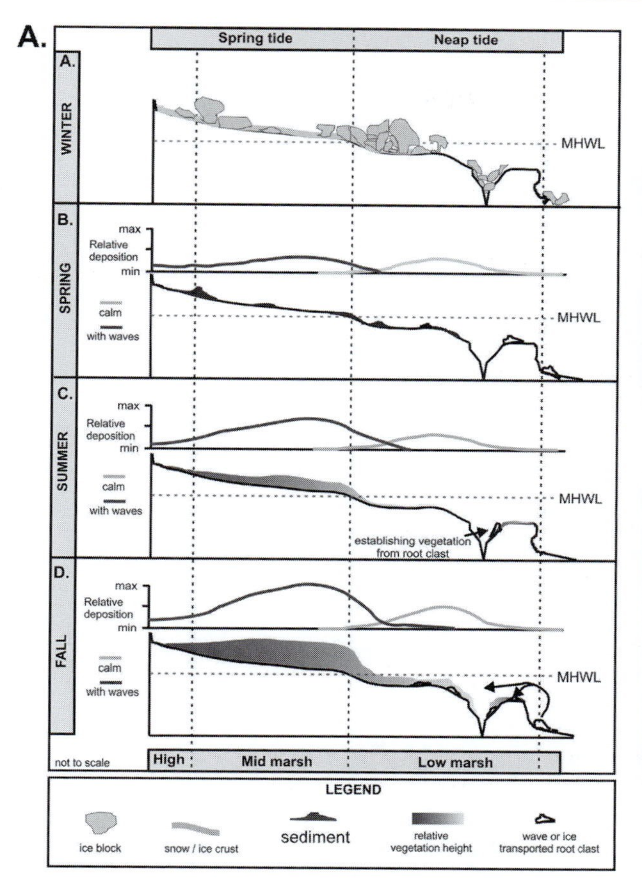

Figure 11.22 Controls on annual deposition on Allen Creek marsh: A. schematic diagram showing seasonal patterns of sediment deposition across Allen Creek marsh and the influence of seasonal vegetation changes and ice (van Proosdij *et al.*, 2006b); B. blocks of ice stranded near the spring high tide line; C. sediment melting out of a remnant ice block at the end of the winter.

At Allen Creek marsh, which lies mostly below the spring high tide line, the decrease in deposition with elevation is much less pronounced and deposition from ice near the landward margin may offset the lower deposition from tidal inundation (Figure 11.24). The average rate of accretion on the marsh surface ranges from about 9 mm a^{-1} to about 1.3 mm a^{-1} and this is consistent with an average rate over about 50 years of about 1.0 mm a^{-1} obtained by Kostaschuk *et al.* (2008).

11.3.5 Decadal Scale Sedimentation, Erosion and Saltmarsh Cycles

There are a few studies where marker horizons have been used to measure sedimentation for periods exceeding a decade (Steers, 1948; Bartholdy *et al.*, 2004; Goodman *et al.*, 2007) but most studies of saltmarsh dynamics over decades to centuries are based on analysis of historical aerial photographs to map changes in areal extent and stratigraphic approaches where suitable dating material is available. In addition to the tradition al methods of stratigraphic dating the accumulation of ^{210}Pb and ^{137}Cs from airborne testing of nuclear weapons has been used to provide a marker horizon in a number of studies (Kearney and Ward, 1986; Marshall *et al.*, 2007). At a decadal to century scale, sedimentation is also influenced by relative sea level rise and particularly its influence on shoreline

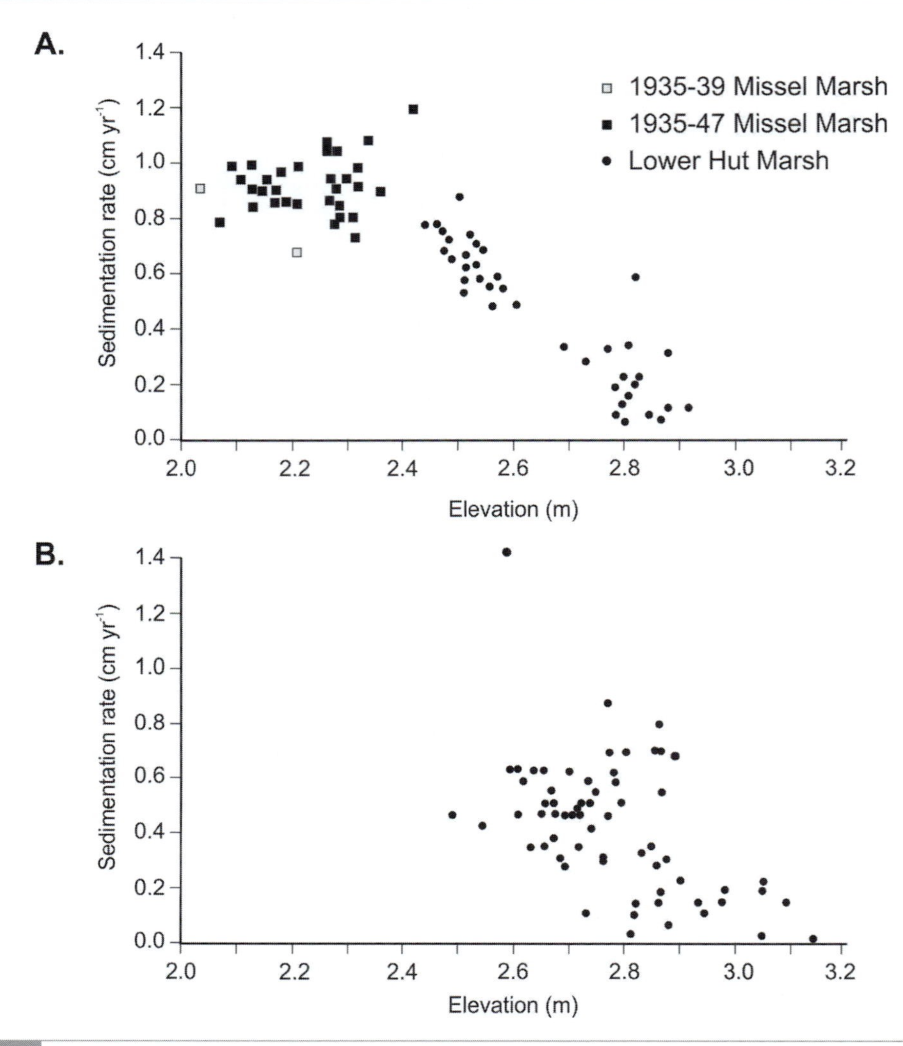

Figure 11.23 Relationship for sedimentation rate and elevation: A. Missel and Hut marshes 1935–47 (data from Steers, 1948); and B. Hut marsh using the same locations as those occupied by Steers for 1983–7 (Stoddart *et al.*, 1989). Note the greater scatter in the data collected over a shorter time period.

displacement and the accommodation space available for sediment deposition. In some regions, such as the eastern Canadian Arctic, northern Scotland and Scandinavia, isostatic uplift results in shoreline retreat and a reduction in the accommodation space at any point on the marsh. Marsh deposits therefore tend to be quite thin but may be preserved over a considerable width normal to the shoreline. In Australia and parts of the Pacific, many marshes (and mangrove swamps) have developed under relatively stable sea level during the late Holocene. In much of central North America and western Europe, marsh development has occurred under rising sea levels and there is considerable preoccupation today with the potential effect on marsh survival of an increasing rate of sea level rise relative to historic accretion rates (Reed, 1995; Kirwan *et al.*, 2016).

11.3.5.1 *Decadal Scale Rates of Accretion*

Accretion rates measured over decades to a few centuries from sites in North America and western Europe typically show accretion rates

Figure 11.24 Erosion of saltmarsh cliffs, Allen Creek marsh: A. photograph of wave action on the cliff on the rising tide; B. view of the marsh margin cliff at low tide showing notching and the accumulation of blocks of eroded material on the erosional platform; C. Minudie marsh, Cumberland Basin, Bay of Fundy showing an old marsh cliff now protected by the establishment of vegetation in front of it and beginning of a progradational cycle. The old cliff will be preserved for a number of years until the developing marsh buries it beneath new deposits.

ranging from < 1 mm a^{-1} to 10–15 mm a^{-1} (Stoddart *et al.*, 1989; Callaway *et al.*, 1997; Temmerman *et al.*, 2004; Alexander *et al.*, 2017; Carey *et al.*, 2017). The wide range of accretion rates reflects the variety of locations for saltmarsh growth and thus differences in the supply of sediment and rate of organic matter accumulation. It also reflects differences in rates between low marsh sites, where deposition is usually highest, and high marsh sites where, as we saw earlier, the reduced number of flooding cycles ultimately reduces the rate of accumulation.

Examination of saltmarsh accretion over centuries to several thousands of years suggests that accretion rates change over time, both in response to changing environmental conditions controlling sediment supply and plant growth, but also in response to fluctuations in sea level and rates of sea level rise (Allen, 2000). In part this may reflect variations in the number of flooding cycles accompanying sea level change and the feedback relationship between flooding frequency and elevation relative to sea level at a point on a marsh. The result is that accretion rates tend to decrease during periods of stable or falling sea level and to increase during periods of rising sea level. In some cases rising sea level and transgression may be accompanied by increased sediment supply to the marsh. There is also some evidence from Louisiana marshes, where subsidence produces very high rates of relative sea level rise, that marshes may be able to accrete at rates up to 10 mm a^{-1} (Reed *et al.*, 2009).

11.3.5.2 *Saltmarsh Cycles and Evolution*

Some saltmarshes develop in areas where they expand to occupy much of the space in a lagoon behind a barrier system or areas protected by levees from wave action in the main bay or lagoon. The marsh margin especially in sandy intertidal systems may be marked by a narrow beach and small dune system that acts to protect the marsh behind from major erosional events or there may be a transitional zone of sparse vegetation that is occasionally pushed landward during storm events. In some lagoon and estuarine situations where fetch lengths are long enough to produce significant wave action the marsh margin may be eroded during storms leading to the formation of a marsh margin cliff that then recedes over time leading to a reduction in marsh area and ultimately to a net loss of sediment from the marsh (Allen, 1989; Bendoni *et al.*, 2016; Leonardi *et al.*, 2016a; Zhao *et al.*, 2017). Recession of the cliff occurs as a result of wave action and erosion of the toe of the cliff which leads to undercutting and eventual toppling of material from the crest of the cliff (Figure 11.24). Cliff erosion takes place through all the mechanisms detailed in Chapter 13, but in this case the cliff may be completely overtopped at high tide as water covers the marsh surface. Direct wave attack may dislodge large blocks from near the crest where the sediments are strongly bound by roots of the vegetation and these can be deposited on the platform in front of the cliff (Figure 11.24B). Blocks of material from the crest can also be deposited on the marsh surface during periods of high waves when the water level is near the level of the cliff top and the process can also occur as a result of ice action under similar water level conditions. As might be expected, wave power reaching the marsh cliff is one of the major controls on the recession rate (Leonardi *et al.*, 2016b). McLoughlin *et al.* (2015) modelled recession rates over over half a century in Hog Island Bay, Virginia using the wave simulation program SWAN. They found that there was a stronger correlation between wave energy and volumetric loss than with the simple recession rate which suggests that cliff height and its interaction with waves and tidal stage may also play a role.

Since the marsh surface is generally an area of net deposition it is possible to determine the mass balance of the marsh on an annual or longer time scale. Where the marsh is mature with few changes in the areal extent, the annual mass balance can be determined from measurement of accretion over the marsh surface area. However, on marshes where there is erosion of the marsh margin erosion of the cliff will produce a loss per unit area equal to the height of the bluff – this may be a metre or more in some marshes. In this circumstance, determination of the mass balance requires careful surveying of the marsh margins to determine areas of retreat and the height and bulk density of sediments making up the marsh cliff. Because the marsh surface continues to accumulate sediment, even while the areal extent is decreasing due to cliff erosion, it is possible for the mass balance to be positive for many years after erosion is initiated. The expansion of vegetation onto previously bare intertidal areas may complicate the assessment of the mass balance (Moreira, 1992; van Proosdij *et al.*, 2006b).

A saltmarsh cycle can be defined as a period of seaward extension of the marsh, followed by development of a marsh cliff and a period of erosion as the cliff retreats landward until it is stabilised by recolonisation of vegetation in front of the cliff (Figure 11.25). They have been recognised and described from a number of saltmarshes worldwide (Yapp *et al.*, 1917; Allen, 1989, 2000; Pringle, 1995; Ollerhead *et al.*, 2005; Van der Wal *et al.*, 2008; Zhao *et al.*, 2017). The cycles of accretion and erosion can be initiated by a number of factors including extrinsic ones such as changes in winds and storm intensity or changes in the position of estuary channels (Pringle, 1995). Zhao *et al.* (2017) report on the formation of cliffs following the introduction of *Spartina alterniflora* which rapidly colonised mudflats along the Jiangsu coast of China. They document saltmarsh cycles that last only a few years because of exposure to large waves. In some areas the marsh cycles may result from intrinsic factors reflecting feedback between vertical growth of the marsh surface and the establishment of a steep section at the outer edge of the marsh that evolves into a cliff and initiates a period of recession while the saltmarsh surface continues to accrete. One of the earliest reports of such an 'intrinsic-effect' cycle comes from Yapp *et al.* (1917) who worked in the Dovey estuary in Wales (cited in Allen, 2000). Cycles reported from the Westerschelde estuary (Van der Wal *et al.*, 2008) and from the Cumberland Basin (Ollerhead *et al.*, 2005) also appear to reflect a form of self-organisational feedback.

In the Cumberland Basin the process of seaward extension of marsh vegetation across tidal

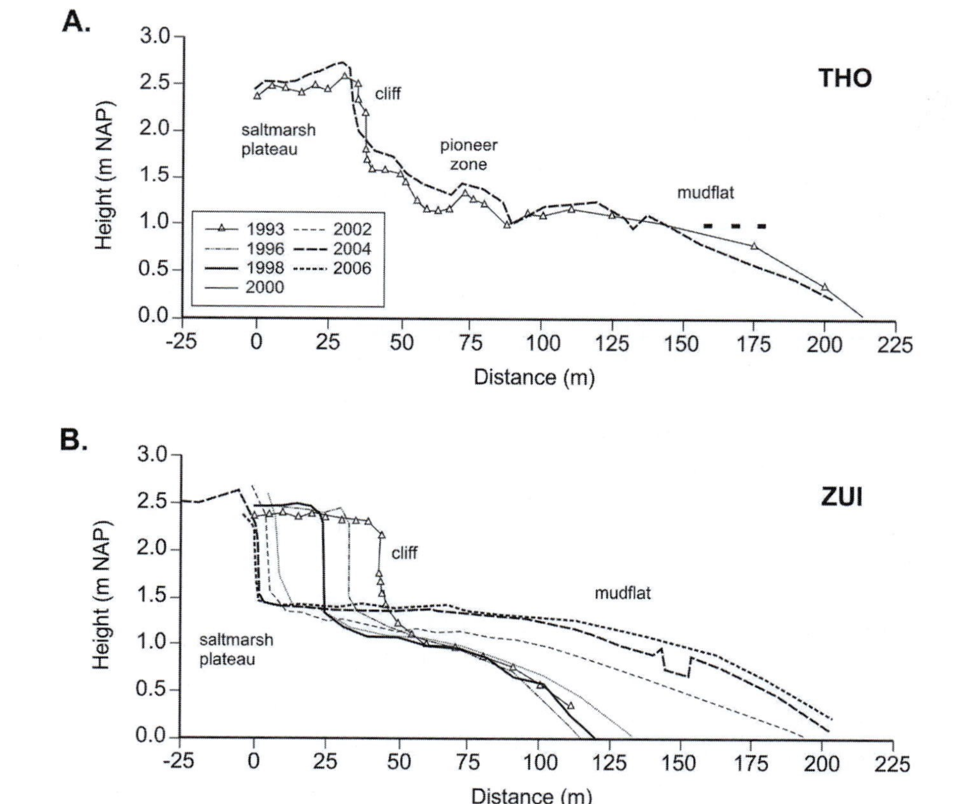

Figure 11.25 Examples of profile evolution on: A. an accreting marsh; and B. an eroding marsh in the Westerschelde, the Netherlands. (After van der Wal *et al.*, 2008; location of the saltmarshes in the Westerschelde, southwest Netherlands: Thomaespolder (THO), Zuidgors (ZUI).)

mudflats is rapid initially but slows as vegetation is established lower in the tidal frame where growth is restricted by longer periods of immersion and greater exposure to wave action (Figure 11.26). A cliff may form as a result of vertical accretion on the marsh over a period of several years, or it may be initiated by a major storm event. Once established, wave reflection from the cliff face prevents the re-establishment of vegetation in front of the cliff until it has

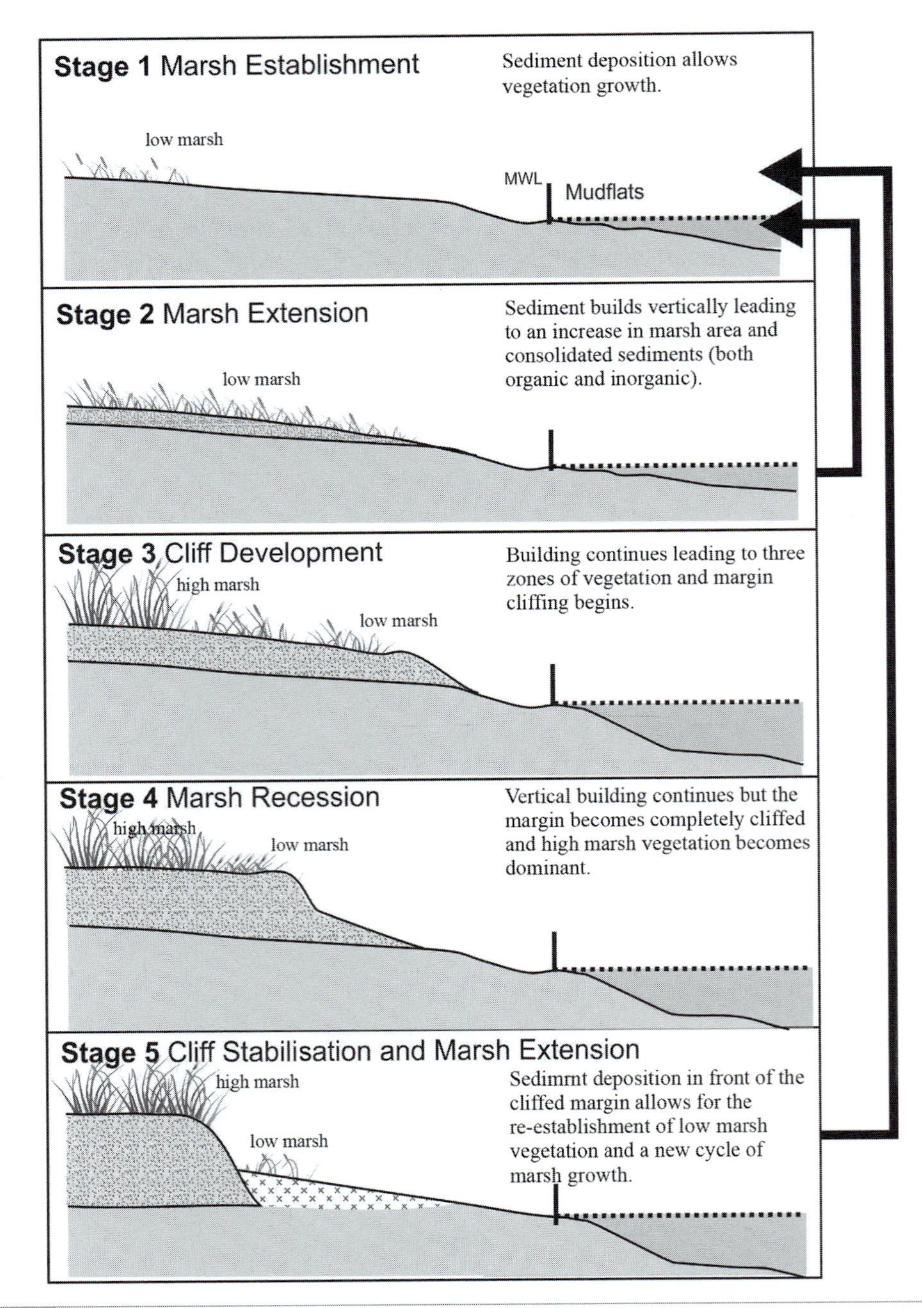

Figure 11.26 Schematic model of the stages in a cycle of accretion and erosion of marshes in the Cumberland Basin. (After Ollerhead *et al.*, 2005.)

retreated to the point where energy dissipation over the erosional mudflat ramp is sufficient to permit revegetation (Figures 11.24C, 11.26). The cycle then begins again, with the residual cliff being preserved and eventually largely buried beneath new saltmarsh deposits (Ollerhead *et al.*, 2005).

Saltmarsh cliffs and sedimentation erosion cycles may also reflect the effects of sea level changes and fluctuations on a scale of hundreds to thousands of years. Marsh evolution on this time scale will reflect the effects of the Holocene transgression as well as local sea level fluctuations. Sea level rise increases the accommodation space for vertical growth of saltmarshes, permitting thick units of sediments, including peat beds, to develop. A range of possible scenarios can be envisaged, reflecting different sea level curves (e.g., fluctuation around a steady level or fluctuation within a general trend for sea level rise) and the local geomorphic setting within which the marsh develops (e.g., transgressive barrier lagoon system, stable barrier lagoon system and estuary). These have been documented extensively in the marshes of the Severn Estuary (see Allen, 2000 for a summary) and for marsh systems in many other parts of the world.

11.3.5.3 *Saltmarshes and Sea Level Rise*

As is the case with coral reefs and mangroves, there has been considerable attention paid to the potential effects of sea level rise on saltmarshes around the world and in particular on whether the rate of sedimentation on marsh surfaces is sufficient to keep pace with present and future rates of sea level rise (Redfield, 1972; Reed, 1995; Crosby et al., 2016). In effect, the first line of defence against rising sea levels is for marshes, and the adjacent tidal flats, to accrete at a rate that equals or exceeds sea level rise so that the shoreline remains in the same position. The problem has stimulated a lot of the research into the controls on saltmarsh sedimentation and on measuring both short-term and decadal-to-century scale rates of saltmarsh accretion. The response of marshes to sea level rise has been explored in a number of simulation models (e.g., French, 1993; Temmerman *et al.*, 2004; Kirwan and Murray, 2008). These offer some insights into the range of potential responses but their predictive ability is generally limited by fairly crude assumptions that make application to a particular site difficult. There are also relatively few attempts to validate models and their applicability to a wide range of marsh environments (e.g., Morgensen and Rogers, 2018). Recent analysis by Kirwan *et al.* (2016) suggests that actual increase in marsh surface elevation is equal to or greater than RSLR for a large proportion of marshes globally and they point particularly to data from the low marsh rather than the high marsh where accretion rates are likely much lower due to infrequent inundation. In particular, they point to decreased vulnerability with increasing suspended sediment supply and tidal range (Figure 11.27). They also note that static models, such as the widely used sea level affecting marshes model (SLAMM), overestimate the rate of submergence because they do not incorporate the dynamic 'ecogeomorphic' feedbacks between marsh plants and increased frequency of inundation (see also Morgensen and Rogers, 2018).

Fixation on the rate of vertical growth of saltmarshes in the face of accelerating RSLR reflects uncertainties over the impact of human activities on sediment supply to estuaries, deltas and sandy coastal systems and the impact this may have on marsh sediment budgets (e.g., Weston, 2014) and the effects of hardening of the coast. The survival of saltmarshes in the face of sea level rise much greater than the present (see Section 3.6.2) does not depend on the ability of marshes to keep pace through vertical accretion. Saltmarsh vegetation can become established within a period of a few years and new saltmarshes develop rapidly whenever new opportunities exist within intertidal areas (Zhao *et al.*, 2017; Bartholdy *et al.*, 2018). Saltmarshes, like coral reefs, did not simply survive the Holocene transgression, but were a significant coastal feature throughout, with new marshes developing as estuaries were drowned and as barriers transgressed or developed anew. As sea level rises, flooding extends farther inland

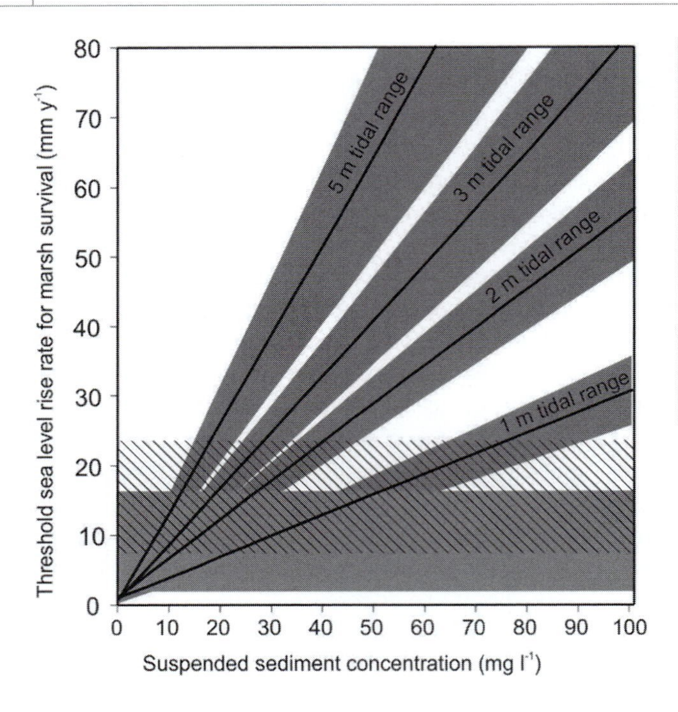

Figure 11.27 The effects of suspended sediment concentration and tidal range on the threshold rate of sea level rise for survival of saltmarshes predicted by an ensemble of five dynamic marsh models (documented in Kirwan *et al.*, 2010). The bands around each of the four tidal range lines represent ± 1 standard error on the predicted threshold rates of sea level rise. The horizontal bands represent the IPCC (Church *et al.*, 2013) projected rate of sea level rise based on process models (solid) and semi-empirical projections (diagonal stripes) with the combined pattern representing the overlap between them. (After Kirwan *et al.*, 2016, figure 3a.)

and in most cases we could expect the same thing to happen in the face of climate change induced sea level rise (Reed, 1995). However, the effects of hardening of the coast could restrict the potential for inland migration of marshes, a concept termed 'coastal squeeze' (Pethick, 2001; Pontee, 2013). The danger here is that hard structures, such as dykes, seawalls and road embankments, may simply prevent transgression of a natural shoreline and thus prevent this form of natural response. There is a need, therefore, to recognise these in coastal management planning, and indeed it is explicitly recognised within programmes of 'managed retreat'. Nevertheless, saltmarsh loss due to infilling, reclamation and a range of other human activities is still a much more significant threat than the effects of sea level rise.

Box 11.2 Coastal Management Application

Managed Realignment and Saltmarsh Restoration

While saltmarshes can adjust to rising sea level in the short term through vertical accretion, in the medium term (decades to centuries), preservation of marshes usually necessitates landward migration into appropriate space made available as a result of inland flooding. However, natural migration is often constrained on developed coasts by hardening of the shoreline – a situation termed coastal squeeze (Pethick, 2001; Pontee, 2013). In addition to hardening through construction of roads and buildings, and their associated shore protection, a significant proportion of coastal squeeze can be attributed to restricted tidal flows in tidal creeks and rivers due to dams, causeways and culverts that are too small to permit the normal tidal flows. Recognition of the need to restore the ecosystem services provided by saltmarshes in the face of further loss due to RSLR has led to efforts to restore natural flows in tidal creeks and rivers (e.g., Burdick *et al.*, 1997; Konisky *et al.*, 2006; Bowron *et al.*, 2009). In the

Box 11.2 (cont.)

Bay of Fundy and Gulf of Maine a major driver was the impact on access to good spawning habitat for wild Atlantic salmon. In Nova Scotia and New Brunswick, Canada, a collaborative study identified all sites where narrow culverts were restricting flows into tidal creeks and rivers and the outcome of this was the development of a long-term management plan to replace them with culverts that would be large enough to restore normal tidal flows (Bowron *et al.*, 2009). One of the first applications of the plan was at the Cheverie Creek, Minas Basin, restoration site described by Bowron *et al.* (2009). There, replacement of an old culvert in December 2005 increased flows into the marsh seven fold and eliminated restrictions on all but 15 per cent of the highest tides. The result was an increase in saltmarsh area from 5 to 43 ha. The project thus provides an example of a relatively simple and inexpensive way to increase saltmarsh area and to permit natural expansion of the saltmarsh as sea level rises.

In many parts of the world reductions in the areal extent of saltmarshes reflects the construction of dykes and the conversion of saltmarshes to freshwater marshes and to agricultural lands (Reed *et al.*, 2018). There is now increased recognition that managed realignment (a form of managed retreat) in dyked areas and the restoration of naturally functioning saltmarshes can not only restore some of the lost ecosystem functions but also have the potential to provide protection against storm surge and flooding in the face of RSLR (French, 2006). Examples of the natural failure of dykes in front of drained marshes in a number of estuaries on the east coast of England provided evidence of the complex controls on the restoration of saltmarshes to these areas and the failure of many sites to develop significant marshes at all (French *et al.*, 2000; Reed *et al.*, 2018). Similar examples have been noted on the coasts of Maine and the Bay of Fundy on the east coast of North America. In particular, it is clear that successful establishment of saltmarsh vegetation is hindered when the depth of flooding is too great because of compaction following dyking, by the size and positioning of the breach or breaches, and by the amount of sediment deposition. Studies of both natural and planned dyke breaches indicate that successful managed realignment projects require the collection of considerable background data on the site and simulation modelling of the flows and sediment inputs under a range of possible breach configurations (Friess *et al.*, 2014; Reed *et al.*, 2018).

Managed realignment through the breaching of two dykes at Aulac in the Cumberland Basin, Bay of Fundy, provides an example of the success of careful planning and modelling in saltmarsh restoration (Boone *et al.*, 2017). The spring tidal range here is 10–12 m and the area is subject to waves with significant wave height (H_s) of 1.5 m and period up to 8 seconds. The topography of the two sites and the dykes was determined from a LiDAR DEM, as well as from ground surveying and aerial photography, and it was noted that the area within the two restoration sites was up to 2.1 m below the level of an adjacent natural marsh. Computer simulation modelling was undertaken using the measured topography, tides and wave climate, to determine the optimum dimensions and number of breaches in the dykes to keep flow speeds $< 1.5 \text{ m s}^{-1}$ so as to limit erosion of the breaches and at the same time maximise the depth and frequency of flooding (Boone *et al.*, 2017; Figure 11.28). A carefully planned monitoring scheme was put in place and included comparison of annual deposition at marker stations within the two sites as well as on an adjacent natural marsh (Figure 11.28). *Spartina alterniflora*, the dominant species of the low marsh in this location (see Figure 11.2B), became established 2 years after the breach and after 5 years was fully established at both sites as a result of both extension of rhizomes and establishment of seedlings (Boone *et al.*, 2017). By 2017, vegetation covers almost all of both sites (Figure 11.29A). At spring high tide the two sites are completely flooded (Figure 11.29B), and the action of waves and tidal currents has led to the dispersal of much of the original dykes at both sites. High marsh vegetation has now been observed along the dyke margins where sediment deposition has created higher topographic points. There are some areas where cumulative deposition is >1.0 m in 2017, but over most of the area of both sites cumulative deposition is <0.4 m (Figure 11.30). The average cumulative increase in elevation during 2006–18 is about 0.65 m at the west site and 0.32 m at the east site (Figure 11.31). In 2018, the data

Box 11.2 (cont.)

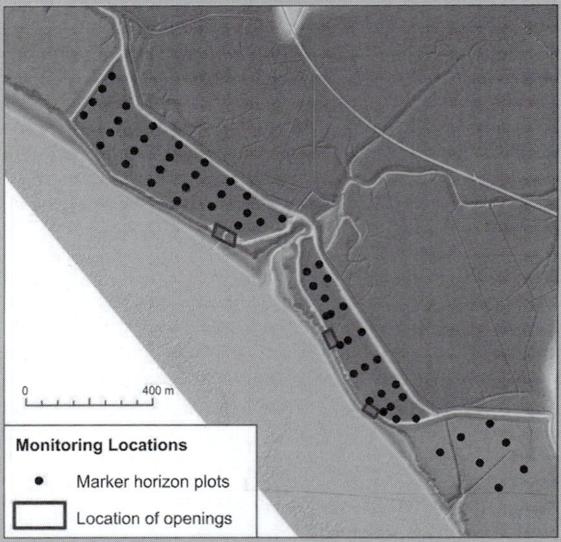

Figure 11.28 Schematic of the two restoration sites at Aulac showing the breaches in the old dyke (black-lined rectangles) and the locations of the marker horizon plots (black dots) used to measure sediment accumulation. The dots to the east of the two cells are marker horizon plots located in the east reference marsh. (After Boone *et al.*, 2017, figure 21.5.)

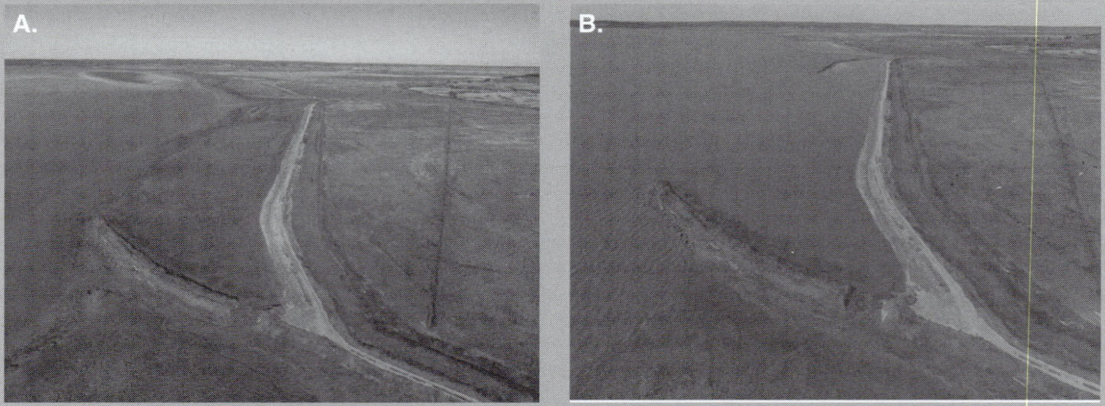

Figure 11.29 Oblique aerial views of the two restoration sites taken on 7 October 2017. A. View looking west at low tide of the east restoration site (foreground) and the west site (top just beyond the tidal creek separating the two sites). Note the presence of vegetation scattered across the surface of the east site and the dispersion of much of the old dyke wall. B. Similar view to A. at spring high tide when both sites, including almost all of the old front dykes, are completely flooded. (Photos courtesy Myriam Barbeau.)

indicate that the restoration of a saltmarsh at both sites has been successful, but it is not clear whether the marsh will extend sufficiently far beyond the original dyke location to provide complete protection to the new dyke landward of it from storm surge and wave action during a severe storm (Jeff Ollerhead, personal communication December, 2018). Recent work by Moller (2019) may provide some guidance on the minimum marsh width needed to reduce wave height to an acceptable level.

Box 11.2 (cont.)

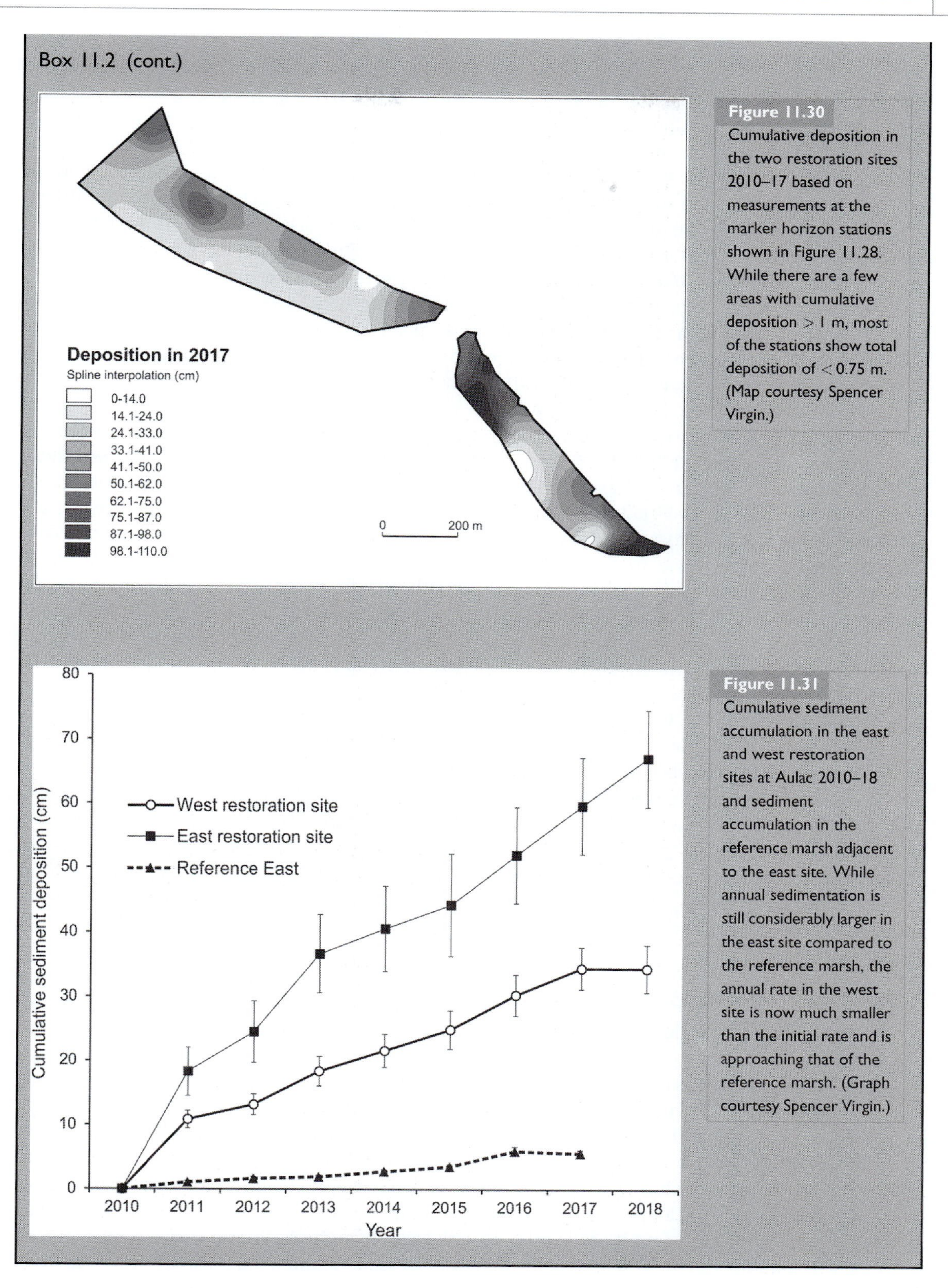

Figure 11.30 Cumulative deposition in the two restoration sites 2010–17 based on measurements at the marker horizon stations shown in Figure 11.28. While there are a few areas with cumulative deposition > 1 m, most of the stations show total deposition of < 0.75 m. (Map courtesy Spencer Virgin.)

Figure 11.31 Cumulative sediment accumulation in the east and west restoration sites at Aulac 2010–18 and sediment accumulation in the reference marsh adjacent to the east site. While annual sedimentation is still considerably larger in the east site compared to the reference marsh, the annual rate in the west site is now much smaller than the initial rate and is approaching that of the reference marsh. (Graph courtesy Spencer Virgin.)

11.4 Mangroves

Mangrove wetlands are the dominant coastal ecosystem of tidal salt and freshwater areas in the tropics and, where they are protected from cold outbreaks, they can extend into the subtropical zone. They differ from saltmarshes in that the vegetation is characterised by trees and some woody shrubs. However, like the grasses and herbs of saltmarshes, they are all adapted to varying levels of flooding and are tolerant of a wide range of salt content. There are about 50 species of mangrove but only 9 that are found in the Americas and West Africa (New World mangroves) with many more Old World species found in East Africa and the Indo-Pacific region. In North America they are well established in southern Florida in the Everglades but further north and along the Gulf coast into Texas their range fluctuates in response to the impact of freezing temperatures during cold air outbreaks (Stevens *et al.*, 2006). Black mangroves can survive lower temperatures than red mangroves and therefore tend to be found at the northern limit (Mitsch and Gosselink, 2000). On the west coast they grow as far north as Baja California. They are found throughout the Caribbean and in South America they extend as far south as 30° on the east coast but on the west coast they are limited to about 5° because of the effects of the cold Humboldt current. Mangroves are not found in the Mediterranean but occur throughout most of the Pacific coast of Asia and along the north and east coasts of Australia and parts of North Island, New Zealand.

11.4.1 Mangrove Swamp: Form, Flooding and Vegetation Zonation

Mangroves can occur as a thin fringe along the borders of bays and lagoons or salt ponds, as well as in a belt hundreds of metres to kilometres in width in prograding deltaic environments such as the muddy deltas of rivers. Mangrove establishment requires shelter from high wave energy conditions and so the geomorphological settings in which they are found are similar to those of saltmarshes (Thom, 1982; Woodroffe, 1992; Augustinus, 1995; Scott *et al.*, 2014; see Figure 11.32A, B). They can be found on open muddy coasts such as French Guiana where wave energy is dissipated over a wide nearshore zone (Allison and Lee, 2004) and on muddy deltaic coasts such as the Red and Mekong Rivers in Vietnam (Van Santen *et al.*, 2007; Fagherazzi *et al.*, 2017), fringing active delta lobe mudflats, behind wave worked cheniers and along the banks of estuaries and distributary channels. They also occur in estuaries and embayments (Woodroffe *et al.*, 1989; Anthony, 2004) and in salt ponds and areas sheltered by sandy barriers (Méndez Linares *et al.*, 2007) and coral reefs (Fujimoto *et al.*, 1996).

Mangrove forests may also be characterised on the basis of the dominant physical processes controlling exchanges of water and sediment (Woodroffe, 1992; see Figure 11.33). Tide dominated mangroves are generally close to the sea or to a tidal creek that permits continuous connection to ocean tides (Figure 10.32A). As a result, flooding occurs daily and salinity is relatively stable and is close to that of sea water. There is also a continuous exchange of organic and inorganic matter between the mangroves and the sea.

River dominated mangroves occur in deltas and estuaries and water level fluctuations due to river flows may be more important than tides. Mangroves may be flooded for periods of weeks or months during the rainy season and there is a strong flow of sediment and organic matter through to the sea. Nevertheless, sedimentation tends to be quite rapid and inorganic sediments generally dominate over organic accumulation.

Interior mangroves, such as the Florida Everglades, occur in low-lying areas some distance from the sea and from tidal influence. They also tend to be cut off from river flooding and so salinity fluctuates with rainfall and evaporation. In the absence of inputs of mineral sediments, accumulation is dominantly organic, leading to the development of thick peat units. Mangroves in salt ponds where there is no permanent connection to the sea (Figure 11.32B), may be subject to similar water level and salinity fluctuations as riverine mangroves in areas with a pronounced dry season. Rainfall and runoff from the surrounding slopes produce high water levels and

Figure 11.32 Examples of mangrove habitats: A. estuarine mangroves, Myall River, NSW, Australia; B. mangroves in a salt pond, South Friar's Beach, St Kitts, West Indies.

reduced salinity during the rainy season, and evaporation leads to reduced water levels and increased salinity during the dry season. Like interior mangroves, they are cut off from regular tidal exchanges with the sea and there is considerable accumulation of organic material. However, runoff from the surrounding hill slopes can bring in considerable amounts of inorganic sediment. In a number of the Caribbean islands storm surge, associated with close passage of hurricanes, can result in overwash of the low barriers and the deposition of sandy beds over the organic-rich silt and clay beds that accumulate most of the time.

There are about 54 species of true mangroves (those found only in wet, saline environments) with about half found in two families, Avicenniaceae and Rhizophoraceae (Hogarth, 1999). Mangroves are not only halophytic but they have adapted to growing in waterlogged soils where

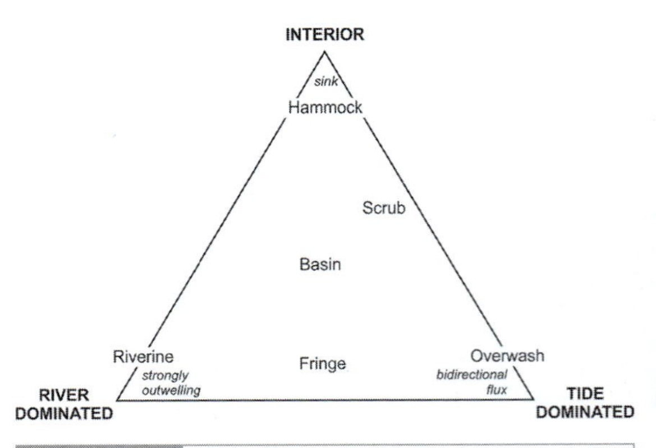

Figure 11.33 The relation between functional type of mangrove forests and the dominant physical processes. Riverine dominated habitats are characterised by strong outflow of sediments and organic material. Tide-dominated mangroves have bidirectional fluxes and interior habitats are sinks largely for organic materials. (After Woodroffe, 1992.)

Figure 11.34 Adaptation by mangroves to waterlogged conditions: red mangrove (*Rhizophora mangle*) in the Buccoo lagoon, Tobago, WI, showing the extensive aerial prop root system permitting it to survive in the water.

oxygen for respiration is difficult to obtain by special developments of the root forms. Roots of Rhizophora diverge from the trunk some distance above the soil forming a branching network of prop roots (Figure 11.34), with air entering the portions of the roots above the water level. Because of this *Rhizophora* can grow down to the low-tide level and even just below, with the result that mangroves can occupy the whole intertidal zone, rather than just the upper half as is the case for saltmarshes. Species of *Avicennia* and *Sonneratia* have roots that spread out just below the surface with pneumatophores that grow upward above the soil in a network of spikes. In a mature forest the pneumatophores can become so dense (Figure 11.35B) that they retard water flows to the extent that the depth of ponded waters is increased (Mullarney *et al.*, 2017a), thus changing the hydrology of the whole area. Mangroves can spread to other sites rapidly because all species can disperse by water and most can survive long periods of transport. They are aided in this by a long period of development of the seed on the tree with the result that what is dispersed on the water is a seedling that can establish itself rapidly once it comes to rest on a suitable substrate.

Like saltmarsh and dune vegetation, there is a tendency for zonation of mangrove species, especially where there is a distinct gradient in flooding frequency or salinity. In the Friar's Bay salt pond (Figure 11.32B), red mangrove occupy most of the portions of the pond that are flooded frequently, with a border of black mangrove (*Avicennia germinans*) in the zone near the upper limit of flooding and some white mangrove (*Laguncularia racemosa*) at the edge of higher ground. Similar zonation is found in much of the Caribbean, Florida and Mexico (Méndez Linares et al., 2007). On relatively stable lagoon and estuarine shorelines, the horizontal zonation reflects primarily adaptation to small differences in elevation and therefore to different flooding levels and disturbance such as hurricanes rather than succession. However, in a study of mangroves in Singapore, Leong *et al.* (2018) found that small differences in elevation did influence the distribution of mangrove types but that there was considerable overlap in the range of different species. In large basins where there is little topographic relief there may be little evidence of zonation and dominance by a single species (Knight *et al.*, 2008). Mangrove zonation is not as sensitive to flooding frequency as are saltmarsh plants. In part this is because of the adaptation of their root system to waterlogged conditions and in part to the fact that the leaf canopy is always above the level of tidal inundation and thus respiration and

Figure 11.35 Photographs of the mangrove fringe, Mekong delta showing A. the transition from mudflats to mangroves. The instrumentation here consists of an ADV close to the bed for measuring turbulence near the bed, and an ADCP for measuring flow and suspended sediment concentration through the water column. B. Close up of pneumatophores of *Sonneratia caseolaris*, the dominant mangrove in this area, with mature trees in the background. (Photographs courtesy University of Waikato/Washington State University Mekong field team; see Bryan *et al.*, 2017, and Mullarney *et al.*, 2017a.)

photosynthesis can occur continuously even if the root system is flooded for days or months. Zonation may also be complicated in areas such as deltas where there are rapid variations in riverine flows and in sediment supply which influence mangrove distribution in addition to tidal effects (Méndez Linares *et al.*, 2007; Fagherazzi et al., 2017; Bullock *et al.*, 2017).

11.4.2 Flows and Sediment Deposition in Mangroves

Conditions for measuring flows, TSS concentration and short-term deposition are not logistically as easy as for saltmarshes and the result is that there are fewer measurements of hydrodynamics and sedimentation than there are for saltmarshes (Augustinus, 1995; Schwarzer *et al.*, 2016; Mullarney *et al.*, 2017a). As is the case with saltmarshes, flow in tidal creeks and through the mangrove forest is driven by tidal water elevation changes and by the tidal prism (Wolanski et al., 1992). The dense prop roots, pneumatophores and tree trunks of the mangrove forest act as very significant roughness elements with much of it extending beyond the water surface through tidal and riverine flows (Furukawa and Wolanski, 1996; Kobashi and Mazda, 2003; Bryan *et al.*, 2017; Mullarney et al., 2017b). The result is that there is a marked reduction in flow speeds and wave action along the edge of tidal creek channels and away from the channels into the forest (Figure 11.35), and similar reductions for flows driven by riverine flooding (Van Santen *et al.*, 2007). Furukawa and Wolanski (1996) report that flows beyond the margin seldom exceed 0.1 m s^{-1}, and similar observations have been found in a number of other studies (Anthony, 2004; Van Santen *et al.*, 2007).

The roots and trunks of mangroves similarly act to produce a rapid dissipation of wave energy away from the margin. The relatively small locally generated waves found in embayments, salt ponds and estuaries are quickly reduced in height to the point where the orbital motion is below that capable of causing sediment entrainment (Quartel *et al.*, 2007). Wave height reduction does depend on tidal stage and on the physical structure of the mangroves. Thus, Mazda *et al.* (2006) found that the rate of reduction of H_s was initially large in *Sonneratia* sp. which possess pneumatophores, but decreased as the tide rose because of the tapering nature of the pneumatophores spikes. Henderson *et al.* (2017) found that, except at very high densities, energy dissipation is rapid leading to most energy being dissipated within 100 m with a water level of 1 m and within 200 m for water depth of 2 m. Energy dissipation may be greater in areas of *Rhizophora* forest, especially for deeper flows, because of the much greater mass of the prop roots (Horstman *et al.*, 2014). Similarly, Massel *et al.* (1999) measured changes in the energy spectra away from the bank of Cocoa Creek, near Townsville, Australia at different tidal stages (Figure 11.36). The first station was located on the mudflat in front of the mangroves and stations 2–4 were located in *Rhizophora* forest with dominant prop roots (Figure 11.36A). Station 5 was only flooded to a shallow depth around high tide and was located in a zone dominated by *Aegiceras* and *Ceriops* mangroves which do not have prop roots.

Similar to tidal flows on saltmarsh surfaces (Figure 11.15), mangroves can act to reduce the speed and direction of tidal currents, encourage deposition of sediment in suspension and to reduce the potential for scour of sediments once deposited. However, on relatively exposed coasts there is a complex interaction between tidal elevation, tidal currents, waves, topography (particularly slope) and the varying drag effects of the roots and trunks of the mangrove canopies (Mullarney *et al.*, 2017b; Bryan *et al.*, 2017). Under some energetic conditions drag around pneumatophores, prop roots and tree trunks can lead to scour, particularly of fine sediments. Flows on the tidal flats and mangrove fringe tend to be flood dominant and become ebb dominant within the forest. However, surface slope, and therefore the magnitude of the tidal prism, is also an important control. Measurements of sediment deposition over tidal cycles or a few days can be carried out in a similar fashion to measurements on marshes, but where there is waterlogging, simple filter paper traps will not work and more robust surfaces such as canvas are deployed (Anthony, 2004; Van Santen *et al.*, 2007). As is the case with saltmarshes, the TSS concentration can be expected to decrease away from the mangrove margin as material is deposited on the surface and, in a simple tidally driven system, deposition can be expected to be at a maximum near the tidal creek bank and to decrease away from it with decreasing frequency of inundation (Furukawa and Wolanski, 1996; Rogers *et al.*, 2005; see Figure 11.37).

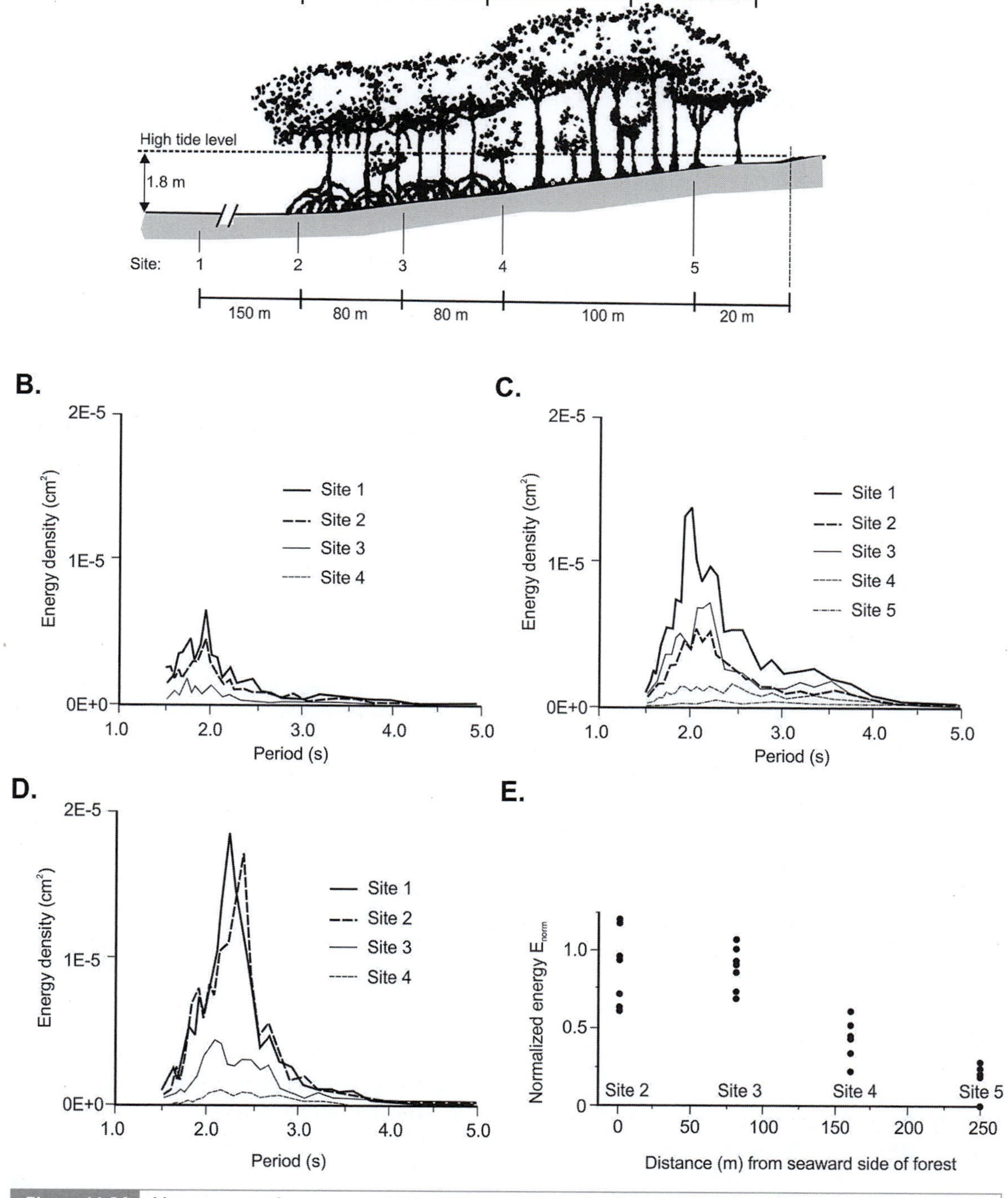

Figure 11.36 Measurements of wave attenuation in a mangrove community along a tidal creek at different tidal stages: A. sketch of the transect showing instrument locations and dominant mangrove species; changes in wave spectra along the transect B. on the rising tide C. at high tide; and D. on the falling tide; E. normalised wave energy versus distance for all measurements (Massel *et al.*, 1999).

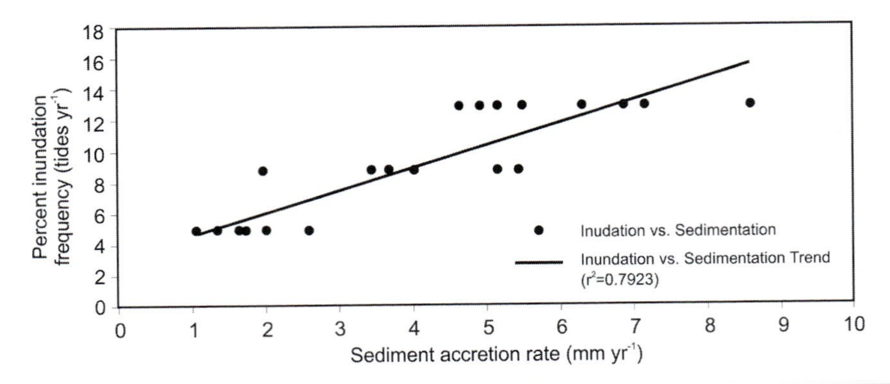

Figure 11.37 Observed and predicted sedimentation rate with distance from the tidal creek for one spring tidal cycle for measurements carried out at Cairns, Queensland, Australia (Furukawa and Wolanski, 1996).

In addition to sediment imported from the bay, estuary or river, there is a steady supply of organic matter from leaf and stem litter and ultimately from the trunks and roots of the mangroves. The result is that mangrove sediments often contain large quantities of organic matter and the bulk density is initially quite low (Sanders *et al.*, 2000). Thus, autocompaction is a problem here in the same way as it is for saltmarshes with the result that long-term estimates of sedimentation must be obtained from stratigraphic coring and dating (Woodroffe, 1981; Ramcharan, 2004). Rates of accretion are similar to those reported for saltmarshes and like them vary with the geomorphic setting.

Box 11.3 Coastal Management Application

Mangroves and Protection Against Tsunamis

Following the 26 December 2004 tsunami there have been several studies that have suggested that destruction in areas protected by a belt of mangroves was much less than in exposed areas (e.g., Danielsen *et al.*, 2005; Kathiresan and Rajendran, 2005; Iverson and Prasad, 2007; Marois and Mitsch, 2015; Blankespoor *et al.*, 2017). The thesis is a simple one based on the known attenuation of wave energy within mangrove forests and it leads to a logical conclusion that planning for rebuilding of settlements in these vulnerable areas should include the potential for developing buffer zones of mangroves. Some of the studies have been criticised for misuse of statistical tests or for drawing conclusions from data that are not justified by the sampling strategy (e.g., Kerr *et al.*, 2006; Baird and Kerr, 2007; Iverson and Prasad, 2007). It is clearly a multivariate problem and it is not easy in retrospect to provide reliable tests of the hypothesis.

Alongi (2008) provides a thoughtful review of some of the discussion, and makes the important point that the hydraulic characteristics of the tsunami waves are quite different from those used to measure wave attenuation (Massel *et al.*, 1999; Mazda *et al.*, 2006; Horstman *et al.*, 2014). There is therefore a need to model the effectiveness of different mangrove species in attenuating tsunami waves and to assess the relative effectiveness of these versus, for example, other tree species or artificial shore protection in highly vulnerable areas.

Further Reading

Mitsch, W.J. and Gosselink, J.G. 2000. *Wetlands*. New York: John Wiley and Sons, **19**, 920 pp.

This is a classic text on the biology and dynamics of all wetlands.

Allen, J.R.L. 2000. Morphodynamics of Holocene saltmarshes: a review sketch from the Atlantic and Southern North Sea coasts of Europe. *Quaternary Science Reviews*, **19**, 1155–231.

This is an excellent review of all aspects of the physical dynamics of saltmarshes.

Mullarney, J.C., Henderson, S.M., Norris, B.K. *et al.* 2017a. A question of scale: how turbulence around aerial roots shapes the seabed morphology in mangrove forests of the Mekong Delta. *Oceanography* 30(3), 34–47, DOI: https://doi.org/10.5670/oceanog.2017.312.

Fagherazzi, S., Bryan, K.R. and Nardin, W. 2017. Buried alive or washed away: the challenging life of mangroves in the Mekong Delta. *Oceanography* 30(3), 48–59, DOI: https://doi.org/10.5670/oceanog.2017.313.

These two papers provide an interesting overview of a series of studies of the interaction between the structure and dynamics of a mangrove forest and sedimentary processes associated with tides, waves and winds.

References

Adam, P. 1990. *Saltmarsh Ecology*. Cambridge: Cambridge University Press, 461 pp.

Adam, P. 1997. Absence of creeks and pans in temperate Australian salt marshes. *Mangroves and Salt Marshes*, **1**, 239–41.

Alexander, C.R., Hodgson, J.Y.S. and Brandes, J.A. 2017. Sedimentary processes and products in a mesotidal salt marsh environment: insights from Groves Creek, Georgia. *Geo-Marine Letters*, **37**, 345–59.

Allen, J.R.L. 1989. Evolution of salt-marsh cliffs in muddy and sandy systems: a qualitative comparison of British west-coast estuaries. *Earth Surface Processes and Landforms*, **14**, 85–92.

Allen, J.R.L. 1990. Salt marsh growth and stratification: a numerical model with special reference to the Severn Estuary, southwest Britain. *Marine Geology*, **95**, 77–96

Allen, J.R.L. 2000. Morphodynamics of Holocene saltmarshes: a review sketch from the Atlantic and Southern North Sea coasts of Europe. *Quaternary Science Reviews*, **19**, 1155–231.

Allen, J.R.L. and Duffy, M.J. 1998. Medium-term sedimentation on high intertidal mudflats and salt marshes in the Severn Estuary, SW Britain: the role of wind and tide. *Marine Geology*, **150**, 1–27.

Allen, J.R.L. and Pye, K. 1992. *Saltmarshes: Morphodynamics, Conservation and Engineering Significance*. Cambridge: Cambridge University Press, 184 pp.

Allison, M.A. and Lee, M.T. 2004. Sediment exchange between Amazon mudbanks and shore-fringing mangroves in French Guiana. *Marine Geology*, **208**, 169–90.

Alongi, D.M. 1998. *Coastal Ecosystem Processes*. Boca Raton, FL: CRC Press, 419 pp.

Alongi, D.M. 2008. Mangrove forests: resilience, protection from tsunamis, and responses to global climate change. *Estuarine, Coastal and Shelf Science*, **76**, 1–13.

Anthony, E.J. 2004. Sediment dynamics and morphological stability of estuarine mangrove swamps in Shebro Bay, West Africa. *Marine Geology*, **208**, 207–24.

Argow, B.A., Hughes, Z.J. and FitzGerald, D.M., 2011. Ice raft formation, sediment load, and theoretical potential for ice-rafted sediment influx on northern coastal wetlands. *Continental Shelf Research*, **31**, 1294–305.

Ashall, L.M., Mulligan, R.P., van Proosdij, D. and Poirier, E. 2016. Application and validation of a three-dimensional hydrodynamic model of a macrotidal salt marsh. *Coastal Engineering*, **114**, 35–46.

Augustinus, P.G.E.F. 1995. Geomorphology and sedimentology of mangroves. In Perillo, G. (ed.), *Geomorphology and Sedimentology of Estuaries*. Amsterdam: Elsevier, pp. 333–57.

Baird, A.H. and Kerr, A.M. 2007. Landscape analysis and tsunami damage in Aceh: comment. *Landscape Ecology*, **23**, 3–5.

Bartholdy, J. 2012. Salt marsh sedimentation. In Davis, R.A. and Dalrymple, R.W. (eds.), *Principles of Tidal Sedimentology*. Berlin: Springer, pp. 151–95.

Bartholdy, J. and Aagaard, T. 2001. Storm surge effects on a back-barrier tidal flat of the Danish Wadden Sea. *Geo-Marine Letters*, **20**, 133–41.

Bartholdy, J., Christiansen, C. and Kunzendorf, H. 2004. Long term variations in backbarrier salt marsh deposition on the Skallingen peninsula: the Danish Wadden Sea. *Marine Geology*, **203**, 1–21.

Bartholdy, A.T., Bartholdy, J. and Kroon, A. 2010a. Salt marsh stability and patterns of sedimentation across a backbarrier platform. *Marine Geology*, **278**, 31–42.

Bartholdy, J., Pedersen, J.B.T.and Bartholdy, A.T. 2010b. Autocompaction in shallow silty salt marsh clay. *Sedimentary Geology*, **223**, 310–19.

Bartholdy, J., Brivio, l., Bartholdy, A., Kim, D. and Fruergaard, M. 2018. The Skallingen spit, Denmark: birth of a back-barrier saltmarsh. *Geo-Marine Letters*, 38(2), 153–66. DOI: 10.1007/s00367–017-0523-5.

Bayliss-Smith, T.P., Healey, R., Lailey, R., Spencer, T. and Stoddart, D.R. 1979. Tidal flows in salt marsh creeks. *Estuarine Coastal and Shelf Science*, **9**, 235–55.

Bendoni, M., Mel, R., Solari, L. *et al.* 2016. Insights into lateral marsh retreat mechanism through localized field measurements. *Water Resources Research*, **52**, 1446–64.

Blankespoor, B., Dasgupta, S. and Lange, G-M. 2017. Mangroves as a protection from storm surges in a changing climate. *Ambio*, **46**, 478–91.

Boone, L.K., Ollerhead, J., Barbeau, M.A. *et al.* 2017. Returning the tide to dikelands in a macrotidal and ice-influenced environment: challenges and lessons learned. In Finkl, C.W. and Makowski C. (eds.), *Coastal Wetlands: Alteration and Remediation*. Cham: Springer, pp. 705–51.

Boumans, R.M. and Day, J.W. Jr. 1993. High precision measurements of sediment elevation in shallow coastal areas using a sedimentation-erosion table. *Estuaries*, **16**, 375–80.

Bowron, T., Neatt, N., van Proosdij, D., Lundholm, J. and Graham, J., 2009. Macro-tidal salt marsh ecosystem response to culvert expansion. *Restoration Ecology*, **19**, 307–22.

Boyd, B.M., Sommerfield, C.K. and Elsey-Quirk, T. 2017. Hydrogeomorphic influences on salt marsh sediment accumulation and accretion in two estuaries of the US Mid-Atlantic coast. *Marine Geology*, **383**, 132–45.

Bryan, K.R., Nardin, W., Mullarney, J.C. and Fagherazzi, S. 2017. The role of cross-shore tidal dynamics in controlling intertidal sediment exchange in mangroves in Cù Lao Dung, Vietnam. *Continental Shelf Research*, **147**, 128–43.

Bullock, E.L., Fagherazzi, S., Nardin, W. *et al.* 2017. Temporal patterns in species zonation in a mangrove forest in the Mekong Delta, Vietnam, using a time series of Landsat imagery. *Continental Shelf Research*, **147**, 144–54.

Burdick, D.M., Dionne, M., Boumans, R.M. and Short, F.T. 1997. Ecological responses to tidal restoration of two northern New England salt marshes. *Wetlands Ecology and Management*, **4**, 129–44.

Cahoon, D.R., Reed, D.J. and Day, J.W. Jr. 1995. Estimating shallow subsidence in microtidal salt marshes of the southeastern United States: Kaye and Bighoorn revisited. *Marine Geology*, **128**, 1–9.

Cahoon, D.R., French, J.R., Spencer, T., Reed, D.J. and Möller, I. 2000. Vertical accretion versus elevational adjustment in UK saltmarshes: an evaluation of alternative methodologies. In Pye, K. and Allen, J.R.L. (eds.), *Coastal and Estuarine Environments. Journal of The Geological Society*, Special Publication, **175**, pp. 223–38.

Callaway, J.C., DeLaune, R.D. and Patrick, W.H. Jr. 1996. Chernobyl 137Cs used to determine sediment accretion rates at selected northern European coastal wetlands. *Limnology and Oceanography*, **41**, 444–50.

Callaway, J.C., DeLaune, R.D. and Patrick, W.H. Jr. 1997. Sediment accretion rates from four coastal wetlands along the Gulf of Mexico. *Journal of Coastal Research*, **13**, 181–91.

Carey, J.C., Moran, S.B., Kelly, R.P., Kolker, A.S. and Fulweiler, R.W. 2017. The declining role of organic matter in New England salt marshes. *Estuaries and Coasts*, **40**, 626–39.

Childers, D.L., Skalar, F.H., Drake, B. and Jordan, T. 1993. Seasonal measurements of sediment elevation in three mid-Atlantic estuaries. *Journal of Coastal Research*, **9**, 986–1003.

Christiansen, T., Wiberg, P.L. and Milligan, T.G. 2000. Flow and sediment transport on a tidal salt marsh. *Estuarine, Coastal and Shelf Science*, **50**, 315–31.

Church, J.A., Clark P.U., Cazenave, A. *et al.*, 2013. Sea level change. In Stocker, T.F., Qin, D., Plattner, G.-K. *et al.* (eds.), *Climate Change 2013: The Physical Science Basis. Contribution of Working Group I to the Fifth Assessment Report of the Intergovernmental Panel on Climate Change*. Cambridge and New York: Cambridge University Press.

Coulombier, T., Neumeier, U. and Bernatchez, P. 2012. Sediment transport in a cold climate salt marsh (St. Lawrence Estuary, Canada), the importance of vegetation and waves. *Estuarine, Coastal and Shelf Science*, **101**, 64–75.

Crosby, S.C., Sax, D.F., Palmer, M.E. *et al.* 2016. Salt marsh persistence is threatened by predicted sea-level rise. *Estuarine, Coastal and Shelf Science*, 181, 93–9.

D'Alpaos, A., Lanzoni, S., Marani, M. *et al.* 2007. Spontaneous tidal network formation within a constructed salt marsh: observations and morphodynamic modelling. *Geomorphology*, **91**, 186–97.

Danielsen, F., Sorensen, M.K., Olwig, M.F. *et al.* 2005. Asian tsunami: a protective role for coastal vegetation. *Science*, **310**, 643.

Davidson-Arnott, R.G.D., van Proosdij, D., Ollerhead, J. and Schostak, L.E. 2002 Hydrodynamics and sedimentation in saltmarshes: examples from a macrotidal marsh, Bay of Fundy. *Geomorphology*, **48**, 209–231.

Fagherazzi, S., Kirwan, M.L., Mudd, S.M. *et al.* 2012. Numerical models of salt marsh evolution: ecological, geomorphic, and climatic factors. *Review of Geophysics*, **50**, RG1002, DOI:10.1029/2011RG000359.

Fagherazzi, S., Bryan, K.R. and Nardin, W. 2017. Buried alive or washed away: The challenging life of mangroves in the Mekong Delta. *Oceanography* 30(3), 48–59, DOI: https://doi.org/10.5670/oceanog.2017.313.

French, C.E., French, J.R., Clifford, N.J. and Watson, C.J. 2000. Sedimentation–erosion dynamics of abandoned reclamations: the role of waves and tides. *Continental Shelf Research*, **20**, 1711–33.

French, J.R. 1993. Numerical simulation of vertical marsh growth and adjustment to accelerated sea-level rise, North Norfolk, UK. *Earth Surface Processes and Landforms*, **18**, 63–81.

French, J.R. and Spencer, T. 1993. Dynamics of sedimentation in a tide dominated backbarrier salt marsh, Norfolk, United Kingdom. *Marine Geology*, 110, 315–31.

French, J.R. and Stoddart, D.R. 1992. Hydrodynamics of salt marsh creek systems: implications for marsh morphodynamic development and matter exchange. *Earth Surface Processes and Landforms*, **17**, 235–52.

French, P.W. 2006. Managed realignment: the developing story of a comparatively new approach to soft engineering. *Estuaries, Coastal and Shelf Science*, **67**, 406–23.

Friess, D.A., Möller, I., Spencer, T. *et al.* 2014. Coastal saltmarsh managed realignment drives rapid breach inlet and external creek evolution, Freiston Shore (UK). *Geomorphology*, **208**, 22–33.

Fujimoto, K., Miyagi, T., Kikuchi, T. and Kawana, T. 1996. Mangrove habitat formation and response to Holocene sea-level changes on Kusrae island, Micronesia. *Mangroves and Salt Marshes*, **1**, 47–57.

Furukawa, K. and Wolanski, E. 1996. Sedimentation in mangrove forests. *Mangroves and Salt Marshes*, **1**, 3–10.

Goodman, J.E., Wood, M.E. and Gehrels, W.R. 2007. A 17-yr record of sediment accretion in the salt marshes of Maine (USA). *Marine Geology*, **242**, 109–21.

Gordon, D.C. and Desplanque, C. 1983. Dynamics and environmental effects of ice in the Cumberland Basin of the Bay of Fundy. *Canadian Journal of Fisheries and Aquatic Sciences*, 40, 1331–42.

Gordon, D.C., Cranford, P.J. and Desplanque, C. 1985. Observations on the ecological importance of salt marshes in the Cumberland Basin, a macrotidal estuary in the Bay of Fundy. *Estuarine, Coastal and Shelf Science*, **20**, 205–27.

Gray, A.J. 1992. Saltmarsh plant ecology: zonation and succession revisited. In Allen, J.R.L. and Pye, K. (eds.), *Saltmarshes, Morphodynamics, Conservation and Engineering Significance*. Cambridge: Cambridge University Press, pp. 63–79.

Healey, R.G., Pye, K., Stoddart, D.R. and Bayliss-Smith T.P. 1981. Velocity variations in salt marsh creeks, Norfolk, England. *Estuarine, Coastal and Shelf Science*, **13**, 535–45.

Henderson, S.M., Norris, B.K., Mullarney, J.C. and Bryan, K.R. 2017. Wave-frequency flows within a near-bed vegetation canopy. *Continental Shelf Research*, **147**, 91–101.

Hogarth, P.J. 1999. *The Biology of Mangroves*. Oxford: Oxford University Press, 228 pp.

Horstman, E.M., Dohmen-Janssen, C.M., Narra, P.M.F. *et al.* 2014. Wave attenuation in mangroves: a quantitative approach to field observations. *Coastal Engineering*, **94**, 47–62.

Houser, C., Trimble, S. and Morales, B. 2015. Influence of blade flexibility on the drag coefficient of aquatic vegetation. *Estuaries and Coasts*, **38**, 569–77.

Iverson, L.R. and Prasad, A.M. 2007. Using landscape analysis to assess and model tsunami damage in Aceh Province, Sumatra. *Landscape Ecology*, **22**, 323–31.

Kakeh, N., Coco, G. and Marani, M. 2016. On the morphodynamic stability of intertidal environments and the role of vegetation. *Advances in Water Resources*, **93**, 303–14.

Kathiresan, K. and Rajendran, N. 2005. Coastal mangrove forests mitigated tsunami. *Estuarine, Coastal and Shelf Science*, 65, 601–6.

Kearney, M.S. and Ward, L.G. 1986. Accretion rates in brackish marshes of a Chesapeake Bay estuarine tributary. *Geological Marine Letters*, **6**, 41–9.

Kerr, A.M., Baird, A.H. and Campbell, S.J. 2006. Comments on 'Coastal mangrove forests mitigated tsunami' by K. Kathiresan and N. Rajendran. *Estuarine, Coastal and Shelf Science*, **67**, 539–41.

Kim, D., Bartholdy, J and Bartholdy, A.T. 2016. Varying patterns of vegetation dynamics across multiple levels of organization in a salt marsh of the Danish Wadden Sea. *Hydrobiologia*, 771, 67–81.

Kirwan, M.L. and Murray, A.B. 2008. Ecological and morphological response of brackish tidal marshland to the next century of sea level rise: Westham Island, British Columbia. *Global and Planetary Change*, **60**, 471–86.

Kirwan, M.L., Guntenspergen, G.R., D'Alpaos, A. *et al.* 2010. Limits on the adaptability of coastal marshes to rising sea level. *Geophysical Research Letters*, 37, L23401.

Kirwan, M.L., Walters, D.C., Reay, W.G. and Carr, J.A. 2016. Sea level driven marsh expansion in a coupled model of marsh erosion and migration. *Geophysical. Research Letters*, **43**, 4366–73.

Knight, J.M., Dale, P.E.R., Dunn, R.J.K., Broadbent, G.J. and Lemckert, C.J. 2008. Patterns of tidal flooding within a mangrove forest: Coombabah Lake, southeast Queensland, Australia. *Estuarine, Coastal and Shelf Science*, **76**, 580–93.

Kobashi, D. and Mazda, Y. 2003. Tidal flow in riverine-type mangroves. *Wetlands Ecology and Management*, **13**, 615–19.

Konisky, R.A., Burdick, D.M., Dionne, M. and Neckles, H.A. 2006. A regional assessment of salt marsh restoration and monitoring in the Gulf of Maine. *Restoration Ecology*, **14**, 516–25.

Kostaschuk, R.A., Chen, Z, Saito, Y. and Wang, Z. 2008. Sedimentation rates and heavy metals in a macrotidal salt marsh: Bay of Fundy, Canada. *Environmental Geology*, **55**, 1291–8.

Leonard, L.A. and Croft, A.L. 2006. The effect of standing biomass on flow velocity and turbulence in *Spartina alterniflora* canopies. *Estuarine, Coastal and Shelf Science*, 69, 325–36.

Leonard, L.A. and Luther, M.E. 1995. Flow hydrodynamics in tidal marsh canopies. *Limnology and Oceanography*, **40**, 1474–84.

Leonard, L.A., Hine, A.C. and Luther, M.E. 1995a. Surficial sediment transport and deposition processes in a *Juncus roemerianus* marsh, west-central Florida. *Journal of Coastal Research*, **11**, 322–36.

Leonard, L.A., Hine, A.C., Luther, M.E., Stumpf, R.P. and Wright, E.E. 1995b. Sediment transport processes in a west-central Florida open marine marsh tidal creek; the role of tides and extra-tropical storms. *Estuarine, Coastal and Shelf Science*, **41**, 225–48.

Leonardi, N., Ganju, N.K. and Fagherazzi, S. 2016a. A linear relationship between wave power and erosion determines salt-marsh resilience to violent storms and hurricanes. *Proceedings of the National Academy of Science*, **113**, 64–8.

Leonardi, N., Defne, Z., Ganju, N.K. and Fagherazzi, S. 2016b. Salt marsh erosion rates and boundary features in a shallow bay. *Journal of Geophysical Research: Earth Surface*, **121**, 1861–75.

Leonardi, N., Carnacina, I., Donatelli, C. *et al.* 2018. Dynamic interactions between coastal storms and salt marshes: a review. *Geomorphology*, **301**, 92–107.

Leong, R.C., Friess, D.A., Crase, B., Lee, W.K. and Webb, E.L. 2018. High-resolution pattern of mangrove species distribution is controlled by surface elevation. *Estuarine, Coastal and Shelf Science*, **202**, 185–92.

Luternauer, J.L., Atkins, R.J., Moody, A.I., Williams, H.F.L. and Gibson, J.W. 1995. *Salt marshes*. In Perillo, G.M.E. (ed.), *Geomorphology and Sedimentology of Estuaries*. Amsterdam: Elsevier, pp. 307–32.

Marois, D.E. and Mitsch, W.J. 2015. Coastal protection from tsunamis and cyclones provided by mangrove wetlands: a review. *International Journal of Biodiversity Science, Ecosystem Services & Management*, **11**, 71–83.

Marshall, W., Gehrels, W.R., Garnett, M. *et al.* 2007. The use of 'bomb spike' calibration and high-precision AMS 14C analyses to date salt-marsh sediments deposited during the past three centuries. *Quaternary Research*, **68**, 325–37.

Massel, S.R., Furukawa, K. and Brinkman, R.M. 1999. Surface wave propagation in mangrove forests. *Fluid Dynamics Research*, **24**, 219–49.

Mazda, Y., Magi. M., Ikeda, Y., Kurokawa, T. and Asano, T. 2006. Wave reduction in a mangrove forest dominated by Sonneratia sp. *Wetlands Ecology and Management*, **14**, 365–78.

McLoughlin, S.M., Wiberg, P.L., Safak, I. and McGlathery, K.J. 2015. Rates and forcing of marsh edge erosion in a shallow coastal bay. *Estuaries and Coasts*, **38**, 620–38.

Méndez Linares, A.P., López-Portillo, J., Hernández-Santana, J.R., Ortiz Pérez, M.A. and Oropeza Orozco, O. 2007. The mangrove communities in the Arroyo Seco deltaic fan, Jalisco, Mexico, and their relation with the geomorphic and physical–geographic zonation. *Catena*, **127**, 127–42.

Mitsch, W.J. and Gosselink, J.G. 2000. *Wetlands*. New York: John Wiley and Sons, 920 pp.

Möller, I. 2006. Quantifying saltmarsh vegetation and its effect on wave height dissipation: results from a UK east coast saltmarsh. *Estuarine, Coastal and Shelf Science*, **69**, 337–51.

Möller, I. 2019. Applying uncertain science to nature-based coastal protection: lessons from shallow wetland-dominated shores. *Frontiers in Earth Science*, **7**, 49. DOI: 10.3389/fenvs.2019.00049.

Möller, I. and Spencer, T. 2002. Wave dissipation over macro-tidal saltmarshes: effects of marsh edge typology and vegetation change. *Journal of Coastal Research*, **SI 36**, 502–21.

Möller, I., Kudella, M., Rupprecht, F. *et al.* 2014. Wave attenuation over coastal salt marshes under storm surge conditions. *Nature: Geoscience*, **7**, 727–31.

Moreira, M.E.S.A. 1992. Recent saltmarsh changes and sedimentation rates in the Sado Estuary, Portugal. *Journal of Coastal Research*, **8**, 631–40.

Morgensen, L.A. and Rogers, K. 2018. Validation and comparison of a model of the effect of sea-level rise on coastal wetlands. *Nature: Scientific Reports*, **8**, 1369. DOI:10.1038/s41598–018-19695-2

Mudd, S.M., D'Alpaos, A. and Morris, J.T. 2010. How does vegetation affect sedimentation on tidal marshes? Investigating particle capture and hydrodynamic controls on biologically mediated sedimentation. *Journal of Geophysical Research: Earth Surface*, **115**(F3), F03029. DOI: http://dx.doi.org/10.1029/2009JF0 01566.

Mueller, P., Jensen, K. and Megonigal, J.P. 2016. Plants mediate soil organic matter decomposition in response to sea level rise. *Global Change Biology*, **22**, 404–14.

Mullarney, J.C., Henderson, S.M., Norris,B.K. *et al.* 2017a. A question of scale: how turbulence around aerial roots shapes the seabed morphology in mangrove forests of the Mekong Delta. *Oceanography* 30(3), 34–47. DOI: https://doi.org/10.5670/oceanog.2017.312.

Mullarney, J.C. Henderson, S.M., Reyns, J.A.H., Norris, B.K. and Bryan, K.R. 2017b. Spatially-varying drag within a wave-exposed mangrove forest and on the adjacent tidal flat. *Continental Shelf Research*, **147**, 102–13.

Murphy, S. and Voulgaris, G. 2006. Identifying the role of tides, rainfall and seasonality in marsh sedimentation using long-term suspended sediment concentration data. *Marine Geology*, **227**, 31–50.

Mwamba, M.J. and Torres, R. 2002. Rainfall effects on marsh sediment redistribution, North Inlet, South Carolina, USA. *Marine Geology*, **189**, 267–87.

Nepf, H.M. 2012. Flow and transport in regions with aquatic vegetation. *Annual Review. of Fluid Mechanics*. **44**, 123–42.

Neubauer, S.C., Anderson, I.C., Constantine, J.A. and Kuehl, S.A. 2002. Sedimentary deposition and accretion in a mid-Atlantic (USA) tidal freshwater marsh. *Estuarine, Coastal and Shelf Science*, **54**, 713–27.

Nolte, S., Koppenaal, E. C., Esselink, P. *et al.* 2013. Measuring sedimentation in tidal marshes: a review on methods and their applicability in biogeomorphological studies. *Journal of Coastal Conservation*, **17**, 301–25.

Nyman, J.A., Walters, R.J., Delaune, R.D. and Patrick, W.H. Jr. 2006. Marsh vertical accretion via vegetative growth. *Estuarine, Coastal and Shelf Science*, **69**, 370–80.

O'Laughlin, C. and van Proosdij, D. 2013. Influence of varying tidal prism on hydrodynamics and sedimentary processes in a hypertidal salt marsh creek. *Earth Surface Processes and Landforms*, **38**, 534–46.

O'Laughlin, C., van Proosdij, D. and Milligan, T.G. 2014. Flocculation and sediment deposition in a hypertidal creek. *Continental Shelf Research*, **82**, 72–84.

Ollerhead, J., Davidson-Arnott, R.G.D. and Scott, A. 2005. Cycles of saltmarsh extension and contraction, Cumberland Basin, Bay of Fundy, Canada. In Sanjaume, E. and Mateu, J.F. (eds.), *Geomorphologia Littoral I Quaternari: Homenatge al Professor V.M. Rossello I Verger*. Valencia: Publicacions Universitat de Valencia, pp. 293–305.

Pethick, J.S. 1981. Long term accretion rates on tidal salt marshes. *Journal of Sedimentary Petrology*. **51**, 571–77.

Pethick, J.S 1992. Saltmarsh geomorphology. In Allen, J.R.L. and Pye, K. (eds.), *Saltmarshes: Morphodynamics, Conservation and Engineering Significance*. Cambridge: Cambridge University Press, pp. 41–62.

Pethick, J.S. 2001. Coastal management and sea level rise. *Catena*, **42**, 307–22.

Perrillo, G.E.M. and Iribarne, O.G. 2003. Processes of tidal channel development in salt and freshwater marshes. *Earth Surface Processes and Landforms*, **28**, 1473–82.

Pontee, N. 2013. Defining coastal squeeze: a discussion. *Ocean and Coastal Management*, **84**, 204–7.

Pringle, A.W. 1995. Erosion of a cyclic saltmarsh in Morecambe Bay, northwest England. *Earth Surface Processes and Landforms*, **20**, 387–405.

Quartel, S., Kroon, A., Augustinus, P.G.E.F., Van Santen, P. and Tri, N.H. 2007. Wave attenuation in coastal mangroves in the Red River Delta, Vietnam. *Journal of Asian Earth Sciences*, **29**, 576–84.

Ramcharan, E.K. 2004. Mid-to-late Holocene sea level influence on coastal wetland development in Trinidad. *Quaternary International*, **120**, 145–51.

Redfield, A.C. 1972. Development of a New England salt marsh. *Ecological Monographes*, **42**, 201–37.

Reed, D.J. 1989. Patterns of sediment deposition in subsiding coastal salt marshes, Terreborne Bay, Louisiana: the role of winter storms. *Estuaries*, **12**, 222–7.

Reed, D.J. 1995. The response of coastal marshes to sea-level rise: survival or submergence? *Earth Surface Processes and Landforms*, **20**, 39–48.

Reed, D.J., Spencer, T., Murray, A.L., French, J.R. and Leonard, L. 1999. Marsh surface sediment deposition and the role of tidal creeks: implications for created and managed coastal marshes. *Journal of Coastal Conservation*, **5**, 81–90.

Reed, D.J., Davidson-Arnott, R.G.D. and Perillo, G.M.E. 2009. The future of coastal systems: from estuaries to dunes. In Slaymaker, O. (ed.), *Landscape Changes in the 21st Century*. Cambridge: Cambridge University Press, pp. 130–57.

Reed, D.J., van Wesenbeeck, B., Herman, P.M.J. and Meselhed, E. 2018. Tidal flat-wetland systems as flood defences: understanding biogeomorphic controls. *Estuarine, Coastal and Shelf Science*, **213**, 269–82.

Robertson, A.I. and Alongi, D.M. (eds.) 1992. *Tropical Mangrove Ecosystems*. Washington, DC: American Geophysical Union, 329 pp.

Rogers, K., Saintilan, N. and Cahoon, D. 2005. Surface elevation dynamics in a regenerating mangrove forest at Homebush Bay, Australia. *Wetlands Ecology and Management*, **13**, 587–98.

Sanders, C.J., Smoak, J.M., Naidu, A.S., Sanders, L.M. and Patchineelam, S.R. 2000. Organic carbon burial in a mangrove forest, margin and intertidal mud flat. *Estuarine, Coastal and Shelf Science*, **90**, 168–72.

Schostak, L.E., Davidson-Arnott, R.G.D., Ollerhead, J. and Kostaschuk, R.A. 2000. Patterns of flow and suspended sediment concentration in a macrotidal saltmarsh creek, Bay of Fundy, Canada. In Pye, K. and Allen, J.R.L. (eds.), *Coastal and Estuarine Environments. Journal of The Geological Society*, Special Publication, **175**, pp. 59–73.

Schwarzer, K., Nguyen, C.T. and Ricklefs, K. 2016. Sediment re-deposition in the mangrove environment of Can Gio, Saigon River estuary (Vietnam). In Vila-Concejo, A., Bruce, E., Kennedy, D.M. and McCarroll, R.J. (eds.), *Proceedings of the Fourteenth International Coastal Symposium* (Sydney, Australia). *Journal of Coastal Research*, **SI 75**, 138–42.

Scott, D.B., Frail-Gauthier, J. and Mudie, P.J. 2014. *Coastal Wetlands of the World: Geology, Ecology, Distribution and Applications*. Cambridge: Cambridge University Press, 351 pp.

Scrimshaw, M.D., Bubb, J.M. and Lester, J.N. 1996. Organochlorine contamination of the UK Essex coast salt marsh sediments. *Journal of Coastal Research*, **12**, 246– 55.

Shi, Z., Hamilton, L.J. and Wolanski, E. 2000. Near-bed currents and suspended sediment transport in saltmarsh canopies. *Journal of Coastal Research*, **16**, 909–14.

Silvestri, S., Defina, A. and Marani, M. 2005. Tidal regime, salinity and salt marsh plant zonation. *Estuarine, Coastal and Shelf Science*, **62**, 119–30.

Steel, T.J. and Pye, K. 1997. The development of saltmarsh tidal creek networks: evidence from the UK. *Proceedings of the Canadian Coastal Conference*, Ottawa: Canadian Coastal Science and Engineering Association, pp. 267–80.

Steers, J.A. 1948. Twelve years' measurement of accretion on Norfolk salt marshes. *Geological Magazine*, **85**, 163–6.

Stevens, P.W., Fox, S.L. and Montague, C.L. 2006. The interplay between mangroves and saltmarshes at the transition between temperate and subtropical climate in Florida. *Wetlands Ecology and Management*, **14**, 435–44.

Stoddart, D.R., Reed, D.J. and French, J.R. 1989. Understanding salt-marsh accretion, Scolt Head Island, Norfolk, England. *Estuaries*, **12**, 228–36.

Stokes, D.J., Healy, T.R. and Cooke, P.J. 2010. Expansion dynamics of monospecific, temperate mangroves and sedimentation in two embayments of a barrier-enclosed lagoon, Tauranga Harbour, New Zealand. *Journal of Coastal Research*, **26**, 113–22.

Temmerman, S., Govers, G., Wartel, S. and Meire, P. 2003. Spatial and temporal factors controlling short-term sedimentation in a salt and freshwater tidal marsh, Scheldt Estuary, Belgium, SW Netherlands. *Earth Surface Processes and Landforms*, **28**, 739–55.

Temmerman, S., Govers, G., Wartel, S. and Meire, P. 2004. Modelling estuarine variations in tidal marsh sedimentation: response to changing sea level and suspended sediment concentration. *Marine Geology*, **212**, 1–19.

Temmerman, S., Bouma, T.J., Govers, G., Wang, Z.B. and De Vries, M.B. 2005. Impact of vegetation on flow routing and sedimentation patterns: three-dimensional modelling for a tidal marsh. *Journal of Geophysical Research*, **110**, F00019–31.

Thom, B.G. 1982. Mangrove ecology: a geomorphological perspective. In Clough, B.F. (ed.), *Mangrove Ecosystems in Australia; Structure, Function and Management*. Canberra: Australian National University Press, pp. 3–17.

Van der Wal, D., Wielemaker-Van den Dool, A. and Herman, P.M.J. 2008. Spatial patterns, rates and mechanisms of saltmarsh cycles (Westerschelde,

The Netherlands). *Estuarine, Coastal and Shelf Science*, **76**, 357–68.

van Proosdij, D. 2001. Spatial and temporal controls on the sediment budget of a macrotidal saltmarsh. PhD Thesis, University of Guelph, 314 pp.

van Proosdij, D., Ollerhead, J., Davidson-Arnott, R.G.D. and Schostak, L. 1999. Allen Creek Marsh, a macrotidal temperate saltmarsh. *Canadian Geographer*, **43**, 316–22.

van Proosdij, D., Ollerhead, J. and Davidson-Arnott, R.G.D. 2000. Sedimentation on a macrotidal saltmarsh. In Pye, K. and Allen, J.R.L. (eds.), *Coastal and Estuarine Environments. Journal of The Geological Society*, Special Publication, **175**, pp. 43–57.

van Proosdij, D., Davidson-Arnott, R.G.D. and Ollerhead, J. 2006a Controls on spatial patterns of sediment deposition across a macro-tidal salt marsh surface over single tidal cycles. *Estuarine, Coastal and Shelf Science*, **69**, 64–86.

van Proosdij, D., Ollerhead, J. and Davidson-Arnott, R.G.D. 2006b Seasonal and annual variations in the volumetric sediment balance of a macro-tidal salt marsh. *Marine Geology*, **225**, 103–27.

Van Santen, P., Augustinus, P.G.E.F., Janssen-Stelder, B.M., Quartel, S. and Tri, N.H. 2007. Sedimentation in an estuarine mangrove system. *Journal of Asian Earth Sciences*, **29**, 566–75.

Weston, N.B. 2014. Declining sediments and rising seas: an unfortunate convergence for tidal wetlands. *Estuaries and Coasts*, 37, 1–23.

Wolanski, E., Mazda, Y. and Ridd, P. 1992. Mangrove hydrodynamics. In Robertson, A.I. and Alongi, D.M. (eds.), *Tropical Mangrove Ecosystems*. Washington, DC: American Geophysical Union, pp. 43–62.

Wood, M.E., Kelley, J.T. and Belknap, D.F. 1989. Patterns of sediment accumulation in the tidal marshes of Maine. *Estuaries*, **12**, 237–46.

Woodroffe, C.D. 1981. Mangrove swamp stratigraphy and Holocene transgression, Grand Cayman Island, West Indies. *Marine Geology*, **41**, 271–94.

Woodroffe, C.D. 1992. Mangrove sediments and geomorphology. In Robertson, A.I. and Alongi, D.M. (eds.), *Tropical Mangrove Ecosystems*. Washington, DC: American Geophysical Union, pp. 7–41.

Woodroffe, C.D., Chappell, J., Thom, B.G. and Wallensky, E. 1989. Depositional model of a macrotidal estuary and floodplain, South Alligator River, northern Australia. *Sedimentology*, **36**, 737–56.

Wu, G., Li, H., Liang, B. *et al.* 2017. Subgrid modeling of salt marsh hydrodynamics with effects of vegetation and vegetation zonation. *Earth Surface Processes and Landforms*, **42**, 1755–68.

Yang, S.L., Shi, B. W., Bouma, T.J., Ysebaert, T. and Luo, X.X. 2012. Wave attenuation at a salt marsh margin: a case study of an exposed coast on the Yangtze estuary. *Estuaries and Coasts*, **35**, 169–82.

Yapp, R.H., Johns, D. and Jones, O.T. 1917. The salt marshes of the Dovey estuary. II. The saltmarshes. *Journal of Ecology*, **5**, 65–103.

Zhao, Y., Yua, Q.,Wang, D. *et al.* 2017. Rapid formation of marsh-edge cliffs, Jiangsu coast, China. *Marine Geology*, **385**, 260–73.

12

Coral Reefs and Atolls

12.1 Synopsis

Coral reefs are the result of simple animals, referred to as polyps, that form large colonies with rich varieties of forms and adaptation to differences in wave energy levels, light and other environmental conditions. Modern coral reefs contain a wide variety of hard and soft corals, but it is the fixing of calcium carbonate ($CaCO_3$) by the scleractinian corals to form a hard external skeleton that provides the basis for reef development as the coral structures grow over decades and centuries. The breakdown of corals, coralline algae and other organisms that live in and on the reef supplies sediment to backreef and lagoon areas. Hard corals that grow fast enough to fix large amounts of calcium carbonate are found primarily in warm, tropical waters and exist in a symbiotic relationship with single cell algae that remove waste and provide a large proportion of the nutrients that the corals need. They also give the corals their distinctive colour. Coral growth is most rapid close to the surface where there is abundant light for the symbiotic algae and food circulated by waves and currents. Living corals reefs have a characteristic zonation that reflects the adaptation of individual species to specific combinations of light, wave energy and food sources. The topography of the reef in a cross section from the land to the open ocean can be divided into three zones: (1) the reef flat, which is a shallow area sheltered from direct wave action and an area of sediment accumulation; (2) the reef crest, is the shallowest part of the reef and is subject to high wave action and frequent damage to coral colonies; and (3) the forereef, which extends seaward into increasing water depths and decreasing light and wave energy. There is rapid coral growth in the portion above 20 m and a wide diversity of species.

Coral colonies are highly dynamic and resilient and because coral larvae are dispersed by waves and currents, they can quickly colonise new substrate or areas damaged by high storm waves. As sea level fluctuated during the Pleistocene, the location of reefs migrated seaward and downward and then landward and upward. During periods of rising sea level, such as the Holocene transgression, there were times and places when the rate of sea level rise exceeded the rate at which corals on many reefs could build upward and so individual reefs were drowned. However, the establishment of coral reef communities always kept pace with sea level change. Most modern reefs were established about 5000 to 7000 years BP as the pace of sea level rise slowed, and many of them are built on a framework of older reef deposits that may have been exposed subaerially during glacial periods.

Coral reefs are dynamic ecosystems that are continually evolving and changing in response to disturbances. These disturbances may be biological, such as those resulting from disease or predation, or they may be physical such as those resulting from sea level change or hurricane impact. Coral reef environments also evolve in response to the redistribution of sediment supplied by the break-up of coral colonies and other reef organisms. These infill lagoons, and extend the shorelines of reef islands, and in places may bury reef substrate. Human activities also disturb reef ecosystems, either directly through activities such as dredging and overfishing, or indirectly through the effects of pollution, suspended sediment runoff due to land clearance or climate change. While these disturbances can have an impact on individual reefs and even reefs worldwide, the reef ecosystem is resilient and persistent, especially when viewed from a geomorphological or geological perspective.

12.2 Corals and Reef Formation

12.2.1 Coral Organisms

Corals are Anthozoans, a class that includes sea fans and anemones, and they form the largest class of organisms within the phylum Cnidaria. Corals may be divided broadly into soft and hard or stony corals (scleractinians), with the hard corals having an external skeleton of calcium carbonate within which the coral polyp lives. Most corals are colonial organisms containing thousands to hundreds of thousands of individuals, termed polyps, and each genera has a characteristic form or structure dictated largely by their growth form and by external forces such as wave energy. Coral polyps are simple organisms and most species consist primarily of a gastrovascular cavity which is open at the top to form the mouth and tentacles which form an extension of the body wall and surround the mouth opening. The tentacles act to capture food in the form of plankton and to clear away debris. The tentacles also contain stinging cells that kill their prey and can give unwary divers and snorkellers a nasty rash.

The hard, scleractinian corals are the chief architects of coral reef formation along with several calcareous algae and minor inputs from other organisms such as molluscs. Many of the scleractinian corals live symbiotically with zooxanthellate single cell algae which inhabit the endothermal tissue of the stomach of the coral polyp. The role of the algae in the symbiotic relationship is to remove metabolic waste and to provide the coral polyp with additional nutrients, in some cases most of its nutrients, while the coral polyp in turn provides a protected environment for the algae to establish and to photosynthesise. The presence of the algae greatly enhances the rate of precipitation of calcium carbonate and hence the rate of reef building. The algae also give the corals distinctive colours, and it is the expulsion of the algae during periods of elevated water temperatures that produces coral bleaching.

While coral reefs can extend for kilometres or hundreds of kilometres along the coast, individual coral polyps are only a few millimetres in diameter. The calcium carbonate secreted by the polyp forms a basal plate and surrounding wall in which the polyp sits and is connected to the other polyps in the colony. This provides protection to the polyp which extends itself to feed, and withdraws completely within the walls for protection. The colonies of polyps making up a coral unit divide asexually and, as older polyps die, their skeletons are incorporated in the coral form – the coral thus consists of an interior calcium carbonate skeleton of dead polyps and an outer layer of living polyps. Corals also reproduce sexually and the resulting coral larvae disperse through the reef and beyond, dictated by wind, tidal currents and wave action. This is the primary source of evolution of the existing reef colonies and the establishment of new colonies.

Colonies of reef building corals take on a wide range of shapes including branching forms such as the large, flattened limbs of elkhorn coral (*Acropora palmata*), the dense finger-like colonies of staghorn coral (*Acropora cervicornis*), ball shaped and massive corals such as brain coral (*Diploria* spp.), and crustose forms which form thin layers over the substrate (Figure 12.1). Other corals such as sea fans have carbonate skeletons that do not contribute directly to the reef structure but do provide a source of carbonate sediment to the reef system. Branching colonies will grow at rates of up to 10 cm a^{-1} while the

Figure 12.1 Examples of warm water corals: A. flat, branching elkhorn coral (*Acropora palmata*) Tobago; B. staghorn coral (*Acropora cervicornis*), Belize; C. pillar coral (*Dendrogyra cylindrus*) centre foreground and mountainous star coral (*Orbicella faveolata*) left foreground, on the right are a variety of gorgonians including sea fans and deadman's fingers, St Barts, 8 m depth; D. black coral and other gorgonians on the reef wall, Grand Turk, 22 m depth.

massive corals grow at half this rate or less. The branching corals are, however, more vulnerable to wave action as they grow larger and the impact of tropical storms can often be seen in the accumulation of large amounts of debris consisting of fragments of branches that have been snapped off. These may be cemented together by coralline algae and provide a substrate for new coral establishment and growth. Corals grow fastest in shallow water near the reef crest where there is plenty of available light for photosynthesis by the zooxanthellates and a continuous supply of nutrients and food. The growth rate declines in deeper water down the front face of the reef, and below 20–25 m depth the main reef building forms are largely replaced by soft corals, sea fans and other scleractinian corals which are adapted to lower light and slower growth. Thus, reef building takes place largely in shallow water close to land or on shallow continental shelves.

Actual rates of calcium carbonate accumulation are highly variable and not easily measured. Extreme rates may range from 1 to 35 kg $CaCO_3$ m^{-2} a^{-1} (Alongi, 1998). Corresponding rates of accretion are 1–7 mm a^{-1} with a mean of about 3 mm a^{-1}. These numbers indicate that coral reefs would not have been able to maintain themselves simply by vertical accretion during periods of rapid sea level rise during the Holocene transgression and that horizontal (landward) transgression through establishment of new colonies would have been the primary process of preservation of coral colonies.

12.2.2 Coral Reef Divisions

There is a wide range of reef planform that we will examine in more detail in the next section. However, most reefs can be divided into a typical set of divisions along a transect from the shore across the reef flat, the reef crest and the forereef to deep water, though there are some differences between reefs in the Indo-Pacific region and those of the Caribbean region (Figure 12.2A, B). The reef flat, or backreef, is the most landward zone and is sheltered from high wave action by the reef crest. It may be only a few metres in width or up to several kilometres. It is generally very shallow with portions close to the reef crest that may be emergent at low tide and often a lagoon that may be a few metres deep between this and the shore. Portions of the reef flat may contain solidified reef material and there will be isolated coral heads consisting of both dead and live corals. Much of the reef flat, however, consists of fragments of coral as well as sand and fine sediments produced by the comminution of dead coral material, molluscs and other organisms that are washed landward across the reef by waves and currents. Bare sand areas will occupy considerable portions of the reef flat and may build above sea level to form beaches and sand cays. The shallow lagoon floor is often covered by extensive area of seagrass accompanied by various algae and this provides a rich habitat for a

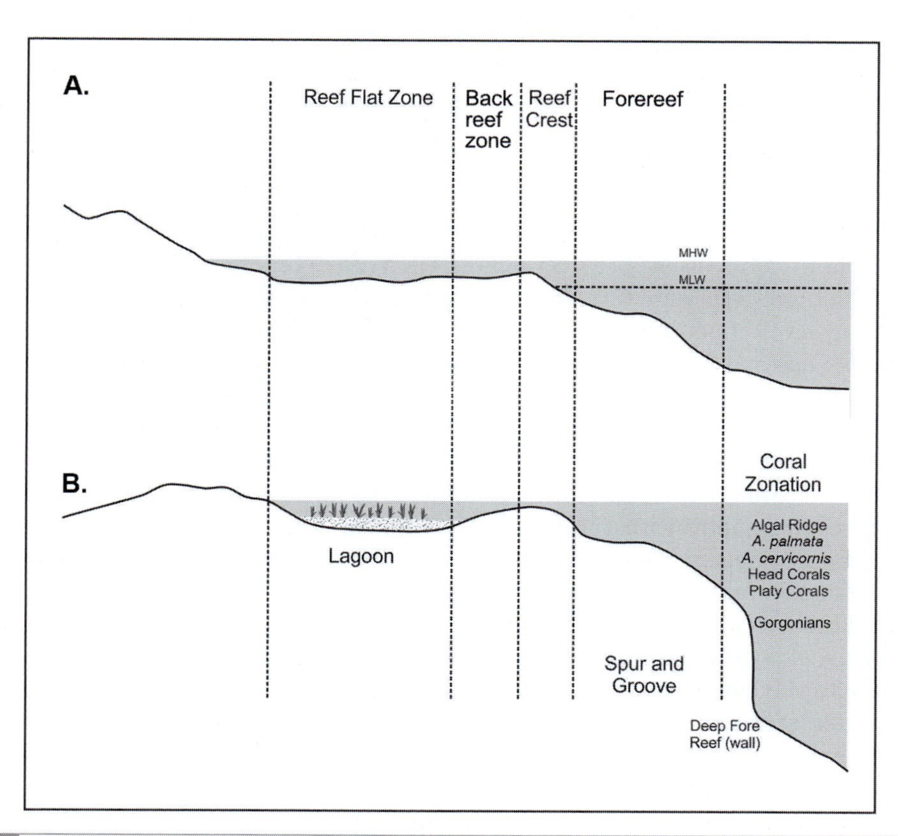

Figure 12.2 Schematic cross sections across Caribbean reefs from the shoreline to deep water showing the four main divisions of the reef (based on MacIntyre, 1988; Hubbard, 1997). Panel A. shows a shallow reef flat which dries at low tide and relatively shallow nearshore with the base of the reef ending in a gently sloping sandy zone. Panel B. shows a lagoon with seagrass beds and a steep nearshore with reef wall grading at depths below 100 m into a carbonate slope. The zonation of corals reflects decreasing wave energy (based on Adey and Burke, 1977).

wide variety of small fish, molluscs and worms and, in deeper lagoons, for larger organisms such as turtles and manatee (Figure 12.2B).

The reef crest is the highest point of the reef and parts of it will be exposed at spring low tide. It is typically a zone of wave shoaling and breaking, and in the most exposed areas only some species of calcareous algae will survive. In slightly deeper or protected areas, the crest will be dominated by strong branching corals such as *Acropora* which grow rapidly in the shallow water because of high rates of photosynthesis and strong circulation, which supplies large quantities of food. Damage to colonies during high wave events is quickly repaired and the dead broken pieces add to the structure of the reef. The reef crest provides protection for a wide range of small fish, crabs and crustaceans and thus is a prime area for snorkelling.

The seaward slope or forereef zone is marked by decreasing wave intensity and decreasing light with increasing distance and depth from the reef crest (Figure 12.2B). The upper portion of the zone down to about 20–25 m is termed the buttress and is characterised by the presence of massive and branching corals as well as coralline algae, all of which are well suited to withstand high wave energy levels. Conditions are ideal for rapid growth because of the shallow, generally clear water which permits maximum light penetration and photosynthesis, and the rapid turnover of water which brings fresh food to the coral polyps. This front face of the reef often develops a spur and groove topography with coral formations dominating the sides and crests of the spurs and narrow, flat floored channels separating them. Sand and broken coral are transported into the bottom of the channels and then work their way downslope to the base of the reef. Reduced wave action often leads to the greatest diversity of coral types at depths of 15–30 m, but below this the reduced light leads to a decline in the growth rate of scleractinian corals and a variety of soft corals, sea whips and sea fans become dominant. Where the drop off is very steep the reef face may be a wall lined with corals to depths greater than 50 m. Alternatively if the slope becomes more gentle sand flats develop and the reef becomes very patchy. A schematic summary of the distribution of dominant reef-building corals and coralline algae for high and moderate energy reefs in the Indo-Pacific tropics is shown in Figure 12.3.

12.2.3 Distribution of Coral Reefs

The reef building hard corals and their symbiotic zooxanthellates flourish in warm waters with temperatures above 18 °C and they generally do not survive when water temperatures exceed 30 °C for lengthy periods. Consequently, they are found primarily in a zone between 25 degrees latitude north and south of the equator. The greatest proportion of reefs worldwide occurs in a band stretching from the east coast of Africa across the Indian and Pacific oceans to Central America, with the highest concentration in the region around Indonesia, Papua New Guinea and Australia (Figure 12.4). This includes the Australian Great Barrier Reef that extends over 15 degrees of latitude northward from Bundaburg in Queensland at about 24.5 degrees south into the Torres Strait between Australia and Papua New Guinea (Hopley et al., 2007). In Australia, the southernmost area of reef development is on Lord Howe Island in the Tasman Sea at latitude 31° 40′ south. The second, much smaller area of reef development is in the Caribbean region extending from the Florida Keys and the Bahamas through Mexico and Central America and the islands of the Caribbean south to Trinidad. The most northern occurrence of reef development here is associated with the islands of Bermuda at about 32° N, where the waters are warmed by the North Atlantic Drift. The second largest barrier reef in the world, the Mesoamerican reef, occurs along the coast of Belize.

Reef development is curtailed on the west coast of South America and to some extent North America by upwelling and the presence of cold currents and this is also a factor in the scarcity of reefs on the west coast of Africa in the Atlantic Ocean. Reef development may be curtailed by high levels of turbidity and/or sedimentation and thus reef development does not occur near

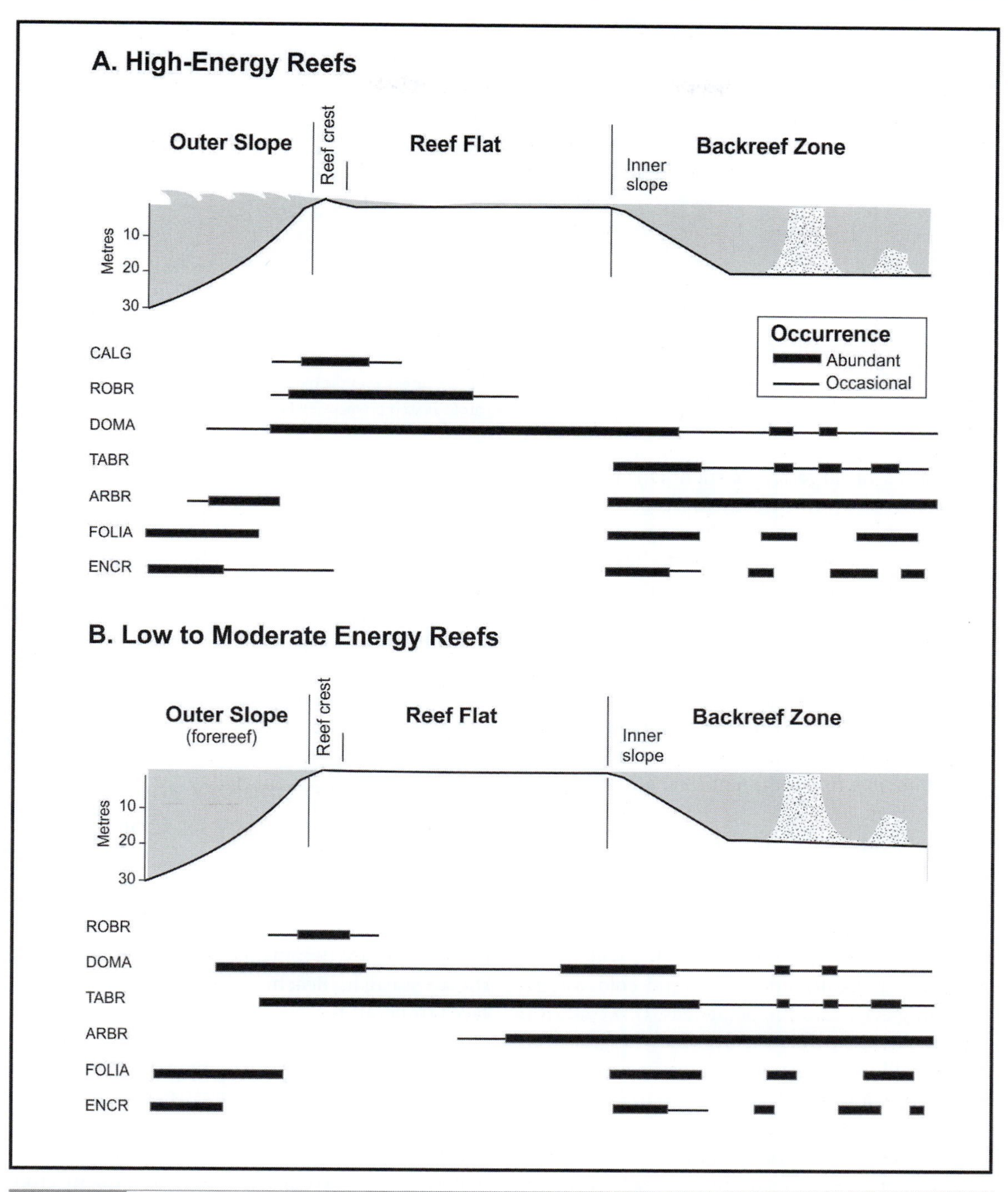

Figure 12.3 Schematic summary of the distribution of dominant reef-building coral (and coralline algal) forms and assemblages in the Indo-Pacific tropics, from deep outer slopes to backreef zones in relation to wave energy and water depth. A. Reef systems located in higher energy (outer shelf) settings. B. Reef systems located in lower energy settings (inner shelf and protected settings). CALG = coralline algae; ROBR = robust, branching corals; DOMA = domal coral; TABR = tabular branching; ARBR = arborescent corals; FOLIA = foliaceous corals; ENCR = encrusting coral (Montaggioni, 2005).

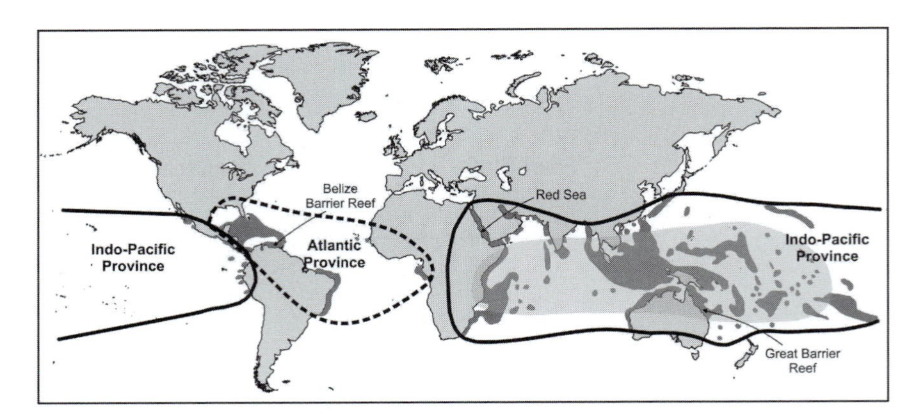

Figure 12.4 Global distribution of coral reefs. The dark shading indicates areas with the most prolific reef development, and the lighter shading indicates the area in the Pacific containing all known atolls (based on maps by Davies, 1980; NOAA, 2017).

the mouths of large rivers such as the Amazon and Orinoco on the northeast coast of South America and other major river systems in the Pacific and Indian Ocean systems (McLaughlin *et al.*, 2003). Coral growth is also inhibited when salinity falls below 27‰ and this is also associated with large rivers reaching the coast. It may also account for gaps in reefs near the mouths of smaller rivers. Finally, coral reefs generally do not establish well on muddy sediments and so development is inhibited also in coastal plains with abundant fine sediment in shallow water.

12.2.4 Cold Water Corals

Most attention over the past 200 years in the study of corals and coral reefs has been on those that are associated with shallow tropical seas. It has been known for some time that a wide range of corals can be found in cool and cold waters, extending even into the polar seas. However, it is only in the last three decades that the great variety of cold water corals (similar in numbers to the warm water corals) has come to be appreciated. These corals include soft corals and sea fans, as well as black corals and varieties of scleractinian corals which can produce large colonies in water depths up to 2000 m which can be compared to the reefs of shallow, tropical seas (Hourigan *et al.*, 2017). The rate of formation is much slower than for tropical reefs because of the absence of the symbiotic zooxanthellates, but it is this absence that permits them to inhabit such cold and deep water. The deep-water reefs are important habitat for a variety of commercial fish and shrimp and they are very vulnerable to destruction by trawl nets used to catch these. Because of the depth at which they tend to occur, these cold-water reefs have little influence on coastal processes so we will focus only on the tropical shallow water reef dynamics in the remainder of this chapter.

12.3 Geomorphology and Sedimentology of Coral Reefs

Tropical coral communities are vulnerable to a variety of natural stresses, such as the effects of overgrazing by organisms such as the crown of thorns starfish, bleaching due to elevated sea surface temperatures, the impacts of hurricanes, and changes in sea level on a geological time scale. In addition, there are impacts due to direct human activities such as harbour construction, dredging, overfishing and recreational diving, as well as indirect effects such as increased fine sediment deposition resulting from deforestation in river valleys. Thus, there is a tendency from an ecological perspective to regard coral reefs as fragile environments. On the other hand, reefs are complex biological, geomorphological and sedimentological features that have existed for hundreds of millions of years and are common features of the sedimentological record. Furthermore, coral reefs worldwide have been subject to large and relatively rapid changes in sea level throughout the Pleistocene, particularly in the past 600 000 years, and have not only survived, but flourished. From this perspective, reefs are robust and highly adaptive features of tropical coasts.

Coral reefs influence the geomorphology and sedimentology of tropical coasts both directly and indirectly, and they are the dominant control on the form and dynamics of many thousands of kilometres of coasts in the two regions where most reefs occur. The upward growth of the coral reef towards the low water line results in the development of a physical barrier which acts to reduce wave energy and thus leads to the development of extensive relatively sheltered areas of reef flat and lagoon, and low-energy shorelines where sediment accumulation and island development can occur. Reef sediments are produced by mechanical disintegration of reef corals and other $CaCO_3$ fixing organisms found in the reef and in the lagoon and reef flats behind the reef front, as well as by the activities of grazers such as parrotfish. Passages through the reef, and complex topography within the reef flat, result in spatially complex wave and current conditions. Temporal variation in energy levels and sediment transport occur because of tidal fluctuations, as well as seasonal variations in wave climate and the impact of major storms such as hurricanes and typhoons. Thus, the primary reef structure, which initiates the building up of the forereef towards sea level, ultimately produces a much larger and more complex coastal form than the initial reef itself. The geomorphology of modern coral coasts is further complicated by the residual effects of the Holocene sea level rise and the relatively stable sea level of the past 5000 years as well as by the inheritance of the effects of fluctuating climate and sea levels over the Pleistocene.

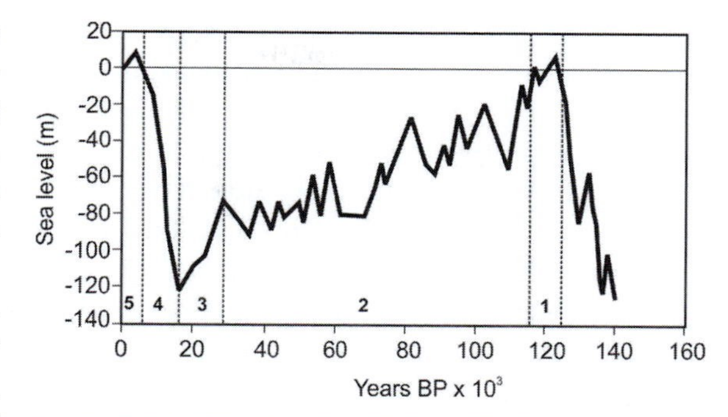

Figure 12.5 Generalised sea level curve for the Pacific over the past 140 000 years encompassing the last interglacial maximum, the last glacial period and the post-glacial transgression (based on Chappell *et al.*, 1996). The curve is based on data from rates of uplift of terraces in the Huon Peninsula, New Guinea and on $\delta^{18}O$ deep sea core data. The original data points have been removed for clarity. The five divisions identified in the time series reflect key changes in sea level affecting coral reef development – see text for more details.

12.3.1 Sea Level Fluctuations and Coral Reef Evolution

Over the past three million years, sea levels have fluctuated through a range of over 100 m in response to the growth and decay of upper latitude glaciers and ice sheets (see Section 3.3.1). Shorelines have migrated seaward during periods of falling sea level and landward during periods of rising sea level and coral communities have adapted to these changes. A generalised sea level curve for past 130 000 years (Figure 12.5), shows five features or periods that are significant for coral reef development: (1) a period of about 12 000 years during the last interglacial that is characterised by stable sea level at roughly the same elevation at present sea level; (2) a period of about 80 000 years during the early and middle glacial period which is characterised by sea level fluctuations on the order of 20–40 m superimposed on a gradual fall in mean sea level to about 60 m below the present level; (3) a rapid decrease in sea level during the last glacial maximum (LGM) to a low level about 120–130 m below the present; (4) a rapid rise in sea level beginning about 20 ka ago and reaching the present sea level about 5000 years ago; and (5) the modern period with sea levels close to their present level. The regions of coral reef growth are outside the areas affected directly by glacial loading and unloading but there will be varying local response to oceanic loading and unloading of the continental shelf during these fluctuations that reflect the morphology and width of the shelf. Relative sea level was also affected by local tectonics during the period, such as the ongoing subsidence of atolls on the Pacific plate, and the well-documented uplift of the Huon Peninsula in New Guinea and the island of Barbados in the Caribbean. What were the changes to the coastal environment during this time and how did reefs change in response to this?

During the last interglacial, we expect reef distribution somewhat like the present though with some differences in the detail – for example, there is evidence that much of the Australian Great Barrier reef only developed at the end of the Holocene transgression (Hopley et al., 2007). It can also be inferred from a study of modern reefs that the pace of new reef development would have slowed considerably by the end of this relatively long period of stable sea level. Reefs would have colonised almost all the places favourable for reef development, and sediment generated by corals and other carbonate producing organisms that make up the reef community would have filled much of the available accommodation space.

The second period is one of gradually declining sea levels, but with fluctuations reflecting alternations in ice sheet growth and decay. Sea level fluctuations are small enough that in many areas we can expect the active coral communities to migrate landward and seaward in response to the changes sea level fluctuations over the reef substrate. Reef flat zones built at high sea level stands were exposed and subject to reworking by subaerial processes, including solution and karstification of the exposed carbonates that form the internal structure of the reef. Initially they might have been re-occupied by active corals during a succeeding high stand but as mean sea level continued to fall they were continuously exposed until the post-glacial sea level rise in period 4. On continental shelves, there were local topographical changes reflecting both the operation of coastal processes as well as the extension of rivers onto the exposed continental shelf bringing changes in the location of freshwater inflows and terrigenous sedimentation. Changing coastal configuration as well as changes in wind patterns and storm tracks influence energy levels and nutrient supply and thus the ability of the reef communities of any location to adapt to the fluctuations in sea level. The impact of these on the fast growing, reef-building corals that are crucial to the rapid development of the reef may have been an important determinant of whether individual reefs maintained themselves or died off.

The third period began around 25 ka BP when there was a rapid sea level drop to the low stand of about 130 m below the present level. On wide continental shelves reef activity was confined to the outer shelf and subaerial drainage extended over much of the shelf. On islands and atolls, reefs likely formed at the base of existing reef fronts. Ongoing solutional processes on the former interglacial reefs led to the formation of caves and collapse features, exemplified by the famous blue hole of the Belize barrier reef. Evidence for these reefs is not widely available to submersibles and deep drilling (Beaman *et al.*, 2008).

The rapid melting of ice sheets beginning around 20 ka BP signals the start of the fourth period – the Holocene transgression. The rapid and nearly continuous rise in sea level to about 5 ka BP at rates of 5–10 mm a^{-1} is punctuated by at least three major meltwater pulses at 19, 14 and 11.5 ka BP with associated rates of accelerated sea level rise of 30, 40+ and 25 mm a^{-1} (Montaggioni, 2005). Each of these pulses is associated with a major period of non-reef construction or reef drowning. However, while there may have been insufficient time for reef building to occur, reef communities continued to exist and they kept pace with rising sea level through the continued recruitment of individuals to newly submerged substrates. Montaggioni (2005) recognised four reef generations beginning with RGO during the low stand, and RGI, RGII and RGIII following each of the meltwater pulses (Figure 12.6). Since about 11 ka BP, the rate of sea level rise has been slow enough for many reef communities to persist to the present time, though some may occur well below the surface because of a slow growth rate (Harris *et al.*, 2008).

The vertical accretion of reefs during sea level rise, particularly during the past 11 ka, has been

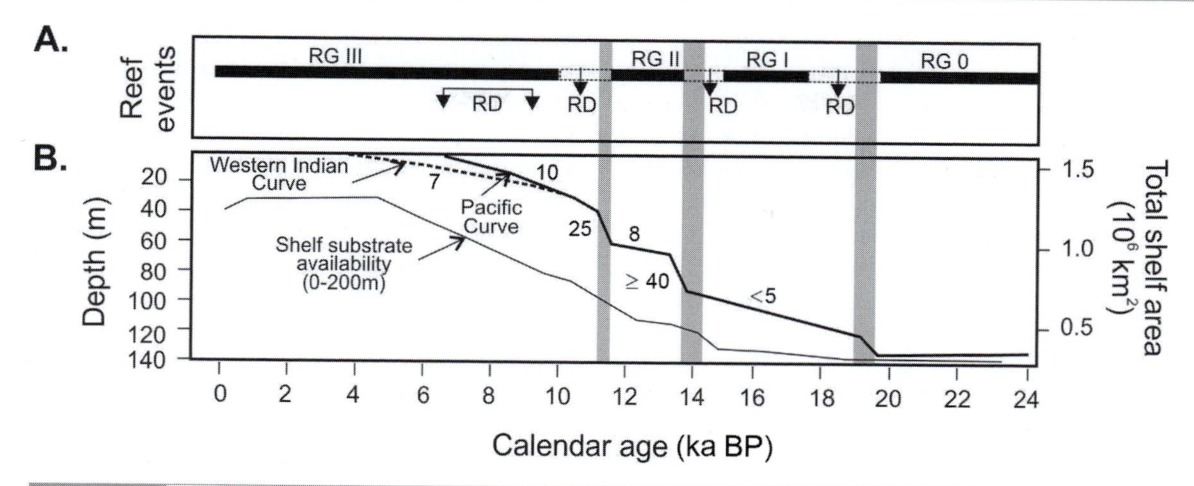

Figure 12.6 Post-glacial reef growth events in the Indo-Pacific province: A. schematic reconstruction of reef evolution showing reef growth phases (RG) and reef drowning (RD) events; B. late glacial and Holocene sea level curve for the western Indian Ocean and Pacific Ocean and associated area of shelf substrate available for reef growth (Montaggioni, 2005).

characterised (rather anthropomorphically) as falling into three categories – keep-up, catch-up and give-up (Neuman and Macintyre, 1985; Woodroffe and Webster, 2014). In the first case the rate of vertical accretion keeps pace with rising sea level. In the second case, accretion is initially slower than sea level rise but catches up, either because of increased carbonate production or a slowing of sea level rise. In the third case, accretion is less than the rate of sea level rise over an extended period, with the result that the reef is submerged to a depth where reef-generating corals can no longer survive and the reef is drowned. Each of these conceptual scenarios is associated with a different facies assemblage that should be recognisable in cores through reef sediments. There is some question as to whether these exist or are an artefact of coring (Blanchon and Blakeway, 2003). Apart from the danger of thinking of these scenarios as reef 'strategies', it probably simplifies too much the complex set of biological and physiological factors that control the development of reef facies assemblages in space and time.

12.3.2 Geomorphic and Sedimentologic Characteristics of Coral Reefs

In the broadest sense, we can distinguish between coral reefs developed on shallow continental shelves in close proximity to major land masses (characterised by fringing and nearshore reefs and complex reef assemblages associated with barrier reefs), and those developed on volcanic platforms or isolated volcanic islands in oceanic settings (notably the coral atolls of the Indo-Pacific region).

However, it is now increasingly recognised that a detailed examination of reefs in any region will show that the simple divisions of fringing reef, barrier reef and atoll, first noted by Darwin, does not encompass all the types of reef form that result from the complex range of topographic and hydrodynamic conditions under which reefs develop (Hopley *et al.*, 2007; Leon and Woodroffe, 2013; Rowlands *et al.*, 2014; Kordi and O'Leary, 2016). Many mainland and large island coasts with a steep nearshore and narrow continental shelf may be characterised by many small patch and planar reefs. Larger fringing reefs may be absent or occupy a comparatively smaller proportion of the total shoreline. On the other hand, other coasts may have fringing reefs that occupy most of the total shoreline (Kordi and O'Leary, 2016). In contrast, wide continental shelves or extensive shoal areas give rise to the development of reefs that are isolated from the mainland coast, and evolve in response to controls such as sea level change, wave climate and coastal currents. Barrier reefs, such as the Great Barrier Reef in Australia and the Belize Barrier Reef, as well as the extensive shoals of the Bahamas Bank, will contain some fringing reefs. In addition, these areas will encompass submerged or patch reefs emerged reefs, and islands which may be located both on the exposed outer zone and in more sheltered central and inshore portions of the continental shelf.

12.3.2.1 *Fringing Reefs*

Fringing reefs are common on mainland coasts as well as the coasts of islands on the continental shelf and on large oceanic islands (Figure 12.7). They are relatively simple in morphology and occur as either shore-attached features or features very close to the shoreline (Kennedy and Woodroffe, 2002; Madden *et al.*, 2013; Kordi and O'Leary, 2016; Ryan *et al.*, 2018). Reefs may develop in several areas, including rocky headlands, bayheads, along narrow linear sandy beaches and as nearshore shoals some distance offshore from a range of coastal forms (Hopley *et al.*, 2007). While there is a tendency to think of reef colonisation as being largely restricted to rock or rubble substrates, there is now much evidence that reefs can colonise sandy substrates equally well. Many fringing reefs have a crest some distance offshore, and a nearly horizontal reef flat that extends right to the shoreline and may be exposed at spring low tide (Figure 12.2A; 12.7A). In much of the Indo-Pacific region, reef flats may be emergent at low tide, perhaps reflecting the small late Holocene fall in sea level postulated for this region. Lagoonal fringing reefs on the other hand have a shallow lagoon between the reef flat and the shoreline that is typically 1–3 m deep (Figure 12.2B; Figure 12.7A, B). This may reflect the establishment of the reef front some distance offshore and limited coral growth close to shore, or the influence of runoff from the land inhibiting coral growth in this location. The shallow lagoon may also result from scour by strong currents, produced by waves coming over the reef front, that act to inhibit deposition of carbonate detritus brought onshore from the reef front or flat.

As with all coral reefs, the evolution of fringing reefs over the past 20 ka was controlled by the Holocene transgression and by local tectonics. Fringing reefs can evolve by growing upward at the same rate as sea level rise, but they can also backstep by colonising available terrestrial substrate as the land surface is drowned. As the rate of sea level rise slows, progradation of the reef front may occur. The result is a complex set of potential scenarios for fringe reef development in the Holocene (Kennedy and Woodroffe,

Figure 12.7 Fringing reefs. A. Fringing reefs on the island of Moorea, Tahiti (see Figure 12.11 for location). On the right the reef flat extends to the shoreline but on the left there is a passage through the reef and this connects to a lagoon between the reef flat and the shoreline. B. Small fringing reef off the coast of Antigua, West Indies. A portion of the reef crest is emergent at low tide and there is a lagoon between the reef and the shoreline.

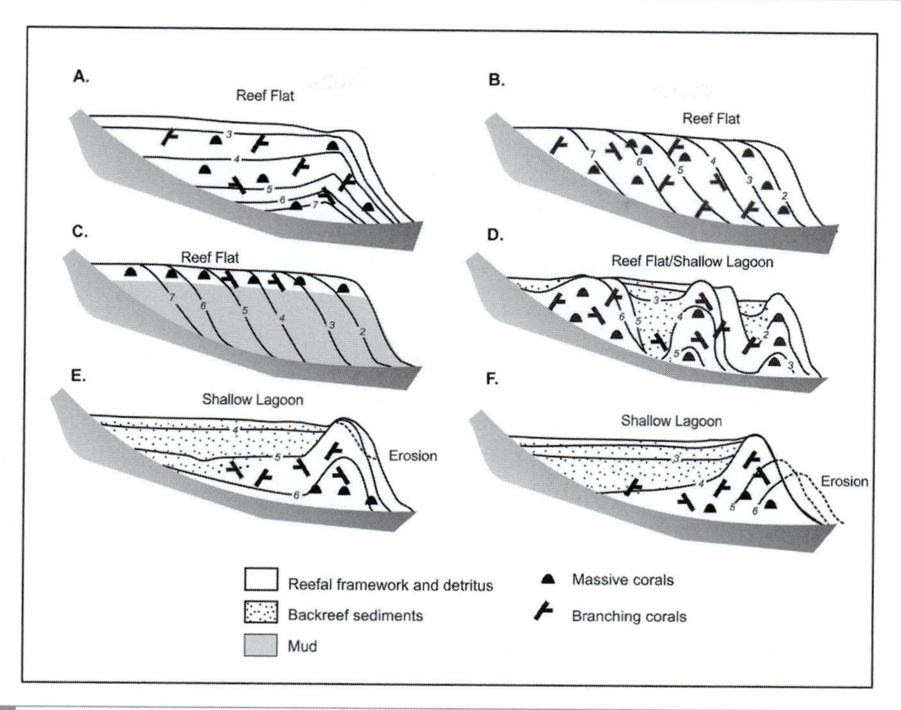

Figure 12.8 Generalised models of fringing reef development. A. Vertical reef accretion keeping up or catching up with sea level rise. B. Lateral (offshore) accretion associated with stable sea level. C. Lateral accretion over non-reefal mud in the forereef zone. D. Episodic progradation model. E. Offshore accretion and gradual infilling of lagoon. F. Offshore reef structure formed primarily as a rubble pile by storm processes which may be reworked landward by large storm events (Kennedy and Woodroffe, 2002).

2002; see Figure 12.8). Coring in fringing reefs of the Great Barrier Reef of Australia indicates that many of them experienced rapid growth in the period around 7000–5000 years BP when sea level rise slowed sufficiently to allow established reef communities to build upward into the available accommodation space (Hopley *et al.*, 2007). This growth slowed markedly after 5000 BP in many fringing reefs, coincident with establishment of a stable sea level and the decreasing areas with room for vertical accretion.

The establishment of a stable sea level marks the end of a continuously increasing vertical accommodation space, but seaward progradation of fringing reefs also appears to have slowed markedly in the last few millennia. In many cases this may reflect the fact that the reef has prograded rapidly over a gently sloping terrigenous surface and the reef front is now over a relatively steep seaward slope or even a 'wall'. Further progradation requires a greater supply of rubble to build the substrate platform up towards the optimum water depth for coral growth and with the steep slope much of this may be carried into considerable depth down the channels on the forereef. In some areas increased turbidity and runoff from mainland coasts due to human activity, including dredging and construction on the coast and on the reef itself, has reduced the optimum conditions for coral growth. Hopley *et al.* (2007) also argue that, with a relatively stable sea level, fringing reefs may now occupy most of the optimum areas for reef colonisation and growth and that the opportunities for establishment of new reefs or rapid growth of established reefs are now very much reduced.

12.3.2.2 *Reefs of Shallow Continental Shelves*

On extensive banks, such as the Bahamas Bank and shallow continental shelves, a wide range of reef forms and associated emerged islands occur.

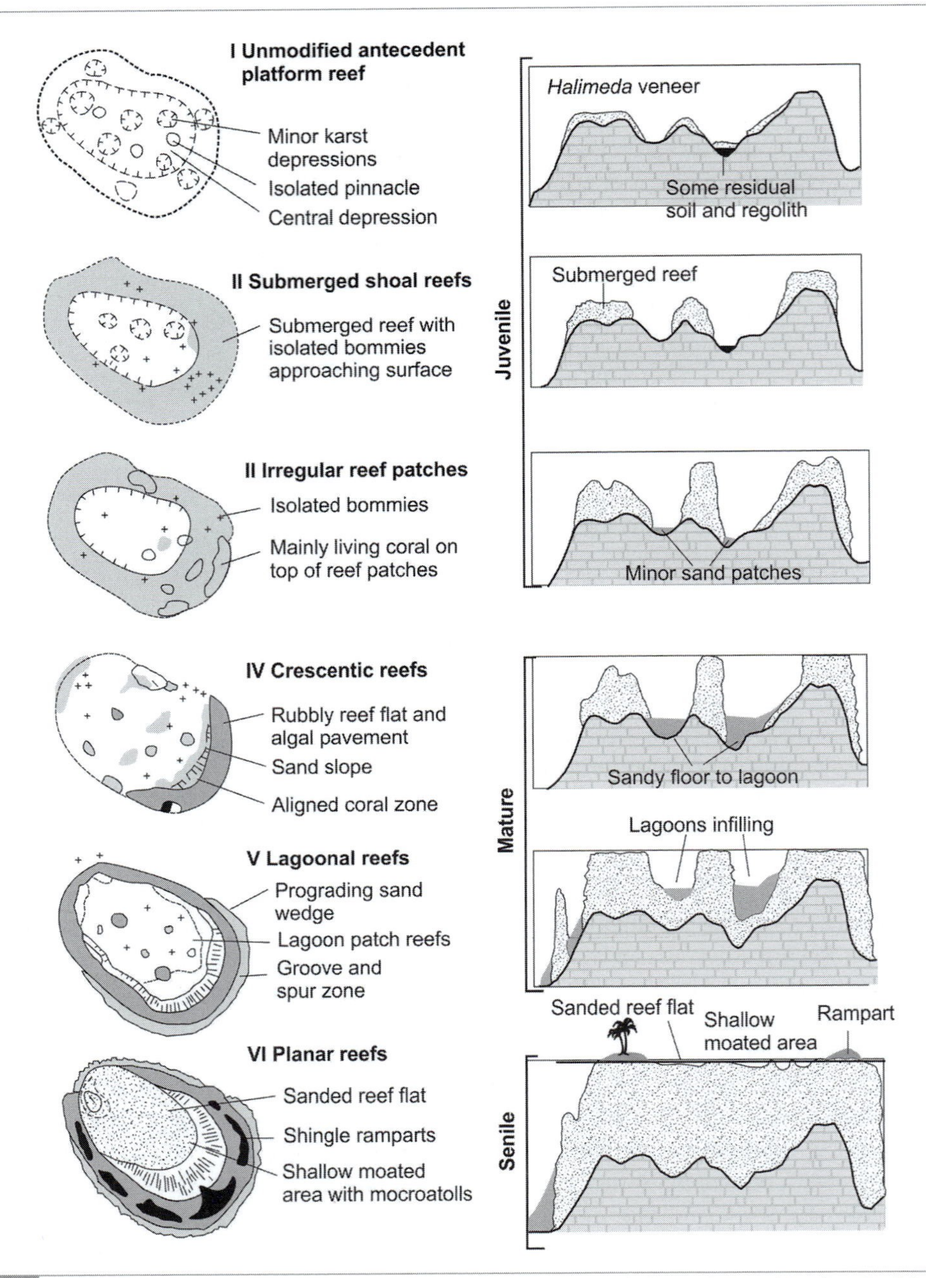

Figure 12.9 Schematic depiction of planform and profile changes associated with the evolution of coral reefs of the Great Barrier Reef, Australia (Hopley *et al.*, 2007).

Because of the considerable control imposed on reef growth and sediment reworking by sea level change since the mid Holocene it is possible in many cases to associate reef form with stages in an evolutionary scheme such as that outlined for reefs on the Great Barrier Reef of Australia (Figure 12.9). Rapid submergence of the shoal or continental shelf platform would have occurred during the early phases of the Holocene transgression with sea level rising too quickly to permit upward reef building to keep pace. However, as the rate of sea level rise slowed after

7000 years BP, vertical accretion of reefs establishing on the recently drowned terrigenous substrate could exceed that of sea level rise and initiate the upward building of the reef relative to sea level. As portions of the reef reach the surface, the form of the reef front and the development of areas of carbonate sediment accumulation are influenced by exposure to waves, by wave refraction and by tidal, wind and wave generated currents. Reef growth tends to be most vigorous on the high-energy windward or exposed side of the reef where there is a constant supply of nutrients and food for the polyps and species of rapidly growing branching corals are most likely to be found.

12.3.2.3 *Coral Atolls*

Atolls are reef complexes found in mid ocean far from the continental shelves and large islands. They are characterised by an annular reef rim that encloses a central lagoon or lagoons and a variety of small islands that are mainly located on the outer margins of the atoll. Almost all atolls are found in the Indo-Pacific zone on the Pacific Plate, and most are associated with volcanoes that form at hot spots within the plate, rather than at convergent or divergent boundaries where volcanic activity is predicted by plate tectonics theory. Whatever the origin of the hotspot, magma reaching the surface of the ocean floor can build up to form a volcano which in many cases will reach the ocean surface and build a volcanic island that may be hundreds of metres above sea level (Figure 12.10A). Movement of the Pacific Plate carries the volcanic island away from its magma source. Gradual subsidence of the plate, at rates of 2–3 m per 10 000 years, as well as subaerial erosion, result in the volcanic island sinking below sea level and the transformation of fringing reefs into an atoll with a central lagoon (Figure 12.10B). If circumstances are favourable, reef growth can keep up with the rate of plate subsidence, leading to the construction of a carbonate platform of increasing thickness over the underlying volcanic core. If the atoll migrates towards higher latitudes, coral growth will become restricted and the reef eventually dies, leaving the top of the seamount to subside further and further below the wave surface

There are many hundreds of atolls in the Pacific, most of which lie close to the equator or south of it (Figure 12.11). Archipelagos such as those of French Polynesia may contain some high islands such as Moorea, where portions of the old volcanic landscape remain, surrounded by fringing reefs (Figure 12.7A, 12.10B). They also include low islands such as Tikehau atoll, where the volcanic core is no longer visible above the surface (Figure 12.10B, 12.12A, B). Atolls form ranges from circular or elliptical to more complex shapes and groupings. Initially this will reflect the shape of the original volcanic shoreline, as well as subaerial processes associated with the development of drainage basins, and the effects of wave erosion and littoral transport (Figure 12.10B). The form will also reflect fluctuations in sea level over the Quaternary and the development of karst landforms during low stands of sea level. Terry and Goff (2013) postulate that many atolls show the presence of what they term arcuate ‘bight-like’ structures which may reflect large submarine failures along the flanks of the volcano. These failures occur in volcanic deposits but may also occur in the reef itself and can be a source of tsunami generation (Keating and McGuire, 2000).

The association of Pacific atolls with subsidence of volcanic islands and continued upbuilding by reef communities goes back to Charles Darwin and the voyage of the *Beagle* in 1836. Partial support for Darwin’s theory came from drilling on Funafuti atoll between 1896 and 1898 which showed the presence of shallow reef carbonate deposits down to a depth >300 m. Subsequently drilling on Bikini and Eniwetok atolls after World War II reached the volcanic basement on which the carbonates were built (Guilcher, 1988; Woodroffe, 2008).

As with all coral reefs, the modern atoll reefs have been greatly influenced by sea level fluctuations in the past few 100 000 years. During the last interglacial, we can expect that there were a range of atoll forms similar to those existing today. As sea levels fell during the glacial period, the active reefs would have migrated down the

A.

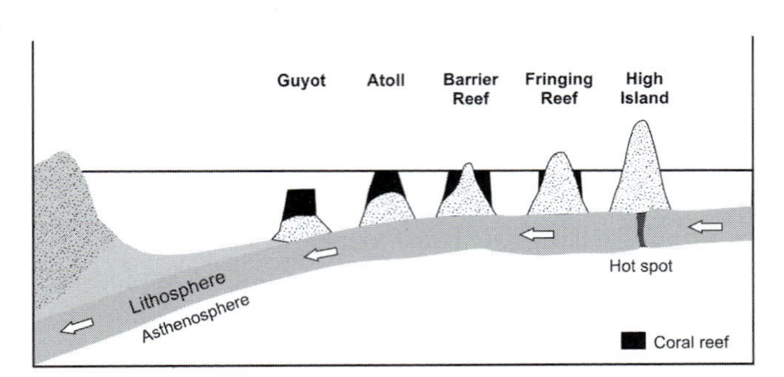

B.

Stage 1

oblique view

fringing reef

cross section view

subsiding volcanic island

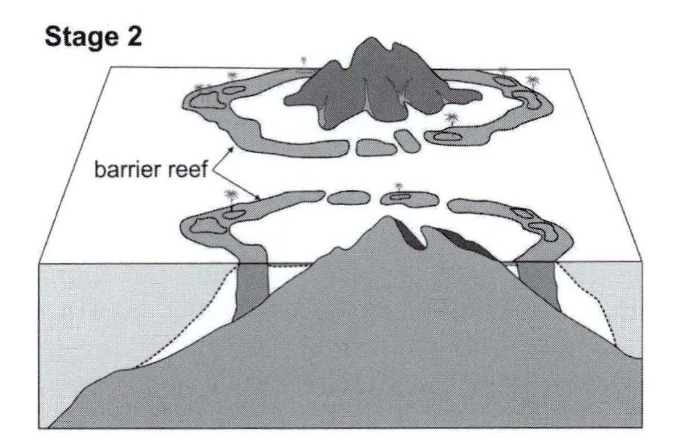

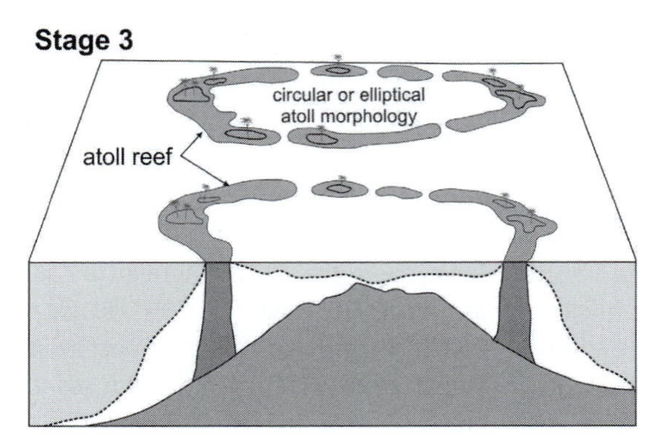

Figure 12.10 Darwin's subsidence theory of atoll formation. A. Schematic depiction of the development of a Pacific Ocean atoll through subsidence of a volcanic island formed over a mid-ocean hotspot. Once the volcanic island moves away from the hotspot, erosion reduces the subaerial relief and ongoing subsidence of the cone leads to the transformation of fringing and barrier reefs into an atoll. B. Classical diagrammatic interpretation leading to the formation of a circular or elliptical atoll morphology. The stages shown represent snapshots during a process that is a continuum through time (after Terry and Goff, 2013).

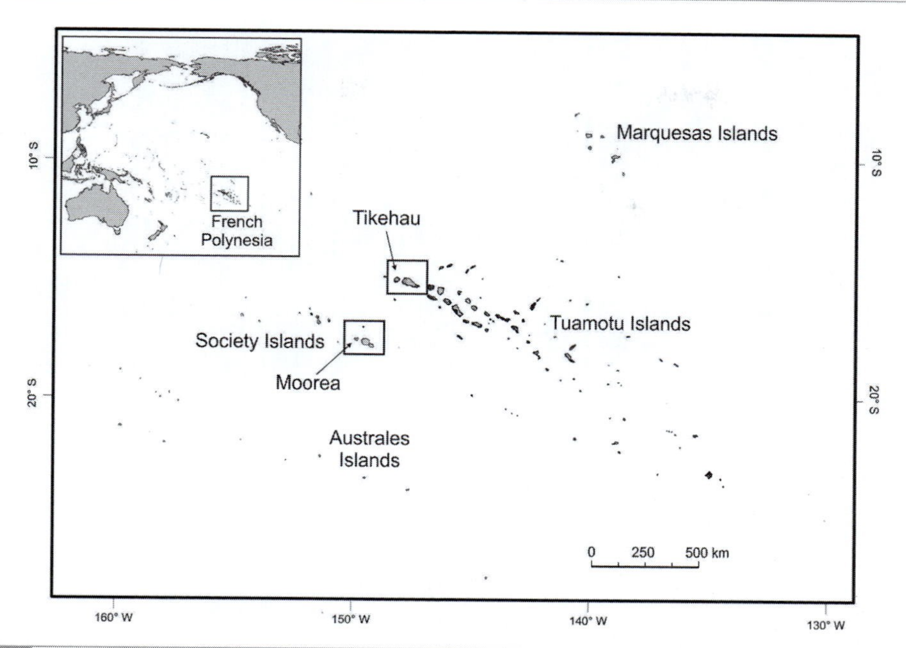

Figure 12.11 Central Pacific islands (including atolls) and archipelagos. The inset shows the area where most atolls are located. The square encompasses the islands of French Polynesia. The main map shows a portion of the Tuamotu Archipelago, which is part of French Polynesia and has some 77 atolls spread over a distance of more than 1500 km (after Duvat and Pillet, 2017, figure 1).

flanks leaving the upper sections exposed to subaerial processes. As sea level rose at the end of the glacial period, the response of the atoll reefs would have been similar to that in other areas, with some of the fringing reefs being drowned during periods of rapid sea level rise, and others building upward along the flanks of the atoll platforms. The final stage came with relatively stable sea level and the building of the modern carbonate cap over underlying Pleistocene limestone (Figure 12.12C) and ongoing development of reef islands and lagoon systems.

12.3.2.4 *Reef Islands*

Atolls, barrier reefs and carbonate platforms are characterised by the development of reef islands which are built up above the mean high tide level (Figures 12.12A, B; 12.13A). The platform on which the island develops usually consists of late Holocene backreef and forereef conglomerates and sandy sediments (Figure 12.12C), and the islands themselves have developed because of accretionary processes driven largely by wave action. In parts of the Pacific, and on the Australian Great Barrier Reef, islands may have an emergent fossil reef platform resulting from a slight fall in sea level. The platform is overlain by beachrock produced by cementation of carbonate sediments on the reef platform. In areas where reefs have built up to sea level more recently, the emergent part of the island may contain rubble and rubble ridges resulting from storm wave activity as well as carbonates sands and finer material deposited as sand sheets by overwash. On atolls, ridges are generally higher on windward coast and build up to 2 or 3 metres above mean sea level (Figure 12.13A, B). Colonisation by vegetation aids the process of island accretion because of the trapping effect of roots and stems. Emergence of the reef crest provides protection for shallow lagoon or lagoons and sandy sediment derived from the reef or from organisms in the lagoons can be reworked by waves to form sand cays. Colonisation by vegetation leads to some aggregation of these islands due to trapping of sediment washed onshore

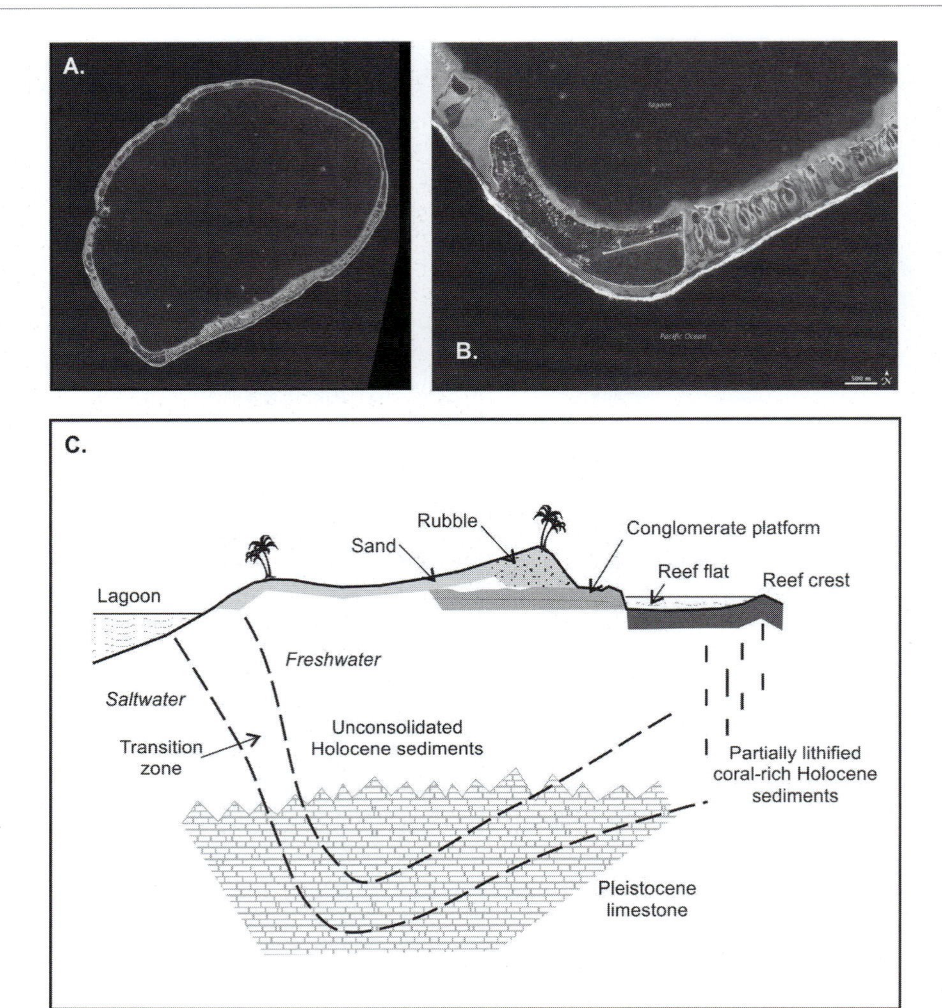

Figure 12.12 Pacific Atolls: A. Tikehau atoll, French Polynesia showing the nearly circular form of the outer reef rim with vegetated islands of various sizes and a single passage on the west side connecting the lagoon to the ocean. B. Close up of Tuherahara Island at the southwest end which contains the only settlement. Note the variation in the continuity of the reef islands between the southwest facing coast and the north coast. C. Idealised cross section of an atoll island rim showing a reef island, its geomorphology, lithology and sediments, and typical configuration of the freshwater lens. (A. and B. NASA images created by Jesse Allen, using EO-1 ALI data provided courtesy of the NASA EO-1 Team; C. after Woodroffe, 2008.)

during storm surges. A portion of the sediment transported onto the island may result from aeolian activity and this is more significant on sand cays where there is an abundance of fine sediment and sufficient fetch for aeolian transport to be effective (Figure 12.13C). Historical aerial photographs and satellite imagery has permitted measurement of changes in island size over decades. On Huvadhoo Atoll, Aslam and Kench (2017) measured changes in island planform for 184 reef islands, including some on lagoonal reef platforms, and found that about 42 per cent showed net erosion, 38 per cent were stable and the remainder accreted over the period. While there was a net loss of island area through natural processes of about 1.5 per cent, the overall land area increased by about 2.4 per cent because of land reclamation on 12 of the larger inhabited islands. Studies of other atoll islands have shown a similar dynamic response

Figure 12.13 Reef islands. A. Oblique aerial photo of a portion of Huvadhoo Atoll, Maldives showing reef islands on the atoll rim. The picture is taken at low tide and the lagoon is on the right side of the picture. B. View along the beach of a reef island on Huvadhoo Atoll showing sediments on which the island is formed. C. Matagorda Island, a small sand cay in the Maldives photographed from a UAV. Photos A. and B. courtesy of Paul Kench, C. courtesy of Mike Hilton.

to processes controlling sediment supply and transfers such as sea level change, waves, currents and the impact of tropical cyclones (e.g., Rankey, 2011; Ford and Kench, 2015; Duvat and Pillet, 2017). On some Caribbean islands (e.g., Anguilla, Grand Turk), lithification of pre-Holocene dunes occurred during periods of lower sea level and this provides greater stability to the island (Figure 12.14). However, islands that are not stabilised, including those along the southern section of the Mesoamerican reef in Belize, have eroded over several decades (1960–2012) and have been reoriented through a loss of sediment supply from the reef (Houser *et al.*, 2014).

Islands generally grow by accretion vertically, and especially seaward and lagoonward. The size and shape of the island will depend on the supply of sediments from the reef, the time since the reef platform reached sea level and the detailed local history of wave and current processes controlling sediment movement. A wide range of possible scenarios for island development can be envisaged (Woodroffe *et al.*, 1999), depending on the relative significance of seaward and lagoonward accretion, vertical accretion and island migration. Woodroffe *et al.* (1999) also document the history of accretion on West Island, Cocos (Keeling) Island with seaward growth dominant in the central section over the period 4000–2000 years BP and more gradual accretion since then with sedimentation focused more on the ends of the island. In general, we should expect island accretion to continue if the biological

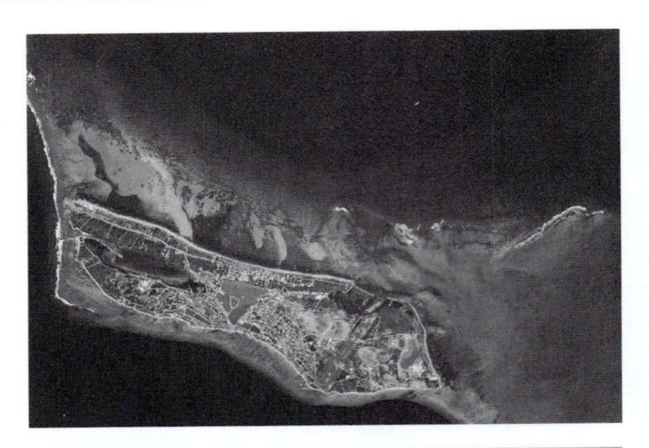

Figure 12.14 Grand Turk Island, one of the Turks and Caicos island groups on the edge of the Bahamas carbonate platform. Most of the island is a few metres above mean sea level but on the north side lithified coastal dunes form a higher ridge. Note the wide reef platform on the north side of the island and the reef wall which is about 150 m offshore on the south side. (Courtesy ISS Crew Earth Observations Facility and the Earth Science and Remote Sensing Unit, Johnson Space Center. Image ISS050-E-41317 taken 12 February 2017 by a member of the Expedition 50 crew.)

productivity of the reef remains high, and the island is the main sink for sediments.

12.3.2.5 *Sedimentary Facies of Coral Reefs*

The general pattern of reef environments and associated sedimentary facies along a transect from reef flat and lagoon, across the reef crest and down the slope of the forereef is given added complexity by the range of reef forms and degree of exposure (Figures 12.2, 12.3). In turn, the sedimentary unit that is formed by accumulation over a period of thousands of years will reflect the influence of changing sea level, wind and wave climate, and terrigenous sediment input as well as changes in the biotic communities. Coring through modern reefs provides an opportunity to examine the interior structure of reefs and, together with dating of materials, to determine the growth history in response to changing environmental conditions and sea levels (e.g., Gischler *et al.*, 2008). This information in turn provides improved understanding of sea level change in the Holocene as well as indirect evidence of changing climate and sea surface temperatures. Finally, we can use information from modern reefs to help in the interpretation of ancient reef deposits.

There were some notable early attempts to core reefs, such as the Royal Society sponsored drilling on Funafuti Atoll in 1896–8 and the first drilling on the Great Barrier Reef in 1926 (Hopley *et al.*, 2007). However, it was the development, in the late 1960s, of lightweight drilling rigs that could be set up on emerged reef flats or mounted on pontoons to drill in shallow lagoons that resulted in an explosion of coring, both in the Caribbean (e.g., MacIntyre *et al.*, 1982) and throughout the Indo-Pacific region. Coring has provided much of the evidence for interpretation of reef evolution in response to sea level rise. Identification of facies in the cores also provides information on the various modes by which reefs are constructed, the nature of the contribution by different coral species as well as species of coralline algae, and the relative importance of secondary sand and muddy deposits in lagoons, reef flats and forereef.

Based on an extensive review of cores from the Indo-Pacific region, Montaggioni (2005) identified ten major sedimentary facies composed of seven forms of growth framework and three detrital facies (Table 12.1). Interpretation of each facies is based on comparison with the modern reef zones shown in Figure 12.3. The framework facies are considered the primary foundation of the reef structure but cores, particularly those from backreef areas, may be dominated by detrital facies. The sequence of facies encountered within an individual core will reflect the location on the reef and the evolutionary history of the reef – for example, vertical growth, seaward progradation or backstepping (Kennedy and Woodroffe, 2002). In areas subject to tropical storms (hurricanes, typhoons), the impact of these storms may lead to destruction of much of the live coral in shallow water on the reef crest and seaward slope so that the reef structure is dominated by coral rubble rather than intact corals (Blanchon *et al.*, 2017).

Coral facies may also be mapped in uplifted coral reefs and the sequence found on the coral

Table 12.1 Facies identified in cores through reefs in the Indo-Pacific region

Facies type	Facies name	Description
Framework	Coralline algal	Reef crests and reef flats – centimetre to decimetre thick laminated crusts binding in place or reworked corals.
Framework	Robust branching coral	Reef crest and proximal reef fronts – open to cavernous interlocking framework formed by branching corals with infilling by rubble and sandy matrix.
Framework	Domal coral	Outer reef slope, crest and reef flats – coral heads encrusted with coralline algae or scattered colonies separated by skeletal sands and silts.
Framework	Tabular branching coral	Semi exposed or sheltered forereef and reef flats – rigid structure of tabular corals either in situ or reworked and mixed with silty sands.
Framework	Arborescent coral	Lower to middle parts of forereef and backreef – framework of encrusted corals or fragments of reworked branches.
Framework	Foliaceous coral	Sheltered areas with high turbidity – platy corals encrusted and often buried in muddy fine sands.
Framework	Encrusting coral	High-energy reef crests less than 10 m – coral components contained within matrix of carbonate and terrigenous mud.
Detrital	Skeletal rubble	Subtidal to intertidal reef flats – coral rubble consisting of broken coral and coralline algae in a matrix of fine to coarse sand. May make up to 60% of a core.
Detrital	Carbonate sand	Inner reef flat and backreef – sand and gravel derived from wide range of corals, coralline algae, mollusks and green algae. Make up 10–50% of cores.
Detrital	Carbonate mud	Sheltered inner reef flats and backreef zones, seagrass beds – may be inorganically precipitated or breakdown of skeletal material.

Note: The distribution of the dominant coral forms and assemblages associated with the framework facies is shown in Figure 12.12.
Source: Based on descriptive summary in Montaggioni (2005), pp. 14–24.

reef terraces of Barbados has been extensively studied, both in terms of the characteristics of the ancient reef sediments, and because dating of the uplifted reefs has made a significant contribution to sea level chronology through the mid to late Pleistocene (Mesolella *et al.*, 1970; Blanchon and Eisenhauer, 2001; Schellmann and Radtke, 2004). Continued uplift of the Barbados ridge provides a sequence of uplifted reefs reaching an elevation of over 300 m and spanning much of the Pleistocene. Portions of the reef platform and reef flats are preserved landward of notches or cliffs cut in the uplifted reefs during high stands of sea level. Data provided by coring and by the study of uplifted reefs can be used to develop predictive models of reef development and internal structure such as ReefSAM (Barrett and Webster, 2017), and in turn these can be used for predicting both past and future reef development under a range of input scenarios.

12.3.3 Wave and Current Dynamics on Reefs

Waves and wave-generated currents influence reefs and reef dynamics in several ways and in turn the presence of fringing and barrier reefs has considerable influence on the amount of wave energy reaching the shoreline. Differences in wave energy and in exposure to intense storms have been used to explain broad differences in reef morphology, while energy gradients with depth down the forereef zone or across the reef crest and reef flat have been recognised as exerting an important control on both the biological zonation of coral reefs and on sediment dispersal and deposition. Wind and wave generated currents influence the transport of sediment over the reef and reef flats, and they are the primary control on the supply of nutrients and food for corals and other organisms of reef communities.

Broad relationships between reef zonation and wave climatology – specifically wave energy levels and the frequency of hurricanes – have been recognised for Caribbean coral reefs (Geister, 1977; Hubbard, 1997). Hubbard (1997) divides the Caribbean into three groups: (1) areas such as the eastern shores of the Windward Islands which are characterised by high wave energy and moderate to high frequency of impacts from hurricanes, either directly or through high swell waves generated by hurricanes passing within 100 km; (2) areas of high wave energy but infrequent disruption by hurricane impacts such as the north shore of St Croix and other protected sites in the Florida Keys; and (3) areas of low wave energy but subject to frequent hurricane impacts such as many parts of the Bahamas. Type 1 reefs are characterised by *Acropora palmata* colonies that are frequently broken down by hurricane wave impacts, with subsequent colonisation of the rubble by coralline algae. Type 2 reefs are dominated by *Acropora palmata* which thrive in the absence of hurricanes. Finally, Type 3 reefs have only isolated *Acropora palmata* colonies and relatively low diversity, but long periods of relatively low-energy conditions encourage grazing which inhibits the growth of thick algal crusts (Hubbard, 1997).

Recent work has also focused on seasonal variations in wind and wave climatology and the effect of this on both physical processes and biological activity. In the Caribbean, in addition to the potential for hurricanes during the period from July through November, the strength of the trade winds, water surface temperatures and runoff from the land also vary seasonally. Reef island shorelines in the Maldives showed large changes in area in response to seasonal reversals in wind and wave conditions associated with west–northwest and northeast monsoons (Kench and Brander, 2006a). Similarly, measurements over a 15-month period by Storlazzi and Jaffe (2008) on a fringing reef on West Maui, Hawaii showed that there were considerable seasonal variations in wave energy levels and in the extent of turbidity in the water column. This had significant effects on mixing, and on the transport of both sediments and larvae in the water column. Because coral spawning occurs primarily during conditions dominated by trade winds (June to September) they speculate that the location of the reef with respect to currents would lead to dispersal of coral larvae to nearby downdrift islands, thus increasing diversity of corals in these areas (Storlazzi and Jaffe, 2008). In contrast, fish communities on reef flats of reefs on Reunion Island in the southwest Indian Ocean exhibit a strong spatial pattern related to geomorphological zones on the fringing reefs and showed little response to seasonal changes in winds and waves (Letourner *et al.*, 2008).

Recognition of the key role played by waves in controlling wave energy levels on the reef crest and currents over the reef flat and within atoll lagoons has led to an increasing number of field measurements as well as to attempts to generate models to predict set-up, wave transmission and currents (Gourlay, 1994; Lugo-Fernàndez *et al.*, 1998; Kench and Brander, 2006b; Harris *et al.*, 2015; Jeanson *et al.*, 2016; Harris *et al.*, 2018). On simple fringing reefs with relatively gentle forereef slope the reef crest acts somewhat like a nearshore bar with the extent of wave set-up and transmission dependant on the crest elevation and tidal stage. Wave breaking on the crest transmits large volumes of water onto and across the reef flat and may drive strong alongshore currents close to the beach. This water then escapes from the lagoon through channels in the reef. Because the reef crest is irregular and because of the existence of channels through the reef, waves may be refracted and focused in places, and the resultant set-up gradients and wave generated currents will determine areas of transport and deposition of sediments.

While both short period wind waves and long period swells are attenuated rapidly over the reef flat, long period swell waves may propagate through channels in the reef and they may also be the dominant waves on the leeward side of reefs (Kench and Brander, 2006b; Lashley *et al.*, 2018). On wide reef flats and shallow lagoons, locally generated wind-waves may become the dominant process influencing island beaches. In general, wave attenuation is greatest for wide, shallow reefs and least for narrow, relatively deep reefs and is greater at high tide than at low tide when more of the reef crest is exposed or close to the surface (Figure 12.15). Kench and Brander (2006b) combine the effects of water depth at spring high tide and reef width to produce a reef energy window index (ψ) which can be used to facilitate comparison of expected wave energy propagation across different reefs:

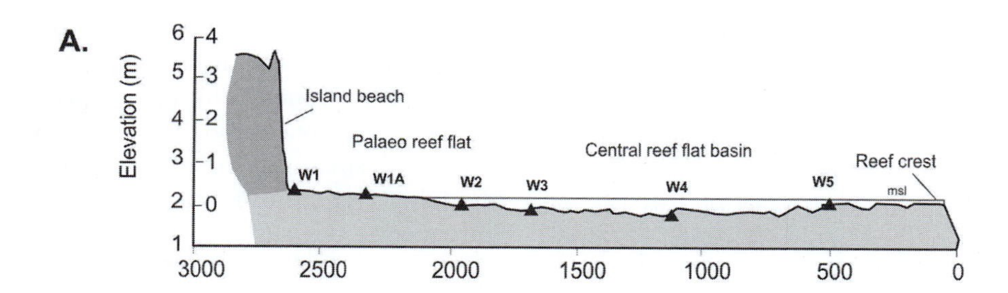

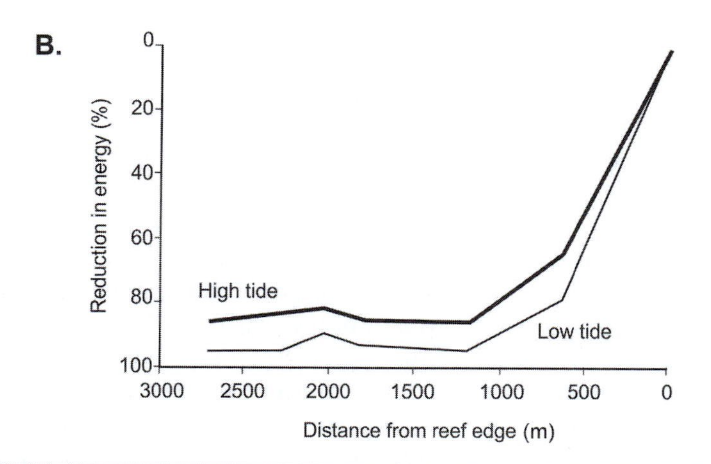

Figure 12.15 Wave energy dissipation across the windward reef flat, Warraber Island, Torres Strait, Australia at high and low tides: A. profile showing reef zones and location of pressure transducers used to measure waves; B. reduction in wave energy relative to offshore significant wave height at high and low tide (Kench and Brander, 2006b). Reproduced with permission from the Coastal Education and Research Foundation, Inc.

$$\psi = \frac{h}{RW}, \tag{12.1}$$

where h is mean water depth over the reef flat at spring high tide; RW is reef width.

This can provide some insight into the relative potential for sediment transport across the reef flat, in combination with the proportion of time over the spring–neap cycle that water depth on the reef flat exceeds a critical depth for wave propagation. It may also reflect the amount of sedimentation that has already occurred.

Wave set-up and subsequent inflow of water across the reef flat is an important source of mass and energy transfer driving transport of nutrients into the backbarrier and lagoon areas and contributes to the overall circulation in the lagoon. On windward coasts the effects of wind stress on the water surface will be added to that of waves and may also be a significant contributor on shallow reef flats. The relative importance of wave set-up and winds will vary with incident wave height and with the tidal range at a site as well as with the depth of the lagoon and the spacing and depth of connections to the sea. Flushing due to tidal action is also very important, but since almost all coral reefs occur in areas of micro- to low-mesotidal range, waves usually are a significant factor and may be the primary factor in microtidal areas. On atolls where wide reef flats and reef islands may greatly constrain direct connection to the ocean through inlet channels, flushing by waves and wind becomes very important for maintaining water quality in the lagoon. In their study of water levels and flows across reef flats on two virtually enclosed microtidal atolls in the South Pacific, Callaghan *et al.* (2006) found that wave breaking produced a nearly continuous superelevation of the water surface on the exposed reef flat which drove an outflow of water across the leeward side of the atoll. In a neat applied study they used analytical modelling of flow over the reef to show that enhanced flushing of the lagoon to reduce the impact of oyster cultivation could be accomplished better by removing obstacles to wave pumping across the reef rim than by tidal pumping through artificial channels dredged through the reef (Callaghan *et al.*, 2006).

12.4 Impacts of Disturbance on Coral Reefs

Coral reefs, like all ecosystems, consist of a biotic component of plants and animals and an abiotic component which includes the reef substrate the surrounding ocean and atmosphere and the physical and chemical processes that take place within the abiotic component. Reef dynamics include both evolutionary processes, such as the upward growth of the reef towards the sea surface, and the response to disturbances produced by both biological and physical causes (Bythell *et al.*, 1993; van Woesik, 1994; Hubbard, 1997; Smith *et al.*, 2008). Disturbances to coral reefs can be grouped into those that result from natural processes, such as bleaching, predation, hurricanes and sea level change, and human activities such as runoff, nutrient loading, dredging and fishing (Brown, 1997). Severe disturbance to the reef has an immediate effect on the fish and shellfish communities that are supported by the reef organisms. Over a period of years to decades it may have an impact on the reef structure itself and thus on the protection it affords to the lagoon and shore zone landward of it from the impact of major storm events (van Zanten *et al.*, 2014; Elliff and Kikuchi, 2017). In the Caribbean, there has been a decline of about 70 per cent in scleractinian corals since the 1980s (Jackson *et al.*, 2014), and similar declines have been reported from the Indo-Pacific region.

12.4.1 Natural Disturbance

As part of the larger reef ecosystem, coral organisms are subject to a wide range of potential natural disturbances that may affect individual organisms over a wide area or many organisms in a more restricted area. These disturbances can be biological in nature or result from physical processes such as wave action or temperature change.

12.4.1.1 *Biological Causes of Disturbance*

Coral organisms are subject to disease and to the effects of invasions by predators in the same way that land-based forests or animal populations are. In the Caribbean, populations of the dominant framework reef builder elkhorn coral (*Acropora palmata*) as well as other corals have declined drastically since the 1980s, primarily because of an outbreak of a bacterial disease termed white-band disease (Aronson and Precht, 2001; Clemens and Brandt, 2015). While this has implications for continued reef building and for the dynamics of the reef crest in the short term, disease-free colonies may establish as a result of recruitment from surviving areas and possibly selection of disease resistant forms in the long term (MacIntyre *et al.*, 2007).

One of the most studied coral reef predators has been the crown-of-thorns starfish (*Acanthaster planci*), which affected large areas of the Australian Great Barrier Reef in the 1960s and 1980s (Moran, 1986; Brown, 1997; De'ath *et al.*, 2012). The starfish, which is 10–15 cm in diameter, has multiple arms and a dense covering of mildly poisonous spines, feed directly on live coral colonies. At low densities, the impact is relatively small and recovery is rapid. During outbreaks the density of starfish in places can be 10–15 adults per m^2 and coral mortality can exceed 90 per cent. Recovery in this case may take 20 to 50 years depending on recruitment patterns and on the growth of algae on dead coral. Human intervention may be feasible on a small scale and several different approaches have been tried (e.g., Boström-Einarsson and Rivera-Posada, 2016).

The dynamics of corals and coral reefs are also affected indirectly by the activities of herbivores such as parrotfish (*Scaridae*) and surgeon fish (*Acanthuridae*), which usually occur in schools on coral reefs. They feed largely on algae and can consume a very large proportion of the total annual growth (Hatcher 1982; O'Farrell *et al.*, 2016). These activities act as an important control on the spread of algae which act to reduce the recruitment of scleractinian corals. Human overfishing of parrotfish is thus a major contributor to the decline of reefs worldwide (Edwards *et al.*, 2014). Large declines in the populations of another herbivore, the sea urchin (*Diadema antillarium*), has led to a large reduction in the abundance of coral and an increase in algae on many Caribbean reefs since the 1980s (Brown, 1997; Bodmer *et al.*, 2015.).

12.4.1.2 *Physical Causes of Disturbance*

Bleaching of corals results from the expulsion of the zooxanthellae and is triggered by thermal stress from high water temperatures or exposure. Bleaching leading to death of corals occurs on the reef crest and backreef area when corals become exposed at low tide, a consequence of the vertical growth of the reef, and may be triggered by seasonal or decadal sea level fluctuations. This is a normal part of reef evolution and dynamics. However, bleaching due to high sea surface temperatures will affect corals on the forereef to depths greater than 10 m. It can be a consequence of climate patterns such as El Niño (Glynn, 1993; Smith *et al.*, 2008; Glynn *et al.*, 2015), or to a general increase in sea surface temperatures worldwide. Bleaching episodes are expected to increase, and likely already are increasing, because of climate change (Hughes *et al.*, 2003; Baker *et al.*, 2008; Li and Reidenbach, 2014). In Australia, mapping of coral bleaching episodes associated with the Great Barrier Reef suggests that the episodes are becoming more frequent and intense, though the variation in spatial patterns between episodes makes the analysis more difficult (Hughes *et al.*, 2017; see Figure 12.16). The 2016 bleaching episode affected much of the northern half of the Great Barrier Reef and a few reefs in northern and western Australia. In 2017, bleaching was of similar magnitude such that about half of the northern reef was thought to have died.

Response to thermal stress varies among individual species of coral and some species show little effect over a wide range of temperatures. The recovery of coral communities from bleaching events may occur through recovery of remnant colonies and through recruitment with the rate and extent to which this occurs being influenced by location as well as the extent to which

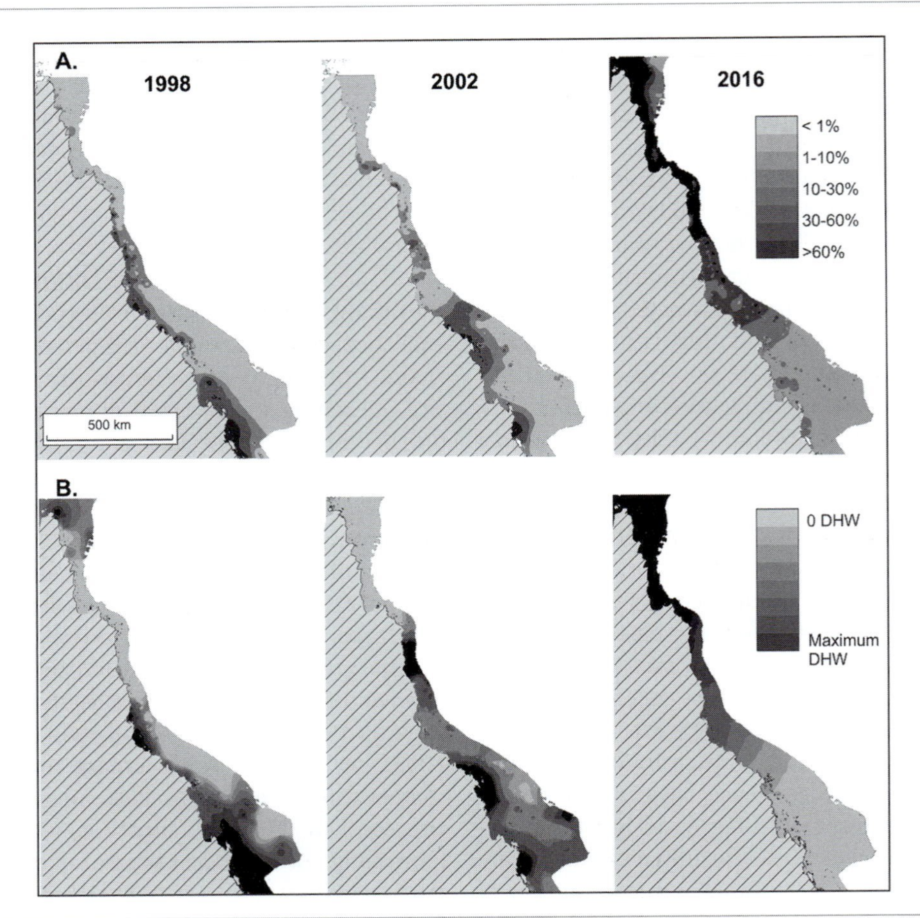

Figure 12.16 Geographic extent and severity of recurrent coral bleaching at a regional scale, Australia. A. The footprint of bleaching on the Great Barrier Reef in 1998, 2002 and 2016, measured by extensive aerial surveys: shading shows percentage of reef exhibiting severe bleaching. The number of reefs surveyed in each year was 638 (1998), 631 (2002), and 1156 (2016). B. Spatial pattern of heat stress based on the number of degree heating weeks (DHWs; °C-weeks) during each mass-bleaching event. Grey scale shows number of weeks ranging from 0 to the maximum for each year (7, 16 and 10 respectively for 1998, 2002 and 2016) (Hughes *et al.*, 2017).

damaged corals are replaced by other species such as algae (Golbuu *et al.*, 2007). Recent work suggests that there is a range of tolerance to thermal stress within a single species and that this may change through time, through genetic adaptation and selection (Woolridge, 2014). As a result, the resilience of coral organisms and coral reefs to climate variation and climate change may be much greater than initial scenarios for climate change effects have portrayed (Hughes *et al.*, 2003). However, knowledge of adaptability and resilience of corals is still very limited and it may be that the vulnerability to global warming is quite high (Mora *et al.*, 2016).

In many areas of the tropics the most significant physical disturbance to reefs is caused by the passage of tropical storms (hurricanes, cyclones, typhoons), which bring with them several different types of disturbance, including the direct mechanical effects of wave action, burial by transported debris, high turbidity due to large waves, and runoff from mainland areas, and changes in salinity (Harmelin-Vivien, 1994). The direct impact of large waves breaking on the reef crest is responsible for the breakage of coral, especially of branching corals such as *Acropora* sp., while domal forms suffer much less (Woodley *et al.*, 1981; Hubbard et al., 1991; Bries *et al.*, 2004). Some damage to coral colonies also occurs because of the impact of large fragments being transported in the waves and abrasion from sand. Damage from tropical cyclones should increase with increasing storm intensity but local sheltering, refraction and shoreline orientation as well as the reef profile generally produces quite a lot of variability in the impact. Most damage to coral colonies occurs in shallow water, and at depths below 8–10m there is generally a rapid decrease in the impacts. However, coral in shallow water landward of the reef crest may survive even very intense storms because of the sheltering effect of rubble ridges. Recovery from hurricane damage is facilitated by three processes (Glynn, 1990; Harmelin-Vivien, 1994): (1) regrowth of surviving coral fragments; (2) regeneration of damaged colonies; and (3) the settlement of coral larvae on damaged reef surfaces. The first two processes appear to be most important where hurricane damage is relatively light. Recruitment of new colonies is most important where damage is severe (Harmelin-Vivien, 1994; Coles and Brown, 2007), and the rate of recovery through this process will depend in part on availability of suitable substrate, on competition with algae and on sources of recruitment.

While hurricane impacts have a damaging effect in the short-term from a biological perspective, they can be regarded as simply part of the dynamics of the reef system which will have influence the overall diversity and structure of the reef ecosystem (Hubbard, 1997). The impact of hurricanes from a geomorphological or geological perspective can be regarded as largely constructive, since these high magnitude events are important for shaping the reef crest, for providing large quantities of rubble for further reef growth and for transporting large quantities of sand and fines landward across the reef flat and seaward down the reef face (Blanchon *et al.*, 2017). Recent work has highlighted the constructive effects of even very intense storms on building reef islands above sea level and adding rubble and sandy sediments to beaches and vegetated areas (Perry *et al.*, 2011; Duvat *et al.* 2017). The characteristics of coral rubble deposition may vary with storm intensity and exposure and this in turn will influence the character of the reef substrate and the process of coral recovery (Perry, 2001). In some cases, extreme storms produce large ridges of cobbles and boulders which may be 1–4 m in height and be built up at the shoreline or on the reef flat (Richmond and Morton, 2007; Reyes *et al.*, 2015).

Similarly, while tsunamis can have a devastating impact on human populations on exposed coral atoll islands, the geomorphological impact appears to be relatively minor (Kench *et al.*, 2008). Kench *et al.* (2008) could repeat pre-tsunami profiles and vegetation line surveys on a number of uninhabited islands on South Maalhosmadulu Atoll in the Maldives atoll chain and were able to examine both erosional and depositional features left by the tsunami impact. At this location, the first wave overtopped all but the highest portions of the ridge along the outer rim of the islands but subsequent waves did not. The first wave flooded across the crest and deposited a relatively thin sheet as a washover fan with sediments ranging in thickness from 0.1–0.3 m near the ridge crest but rapidly tapering to a few centimetres inland. Subsequent waves produced small scarps and eroded the vegetation line. In most areas erosion of beach sediments and of the vegetated portion of the island was small (<4 per cent by area). The tsunami impacts varied spatially depending on local relief, sources of sediment and configuration of the reefs and channels. Kench *et al.* (2008) concluded that most evidence of the tsunami would quickly disappear because of ongoing coastal processes and the most significant effect on island evolution was probably the addition of sediments to the central portion of the islands. Thus, like tropical cyclones, tsunamis may be viewed geomorphologically as generally having a positive impact on reef and atoll islands.

12.4.2 Human Causes of Disturbance

There is considerable evidence of a worldwide decline in the extent and biodiversity of coral reefs (Gardner *et al.*, 2003; Mora, 2008). Much of this has been attributed to the direct and indirect impact of human activities, though there is still some difficulty in determining the impact of individual stresses in a system that is increasingly subject to multiple stressors (Hughes *et al.*, 2003; Aronson and Precht, 2006). Locally, human impacts on coral reefs include (Brown, 1997): sedimentation and high turbidity resulting from increases in soil erosion inland and from dredging and construction activities in the waters around coral reefs. Often this may be associated with tourism infrastructure; oil pollution from oil spills or chronic leakage in areas around oil refineries and harbours; and from mining of coral reefs for building materials and for making lime. Tourism development is often a major source of local threats both from the construction of facilities near the shoreline or in shallow water, and from the accompanying pollution from untreated or poorly treated sewage and runoff from fertilisers and other chemicals used to maintain hotel grounds and golf courses. Coastal development may result in destruction of mangroves and saltmarshes and the infilling of small lagoons and salt ponds leading to reduced trapping of fines before they reach coastal waters. Anchor dragging (Maynard, 2008) and tourism diving and snorkelling can have severe local physical impacts on reefs, though recognition of this has led to the training of dive masters and the adaptation of simple solutions such as the deployment of buoys along highly frequented reefs. Physical damage may also result from destructive fishing techniques such as blasting or the use of cyanide (Souter and Linden, 2000).

There is considerable concern also over the potential impact of human-induced climate change on coral reefs, particularly through higher sea surface temperatures, increased rates of sea level rise (see Box 12.1) and possibly increased intensity of tropical storms. As noted earlier, coral reef communities show considerable resilience in the face of these types of stresses and the impact of climate change may be to produce changes in the diversity and make-up of reef communities rather than a decline in the reefs themselves (Hughes *et al.*, 2003). It seems that the more immediate destructive threats are the local and regional ones, and these will require a diverse suite of approaches and measures to produce a significant reduction in them (Souter and Linden, 2000). It may also require a much more concerted effort on the part of scientists and social scientists to develop a robust framework for addressing the crisis and a convincing public relations effort to get it implemented (Risk, 1999).

Box 12.1 Coastal Management Application

Vulnerability of Ocean Atoll Reef Islands to Sea Level Rise

Over the past two or three decades there has been considerable speculation that human-induced climate change is threatening the existence of Pacific atoll island communities such as the Maldives and Vanuatu. The major threats are seen as vulnerability to salt water intrusion into the freshwater lenses under the islands, increased erosion and flooding during cyclones and possibly increased frequency of large magnitude cyclones. As is the case with any natural hazard, there are two components to consider – the nature of the physical process or processes that forms the hazard, and the ability of humans to adjust/adapt to the threat. Given our understanding of processes operating on coral reefs and reef islands, how real is the threat? How will reef and island processes change/adjust to increased sea level rise and possibly greater frequency of storms? How can our understanding of these processes aid in the development of adaptive measures that will mitigate the potential impact on human populations?

Most Pacific atolls have a ring of islands formed on reef flats that surround some form of central lagoon. Cross sections of islands generally show a high ridge on the exposed side of the island that is generally 2–4 m above present sea level (Woodroffe, 2008), though some islands may have higher spots because of dune building where

Box 12.1 (cont.)

there is abundant sand. Less prominent ridges may occur on lagoon shores where there is sufficient fetch for moderate wave generation or ridges may be the result of agricultural activities in swampy areas. Some portion of the present height above sea level may reflect slight sea level fall since the Holocene maximum (Dickinson, 2001), but the significance of this varies regionally across the Pacific and will also vary locally depending on the history of the accretion of individual atolls and islands.

In most instances much of the accretion above sea level results from the impact of large storms which produce storm surge and wave run-up that leads to sediment being washed inland – as we have seen this may be in the form of rubble ridges where this reefal material is available Figure 12.13A, B), and these form the most stable ridges (Perry *et al.*, 2014; Duvat *et al.* 2017). We can expect that the frequency of water levels above a critical height will increase a result of rising sea level, but this will be superimposed slowly on higher frequency changes due to tides and seasonal and decadal sea level oscillations. Increased water depth will provide additional space for upward reef growth and continued supply of sediment to the reef flat (Woodroffe, 2008). Geomorphologically, sea level rise due to global warming will have a limited impact on coral atolls and their islands and the ongoing process of reef building and sediment dispersal by waves and currents will continue to be the primary control.

As is the case with barrier islands, humans occupying atoll islands have chosen to live in a dynamic environment and may need to find a way to permit processes that lead to building up of the islands to continue while adapting human activities to minimise the risk posed by storm overwash and flooding. In cases such as Funafuti Atoll, Tuvalu, human vulnerability to the threats posed by sea level rise has been increased by building on low-lying lands that are most threatened by flooding (Yamano *et al.*, 2007). Adaptation to the threat of increased overwash and flooding therefore requires an understanding of the physical process posing the threat and the full range of adaptive strategies that are available (Solomon and Forbes, 1999; Ford and Kench, 2015). A key component of any management plan must, however, be the recognition of the significance of all the processes controlling island evolution and adaptation to sea level rise must be seen within the context of a much broader management and environmental strategy that must include the threats posed by human activities as well. In this respect, integrated coastal management for coral atoll islands is no different than for any dynamic coastal environment.

Further Reading

Barnes, R.S.K. and Hughes, R.N. 1999. *An Introduction to Marine Ecology*, 3rd Edition. Oxford: Blackwell Science Ltd, pp. 117–41.

One of several texts that provide useful background on the biology of reefs.

Birkeland, C. (ed.) 1997. *The Life and Death of Coral Reefs*. New York: Chapman and Hall, 536 pp.

This is a good compilation of chapters by some of the experts on coral reefs.

Hopley, D., Smithers, S.G. and Parnell, K.E. 2007. *The Geomorphology of the Great Barrier Reef*. Cambridge: Cambridge University Press, 532 pp.

This is an excellent book that brings together a wealth of information on the barrier reef and provides a strong geomorphological perspective.

Montaggioni, L.F. 2005. History of Indo-Pacific coral reef systems since the last glaciation: development patterns and controlling factors. *Earth-Science Reviews*, **71**, 1–75.

An excellent review of the stratigraphic history of coral reefs in the region in response to post-glacial sea level changes.

References

Adey, W.H. and Burke, R.B. 1977. Holocene bioherms of Lesser Antilles: geographic control of development. In Frost, S.H., Weiss, M.P. and Saunders, J.B. (eds.), *Reefs and Related Carbonates: Ecology and Sedimentology*, Studies in Ecology **4**. Tulsa, OK: American Association of Petroleum Geologists, pp. 67–81.

Alongi, D.M. 1998. *Coastal Ecosystem Processes*. Boca Raton, FL: CRC Press, pp. 139–82

Aronson, R.B. and Precht, W.B. 2001. White band disease and the changing face of Caribbean reefs. *Hydrobiologia*, **460**, 25–38.

Aronson, R.B. and Precht, W.B. 2006. Conservation, precaution and Caribbean reefs. *Coral Reefs*, **25**, 441–50.

Aslam, M. and Kench, P.S. 2017. Reef island dynamics and mechanisms of change in Huvadhoo Atoll, Republic of Maldives, Indian Ocean. *Anthropocene*, **18**, 57–68.

Baker, A.C., Glynn, P.W. and Riegl, B. 2008. Climate change and coral reef bleaching: an ecological assessment of long-term impacts, recovery trends and future outlook. *Estuarine, Coastal and Shelf Science*, **80**, 435–71.

Barrett, S.J. and Webster, J.M, 2017. Reef Sedimentary Accretion Model (ReefSAM): understanding coral reef evolution on Holocene time scales using 3D stratigraphic forward modelling. *Marine Geology*, **391**, 108–26.

Beaman, R.J., Webster, J.M. and Wust, R.A.J. 2008. New evidence for drowned shelf edge reefs in the Great Barrier Reef, Australia. *Marine Geology*, **247**, 17–34.

Blanchon, P. and Eisenhauer, A. 2001. Multi-stage reef development on Barbados during the last interglacial. *Quaternary Science Review*, **20**, 1093–112.

Blanchon, P. and Blakeaway, D.R. 2003. Are catch-up reefs an artefact of coring? *Sedimentology*, **50**, 1271–82.

Blanchon, P., Richards, S., Bernal, J.P. *et al.* 2017. Retrograde accretion of a Caribbean fringing reef controlled by hurricanes and sea-level rise. *Frontiers in Earth Science*, 5, 78.

Bodmer, M.D.V., Rogers, A.D., Speight, M.R., Lubbock, N and Exton, D.A. 2015. Using an isolated population boom to explore barriers to recovery in the keystone Caribbean coral reef herbivore *Diadema antillarum*. *Coral Reefs*, **34**, 1011–21.

Boström-Einarsson, L. and Rivera-Posada, J. 2016. Controlling outbreaks of the coral-eating crown-of-thorns starfish using a single injection of common household vinegar. *Coral Reefs*, **35**, 223–8.

Bries, J.M., Debrot, A.O. and Meyer, D.L. 2004. Damage to the leeward reefs of Curaço and Bonaire, Netherlands Antilles from a rare storm event: Hurricane Lenny, November, 1999. *Coral Reefs*, **23**, 297–307.

Brown, B.E. 1997. Disturbances to reefs in recent times. In Birkeland, C. (ed.), *Life and Death of Coral Reefs*. New York: Chapman and Hall, pp. 354–85.

Bythell, J.C., Gladfelter, E.H. and Bythell, M. 1993. Chronic and catastrophic natural mortality of three common Caribbean reef corals. *Coral Reefs*, **12**, 143–52.

Callaghan, D.P., Nielsen, P., Cartwright, N., Gourlay, M.R. and Baldock, T.E. 2006. Atoll lagoon flushing forced by waves. *Coastal Engineering*, **53**, 691–704.

Chappell, J., Omura, A., Esat, T. *et al.* 1996. Reconciliation of late Quaternary sea levels derived from coastal terraces at Huon Peninsula with deep sea oxygen isotope records. *Earth and Planetary Science Letters*, **141**, 227–36.

Clemens, E. and Brandt, M.E. 2015. Multiple mechanisms of transmission of the Caribbean coral disease white plague. *Coral Reefs*, **34**, 1179–88.

Coles, S.L. and Brown, E.K. 2007. Twenty-five years of change in coral coverage on a hurricane impacted reef in Hawai'i: the importance of recruitment. *Coral Reefs*, **26**, 705–17.

Davies, J.L. 1980. *Geographic Variation in Coastal Development*, 2nd edition. London: Longman, 212 pp.

De'ath, G., Fabricius, K.E., Sweatman, H. and Puotinen, M. 2012. The 27-year decline of coral cover on the Great Barrier Reef and its causes. *Proceedings National Academy of Sciences*, **109**, 17995–9.

Dickinson, W.R. 2001. Paleoshoreline record of relative Holocene sea levels on Pacific islands. *Earth-Science Reviews*, **55**, 191–234.

Duvat, V.K.E. and Pillet, V. 2017. Shoreline changes in reef islands of the Central Pacific: Takapoto Atoll, Northern Tuamotu, French Polynesia. *Geomorphology*, **282**, 96–118.

Edwards, C.B., Friedlander, A.M., Green, A.G. *et al.* 2014. Global assessment of the status of coral reef herbivorous fishes: evidence for fishing effects. *Proceedings Royal Society of London, B, Biological Sciences*, **281**(1774), 20131835.

Elliff, C.I. and Kikuchi, R.K.P. 2017. Ecosystem services provided by coral reefs in a Southwestern Atlantic

Archipelago. *Ocean and Coastal Management*, **136**, 49–55.

Ford, M.R. and Kench, P.S. 2015. Multi-decadal shoreline changes in response to sea level rise in the Marshall Islands. *Anthropocene*, **11**, 14–24.

Gardner, T.A., Cote, I.M., Gill, J.A., Grant A. and Watkinson, A.R. 2003. Long-term region-wide declines in Caribbean corals. *Science*, **301**, 958–60.

Geister, J. 1977. The influence of wave exposure on the ecological zonation of Caribbean reefs. *Proceedings Third International Coral Reef Symposium, Miami*, **1**, 23–9.

Gischler, E., Hudson, J.H. and Pisera, A. 2008. Late Quaternary reef growth and sea level in the Maldives (Indian Ocean). *Marine Geology*, **250**, 104–13.

Glynn, P.W. 1990. Coral mortality and disturbances to coral reefs in the tropical eastern Pacific. In Glynn, P.W. (ed.), *Global Ecological Consequences of the 1982-83 El Nino–Southern Oscillation*. Amsterdam: Elsevier Oceanography Series, **52**, pp. 55–126.

Glynn, P.W. 1993. Coral reef bleaching: ecological perspectives. *Coral Reefs*, **12**, 1–17.

Glynn, P.W., Riegl, B., Purkis, S., Kerr, J.M. and Smith, T.B. 2015. Coral reef recovery in the Galápagos Islands: the northernmost islands (Darwin and Wenman). *Coral Reefs*, **34**, 421–36.

Golbuu, Y., Victor, S., Penland, L. *et al.* 2007. Palau's coral reefs show differential habitat recovery following the 1998-bleaching event. *Coral Reefs*, **26**, 319–32.

Gourlay, M.R. 1994. Wave transformation on a coral reef. *Coastal Engineering*, **23**, 17–42.

Guilcher, A. 1988. *Coral Reef Geomorphology*. Chichester: Wiley, 228 pp.

Harmelin-Vivien, M.L. 1994. The effects of storms and cyclones on coral reefs: a review. *Journal of Coastal Research*, **SI 12**, 211–31.

Harris, P.T., Heap, A.D., Marshall, J.F. and McCulloch, M. 2008. A new coral reef province in the Gulf of Carpentaria, Australia: colonisation, growth and submergence during the early Holocene. *Marine Geology*, **251**, 85–97.

Harris, D.L., Vila-Concejo, A., Webster, J.M. and Power, H.E. 2015. Spatial variations in wave transformation and sediment entrainment on a coral reef sand apron. *Marine Geology*, **363**, 220–29.

Harris, D.L., Power, H.E., Kinsela, A.E., Webster, J.M. and Vila-Concejo, A. 2018. Variability of depth-limited waves in coral reef surf zones. *Estuarine, Coastal and Shelf Science*, **211**, 36–44.

Hatcher, B.G. 1982. The interaction between grazing organisms and the epilithic algal community of a coral reef: a quantitative assessment. *Proceedings Fourth International Coral Reef Symposium, Manilla*, **2**, 515–24.

Hopley, D., Smithers, S.G. and Parnell, K.E. 2007. *The Geomorphology of the Great Barrier Reef*. Cambridge: Cambridge University Press, 532 pp.

Hourigan, T.F., Etnoyer, P.J. and Cairns, S.D. 2017. The state of deep-sea coral and sponge ecosystems of the United States. NOAA Technical Memorandum NMFS-OHC-4. Silver Spring, MD. 467 pp.

Houser, C., D'Ambrosio, T., Bouchard, C. *et al.* 2014. Erosion and reorientation of the Sapodilla Cays, Mesoamerican Reef Belize from 1960 to 2012. *Physical Geography*, **35**(4), 335–54.

Hubbard, D.K. 1997. Reefs as dynamic systems. In Birkeland, C. (ed.), *Life and Death of Coral Reefs*. New York: Chapman and Hall, pp. 43–67.

Hubbard, D.K., Parsons, K.M., Bythell, J.C. and Walker, N.D. 1991. The effects of Hurricane Hugo on the reefs and associated environments of St. Croix, US Virgin Islands: a preliminary assessment. *Journal of Coastal Research*, **8**, 33–48.

Hughes, T.P., Baird, A.H., Bellwood, D.R. *et al.* 2003. Climate change, human impacts, and the resilience of coral reefs. *Science*, **301**, 929–33.

Hughes, T.P, Kerry, J.P., Álvarez-Noriega, M. *et al.* 2017. Global warming and recurrent mass bleaching of corals. *Nature*, **543**, 373–7. DOI:10.1038/nature21707.

Jackson, J.B.C., Donovan, M.K., Cramer, K. and Lam, V. 2014. Status and trends of Caribbean coral reefs: 1970–2012. *Global Coral Reef Monitoring Network*. Switzerland: IUCN, Gland.

Jeanson, M., Dolique, F., Sedrati, M. *et al.* 2016. Wave modification across a coral reef: Cap Chevalier, Martinique Island. *Journal of Coastal Research*, SI 75(1), 582–6.

Keating, B.H. and McGuire, W.J. 2000. Island edifice failures and associated tsunami hazards. *Pure and Applied Geophysics*, **157**, 899–955.

Kench, P.S. and Brander, R.W. 2006a. Response of reef island shorelines to seasonal climate oscillations: South Maalhosmadulu atoll, Maldives. *Journal of Geophysical Research*, **111**, F01001, 1–12.

Kench, P.S. and Brander, R.W. 2006b. Wave processes on coral reef flats: implications for reef geomorphology using Australian case studies. *Journal of Coastal Research*, **22**, 209–23.

Kench, P.S., Nichols, S.L., Smithers, S.G., McLean, R.F. and Brander, R.W. 2008. Tsunamis as agents of geomorphic change in mid-ocean reef islands. *Geomorphology*, **95**, 361–83.

Kennedy, D.M. and Woodroffe, C.D. 2002. Fringing reef growth and morphology: a review. *Earth-Science Reviews*, **57**, 255–77.

Kordi, M.N. and O'Leary, M. 2016. Geomorphic classification of coral reefs in the north western Australian shelf. *Regional Studies in Marine Science*, **7**, 100–10.

Lashley, C.H., Roelvink, D., van Dongeren, A., Buckley, M.L. and Lowe, R.J. 2018. Nonhydrostatic and surfbeat model predictions of extreme wave run-up in fringing reef environments. *Coastal Engineering*, **137**, 11–27.

Leon, J.X. and Woodroffe, C.D. 2013. Morphological characterisation of reef types in Torres Strait and an assessment of their carbonate production. *Marine Geology*, **338**, 64–75.

Letourner, Y., Gaertner, J-C., Durbec, J-P. and Jessu, M.E. 2008. Effects of geomorphological zones, reefs and seasons on coral reef fish communities of Réunion Island, Mascarene Archipelago, SW Indian Ocean. *Estuarine, Coastal and Shelf Science*, **77**, 697–709.

Li, A. and Reidenbach, M.A. 2014. Forecasting decadal changes in sea surface temperatures and coral bleaching within a Caribbean coral reef. *Coral Reefs*, **33**, 847–61.

Lugo-Fernàndez, A., Roberts, H.H. and Wiseman, J.W. 1998. Tide effects on wave attenuation and wave setup on a Caribbean coral reef. *Estuarine, Coastal and Shelf Science*, **47**, 385–93.

MacIntyre, I.G. 1988. Modern coral reefs of western Atlantic: new geological perspectives. *American Association of Petroleum Geologists Bulletin*, **72**, 1360–9.

MacIntyre, I.G., Burke, R.B. and Stuckenrath, R. 1982. Core holes in the outer forereef off Carrie Bow Cay, Belize: a key to the Holocene history of the Belizean barrier reef complex. In *Proceedings Fourth International Coral Reef Symposium*, Manilla: University of Philippines, **1**, pp. 567–74.

MacIntyre, I.G., Glynn, P.W. and Toscano, M.A. 2007. The demise of a major *Acropora palmata* bank-barrier reef off the southeast coast of Barbados. *Coral Reefs*, **26**, 765–73.

Madden, R.H., Wilson, M.E., O'Shea, M. 2013. Modern fringing reef carbonates from equatorial SE Asia: An integrated environmental, sediment and satellite characterisation study. *Marine Geology*, **344**, 163–85.

Maynard, J.A. 2008. Severe anchor damage to *Lobophyllia variegata* colonies on the Fujikawa Maru, Truk Lagoon, Micronesia. *Coral Reefs*, **27**, 273.

McLaughlin, C.J., Smith, C.A., Buddemeier, R.W., Bartley, J.D. and Maxwell, B.A. 2003. Rivers, runoff, and reefs. *Global Planetary Change*, **39**, 191–9.

Mesolella, K.J., Sealy, H.A. and Mathews, R.K. 1970. Facies geometries within Pleistocene reefs of Barbados, West Indies. *American Association of Petroleum Geologists Bulletin*, **54**, 1899–917.

Montaggioni, L.F. 2005. History of Indo-Pacific coral reef systems since the last glaciation: development patterns and controlling factors. *Earth-Science Reviews*, **71**, 1–75.

Mora, C. 2007. A clear human footprint in the coral reefs of the Caribbean. *Proceedings of the Royal Society B*, **275**(1636), 767–73. DOI: https://doi.org/10.1098/rspb.2007.1472.

Mora, C., Graham, N.A.J. and Nyström, M. 2016. Ecological limitations to the resilience of coral reefs. *Coral Reefs*, **35**, 1271–80.

Moran, P.J. 1986. The *Acanthaster* phenomenon. *Oceanography and Marine Biology Annual Review*, **24**, 379–480.

Neuman, A.C. and MacIntyre, A.G. 1985. Reef response to sea level rise: keep up, catch up or give up. *Proceedings Fifth International Coral Reef Congress, Tahiti*, **3**, 105–10.

National Oceanic and Atmospheric Administration. 2017. Distribution of Coral Reefs. Available online at https://oceanservice.noaa.gov/education/kits/corals/media/supp_coral05a.html.

O'Farrell, S., Luckhurst, B.E., Box, S.J. and Mumby, P.J. 2016. Parrotfish sex ratios recover rapidly in Bermuda following a fishing ban. *Coral Reefs*, **35**, 421–5.

Perry, C.T. 2001. Storm-induced coral rubble deposition: Pleistocene records of natural reef disturbance and community response. *Coral Reefs*, **20**, 171–83.

Perry, C.T., Kench, P.S., Smithers, S.G. *et al.* 2011. Implications of reef ecosystem change for the stability and maintenance of coral reef islands. *Global Change Biology*, **17**, 3679–96.

Perry, C.T., Smithers, S.G., Kench, P.S. and Pears, B. 2014. Impacts of Cyclone Yasi on nearshore, terrigenous sediment-dominated reefs of the central Great Barrier Reef, Australia. *Geomorphology*, **222**, 92–105.

Rankey, E.C. 2011. Nature and stability of atoll island shorelines: Gilbert Island chain, Kiribati, equatorial Pacific. *Sedimentology*, 58, 1831–59.

Reyes, M., Engel, M., May, S.M., Brill, D. and Brueckner, H. 2015. Life and death after super typhoon Haiyan. *Coral Reefs*, **34**, 419.

Richmond, B.M. and Morton, R.A. 2007. Coral-gravel storm ridges: examples from the tropical Pacific and Caribbean. In *The Proceedings of the Coastal Sediments '07*, Reston, VA: American Society of Civil Engineers, pp. 572–83.

Risk, M.J. 1999. Paradise lost: how marine science failed the world's coral reefs. *Marine and Freshwater Research*, **50**, 831–37.

Rowlands, G., Purkis, S. and Bruckner, A. 2014. Diversity in the geomorphology of shallow-water carbonate depositional systems in the Saudi Arabian Red Sea. *Geomorphology*, **222**, 3–13.

Ryan, E.J, Smithers, S.G., Lewis, S.E. *et al.* 2018. Fringing reef growth over a shallow last interglacial reef foundation at a mid-shelf high island: Holbourne Island, central Great Barrier Reef. *Marine Geology*, **398**, 137–50.

Schellmann, G. and Radtke, U. 2004. A revised morpho- and chronostratigraphy of the Late and Middle Pleistocene coral reef terraces on Southern Barbados (West Indies). *Earth-Science Reviews*, **64**, 157–87.

Smith, L.D., Gilmour, J.P. and Heyward, A.J. 2008. Resilience of coral communities on an isolated system of reefs following catastrophic mass-bleaching. *Coral Reefs*, **27**, 197–205.

Solomon, S.M. and Forbes, D.L. 1999. Coastal hazards and associated management issues on South Pacific Islands. *Ocean and Coastal Management*, **42**, 523–54.

Souter, D.W. and Linden, O. 2000. The health and future of coral reef systems. *Ocean and Coastal Management*, **43**, 657–88.

Storlazzi, C.D. and Jaffe, B.E. 2008. The relative contribution of processes driving variability in flow, shear, and turbidity over a fringing coral reef: West Maui, Hawaii. *Estuarine, Coastal and Shelf Science*, **77**, 549–64.

Terry, J.P. and Goff, J. 2013. One hundred and thirty years since Darwin: 'Reshaping' the theory of atoll formation. *The Holocene*, **23**, 615–19.

van Woesik, R. 1994. Contemporary disturbances to coral communities of the Great Barrier Reef. *Journal of Coastal Research*: *Coastal Hazards* **SI 12**, 233–52.

van Zanten, B.T., van Beukering, P.J.H. and Wagtendonk, A.J. 2014. Coastal protection by coral reefs: a framework for spatial assessment and economic valuation. *Ocean and Coastal Management* 96, 94–103.

Woodley, J.D., Chornesky, E.A., Clifford, P.A. *et al.* 1981. Hurricane Allen's impact on Jamaican coral reefs. *Science*, **214**, 749–55.

Woodroffe, C.D. 2008. Reef-island topography and the vulnerability of atolls to sea level rise. *Global and Planetary Change*, **62**, 77–96.

Woodroffe, C.D. and Webster, J.M., 2014. Coral reefs and sea-level change. *Marine Geology* **352**, 248–67.

Woodroffe, C.D., Mclean, R.F., Smithers, S.G. and Lawson, E.M. 1999. Atoll reef-island formation and response to sea-level change: West Island, Cocos (Keeling) Islands. *Marine Geology*, **160**, 85–104.

Wooldridge, S.A. 2014. Differential thermal bleaching susceptibilities amongst coral taxa: re-posing the role of the host. *Coral Reefs*, **33**, 15–27.

Yamano, H., Kayanne, H., Yamaguchi, T. *et al.* 2007. Atoll island vulnerability to flooding and inundation revealed by historical reconstruction: Fongafale islet, Funafuti Atoll, Tuvalu. *Global and Planetary Change*, **57**, 407–16.

13

Cliffed and Rocky Coasts

13.1 Synopsis

The focus of this chapter is on coasts where the shoreline is largely developed in rocks or sediments that possess strength due to cohesion and thus, are able to offer resistance to wave action. Beach-forming materials such as sand and gravel are generally scarce and are found primarily as a thin layer fronting the cliff toe or in isolated pocket beaches. The term cliff is used where the slope angle >40°, and cliffed shorelines are those characterised by steep slopes rising abruptly from the water or from the back of a platform that is narrow enough for the toe of the slope to be affected by wave action during storms. Cliffed shorelines may develop in sedimentary rocks, ranging from recent deposits with some cohesion (due to the presence of clays or to overconsolidation due to glacial loading) through weakly cemented shale and sandstone. The most resistant cliffs are found in rocks such as limestone, where chemical bonding is important, and in massive igneous and metamorphic rocks such as basalt or granite that possess strength due to crystallisation from melt and high pressures. Unlike sandy coasts, where progressive erosion of the shoreline may be halted by phases of deposition and progradation, erosion of bedrock destroys the bonding that provides the material strength needed to sustain the cliffed coast, and thus, there is no reversal of the erosion process. Cliffed coasts remain stationary or they recede.

On very hard, rocky coasts (Figure 13.1A), rock strength greatly exceeds the erosional forces of individual waves and erosion takes place very slowly – perhaps millimetres to a few centimetres per century. Since sea level has been at the present level in most parts of the world for <5000 years (see Chapter 3), cliffs in these materials reflect the operation of both coastal and subaerial processes operating over 10^3–10^5 years, which includes varying positions of sea level relative to the coast. Thus, the present-day shoreline reflects the position of the sea against a pre-existing topography rather than an equilibrium morphology that is fully adjusted to wave–current forces today. The spectacular cliffs along fjords, associated with glaciated highlands, are a good example of this. There, contemporary coastal processes act to modify the shoreline, but most of the coastal morphology is inherited.

On soft coasts, such as those developed in glacial till, shale and weak sandstones, wave action is able to erode the toe of the cliff relatively rapidly and to remove the eroded debris (Figure 13.1B). Recession of the cliffs can be on the order of decimetres to metres per year, and thus the cliff form is controlled by modern coastal processes and is not an inherited feature. Because of the relatively rapid rates of recession, these coasts offer an opportunity to study and model the controls on cliff erosion and profile evolution, and to extrapolate the results to hard rock coasts. Studies of soft rock coasts are also timely because the rapid recession rates along

Figure 13.1 Cliff types. A. Cliff developed in 'hard' bedrock which is resistant to erosion and changes only over hundreds to thousands of years Saguenay Fjord, Québec, Canada. B. Cliff or bluff developed in 'soft' glacial till which erodes rapidly, with noticeable change within several years (Lake Erie north shore 10 km east of Port Maitland). Coasts such as this where bluffs are developed in sediments, rather than weak bedrock, are termed cohesive coasts.

many Arctic coasts associated with climate warming are a major concern to northern nations such as Canada and Russia. Also, rapid erosion of soft cliffs may result in threats to houses and infrastructure located on the cliff top, leading to calls for intervention in the form of shore protection.

13.2 Cliffed Coast Morphology

13.2.1 Cliff Form and Occurrence

While cliffs can be formed in cohesionless sands where plant roots and soil moisture provide some strength, most cliffed shorelines develop in material that possesses strength due to cohesion provided by the bonding of clay minerals, cementation by chemical precipitates or the crystal bonding of igneous and metamorphic rocks. The term cliff is used here for all shorelines with a steep subaerial slope, with some portion of the slope exceeding 40°. The height of the cliff should exceed the maximum height of wave run-up and overtopping (although wave spray may reach the top of the cliff). If wave overtopping of the cliff top occurs then the shoreline feature is termed a bank. The term bluff is often used interchangeably with cliff, but here its use is restricted to describing cliffs formed on cohesive coasts in unconsolidated or weakly consolidated sediments, such as sand, silt, clay and till. The presence of a cliffed shoreline reflects the existence of relatively high ground near the coast, and this, in turn, may result from tectonic forces on a continental scale, local folding and faulting, fluvial or glacial erosion, water level change, or simply recession due to coastal erosion on a gently sloping plain. Thus, the height of the coastal cliff is controlled primarily by the relief of the coastal zone and secondarily due to the operation of coastal processes.

In examining the processes and components that make up the coastal cliff recession system it is useful to begin with a description of the morphological components of the system in a profile normal to the shoreline (Figure 13.2A). Not all features need be present on all cliffed coasts. The major components are:

1. The **tableland** or area inland from the cliff top.
2. The cliff top or **crest** marks the change in slope from the tableland to the cliff face.
3. The subaerial **cliff face** extends from the cliff top to the toe of the slope where it intersects the beach or platform. This area is dominated by erosion due to processes resulting from mass wasting and overland flow. In the case of plunging cliffs the face extends below the water level.

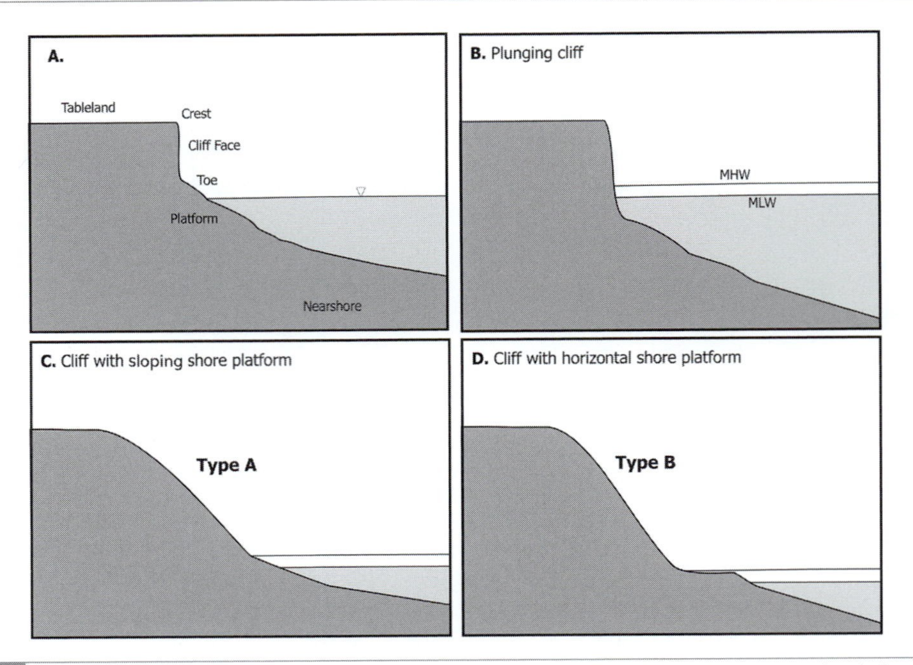

Figure 13.2 Coastal cliffs: A. major components of the coastal cliff system; B., C. and D. profiles associated with the three major types of coastal cliffs; C. and D. refer to the nomenclature of Sunamura (1992).

4. The **cliff toe** forms the transition from the subaerial cliff to the beach and shore platform. The upper limit of the cliff toe is marked by the height to which wave action (not including spray) can reach, whereas the lower limit is marked by the junction with the more gently sloping shore platform. It also marks the transition from the cliff face, which is dominated by subaerial erosional processes leading to horizontal recession, and the shore platform, which is dominated by processes resulting in vertical lowering.
5. The **shore platform** extends from the base of the cliff offshore to a point at, or just below, spring low tide. The shore platform may be overlain by varying amounts of surficial sediments that are critical to mechanical abrasion. The platform is subject to wave action as well as weathering processes during subaerial exposure.
6. The **nearshore slope** forms the subaqueous extension of the intertidal platform, and is a zone of shoaling and breaking waves extending offshore to the limit of wave erosion and transport of sediment.

13.2.2 Cliff Coast Classification

Three major types of cliffed coast can be recognised on the basis of the profile form in a direction normal to the shore and the relative position of sea level on the profile (Figure 13.2B, C, D). Plunging cliffs occur where the cliff extends below the water line to some considerable depth (Figures 13.1A; 13.3A, B). Waves break directly against the cliff face, and there is no beach, ramp or platform that would lead to wave breaking offshore or the accumulation of sediments at the cliff toe. Plunging cliffs usually occur in resistant bedrock where the slope and relief are determined by tectonic events (e.g., folding or block faulting) or where erosion by glaciers or rivers, and subsequent drowning due to sea level rise, has produced a steep cliff with deep water at the cliff base. Over time, jointing and vertical weaknesses may be exploited by wave action to produce chimneys and small pocket beaches, and irregularities in the cliff face may be enhanced by runoff and gully development. However, erosional processes are generally very slow on these coasts, and so it takes a long time for these features to emerge. Plunging cliffs are not generally found on soft rock coasts because the material forming the cliff material is too weak to withstand direct wave attack for very long, and cliff recession soon leads to the formation of a sloping platform and beach.

Where the overall slope of the inherited coastal morphology is less steep, the toe of the cliff face will be located in or above the intertidal zone. Erosion of the cliff toe will occur, leading to recession and the generation of a platform as the cliff face recedes. Weaknesses such as joints and bedding planes of varying lithology and strength in the rock making up the lower part of the cliff will lead to spatially uneven rates of erosion and the development of a variety of erosional forms such as notches, blowholes, caves, arches and stacks (Figures 13.2C, D; 13.3C, D). The coastline tends to become highly irregular and the inner nearshore is often rocky with a variety of shallow reefs and emergent boulders. These features are absent on cohesive coasts where rapid erosion on the beach and shallow nearshore quickly removes any irregularities.

Horizontal erosion on cliffed coasts is focused at the toe of the cliff, and recession of the shoreline will tend to produce a quasi-horizontal erosion surface or platform. However, erosional processes in the intertidal and subtidal zones also act on the platform leading to vertical lowering of the surface. This in turn generates a cliff with a Type A platform (Sunamura, 1992) with an intertidal zone that slopes away from the base of the cliff and grades into the underwater nearshore profile without any abrupt transition (Figure 13.1C). Type A platforms are the most common form of platform on cliffed shorelines, particularly in rocks of moderate to low strength and in areas where sand and gravel are present in the intertidal zone (Figures 13.1B; 13.3C). However, vertical erosion of the platform in the

Figure 13.3 Photographs of cliffed and rocky coasts: A. plunging cliffs along a fjord coast in Labrador; B. plunging cliffs on the west coast of Ireland; C. cliffed coast in northwest Scotland with Type A platform, stacks and caves; D. cliffed coast with Type B shore platform near Wollongong, NSW, Australia. A. courtesy Don Forbes.

intertidal and shallow subtidal zone is sometimes very slow compared to horizontal recession of the cliff toe. This leads to the development of a cliff with a Type B shore platform that has a nearly horizontal surface in front of the base of the cliff, and that often terminates abruptly in a seaward drop to the nearshore (Figures 13.2D; 13.3D). The elevation of the platform may be close to the high-tide level, the low-tide level, or somewhere in between. The quasi horizontal Type B platforms have generated much interest and there is considerable debate about the processes operating on them and the controls on their origin. We will examine this problem at the end of the chapter after we look at cliffed coast processes generally, and at processes operating on cohesive and rock coasts.

13.3 Cliffed Coast Erosion System

Unlike coasts dominated by sedimentary deposits, where erosion of the beach may be balanced by subsequent accretion, the dominant controls on the evolution of cliffed coasts are related to erosional processes. Attention is focused, therefore, primarily on rates of recession (retreat) and the processes leading to this. As is the case for clastic shorelines, it is useful to distinguish clearly between recession and erosion. Recession is used here as a measure of the horizontal, landward retreat of a reference point on the cliff face – for example, the cliff toe or the cliff top. It can be expressed in absolute terms over a defined time period, or as a rate such as m a^{-1}. Erosion is a more general term and, strictly speaking, it refers to the mass of rock material removed from the cliff face or platform either by subaerial processes or by waves and currents. Ideally, erosion is expressed as a mass or mass per unit length alongshore (e.g., kg m^{-1}), though for some purposes it is useful to report it as a volume of material removed per unit area. Vertical lowering of the platform and nearshore through erosion is a linear rate term that has the same units as the horizontal recession rate, and it too can be used to derive a volumetric erosion rate.

13.3.1 Coastal Cliff Recession System

Sunamura (1983, 1992) summarises the major controls on recession of coastal cliffs (Figure 13.4). The primary processes result from the action of waves reaching the bluff toe which lead directly to erosion and recession of the toe. Toe erosion in turn increases the slope angle of the cliff and is an important control on the relative importance and significance of a suite of subaerial processes acting on the cliff slope. These include mass movement processes (creep, falls, slumps, slides and flows) as well as transport processes resulting from unchannelised and channelised water on the slopes (wave splash, rain splash, overland flow, rills, gullies). Material eroded from the cliff face by these processes will be transported by gravity down slope and accumulate temporarily at the cliff toe. The material brought to the cliff toe provides protection to the base of the cliff from wave action and if sufficient debris accumulates then toe erosion may cease altogether. Thus, continued recession requires that the debris is removed offshore

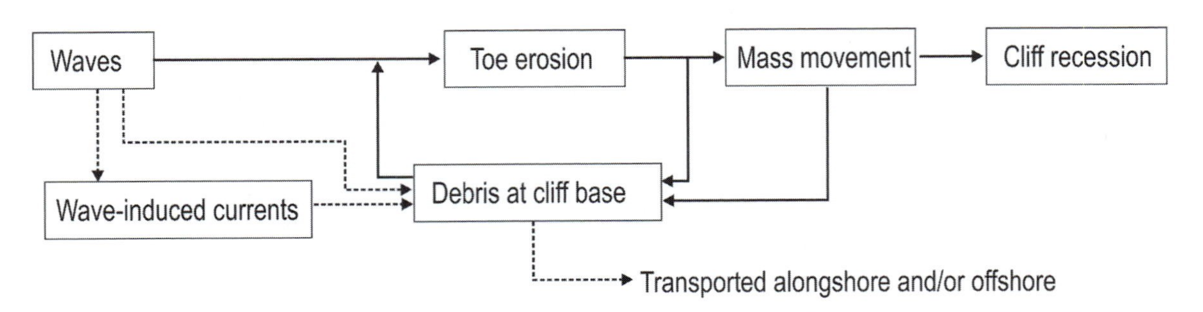

Figure 13.4 Coastal cliff recession system (Sunamura, 1983, 1992).

and/or alongshore by waves and wave-generated currents – hence the feedback loop shown in Figure 13.4. Where this removal does not occur, or where sediment is brought in from updrift, the toe eventually becomes protected and recession of the cliff by subaerial processes will continue only until a stable angle has been achieved. It explains why, for example, cliffs are so prominent at headlands, because net sediment transport will almost always be into the neighbouring bays.

Coastal cliffs and bluffs exhibit the full range of mass wasting and slope removal processes from the movement of individual particles, to shallow slumps and slides, to spectacular deep-seated failures (Figure 13.1B). The relative importance of a particular process will depend on the nature of the material making up the cliff (lithology, stratigraphy, jointing), climatic factors such as precipitation and temperature, vegetation cover, groundwater hydrology, and the stability of the cliff face as dictated by the slope angle. The form of coastal cliffs is directly related to the interplay between these factors as well as to the regional geomorphological or tectonic context. The rich variety of coastal cliff scenery is a reflection of this complexity (Figure 13.3). It can be argued that erosion and removal of sediment at the cliff toe are the primary controls on the rate of coastal cliff recession over the long term. While the subaerial processes of mass wasting and erosion by water acting on the cliff face are interesting, they are generally only a minor determinant of the long-term (century scale) recession rate. However, at the decadal scale, recent work has demonstrated that there is considerable spatial and temporal variation in cliff-top recession and in erosion of the cliff face (Brooks and Spencer, 2014; Pye and Blott, 2015). Some of this arises because of decadal scale variation in the drivers of toe and slope erosion, including sea level and storminess, the small scale variability in beach and nearshore topography, and the lithology of material in the cliff face. Because of the demand by coastal managers for prediction of recession rates over a few years or decades there has been an increase in interest in modelling and predicting short-term recession rates and the location of potential cliff failure (Bernatchez *et al.*, 2011; Castedo *et al.*, 2013). However, these models are not useful for predicting long-term coastal evolution.

13.3.2 Wave-Induced Cliff Erosion Model

The schematic cliff system model shown in Figure 13.3 can be expanded to incorporate a more detailed examination of the interrelationship among the processes and factors controlling erosion of the cliff toe (Sunamura 1983, 1992; see Figure 13.5). The approach here is a mechanical force balance one with toe erosion being controlled by the relative magnitude of two groups of factors: (1) those determining the erosional forces produced by waves reaching the toe of the cliff – the assailing force; and (2) those controlling the strength of the material forming the lower cliff – the resisting force. Conceptually this is nice and relatively simple; and it does point to the need to quantify the processes of erosion in terms of forces. Note that in this model there is no consideration of subaerial processes on the cliff face and that there is the explicit assumption that material is removed alongshore and therefore does not provide any substantial protective role. This is not unreasonable for many coasts, especially when considering long-term shoreline evolution.

The schematic model outlined by Sunamura shown in Figure 13.5 is divided into two components: the wave system and the cliff system. Offshore wave energy (determined by the wave climate) is transformed through shoaling, refraction and breaking over the nearshore profile towards the beach/platform (if present) and cliff toe. The characteristics of the waves at the cliff toe then depend on the nearshore and intertidal morphology, mean sea level and short-term fluctuations due to tides and storm surge, and the effects of these operating together on wave shoaling and breaking transformation (see Chapter 5). The wave system is examined next and the cliff system will be described immediately after.

13.3.3 Cliff Toe Erosion Processes

Wave action at the cliff toe produces both hydraulic and mechanical forces, which make

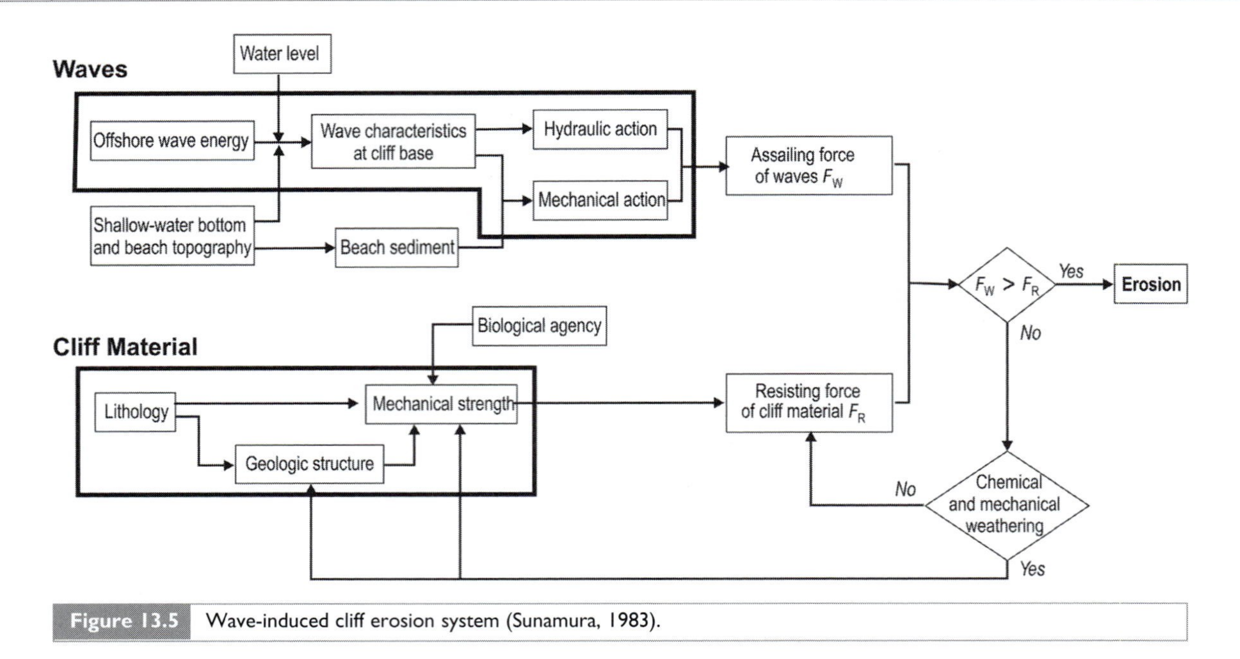

Figure 13.5 Wave-induced cliff erosion system (Sunamura, 1983).

up the assailing force of waves, F_W (Sunamura, 1992; Stephenson *et al.*, 2013). Hydraulic forces include those resulting from compression due to the collision of the wave with the face of the cliff, tension as the water recedes from the cliff, and shearing from water moving upward or downward over the cliff face. The compressive force is greatest when the wave breaks directly against the cliff face with the plunging jet producing a nearly horizontal force into the cliff. This direct impact is termed water hammer and the shock of large waves breaking against a rock cliff can produce low-frequency oscillations or microseismic motions throughout the cliff (Adams *et al.*, 2005; Young *et al.*, 2016). Much of the time a small layer of air is trapped between the breaking wave and the cliff face and the rapid compression of the air pocket develops high pressures within joints and cracks in the rock. Repeated impacts from waves and the forcing of water or air into spaces in the rocks results in the weakening and expansion of cracks, dislodgement of rock material and can produce blowholes that spurt a mixture of air and water high up onto the cliff. The recession of water in the wave trough and the downward drainage of water generate tension on the face. As material is loosened by repeated wave impacts it may be pulled or pushed off the face in a series of processes termed plucking or quarrying. Most of the dislodged rock material ends up at the base of the cliff, but some can be transported right onto the cliff top by the vertical component of wave breaking to produce cliff-top storm deposits (Hall *et al.*, 2006).

Where the mean water level is quite high up the cliff face, hydrostatic pressures are exerted on the cliff materials that lead to buoyancy forces. The hydrostatic pressures are greatest near the foot of the cliff while pressures generated by wave breaking tend to be focused at, or just above, the mean water level. Wave impact forces against vertical or sloping structures as well as hydrostatic forces can be measured in laboratory studies and modelled semi-theoretically. However, it is much more difficult to apply this to real cliffs where wave breaking is often chaotic, the cliff face is irregular and the angle of wave approach often varies significantly from cliff perpendicular.

Shearing forces are produced by two different mechanisms. Waves may break on the beach or platform, generating swash that can run up the toe of the cliff for some distance, and producing a shearing force at the toe. Alternatively, where the mean water level is above the cliff toe a portion of the wave breaking on the cliff is usually directed upwards, producing shearing and tensional forces on the cliff face.

In addition to the hydraulic forces generated by the waves, mechanical forces are generated by entrained sand, gravel and rock particles. The point impacts of rocks and boulders hurled or rolled against the cliff face can be substantial. However, abrasion by sand and gravel is probably more significant because it operates over a greater area and for a greater proportion of time. Abrasion and rock impact are most important at mean water levels ranging from just below the toe, where swash bores frequently reach the toe, to just above the toe where turbulence from wave breaking and orbital velocities at the bed can entrain material. As the water depth at the toe gets large there is less mobilisation of sediment and hydraulic forces become dominant. Thus, wave-induced forces will vary with the form of the platform and nearshore, as well as with the depth of water at the cliff toe and the availability of clastic material for impact and abrasion. They will also vary temporally as incident wave conditions and water levels change.

The significance of each of the hydraulic and mechanical forces associated with wave action can be related to four scenarios that reflect increasing water depth at the cliff toe (Figure 13.6):

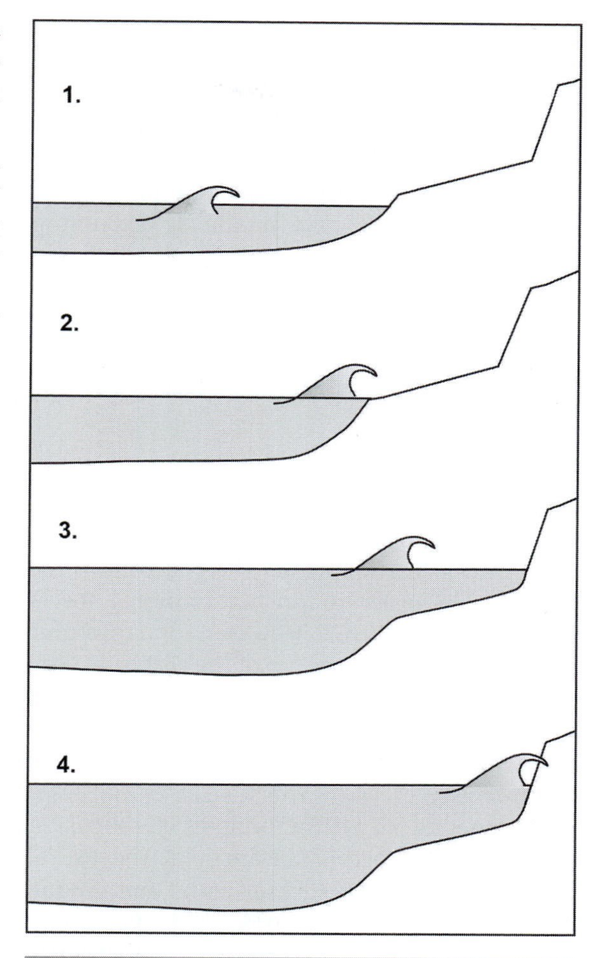

Figure 13.6 Effects of changes in mean water level relative to the elevation of the cliff toe on the relative significance of hydraulic and abrasion forces leading to toe erosion: 1. mean water level far below cliff/platform junction – waves do not reach cliff toe; 2. mean water level below cliff/platform junction – swash run-up reaches the cliff toe; 3. mean water level just above cliff/platform junction – broken waves reach cliff toe; and 4. mean water level considerably above cliff/platform junction – waves break against the cliff face.

1. In the first scenario the mean water level is well below the cliff toe. Waves shoal and break across the nearshore profile, and platform or beach, thus dissipating their energy before the cliff toe is reached. Erosion by waves is focused on the platform and nearshore, while subaerial processes dominate on the cliff.
2. Where water level is below the cliff toe, but wave run-up can reach the toe as swash bores, the dominant erosional processes result from turbulence in the bore, shearing as the wave runs up on the toe, and abrasion from sediments entrained by the swash bore. Unconsolidated material delivered to the toe due to mass wasting of the cliff face can be removed by the swash or may accumulate there depending on debris size and swash energetics.
3. Where the water level is above the toe of the cliff, but water depth is relatively shallow, wave compressional forces are produced from wave impacts. Most waves will begin breaking seaward of the cliff face, so impact forces are moderate. Hydrostatic forces are also

significant, but still relatively small because of the shallow water depths. Abrasion remains significant where there is sand and gravel present on the platform.

4. When the water at the cliff toe is relatively deep, wave breaking frequently takes place right at the cliff face producing maximum water hammer and compressive forces. Hydrostatic forces also become more significant because of the greater water depth, while abrasion is greatly reduced because of the reduction in turbulence and wave orbital velocities at depth.

Direct measurements of wave forces at the cliff toe in the field is extremely difficult and there has been very limited success in modelling wave transformation across the platform to the cliff face in order to predict erosional forces (Young *et al.*, 2016). However, microseismic measurements from the top of the cliff have been shown to be correlated with observed wave and water level conditions at the bluff toe and therefore provide a means of continuously monitoring the intensity of wave attack (Lim *et al.*, 2011; Vann Jones *et al.*, 2015; Young *et al.*, 2016). At the same time, recent advances in the use of remote sensing tools such as aerial and ground based LiDAR (Joyal *et al.*, 2015; Earlie *et al.*, 2015; Obu *et al.*, 2017; Turowski and Cook, 2017; Young, 2018) and structure from motion photography (e.g., Klemas, 2015; Warrick *et al.*, 2017) have made it much easier to monitor cliff failures in both soft and hard materials and thus to correlate the rate and timing of these with offshore wave measurements and with microseismic activity.

13.3.4 Resistance of Cliffs to Toe Erosion

The strength of the cliff system is designated the resisting force, F_R (Figure 13.5). This is determined by intrinsic properties such as lithology, which determines the hardness of the rock material, and by the structure of the rock, including bed thickness and dip angle, joints and fractures, and the relative strength and position of interbedded units. Qualitatively it is easily understood that relatively hard rocks such as granite, basalt and limestone are much more resistant to wave erosion than soft shale or cohesive till. Similarly, massive sandstone and chalk are likely to be much more resistant than thinly bedded sandstones, because there are few weaknesses that can be exploited by water hammer, and compression of air by waves breaking against the cliff face. However, where resistant rock units are interbedded with weaker units such as shale, failure usually results from exploitation of the weaker beds, so the stratigraphic position of beds relative to mean sea level is important (Figure 13.7). Despite such qualitative understanding, it is very difficult to quantify rock strength (see Box 13.1), especially in the field.

Figure 13.7 Erosion of shale at the cliff toe leading to undercutting and failure of blocks of relatively resistant sandy dolomite, Cape Dundas, Georgian Bay, Ontario. The shale is one of the lower units in the Niagara Escarpment. Note the rectangular jointing pattern in the overlying rocks which controls the size of the failure blocks.

In the short term (i.e., seconds to days), if $F_R > F_W$ then no erosion will occur. However, over a longer period it is likely that the value of F_R at points on the cliff toe will decrease as a result of weathering, thus producing the feedback loop in Figure 13.5. In this model, the term weathering is used to include all forms of: (1) chemical weathering, for example, oxidation and hydrolysis; and (2) mechanical weathering, for example, wetting and drying, and freeze-thaw. It also includes all the chemical and mechanical effects of a variety of biological agents such as plants and marine boring organisms

Box 13.1 Measurement of Rock Strength

Intuitively from Figure 13.5 it is evident that the rate of erosion of the cliff toe (and by extension, of the beach and nearshore platform) should be controlled by the resistance of the material making up the lower cliff – that is, some measure of its strength in relation to an applied force or stress. More importantly, if we are to be able to develop a numerical model for predicting toe erosion and cliff recession it is important to be able to quantify the material strength. Since we have a variety of techniques for measuring the strength of rock and cohesive materials and quite an extensive set of theory and empirical data related to soil and rock mechanics, it would seem relatively simple to produce good estimates of the strength of materials making up a cliff. However, the task is complicated by two sets of factors: first, the strength of material must be measured as a response to an applied force; however, erosion processes at on the cliff coast profile, and particularly at the cliff toe produce several types of applied force (e.g., compressive forces, tensional forces, shear forces and forces due to impact and abrasion). Thus, ideally we should measure the strength in response to all of these forces working in combination. The task is further complicated by the effects of structural weaknesses from bedding planes and jointing, which are difficult to quantify but which may be extremely important where wave breaking produces cavitation and compression of entrapped air. Second, the intrinsic strength of the material may change over time as a result of a variety of weathering processes, which means that any model must account for these effects.

The mechanical strength of materials can be evaluated using tests from soil and rock mechanics. These are described in standard engineering soil mechanics and rock mechanics texts, and in documents from the American Society for Testing and Materials (ASTM) and ASTM International. For cohesive materials compressive strength may be measured using an unconfined compression test. Shear strength and compressive strength may be measured in a triaxial compression test with the results plotted as a family of Mohr's circles. Shear strength may also be measured using a shear box or directly in the field using a shear vane. The unconfined compressive strength of rocks can also be estimated using a Schmidt hammer (Aydin and Basu, 2005). When examining erosion of the platform and nearshore profile, resistance to abrasion may be more important. Ultimately, many of the strength properties may be correlated with compressive strength (Sunamura, 1992; Budetta *et al.*, 2000), and thus much of the work to date has attempted to relate erosion rates to some measure of this. Moses *et al.* (2014) provide a good review of methods for measuring changes in subsurface rock properties due to weathering.

(Naylor *et al.*, 2012). In the broader sense it also includes various forms of mechanical weakening due to loading and unloading or fatigue failure. This may result from repetitive flexing due to impact pressure fluctuations, the effects of air compression in joints, and hydrostatic pressure fluctuations produced by fluctuating water depths.

Erosion of the cliff occurs if $F_W > F_R$. This should not be envisioned as a threshold that applies to the cliff as a whole but rather it may be better to think of it as applying to a point on the cliff toe. There will be an infinite number of points on the cliff toe each with its own F_R value which together produce a characteristic probability distribution. In turn, because we are dealing with random seas, complex nearshore topography and varying breaker characteristics, there will be a range of instantaneous F_W values also characterised by a probability distribution. Over time F_W may vary systematically with tidal stage and less predictably with changing incident wave and wind conditions. At an instant in time, wherever $F_W > F_R$, erosion will occur. If the mean value of F_W is small and the mean value of F_R is large, there may be long periods of time where there is no overlap between the two probability distributions or only a small area (Figure 13.8).

In the short term an increase in the probability of erosion is likely to be a function of

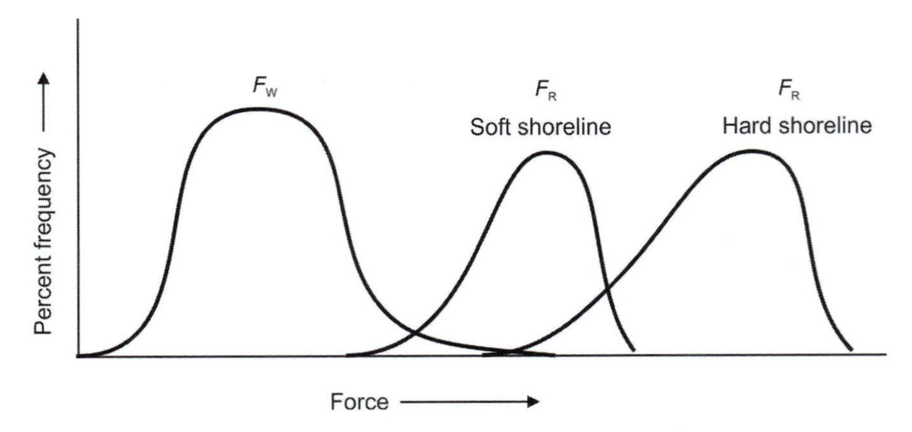

Figure 13.8 Schematic depiction of hypothetical average probability distribution functions for F_W and F_R. Erosion occurs only for the area of overlap between the two distributions and the greater the extent of overlap the more frequent is erosion.

increasing F_W because of increasing storm intensity and/or tidal height. Thus the upper end of the F_W distribution represents conditions that exist for some time in an intense storm, perhaps coinciding with spring high tides, but then do not recur for several months or years. The lower end of the F_R distribution reflects the effects of weathering and fatigue failure on rock strength and it may be truncated after a severe storm when much of the weathered material is removed. The strength of hard rock coasts made up of massive igneous or metamorphic rocks is much greater than wave impact forces, even from very large waves, and there is almost no overlap between the two probability distributions – erosion here proceeds very slowly. Such rocks are usually also very resistant to weathering and so cliff retreat over many centuries may be a metre or less. In softer rocks, especially those where thin bedding or jointing provides weak points to be exploited by wave pressure forces, the overlap of the two distributions will be greater and recession may be measured in centimetres or even metres per year.

There are numerous published cliff recession rates from individual sites and studies and some attempts to compile these systematically for regions of the world (Sunamura, 1983; Hampton and Griggs, 2004; Irrgang *et al*, 2018). However, direct measurement of soft cliff recession through surveying is often difficult, especially where it is desirable to collect the data over some alongshore length rather than at a few profiles. A relatively simple measurement of bluff crest recession over a small area can be carried out by setting up stakes on the bluff top and measuring the distance to the edge on a monthly, seasonal or annual basis (Bernatchez and Dubois, 2008). However, over the short term this is driven primarily by subaerial weathering and mass wasting processes and there is a disconnect from the processes controlling toe erosion. The compilation of a data set involves carrying out measurements over a number of years and there are few programmes anywhere in the world that do this systematically over periods of decades. Most measurements of long-term recession rates have made use of rectified aerial photographs and measured change over the interval between photographs (Buckler and Winters, 1983). Recent advances in digital photogrammetry and the use of DGPS to collect ground control points have made the task easier (Lantuit and Pollard, 2008) and likely improved accuracy. Recession of top of bluff is generally the easiest to measure on aerial photographs, but it is subject to much greater temporal variability than is the toe. However, the bluff toe, which provides a better estimate of cliff recession rates related to wave energy, may be difficult to discern because of overhanging vegetation, slumps and shadow. This is especially true for high cliffs. Errors associated with mapping the cliff toe from aerial photographs are also greater because it is generally difficult to make use of ground control points below the bluff top. As noted earlier, advances in remote sensing, especially airborne and land based LiDAR, structure from motion photography and the deployment of drones has made it much easier to collect areal data from the cliff face over many kilometres (Turowski and Cook, 2017; Young, 2018). However, on coasts where recession rates are small it will take many decades before a robust data set can be acquired.

In order to operationalise and test the predictive capability of the schematic model illustrated in Figure 13.5, we would need to have data for F_W, F_R, and recession rates from a variety of sites. Clearly, the difficulties noted above in measuring all three of these parameters make this task a daunting one and there are as yet relatively few data sets with measurements of both recession rates and wave processes. Nevertheless, empirical studies do provide insights into the operation of some of the parameters discussed above. Thus, they allow us to get a sense of how the parameters control the type of equilibrium form, and the long-term recession rate that might be predicted to develop at a particular location. Before we examine this, it is useful to consider the time scale over which modern coastal cliffs have evolved in relation to cliff recession rates.

Coastal cliffs can be divided into three groups on the basis of the cliff recession rate: (1) soft cliffs with recession rate on the order of $R > 0.05\,\mathrm{m\,a^{-1}}$; (2) cliffs developed in rocks of intermediate resistance with recession rate $0.05 < R > 0.005\,\mathrm{m\,a^{-1}}$; and (3) strongly resistant cliffs with $R < 0.005\,\mathrm{m\,a^{-1}}$. Recession of soft cliff shorelines over the past 5000 years (the period when sea level was relatively stable) will range from a few tens of metres to several kilometres, and thus, the form of the cliff, shore platform and nearshore is likely to reflect modern sea level and wave conditions. Cliffs in rocks with intermediate resistance will have receded a few metres to 15–25 metres, and thus, the profile will reflect a mixture of contemporary erosion processes and inherited forms from prior sea-level stands. Finally, cliffs developed in hard rock will show little measurable recession since sea levels stabilised, and thus the form is largely an inherited one. All that can be said about these coasts is that $F_R >> F_W$.

As was noted earlier, while it is evident from Figures 13.4 and 13.5 that the immediate control on cliff recession is erosion at the toe, consideration of the long-term evolution of cliffed profiles where there is a platform present suggests that the ultimate control on the cliff recession is the rate of lowering of the platform and nearshore profile. This is shown in the process-response model for cohesive shores (Davidson-Arnott, 1990; see Figure 13.9) which is modified from the original system of Sunamura (1983; Figure 13.5). This model introduces the effects of wave action and water level fluctuation on vertical erosion of the nearshore profile, in addition to consideration of wave-induced forces at the bluff toe. Of particular importance are the feedback relationships shown by dashed lines. If the rate of horizontal retreat of the bluff toe exceeds the vertical lowering of the nearshore, this leads to an extension of the platform profile and a reduction in the amount of wave energy reaching the bluff toe. This may be aided by the protective effect of sediment supplied from the upper bluff. Conversely, if bluff recession lags behind vertical lowering of the profile, then the nearshore profile deepens, permitting more wave energy to reach the bluff toe, rather than being dissipated by breaking and bottom friction. It should be noted also that water level changes act in the opposite direction in the nearshore system compared to that at the bluff toe. Thus, a rise in water level tends to increase wave forces at the bluff toe and the platform immediately adjacent to the bluff. Conversely, a fall in water level tends to increase the wave forces on the lower platform and on the nearshore profile. In the Great Lakes, to which the model was initially applied, water level fluctuations are driven by

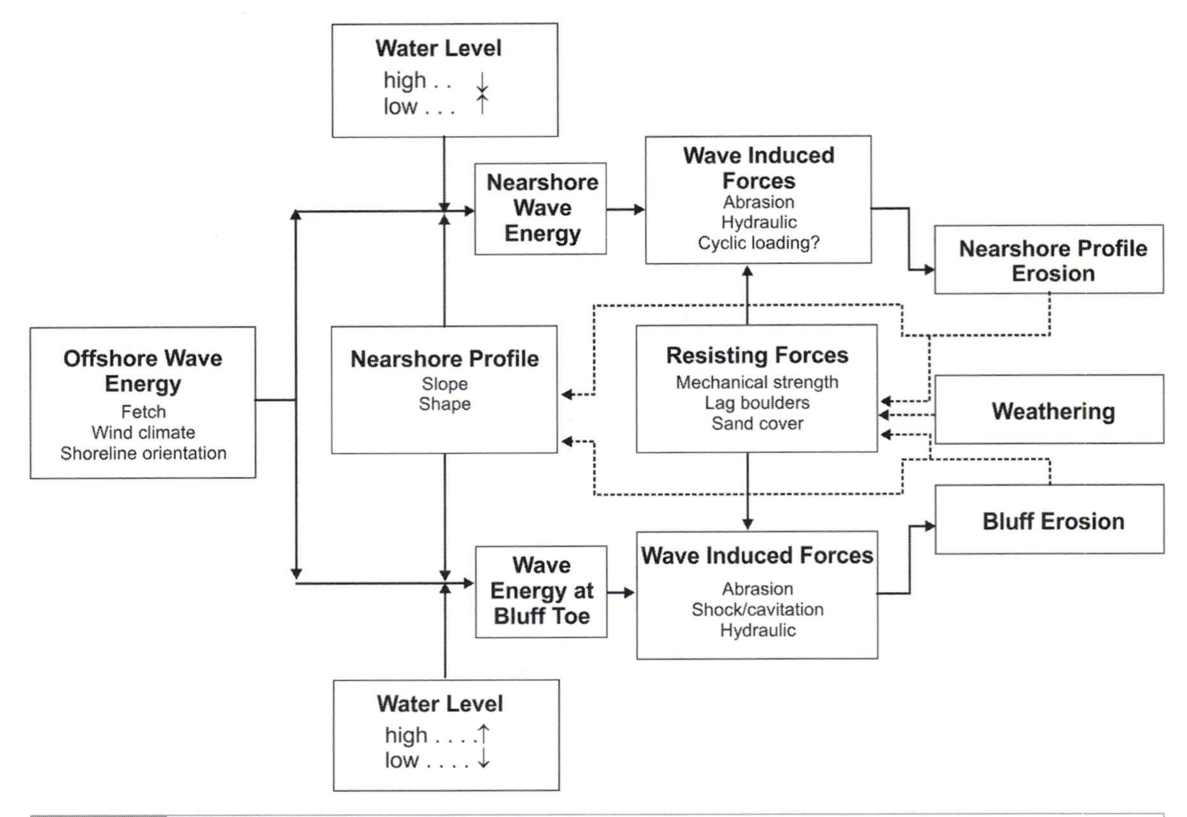

Figure 13.9 Cohesive shoreline process response model. Note the explicit inclusion of erosion of the nearshore profile as well as the inverse effects of water level fluctuations on nearshore and toe erosion (Davidson-Arnott, 1990).

seasonal and multi-year variations in precipitation in the basins, as well as short-term fluctuations due to storm surge and seiching. On oceanic coasts daily and fortnightly tidal fluctuations (Chapter 3) act in a similar fashion in addition to the meteorological and wave induced changes.

The focus of this section has been largely on establishing a conceptual basis for measurement and modelling of coastal cliff erosion and retreat. We can now turn to examining the results of field and laboratory studies of cliffed coast morphodynamics. The next section details the results of studies conducted on weak or soft coasts – coasts where the recession rate is typically tens of centimetres per year. The reasons for choosing to examine this before resistant rock coasts is that the large rates of recession makes it much easier to measure recession rates over short time periods, and because the cliff coast profile is much more likely to be in equilibrium with contemporary water levels and coastal processes. Following this we can see how an understanding of processes on these rapidly eroding soft coasts can be applied to studies of resistant rock coasts and thereby inform the debate over shore platforms.

13.4 Erosion of Soft Rock Cliff Coasts

13.4.1 Soft Rock Coast Characteristics

The term soft or soft rock shoreline is used to describe cliffed coastlines in which the profile is developed in relatively non-resistant rock characterised by cliff recession rates $R > 0.05\ \mathrm{m\,a^{-1}}$ and a uniaxial compressive strength generally < 5 MPa (Sunamura, 2015). Included in this are cohesive bluff coasts formed in sediments with large silt and clay content (Hutchinson, 1973, 1986; Quigley *et al.*, 1977; Carter and Guy, 1988; Brew, 2004) as well as weakly lithified shales and sandstones (Komar and Shih, 1993; Moon and Healey, 1994; Brooks and Spencer, 2010). Cohesive bluff shorelines are found predominantly in mid and high latitudes where they are formed in glacial till, glaciofluvial and glaciolacustrine sand, silt and clay, and Holocene mud. These shorelines are characterised by steep, subaerial cliffs, narrow beaches of mixed sand and gravel, and a moderate to steep, intertidal and nearshore profile that may be concave in form. In western Europe they are common on many stretches of the east and south coast of England (e.g., the Holderness coast of Yorkshire), parts of Ireland, and on the Baltic coasts of Denmark and Germany. They are also found on all three marine coasts of Canada (the Pacific, the Arctic and the Atlantic), on the northeast coast of the US, and the southwest coast of the South Island of New Zealand. Soft cliff coasts in shale and sandstone are not restricted to any particular latitude or part of the world (Brooks and Spencer, 2010; Young, 2018).

Because of the high rates of recession, the modern cliff, platform and nearshore profile in many areas has often evolved over a period of a few thousand years and thus under a relatively stable water level and wave climate. Under these conditions the rate of century-scale horizontal recession of the cliff toe is probably in dynamic equilibrium with the rate of vertical lowering of the nearshore profile (Davidson-Arnott and Askin, 1980; Davidson-Arnott and Ollerhead, 1995; see Figure 13.10). This forms the conceptual basis for understanding the controls on soft coast erosion and the evolution of the coastal profile. At the same time field measurements can provide data that can be used to test models of the evolution of cliff coast profiles (Moses *et al.*, 2006; Trenhaile, 2009). The key processes associated with various locations on the profile are summarised in Figure 13.11.

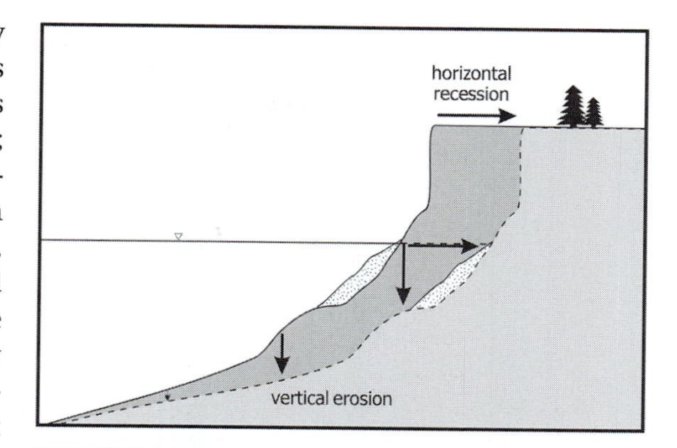

Figure 13.10 Equilibrium profile development on a cohesive coast (Davidson-Arnott and Ollerhead, 1995). The presence of some sand and gravel deposits overlying the till is indicated by the stippled area on the profile.

13.4.2 Vertical Erosion of the Platform and Subaqueous Profile

Recession of soft cliffs, and particularly bluffs on cohesive shorelines is driven by erosion of the cliff toe and this is accompanied by vertical lowering of the profile seaward of the base of the bluff – that is, the platform and the nearshore zone. If the assumption of an equilibrium profile similar to that shown in Figure 13.10 is valid, then it is possible to predict the rate of vertical lowering of the nearshore from the recession rate of the bluff toe and the local slope using equation 13.1 (Zenkovitch, 1967; Philpott, 1986; Sunamura, 1992):

$$\frac{\partial y}{\partial t} = \frac{\partial x}{\partial t} \tan \alpha, \qquad (13.1)$$

where x and y are the horizontal and vertical coordinates of any point on the profile and tan α is the nearshore profile slope at y.

Evidence from a number of soft rock shores provides empirical support for the validity of Equation 13.1. Evaluation of bluff recession and nearshore profile change along the north shore of Lake Erie over an 80-year period (Philpott, 1986) showed average downcutting rates ranging from about 0.5 cm a^{-1} in a water depth of 6 m to about 5–6 cm a^{-1} in a water depth of 1 m. These were associated with recession rates ranging from 1.5 to 3 m a^{-1}. Similar magnitudes were determined for profile erosion at a site on southwest Lake Ontario based on recession over a 35-year period (Davidson-Arnott, 1986). Healy *et al.* (1987) estimated the average long-term

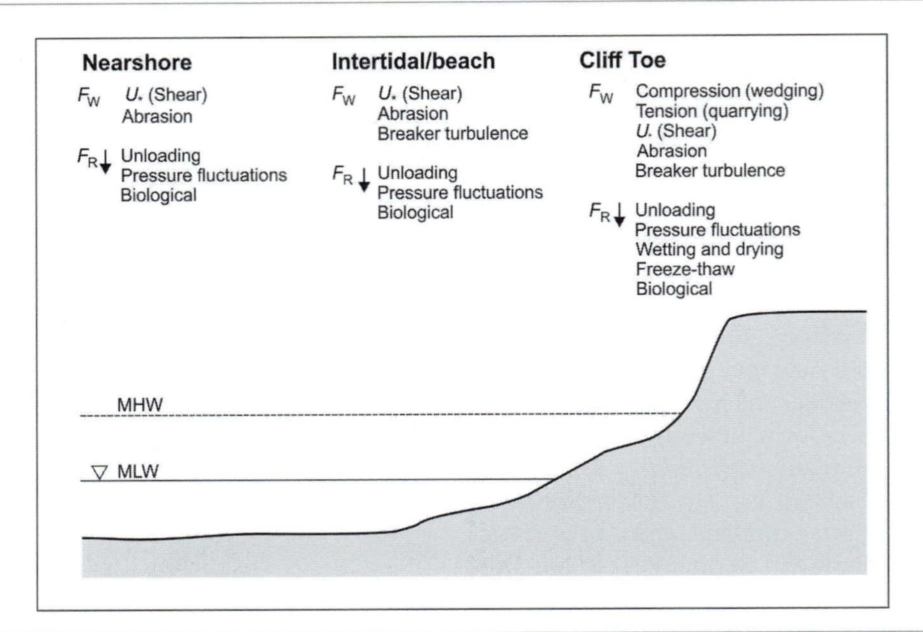

Figure 13.11 Generalised profile of a soft cliff coast with the primary erosional forces and weathering processes associated with each of the major components.

erosion rate for Kiel Bay over the past 5800 years to be 0.06–0.15 cm a^{-1} out to a depth of about 10 m. The much smaller average rates here reflect much slower rates of erosion in deep water compared to shallow water and indicate that erosion in depths below about 6 m is quite small because of the very large waves required to act on the bed at greater water depths. On the Holderness coast of England recent work by Pye and Blott (2015) has shown that much of approximately 60 km of the cohesive shoreline between the south side of Flamborough Head in the north and Kilnsea in the south has retreated between 150 m and 300 m since 1852 (Figure 13.12). As a result, the 1852 cliff line now lies close to, or seaward of the spring low tide line. In turn this means that the intertidal platform has been subject to vertical erosion on the order of 3–5 m over that period. These studies thus show that Equation 13.1 is a good predictor of platform and nearshore erosion, and that the profile evolution at these locations is in dynamic equilibrium with contemporary waves and sea-level position.

The shape of many cohesive profiles is concave, with steeper slopes close to shore and the gradient decreasing into deeper water (Figure 13.13A). This shows that erosion is greater close to shore and that the rate of erosion decreases exponentially offshore (Philpott, 1986; Sunamura, 1992). The actual processes that control the erosive forces, F_W, and resisting forces F_R on the platform and nearshore profile are complex and only poorly understood. However, some insights can be gained from field measurements carried out on nearshore profiles on cohesive coasts in the Great Lakes and on chalk platforms on the coast of England. In both areas erosion is activated by wave action and the erosion rate is determined in part by a reduction in strength of the profile substrate as a result of weathering. However, the relative significance of particular processes and forms of weathering varies between the nearshore profile which is always submerged and the intertidal area which is regularly exposed subaerially (Figure 13.11).

Short-term field measurements of erosion of the nearshore profile in till in Lake Ontario and Lake Huron provide some insight into the erosional mechanisms and the factors controlling their temporal and spatial pattern (Davidson-Arnott 1986; Davidson-Arnott and Ollerhead, 1995; Davidson-Arnott *et al.*, 1999; Davidson-

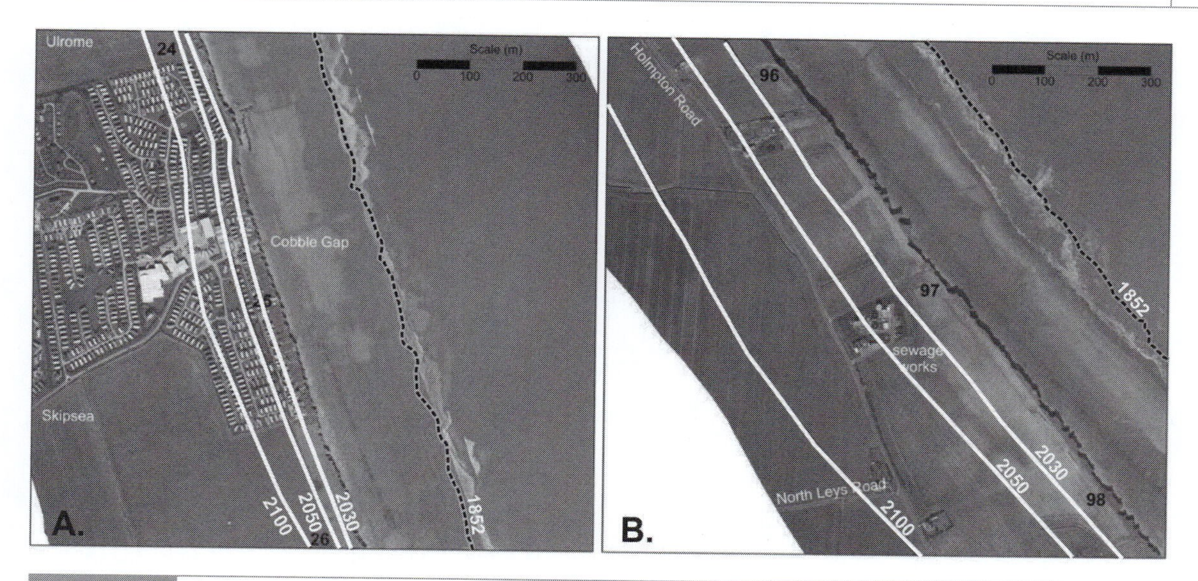

Figure 13.12 Recession of coastal cliffs in glacial till at two locations on the Holderness coast of Yorkshire, England: A. Skipsea and B. south of Withernsea (from Pye and Blott, 2015, figure 17). The position of the cliff top in 1852 can be seen between 200 and 250 m seaward of the current cliff edge at these locations. The future position of the cliff edge in 2030, 2050 and 2100 extrapolated from the measured recession rates are shown inland from the present cliff edge.

Arnott and Langham, 2000). The measurements were made using a modified micro-erosion meter (MEM) which measures the distance to the till surface at stations consisting of three pins set into the till (Askin and Davidson-Arnott, 1981; see Box 13.2). Measurements made along two profiles on Lake Ontario between 1980 and 1984 in water depths ranging from about 1 m to just over 6 m showed average rates of erosion of 30–70 mm a^{-1} in water depths < 2 m with rates decreasing to about 10 mm a^{-1} in depths greater than 6 m (Davidson-Arnott, 1986; see Figure 13.13B). These weighted average annual values conceal significant variability from year to year (Figure 13.13C). There is considerable variation in the length of time over which the data were collected (from a few months to 3 years) and this accounts for some of the variability. Measurements made in shallow water are under-represented because of loss of stations due to the large erosion rate, or because the steel pins were damaged or plucked from the surface by ice action over the winter. Vertical erosion rates predicted from the slope of the nearshore profile at Grimsby (Figure 13.13A) using Equation 13.1 with a long-term cliff recession rate of 1.1 m a^{-1} are very similar to the short-term MEM measurements (Figure 13.14) and provide support for the equilibrium profile concept. Note that Figure 13.12 is similar to one produced by Sunamura (1992, figure 6.2) but he inadvertently used an incorrect recession rate of 1.4 m a^{-1} that produced an offset between measured and predicted data.

Measurements of vertical erosion at intervals of 2–4 weeks were made in 1992 at 16 stations in water depths of 0.5–3 m at St Catharines on Lake Ontario (Davidson-Arnott and Ollerhead, 1995). A total of 94 MEM measurements were collected over six intervals. At some stations erosion was recorded during each interval and altogether erosion was measured 80 per cent of the time. However, only four records had erosion during an interval that was > 1.0 cm. These data show that even modest wave events are able to produce measurable erosion along this shoreline and that erosion of the platform is nearly continuous, in contrast to the episodic events that are characteristic of bluff recession. Short-term erosion rates showed a positive correlation with measured offshore wave energy (Figure 13.16).

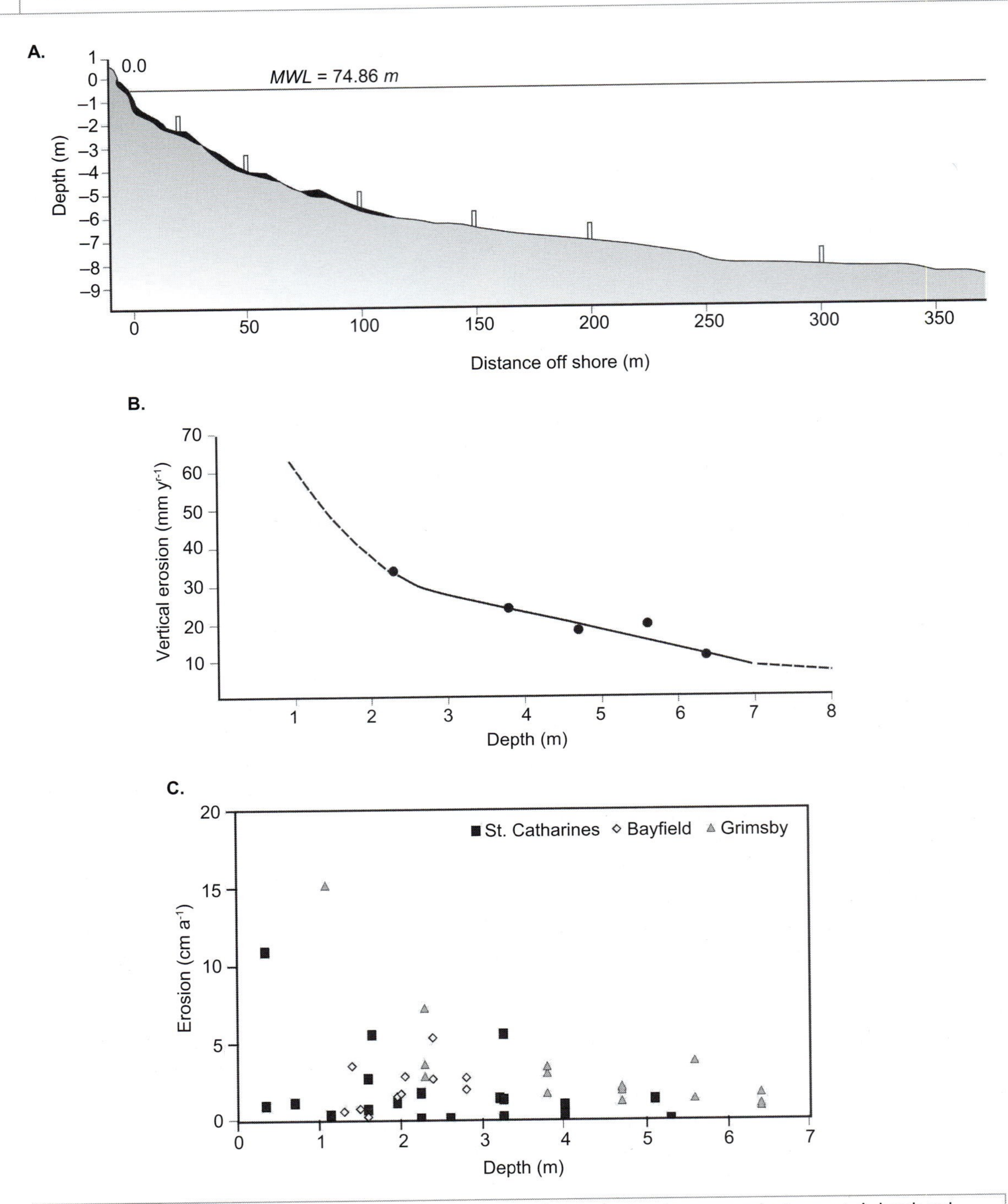

Figure 13.13 Variation of vertical erosion rates with water depth over the nearshore on a cohesive coast: A. beach and nearshore profile on a cohesive coast, Grimsby, Lake Ontario (from Davidson-Arnott, 1986); B. weighted average annual measurements at Grimsby based on MEM data from two profiles 1980–4 (from Davidson-Arnott, 1986); C. annual erosion for individual stations at Grimsby, St Catharines (Lake Ontario) and Bayfield (Lake Huron). Grimsby data are from Davidson-Arnott (1986) and data from the other two sites from Davidson-Arnott *et al.* (1999).

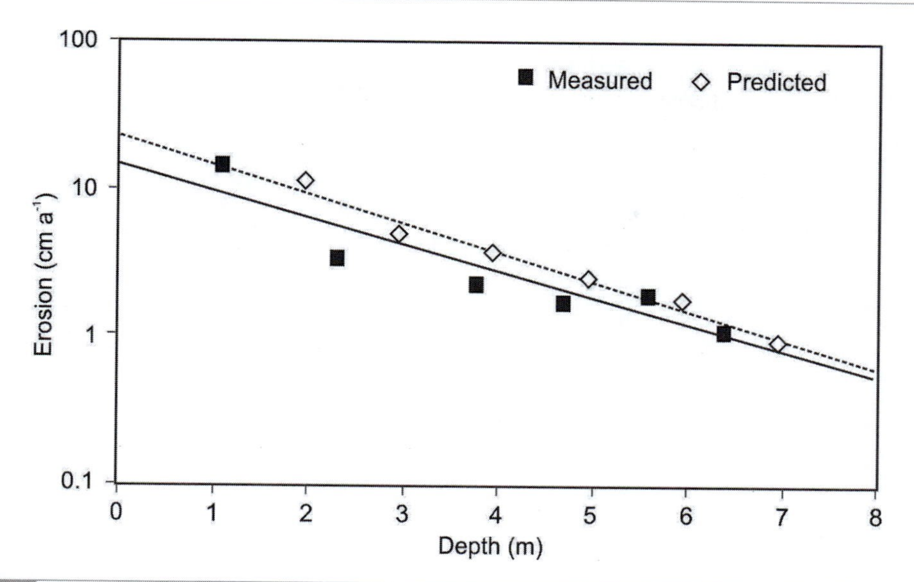

Figure 13.14 Predicted vertical erosion rate based on Equation 13.1 and bluff recession rate of $1.1\ \mathrm{m\,a^{-1}}$ for Grimsby, Lake Ontario compared to measured erosion rates (Davidson-Arnott, 1986).The lines are best fit curves through the six data points.

Erosion of soft cohesive sediments (vane shear strengths 4–25 kPa) can be modelled simply (Parthenaides, 1965; deVries, 1992) by an excess shear stress of the following form:

$$E = M\left(\frac{\tau_b - \tau_c}{\tau_c}\right), \quad (13.2)$$

where E is the rate of surface erosion ($\mathrm{kg\,m^{-2}\,s^{-1}}$); M is the erosion rate coefficient ($\mathrm{kg\,m^{-2}\,s^{-1}}$); τ_b is the bed shear stress ($\mathrm{N\,m^{-2}}$); and τ_c is the critical shear stress ($\mathrm{N\,m^{-2}}$).

The critical shear stress τ_c is a complex function of the shear strength, clay content, structure and other geotechnical properties of the bed materials (Kamphuis and Hall, 1983; Pachure and Mehta, 1985). Actual erosion results from fluid forces associated with wave orbital motion under shoaling waves. Laboratory experiments by Skafel and Bishop(1994) and Skafel (1995) in a 100 m long wave tank showed erosion in a narrow zone associated with plunging breakers, while erosion was much reduced with spilling breakers. However, direct measurements of vane shear strength for unweathered till in Lake Ontario are on the order of 40–80 kPa which is much greater than the shear stress associated with wave orbital motion alone. A number of laboratory (Sunamura, 1977; Kamphuis and Hall, 1983; Skafel and Bishop, 1994; Skafel, 1995) and field studies (Healy and Wefer, 1980; Davidson-Arnott and Askin, 1980) have indicated that abrasion by sand and gravel under wave action is likely to be an important control on the rate of erosion of the substrate on soft rock coasts. Field measurements on Lake Ontario by Davidson-Arnott and Ollerhead (1995) provided evidence to support the importance of abrasion and further work at the same location (Davidson-Arnott and Langham, 2000) using a series of isolated plots (Figure 13.12B) provided a further demonstration of this.

Davidson-Arnott and Langham (2000) also provided evidence of the important role of weathering of the till in the intervals between significant wave events. Observations of the till surface and measurements of the vane shear strength showed that a thin surface layer of reduced shear strength was present over the whole profile with the shear strength decreasing with distance offshore (greater water depth) and reduced frequency and rate of erosion (Figure 13.17). Shear strength and moisture content was measured at 2–5 cm intervals to a depth

Box 13.2 Measurement of Vertical Erosion

Successful modelling of the morphodynamics of erosion of the intertidal and nearshore platforms on cohesive or bedrock coasts requires measurements of profile change or rates of vertical lowering (Turowski and Cook, 2017). Because the rate of erosion may be < 10 mm per decade, standard surveying along profiles is not practical. Instead, this requires repeated high precision measurements be taken at points along the profile. This has been done almost exclusively through the use of some form of micro-erosion meter (MEM). The original design was used first for monitoring solution rates in limestone bedrock (High and Hanna, 1970) and was modified by Robinson (1976) for use on intertidal platforms. It has been used successfully for measuring erosion of intertidal rock platforms (Kirk, 1977; Robinson, 1977a, b; Stephenson and Kirk, 1996). The instrument consists of a triangular base that is constructed to fit onto three metal pins or studs that are drilled into the rock surface and the tops levelled. A high precision engineer's dial gauge is mounted vertically on a platform that is secured to the base. In making measurements the MEM is brought to the measurement point, fitted over the metal studs and measurements of the distance to the surface are made in the middle of the three sides of the triangle to a precision of < 0.1 mm. Improvements made by Trudgill *et al.* (1981) produced the traversing micro-erosion meter or TMEM, which could measure a much greater number of points within the triangle and Stephenson (1997) replaced the dial gauge with a digital one that could be linked to a field computer for recording.

A simpler version of this was used to measure erosion underwater on cohesive profiles (Askin and Davidson-Arnott, 1981). Because erosion rates on the cohesive profile are millimetre to centimetre per year, the high precision required for measurements in rock was unnecessary. The engineers gauge was replaced by a simple metal ruler with a millimetre scale and the rotating assembly was dispensed with in favour of simply lifting the instrument off the pins and rotating it 120 degrees (Figure 13.15A). Metal pins were hammered into the relatively soft till and the surfaces levelled to provide a horizontal surface on which to place this modified MEM. The instrument proved robust and easy to use underwater often under low visibility (Figure 13.15B).

Figure 13.15 The modified micro erosion meter developed for measurement erosion underwater on a cohesive coast. A. Photograph of the MEM taken on the foreshore sitting on the aluminium nails which are driven into the till at each measurement point. The rod used to measure distance to the bed slides in a vertical tube and, in contrast to the point measurement used for subaerial bedrock measurements, has a 2 cm diameter foot on it to prevent penetration in the softened upper layer of till. B. Installation of a MEM station underwater (Davidson-Arnott and Langham, 2000). The three pins that mark the station have been hammered into the till surface using a template. The diver at the top of the photograph is about to use the carpenter's level floating below him to check that the MEM is level in all directions and the diver at the left is waiting to make the initial set of measurements.

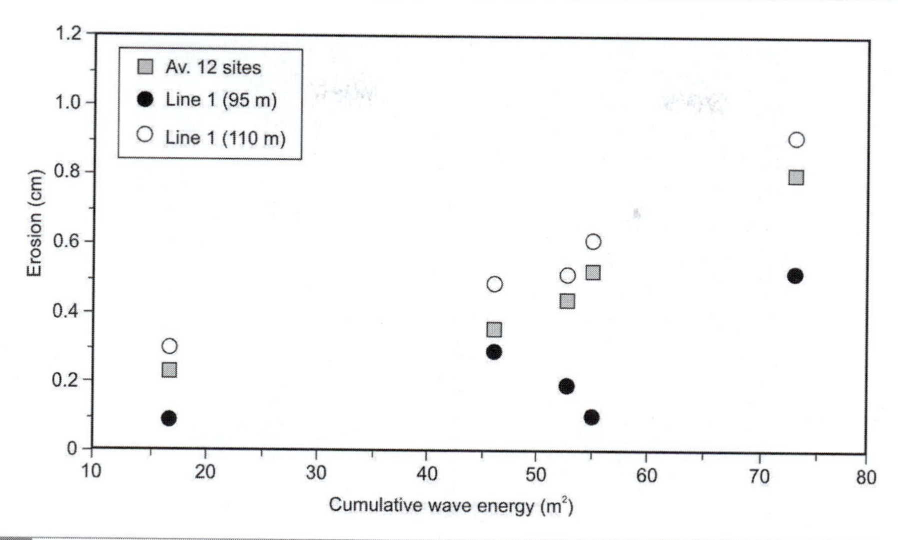

Figure 13.16 Relationship between wave energy and short-term erosion of the nearshore platform at St Catharines, Lake Ontario (Davidson-Arnott and Ollerhead, 1995).

of about 40 cm in cores taken at 70 m, 150 m and 220 m offshore (water depths of about 2, 4 and 5 m). These showed an inverse relationship between shear strength and moisture content with depth below the surface (Figure 3.18A,B). Below 20 cm the values for both were essentially constant with moisture content readings of about 17 per cent and shear strength of about 40–80 kPa, which are typical of the unweathered, overconsolidated till. Between 20 cm and the till surface shear strength decreased rapidly to < 15 kPa and moisture content increased to about 25 per cent. The reduction in strength reflects expansion of the till toward the unconfined surface and is likely aided by cyclic loading and unloading due to wave-induced pressure fluctuations (Davidson-Arnott and Askin, 1980; Davidson-Arnott and Langham, 2000). Weathering serves to reduce F_R and thus enhances the potential rate of erosion by abrasion where sand and gravel are available. In deeper water where surficial sand and gravel may not be present, softening of the till surface over weeks and months may permit erosion by waves large enough to affect the bed in depths of 5–10 m.

On marine coasts, soft cliffs will usually be fronted by an intertidal platform whose width increases with increasing tidal range and decreasing slope (Figure 13.12). This brings about some changes to the dominant erosion and weathering processes compared to the subaqueous section of the profile (Figure 13.11). In addition to the direct movement of sediment under shoaling waves, abrasion will be enhanced by the migration of the breaker zone over the intertidal profile with the rise and fall of tides and because of storm surge. Weathering due to pressure fluctuations may be less effective, but exposure to air permits the operation of weathering due to wetting and drying and potentially salt weathering. In temperate climates frost weathering may also be a significant factor (Jerwood *et al.*, 1990a, b; Foote *et al.*, 2006; Moses and Robinson, 2011). A summary of measurements from chalk cliff platforms in England (Moses and Robinson, 2011) shows that vertical erosion rates range about 1–12 mm a^{-1} with the majority of measurements being in the range of 2–5 mm a^{-1}. This is an order of magnitude lower than for cohesive bluff platforms but still much greater than rates on hard rock platforms.

13.4.3 Cliff Toe Erosion

Recession of the bluff is initiated by wave action at the base (Figure 13.4; 13.19) which leads to notch formation (Figure 13.19B) and an increase in the angle of the lower slope. Waves also act to

A.

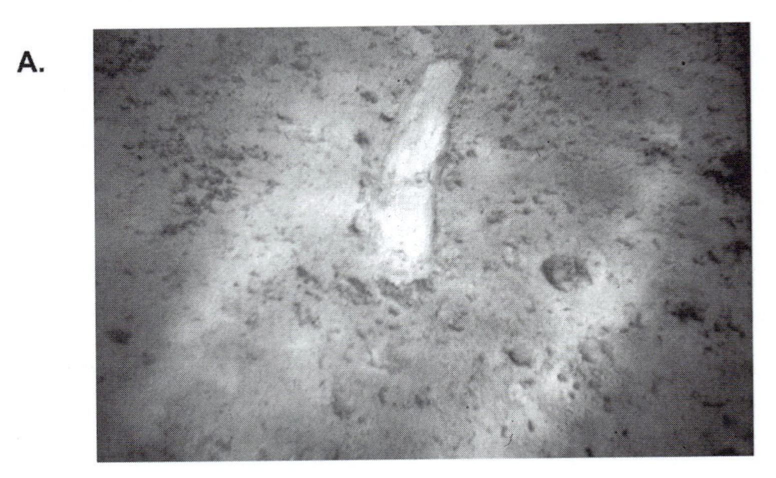

B.

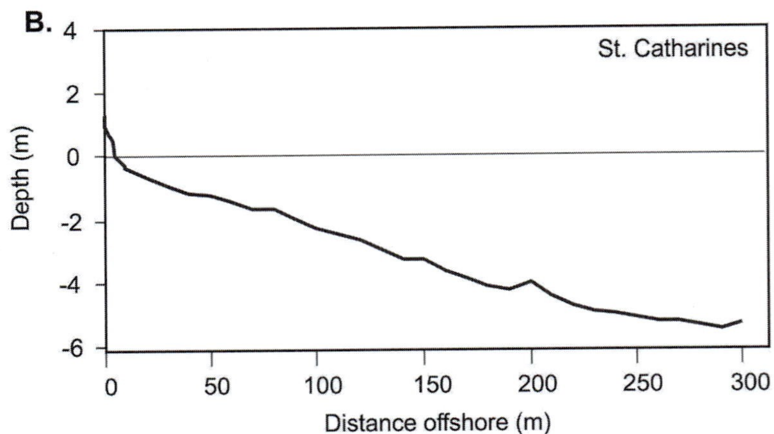

C.

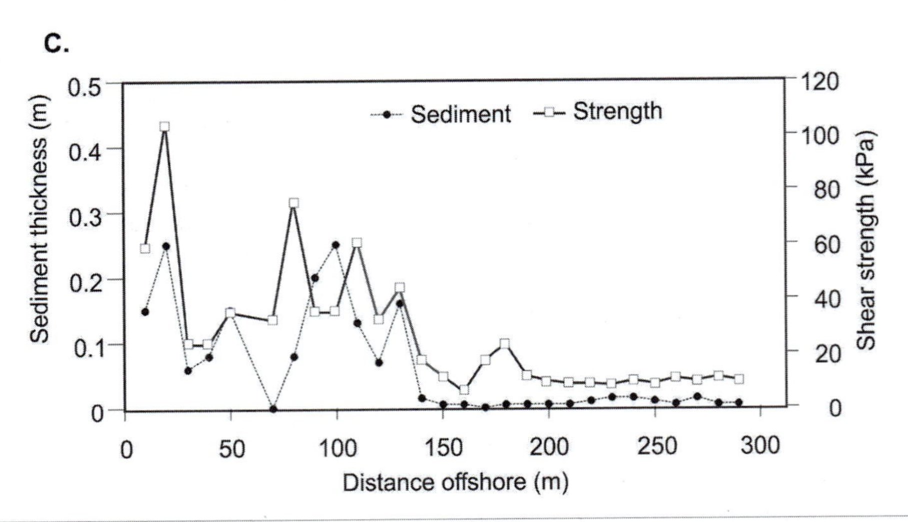

Figure 13.17 Characteristics of the nearshore profile at St Catharines, Lake Ontario: A. till surface underwater showing removal of a thin layer by dragging a finger along the surface; B. topographic profile; C. variations in sediment cover and shear strength of the till surface (Davidson-Arnott and Langham, 2000).

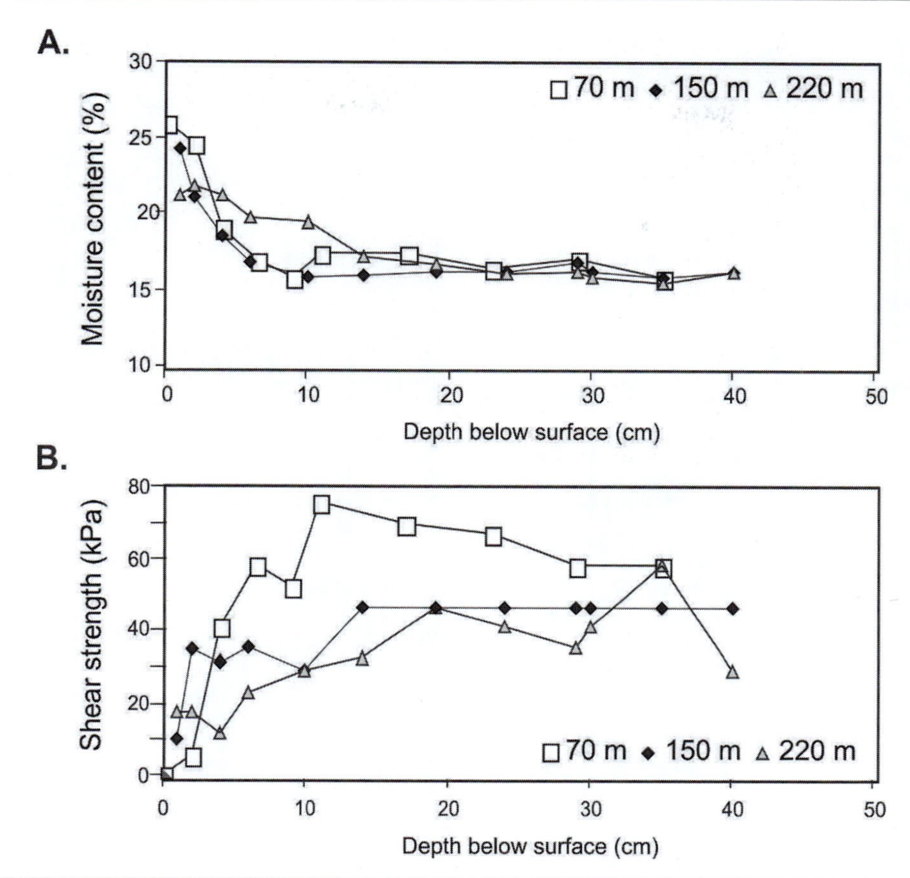

Figure 13.18 Variations in shear strength and moisture content with depth below the till surface in 5 cm diameter cores taken at St Catharines, Lake Ontario: A. variations in moisture content for cores taken at 70 m, 150 m and 220 m offshore; B. variations in shear strength for the same cores. The profile is shown in Figure 13.16B.

remove the products released from the cliff both by direct wave erosion and by mass wasting and fluid processes operating on the cliff face above the toe. In low-energy environments, cliff recession is slow and erosion of the base may result primarily from mass wasting aided by weathering. In these environments waves act primarily to remove the material offshore and alongshore, and subaerial erosion processes dominate (Wilcock *et al.*, 1998; Greenwood and Orford, 2008). Fine materials tend to be dispersed offshore into deep water, while sand and gravel remain on the beach and inner nearshore, and are transported alongshore. The slumped material may protect the toe of the bluff for a short while, but on soft rock coasts it has little strength and is generally removed very quickly (Figure 13.19B, C). On exposed coasts recession rates are generally $> 0.3\ \text{m a}^{-1}$ and direct wave attack at the toe occurs frequently – often several times a month. On marine coasts this will be determined by the tidal range as well as by seasonal and decadal scale variations in sea level and short-term fluctuations related to storm surge. In the North American Great Lakes, where cohesive cliffs comprise more than 1200 km of shoreline, there are no tides but erosion varies with seasonal and decadal scale fluctuations in lake level.

13.4.3.1 *Erosion Rate Measurement*

Wave erosion of the toe takes place by both hydraulic forces (impact, compression and

Figure 13.19 Bluff erosion. A. Waves breaking at the bluff toe, south shore Lake Erie (Amin and Davidson-Arnott, 1995) Note the rods sticking out of the face of the bluff. In this study they were later replaced with smaller pins. B. Notch formed in till from toe erosion. C. Erosion of till in a drumlin, Hirtles Beach, southwest Nova Scotia. The cobble beach has accumulated from erosion of this drumlin and one at the far end of this headland bay beach.

cavitation, shearing) and by abrasion, as detailed in Section 13.3 and Figure 13.11. This complexity makes prediction of F_W difficult. Likewise determination of F_R is also difficult (Budetta *et al.*, 2000; Sunamura, 2015). At this location on the profile the compressive strength is probably a better measure than shear strength but account also has to be taken of the presence of joints in rock and clay tills (Budetta *et al.*, 2000). Material here is also subject to wetting and drying, salt weathering, frost action in temperate climates and the effects of positive pore water pressure and water table fluctuations.

Toe erosion associated with individual storms or periods of a few weeks, has been measured in several studies (Carter and Guy, 1988; Amin and Davidson-Arnott, 1995; Greenwood and Orford, 2008). Carter and Guy (1988) measured the distance to the cliff face at a single height from a rope stretched alongshore and fixed to pipes hammered into the beach foreshore at five sites at the west end of Lake Erie – four sites in till or clay and one in weak shale. The other three studies were all in glacial till and used pins hammered into the bluff face at various heights above the beach (Figure 13.19A). The frequency of wave attack varied with site exposure (fetch length) and the number of erosional events depended on the magnitude of storms and storm surge. In the Great Lakes, the number of events was also closely related to long-term lake level fluctuations (Carter and Guy, 1988).

Amin and Davidson-Arnott measured toe erosion at four sites near Erie, Pennsylvania during 1986, at the peak of a high water phase when waves reached the bluff toe frequently. Erosion of the toe during individual storms produced recession ranging from < 1 cm to as much as 10 cm. (Figure 13.20A, B, C), but differences between measurement lines and between different elevations above the bed were much smaller when averaged over six months (Figure 13.21). Wave erosion reached as high as 4 m up the cliff face during intense storms (Amin and Davidson-Arnott, 1995), however, measured recession was not much greater for intense storms than for moderate ones. This can be attributed to two factors. First, the impact of larger waves is spread over a greater height and thus the actual intensity of attack is dissipated over a much larger area. When water level is quite high and waves break directly against the bluff face, there is a partitioning of the erosional forces such that the upper section is subject primarily to wave impact forces while the lower section is affected most by abrasion. Second, the unweathered till has a compressive strength of about 16 MPa and is thus relatively resistant to compressive forces. During the intervals between storms, a shallow layer a few centimetres thick weathers through wetting and drying or freezing and thawing, and this is manifested in an increase in the moisture content from about 8 per cent to more than 12 per cent. The reduced strength of this layer means that it can be removed readily by even moderate wave action (Figure 13.19A). As this weathered layer is stripped away, the underlying unweathered till is exposed and the erosion rate is slowed considerably. Thus, as is the case for erosion of the intertidal platform and the nearshore zone, weathering during inter-storm periods may be as significant a control on the recession associated with an individual storm event as the absolute wave energy.

13.4.3.2 *Prediction of Erosion*

While we cannot test a numerical model based on the physical processes thought to control toe erosion, because of the dearth of such measurements at the toe during a storm. However, several studies have used correlation and regression to explore the contribution of factors related to the wave energy reaching the cliff toe with measured recession rates (McGreal, 1978; Carter and Guy, 1988; Amin and Davidson-Arnott, 1997; Greenwood and Orford, 2008; Bezerra *et al.*, 2011). The results of these studies all show a significant correlation between some measure of wave energy at the bluff toe and the rate of recession of the toe. However, there is little agreement between studies on what is the most effective measure and the significance of, e.g., the protective beach or elevated water level.

13.4.3.3 *Role of Sediment Cover and Beach Width*

On many soft rock coasts abrasion by surficial sand and gravel is clearly a significant contributor to erosion of the nearshore profile, platform and toe erosion. The thickness of sand on the beach, platform and nearshore profile can vary considerably. The measured thickness shown in Figure 13.17C, for example, ranges from 0 to over 30 cm and on other cohesive coast profiles larger values are found as well as accumulations of lag boulder deposits which act to shield the underlying till surface. This raises the question of how does the efficacy of abrasion vary with the thickness of the surficial sediment cover?

Conceptually we can expect that abrasion under given wave conditions should increase rapidly as surficial sediments become available because the greater mass and larger number of contacts should speed up the process. However, there should be an optimal concentration (or bed thickness) beyond which erosion should decrease because the sediment layer begins to protect the underlying bed surface. Greater wave activity is required to mobilise the surficial cover to the extent that bed material can be abraded. Ultimately, with increasing thickness of surficial sediments all the sand movement takes place in the sand layer above the contact with the cohesive platform. Skafel and Bishop (1994) found that this occurred in the wave tank with a sand thickness greater than about 1 cm. In the field, with larger waves and the development of bedforms, a more realistic cut-off is likely to be on the order of 5–10 cm. On a larger scale, areas of accumulation, such as bars in the nearshore and berms on

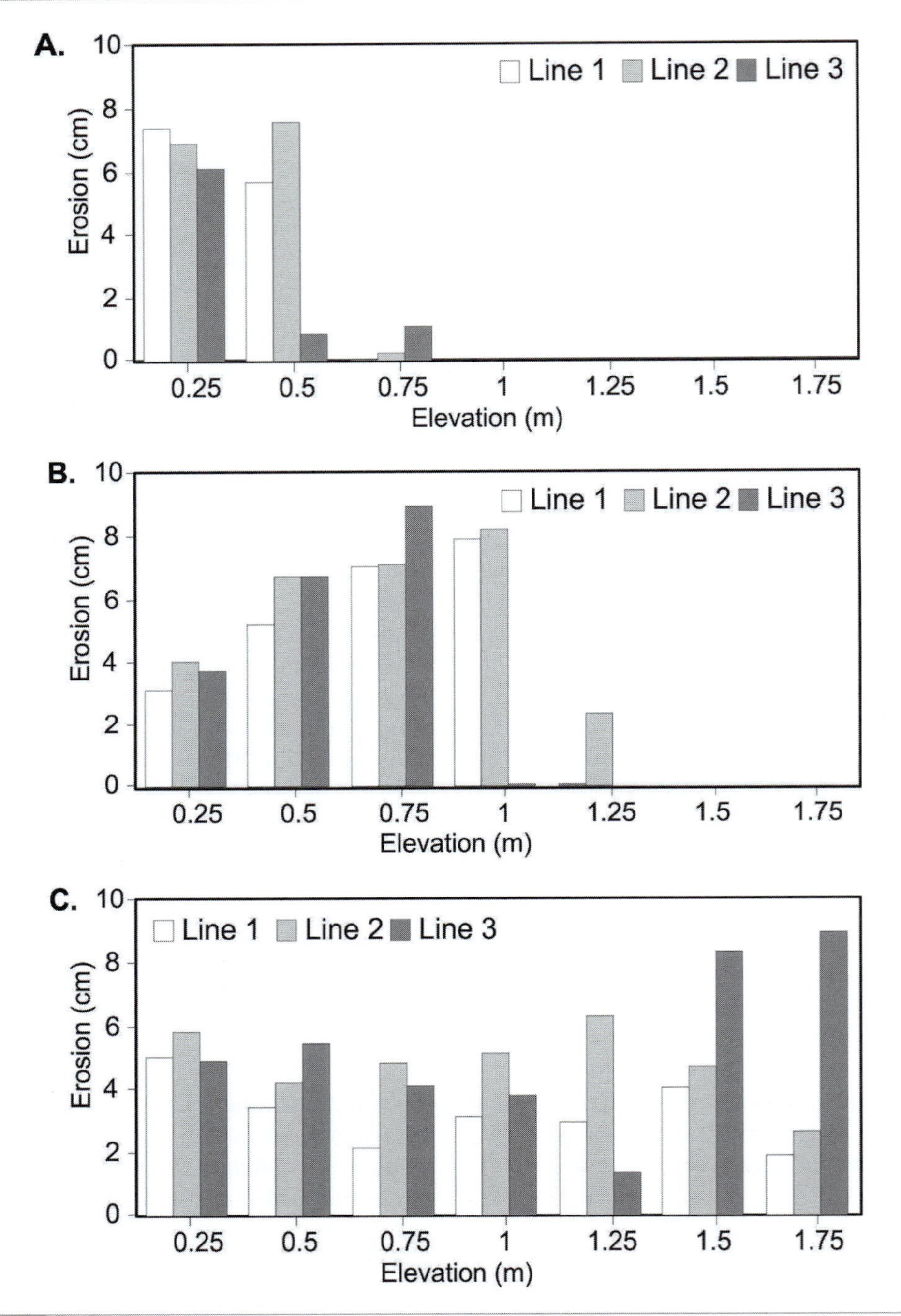

Figure 13.20 Variations in horizontal erosion during a single storm with height above the bluff/toe junction for a site (site 3) on the south shore of Lake Erie (based on data from Amin, 1991). The periods are chosen to illustrate the effects of increasing storm intensity (significant wave height and maximum water level) on the amount of erosion and height above the beach up to which erosion occurs: A. low magnitude storm 24 July–5 August 1986; B. moderate magnitude storm 17–28 August; and C. intense storm 2–16 November. Data are for three lines of erosion pins spaced 2 m apart and show that there is considerable variation for each individual storm and for closely spaced points on the bluff face.

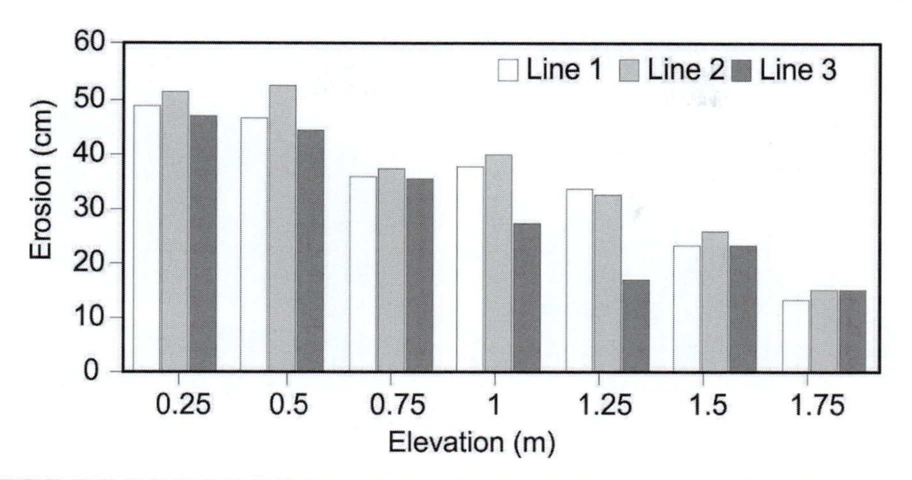

Figure 13.21 Variations in horizontal erosion of the bluff toe over the period 15 July–9 December 1986 at site 3 (based on data from Amin, 1991). Note that maximum erosion here occurs close to the toe and that the differences between the three lines are not significant when averaged over the whole period compared to individual storm events shown in Figure 13.19.

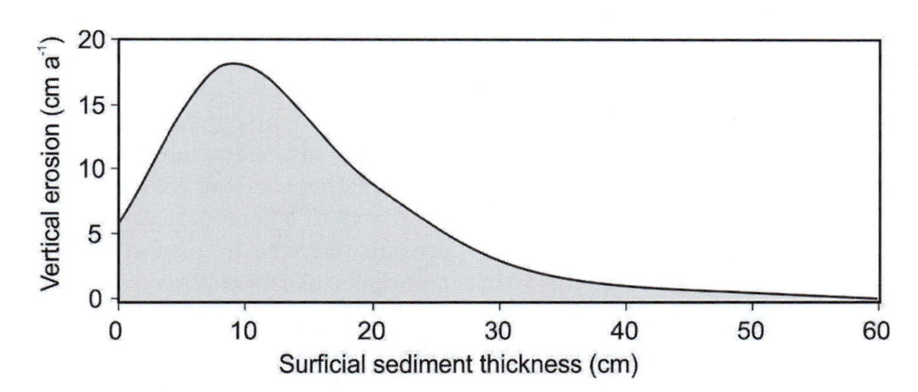

Figure 13.22 Sketch of the hypothetical annual erosion rate versus sand cover thickness for the beach and nearshore area on a cohesive coast. The scales on the *x* and *y* axes are intended to provide an indication of the likely values and are not based on actual data.

the beach, migrate onshore or offshore as wave conditions and water level varies and this movement exposes new areas of cohesive substrate (e.g., in nearshore troughs) while other areas become protected. Some direct evidence of the effect of the mobility of sediment close to shore on abrasion is provided by Davidson-Arnott and Ollerhead (1995) and this work showed that erosion took place during periods with large waves even in areas where the surficial sediment cover during events with small waves was 30 cm or more. As sediment cover increases there are fewer locations and times when exposed areas are subject to abrasion and thus the overall rate of vertical lowering of the platform is reduced. Averaged over time and along the profile, the relationship between abrasion rate and surficial sediment thickness looks something like that shown schematically in Figure 13.22.

Support for the protective role of the surficial cover, particularly sand, comes from observations of the association of wide beaches and slow recession rates (Lawrence and Davidson-Arnott, 1997; Sallenger *et al.*, 2002; Lee, 2008; Pye and Blott, 2015). An example of this can be seen in the oblique aerial photographs taken on 30 November 2015 of two sections along the east coast of Lake Huron south of Goderich (Figure 13.23). The

Figure 13.23 Oblique aerial photographs taken on 30 November 2015 of sections of the Lake Huron shoreline south of the town of Goderich: A. section with narrow beach and average recession rate around 1.0 m a^{-1}; B. section with a wide beach and average recession rate around 0.2 m a^{-1}. The bluffs here are developed in the massive St Josephs till which supports very steep cliffs about 40 m high. Recession of the bluff face takes place primarily through shallow slides, overland flow and gulley development. Note that in both areas, in addition to the gullies on the bluff slope, there are larger ones that drain the interior tableland.

photos are taken after more than 15 years of below-average lake level followed by a rapid rise by about 0.75 m over 18 months. At the northern section (Figure 13.23A) the long-term average recession rate is 1–1.2 m a^{-1} and erosion has been active even during the period of lower lake level. The beach is narrow or non-existent and wave action has resulted in undercutting of the 30 m high till bluff producing block falls controlled by the development of vertical tension cracks in the till. At the extreme left of the photo a large amount of material has been delivered to the beach through shallow slides and overland flows and is in the process of being eroded by wave action. The bluff is either bare or covered by grasses and shrubs because mass movement is so active. At the second section, about 3.5 km south (Figure 13.23B), the average recession rate is 0.2–0.25 m a^{-1}. Here the beach is much wider and the bluffs are well vegetated, mostly by cedars which grow rapidly and are able to withstand some instability. The toe of the bluff has been protected from erosion by the wide beach over the previous years of lower lake level and has only just been subject to wave action in a few places.

On long stretches of coast where erosion provides a substantial input of sand and gravel, collective movement of sediment in the form of longshore sandwaves (see Box 7.1) is characterised by exposure of the substrate at the updrift embayment and nearly complete protection at the downdrift embayment. Migration of the sandwave along the coast then gives rise to a temporal variation in enhanced erosion followed by protection of the cliff toe and upper platform. This has been documented in detail on the Holderness coast by Pringle (1981, 1985) who provided some of the earliest descriptions of longshore sandwaves (locally termed ords). The presence of ords is also brought out in the detailed mapping of Pye and Blott (2015; see Figure 13.12). Similar features are found along some cohesive shorelines on the east coast of Lake Huron and the north coast of Lake Erie.

13.4.4 Subaerial Bluff Processes

On soft rock cliffs recession rates are relatively large and this is associated with frequent wave erosion of the toe and steep cliff profiles. As a result of this, mass movement and water erosion also occurs frequently. All forms of mass wasting failures may occur on cohesive bluffs (Hutchinson, 1973, 1986; Wilcock *et al.*, 1998; Hampton and Griggs, 2004, Collins and Sitar, 2008). On cohesive bluffs with a large clay content and simple stratigraphy, most of the retreat results from shallow slides, slumps and mudflows especially where there is ongoing wave erosion at the toe (Figure 13.23). Because of the rapid recession and frequent activity there is generally very limited protective vegetation cover, which enhances the role of processes such as rain splash, overland flow and the development of shallow rills. On high cohesive bluffs (> 6m) the path length is long enough for rills and gullies to develop, and headward erosion of gullies may complicate crest line retreat and the delivery of sediment to the bluff toe (Figure 13.23B). On most of these coasts bluff recession by shallow slides and running water is usually so rapid that deep-seated failures are rare. High bluffs are more likely to have complex stratigraphy with interbedded units of sand and clays. Failure here is often associated with seepage along the junctions of sand units with underlying units of lower permeability, and complex failures may occur along these shear surfaces (Hutchinson, 1973; Quigley *et al.*, 1977).

While block failure of undercut cliffs can yield immediate transfers of talus material to the cliff toe (Figure 13.6), most sediment on cohesive bluffs is brought to the base of the cliff by mass wasting processes active on the cliff face. Thus, the timing of sediment delivery to the beach may not correlate well with actual erosion of the toe. Instead it may be better correlated with seasonal factors such as snow melt, rainfall events, groundwater table and vegetation cover (Pierre and Lahousse, 2006; Young *et al.*, 2009; Kuhn and Prüfer, 2014; Zoet and Rawling, 2017). Cohesive cliff materials with silt and clay are rapidly broken down and transported alongshore or offshore and deposited in deep water or in sheltered embayments. Sand and gravel is generally transported on the beach or shallow nearshore. On weak shales and sandstones, initial rockfalls may trigger a series of subsequent landslides that work their way to the top of the cliff (Figure 13.7), while the talus is removed from the base of the cliff by wave action (Young *et al.*, 2009; Collins and Sitar, 2008; Kuhn and Prüfer, 2014; Joyal *et al.*, 2015). Because of the greater structural integrity of these materials, removal of the talus may take months to a year or more with the result that the cliff toe is protected from direct attack for some time and subsequent failures are separated by periods with little activity on the slope.

13.4.5 Modelling Soft Cliff Erosion and Evolution

While there are now a number of 2D and 3D models that are used regularly to predict profile change under wave action on sandy coasts, only recently have numerical models been developed to predict profile and shoreline evolution or soft rock coasts. Nairn (1986) developed a simple numerical model to simulate erosion of a cohesive profile and the approach was extended in the COSMOS 2D numerical model (Nairn and Southgate, 1993). The latter included a module that permitted simulation of downcutting of the nearshore profile on a cohesive shoreline and this has been incorporated in various ways to assess the effects of a number of different scenarios, including water level variations, on the erosion of cohesive shorelines in the Great Lakes (e.g., Nairn *et al.*, 1999) and by extrapolation to predict the effects on bluff recession rates. However the results rely heavily on calibrating the model through the use of historical recession data.

Walkden and Hall (2005) developed a 3D simulation model SCAPE (soft-cliff and platform erosion) for the erosion of soft rock shores and it has been applied to evaluating the effects of sea level rise (Dickson *et al.*, 2007; Walkden and Dickson, 2008). Walkden and Hall (2011) have also expanded it to model sediment input to the littoral system from cliff and platform erosion and alongshore littoral transfers. The model simulates erosion primarily as a function of the relationship between breaking wave characteristics and rock resistance based on an equation from Kamphuis (1987):

$$E = \frac{H_b^{13/4} T^{\frac{3}{2}} \tan \alpha}{R_r} \tag{13.3}$$

where: E is the erosion rate and R_r is a function of the rock resistance and other factors that require calibration.

The inclusion of the slope angle is a bit problematic (as Walkden and Hall discuss in their paper). There does not seem to be any obvious physical rationale for erosion rate to depend on slope angle, especially for the small slope angles found in the intertidal or shallow nearshore zones. Rather, it reflects an empirical association between high recession rates and a steeper inner nearshore slope (concavity) on profiles in the Great Lakes. Walkden and Hall's model is also a bit restricted since it does not simulate erosion of the profile seaward of the breaker zone. Nevertheless the modelling efforts provide some useful insights into the complexity of factors controlling recession of soft rock coasts and therefore the difficulty of isolating the effects of sea level rise from other controls.

Trenhaile (2009) describes a simulation model that makes use of a version of the excess stress approach (Equation 13.2) to predict erosion by waves across the nearshore and intertidal zone and also accounts for the presence of cohesionless beach material. It explicitly recognises the link between nearshore profile erosion and horizontal bluff recession and seems to offer the potential to explore a number of scenarios related to varying lithology, thickness of beach sediments and sea level rise. The model was extended to examine rates of erosion in the upper Great Lakes under varying lake level scenarios (Geomorphic Solutions, 2011a, b).

13.5 Hard Rock Coasts

The processes and controls of erosion of hard rock coasts are the same as for soft rock shorelines and the cliff profile forms. The schematic models of cliff systems shown in Figures 13.2, 13.4 and 13.5 are generally applicable. Differences in the rate of toe erosion, the characteristics of subaerial mass wasting features and the form of the cliff and platform therefore can be attributed largely to structural and strength differences between cliffs in hard rock and those developed in cohesive sediments or weakly lithified shale and sandstone. Based on some measure of rock strength, we can expect to see a decrease in rock cliff recession rates, measured over periods of decades or longer, with increasing rock strength. There are only a few studies that have done this, but the data plotted in Figure 13.24 give an indication of how these rates may vary (Sunamura, 2004). These data show that weakly cemented sedimentary rocks such as sandstones and shales can have recession rates that are comparable to those of cohesive coasts, putting them into the category of soft coasts. Cliffs in andesite, basalt and dolerite, which have a high compressive strength and relatively few planes of weakness, have recession rates that are too small to measure over a few decades, as is the case for cliffs formed in massive limestone and crystalline rocks of shield areas.

13.5.1 Weathering, Toe Erosion and Notch Formation

On hard rock cliffs and platforms, direct erosion by fluid forces such as turbulence on the bed and breaking wave impacts on the cliff toe is focused on the exploitation of any form of pre-existing planes of weakness such as bedding planes, joints and fractures. In some areas large-scale patterns of varying lithology produce a highly irregular shoreline and rapid variations in character, reflecting varying response of differing lithology and structure (Figure 13.25A). Where sand and coarser material is available abrasion may be significant but it will be much less effective than on soft rock coasts. Erosion of soft rock coasts often contributes large amounts of material for abrasion, but on hard rock coasts the rate of supply is relatively small and much of the available material tends to get concentrated in pocket beaches with little remaining on rock platforms to erode the cliff toe. As a result of this, the significance of chemical and biological weathering often becomes more important in determining the rate of toe erosion and platform lowering.

The base of cliffs and the portions of platforms, isolated rocks and stacks are all subject to wetting and drying as a result of tides, waves and spray from breaking waves. They are therefore subject to conditions that tend to enhance the operation of various forms of both mechanical and chemical weathering. Stephenson *et al.*

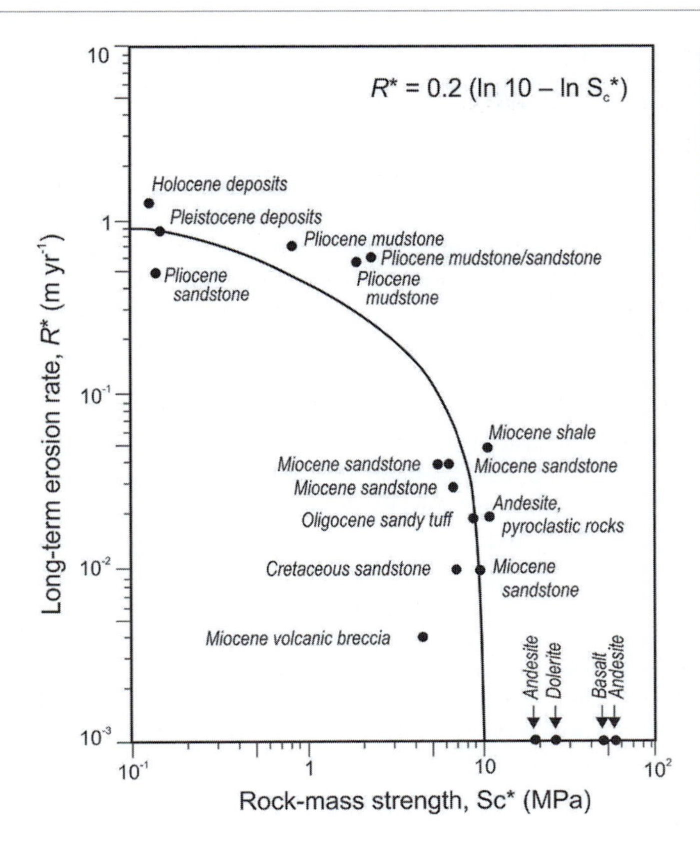

Figure 13.24 Relationship between long-term cliff recession rate and cliff-forming rock-mass strength of cliffs on open coasts of Japan. Rocks with recession rates that are too small to be measured are indicated by arrows (Sunamura, 2004).

(2013) identify salt weathering and wetting and drying as the most important mechanical weathering processes, with frost weathering becoming increasingly important at higher latitudes. While there are some qualitative observations from the field of the role of salt weathering, particularly in warm climates, little quantitative experimentation has been carried out on coasts. Following the pioneering work of Cooke and Smalley (1968) on salt weathering in desert environments there have been a number of laboratory experiments designed to determine the effects of different salt solutions and simple wetting and drying on samples taken from rock coasts (e.g., Jerwood *et al.*, 1990a, b; Kanyaya and Trenhaile, 2005; Trenhaile *et al.*, 2015). These suggest that both salt weathering, and wetting and drying can operate, but that their efficacy will vary depending on rock type, micro structure and on the environmental conditions. Thus, for example, Trenhaile *et al.* (2015) carried out laboratory experiments on rock samples from notches on a cliff coast in Baja California composed of andesitic lahar deposits. They found that wetting and drying with salt water was more effective at causing rock break down than simple wetting and drying in freshwater. However, because salt weathering requires alternate wetting and drying, it is difficult to isolate the relative effectiveness of salt weathering in these types of experiments. Similarly, while there are lots of observations that suggest that frost weathering should be an important control on erosion of high latitude coasts, direct measurements are few (Hansom *et al.*, 2014).

On irregular, and particularly on subhorizontal surfaces, pooling of water for many hours enhances the role of solution weathering. Solution weathering can produce small-scale features in a variety of rock types but it becomes particularly important in the break-down of carbonate cliffs and platforms (Trudgill, 1976; Moses, 2013). On headlands where water is

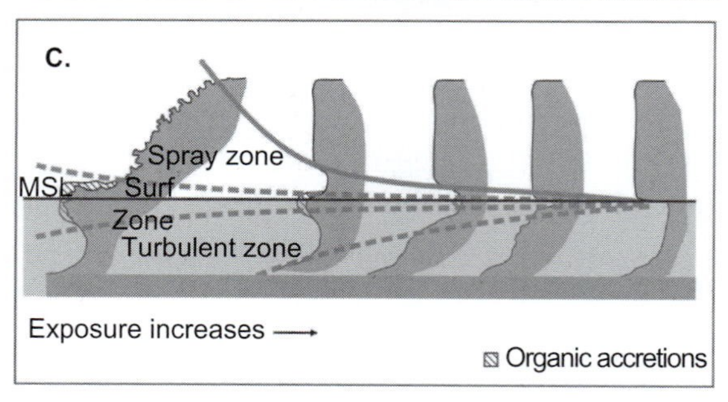

Figure 13.25 Features of erosion of bedrock coastal cliffs: A. structural and lithological control on bedrock erosion near Kaka Point, South Island, New Zealand; B. notch formation in carbonate rock, New Zealand with rough (karren) surface on the platform; C. forms of notch development in limestone, Curacao (after Focke, 1978).

always above the cliff toe, solution weathering can produce a characteristic notch with an overhang or visor (Figure 13.25B). Enlargement and deepening of the notch eventually leads to failure and results in large blocks falling onto the sea bed. Solution is enhanced in warm waters and temperatures of the tropics and the form may also vary as a response to wave exposure (Focke, 1978; see Figure 13.25C). On the cliff top, solution weathering may produce a rough surface (karren) though this may be at least partly inherited from beneath a thin soil layer which is removed near the cliff edge by runoff from rain and wave spray (Figure 13.25B).

On soft rock coasts biological processes that may lead to weathering and erosion tend to be much less significant than for hard rock coasts because of the greater effectiveness of hydraulic action by waves and abrasion and relatively rapid removal of surface material. However, on hard rock surfaces where erosion is very slow, the role of biological processes is much more significant both in direct erosion of the surface and in facilitating the action of other erosional processes (Naylor *et al.*, 2012). Direct erosion can take place at differing scales dependent on the nature of the organism carrying this out. Naylor *et al.* (2012) recognise three groups of boring organisms based on the scale of their action: (1) cyanobacteria (μm); (2) spionid polychaetes (mm); and (3) molluscs such as piddocks (cm) (Figure 13.26). In addition, grazers such as limpets and chitons will erode the rock surface as they track across it. Holes and tracks made by these organisms can be recognised on rocks, on platforms and in tidal notches. The difficulty is to determine how much erosion of the rock material they are responsible for on an annual basis and what portion of the total erosion this might represent. Based on the amount of chalk in faeces of captive limpets, Andrews and Williams (2000) estimated that erosion by limpets on a chalk platform in southeast England contributed on the order of 0.15 mm of downwearing or roughly 12 per cent of the total annual rate.

Where there is a well-defined cliff and platform, the toe of the cliff in hard rock is often undercut to produce a notch and visor form because the hard rock is able to maintain an overhang (Figure 13.25C; 13.27A, B). In some cases, especially on carbonate coasts, the notch is produced primarily by solution weathering and wave action serves to remove the eroded

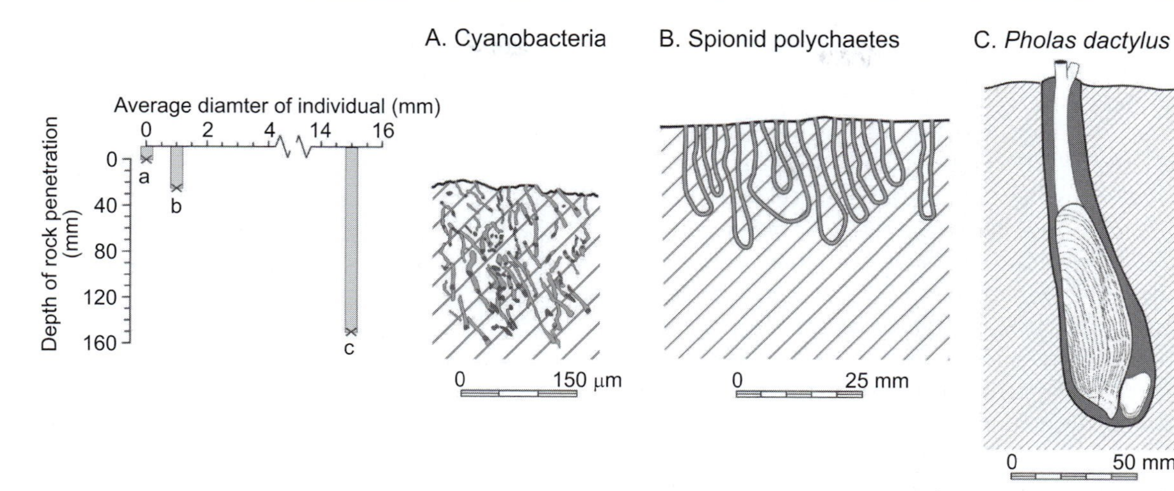

Figure 13.26 Bioerosion (boring) at three different scales: A. micro-scale (μm) cyanobacteria; B. macro-scale (mm) spionid polychaetes; and C. macro-scale (cm) *Pholas dactylus*. The inset graph demonstrates relationships between average width and depth of boreholes (after Naylor *et al.*, 2012, figure 4).

Figure 13.27 Features of erosion of hard coastal cliffs: A. notch and visor in resistant sandstone, Scotland; B. platform and notch giving rise to 'old hat' form in limestone boulders, east coast Barbados; C. erosion of weakly lithified sandstones, Antigua (note the recent rockfalls which act to protect the toe); D. cliff recession leading to the isolation of stacks – Twelve Apostles Marine Park, Victoria, Australia.

material in solution and to hasten collapse when fracturing occurs. Where gravel and cobbles are present on the coast they tend to be concentrated near the high tide line at the base of the cliff (Figure 13.27A) and so abrasion may be focussed at the base of the notch. While erosion due to abrasion may be relatively slow compared to a soft rock coast, it is enhanced by turbulence around the irregular rock surface and within the notch. It is also likely hastened by the presence of particles with a relatively large mass (pebbles and cobbles) compared to sand. The notch development in Paleogene/Neogene (Tertiary) limestone boulders at Bathsheba (Figure 13.27B), on the east coast of Barbados is likely produced by solution weathering, abrasion and possibly bioerosion by chitons and other organisms (Donovan and Harper, 2005).

Wave action may eventually lead to collapse of a portion of the roof of the notch of rock and this in turn may trigger failures farther up the cliff face. In some places massive failures occur through deep seated slides and the effect of this is to push the shoreline out for some distance. Such large-scale movements may lead to protection of the toe for years and even decades (Komar, 1998). However, most hard rock coasts are dominated by small failures, ranging from the detachment of individual blocks to topples, slumps and slides (Figure 13.27C). These may cover only a portion of the cliff toe and extend alongshore for metres or a few tens of metres (Andriani and Walsh, 2007). In contrast to soft rock coasts where material reaching the base of the cliff is generally broken down and removed quickly, on hard coasts slides and rock avalanches may bring large blocks of hard material which act to protect the toe from further erosion over periods of decades to centuries for very resistant material. Thus while cliff height is not generally correlated (inversely) with cliff recession rates for soft rock coasts, it can play an important role for hard rock coasts. Some evidence for this comes from the work of Mano and Suzuki (1999) who found that cliff recession on the Fukushima coast in Japan was correlated with a measure of wave energy and inversely related to rock strength and cliff height.

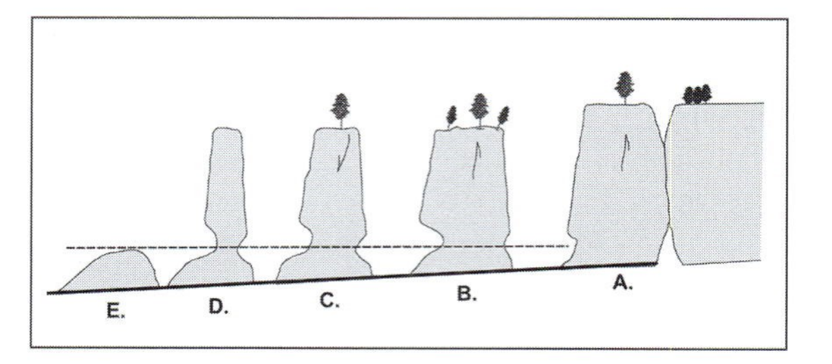

Figure 13.28 Schematic model for the evolution of stacks, Hopewell Rocks, Bay of Fundy, New Brunswick, Canada (Trenhaile *et al.*, 1998).

On a larger scale, joint patterns or alternating hard and soft rocks may result in the isolation of parts of the cliff producing stacks as the main shoreline cliff recedes (Figure 13.27D). Trenhaile *et al.* (1998) have documented the relationship of stacks at Hopewell Cape, Bay of Fundy, which are developed in arkosic sandstones and poorly sorted conglomerates of Carboniferous age, to the joint pattern in this location and produced a simple evolutionary model that probably can be extended to stacks in many other areas (Figure 13.28). Limber and Murray (2015) explored a number of the possible controls on stack formation and evolution using a computer simulation model.

13.5.2 Shore Platforms

There is a rich literature on shore platforms and no shortage of recent research (Bartrum, 1916; Trenhaile, 1972; Stephenson *et al.*, 2013; Kennedy *et al.*, 2017). There is also no shortage of confusion over definition of the term and some controversy over the perceived importance of weathering versus marine processes in platform development (e.g., Stephenson, 2000). The task of determining the relative importance of key processes is not made easier by the relatively slow rate of erosion compared to soft rock coasts, by the effects of structural controls in some areas (Kennedy, 2010), and by the role of inheritance of features from past interglacial sea levels (Brooke *et al.*, 1994). In this section we can make use of insights from the study of soft coast evolution to shed some light on the mechanisms that may be at work on hard rock coasts and which control the evolution of the profile following recession of the cliff toe.

As we saw in Section 13.2.1, the shore platform is defined as the erosional surface in bedrock or other resistant material that extends from the cliff toe to the seaward limit of the intertidal zone. The platform is considered to be formed by recession of the coastal cliff toe and subsequently modified by weathering and wave action. In terms of morphodynamics, it is connected on the landward side to the subaerial processes acting on the cliff face and on the seaward side to subaqueous processes operating on the nearshore slope. On many soft rock coasts formation of the platform due to cliff recession may have begun at a position that is now at the outer edge of the nearshore profile. However, because of the slow rate of cliff recession in hard rock it may not lie much beyond the low tide line and in some cases cliff retreat will only be a few metres at the present sea level.

In contrast to the steep face of the subaerial cliff, the platform slope is relatively gentle and parts of it may be nearly horizontal – hence the term platform. Shore platforms are usually portrayed as consisting of two types on the basis of the profile normal to the shoreline (Figure 13.2 C, D), and much of the attention has been focused on Type B platforms that may be marked by a scarp or low cliff separating the nearly horizontal platform from the nearshore profile. The platform may have varying amounts of surficial sediments on it, but the thickness should not be great enough to prevent exposure of the bedrock at least intermittently.

As we noted in Section 13.2, Type A platforms reflect conditions where vertical lowering of the platform in the intertidal area is similar to that in the inner nearshore and keeps pace with horizontal retreat of the cliff toe. The cohesive profiles described in Section 13.4 fall into this category and the conceptual model of a dynamic equilibrium between horizontal retreat of the cliff toe and vertical lowering of the platform should be applicable to Type A profiles in both weak and hard bedrock. Where cliff retreat rates are comparatively rapid there has been sufficient time for an equilibrium to be established and we can expect platform profiles in areas with relatively weak bedrock such as the Bay of Fundy (e.g., Trenhaile *et al.*, 1998) and the California coast (Griggs and Patsch, 2004) to fit this model. There is evidence that only Type A platforms are found on high macrotidal and hypertidal coasts and thus that Type B platforms are restricted to coasts with a tidal range of < 3 m (Stephenson *et al.*, 2013). In addition, the existence of a substantial Type B platform generally requires relatively hard bedrock (Figure 13.29).

Figure 13.29 Photographs of the intertidal components of Type B shore platforms: A. a small rock platform on a headland, St. Lucia, West Indies; B. rock platform Catlins, New Zealand. The boulders on the surface suggest that erosion is dominated by the quarrying of individual blocks.

Profile and MEM measurements have been carried out on shore platforms on the Kaikoura Peninsula, South Island, New Zealand beginning in 1973 (Kirk, 1977). Kirk, and later Stephenson and Kirk, have updated profile and MEM measurements for several decades (e.g., Stephenson and Kirk, 1996, Stephenson *et al.* 2010) so that it now has one of the longest record of erosion measurements and surveyed profiles. Profile KM2 (shown before and after the 2016 earthquake in Figure 3.4) has the classic Type B raised rim at the seaward margin which is succeeded by a low tide cliff which ends at the nearshore profile slope. However, recent work has pointed to the existence of a wide range of profile morphology here (Stephenson *et al.*, 2013). The raised rim is found only in a few places, and many of the profiles have a sloping ramp rather than a cliff at the seaward margin.). Indeed Stephenson and Kirk (2000a, b) describe three of the seven profiles measured at Kaikoura as being Type B and four as being Type A.

Recent measurements using LiDAR by Kennedy (2016) of an 85 km long stretch of the Victoria, Australia coast provide a good picture of the range of both intertidal platform morphology and also of the subtidal profile and it provides evidence for viewing Type A and B profiles as end members of a considerable range of profile form. Based on an analysis of 47 complete profiles Kennedy found that a simple vertical cliff is uncommon at the seaward margin of the intertidal platform and he recognised four groups of characteristic profile shapes (Figure 13.30; Kennedy, 2015).

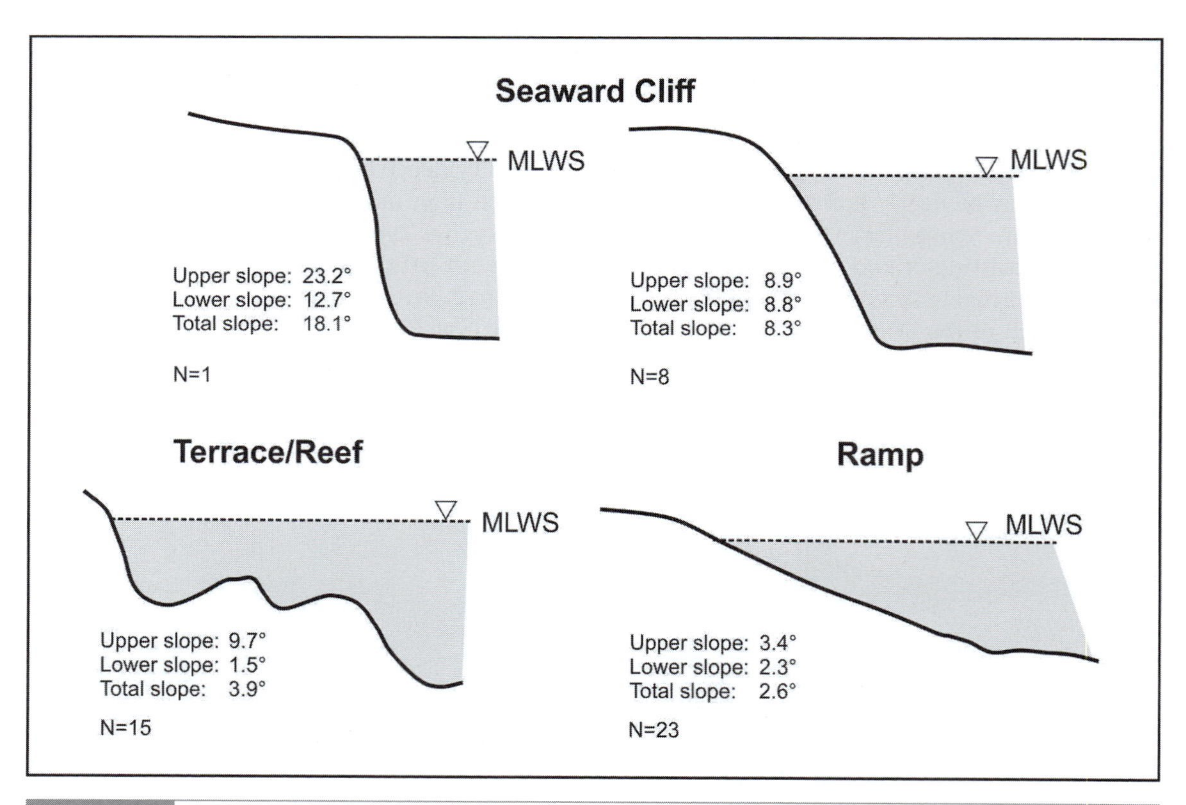

Figure 13.30 A morphological model of the three types of subtidal shore platform morphology and their average profile slope based on an analysis of 47 profiles along an 85 km section of the Victoria coast, Australia (Kennedy, 2015). The seaward cliff (defined by clusters 1 and 2 is most similar to the traditional classifications of microtidal shore platforms. Note that Cluster 1 contains only 1 profile.

Kennedy (2015) has argued that on a rock coast the seaward limit of the platform should be defined by the depth at which wave action first begins to erode the profile – that is, at the wave base. In part this can be defended based on the fact that the characteristics, and therefore the erosive ability, of waves reaching the horizontal (intertidal) platform will be controlled by the deepwater characteristics and the nature of the nearshore profile over which wave transformation takes place. Additionally, based on the material presented for soft rock coasts in Section 13.4.2, erosion of the subaqueous profile over time will lead to greater wave energy reaching the seaward edge of the intertidal platform. While there is a good argument for thinking in terms of the whole profile from the seaward limit of wave erosion to the top of the beach or cliff toe, processes in the intertidal zone (the beach of sandy shores) are sufficiently different from those in the subaqueous zone that we can make a distinction between the two components.

Since the time available for vertical lowering of the platform increases away from the cliff toe, the presence of the nearly horizontal intertidal portion of the platform associated with the Type B side of the spectrum must therefore reflect relatively slow rates of vertical erosion in that zone compared to similar portions of Type A platforms. Measurements of vertical lowering of the intertidal areas of Type B platforms generally range between 0.5 and 1.5 mm a^{-1} (e.g., table 1 in Stephenson, 2000; Stephenson *et al.*, 2010). These are at least an order of magnitude less than those measured on soft coasts. Using the terminology of Sunamura, is vertical lowering here slow because F_W is small, or is it slow because F_R is relatively large? At the moment there seems to be no clear answer to this and especially no explanation of the causal relation. There is now some progress being made in quantifying the spatial variability in rock profile form and there are some more measurements of vertical erosion rates. There is also some support for the key role that is likely played by abrasion and by the availability of gravel and cobbles to carry this out. The pioneering work of Robinson (1977a, b) on the northeast Yorkshire coast showed that areas where there was a substantial accumulation of sand and pebbles in front of the cliff had high rates of vertical erosion producing a ramp with a slope of 2.5°– 15°. In contrast, areas where there was no surficial sediment over the shale bedrock had relatively low rates of vertical erosion producing a 'plane' with a slope of about 1°. Several other studies have associated larger lowering rates with the presence of sediment but there are no data to link think the work done to the magnitude and frequency of wave processes over decades. A recent study, by Cullen and Burke, 2018, points a way towards identifying the nature of clast abrasion during individual storm events and may provide a means of quantifying vertical lowering over a significant area.

Further reading

Kennedy, D.A., Stephenson, W.J. and Naylor, L.A. (eds.), 2014. *Rock Coast Geomorphology: A Global Synthesis.* Memoir 40. London: Geological Society of London Books, 286 pp.

This provides an up-to-date review of our understanding of erosional processes and evolution on rock coasts.

Hampton, M.A. and Griggs, G.B. (eds.), 2004. Formation, Evolution and Stability of coastal Cliffs-Status and Trends. United States Geological Survey, Professional Paper 1693, 123 pp.

This contains a good review of cliff processes as well as regional reports from the US west coast and Great Lakes. It is available online as a PDF.

Sunamura, T. 1992. *Geomorphology of Rocky Coasts.* Chichester, UK: John Wiley & Sons, 302 pp.

Although getting a bit old now, this is a classic text on the study of rock coasts.

References

Adams, P.N., Storlazzi, C.D. and Anderson, R.S. 2005. Nearshore wave-induced cyclical flexing of sea cliffs. *Journal of Geophysical Research*, **110**, F02002, 1–19.

Amin, S.M.N. 1991. Bluff Toe Erosion: Magnitude, Processes and Factors Along a Section of Lake Erie South Shore. PhD Thesis, Kent State University, 235 pp.

Amin, S.M.N. and Davidson-Arnott, R.G.D. 1995. Toe erosion of glacial till bluffs, Lake Erie south shore. *Canadian Journal of Earth Sciences*, **32**, 829–37.

Amin, S.M.N. and Davidson-Arnott, R.G.D. 1997. A statistical analysis of the controls on shoreline erosion rates, Lake Ontario. *Journal of Coastal Research*, **13**, 1093–101.

Andrews, C. and Williams, R.B.G. 2000. Limpet erosion of chalk shore platforms in southeast England. *Earth Surface Processes and Landforms*, **25**, 1371–81.

Andriani, G.F. and Walsh, N. 2007. Rocky coast geomorphology and erosional processes: a case study along the Murgia coastline south of Bari, Apulia – SE Italy. *Geomorphology*, **87**, 224–38.

Askin, R.W. and Davidson-Arnott, R.G.D. 1981. Micro-erosion meter modified for use underwater. *Marine Geology*, **40**, M45–M48.

Aydin, A. and Basu, A. 2005. The Schmidt hammer in rock material characterisation. *Engineering Geology*, **81**, 1–14.

Bartrum, J.A., 1916. High-water rock-platforms: a phase of shore-line erosion. *Transactions and Proceedings of the New Zealand Institute*, **48**, 132–34.

Bernatchez, P. and Dubois, J-M. 2008. Seasonal quantification of coastal processes and cliff erosion on fine sediment shorelines in a cold temperate climate, north shore of the St Lawrence Maritime Estuary, Québec. *Journal of Coastal Research*, **24**, 169–80.

Bernatchez, P., Jolivet, Y. and Corriveau, M. 2011. Development of an automated method for continuous detection and quantification of cliff erosion events. *Earth Surface Processes and Landforms*, **36**, 347–62.

Bezerra, M.M., Moura, D., Ferreira, O. and Taborda, R. 2011. Influence of wave action and lithology on sea cliff mass movements in Central Algarve coast, Portugal. *Journal of Coastal Research*, **27**, 162–71.

Brew, D. 2004. Understanding and Predicting Beach Morphological Change Processes Associated with the Erosion of Cohesive Foreshores Scoping Report. London: Technical Report FD1915DEFRA/Environment Agency,, 73pp.

Brooke, B.P., Young, R.W., Bryant, E.A., Murray-Wallace, C.V. and Price, D.M. 1994. A Pleistocene origin for shore platforms along the northern Illawarra coast, New South Wales. *Australian Geographer*, **25**, 178–85.

Brooks, S.M. and Spencer, T. 2010. Temporal and spatial variations in recession rates and sediment release from soft rock cliffs, Suffolk coast, UK. *Geomorphology*, **124**, 26–41.

Brooks, S.M. and Spencer, T. 2014. Importance of decadal scale variability in shoreline response: examples from soft rock cliffs, East Anglian coast, UK. *Journal of Coastal Conservation*, 18, 581–93.

Buckler, W.R. and Winters, H.A. 1983. Lake Michigan bluff recession. *Annals of the Association of American Geographers*, **73**, 89–110.

Budetta, P., Galieta, G. and Santo, A. 2000. A methodology for the study of the relation between coastal cliff erosion and the mechanical strength of soils and rock masses. *Engineering Geology*, **56**, 243–56.

Carter, C.H. and Guy, D.E. Jr. 1988. Coastal erosion: processes, timing and magnitudes at the bluff toe. *Marine Geology*, **84**, 1–17.

Castedo, R., Fernández, M., Trenhaile, A. and Paredes, C. 2013. Modelling cyclic recession of cohesive clay coasts: Effects of wave erosion and bluff stability. *Marine Geology*, **335**, 162–76.

Collins, B.D. and Sitar, N. 2008. Processes of coastal bluff erosion in weakly lithified sands, Pacifica, California, USA. *Geomorphology*, **97**, 483–501.

Cooke, R.U. and Smalley, I.J. 1968. Salt weathering in deserts. *Nature*, **220**, 1226–7.

Cullen, N.D. and Bourke, M.C. 2018. Clast abrasion of a rock shore platform on the Atlantic coast of Ireland. *Earth Surface Processes and Landforms*, **43**, 2627–41.

Davidson-Arnott, R.G.D. 1986. Rates of erosion of till in the nearshore zone. *Earth Surface Processes and Landforms*, **11**, 53–8.

Davidson-Arnott, R.G.D. 1990. The effects of water level fluctuations on coastal erosion in the Great Lakes. *Ontario Geographer*, **10**, 12–25.

Davidson-Arnott, R.G.D. and Askin, R.W. 1980. Factors controlling erosion of the nearshore profile in overconsolidated till, Grimsby, Lake Ontario. *Proceedings*

Canadian Coastal Conference. Ottawa: National Research Council of Canada, pp. 185–99.

Davidson-Arnott, R.G.D. and Ollerhead, J. 1995. Nearshore erosion on a cohesive shoreline. *Marine Geology*, **122**, 349–65.

Davidson-Arnott, R.G.D. and Langham, D.R.J. 2000. The effects of softening on nearshore erosion on a cohesive shoreline. *Marine Geology*, **166**, 145–62.

Davidson-Arnott, R.G.D., van Proosdij, D., Ollerhead, J. and Langham, D. 1999. Rates of erosion of till in the nearshore zone on Lakes Huron and Ontario. *Proceedings Canadian Coastal Conference*, Ottawa: Canadian Coastal Science and Engineering Association, pp. 627–36.

deVries, J.W. 1992. Field measurements of the erosion of cohesive sediments. *Journal of Coastal Research*, **8**, 312–18.

Dickson, M.E., Walkden, M.J.A., and Hall, J.W. 2007. Systemic impacts of climate change on an eroding coastal region over the twenty-first century. *Climatic Change*, **84**, 141–66.

Donovan, S.K. and Harper, D.A.T. 2005. The geology of Barbados: a field guide. *Caribbean Journal of Earth Science*, **38**, 21–33.

Earlie, C.S., Masselink, G., Russell, P.E. and Shail, R.K. 2015. Application of airborne LiDAR to investigate rates of recession in rocky coast environments. *Journal of Coastal Conservation*, **19**, 831–45.

Föcke, J.W. 1978. Limestone cliff morphology on Curaçao (Netherlands Antilles) with special attention to the origin of notches and vermetid/coralline algal surf benches ('cornices', 'trottoirs'). *Zeitschrift für Geomorphologie*, **22**, 329–49.

Foote, Y., Plessis, E., Robinson, D.A., Hénaff, A. and Costa, S. 2006. Rates and patterns of downwearing of chalk shore platforms of the Channel: comparisons between France and England. *Zeitschrift für Geomorphologie*, **144**, 93–115.

Geomorphic Solutions, 2011a. Modelling bluff erosion in the upper Great Lakes: model description. International Joint Commission, Upper Great Lakes Study, Report No. CZN08–R10, 9 pp.

Geomorphic Solutions, 2011b. Cohesive shoreline erosion modelling results. International Joint Commission, Upper Great Lakes Study, Report No. CZN08–R09, 26 pp.

Greenwood, R.O. and Orford, J.D. 2008. Temporal patterns and processes of retreat of drumlin coastal cliffs: Strangford Lough, Northern Ireland. *Geomorphology*, **94**, 153–69.

Griggs, G.B. and Patsch, K.B. 2004. California's coastal cliffs and bluffs. In Hampton, M.A. and Griggs, G.B. (eds.), *Formation, Evolution and Stability of coastal Cliffs-Status and Trends*. United States Geological Survey, Professional Paper 1693, pp. 53-64.

Hall, A.M., Hansom, J.D., Williams, D.M. and Jarvis, J. 2006. Distribution, geomorphology and lithofacies of cliff-top storm deposits: examples from high energy coasts of Scotland and Ireland. *Marine Geology*, **232**, 131–55.

Hampton, M.A. and Griggs, G.B. (eds.), 2004. *Formation, Evolution and Stability of coastal Cliffs-Status and Trends*. United States Geological Survey, Professional Paper 1693, 123 pp.

Hansom,J.D., Forbes, D.L. and Etienne, S. 2014. The rock coasts of polar and sub-polar regions. In Kennedy, D.A., Stephenson, W.J. and Naylor, L.A. (eds.), *Rock Coast Geomorphology: A Global Synthesis*, vol. 40, Geological Society, London, Memoirs, pp. 263–81.

Healy, T.R. and Wefer, G. 1980. The efficacy of submarine erosion versus cliff retreat as a supplier of marine sediment in the Kieler Bucht, Western Baltic. *Meyniana*, **32**, 89–96.

Healy, T.R., Sneyd, A.D. and Werner, F. 1987. First approximation sea-level dependent mathematical model for volume eroded and submarine profile development in a semi-enclosed sea: Kiel Bay, Western Baltic. *Mathematical Geology*, **19**, 41–50.

High, C.J. and Hanna, F.K. 1970. A method for direct measurement of erosion on rock surfaces. *British Geomorphology Research Group, Technical Bulletin* **5**, 1–25.

Hutchinson, J.N. 1973. The response of London Clay cliffs to differing rates of toe erosion. *Geologia Applicata e Idrogeologia*, **8**, 221–39.

Hutchinson, J.N. 1986. Cliffs and shores in cohesive materials: geotechnical and engineering aspects. *Proceedings Symposium on Cohesive Shores*, Ottawa: National Research Council of Canada, pp. 1–44.

Irrgang, A.M., Lantuit, H., Manson, G.K. *et al.* 2018. Variability in rates of coastal change along the Yukon coast, 1951 to 2015. *Journal of Geophysical Research: Earth Surface*, **123**, 779–800.

Jerwood, L.C., Robinson, D.A. and Williams, R.B.G. 1990a. Experimental frost and salt weathering of chalk I. *Earth Surface Processes and Landforms*, **15**(7), 611–24.

Jerwood, L.C., Robinson, D.A. and Williams, R.B.G. 1990b. Experimental frost and salt weathering of chalk II. *Earth Surface Processes and Landforms*, 15(8), 699–708.

Joyal, G., Lajeunesse, P., Morissette, A. and Bernatchez, P. 2016. Influence of lithostratigraphy on the retreat of an unconsolidated sedimentary coastal cliff (St Lawrence estuary, eastern Canada). *Earth Surface Processes and Landforms*, **41**, 1055–72.

Kamphuis, J.W. 1987. Recession rate of glacial till bluffs. *Journal of Waterways, Port, Coastal and Ocean Engineering*, **113**, 60–73.

Kamphuis J.W. and Hall, K.R. 1983. Cohesive material erosion by unidirectional current. *Journal Hydraulics Division, ASCE*, **109**, 49–61.

Kanyaya, J.I. and Trenhaile, A.S. 2005. Tidal wetting and drying on shore platforms: an experimental assessment. *Geomorphology*, **70**, 129–46.

Kennedy, D.M. 2010. Geological control on the morphology of estuarine shore platforms: Middle Harbour, Sydney, Australia. *Geomorphology*, **114**, 71–7.

Kennedy, D.M. 2015. Where is the seaward edge? A review and definition of shore platform morphology. *Earth Science Reviews*, **147**, 99–108.

Kennedy, D.M. 2016. The subtidal morphology of microtidal shore platforms and its implication for wave dynamics on rocky coasts. *Geomorphology*, **268**, 146–58.

Kennedy, D.M., Coombes, M.A and Mottershead, D.N. 2017. The temporal and spatial scales of rocky coast geomorphology: a commentary. *Earth Surface Processes and Landforms*, **42**, 1597–600.

Kirk, R.M. 1977. Rates and forms of erosion on intertidal platforms at Kaikoura Peninsula, South Island, New Zealand. *New Zealand Journal of Geology and Geophysics*, **20**, 571–613.

Klemas,V.V. 2015. Coastal and environmental remote sensing from unmanned aerial vehicles: An overview. *Journal of Coastal Research*, **31**, 1260–7.

Komar, P.D. 1998. Wave erosion of a massive artificial coastal landslide. *Earth Surface Processes and Landforms*, **23**, 415–28.

Komar, P.D. and Shih, S.M. 1993. Cliff erosion along the Oregon coast: a tectonic-sea level imprint plus local controls by beach processes. *Journal of Coastal Research*. **9**, 746–65.

Kuhn, D. and Prüfer, S. 2014. Coastal cliff monitoring and analysis of mass wasting processes with the application of terrestrial laser scanning: a case study of Rügen, Germany. *Geomorphology*, **213**, 153–65.

Lantuit, H. and Pollard, W.H. 2008. Fifty years of coastal erosion and retrogressive thaw slump activity on Herschel Island, southern Beaufort Sea, Yukon Territory, Canada. *Geomorphology*, **95**, 84–102.

Lawrence, P.L. and Davidson-Arnott, R.G.D. 1997. Alongshore wave energy and sediment transport on southeastern Lake Huron. *Journal of Coastal Research*, **13**, 1004–15.

Lee, E.M. 2008. Coastal cliff behaviour: observations on the relationship between beach levels and recession rates. *Geomorphology*, **101**, 558–71.

Lim, M., Rosser, N.J., Petley, D.N. and Keen, M. 2011. Quantifying the controls and influence of tide and wave impacts on coastal rock cliff erosion. *Journal of Coastal Research*, **27**, 46–56.

Limber, P.W. and Murray, A.B. 2015. Sea stack formation and the role of abrasion on beach-mantled headlands. *Earth Surface Processes and Landforms*, **40**, 559–68.

Mano, A. and Suzuki, S. 1999. Erosion characteristics of sea cliff on the Fukushima coast. *Coastal Engineering Journal*, **41**, 43–63.

McGreal, W.S. 1978. Marine erosion of glacial sediments from a low-energy cliffline environment near Kilkeel, Northern Ireland. *Marine Geology*, **32**, 89–103.

Moon, V.G. and Healey, T.R. 1994. Mechanisms of coastal cliff retreat and hazard delineation in soft flysch deposits. *Journal of Coastal Research*, 10, 663–80.

Moses, C.A., 2013. Tropical rock coasts Cliff, notch and platform erosion dynamics. *Progress in Physical Geography*, **37**, 206–26.

Moses C. and Robinson, D. 2011. Chalk coast dynamics: implications for understanding rock coast evolution. *Earth-Science Reviews*, **109**, 63–73.

Moses, C.A., Robinson, D.A., Williams, R.B.G. and Marques, F.M.S.F. 2006. Predicting rates of shore platform downwearing from rock geotechnical properties and laboratory simulation of weathering and erosion processes. *Zeitschrift für Geomorphologie*, **144**, 19–37.

Moses, C., Robinson, D. and Barlow J. 2014. Methods for measuring rock surface weathering and erosion: a critical review. *Earth-Science Reviews*, **135**, 141–61.

Nairn, R.B. 1986. Physical modelling of erosion on cohesive profiles. *Proceedings Symposium on Cohesive Shores* Ottawa: National Research Council of Canada, pp. 210–25.

Nairn, R.B. and Southgate, H.N. 1993. Deterministic profile modelling of nearshore processes. Part 2. Sediment transport and beach profile development. *Coastal Engineering*, **19**, 57–96.

Nairn, R.B., Davis, J.E. and Thieme, S.J. 1999. A GIS-linked flood and erosion prediction system for Lake Michigan. In *The Proceedings of the Coastal Sediments '99*. Reston, VA: American Society of Civil Engineers, pp. 1978–93.

Naylor, L.A., Coombes, M.A. and Viles, H.A. 2012. Reconceptualising the role of organisms in the erosion of rock coasts: a new model. *Geomorphology*, **157–8**, 17–30.

Obu, J., Lantuit, H., Grosse, G. *et al.* 2017. Coastal erosion and mass wasting along the Canadian Beaufort Sea based on annual airborne LiDAR elevation data. *Geomorphology*, **293**, 331–46.

Pachure, T.M. and Mehta, A.J. 1985. Erosion of soft cohesive sediment deposits. *Journal of Hydraulic Engineering*, **110**, 1308–26.

Parthenaides, E. 1965. Erosion and deposition of cohesive soils. *Journal of the Hydraulics Division, ASCE*, **91**, 105–39.

Philpott, K.L. 1986. Coastal engineering aspects of the Port Burwell shore erosion damage litigation. *Proceedings Symposium on Cohesive Shores*, Ottawa: National Research Council of Canada, pp. 309–38.

Pierre, G. and Lahousse, P. 2006. The role of groundwater in cliff instability: An example at Cape Blanc-Nez; Pas-de-Calais, France. *Earth Surface Processes and Landforms*, **31**, 31–45.

Pringle, A.W. 1981 Beach development and coastal erosion in Holderness, North Humberside. In Flenley, J. (ed.), *The Quaternary in Britain*. Oxford: Pergamon Press, pp. 194–205.

Pringle, A.W. 1985. Holderness coast erosion and the significance of ords. *Earth Surface Processes and Landforms*, **10**, 107–24.

Pye, K. and Blott, S.J. 2015. Spatial and temporal variations in soft-cliff erosion along the Holderness coast, East Riding of Yorkshire, UK. *Journal of Coastal Conservation*, **19**, 785–808.

Quigley, R.M., Gelinas, P.J., Bou, W.T. and Packer, R.W. 1977. Cyclic erosion-instability relationships: Lake Erie north shore bluffs. *Canadian Geotechnical Journal*, **14**, 301–23.

Robinson, L.A. 1976. The micro-erosion meter technique in a littoral environment. *Marine Geology*, **22**, M51–M58.

Robinson, L.A. 1977a. Marine erosive processes at the cliff foot. *Marine Geology*, **23**, 257–71.

Robinson, L.A. 1977b. Erosive processes on the shore platform of northeast Yorkshire, England. *Marine Geology*, **23**, 339–61.

Sallenger, A.H. Jr, Krabill, W., Brock, J. *et al.* 2002. Sea-cliff erosion as a function of beach changes and extreme wave runup during the 1997–1998 El Niño. *Marine Geology*, **187**, 279–97.

Skafel, M.G. 1995. Laboratory measurements of near-shore velocities and erosion of cohesive sediment (till) shorelines. *Coastal Engineering*, **24**, 343–9.

Skafel, M.G. and Bishop, C.T. 1994. Flume experiments on the erosion of till shores by waves. *Coastal Engineering*, **23**, 329–48.

Stephenson WJ. 1997. Improving the traversing micro-erosion meter. *Journal of Coastal Research*, **13**(1), 236–41.

Stephenson, W.J. 2000. Shore platforms: a neglected coastal feature? *Progress in Physical Geography*, **24**, 311–27.

Stephenson, W.J. and Kirk, R.M. 1996. Measuring erosion rates using the micro-erosion meter: 20 years of data from shore platforms, Kaikoura Peninsula, South Island New Zealand. *Marine Geology*, **131**, 209–18.

Stephenson, W.J. and Kirk, R.M. 2000a. Development of shore platforms on Kaikoura Peninsula, South Island, New Zealand: part 1, the role of waves. *Geomorphology*, **32**, 21–41.

Stephenson, W.J. and Kirk, R.M. 2000b. Development of shore platforms on Kaikoura Peninsula, South Island, New Zealand: part 2, the role of subaerial weathering. *Geomorphology*, **32**, 43–56.

Stephenson, W.J., Kirk, R.M., Hemmingsen, S.A. and Hemmingsen, M.A. 2010. Decadal scale micro erosion rates on shore platforms. *Geomorphology*, **114**, 22–9.

Stephenson W.J., Dickson, M.E. and Trenhaile, A.S. 2013. Rock coasts. In Shroder J. (editor-in-chief), Sherman D.J. (ed.), *Treatise on Geomorphology, Volume 10, Coastal Geomorphology*. San Diego, CA: Academic Press, pp. 289–307.

Sunamura, T. 1977. A relationship between wave-induced cliff erosion and erosive force of waves. *Journal of Geology*, **85**, 613–18.

Sunamura, T. 1983. Processes of sea cliff and platform erosion. In Komar, P.D. (ed.), *Handbook of Coastal Processes and Erosion*. Boca Raton, FL: CRC Press, pp. 233–65.

Sunamura, T. 1992. *Geomorphology of Rocky Coasts*. Chichester, UK: John Wiley & Sons, 302 pp.

Sunamura, T. 2004. Cliffs, lithology versus erosion rates. In Schwartz, M. (ed.), *Encyclopedia of Coastal Sciences*. Dordrecht: Kluwer Academic Publishers, pp. 241–3.

Sunamura, T. 2015. Rocky coast processes: with special reference to the recession of soft rock cliffs. *Proceedings of the Japan Academy, Serial B Physical and Biological Sciences*, **91**, 481–500.

Trenhaile, A.S. 1972. The shore platforms of the Vale of Glamorgan, Wales. *Transactions of the Institute of British Geographers*, **56**, 127–44.

Trenhaile, A.S. 2009. Modelling the erosion of cohesive clay coasts. *Coastal Engineering*, **56**, 59–72.

Trenhaile, A.S., Pepper, D.A., Trenhaile, R.W. and Dalimonte, M. 1998. Stacks and notches at Hopewell Rocks, New Brunswick, Canada. *Earth Surface Processes and Landforms*, **23**, 975–86.

Trenhaile, A.S., Gagnon, J., Prestanski, K.J. and Porter, N.I. 2015. Cliff notch formation along the subtropical La Paz Peninsula, Southern Baja California, Mexico. *Marine Geology*, **368**, 15–24.

Trudgill, S.T. 1976. The marine erosion of limestones on Aldabra Atoll, Indian Ocean. *Zeitschrift für Geomorphologie*, **26**, 164–200.

Trudgill, S., High, C.J. and Hanna, F.K. 1981. Improvements to the microerosion meter. *British Geomorphological Research Group, Technical Bulletin*, **29**, 3–17.

Turowski, J.M. and Cook, K.L. 2017. Field techniques for measuring bedrock erosion and denudation. *Earth Surface Processes and Landforms*, **42**, 109–27.

Vann Jones, E.C., Rosser, N.J., M.J. Brain, M.J. and Petley, D.N. 2015. Quantifying the environmental controls on erosion of a hard rock cliff. *Marine Geology*, **363**, 230–42.

Walkden, M. and Dickson, M. 2008. Equilibrium erosion of soft rock shores with a shallow or absent beach under increased sea level rise. *Marine Geology*, **251**, 75–84.

Walkden, M.J.A. and Hall, J.W. 2005. A predictive mesoscale model of the erosion and profile development of soft rock shores. *Coastal Engineering*, **52**, 535–63.

Walkden, M.J.A. and Hall, J.W. 2011. A mesoscale predictive model of the evolution and management of a soft-rock coast. *Journal of Coastal Research*, **27**, 529–43.

Warrick, J.A., Ritchie, A.C., Adelman, G., Adelman, K. and Limber, P.W., 2017. New techniques to measure cliff change from historical oblique aerial photographs and structure-from-motion photogrammetry. *Journal of Coastal Research*, **33**, 39–55.

Wilcock, P.R., Miller, D.S., Shea, R.H. and Kerkin, R.T. 1998. Frequency of effective wave activity and the recession of coastal bluffs: Calvert Cliffs, Maryland. *Journal of Coastal Research*, **14**, 256–58.

Young, A.P. 2018. Decadal-scale coastal cliff retreat in southern and central California. *Geomorphology*, **300**, 164–75.

Young, A.P., Guza, R.T., Flick, R.E., O'Reilly, W.C., Gutierrez, R. 2009. Rain, waves, and short-term sea-cliff evolution. *Marine Geology*, **267**, 1–7.

Young, A.P., Guza, R.T., O'Reilly, W.C., Burvingt, O. and Flick, R.E. 2016. Observations of coastal cliff base waves, sand levels, and cliff top shaking. *Earth Surface Processes and Landforms*, **41**, 1564–73.

Zenkovitch, V.P. 1967. *Processes of Coastal Development*. Edinburgh: Oliver and Boyd, 738 pp.

Zoet, L.K. and Rawling, J.E. 2017. Analysis of a sudden bluff failure along the southwest Lake Michigan shoreline. *Journal of Great Lakes Research*, **43**, 999–1004.

Index

acoustic Doppler velicometers, 160
acoustic Doppler current profilers, 160
ADCP, 92, 204, *see* acoustic Doppler current profiler
ADV, 92, *see* acoustic Doppler velocimeter
aeolian sediment supply
  effect of tidal cycles, 312
  seasonal effects, 328
aeolian transport
  activity parameter, 322
  beach moisture content, 312
  bedload, 308
  concentration profile, 309
  cosine effect, 325
  critical fetch distance, 324
  effect of salts, 314
  effects of flow unsteadiness, 321
  effects of moisture, 312
  effects of pebble lag, 315
  effects of profile slope, 316
  equilibrium transport rate, 311
  fetch effect, 324
  fluid threshold, 307
  impact threshold, 308
  instantaneous transport prediction, 310
  intermittency, 321
  non-equilibrium conditions, 311
  potential sediment supply, 327
  probabilitic threshold, 322
  saltation, 308
  saltation cascade, 324
  saltation load, 308
  sediment flurries, 321
  Shields parameter, 307
  supply limited, 311
  suspension load, 308
  transport events, 327
  wind events, 327
atoll, *see* coral reef geomorphology: atoll

barrier dynamics
  barrier drowning, 381
  barrier evolution, 379
  Great Lakes overwash cycles, 360
  inlet formation, 350
  inundation overwash, 351, 357
  overstepping, 375
  overwash, 350, 354
  overwash fan, 355
  overwash regime, 351–2
  progradation, 379
  response to sea level change, 379
  response to sea level rise, 380
  run-up overwash, 351
  spit morphodynamics, 370
  stages of dune recovery, 358
  tidal inlet formation, 366
  tidal inlet stability, 367
  vegetation and washover healing, 357
  washover healing, 356, 362
  washover terrace, 357
barriers
  backbarrier sediments, 343
  barrier components, 347
  barrier island form, 364
  barrier island formation, 376
    mainland beach detachment, 378
    shoal aggradation, 376
    spit detachment, 376
  barrier islands, 280, 347, 375
    natural hazards, 381
  cuspate forelands, 346
  ebb tidal delta, 203
  flood tidal delta, 366
  flying spits, 203
  human impacts, 381–2
  lagoon types, 368
  lagoons, 359, 368
    choked, 368–9
    leaky, 368
    restricted, 368
  littoral sediment budget, 379
  management of transgressive barriers, 383
  morphological classification
    no free ends, 344
    one free end, 345
    two free ends, 347
  spit evolution, 371
  spit platform, 371
  spits, 347
    distal prograding end, 374
    dune ridge curvature, 371, 373
    spit extension, 374
    transgressive proximal end, 374
    transition zone, 374
  tidal inlet channel, 364
  tidal inlet prism, 359
  tidal inlet spacing, 363
  tidal inlets, 359
beach, 230
  beach plan form, 246
  berm, 201
  coarse sand beach, 201
  cobble, 230
  cobble beach, 201, 239, *see* beach: gravel beach

beach (cont.)
  fine sand beach, 202
  gravel beach, 236
  log spiral, 248
  mixed sediment, 236
  sandy beach, 230, 236
  swash aligned, 246
beach berm, 265
  height, 266
beach cusps, 269
  formation, 270
  giant cusps, 269
beach morphodynamics, 249
  dimensionless fall velocity parameter, 245, 257
  sandy beaches, 263
  temporal scales, 249
beach morphology, 232, 234
beach nourishment, 266
  mega-nourishment, 268
  Sandmotor, 268
beach profile
  barred, 239, 243, 245
  barred profile controls, 246
  equilibrium profile form, 242–3
  measurement, 233
  planar, 243
  sediment size, 242
  sweep zone, 239
  sweep zone and oil spill, 242
beach sediment budget, *see* littoral sediment budget
beach slope, 236
beach stage model, 257, 260, 266
beach system, 232
beach video monitoring, 233
  Argus system, 233, 260
beach water table, 264
bed elevation measurements, 235
  ultrasonic distance sensor, 235
bed shear stress waves, 186, 189
bedforms due to waves, 187
  3D ripples, 189–90
  asymetric flow regime, 189
  asymmetric ripples, 190
  barred profile, 190
  cross ripples, 190
  lunate megaripples, 189–90
  oscillatory flow regime, 187
  planar profile, 190
  plane bed, 190
  ripples, 187
  rolling grain ripples, 187
  sheet flow, 193
  surface roughness, 191
  vortex ripples, 187
bluff, *see* cohesive bluffs
boundary layer
  oscillatory flow, 185
  unidirectional flow, 184
breaker zone, 15
cliffed coast
  classification, 478
  definition, 477
  erosion modelling
    COSMOS 2-D, 503
  horizontal platform, 480
  morphological components, 477
  plunging cliffs, 478
  process-response model, 487
  shore platform, 478, 508
    range in form, 510
  shore platform types, 509
  shore platforms
    abrasion effects, 511
  sloping platform, 479
  soft coast erosion, 503
  soft coast erosion modelling
    SCAPE, 503
    Trenhaile, 504
  soft rock shoreline definition, 488
  toe erosion of hard rock, 504
cliffed coast classification
  recession rate, 487
cliffed coast erosion, 480
  effect of water level, 483
  effects of rock weathering, 484
  erosion probability, 485
  hydraulic forces, 482
  hydrostatic forces, 482
  measurement of rock strength, 485
  notch and visor formation, 506
  rock resistance, 484
  role of biological agents, 506
  role of cliff height, 508
  Sunamura model, 481
  toe erosion processes, 481
  weathering of hard rock, 504
  weathering rate, 484
cliffed coast recession
  cliff recession system, 480
  decadal scale, 481
  effect of platform lowering, 487
  hard rock cliffs, 504
  measurement techniques, 486
  recession models, 481
  recession rate, 486
coastal classification, 15–18
  Davies, 19
  Fairbridge, 18
  Inman and Nordstrom, 17
  Shepard, 17
  Valentin, 18
coastal dune vegetation
  dune rebuilding, 299
  invasive species, 285
  pioneer plant species, 300

plant zonation, 300
vegetation and stability, 284
vegetation cover, 299
coastal dunes, 280
beach–dune interaction, 330, 333
blowouts, 290
disturbance, 290
dune fields, 203, 290, 293
dune sediment budget, 212, 282, 302
embryo dune, *see* incipient dune
foredune equilibrium, 298
foredune erosion by storms, 297
foredune evolution
beach sediment budget, 331
beach stage model, 331
conceptual models, 330
Hesp model, 333
large-scale controls, 333
foredune lee slope, 288
foredune scarps, 297
foredune stoss slope, 288
foredunes, 280, 283, 288
impeded, 283
incipient dune, 289, 301
mobile, 283
nebkha, 294, 357
parabolic dunes, 283, 291
sand ramp, 298
secondary dunes, 290
sediment size distribution, 282
shadow dunes, 294
transgressive dune fields, 293
transgressive dunes, 302
coastal evolution, 18, 25
coastal geomorphology, 9
coastal management
sandy beach setbacks, 385
setbacks, 384
stable beach setback, 385
transgressive sandy beaches, 386
coastal sediment compartments, *see* littoral cells
coastal systems, 9
coastal zone, 12
coastal zone description
backshore, 14
beach, 13
coastal zone, 12
foreshore, 14
littoral zone, 13
nearshore zone, 13
offshore zone, 13
shoreface, 13
shoreline, 12
cohesive bluff erosion
abrasion by sand, 493
equilibrium profile, 489
mass wasting processes, 502
measurement of toe erosion, 497
measurement of vertical lowering, 494
nearshore erosion, 490–1
nearshore erosion rate, 491
shear strength of substrate, 493
toe erosion, 495, 499
vertical lowering, 489–90
weathering, 495
weathering of substrate, 493
cohesive bluff recession, 489
cohesive bluff recession rate
effect of beach width, 501
surficial sediment cover, 499
cohesive bluffs, 477, 488
distribution, 489
computational fluid dynamics, 291, 318
coral organisms, 445
calcium carbonate secretion, 445
cold water corals, 450
rates of carbonate accumulation, 446
reef-building corals, 445
scleractinian hard corals, 445
soft corals, 446
coral reef disturbance, 466
biological causes, 467
coral bleaching, 467
human causes, 470
natural causes, 466
physical causes, 467
tropical cyclones, 468
tsunamis, 469
coral reef geomorphology, 453
atoll response to RSL change, 457
atolls, 457
fringing reefs, 454
Pacific atolls, 457
reef island accretion, 461
reef islands, 459
shallow continental shelves, 455
volcanic islands, 457
coral reef hydrodynamics, 464
inlets and flushing, 466
seasonal wave climate, 464
wave attenuation, 464
wave energy and zonation, 464
coral reefs, 447
coastal protection, 451
coring, 462
distribution, 448
facies in uplifted reefs, 462
forereef slope, 448
framework facies, 462
geomorphic resilience, 450
Great Barrier Reef, 448
physical zonation, 447
reef crest, 448
response to sea level change, 451
response to sea level fall, 452
response to sea level rise, 452, 471

coral reefs (cont.)
  sedimentary facies, 462
  vertical accretion, 452
  vulnerability to climate change, 470

DEM, *see* digital elevation model
digital elevation model, 21, 23
drone, *see* UAV

echo sounding, 233
electromagnetic current meter, 160, 195
equatorial siphoning, 39
ETRS89, (European Terrestrial Reference System 1989), 32
Exner equation, 24
extra tropical storms, *see* mid-latitude cyclones

fluid properties of air, 303
forebulge, 41

geographic information system, 21
geoid, 39
GIS, *see* geographic information system
global climate change, 25
GLOSS, (Global Sea Level Observing System), 35
ground penetrating radar, 370

headland bay beach, 246
Holocene transgression, 379
hurricanes, *see* tropical cyclone
hydrodynamic models, 24

incipient dunes
  evolution model, 289
intertidal zone, 15, 44
isostatic adjustment, *see* sea level change
isostatic loading, *see* isostasy

Lanphere Dunes, 285
LiDAR, 20, 234, 352, 484, 510
littoral cells, 201, 213
  boundaries, 213
  interruptive boundary, 216
  Lake Erie, 214
littoral sediment budget, 211, 213, 371
  coastal erosion source, 213
  modelling, 371
  river source, 212
  sink, 211, 213
  source, 211, 213
littoral sediment gradient, 217, 371, 374
  ABC model, 217
  modelling, 218
littoral sediment transport
  gross littoral transport, 202
  net littoral transport, 202, 206
littoral zone, 231
longshore sandwaves, 209, 360, 502

mangrove hydrodynamics, 434
  effects of vegetation, 434
  effects on flows, 434
  reduction of wave energy, 434
mangrove sedimentation, 434
  autocompaction, 436
  organic deposition, 436
mangroves, 395, 430
  adaptation to waterlogging, 431
  classification of systems, 430
  primary productivity, 398
  protection against tsunamis, 436
  river dominated, 430
  salt ponds, 430
  species distribution, 430
  species zonation, 432
  TSS concentation, 434
  vegetation, 396
Mason Bay Dunes, 287
MEM, 494
micro-erosion meter, *see* MEM
mid-latitude cyclones, 59
Milankovitch cycles, 38
morphodynamics, 11, 22
  beach and nearshore, 232

NAD83, 32
nearshore and intertidal bars
  classification, 249
  controls on dynamic range, 256
  ridge and runnel, 250
nearshore bars
  bar switching, 253
  bar welding, 260, 266
  formation, 253
  formation by cross-shore flows, 254
  formation by infragravity waves, 253–4
  inner bars, 252, 256
  multiple parallel, 250
  tidal range control, 262
nearshore morphology, 233
nearshore sediment size, 238
nearshore slope, 238
numerical model, 22

OBS, 160, 195, 204, 206, *see* optical backscatterance sensors
oil spill, 241
optical backscatterance sensor, *see* OBS
optical luminescence, 36

particle image velocimetry, 160
pressure bulge, *see* forebulge

reductionist approach, 21–2
rips, *see* surf zone circulation

saltmarsh erosion
  erosion cycle model, 423
  erosion cycles, 422–3

modelliing, 422
wave attack, 422
saltmarsh hydrodynamics, 406
effects of vegetation, 408
flow modelling, 409
flows in tidal creeks, 406
flows over marsh surface, 408
wave reduction, 409
saltmarsh sedimentation, 410
annual, 417
autocompaction, 411
decadal, 420
decadal scale, 419
effects of winter ice, 417
measurement of accretion, 413
models, 415
organic accretion, 411
rate of deposition, 414
sediment budget, 410–11
sediment erosion table, 413
spatial patterns, 416, 418
tidal cycles, 413
trap measurements, 414
TSS concentration, 413, 417
saltmarshes, 395
annual mass balance, 422
distribution, 398
effects of sea level rise, 425
evolution, 400, 404
high marsh, 398
low marsh, 400
managed realignment, 426
Cumberland Basin, 427
minerogenic, 396
morphology, 398
organogenic, 396
primary productivity, 398
RSLR and coastal squeeze, 426
sediment sources, 404
tidal creek evolution, 405
tidal creek network, 405
vegetation, 396
vegetation zonation, 398
satellite mapping, 234
SBEACH, 23
sea level, 32
ellipsoid model, 32
geoid model, 32, 34
mean sea level, 32, 36
measurement, 35
satellite altimetry, 35, 65
sea level change, 31–2
Bruun rule, 25, 67
decadal, 56
eustatic, 31, 36, 38
forecast sea level rise, 68
historic, 36
Holocene transgression, 31, 38–9, 41
hydro-isostasy, 40–1
isostatic, 36, 42
regression, 31
sea level rise, 63
seiche, 61
shoreline recession, 67
storm surge, 56, 358
storm surge models, 58
tectonic, 40
transgression, 31
sediment transport, 24
cross-shore, 184, 194
direction indicators, 202
energetics approach, 199
fluorescent tracers, 207
longshore transport, 184, 200, 202
beach drifting, 200–1
longshore currents, 200–2
rip current effects, 200
suspended sediment concentration profile, 195, 198
suspended sediment transport, 195
threshold of motion, 186
transport rate, 198
sediment transport by waves, 184
sediment transport measurement
fluorescent tracers, 204
radioactive tracers, 204
RFID tags, 205
sediment tracers, 204
streamer traps, 205
sediment transport models
Baillard, 199–200
Bowen, 199
Delft3D, 218
SBEACH, 199
wave tanks, 200
sediment transport models, 199
sediment transport rate, 200
self-organisational models, 170
shoaling zone, 14
shore platform, *see* cliffed coast: shore platform
shoreface, 184
inner, 231
outer, 231
shoreline change, 11
shoreline mapping, 19
stilling well, 35
structure from motion, 484
surf zone, 15, 130, 158
barred, 161
planar, 158
surf zone circulation
boundary rips, 169
channelised rips, 167
drogue tracking, 160
feeder currents, 165
longshore current, 174
barred, 175
longshore current speed, 174
mass transport, 158

surf zone circulation (cont.)
- measurement, 159
- planar, 161
- planar rips, 167
- rip cell, 164
- rip cell spacing, 170
- rip channel, 164
- rip current, 164
- rip current hazards, 172
- rip head, 164
- rip neck, 164
- rip speed, 169
- rips, 164
- shear waves, 177
- undertow, 158, 161, 171, 184, 195, 256
- undertow speed, 158, 163
- wind-driven currents, 177, 184

swash zone, 15
- backwash, 143
- run-up, 143
- swash, 143

template models, 170
thermo-luminescence, 36
tide gauges, 35
tides
- amphidrome, 53
- astronomical tides, 44
- diurnal, 45
- dynamic theory, 52
- equilibrium theory, 48
- harmonic analysis, 51
- hypertidal, 47
- meteorological tides, 44, 55
- neap tide, 49
- semi-diurnal, 45, 49
- spring–neap cycle, 45
- spring tide, 46, 49
- tidal bore, 55
- tidal currents, 177
- tidal range, 44–6, 54
- tidal type, 45
- tidal wave, 48, 52

total suspended solids, *see* saltmarsh sedimentation:TSS
tropical cyclone, 59
typhoons, *see* tropical cyclone

UAV, 235
undertow, *see* surf zone circulation

washover fan, *see* barrier dynamics: overwash fan
wave analysis, 93
- energy density spectrum, 98
- frequency domain, 96
- mean wave height, 95
- RMS height, 96
- significant wave height, 96
- spectral analysis, 96
- time domain, 93
- variance spectrum, 98
- wave climate, 106
- wavelet analysis, 101
- zero crossing, 93

wave breaking, 128, 183
- breaker height, 129
- breaker indices, 135
- breaker types, 131
- breaker zone, 129
- collapsing breaker, 134
- deep water, 128
- edge waves, 151
- harmonic waves, 140
- infragravity waves, 149
- plunging breaker, 132, 158
- radiation stress, 174
- reflection, 140
- saturated surf zone, 139
- shallow water, 129
- spilling breaker, 131, 158
- surf bore, 137
- surf scaling parameter, 135
- surging breaker, 134
- vortical motion, 137
- water depth, 129
- wave set-up, 148

wave description
- amplitude, 76
- capillary waves, 79–80
- celerity, 76
- deep water length, 113
- deep water wave, 78
- definition, 76
- edge wave, 170
- energy density, 117
- frequency, 76
- group celerity, 117
- height, 76
- infragravity waves, 80, 195
- length, 76
- linear wave profile, 111
- maximum orbital velocity, 115
- mean water level, 76
- orbital diameters, 115
- orbital motion, 76
- period, 76
- power, 117
- pressure fluctuations, 116
- sea, 79
- shallow water L, 113
- still water level, 76
- Stokes drift, 119, 158, 184
- surface gravity waves, 76, 81
- swell, 79
- tsunami, 144

wave energy flux
- longshore component P_L, 207

wave generation
  dispersion, 84, 112
  energy spectrum, 82
  event duration, 81
  fetch length, 81, 86
  fully arisen sea, 85
  spectrum, 80
  wave growth, 82
  wave spectrum, 91
  whitecapping, 84, 129
  wind speed, 81
wave measurement, 86
  direction, 91
  pressure transducers, 89
  remote sensing, 93
  wave staffs, 89
  wave-rider buoy, 92
wave prediction, 101
  direction, 104
  forecasting, 101
  hndcasting, 101
  JONSWAP, 102
  SMB method, 102
  SWAN, 103
wave shoaling, 122–3
  diffraction, 126
  MIKE 21, 128
  orthogonals, 122
  radiation stress, 147
  refraction, 122, 125, 371
  Snell's law, 127
  STWAVE, 128
  wave refraction, 373
  wave set down, 147
wave theory, 110
  Airy (linear), 110
  applicability, 120
  cnoidal, 120
  Stokes second-order, 118
  trochoidal, 120
  wave potential function, 111
wetland ecosystem, 395
WGS84, 32
wind flow
  bed shear velocity, 305
  boundary layer profile, 303
  coherent flow structures, 322
  drag by plants, 317
  flow separation, 319
  flow unsteadiness, 319
  Reynolds shear stress, 323
  roughness elements, 306
  roughness length, 303
  sonic anemometers, 323
  speed up, 317
  topographic effects, 317
  topographic steering, 318

XBEACH, 23